# Frequently Used Formulas

Slope of a Line Containing the Points $(x_1, y_1)$ and $(x_2, y_2)$:

$$m = \frac{y_2 - y_1}{x_2 - x_1}, \text{ where } x_2 - x_1 \neq 0.$$

Slope—Intercept Form of a Line: $y = mx + b$

Standard Form of a Line:

$$Ax + By = C \text{ where } A, B, \text{ and } C \text{ are integers and } A \text{ is positive.}$$

Point-Slope Formula: If $(x_1, y_1)$ is a point on line $L$ and $m$ is the slope of line $L$, then the equation of $L$ is given by

$$y - y_1 = m(x - x_1).$$

Factoring Formulas

1. Difference of Two Squares: $a^2 - b^2 = (a + b)(a - b)$
2. Sum and Difference of Two Cubes:

$$a^3 + b^3 = (a + b)(a^2 - ab + b^2)$$
$$a^3 - b^3 = (a - b)(a^2 + ab + b^2)$$

Distance Formula: The distance, $d$, between the points $(x_1, y_1)$ and $(x_2, y_2)$ is given by $d = \sqrt{(x_2 - x_1)^2 + (y_2 - y_1)^2}$.

Quadratic Formula: The solutions of any quadratic equation of the form $ax^2 + bx + c = 0$ $(a \neq 0)$ are

$$x = \frac{-b \pm \sqrt{b^2 - 4ac}}{2a}.$$

# Beginning Algebra

## Sherri Messersmith
*College of DuPage*

McGraw Hill

*Connect
Learn
Succeed*™

BEGINNING ALGEBRA

Published by McGraw-Hill, a business unit of The McGraw-Hill Companies, Inc., 1221 Avenue of the
Americas, New York, NY 10020. Copyright © 2012 by The McGraw-Hill Companies, Inc. All rights reserved.
No part of this publication may be reproduced or distributed in any form or by any means, or stored in a
database or retrieval system, without the prior written consent of The McGraw-Hill Companies, Inc.,
including, but not limited to, in any network or other electronic storage or transmission, or broadcast for
distance learning.

Some ancillaries, including electronic and print components, may not be available to customers outside the
United States.

This book is printed on acid-free paper.

1 2 3 4 5 6 7 8 9 0 DOW/DOW 1 0 9 8 7 6 5 4 3 2 1

ISBN 978–0–07–340616–9
MHID 0–07–340616–3

ISBN 978–0–07–329710–1 (Annotated Instructor's Edition)
MHID 0–07–329710–0

Vice President, Editor-in-Chief: *Marty Lange*
Vice President, EDP: *Kimberly Meriwether David*
Vice-President New Product Launches: *Michael Lange*
Editorial Director: *Stewart K. Mattson*
Executive Editor: *Dawn R. Bercier*
Developmental Editor: *Emily Williams*
Director of Digital Content Development: *Emilie J. Berglund*
Marketing Manager: *Peter A. Vanaria*
Lead Project Manager: *Peggy J. Selle*
Buyer II: *Sherry L. Kane*
Senior Media Project Manager: *Sandra M. Schnee*
Senior Designer: *David W. Hash*
Cover Designer: *Greg Nettles/Squarecrow Creative*
Cover Image: *© Bruce Johnston*
Lead Photo Research Coordinator: *Carrie K. Burger*
Compositor: *Aptara, Inc.*
Typeface: *10.5/12 Times New Roman*
Printer: *R. R. Donnelley*

All credits appearing on page or at the end of the book are considered to be an extension of the copyright page.

**Library of Congress Cataloging-in-Publication Data**

Messersmith, Sherri.
    Beginning algebra / Sherri Messersmith. — 1st ed.
        p. cm.
    Includes index.
    ISBN 978–0–07–340616–9 — ISBN 0–07–340616–3 (hard copy : alk. paper) 1. Algebra—Textbooks. I. Title.
    QA152.3.M465 2012
    512.9—dc22

                                                                                            2010044510

www.mhhe.com

# Message from the Author

Dear Colleagues,

Students constantly change—and over the last 10 years, they have changed a lot, therefore this book was written for today's students. I have adapted much of what I had been doing in the past to what is more appropriate for today's students. This textbook has evolved from the notes, worksheets, and teaching strategies I have developed throughout my 25-year teaching career in the hopes of sharing with others the successful approach I have developed.

To help my students learn algebra, I meet them where they are by helping them improve their basic skills and then showing them the connections between arithmetic and algebra. Only then can they learn the algebra that is the course. Throughout the book, concepts are presented in **bite-size pieces** because developmental students learn best when they have to digest fewer new concepts at one time. The **Basic Skills Worksheets** are quick, effective tools that can be used in the classroom to help strengthen students' arithmetic skills, and most of these worksheets can be done in 5 minutes or less. The **You Try** exercises follow the examples in the book so that students can practice concepts immediately. The **Fill-It-In exercises** take students through a step-by-step process of working multistep problems, asking them to fill in a step or a reason for a given step to prepare them to work through exercises on their own and to reinforce mathematical vocabulary. **Modern applications** are written with student interests in mind; students frequently comment that they have never seen "fun" word problems in a math book before this one. **Connect Mathematics hosted by ALEKS** is an online homework manager that will identify students' strengths and weaknesses and take the necessary steps to help them master key concepts.

The **writing style** is friendlier than that of most textbooks. Without sacrificing mathematical rigor, this book uses language that is mathematically sound yet easy for students to understand. Instructors and students appreciate its conversational tone because it sounds like a teacher talking to a class. The **use of questions** throughout the prose contributes to the conversational style while teaching students how to ask themselves the questions we ask ourselves when solving a problem. This friendly, less intimidating writing style is especially important because many of today's developmental math students are enrolled in developmental reading as well.

Beginning Algebra, is a compilation of what I have learned in the classroom, from faculty members nationwide, from the national conferences and faculty forums I have attended, and from the extensive review process. Thank you to everyone who has helped me to develop this textbook. My commitment has been to write the most mathematically sound, readable, student-friendly, and up-to-date text with unparalleled resources available for both students and instructors. To share your comments, ideas, or suggestions for making the text even better, please contact me at sherri.messersmith@gmail.com. I would love to hear from you.

Sherri Messersmith

# About the Author

**Sherri Messersmith** has been teaching at College of DuPage in Glen Ellyn, Illinois, since 1994. She has over 25 years of experience teaching many different courses from developmental mathematics through calculus. She earned a bachelor of science degree in the teaching of mathematics at the University of Illinois at Urbana-Champaign and went on to teach at the high school level for two years. Sherri returned to UIUC and earned a master of science in applied mathematics and stayed on at the university to teach and coordinate large sections of undergraduate math courses. Sherri has authored several textbooks, and she has also appeared in videos accompanying several McGraw-Hill texts.

Sherri lives outside of Chicago with her husband, Phil, and their daughters, Alex and Cailen. In her precious free time, she likes to read, play the guitar, and travel—the manuscripts for this and her previous books have accompanied her from Spain to Greece and many points in between.

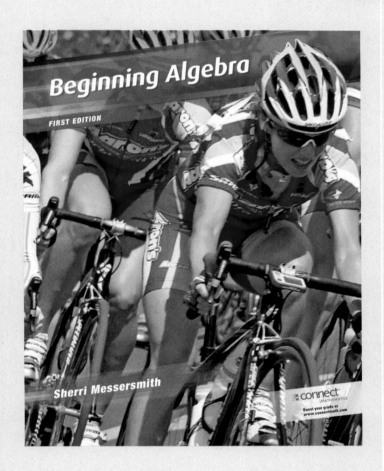

In order to be successful, a cyclist must follow a strict training regimen. Instead of attempting to compete immediately, the athlete must practice furiously in smaller intervals to build up endurance, skill, and speed. A true competitor sees the connection between the smaller steps of training and final accomplishment. Similarly, after years of teaching, it became clear to Sherri Messersmith that mastering math for most students is less about the memorization of facts and more of a journey of studying and understanding what may seem to be complex topics. Like a cyclist training for a long race, as pictured on the front cover, students must build their knowledge of mathematical concepts by connecting and applying concepts they already know to more challenging ones, just as a cyclist uses training and hard work to successfully work up to longer, more challenging rides. After following the methodology applied in this text, like a cyclist following a training program, students will be able to succeed in their course.

# Brief Contents

# Preface

## Building a Better Path to Success

*Beginning Algebra*, helps students build a better path to success by providing the tools and building blocks necessary for success in their mathematics course. The author, Sherri Messersmith, learned in her many years of teaching that students had a better rate of success when they were connecting their knowledge of arithmetic with their study of algebra. By **making these connections between arithmetic and algebra** and **presenting the concepts in more manageable, "bite-size" pieces,** Sherri better equips her students to learn new concepts and strengthen their skills. In this process, students practice and build on what they already know so that they can understand and **master new concepts** more easily. These practices are integrated throughout the text and the supplemental materials, as evidenced below:

### Connecting Knowledge

- **Examples** draw upon students' current knowledge while connecting to concepts they are about to learn with the positioning of arithmetic examples before corresponding algebraic examples.

- Inclusion of a **geometry review** in Section 1.3 allows students to practice several geometry concepts without variables *before* they have to write algebraic equations to solve geometry problems.

- The very popular author-created **worksheets** that accompany the text provide instructors with additional exercises to assist with overcoming potential stumbling blocks in student knowledge and to help students see the connections among multiple mathematical concepts. The worksheets fall into three categories:
  - Worksheets to Strengthen Basic Skills
  - Worksheets to Help Teach New Concepts
  - Worksheets to Tie Concepts Together

### Presenting Concepts in Bite-Size Pieces

- The **chapter organization** help break down algebraic concepts into more easily learned, more manageable pieces.

- New **Fill-It-In exercises** take a student through the process of working a problem step-by-step so that students have to provide the reason for each mathematical step to solve the problem, much like a geometry proof.

- New **Guided Student Notes** are an amazing resource for instructors to help their students become better note-takers. They contain in-class examples provided in the margin of the text along with additional examples not found in the book to emphasize the given topic so that students spend less time copying down information and more time engaging within the classroom.

- **In-Class Examples** give instructors an additional tool that exactly mirrors the corresponding examples in the book for classroom use.

### Mastering Concepts:

- **You Try** problems follow almost every example in the text to provide students the opportunity to immediately apply their knowledge of the concept being presented.

- **Putting It All Together** sections allow the students to synthesize important concepts presented in the chapter sooner rather than later, which helps in their overall mastery of the material.

- **Connect Math hosted by ALEKS** is the combination of an online homework manager with an artificial-intelligent, diagnostic assessment. It allows students to identify their strengths and weaknesses and to take the necessary steps to master those concepts. Instructors will have a platform that was designed through a comprehensive market development process involving full-time and adjunct math faculty to better meet their needs.

# Connecting Knowledge

**Examples** The **examples** in each section begin with an arithmetic equation that mirrors the algebraic equation for the concept being presented. This positioning allows students to apply their knowledge of arithmetic to the algebraic problem, making the concept more easily understandable.

---

**1. Add and Subtract Rational Expressions with a Common Denominator**

Let's first look at fractions and rational expressions with common denominators.

> **Example 1**
>
> Add or subtract.
>
> a) $\dfrac{8}{11} - \dfrac{5}{11}$    b) $\dfrac{2x}{4x - 9} + \dfrac{5x + 3}{4x - 9}$
>
> **Solution**
>
> a) Since the fractions have the same denominator, subtract the terms in the numerator and keep the common denominator.
>
> $$\frac{8}{11} - \frac{5}{11} = \frac{8 - 5}{11} = \frac{3}{11} \qquad \text{Subtract terms in the numerator.}$$
>
> b) Since the rational expressions have the same denominator, add the terms in the numerator and keep the common denominator.
>
> $$\frac{2x}{4x - 9} + \frac{5x + 3}{4x - 9} = \frac{2x + (5x + 3)}{4x - 9} \qquad \text{Add terms in the numerator.}$$
> $$= \frac{7x + 3}{4x - 9} \qquad \text{Combine like terms.}$$

"Messersmith does a great job of addressing students' abilities with the examples and explanations provided, and the thoroughness with which the topic is addressed is excellent." Tina Evans, Shelton State Community College

"The author is straightforward, using language that is accessible to students of all levels of ability. The author does an excellent job explaining difficult concepts and working from easier to more difficult problems." Lisa Christman, University of Central Arkansas

**Geometry Review** A **geometry review** in Chapter 1 Section 1.3, provides the material necessary for students to revisit the geometry concepts they will need later in the course. Reviewing the geometry early, rather than in an appendix or not at all, removes a common stumbling block for students. The book also includes geometry applications, where appropriate, throughout.

"The Geometry Review is excellent for this level." Abraham Biggs, Broward College-South

---

> **Example 3**
>
> Find the perimeter and area of each figure.
>
> a) 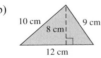 7 in. / 9 in.    b) 10 cm / 8 cm / 9 cm / 12 cm
>
> **Solution**
>
> a) This figure is a rectangle.
>
> Perimeter: $P = 2l + 2w$
> $P = 2(9 \text{ in.}) + 2(7 \text{ in.})$
> $P = 18 \text{ in.} + 14 \text{ in.}$
> $P = 32 \text{ in.}$
>
> Area: $A = lw$
> $A = (9 \text{ in.})(7 \text{ in.})$
> $A = 63 \text{ in}^2$ or 63 square inches

**Worksheets** Supplemental **worksheets** for *every* section are available online through Connect. They fall into three categories: worksheets to strengthen basic skills, worksheets to help teach new concepts, and worksheets to tie concepts together. These worksheets provide a quick, engaging way for students to work on key concepts. They save instructors from having to create their own supplemental material, and they address potential stumbling blocks. They are also a great resource for standardizing instruction across a mathematics department.

---

Worksheet 2.1 A        Name: _____

Messersmith–Beginning Algebra

*Rewrite each using exponents.*

1) $5 \cdot 5 \cdot 5 \cdot 5$            3) $2 \cdot p \cdot p \cdot p$

2) $(-3)(-3)(-3)(-3)(-3)$     4) $(2p)(2p)(2p)$

*Evaluate.*

5)    $3^4$                7) $-2^6$

6)    $\left(\dfrac{2}{5}\right)^3$          8) $(-2)^6$

*Simplify the expression using the product rule.*

9) $y^5 \cdot y^4$            12) $(6a^4)(4a^3)$

10) $w^7 \cdot w^4$           13) $(5t^2)(-9t^3)$

11) $p^3 \cdot p \cdot p^8$       14) $\left(\dfrac{3}{4}c\right)\left(-\dfrac{2}{3}c^5\right)(-10c^8)$

*Simplify the expression using one of the power rules.*

15) $(y^2)^3$              18) $(2h)^4$

16) $\left(\dfrac{x}{3}\right)^2$          19) $(-3d)^3$

17) $\left(\dfrac{a}{b}\right)^{12}$        20) $(-12uv)^2$

---

Worksheet 3C        Name: _____

Messersmith–Beginning Algebra

Find 2 numbers that . . .

| MULTIPLY TO | *and* ADD TO | ANSWER |
|---|---|---|
| −27 | −6 | −9 and 3 |
| 72 | 18 | |
| 24 | −11 | |
| −4 | 3 | |
| 10 | −7 | |
| 121 | 22 | |
| −54 | −3 | |
| 54 | 29 | |
| 16 | −10 | |
| 30 | 17 | |
| 9 | −6 | |
| −8 | −2 | |
| 21 | 10 | |
| 60 | −19 | |
| 56 | 15 | |
| −28 | 3 | |
| −72 | −6 | |
| 100 | 25 | |
| −40 | 6 | |
| 11 | −12 | |
| 20 | 12 | |
| −35 | −2 | |
| 77 | 18 | |
| 108 | 21 | |
| −3 | −2 | |

# Presenting Concepts in Bite-Size Pieces

**Chapter Organization**  The **chapter organization** is designed to present the context in "bite-size" pieces, focusing not only on the mathematical concepts but also on the "why" behind those concepts. By breaking down the sections into manageable chunks, the author has identified the core places where students traditionally struggle.

"The material is presented in a very understandable manner, in that it approaches all topics in bite-sized pieces and explains each step thoroughly as it proceeds through the examples."
Lee Ann Spahr,
Durham Technical Community College

CHAPTER 2

## The Rules of Exponents

### Algebra at Work: Custom Motorcycle Shop

The people who build custom motorcycles use a lot of mathematics to do their jobs. Mark is building a chopper frame and needs to make the supports for the axle. He has to punch holes in the plates that will be welded to the frame.

Mark has to punch holes with a diameter of 1 in. in mild steel that is $\frac{3}{8}$ in. thick. The press punches two holes at a time. To determine how much power is needed to do this job, he uses a formula containing an exponent, $P = \dfrac{t^2 dN}{3.78}$. After substituting the numbers into the expression, he calculates that the power needed to punch these holes is 0.07 hp.

**In-Class Examples**  To give instructors additional material to use in the class-room, a matching ***In-Class Example*** is provided in the margin of the Annotated Instructor's Edition for every example in the book. The more examples a student reviews, the better chance he or she will have to understand the related concept.

---

**Example 11**

Write an equation and solve.

Alex and Jenny are taking a cross-country road trip on their motorcycles. Jenny leaves a rest area first traveling at 60 mph. Alex leaves 30 minutes later, traveling on the same highway, at 70 mph. How long will it take Alex to catch Jenny?

**Solution**

*Step 1:*  **Read** the problem carefully, and identify what we are being asked to find.

We must determine how long it takes Alex to catch Jenny.

We will use a picture to help us see what is happening in this problem.

Since both girls leave the same rest area and travel on the same highway, when Alex catches Jenny they have driven the *same* distance.

---

"I like that the teacher's edition gives in-class examples to use so the teacher doesn't have to spend prep time looking for good examples or using potential homework/exam questions for in-class examples."
Judith Atkinson,
University of Alaska–Fairbanks

**Fill-It-In** **Fill-It-In exercises** take a student through the process of working a problem step-by-step so that students either have to provide the reason for each mathematical step or fill in a mathematical step when the reason is given. These types of exercises are unique to this text and appear throughout.

---

**Fill It In**

Fill in the blanks with either the missing mathematical step or reason for the given step.

3) $y^2 + 18y$

_____     Find half of the coefficient of $y$.

_____     Square the result.

_____     Add the constant to the expression.

The perfect square trinomial is _____ .

The factored form of the trinomial is _____ .

4) $c^2 - 5c$

$\dfrac{1}{2}(-5) = -\dfrac{5}{2}$    _____ .

$\left(-\dfrac{5}{2}\right)^2 = \dfrac{25}{4}$    _____ .

$c^2 - 5c + \dfrac{25}{4}$    _____ .

The perfect square trinomial is _____ .

The factored form of the trinomial is _____ .

---

---

**Fill It In**

Fill in the blanks with either the missing mathematical step or reason for the given step.

3) $y^2 + 18y$

$\dfrac{1}{2}(18) = 9$     Find half of the coefficient of $y$.

$9^2 = 81$     Square the result.

$y^2 + 18y + 81$     Add the constant to the expression.

The perfect square trinomial is $y^2 + 18y + 81$.

The factored form of the trinomial is $(y + 9)^2$.

4) $c^2 - 5c$

$\dfrac{1}{2}(-5) = -\dfrac{5}{2}$    Find half of the coefficient of $c$.

$\left(-\dfrac{5}{2}\right)^2 = \dfrac{25}{4}$    Square the result.

$c^2 - 5c + \dfrac{25}{4}$    Add the constant to the expression.

The perfect square trinomial is $c^2 - 5c + \dfrac{25}{4}$.

The factored form of the trinomial is $\left(c - \dfrac{5}{2}\right)^2$.

Guided Student Notes                    Name: _____

Messersmith–Beginning Algebra

### 1.1 Review of Fractions

**Definition of Fraction**

*What part of the figure is shaded?*

1)

**Definition of Lowest Terms**

**Factors of a Number**

2) Find all factors of 18.          3) Find all factors of 54.

---

---

Guided Student Notes                    Name: _____

Messersmith–Beginning Algebra

### 2.1a Basic Rules of Exponents Product Rule and Power Rule

**Base**                              **Exponent**

*Identify the base and the exponent in each expression and evaluate.*

1) $3^4$                             5) $(-5)^3$

2) $(-3)^4$                          6) $2(5)^2$

3) $-3^4$                            7) $4a(-3)^2$

4) $-5^2$                            8) $-(2)^4$

**Product Rule**                     **Power Rule**

*Find each product.*                 *Simplify using the power rule.*

 9) $5^2 \cdot 5$                    13) $(4^6)^3$

10) $y^4 y^9$                        14) $(m^2)^5$

11) $-4x^5 \cdot (-10x^8)$           15) $(q^8)^7$

12) $d \cdot d^7 \cdot d^4$          16) $(df^2)^3$

# Mastering Concepts

**You Try** After almost every example, there is a *You Try* problem that mirrors the example. This provides students the opportunity to practice a problem similar to what the instructor has presented before moving on to the next concept.

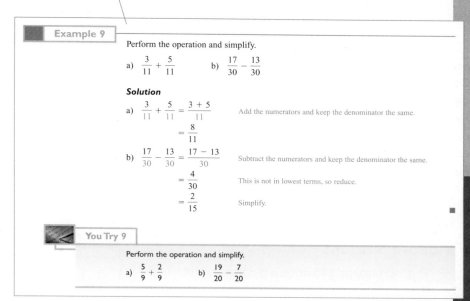

**Example 9**

Perform the operation and simplify.

a) $\dfrac{3}{11} + \dfrac{5}{11}$    b) $\dfrac{17}{30} - \dfrac{13}{30}$

**Solution**

a) $\dfrac{3}{11} + \dfrac{5}{11} = \dfrac{3 + 5}{11}$    Add the numerators and keep the denominator the same.

$= \dfrac{8}{11}$

b) $\dfrac{17}{30} - \dfrac{13}{30} = \dfrac{17 - 13}{30}$    Subtract the numerators and keep the denominator the same.

$= \dfrac{4}{30}$    This is not in lowest terms, so reduce.

$= \dfrac{2}{15}$    Simplify.

**You Try 9**

Perform the operation and simplify.

a) $\dfrac{5}{9} + \dfrac{2}{9}$    b) $\dfrac{19}{20} - \dfrac{7}{20}$

**Putting It All Together** Several chapter include a *Putting It All Together* section, in keeping with the author's philosophy of breaking sections into manageable chunks to increase student comprehension. These sections help students synthesize key concepts before moving on to the rest of the chapter.

## Putting It All Together

**Objective**
1. Combine the Rules of Exponents

### 1. Combine the Rules of Exponents

Let's review all the rules for simplifying exponential expressions and then see how we can combine the rules to simplify expressions.

**Summary** Rules of Exponents

In the rules stated here, $a$ and $b$ are any real numbers and $m$ and $n$ are positive integers.

| | |
|---|---|
| Product rule | $a^m \cdot a^n = a^{m+n}$ |
| Basic power rule | $(a^m)^n = a^{mn}$ |
| Power rule for a product | $(ab)^n = a^n b^n$ |
| Power rule for a quotient | $\left(\dfrac{a}{b}\right)^n = \dfrac{a^n}{b^n}, \quad (b \neq 0)$ |
| Quotient rule | $\dfrac{a^m}{a^n} = a^{m-n}, \quad (a \neq 0)$ |

Changing from negative to positive exponents, where $a \neq 0$, $b \neq 0$, and $m$ and $n$ are any integers:

$$\frac{a^{-m}}{b^{-n}} = \frac{b^n}{a^m} \qquad \left(\frac{a}{b}\right)^{-m} = \left(\frac{b}{a}\right)^m$$

In the following definitions, $a \neq 0$, and $n$ is any integer.

| | |
|---|---|
| Zero as an exponent | $a^0 = 1$ |
| Negative number as an exponent | $a^{-n} = \dfrac{1}{a^n}$ |

**|MATH**

Hosted by **ALEKS Corp.**

**Connect Math Hosted by ALEKS Corporation** is an exciting, new assignment and assessment platform combining the strengths of McGraw-Hill Higher Education and ALEKS Corporation. Connect Math Hosted by ALEKS is the first platform on the market to combine an artificially-intelligent, diagnostic assessment with an intuitive ehomework platform designed to meet your needs.

Connect Math Hosted by ALEKS Corporation is the culmination of a one-of-a-kind market development process involving math full-time and adjunct Math faculty at every step of the process. This process enables us to provide you with a solution that best meets your needs.

Connect Math Hosted by ALEKS Corporation is built by Math educators for Math educators!

**1** *Your students want a well-organized homepage where key information is easily viewable.*

### Modern Student Homepage

▶ This homepage provides a dashboard for students to immediately view their assignments, grades, and announcements for their course. (Assignments include HW, quizzes, and tests.)

▶ Students can access their assignments through the course Calendar to stay up-to-date and organized for their class.

*Modern, intuitive, and simple interface.*

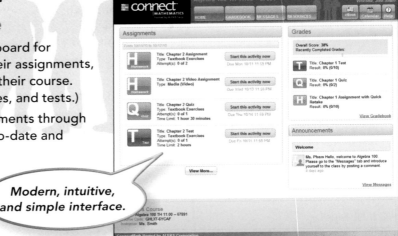

**2** *You want a way to identify the strengths and weaknesses of your class at the beginning of the term rather than after the first exam.*

### Integrated ALEKS® Assessment

▶ This artificially-intelligent (AI), diagnostic assessment identifies precisely what a student knows and is ready to learn next.

▶ Detailed assessment reports provide instructors with specific information about where students are struggling most.

▶ This AI-driven assessment is the only one of its kind in an online homework platform.

*Recommended to be used as the first assignment in any course.*

ALEKS is a registered trademark of ALEKS Corporation.

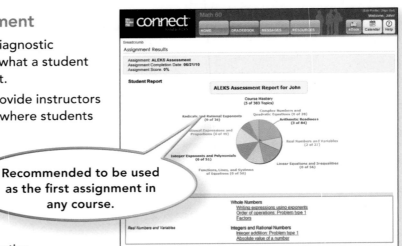

# Built by Math Educators for Math Educators

**③ Your students want an assignment page that is easy to use and includes lots of extra help resources.**

## Efficient Assignment Navigation

▶ Students have access to immediate feedback and help while working through assignments.

▶ Students have direct access to a media-rich eBook for easy referencing.

▶ Students can view detailed, step-by-step solutions written by instructors who teach the course, providing a unique solution to each and every exercise.

Students can easily monitor and track their progress on a given assignment.

**④ You want a more intuitive and efficient assignment creation process because of your busy schedule.**

## Assignment Creation Process

▶ Instructors can select textbook-specific questions organized by chapter, section, and objective.

▶ Drag-and-drop functionality makes creating an assignment quick and easy.

▶ Instructors can preview their assignments for efficient editing.

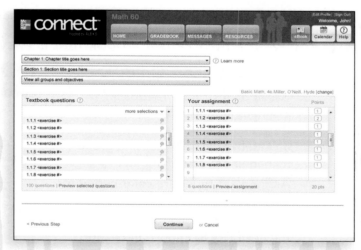

*Connect Learn Succeed*™

# www.connectmath.com

**5** *Your students want an interactive eBook with rich functionality integrated into the product.*

 **McGraw Hill** connect™ (plus+)
|MATH

Hosted by **ALEKS Corp.**

### Integrated Media-Rich eBook

▶ A Web-optimized eBook is seamlessly integrated within ConnectPlus Math Hosted by ALEKS Corp for ease of use.

▶ Students can access videos, images, and other media in context within each chapter or subject area to enhance their learning experience.

▶ Students can highlight, take notes, or even access shared instructor highlights/notes to learn the course material.

▶ The integrated eBook provides students with a cost-saving alternative to traditional textbooks.

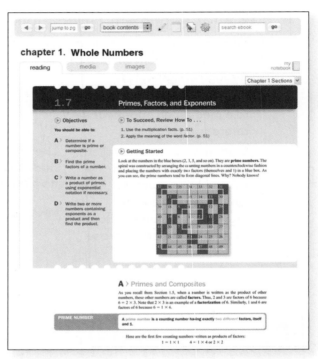

**6** *You want a flexible gradebook that is easy to use.*

### Flexible Instructor Gradebook

▶ Based on instructor feedback, Connect Math Hosted by ALEKS Corp's straightforward design creates an intuitive, visually pleasing grade management environment.

▶ Assignment types are color-coded for easy viewing.

▶ The gradebook allows instructors the flexibility to import and export additional grades.

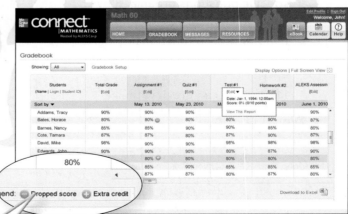

Instructors have the ability to drop grades as well as assign extra credit.

# Built by Math Educators for Math Educators

 **7** You want algorithmic content that was developed by math faculty to ensure the content is pedagogically sound and accurate.

## Digital Content Development Story

The development of McGraw-Hill's Connect Math Hosted by ALEKS Corp. content involved collaboration between McGraw-Hill, experienced instructors, and ALEKS, a company known for its high-quality digital content. The result of this process, outlined below, is accurate content created with your students in mind. It is available in a simple-to-use interface with all the functionality tools needed to manage your course.

1. McGraw-Hill selected experienced instructors to work as Digital Contributors.
2. The Digital Contributors selected the textbook exercises to be included in the algorithmic content to ensure appropriate coverage of the textbook content.
3. The Digital Contributors created detailed, stepped-out solutions for use in the Guided Solution and Show Me features.
4. The Digital Contributors provided detailed instructions for authoring the algorithm specific to each exercise to maintain the original intent and integrity of each unique exercise.
5. Each algorithm was reviewed by the Contributor, went through a detailed quality control process by ALEKS Corporation, and was copyedited prior to being posted live.

## Connect Math Hosted by ALEKS Corp.
### Built by Math Educators for Math Educators

**Lead Digital Contributors**

**Tim Chappell**
*Metropolitan Community College, Penn Valley*

**Jeremy Coffelt**
*Blinn College*

**Nancy Ikeda**
*Fullerton College*

**Amy Naughten**

### Digital Contributors

Al Bluman, *Community College of Allegheny County*

John Coburn, *St. Louis Community College, Florissant Valley*

Vanessa Coffelt, *Blinn College*

Donna Gerken, *Miami-Dade College*

Kimberly Graham

J.D. Herdlick, *St. Louis Community College, Meramec*

Vickie Flanders, *Baton Rouge Community College*

Nic LaHue, *Metropolitan Community College, Penn Valley*

Nicole Lloyd, *Lansing Community College*

Jackie Miller, *The Ohio State University*

Anne Marie Mosher, *St. Louis Community College, Florissant Valley*

Reva Narasimhan, *Kean University*

David Ray, *University of Tennessee, Martin*

Kristin Stoley, *Blinn College*

Stephen Toner, *Victor Valley College*

Paul Vroman, *St. Louis Community College, Florissant Valley*

Michelle Whitmer, *Lansing Community College*

# Build a Better Path to Success with ALEKS®

## Experience Student Success!

**ALEKS** Aleks is a unique online math tool that uses adaptive questioning and artificial intelligence to correctly place, prepare, and remediate students . . . all in one product! Institutional case studies have shown that **ALEKS has improved pass rates by over 20% versus traditional online homework, and by over 30% compared to using a text alone.**

By offering each student an individualized learning path, ALEKS directs students to work on the math topics that they are ready to learn, Also, to help students keep pace in their course, instructors can correlate ALEKS to their textbook or syllabus in seconds.

To learn more about how ALEKS can be used to boost student performance, please visit www.aleks.com/highered/math or contact your McGraw-Hill representative.

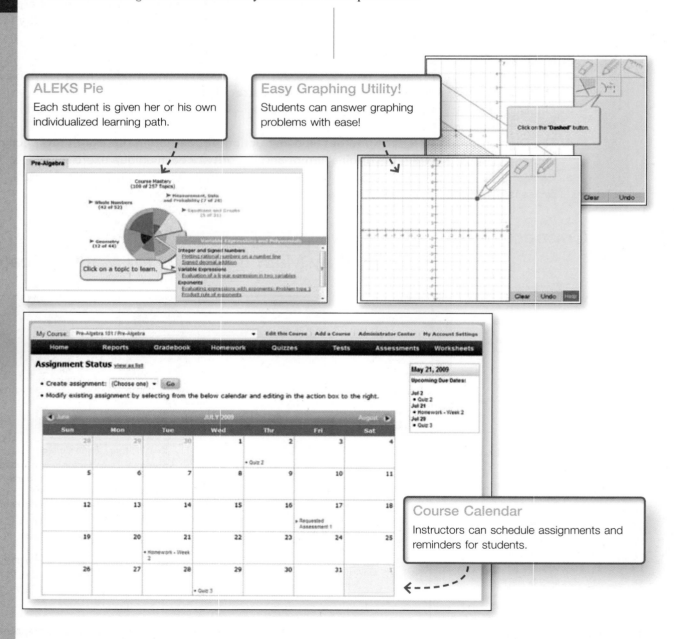

**ALEKS Pie**
Each student is given her or his own individualized learning path.

**Easy Graphing Utility!**
Students can answer graphing problems with ease!

**Course Calendar**
Instructors can schedule assignments and reminders for students.

# New ALEKS Instructor Module

**Enhanced Functionality and Streamlined Interface Help to Save Instructor Time**

**ALEKS** The new ALEKS Instructor Module features enhanced functionality and a streamlined interface based on research with ALEKS instructors and homework management instructors. Paired with powerful assignment-driven features, textbook integration, and extensive content flexibility, the new ALEKS instructor Module simplifies administrative tasks and makes ALEKS more powerful than ever.

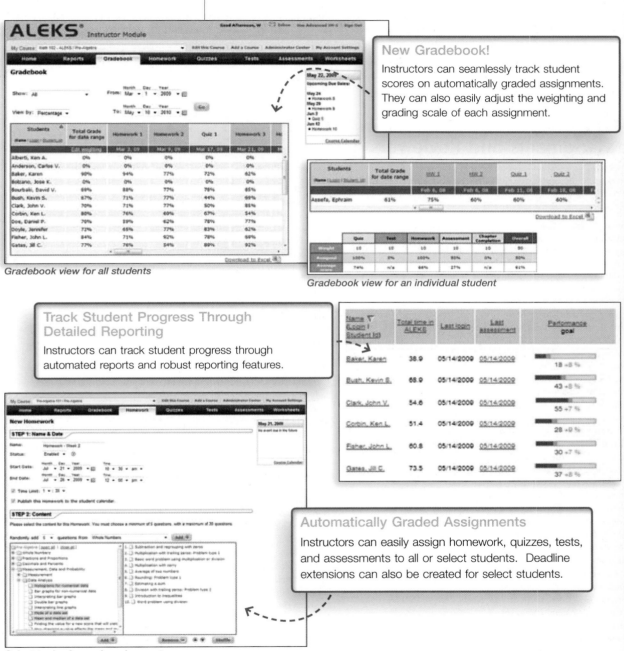

*Gradebook view for all students*

## New Gradebook!

Instructors can seamlessly track student scores on automatically graded assignments. They can also easily adjust the weighting and grading scale of each assignment.

*Gradebook view for an individual student*

## Track Student Progress Through Detailed Reporting

Instructors can track student progress through automated reports and robust reporting features.

## Automatically Graded Assignments

Instructors can easily assign homework, quizzes, tests, and assessments to all or select students. Deadline extensions can also be created for select students.

*Select topics for each assignment*

**Learn more about ALEKS by visiting** www.aleks.com/highered/math **or contact your McGraw-Hill representative.**

# 360° Development Process

McGraw-Hill's 360° Development Process is an ongoing, never ending, market-oriented approach to building accurate and innovative print and digital products. It is dedicated to continual large scale and incremental improvement driven by multiple customer feedback loops and checkpoints. This is initiated during the early planning stages of our new products, and intensifies during the development and production stages, then begins again upon publication, in anticipation of the next edition.

A key principle in the development of any mathematics text is its ability to adapt to teaching specifications in a universal way. The only way to do so is by contacting those universal voices—and learning from their suggestions. We are confident that our book has the most current content the industry has to offer, thus pushing our desire for accuracy to the highest standard possible. In order to accomplish this, we have moved through an arduous road to production. Extensive and open-minded advice is critical in the production of a superior text.

## Acknowledgments and Reviewers

The development of this textbook series would never have been possible without the creative ideas and feedback offered by many reviewers. We are especially thankful to the following instructors for their careful review of the manuscript.

### Manuscript Reviewers

Kent Aeschliman, *Oakland Community College Highland Lakes*
Froozan Afiat, *College of Southern Nevada–Henderson*
Carlos Amaya, *California State University–Los Angeles*
Judy Atkinson, *University of Alaska–Fairbanks*
Rajalakshmi Baradwai, *University of Maryland Baltimore County*
Carlos Barron, *Mountain View College*
Jon Becker, *Indiana University–Northwest–Gary*
Abraham Biggs, *Broward College–South*
Donald Bigwood, *Bismarck State College*
Lee Brendel, *Southwestern Illinois College*
Joan Brown, *Eastern New Mexico University*
Shirley Brown, *Weatherford College*
Debra Bryant, *Tennessee Tech University*
Gail Butler, *Erie Community College North Campus–Williamsville*
Kim Cain, *Miami University–Hamilton*
Ernest Canonigo, *California State University–Los Angeles*
Douglas Carbone, *Central Maine Community College*
Randall Casleton, *Columbus State University*
Jose Castillo, *Broward College–South*
Dwane Christensen, *California State University–Los Angeles*
Lisa Christman, *University of Central Arkansas*
William Clarke, *Pikes Peak Community College*
Delaine Cochran, *Indiana University Southeast*
Wendy Conway, *Oakland Community College Highland Lakes*
Charyl Craddock, *University of Tennessee–Martin*
Greg Cripe, *Spokane Falls Community College*
Joseph De Guzman, *Riverside Community College*
Robert Diaz, *Fullerton College*
Paul Diehl, *Indiana University Southeast*
Deborah Doucette, *Erie Community College North Campus–Williamsville*
Scott Dunn, *Central Michigan University*
Angela Earnhart, *North Idaho College*
Hussain Elalaoui-Talibi, *Tuskegee University*
Joseph Estephan, *California State University–Dominguez Hills*

Tina Evans, *Shelton State Community College*
Angela Everett, *Chattanooga State Tech Community College (West)*
Christopher Farmer, *Southwestern Illinois College*
Steve Felzer, *Lenoir Community College*
Angela Fisher, *Durham Tech Community College*
Marion Foster, *Houston Community College–Southeast College*
Mitzi Fulwood, *Broward College–North*
Scott Garvey, *Suny Agriculture & Tech College–Cobleskille*
Antonnette Gibbs, *Broward College–North*
Sharon Giles, *Grossmont College*
Susan Grody, *Broward College–North*
Kathy Gross, *Cayuga Community College*
Margaret Gruenwald, *University of Southern Indiana*
Kelli Hammer, *Broward College–South*
Pamela Harden, *Tennessee Tech University*
Jody Harris, *Broward College–Central*
Terri Hightower Martin, *Elgin Community College*
Michelle Hollis, *Bowling Green Community College at Western Kentucky University*
Joe Howe, *Saint Charles County Community College*
Barbara Hughes, *San Jacinto College-Pasadena*
Michelle Jackson, *Bowling Green Community College at Western Kentucky University*
Pamela Jackson, *Oakland Community College–Farmington Hills*
Nancy Johnson, *Broward College–North*
Tina Johnson, *Midwestern State University*
Maryann Justinger, *Erie Community College South Campus–Orchard Park*
Cheryl Kane, *University of Nebraska–Lincoln*
Avi Kar, *Abraham Baldwin Agricultural College*
Ryan Kasha, *Valencia Community College–West Campus*
Joe Kemble, *Lamar University–Beaumont*
Pat Kier, *Southwest Texas Junior College–Uvalde*
Heidi Kilthau-Kiley, *Suffolk County Community College*
Jong Kim, *Long Beach City College*
Lynette King, *Gadsden State Community College*
Edward Koslowska, *Southwest Texas Junior College–Uvalde*

Randa Kress, *Idaho State University*
Debra Landre, *San Joaquin Delta Community College*
Cynthia Landrigan, *Erie Community College South Campus–Orchard Park*
Richard Leedy, *Polk Community College*
Janna Liberant, *Rockland Community College*
Shawna Mahan, *Pikes Peak Community College*
Aldo Maldonado, *Park University–Parkville*
Rogers Martin, *Louisiana State University–Shreveport*
Carol Mcavoy, *South Puget Sound Community College*
Peter Mccandless, *Park University–Parkville*
Gary Mccracken, *Shelton State Community College*
Margaret Messinger, *Southwest Texas Junior College–Uvalde*
Kris Mudunuri, *Long Beach City College*
Amy Naughten, *Middle Georgia College*
Elsie Newman, *Owens Community College*
Paulette Nicholson, *South Carolina State University*
Ken Nickels, *Black Hawk College*
Rhoda Oden, *Gadsden State Community College*
Charles Odion, *Houston Community College–Southwest*
Karen Orr, *Roane State Community College*
Keith Pachlhofer, *University of Central Arkansas*
Charles Patterson, *Louisiana Tech University*
Mark Pavitch, *California State University–Los Angeles*
Jean Peterson, *University of Wisconsin–Oshkosh*
Novita Phua, *California State University–Los Angeles*
Mohammed Qazi, *Tuskegee University*
L. Gail Queen, *Shelton State Community College*
William Radulovich, *Florida Community College*
Kumars Ranjbaran, *Mountain View College*
Gary Rattray, *Central Maine Community College*
David Ray, *University of Tennessee–Martin*
Janice Reach, *University of Nebraska at Omaha*
Tracy Romesser, *Erie Community College North Campus–Williamsville*
Steve Rummel, *Heartland Community College*
John Rusnak, *Central Michigan University*
E. Jennell Sargent, *Tennessee State University*
Jane Serbousek, *Noth Virginia Community College–Loudoun Campus*
Brian Shay, *Grossmont College*
Azzam Shihabi, *Long Beach City College*
Mohsen Shirani, *Tennessee State University*
Joy Shurley, *Abraham Baldwin Agricultural College*
Nirmal Sohi, *San Joaquin Delta Community College*
Lee Ann Spahr, *Durham Technical Community College*
Joel Spring, *Broward College–South*
Sean Stewart, *Owens Community College*
David Stumpf, *Lakeland Community College*
Sara Taylor, *Dutchess Community College*
Roland Trevino, *San Antonio College*
Bill Tusang, *Suny Agriculture & Technical College–Cobleskille*
Mildred Vernia, *Indiana University Southeast*
Laura Villarreal, *University of Texas at Brownsville*
James Wan, *Long Beach City College*
Terrence Ward, *Mohawk Valley Community College*
Robert White, *Allan Hancock College*
Darren White, *Kennedy-King College*
Marjorie Whitmore, *Northwest Arkansas Community College*
John Wilkins, *California State University–Dominguez Hills*
Henry Wyzinski, *Indiana University–Northwest-Gary*
Mina Yavari, *Allan Hancock College*
Diane Zych, *Erie Community College North Campus–Williamsville*

## Instructor Focus Groups

Lane Andrew, *Arapahoe Community College*
Chris Bendixen, *Lake Michigan College*
Terry Bordewick, *John Wood Community College*
Jim Bradley, *College of DuPage*
Jan Butler, *Colorado Community Colleges Online*
Robert Cappetta, *College of DuPage*
Margaret Colucci, *College of DuPage*
Anne Conte, *College of DuPage*
Gudryn Doherty, *Community College of Denver*
Eric Egizio, *Joliet Junior College*
Mimi Elwell, *Lake Michigan College*
Vicki Garringer, *College of DuPage*
Margaret Gruenwald, *University of Southern Indiana*
Patricia Hearn, *College of DuPage*
Mary Hill, *College of DuPage*
Maryann Justinger, *Eric Community College–South*
Donna Katula, *Joliet Junior College*
Elizabeth Kiedaisch, *College of DuPage*
Geoffrey Krader, *Morton College*
Riki Kucheck, *Orange Coast College*
James Larson, *Lake Michigan College*
Gail Laurent, *College of DuPage*
Richard Leedy, *Polk State College*
Anthony Lenard, *College of DuPage*
Zia Mahmood, *College of DuPage*
Christopher Mansfield, *Durham Technical Community College*
Terri Martin, *Elgin Community College*
Paul Mccombs, *Rock Valley College*
Kathleen Michalski, *College of DuPage*
Kris Mudunuri, *Long Beach City College*
Michael Neill, *Carl Sandburg College*
Catherine Pellish, *Front Range Community College*
Larry Perez, *Saddleback College*
Christy Peterson, *College of DuPage*
David Platt, *Front Range Community College*
Jack Pripusich, *College of DuPage*
Patrick Quigley, *Saddleback College*
Eleanor Storey, *Front Range Community College*
Greg Wheaton, *Kishwaukee College*
Steve Zuro, *Joliet Junior College*
Carol Schmidt *Lincoln Land Community College*
James Carr *Normandale Community College*
Kay Cornelius *Sinclair Community College*
Thomas Pulver *Waubonsee Community College*
Angie Matthews *Broward Community College*
Sondra Braesek *Broward Community College*
Katerina Vishnyakova *Colin County Community College*
Eileen Dahl *Hennepin Technical College*
Stacy Jurgens *Mesabi Range Community and Technical College*
John Collado *South Suburban College*
Barry Trippett *Sinclair Community College*
Abbas Meigooni *Lincoln Land Community College*
Thomas Sundquist *Normandale Community College*
Diane Krasnewich *Muskegon Community College*
Marshall Dean *El Paso Community College*
Elsa Lopez *El Paso Community College*
Bruce Folmar *El Paso Community College*
Pilar Gimbel *El Paso Community College*
Ivette Chuca *El Paso Community College*
Kaat Higham *Bergen Community College*
Joanne Peeoples *El Paso Community College*

Diana Orrantia *El Paso Community College*
Andrew Stephan *Saint Charles County Community College*
Joe Howe *Saint Charles County Community College*
Wanda Long *Saint Charles County Community College*
Staci Osborn *Cuyahoga Community College*

Kristine Glasener *Cuyahoga Community College*
Derek Hiley *Cuyahoga Community College*
Penny Morries *Polk State College*
Nerissa Felder *Polk State College*

## Digital Contributors

Donna Gerken, Miami–Dade College
Kimberly Graham
Nicole Lloyd, Lansing Community College
Reva Narasimhan, Kean University
Amy Naughten
Michelle Whitmer, Lansing Community College

Additionally, I would like to thank my husband, Phil, and my daughters, Alex and Cailen, for their patience, support, and understanding when things get crazy. A big high five goes out to Sue Xander and Mary Hill for their great friendship and support.

Thank you to all of my colleagues at College of DuPage, especially Betsy Kiedaisch, Christy Peterson, Caroline Soo, and Vicki Garringer for their contributions on supplements. And to the best Associate Dean ever, Jerry Krusinski: your support made it possible for me to teach and write at the same time.

Thanks to Larry Perez for his video work and to David Platt for his work on the Using Technology boxes. Thanks also go out to Kris Mudunuri, Diana Orrantia, Denise Lujan, K.S. Ravindhran, Susan Reiland, Pat Steele, and Lenore Parens for their contributions.

To all of the baristas at my two favorite Starbucks: thanks for having my high-maintenance drink ready before I even get to the register and for letting me sit at the same table for hours on end.

There are so many people to thank at McGraw-Hill: my fellow Bengal Rich Kolasa, Michelle Flomenhoft, Torie Anderson, Emilie Berglund, Emily Williams, Pete Vanaria, and Dawn Bercier. I would also like to thank Stewart Mattson, Marty Lange, and Peggy Selle for everything they have done.

To Bill Mulford, who has been with me from the beginning: thanks for your hard work, ability to multitask and organize everything we do, and for your sense of humor through it all. You are the best.

*Sherri Messersmith*

**Sherri Messersmith**
*College of DuPage*

## Supplements for the Student

### Connect

Connect Math hosted by ALEKS is an exciting, new assignment and assessment ehomework platform. Starting with an easily viewable, intuitive interface, students will be able to access key information, complete homework assignments, and utilize an integrated, media–rich eBook.

***ALEKS Prep for Developmental Mathematics***   ALEKS Prep for Beginning Algebra and Prep for Intermediate Algebra focus on prerequisite and introductory material for Beginning Algebra and Intermediate Algebra. These prep products can be used during the first 3 weeks of a course to prepare students for future success in the course and to increase retention and pass rates. Backed by two decades of National Science Foundation funded research, ALEKS interacts with student much like a human tutor, with the ability to precisely assess a student preparedness and provide instruction on the topics the student is most likely to learn.

ALEKS Prep Course Products Feature:

Artificial intelligence targets gaps in individual student knowledge

Assessment and learning directed toward individual students' needs

Open response environment with realistic input tools
Unlimited online access—PC and Mac compatible Free trial at www.aleks.com/free_trial/instructor

***Student Solution Manual***   The student's solution manual provides comprehensive, worked-out solutions to the odd-numbered exercises in the section exercises, summary exercises, self-test, and the cumulative review. The steps shown in the solutions match the style of solved examples in the textbook.

***Online Videos***   In the online exercise videos, the author, Sherri Messersmith, works through selected exercises using the solution methodology employed in her text. Each video is available online as part of Connect and is indicated by an icon next to a corresponding exercise in the text. Other supplemental videos include eProfessor videos, which are animations based on examples in the book, exercise videos by Larry Perez, and Connect2Developmental Mathematics videos, which use 3D animations and lectures to teach algebra concepts by placing them in a real-world setting. The videos are closed-captioned for the hearing impaired, are subtitled in Spanish, and meet the Americans with Disabilities Act Standards for Accessible Design.

# Supplements for the Instructor

## Connect

Connect Math hosted by ALEKS is an exciting, new assignment and assessment ehomework platform. Instructors can assign an AI-driven assessment from the ALEKS corporation to identify the strengths and weaknesses of the class at the beginning of the term rather than after the first exam. Assignment creation and navigation is efficient and intuitive. The gradebook, based on instructor feedback, has a straightforward design and allows flexibility to import and export additional grades.

***Instructor's Testing and Resource Online***   Provides wealth of resources for the instructor. Among the supplements is a computerized test bank utilizing Brownstone Diploma algorithm-based testing software to create customized exams quickly. This user-friendly program enables instructors to search for questions by topic, format, or difficulty level; to edit existing questions, or to add new ones; and to scramble questions and answer keys for multiple versions of a single test. Hundreds of text-specific, open-ended, and multiple-choice questions are included in the question bank. Sample chapter tests are also provided. CD available upon request.

***Annotated Instructor's Edition***   In the Annotated Instructor's Edition (AIE), answers to exercises and tests appear adjacent to each exercise set, in a color used only for annotations. Also found in the AIE are icons with the practice exercises that serve to guide instructors in their preparation of homework assignments and lessons.

***Instructor's Solution Manual***   The instructor's solution manual provides comprehensive, worked-out solutions to all exercises in the section exercises, summary exercises, self-test, and the cumulative review. The steps shown in the solutions match the style of solved examples in the textbook.

***Worksheets***   The very popular author-created worksheets that accompany the text provide instructors with additional exercises to assist with overcoming potential stumbling blocks in student knowledge and to help students see the connection among multiple mathematical concepts. The worksheets fall into three categories: Worksheets to Strengthen Basic Skills, Worksheets to Help Teach New Concepts, Worksheets to Tie Concepts Together.

***Guided Student Notes***   Guided Student Notes are an amazing resource for instructors to help their students become better note-takers. They are similar to "fill-in-the-blank" notes where certain topics or definitions are prompted and spaces are left for the students to write down what the instructor says or writes on the board. The Guided Student Notes contain in-class examples provided in the margin of the text along with additional examples not found in the book to emphasize the given topic. This allows students to spend less time copying down from the board and more time thinking and learning about the solutions and concepts. The notes are specific to the Messersmith textbook and offer ready-made lesson plans for teachers to either "print and go" or require students to print and bring to class.

***Powerpoints***   These powerpoints will present key concepts and definitions with fully editable slides that follow the textbook. Project in class or post to a website in an online course.

# Table of Contents

 **create**

## the future of custom publishing is here.

Introducing McGraw-Hill Create™ –a new, self-service website that allows you to create custom course materials by drawing upon McGraw-Hill's comprehensive, cross-disciplinary content and other third party resources. Select, then arrange the content in a way that makes the most sense for your course. Even personalize your book with your course information and choose the best format for your students–color print, black-and-white print, or eBook.

- **Build custom print and eBooks easily**

- **Combine material from different sources and even upload your own content**

- **Receive a PDF review copy in minutes**

## begin creating now: www.mcgrawhillcreate.com

Start by registering for a free Create account. If you already have a McGraw-Hill account with one of our other products, you will not need to register on Create. Simply Sign In and enter your username and password to begin using the site.

**Sign In**

Please sign in to access your Create projects. If you already have a McGraw-Hill Account, you can use your existing username and password.

* Email
Your email address here

* Password
•••••••••

Forgot Your Password?

**New to Create?**

Register online to set up a new account. Then you can take full advantage of Create to build custom books tailored to your course.

Register Now

* Required Field

Sign In

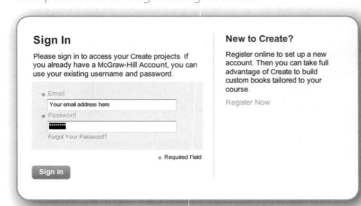

# what you've only imagined.

## edit, share and approve like never before.

After you've completed your project, simply go to the My Projects tab at the top right corner of the page to access and manage all of your McGraw-Hill Create™ projects. Here you will be able to edit your projects and share them with colleagues. An ISBN will be assigned once your review copy has been ordered. Click Approve to make your eBook available for student purchase or print book available for your school or bookstore to order. At any time you can modify your projects and are free to create as many projects as needed.

## receive your pdf review copy in minutes!

Request an eBook review copy and receive a free PDF sample in minutes! Print review copies are also available and arrive in just a few days.

Finally–a way to quickly and easily create the course materials you've always wanted. *Imagine that.*

## questions?

Please visit www.**mcgrawhill**create.com/createhelp for more information.

McGrawHillCreate.com

# Applications Index

# The Real Number System and Geometry

## Algebra at Work: Landscape Architecture

Jill is a landscape architect and uses multiplication, division, and geometry formulas on a daily basis. Here is an example of the type of landscaping she designs. When Jill is asked to create the landscape for a new house, her first job is to draw the plans.

The ground in front of the house will be dug out into shapes that include rectangles and circles, shrubs and flowers will be planted, and mulch will cover the ground. To determine the volume of mulch that will be needed, Jill must use the formulas for the area of a rectangle and a circle and then multiply by the depth of the mulch. She will calculate the total cost of this landscaping job only after determining the cost of the plants, the mulch, and the labor.

It is important that her numbers are accurate. Her company and her clients must have an accurate estimate of the cost of the job. If the estimate is too high, the customer might choose another, less expensive landscaper to do the job. If the estimate is too low, either the client will have to pay more money at the end or the company will not earn as much profit on the job.

In this chapter, we will review formulas from geometry as well as some concepts from arithmetic.

## Section 1.1 Review of Fractions

### Objectives

1. **Understand What a Fraction Represents**
2. **Write Fractions in Lowest Terms**
3. **Multiply and Divide Fractions**
4. **Add and Subtract Fractions**

Why review fractions and arithmetic skills? Because the manipulations done in arithmetic and with fractions are precisely the same skills needed to learn algebra.

Let's begin by defining some numbers used in arithmetic:

Natural numbers: 1, 2, 3, 4, 5, . . .

Whole numbers: 0, 1, 2, 3, 4, 5, . . .

**Natural numbers** are often thought of as the counting numbers. **Whole numbers** consist of the natural numbers and zero.

Natural and whole numbers are used to represent complete quantities. To represent a part of a quantity, we can use a fraction.

### 1. Understand What a Fraction Represents

What is a fraction?

**Definition**

A **fraction** is a number in the form $\dfrac{a}{b}$ where $b \neq 0$, $a$ is called the **numerator**, and $b$ is the **denominator**.

**Note**

1) A fraction describes a part of a whole quantity.

2) $\dfrac{a}{b}$ means $a \div b$.

---

**Example 1**

What part of the figure is shaded?

**Solution**

The whole figure is divided into three equal parts. Two of the parts are shaded. Therefore, the part of the figure that is shaded is $\dfrac{2}{3}$.

$\dfrac{2}{3}$  $\begin{array}{l}\rightarrow \text{ Number of shaded parts}\\ \rightarrow \text{ Total number of equal parts in the figure}\end{array}$

---

**You Try 1**

What part of the figure is shaded?

### 2. Write Fractions in Lowest Terms

A fraction is in **lowest terms** when the numerator and denominator have no common factors except 1. Before discussing how to write a fraction in lowest terms, we need to know about factors.

Consider the number 12.        12  =  3  ·  4

$$\uparrow \qquad\qquad \uparrow \qquad\quad \uparrow$$

Product        Factor        Factor

*3* and *4* are *factors* of 12. (When we use the term **factors**, we mean natural numbers.) Multiplying *3* and *4* results in 12. *12* is the **product**.

Does 12 have any other factors?

 **Example 2**

Find all factors of 12.

**Solution**

$12 = 3 \cdot 4$     Factors are 3 and 4.
$12 = 2 \cdot 6$     Factors are 2 and 6.
$12 = 1 \cdot 12$    Factors are 1 and 12.

These are all of the ways to write *12* as the product of two factors. The factors of 12 are 1, 2, 3, 4, 6, and 12.

 **You Try 2**

Find all factors of 30.

We can also write 12 as a product of *prime numbers.*

**Definition**

A **prime number** is a natural number whose only factors are 1 and itself. (The factors are natural numbers.)

 **Example 3**

Is 7 a prime number?

**Solution**

Yes. The only way to write 7 as a product of natural numbers is $1 \cdot 7$.

**You Try 3**

Is 19 a prime number?

**Definition**

A **composite number** is a natural number with factors other than 1 and itself. Therefore, if a natural number is not prime, it is composite.

**Note**

The number 1 is neither prime nor composite.

**You Try 4**

a)   What are the first six prime numbers?

b)   What are the first six composite numbers?

To perform various operations in arithmetic and algebra, it is helpful to write a number as the product of its **prime factors**. This is called finding the **prime factorization** of a number. We can use a **factor tree** to help us find the prime factorization of a number.

## Example 4

Write 12 as the product of its prime factors.

### Solution

Use a factor tree.

$$12$$
$$/ \ \backslash$$
③ · 4          Think of *any* two natural numbers that multiply to 12.
$$/ \ \backslash$$
② · ②          4 is not prime, so break it down into the product of two factors, 2 × 2.

When a factor is a prime number, circle it, and that part of the factor tree is complete. When all of the numbers at the end of the tree are primes, you have found the *prime factorization* of the number.

Therefore, $12 = 2 \cdot 2 \cdot 3$. Write the prime factorization from the smallest factor to the largest. ■

## Example 5

Write 120 as the product of its prime factors.

### Solution

$$120$$
$$/ \qquad \backslash$$
10  ·  12          Think of *any* two natural numbers that multiply to 120.
$$/ \backslash \ / \backslash$$
② · ⑤ ② · 6          10 and 12 are not prime, so write them as the product of two factors.
$$/ \backslash$$          Circle the primes.
② · ③          6 is not prime, so write it as the product of two factors. The factors are primes. Circle them.

Prime factorization: $120 = 2 \cdot 2 \cdot 2 \cdot 3 \cdot 5$. ■

**You Try 5**

Use a factor tree to write each number as the product of its prime factors.

a)   20          b)   36          c)   90

Let's return to writing a fraction in lowest terms.

**Example 6**

Write each fraction in lowest terms.

a) $\dfrac{4}{6}$     b) $\dfrac{48}{42}$

### Solution

a) $\dfrac{4}{6}$     There are two ways to approach this problem:

**Method 1**
Write 4 and 6 as the product of their primes, and divide out common factors.

$$\frac{4}{6} = \frac{2 \cdot 2}{2 \cdot 3}$$     Write 4 and 6 as the product of their prime factors.

$$= \frac{\overset{1}{\cancel{2}} \cdot 2}{\underset{1}{\cancel{2}} \cdot 3}$$     Divide out common factor.

$$= \frac{2}{3}$$     Since 2 and 3 have no common factors other than 1, the fraction is in lowest terms.

**Method 2**
Divide 4 and 6 by a common factor.

$$\frac{4}{6} = \frac{4 \div 2}{6 \div 2} = \frac{2}{3}$$     $\dfrac{4}{6}$ and $\dfrac{2}{3}$ are **equivalent fractions** since $\dfrac{4}{6}$ simplifies to $\dfrac{2}{3}$.

b) $\dfrac{48}{42}$ is an **improper fraction**. A fraction is *improper* if its numerator is greater than or equal to its denominator. We will use two methods to express this fraction in lowest terms.

**Method 1**
Using a factor tree to get the prime factorizations of 48 and 42 and then dividing out common factors, we have

$$\frac{48}{42} = \frac{\overset{1}{\cancel{2}} \cdot 2 \cdot 2 \cdot 2 \cdot \overset{1}{\cancel{2}}}{\underset{1}{\cancel{2}} \cdot \underset{1}{\cancel{3}} \cdot 7} = \frac{2 \cdot 2 \cdot 2}{7} = \frac{8}{7} \text{ or } 1\frac{1}{7}$$

The answer may be expressed as an improper fraction, $\dfrac{8}{7}$, or as a **mixed number**, $1\dfrac{1}{7}$, as long as each is in lowest terms.

**Method 2**
48 and 42 are each divisible by 6, so we can divide each by 6.

$$\frac{48}{42} = \frac{48 \div 6}{42 \div 6} = \frac{8}{7} \text{ or } 1\frac{1}{7}$$

■

**You Try 6**

Write each fraction in lowest terms.

a) $\dfrac{8}{14}$     b) $\dfrac{63}{36}$

## 3. Multiply and Divide Fractions

### Procedure    Multiplying Fractions

To multiply fractions, $\dfrac{a}{b} \cdot \dfrac{c}{d}$, we multiply the numerators and multiply the denominators. That is,

$$\frac{a}{b} \cdot \frac{c}{d} = \frac{a \cdot c}{b \cdot d} \text{ if } b \neq 0 \text{ and } d \neq 0.$$

### Example 7

Multiply. Write each answer in lowest terms.

a) $\dfrac{3}{8} \cdot \dfrac{7}{4}$     b) $\dfrac{10}{21} \cdot \dfrac{21}{25}$     c) $4\dfrac{2}{5} \cdot 1\dfrac{7}{8}$

**Solution**

a) $\dfrac{3}{8} \cdot \dfrac{7}{4} = \dfrac{3 \cdot 7}{8 \cdot 4}$     Multiply numerators; multiply denominators.

$= \dfrac{21}{32}$     21 and 32 contain no common factors, so $\dfrac{21}{32}$ is in lowest terms.

b) $\dfrac{10}{21} \cdot \dfrac{21}{25}$

If we follow the procedure in the previous example, we get

$$\frac{10}{21} \cdot \frac{21}{25} = \frac{10 \cdot 21}{21 \cdot 25}$$

$$= \frac{210}{525} \qquad \frac{210}{525} \text{ is not in lowest terms.}$$

We must reduce $\dfrac{210}{525}$ to lowest terms:

$\dfrac{210 \div 5}{525 \div 5} = \dfrac{42}{105}$     $\dfrac{42}{105}$ is not in lowest terms. Each number is divisible by 3.

$= \dfrac{42 \div 3}{105 \div 3} = \dfrac{14}{35}$     14 and 35 have a common factor of 7.

$= \dfrac{14 \div 7}{35 \div 7} = \dfrac{2}{5}$     $\dfrac{2}{5}$ is in lowest terms.

Therefore, $\dfrac{10}{21} \cdot \dfrac{21}{25} = \dfrac{2}{5}$.

However, we can take out the common factors before we multiply to avoid all of the reducing in the steps above.

**5** is the greatest common factor of 10 and 25. Divide 10 and 25 by **5**.

$$\frac{\overset{2}{\cancel{10}}}{\underset{1}{\cancel{21}}} \times \frac{\overset{1}{\cancel{21}}}{\underset{5}{\cancel{25}}} = \frac{2}{1} \times \frac{1}{5} = \frac{2 \times 1}{1 \times 5} = \boxed{\frac{2}{5}}$$

**21** is the greatest common factor of 21 and 21. Divide each 21 by **21**.

**Note**

Usually, it is easier to remove the common factors before multiplying rather than after finding the product.

c)  $4\dfrac{2}{5} \cdot 1\dfrac{7}{8}$

Before multiplying mixed numbers, we must change them to improper fractions.

Recall that $4\dfrac{2}{5}$ is the same as $4 + \dfrac{2}{5}$. Here is one way to rewrite $4\dfrac{2}{5}$ as an improper fraction:

1)  Multiply the denominator and the whole number:   $5 \cdot 4 = 20$.

2)  Add the numerator:   $20 + 2 = 22$.

3)  Put the sum over the denominator:   $\dfrac{22}{5}$

To summarize, $4\dfrac{2}{5} = \dfrac{(5 \cdot 4) + 2}{5} = \dfrac{20 + 2}{5} = \dfrac{22}{5}$.

Then, $1\dfrac{7}{8} = \dfrac{(8 \cdot 1) + 7}{8} = \dfrac{8 + 7}{8} = \dfrac{15}{8}$.

$$4\dfrac{2}{5} \cdot 1\dfrac{7}{8} = \dfrac{22}{5} \cdot \dfrac{15}{8}$$

$$= \dfrac{\overset{11}{\cancel{22}}}{\cancel{5}} \cdot \dfrac{\overset{3}{\cancel{15}}}{\cancel{8}}$$    ↙ 5 and 15 each divide by **5**.

$$= \dfrac{11}{1} \cdot \dfrac{3}{4}$$    ↖ 8 and 22 each divide by **2**.

$$= \dfrac{33}{4} \text{ or } 8\dfrac{1}{4}$$    Express the result as an improper fraction or as a mixed number.  ∎

**You Try 7**

Multiply. Write the answer in lowest terms.

a)  $\dfrac{1}{5} \cdot \dfrac{4}{9}$        b)  $\dfrac{8}{25} \cdot \dfrac{15}{32}$        c)  $3\dfrac{3}{4} \cdot 2\dfrac{2}{3}$

### Dividing Fractions

To divide fractions, we must define a reciprocal.

**Definition**

The **reciprocal** of a number, $\dfrac{a}{b}$, is $\dfrac{b}{a}$ since $\dfrac{a}{b} \cdot \dfrac{b}{a} = 1$. That is, a nonzero number times its reciprocal equals 1.

For example, the reciprocal of $\dfrac{5}{9}$ is $\dfrac{9}{5}$ since $\dfrac{\cancel{5}}{\cancel{9}} \cdot \dfrac{\cancel{9}}{\cancel{5}} = \dfrac{1}{1} = 1$.

**Definition**

**Division of fractions:** Let $a, b, c,$ and $d$ represent numbers so that $b, c,$ and $d$ do not equal zero. Then,

$$\frac{a}{b} \div \frac{c}{d} = \frac{a}{b} \cdot \frac{d}{c}.$$

**Note**

To perform division involving fractions, multiply the first fraction by the reciprocal of the second.

---

**Example 8**

Divide. Write the answer in lowest terms.

a) $\dfrac{3}{8} \div \dfrac{10}{11}$     b) $\dfrac{3}{2} \div 9$     c) $5\dfrac{1}{4} \div 1\dfrac{1}{13}$

**Solution**

a) $\dfrac{3}{8} \div \dfrac{10}{11} = \dfrac{3}{8} \cdot \dfrac{11}{10}$     Multiply $\dfrac{3}{8}$ by the reciprocal of $\dfrac{10}{11}$.

$= \dfrac{33}{80}$     Multiply.

b) $\dfrac{3}{2} \div 9 = \dfrac{3}{2} \cdot \dfrac{1}{9}$     The reciprocal of 9 is $\dfrac{1}{9}$.

$= \dfrac{\overset{1}{\cancel{3}}}{2} \cdot \dfrac{1}{\underset{3}{\cancel{9}}}$     Divide out a common factor of 3.

$= \dfrac{1}{6}$     Multiply.

c) $5\dfrac{1}{4} \div 1\dfrac{1}{13} = \dfrac{21}{4} \div \dfrac{14}{13}$     Change the mixed numbers to improper fractions.

$= \dfrac{21}{4} \cdot \dfrac{13}{14}$     Multiply $\dfrac{21}{4}$ by the reciprocal of $\dfrac{14}{13}$.

$= \dfrac{\overset{3}{\cancel{21}}}{4} \cdot \dfrac{13}{\underset{2}{\cancel{14}}}$     Divide out a common factor of 7.

$= \dfrac{39}{8}$ or $4\dfrac{7}{8}$     Express the answer as an improper fraction or mixed number.

---

**You Try 8**

Divide. Write the answer in lowest terms.

a) $\dfrac{2}{7} \div \dfrac{3}{5}$     b) $\dfrac{3}{10} \div \dfrac{9}{16}$     c) $9\dfrac{1}{6} \div 5$

## 4. Add and Subtract Fractions

The pizza on top is cut into eight equal slices. If you eat two pieces and your friend eats three pieces, what fraction of the pizza was eaten?

Five out of the eight pieces were eaten. As a fraction, we can say that you and your friend ate $\frac{5}{8}$ of the pizza.

Let's set up this problem as the sum of two fractions.

Fraction you ate + Fraction your friend ate = Fraction of the pizza eaten

$$\frac{2}{8} \qquad + \qquad \frac{3}{8} \qquad = \qquad \frac{5}{8}$$

To add $\frac{2}{8} + \frac{3}{8}$, we added the numerators and kept the denominator the same. Notice that these fractions have the same denominator.

---

### Definition

Let $a$, $b$, and $c$ be numbers such that $c \neq 0$.

$$\frac{a}{c} + \frac{b}{c} = \frac{a+b}{c} \quad \text{and} \quad \frac{a}{c} - \frac{b}{c} = \frac{a-b}{c}$$

To add or subtract fractions, the denominators must be the same. (This is called a **common denominator**.) Then, add (or subtract) the numerators and keep the same denominator.

---

**Example 9**

Perform the operation and simplify.

a) $\dfrac{3}{11} + \dfrac{5}{11}$ 　　　 b) $\dfrac{17}{30} - \dfrac{13}{30}$

**Solution**

a) $\dfrac{3}{11} + \dfrac{5}{11} = \dfrac{3+5}{11}$ 　　　　Add the numerators and keep the denominator the same.

$= \dfrac{8}{11}$

b) $\dfrac{17}{30} - \dfrac{13}{30} = \dfrac{17-13}{30}$ 　　　　Subtract the numerators and keep the denominator the same.

$= \dfrac{4}{30}$ 　　　　This is not in lowest terms, so reduce.

$= \dfrac{2}{15}$ 　　　　Simplify. ∎

---

**You Try 9**

Perform the operation and simplify.

a) $\dfrac{5}{9} + \dfrac{2}{9}$ 　　　 b) $\dfrac{19}{20} - \dfrac{7}{20}$

When adding or subtracting mixed numbers, either work with them as mixed numbers or change them to improper fractions first.

## Example 10

Add $2\frac{4}{15} + 1\frac{7}{15}$.

### Solution

**Method 1**
To add these numbers while keeping them in mixed number form, add the whole number parts and add the fractional parts.

$$2\frac{4}{15} + 1\frac{7}{15} = (2+1) + \left(\frac{4}{15} + \frac{7}{15}\right) = 3\frac{11}{15}$$

**Method 2**
Change each mixed number to an improper fraction, then add.

$$2\frac{4}{15} + 1\frac{7}{15} = \frac{34}{15} + \frac{22}{15} = \frac{34+22}{15} = \frac{56}{15} \text{ or } 3\frac{11}{15}$$

■

 **You Try 10**

Add $4\frac{3}{7} + 5\frac{1}{7}$.

The examples given so far contain common denominators. How do we add or subtract fractions that do not have common denominators? We find the least common denominator for the fractions and rewrite each fraction with this denominator.

The **least common denominator (LCD)** of two fractions is the least common multiple of the numbers in the denominators.

## Example 11

Find the LCD for $\frac{3}{4}$ and $\frac{1}{6}$.

### Solution

**Method 1**
List some multiples of 4 and 6.

4:   4, 8, $\boxed{12}$, 16, 20, *24*, . . .
6:   6, $\boxed{12}$, 18, *24*, 30, . . .

Although 24 is a multiple of 6 and of 4, the *least* common multiple, and therefore the least common denominator, is 12.

**Method 2**
We can also use the prime factorizations of 4 and 6 to find the LCD.

To find the LCD:

1)   Find the prime factorization of each number.
2)   The least common denominator will include each different factor appearing in the factorizations.
3)   If a factor appears more than once in any prime factorization, use it in the LCD the *maximum number of times* it appears in any single factorization. Multiply the factors.

$$4 = 2 \cdot 2$$
$$6 = 2 \cdot 3$$

The least common multiple of 4 and 6 is

$$\underbrace{2 \cdot 2}_{\substack{\text{2 appears at} \\ \text{most twice in} \\ \text{any single} \\ \text{factorization.}}} \quad \cdot \quad \underbrace{3}_{\substack{\text{3 appears} \\ \text{once in a} \\ \text{factorization.}}} \quad = 12$$

The LCD of $\dfrac{3}{4}$ and $\dfrac{1}{6}$ is 12. ∎

**You Try 11**

Find the LCD for $\dfrac{5}{6}$ and $\dfrac{4}{9}$.

To add or subtract fractions with unlike denominators, begin by identifying the least common denominator. Then, we must rewrite each fraction with this LCD. This will not change the value of the fraction; we will obtain an *equivalent* fraction.

**Example 12**

Rewrite $\dfrac{3}{4}$ with a denominator of 12.

**Solution**

We want to find a fraction that is equivalent to $\dfrac{3}{4}$ so that $\dfrac{3}{4} = \dfrac{?}{12}$.

To obtain the new denominator of 12, the "old" denominator, 4, must be multiplied by 3. But, if the denominator is multiplied by 3, the numerator must be multiplied by 3 as well. When we multiply $\dfrac{3}{4}$ by $\dfrac{3}{3}$, we have multiplied by 1 since $\dfrac{3}{3} = 1$. This is why the fractions are equivalent.

$$\frac{3}{4} \cdot \frac{3}{3} = \frac{9}{12} \qquad \text{So, } \frac{3}{4} = \frac{9}{12}. \qquad ∎$$

> **Procedure**   Adding or Subtracting Fractions with Unlike Denominators
>
> To add or subtract fractions with unlike denominators:
>
> 1) Determine, and write down, the least common denominator (LCD).
>
> 2) Rewrite each fraction with the LCD.
>
> 3) Add or subtract.
>
> 4) Express the answer in lowest terms.

### You Try 12

Rewrite $\dfrac{5}{6}$ with a denominator of 42.

## Example 13

Add or subtract.

a)  $\dfrac{2}{9} + \dfrac{1}{6}$     b)  $6\dfrac{7}{8} - 3\dfrac{1}{2}$

**Solution**

a)  $\dfrac{2}{9} + \dfrac{1}{6}$                 LCD = 18          Identify the least common denominator.

$\dfrac{2}{9} \cdot \dfrac{2}{2} = \dfrac{4}{18}$          $\dfrac{1}{6} \cdot \dfrac{3}{3} = \dfrac{3}{18}$          Rewrite each fraction with a denominator of 18.

$\dfrac{2}{9} + \dfrac{1}{6} = \dfrac{4}{18} + \dfrac{3}{18} = \dfrac{7}{18}$

b)  $6\dfrac{7}{8} - 3\dfrac{1}{2}$

**Method 1**

Keep the numbers in mixed number form. Subtract the whole number parts and subtract the fractional parts. Get a common denominator for the fractional parts.

LCD = 8                              Identify the least common denominator.

$6\dfrac{7}{8}$:   $\dfrac{7}{8}$ has the LCD of 8.

$3\dfrac{1}{2}$:   $\dfrac{1}{2} \cdot \dfrac{4}{4} = \dfrac{4}{8}$.     So, $3\dfrac{1}{2} = 3\dfrac{4}{8}$.     Rewrite $\dfrac{1}{2}$ with a denominator of 8.

$6\dfrac{7}{8} - 3\dfrac{1}{2} = 6\dfrac{7}{8} - 3\dfrac{4}{8}$

$= 3\dfrac{3}{8}$                    Subtract whole number parts and subtract fractional parts.

**Method 2**
Rewrite each mixed number as an improper fraction, get a common denominator, then subtract.

$$6\frac{7}{8} - 3\frac{1}{2} = \frac{55}{8} - \frac{7}{2} \qquad LCD = 8 \qquad \frac{55}{8} \text{ already has a denominator of 8.}$$

$$\frac{7}{2} \cdot \frac{4}{4} = \frac{28}{8} \qquad \text{Rewrite } \frac{7}{2} \text{ with a denominator of 8.}$$

$$6\frac{7}{8} - 3\frac{1}{2} = \frac{55}{8} - \frac{7}{2} = \frac{55}{8} - \frac{28}{8} = \frac{27}{8} \text{ or } 3\frac{3}{8}$$

 **You Try 13**

Perform the operations and simplify.

a) $\dfrac{11}{12} - \dfrac{5}{8}$ 

b) $\dfrac{1}{3} + \dfrac{5}{6} + \dfrac{3}{4}$ 

c) $4\dfrac{2}{5} + 1\dfrac{7}{15}$

---

**Answers to You Try Exercises**

1) $\dfrac{3}{5}$   2) 1, 2, 3, 5, 6, 10, 15, 30   3) yes   4) a) 2, 3, 5, 7, 11, 13   b) 4, 6, 8, 9, 10, 12

5) a) $2 \cdot 2 \cdot 5$   b) $2 \cdot 2 \cdot 3 \cdot 3$   c) $2 \cdot 3 \cdot 3 \cdot 5$   6) a) $\dfrac{4}{7}$   b) $\dfrac{7}{4}$ or $1\dfrac{3}{4}$   7) a) $\dfrac{4}{45}$   b) $\dfrac{3}{20}$   c) 10

8) a) $\dfrac{10}{21}$   b) $\dfrac{8}{15}$   c) $\dfrac{11}{6}$ or $1\dfrac{5}{6}$   9) a) $\dfrac{7}{9}$   b) $\dfrac{3}{5}$   10) $9\dfrac{4}{7}$   11) 18   12) $\dfrac{35}{42}$

13) a) $\dfrac{7}{24}$   b) $\dfrac{23}{12}$ or $1\dfrac{11}{12}$   c) $\dfrac{88}{15}$ or $5\dfrac{13}{15}$

---

## 1.1 Exercises

**Objective 1: Understand What a Fraction Represents**

1) What fraction of each figure is shaded? If the fraction is not in lowest terms, reduce it.

a)

b)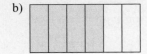

c)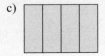

2) What fraction of each figure is *not* shaded? If the fraction is not in lowest terms, reduce it.

a)           b)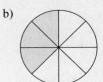

c)

3) Draw a rectangle divided into 8 equal parts. Shade in $\frac{4}{8}$ of the rectangle. Write another fraction to represent how much of the rectangle is shaded.

4) Draw a rectangle divided into 6 equal parts. Shade in $\frac{2}{6}$ of the rectangle. Write another fraction to represent how much of the rectangle is shaded.

## Objective 2: Write Fractions in Lowest Terms

5) Find all factors of each number.

   a) 18

   b) 40

   c) 23

6) Find all factors of each number.

   a) 20

   b) 17

   c) 60

7) Identify each number as prime or composite.

   a) 27

   b) 34

   c) 11

8) Identify each number as prime or composite.

   a) 2

   b) 57

   c) 90

9) Is 3072 prime or composite? Explain your answer.

10) Is 4185 prime or composite? Explain your answer.

11) Use a factor tree to find the prime factorization of each number.

   a) 18          b) 54

   c) 42          d) 150

12) Explain, in words, how to use a factor tree to find the prime factorization of 72.

13) Write each fraction in lowest terms.

   a) $\frac{9}{12}$          b) $\frac{54}{72}$

   c) $\frac{84}{35}$          d) $\frac{120}{280}$

14) Write each fraction in lowest terms.

   a) $\frac{21}{35}$          b) $\frac{48}{80}$

   c) $\frac{125}{500}$          d) $\frac{900}{450}$

## Objective 3: Multiply and Divide Fractions

15) Multiply. Write the answer in lowest terms.

   a) $\frac{2}{7} \cdot \frac{3}{5}$          b) $\frac{15}{26} \cdot \frac{4}{9}$

   c) $\frac{1}{2} \cdot \frac{14}{15}$          d) $\frac{42}{55} \cdot \frac{22}{35}$

   e) $4 \cdot \frac{1}{8}$          f) $6\frac{1}{8} \cdot \frac{2}{7}$

16) Multiply. Write the answer in lowest terms.

   a) $\frac{1}{6} \cdot \frac{5}{9}$          b) $\frac{9}{20} \cdot \frac{6}{7}$

   c) $\frac{12}{25} \cdot \frac{25}{36}$          d) $\frac{30}{49} \cdot \frac{21}{100}$

   e) $\frac{7}{15} \cdot 10$          f) $7\frac{5}{7} \cdot 1\frac{5}{9}$

17) When Elizabeth multiplies $5\frac{1}{2} \cdot 2\frac{1}{3}$, she gets $10\frac{1}{6}$. What was her mistake? What is the correct answer?

18) Explain how to multiply mixed numbers.

19) Divide. Write the answer in lowest terms.

   a) $\frac{1}{42} \div \frac{2}{7}$          b) $\frac{3}{11} \div \frac{4}{5}$

   c) $\frac{18}{35} \div \frac{9}{10}$          d) $\frac{14}{15} \div \frac{2}{15}$

   e) $6\frac{2}{5} \div 1\frac{13}{15}$          f) $\frac{4}{7} \div 8$

20) Explain how to divide mixed numbers.

## Objective 4: Add and Subtract Fractions

21) Find the least common multiple of 10 and 15.

22) Find the least common multiple of 12 and 9.

23) Find the least common denominator for each group of fractions.

   a) $\frac{9}{10}, \frac{11}{30}$          b) $\frac{7}{8}, \frac{5}{12}$

   c) $\frac{4}{9}, \frac{1}{6}, \frac{3}{4}$

24) Find the least common denominator for each group of fractions.

   a) $\frac{3}{14}, \frac{2}{7}$          b) $\frac{17}{25}, \frac{3}{10}$

   c) $\frac{29}{30}, \frac{3}{4}, \frac{9}{20}$

25) Add or subtract. Write the answer in lowest terms.

a) $\dfrac{6}{11} + \dfrac{2}{11}$

b) $\dfrac{19}{20} - \dfrac{7}{20}$

c) $\dfrac{4}{25} + \dfrac{2}{25} + \dfrac{9}{25}$

d) $\dfrac{2}{9} + \dfrac{1}{6}$

e) $\dfrac{3}{5} + \dfrac{11}{30}$

f) $\dfrac{13}{18} - \dfrac{2}{3}$

g) $\dfrac{4}{7} + \dfrac{5}{9}$

VIDEO h) $\dfrac{5}{6} - \dfrac{1}{4}$

i) $\dfrac{3}{10} + \dfrac{7}{20} + \dfrac{3}{4}$

j) $\dfrac{1}{6} + \dfrac{2}{9} + \dfrac{10}{27}$

26) Add or subtract. Write the answer in lowest terms.

a) $\dfrac{8}{9} - \dfrac{5}{9}$

b) $\dfrac{14}{15} - \dfrac{2}{15}$

c) $\dfrac{11}{36} + \dfrac{13}{36}$

d) $\dfrac{16}{45} + \dfrac{8}{45} + \dfrac{11}{45}$

e) $\dfrac{15}{16} - \dfrac{3}{4}$

f) $\dfrac{1}{8} + \dfrac{1}{6}$

g) $\dfrac{5}{8} - \dfrac{2}{9}$

h) $\dfrac{23}{30} - \dfrac{19}{90}$

i) $\dfrac{1}{6} + \dfrac{1}{4} + \dfrac{2}{3}$

j) $\dfrac{3}{10} + \dfrac{2}{5} + \dfrac{4}{15}$

27) Add or subtract. Write the answer in lowest terms.

a) $8\dfrac{5}{11} + 6\dfrac{2}{11}$

b) $2\dfrac{1}{10} + 9\dfrac{3}{10}$

c) $7\dfrac{11}{12} - 1\dfrac{5}{12}$

d) $3\dfrac{1}{5} + 2\dfrac{1}{4}$

e) $5\dfrac{2}{3} - 4\dfrac{4}{15}$

VIDEO f) $9\dfrac{5}{8} - 5\dfrac{3}{10}$

g) $4\dfrac{3}{7} + 6\dfrac{3}{4}$

h) $7\dfrac{13}{20} + \dfrac{4}{5}$

28) Add or subtract. Write the answer in lowest terms.

a) $3\dfrac{2}{7} + 1\dfrac{3}{7}$

b) $8\dfrac{5}{16} + 7\dfrac{3}{16}$

c) $5\dfrac{13}{20} - 3\dfrac{5}{20}$

d) $10\dfrac{8}{9} - 2\dfrac{1}{3}$

e) $1\dfrac{5}{12} + 2\dfrac{3}{8}$

f) $4\dfrac{1}{9} + 7\dfrac{2}{5}$

g) $1\dfrac{5}{6} + 4\dfrac{11}{18}$

h) $3\dfrac{7}{8} + 4\dfrac{2}{5}$

**Mixed Exercises: Objectives 3 and 4**

29) For Valentine's Day, Alex wants to sew teddy bears for her friends. Each bear requires $1\dfrac{2}{3}$ yd of fabric. If she has 7 yd of material, how many bears can Alex make? How much fabric will be left over?

30) A chocolate chip cookie recipe that makes 24 cookies uses $\dfrac{3}{4}$ cup of brown sugar. If Raphael wants to make 48 cookies, how much brown sugar does he need?

31) Nine weeks into the 2009 Major League Baseball season, Nyjer Morgan of the Pittsburgh Pirates had been up to bat 175 times. He got a hit $\dfrac{2}{7}$ of the time. How many hits did Nyjer have?

32) When all children are present, Ms. Yamoto has 30 children in her fifth-grade class. One day during flu season, $\dfrac{3}{5}$ of them were absent. How many children were absent on this day?

33) Mr. Burnett plans to have a picture measuring $18\dfrac{3}{8}$" by $12\dfrac{1}{4}$" custom framed. The frame he chose is $2\dfrac{1}{8}$" wide. What will be the new length and width of the picture plus the frame?

34) Andre is building a table in his workshop. For the legs, he bought wood that is 30 in. long. If the legs are to be $26\dfrac{3}{4}$ in. tall, how many inches must he cut off to get the desired height?

35) When Rosa opens the kitchen cabinet, she finds three partially filled bags of flour. One contains $\frac{2}{3}$ cup, another contains $1\frac{1}{4}$ cups, and the third contains $1\frac{1}{2}$ cups. How much flour does she have all together?

36) Tamika takes the same route to school every day. (See the figure.) How far does she walk to school?

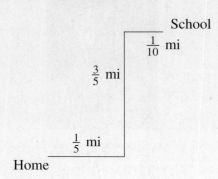

School

$\frac{1}{10}$ mi

$\frac{3}{5}$ mi

$\frac{1}{5}$ mi

Home

37) The gas tank of Jenny's car holds $11\frac{3}{5}$ gal, while Scott's car holds $16\frac{3}{4}$ gal. How much more gasoline does Scott's car hold?

38) Mr. Johnston is building a brick wall along his driveway. He estimates that one row of brick plus mortar will be $4\frac{1}{4}$ in. high. How many rows will he need to construct a wall that is 34 in. high?

39) For homework, Bill's math teacher assigned 42 problems. He finished $\frac{5}{6}$ of them. How many problems did he do?

40) Clarice's parents tell her that she must deposit $\frac{1}{3}$ of the money she earns from babysitting into her savings account, but she can keep the rest. If she earns $117 in 1 week during the summer, how much does she deposit, and how much does she keep?

41) A welder must construct a beam with a total length of $32\frac{7}{8}$ in. If he has already joined a $14\frac{1}{6}$-in. beam with a $10\frac{3}{4}$-in. beam, find the length of a third beam needed to reach the total length.

42) Telephia, a market research company, surveyed 1500 teenage cell phone owners. The company learned that $\frac{2}{3}$ of them use cell phone data services. How many teenagers surveyed use cell phone data services? (*American Demographics,* May 2004, Vol. 26, Issue 4, p. 10)

43) A study conducted in 2000 indicated that about $\frac{3}{5}$ of the full-time college students surveyed had consumed alcohol sometime during the 30 days preceding the survey. If 400 students were surveyed, how many of them drank alcohol within the 30 days before the survey? (*Alcohol Research & Health,* The Journal of the Nat'l Institute on Alcohol Abuse & Alcoholism, Vol. 27, No. 1, 2003)

## Section 1.2 Exponents and Order of Operations

### Objectives
1. Use Exponents
2. Use the Order of Operations

### 1. Use Exponents

In Section 1.1, we discussed the prime factorization of a number. Let's find the prime factorization of 8.

```
      8                    8 = 2 · 2 · 2
     / \
    4 · ②
   / \
  ② · ②
```

We can write 2 · 2 · 2 another way, by using an *exponent*.

$$2 \cdot 2 \cdot 2 = 2^3 \leftarrow \text{exponent (or power)}$$
$$\uparrow$$
$$\text{base}$$

2 is the *base*. 2 is a *factor* that appears three times. 3 is the *exponent* or *power*. An **exponent** represents repeated multiplication. We read $2^3$ as "2 to the third power" or "2 cubed." $2^3$ is called an **exponential expression**.

## Example 1

Rewrite each product in exponential form.

a)   $9 \cdot 9 \cdot 9 \cdot 9$        b)   $7 \cdot 7$

**Solution**

a)   $9 \cdot 9 \cdot 9 \cdot 9 = 9^4$        9 is the base. It appears as a factor 4 times. So, 4 is the exponent.

b)   $7 \cdot 7 = 7^2$        7 is the base. 2 is the exponent.
      This is read as "7 squared."

## You Try 1

Rewrite each product in exponential form.

a)   $8 \cdot 8 \cdot 8 \cdot 8 \cdot 8$        b)   $\dfrac{3}{2} \cdot \dfrac{3}{2} \cdot \dfrac{3}{2} \cdot \dfrac{3}{2}$

We can also evaluate an exponential expression.

## Example 2

Evaluate.

a)   $2^5$        b)   $5^3$        c)   $\left(\dfrac{4}{7}\right)^2$        d)   $8^1$        e)   $1^4$

**Solution**

a)   $2^5 = 2 \cdot 2 \cdot 2 \cdot 2 \cdot 2 = 32$        2 appears as a factor 5 times.

b)   $5^3 = 5 \cdot 5 \cdot 5 = 125$        5 appears as a factor 3 times.

c)   $\left(\dfrac{4}{7}\right)^2 = \dfrac{4}{7} \cdot \dfrac{4}{7} = \dfrac{16}{49}$        $\dfrac{4}{7}$ appears as a factor 2 times.

d)   $8^1 = 8$        8 is a factor only once.

e)   $1^4 = 1 \cdot 1 \cdot 1 \cdot 1 = 1$        1 appears as a factor 4 times.

### Note
1 raised to any natural number power is 1 since 1 multiplied by itself equals 1.

## You Try 2

Evaluate.

a)   $3^4$        b)   $8^2$        c)   $\left(\dfrac{3}{4}\right)^3$

It is generally agreed that there are some skills in arithmetic that everyone should have in order to be able to acquire other math skills. Knowing the basic multiplication facts, for example, is essential for learning how to add, subtract, multiply, and divide fractions as well as how to perform many other operations in arithmetic and algebra. Similarly, memorizing powers of certain bases is necessary for learning how to apply the rules of exponents (Chapter 2) and for working with radicals (Chapter 10). Therefore, the powers listed here must be memorized in order to be successful in the previously mentioned, as well as other, topics. Throughout this book, it is assumed that students know these powers:

| Powers to Memorize | | | | | | |
|---|---|---|---|---|---|---|
| $2^1 = 2$ | $3^1 = 3$ | $4^1 = 4$ | $5^1 = 5$ | $6^1 = 6$ | $8^1 = 8$ | $10^1 = 10$ |
| $2^2 = 4$ | $3^2 = 9$ | $4^2 = 16$ | $5^2 = 25$ | $6^2 = 36$ | $8^2 = 64$ | $10^2 = 100$ |
| $2^3 = 8$ | $3^3 = 27$ | $4^3 = 64$ | $5^3 = 125$ | | | $10^3 = 1000$ |
| $2^4 = 16$ | $3^4 = 81$ | | | | | |
| $2^5 = 32$ | | | | $7^1 = 7$ | $9^1 = 9$ | $11^1 = 11$ |
| $2^6 = 64$ | | | | $7^2 = 49$ | $9^2 = 81$ | $11^2 = 121$ |
| | | | | | | $12^1 = 12$ |
| | | | | | | $12^2 = 144$ |
| | | | | | | $13^1 = 13$ |
| | | | | | | $13^2 = 169$ |

(Hint: Making flashcards might help you learn these facts.)

## 2. Use the Order of Operations

We will begin this topic with a problem for the student:

**You Try 3**

Evaluate $40 - 24 \div 8 + (5 - 3)^2$.

What answer did you get? 41? or 6? or 33? Or, did you get another result?

Most likely you obtained one of the three answers just given. Only one is correct, however. If we do not have rules to guide us in evaluating expressions, it is easy to get the incorrect answer.

Therefore, here are the rules we follow. This is called the **order of operations**.

**Procedure    The Order of Operations**

Simplify expressions in the following order:

1) If parentheses or other grouping symbols appear in an expression, simplify what is in these grouping symbols first.

2) Simplify expressions with exponents.

3) Perform multiplication and division from left to right.

4) Perform addition and subtraction from left to right.

Think about the "You Try" problem. Did you evaluate it using the order of operations? Let's look at that expression:

### Example 3

Evaluate $40 - 24 \div 8 + (5 - 3)^2$.

**Solution**

| | |
|---|---|
| $40 - 24 \div 8 + (5 - 3)^2$ | First, perform the operation in the parentheses. |
| $40 - 24 \div 8 + 2^2$ | Exponents are done before division, addition, and subtraction. |
| $40 - 24 \div 8 + 4$ | Perform division before addition and subtraction. |
| $40 - 3 + 4$ | When an expression contains only addition and subtraction, |
| $37 + 4$ | perform the operations starting at the left and moving to the right. |
| $41$ | |

### You Try 4

Evaluate: $12 \cdot 3 - (2 + 1)^2 \div 9$.

A good way to remember the order of operations is to remember the sentence, "**P**lease **E**xcuse **M**y **D**ear **A**unt **S**ally" (**P**arentheses, **E**xponents, **M**ultiplication and **D**ivision from left to right, **A**ddition and **S**ubtraction from left to right). Don't forget that multiplication and division are at the same "level" in the process of performing operations and that addition and subtraction are at the same "level."

### Example 4

Evaluate.

a)   $9 + 20 - 5 \cdot 3$

b)   $5(8 - 2) + 3^2$

c)   $4[3 + (10 \div 2)] - 11$

d)   $\dfrac{(9 - 6)^3 \cdot 2}{26 - 4 \cdot 5}$

**Solution**

a)
| | |
|---|---|
| $9 + 20 - 5 \cdot 3 = 9 + 20 - 15$ | Perform multiplication before addition and subtraction. |
| $= 29 - 15$ | When an expression contains only addition and subtraction, work from left to right. |
| $= 14$ | Subtract. |

b)
| | |
|---|---|
| $5(8 - 2) + 3^2 = 5(6) + 3^2$ | Parentheses |
| $= 5(6) + 9$ | Exponent |
| $= 30 + 9$ | Multiply. |
| $= 39$ | Add. |

c)   $4[3 + (10 \div 2)] - 11$

This expression contains two sets of grouping symbols: **brackets** [ ] and **parentheses** ( ). Perform the operation in the **innermost** grouping symbol first which is the parentheses in this case.

| | |
|---|---|
| $4[3 + (10 \div 2)] - 11 = 4[3 + 5] - 11$ | Innermost grouping symbol |
| $= 4[8] - 11$ | Brackets |
| $= 32 - 11$ | Perform multiplication before subtraction. |
| $= 21$ | Add. |

d) $\dfrac{(9-6)^3 \cdot 2}{26 - 4 \cdot 5}$

The fraction bar in this expression acts as a grouping symbol. Therefore, simplify the numerator, simplify the denominator, then simplify the resulting fraction, if possible.

$$\frac{(9-6)^3 \cdot 2}{26 - 4 \cdot 5} = \frac{3^3 \cdot 2}{26 - 20} \qquad \text{Parentheses} \atop \text{Multiply.}$$

$$= \frac{27 \cdot 2}{6} \qquad \text{Exponent} \atop \text{Subtract.}$$

$$= \frac{54}{6} \qquad \text{Multiply.}$$

$$= 9$$

**You Try 5**

Evaluate:

a)   $35 - 2 \cdot 6 + 1$

b)   $3 \cdot 12 - (7 - 4)^3 \div 9$

c)   $9 + 2[23 - 4(1 + 2)]$

d)   $\dfrac{11^2 - 7 \cdot 3}{20(9 - 4)}$

**Using Technology**

We can use a graphing calculator to check our answer when we evaluate an expression by hand. The order of operations is built into the calculator. For example, evaluate the expression $\dfrac{2(3 + 7)}{13 - 2 \cdot 4}$.

To evaluate the expression using a graphing calculator, enter the following on the home screen: $(2(3+7))/(13-2*4)$ and then press ENTER. The result is 4, as shown on the screen.

Notice that it is important to enclose the numerator and denominator in parentheses since the fraction bar acts as both a division and a grouping symbol.

Evaluate each expression by hand, and then verify your answer using a graphing calculator.

1)   $45 - 3 \cdot 2 + 7$

2)   $24 \div \dfrac{6}{7} - 5 \cdot 4$

3)   $5 + 2(9 - 6)^2$

4)   $3 + 2[37 - (4 + 1)^2 - 2 \cdot 6]$

5)   $\dfrac{5(7 - 3)}{50 - 3^2 \cdot 4}$

6)   $\dfrac{25 - (1 + 3)^2}{6 + 14 \div 2 - 8}$

**Answers to You Try Exercises**

1) a) $8^5$   b) $\left(\dfrac{3}{2}\right)^4$   2) a) 81   b) 64   c) $\dfrac{27}{64}$   3) 41   4) 35   5) a) 24   b) 33   c) 31   d) 1

**Answers to Technology Exercises**

1) 46   2) 8   3) 23   4) 3   5) $\dfrac{10}{7}$ or $1\dfrac{3}{7}$   6) $\dfrac{9}{5}$ or $1\dfrac{4}{5}$

## 1.2 Exercises

**Objective 1: Use Exponents**

1) Identify the base and the exponent.

   a) $6^4$

   b) $2^3$

   c) $\left(\dfrac{9}{8}\right)^5$

2) Identify the base and the exponent.

   a) $5^1$

   b) $1^8$

   c) $\left(\dfrac{3}{7}\right)^2$

3) Write in exponential form.

   a) $9 \cdot 9 \cdot 9 \cdot 9$

   b) $2 \cdot 2 \cdot 2 \cdot 2 \cdot 2 \cdot 2 \cdot 2 \cdot 2$

   c) $\dfrac{1}{4} \cdot \dfrac{1}{4} \cdot \dfrac{1}{4}$

4) Explain, in words, why $7 \cdot 7 \cdot 7 \cdot 7 \cdot 7 = 7^5$.

5) Evaluate.

   a) $8^2$

   b) $11^2$

   c) $2^4$

   d) $5^3$

   e) $3^4$

   f) $12^2$

   g) $1^2$

   h) $\left(\dfrac{3}{10}\right)^2$

   i) $\left(\dfrac{1}{2}\right)^6$

   j) $(0.3)^2$

6) Evaluate.

   a) $9^2$

   b) $13^2$

   c) $3^3$

   d) $2^5$

   e) $4^3$

   f) $1^4$

   g) $6^2$

   h) $\left(\dfrac{7}{5}\right)^2$

   i) $\left(\dfrac{2}{3}\right)^4$

   j) $(0.02)^2$

7) Evaluate $(0.5)^2$ two different ways.

8) Explain why $1^{200} = 1$.

**Objective 2: Use the Order of Operations**

9) In your own words, summarize the order of operations.

Evaluate.

10) $20 + 12 - 5$

11) $17 - 2 + 4$

12) $51 - 18 + 2 - 11$

13) $48 \div 2 + 14$

14) $15 \cdot 2 - 1$

15) $20 - 3 \cdot 2 + 9$

16) $28 + 21 \div 7 - 4$

17) $8 + 12 \cdot \dfrac{3}{4}$

18) $27 \div \dfrac{9}{5} - 1$

19) $\dfrac{2}{5} \cdot \dfrac{1}{8} + \dfrac{2}{3} \cdot \dfrac{9}{10}$

20) $\dfrac{4}{9} \cdot \dfrac{5}{6} - \dfrac{1}{6} \cdot \dfrac{2}{3}$

21) $2 \cdot \dfrac{3}{4} - \left(\dfrac{2}{3}\right)^2$

22) $\left(\dfrac{3}{2}\right)^2 - \left(\dfrac{5}{4}\right)^2$

23) $25 - 11 \cdot 2 + 1$

24) $2 + 16 + 14 \div 2$

25) $39 - 3(9 - 7)^3$

26) $1 + 2(7 - 1)^2$

27) $60 \div 15 + 5 \cdot 3$

28) $27 \div (10 - 7)^2 + 8 \cdot 3$

29) $7[45 \div (19 - 10)] + 2$

30) $6[3 + (14 - 12)^3] - 10$

31) $1 + 2[(3 + 2)^3 \div (11 - 6)^2]$

32) $(4 + 7)^2 - 3[5(6 + 2) - 4^2]$

33) $\dfrac{4(7 - 2)^2}{12^2 - 8 \cdot 3}$

34) $\dfrac{(8 + 4)^2 - 2^6}{7 \cdot 8 - 6 \cdot 9}$

35) $\dfrac{4(9 - 6)^3}{2^2 + 3 \cdot 8}$

36) $\dfrac{7 + 3(10 - 8)^4}{6 + 10 \div 2 + 11}$

## Section 1.3 Geometry Review

### Objectives

1. **Identify Angles and Parallel and Perpendicular Lines**
2. **Identify Triangles**
3. **Use Area, Perimeter, and Circumference Formulas**
4. **Use Volume Formulas**

Thousands of years ago, the Egyptians collected taxes based on how much land a person owned. They developed measuring techniques to accomplish such a task. Later the Greeks formalized the process of measurements such as this into a branch of mathematics we call geometry. "Geometry" comes from the Greek words for "earth measurement." In this section, we will review some basic geometric concepts that we will need in the study of algebra.

Let's begin by looking at angles. An angle can be measured in **degrees**. For example, 45° is read as "45 degrees."

### 1. Identify Angles and Parallel and Perpendicular Lines

#### Angles

An **acute angle** is an angle whose measure is greater than 0° and less than 90°.

A **right angle** is an angle whose measure is 90°, indicated by the └ symbol.

An **obtuse angle** is an angle whose measure is greater than 90° and less than 180°.

A **straight angle** is an angle whose measure is 180°.

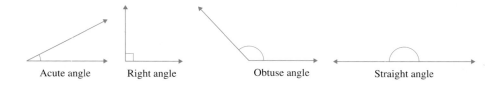

| Acute angle | Right angle | Obtuse angle | Straight angle |

Two angles are **complementary** if their measures add to 90°.

Two angles are **supplementary** if their measures add to 180°.

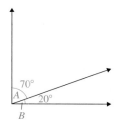

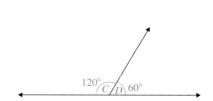

$A$ and $B$ are **complementary angles** since $m\angle A + m\angle B = 70° + 20° = 90°$.

$C$ and $D$ are **supplementary angles** since $m\angle C + m\angle D = 120° + 60° = 180°$.

**Note**

The measure of angle $A$ is denoted by $m\angle A$.

---

**Example 1**

$m\angle A = 41°$. Find its complement.

**Solution**

$$\text{Complement} = 90° - 41° = 49°$$

Since the sum of two complementary angles is 90°, if one angle measures 41°, its complement has a measure of $90° - 41° = 49°$.

**You Try 1**

$m\angle A = 62°$. Find its supplement.

Next, we will explore some relationships between lines and angles.

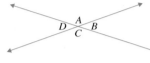

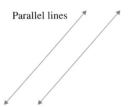

Figure 1.1

### Vertical Angles

When two lines intersect, four angles are formed (see Figure 1.1). The pair of opposite angles are called **vertical angles**. Angles $A$ and $C$ are *vertical angles,* and angles $B$ and $D$ are *vertical angles. The measures of vertical angles are equal.* Therefore, $m\angle A = m\angle C$ and $m\angle B = m\angle D$.

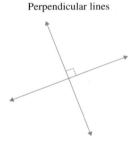

Figure 1.2

### Parallel and Perpendicular Lines

**Parallel lines** are lines in the same plane that do not intersect (Figure 1.2). **Perpendicular lines** are lines that intersect at right angles (Figure 1.3).

### 2. Identify Triangles

We can classify triangles by their angles and by their sides.

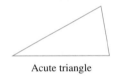   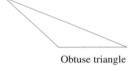   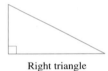

  Acute triangle                Obtuse triangle              Right triangle

An **acute triangle** is one in which all three angles are acute.

An **obtuse triangle** contains one obtuse angle.

A **right triangle** contains one right angle.

Figure 1.3

---

**Property**

The sum of the measures of the angles of any triangle is 180°.

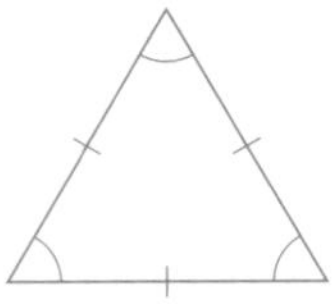

---

  Equilateral triangle 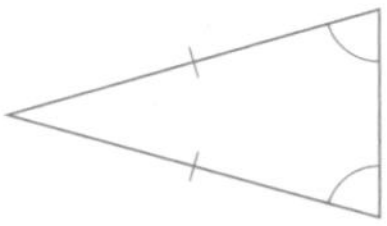          Isosceles triangle 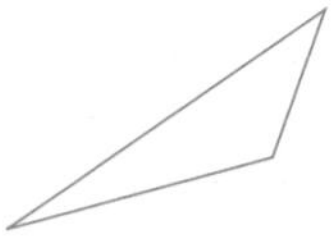            Scalene triangle

If a triangle has three sides of equal length, it is an **equilateral triangle**. (Each angle measure of an equilateral triangle is 60°.)

If a triangle has two sides of equal length, it is an **isosceles triangle**. (The angles opposite the equal sides have the same measure.)

If a triangle has no sides of equal length, it is a **scalene triangle**. (No angles have the same measure.)

**Example 2**

Find the measures of angles $A$ and $B$ in this isosceles triangle.

**Solution**

The single hash marks on the two sides of the triangle mean that those sides are of equal length.

$$m\angle B = 39° \qquad \text{Angle measures opposite sides of equal length are the same.}$$
$$39° + m\angle B = 39° + 39° = 78°.$$

We have found that the sum of two of the angles is 78°. Since all of the angle measures add up to 180°,

$$m\angle A = 180° - 78° = 102°$$

 **You Try 2**

Find the measures of angles $A$ and $B$ in this isosceles triangle.

### 3. Use Area, Perimeter, and Circumference Formulas

The **perimeter** of a figure is the distance around the figure, while the **area** of a figure is the number of square units enclosed within the figure. For some familiar shapes, we have the following formulas:

| Figure | | Perimeter | Area |
|---|---|---|---|
| Rectangle: | | $P = 2l + 2w$ | $A = lw$ |
| Square: | | $P = 4s$ | $A = s^2$ |
| Triangle: $h$ = height | | $P = a + b + c$ | $A = \dfrac{1}{2}bh$ |
| Parallelogram: $h$ = height | | $P = 2a + 2b$ | $A = bh$ |
| Trapezoid: $h$ = height | | $P = a + c + b_1 + b_2$ | $A = \dfrac{1}{2}h(b_1 + b_2)$ |

The perimeter of a circle is called the **circumference**. The **radius**, $r$, is the distance from the center of the circle to a point on the circle. A line segment that passes through the center of the circle and has its endpoints on the circle is called a **diameter**.

Pi, $\pi$, is the ratio of the circumference of any circle to its diameter. $\pi \approx 3.14159265 \ldots$, but we will use 3.14 as an approximation for $\pi$. The symbol $\approx$ is read as "approximately equal to."

**Circumference**  **Area**

$C = 2\pi r$    $A = \pi r^2$

## Example 3

Find the perimeter and area of each figure.

a)

7 in.

9 in.

b)

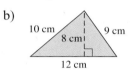

10 cm   9 cm

8 cm

12 cm

### Solution

a)  This figure is a rectangle.

Perimeter: $P = 2l + 2w$
$P = 2(9 \text{ in.}) + 2(7 \text{ in.})$
$P = 18 \text{ in.} + 14 \text{ in.}$
$P = 32 \text{ in.}$

Area: $A = lw$
$A = (9 \text{ in.})(7 \text{ in.})$
$A = 63 \text{ in}^2$ or 63 square inches

b)  This figure is a triangle.

Perimeter: $P = a + b + c$
$P = 9 \text{ cm} + 12 \text{ cm} + 10 \text{ cm}$
$P = 31 \text{ cm}$

Area: $A = \dfrac{1}{2} bh$

$A = \dfrac{1}{2}(12 \text{ cm})(8 \text{ cm})$

$A = 48 \text{ cm}^2$ or 48 square centimeters

## You Try 3

**Find the perimeter and area of the figure.**

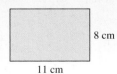

8 cm

11 cm

**Example 4**

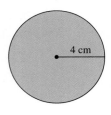

Find (a) the circumference and (b) the area of the circle shown at left. Give an exact answer for each and give an approximation using 3.14 for $\pi$.

*4 cm*

### Solution

a)  The formula for the circumference of a circle is $C = 2\pi r$. The radius of the given circle is 4 cm. Replace $r$ with 4 cm.

$$C = 2\pi r$$
$$= 2\pi(4 \text{ cm}) \qquad \text{Replace } r \text{ with 4 cm.}$$
$$= 8\pi \text{ cm} \qquad \text{Multiply.}$$

Leaving the answer in terms of $\pi$ gives us the exact circumference of the circle, $8\pi$ cm.
To find an approximation for the circumference, substitute 3.14 for $\pi$ and simplify.

$$C = 8\pi \text{ cm}$$
$$\approx 8(3.14) \text{ cm} = 25.12 \text{ cm}$$

b)  The formula for the area of a circle is $A = \pi r^2$. Replace $r$ with 4 cm.

$$A = \pi r^2$$
$$= \pi(4 \text{ cm})^2 \qquad \text{Replace } r \text{ with 4 cm.}$$
$$= 16\pi \text{ cm}^2 \qquad 4^2 = 16$$

Leaving the answer in terms of $\pi$ gives us the exact area of the circle, $16\pi$ cm$^2$.
To find an approximation for the area, substitute 3.14 for $\pi$ and simplify.

$$A = 16\pi \text{ cm}^2$$
$$\approx 16(3.14) \text{ cm}^2$$
$$= 50.24 \text{ cm}^2$$

■

 **You Try 4**

Find (a) the circumference and (b) the area of the circle. Give an exact answer for each and give an approximation using 3.14 for $\pi$.

*5 in.*

A **polygon** is a closed figure consisting of three or more line segments. (See the figure.) We can extend our knowledge of perimeter and area to determine the area and perimeter of a polygon.

Polygons:

**Example 5**

Find the perimeter and area of the figure shown here.

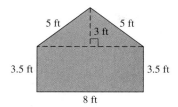

**Solution**

*Perimeter*: The perimeter is the distance around the figure.

$$P = 5 \text{ ft} + 5 \text{ ft} + 3.5 \text{ ft} + 8 \text{ ft} + 3.5 \text{ ft}$$
$$P = 25 \text{ ft}$$

*Area*: To find the area of this figure, think of it as two regions: a triangle and a rectangle.

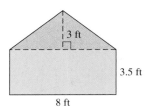

Total area = Area of triangle + Area of rectangle

$$= \frac{1}{2} bh + lw$$

$$= \frac{1}{2} (8 \text{ ft})(3 \text{ ft}) + (8 \text{ ft})(3.5 \text{ ft})$$

$$= 12 \text{ ft}^2 + 28 \text{ ft}^2$$

$$= 40 \text{ ft}^2$$

■

 **You Try 5**

Find the perimeter and area of the figure.

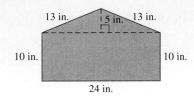

## 4. Use Volume Formulas

The **volume** of a three-dimensional object is the amount of space occupied by the object. Volume is measured in cubic units such as cubic inches (in$^3$), cubic centimeters (cm$^3$), cubic feet (ft$^3$), and so on. Volume also describes the amount of a substance that can be enclosed within a three-dimensional object. Therefore, volume can also be measured in quarts, liters, gallons, and so on. In the figures, $l$ = length, $w$ = width, $h$ = height, $s$ = length of a side, and $r$ = radius.

## Volumes of Three-Dimensional Figures

| | | |
|---|---|---|
| Rectangular solid | | $V = lwh$ |
| Cube | | $V = s^3$ |
| Right circular cylinder | | $V = \pi r^2 h$ |
| Sphere | | $V = \dfrac{4}{3}\pi r^3$ |
| Right circular cone | | $V = \dfrac{1}{3}\pi r^2 h$ |

**Example 6**

Find the volume of each. In (b) give the answer in terms of $\pi$.

a)

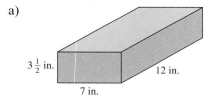

$3\frac{1}{2}$ in.    7 in.    12 in.

b)

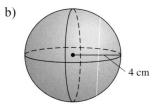

4 cm

### Solution

a)  $V = lwh$      Volume of a rectangular solid

$\quad = (12 \text{ in.})(7 \text{ in.})\left(3\dfrac{1}{2} \text{ in.}\right)$      Substitute values.

$\quad = (12 \text{ in.})(7 \text{ in.})\left(\dfrac{7}{2} \text{ in.}\right)$      Change to an improper fraction.

$\quad = \left(84 \cdot \dfrac{7}{2}\right) \text{in}^3$      Multiply.

$\quad = 294 \text{ in}^3 \text{ or } 294 \text{ cubic inches}$

b)   $V = \dfrac{4}{3}\pi r^3$        Volume of a sphere

$= \dfrac{4}{3}\pi(4\text{ cm})^3$        Replace $r$ with 4 cm.

$= \dfrac{4}{3}\pi(64\text{ cm}^3)$        $4^3 = 64$

$= \dfrac{256}{3}\pi\text{ cm}^3$        Multiply.

**You Try 6**

Find the volume of each figure. In (b) give the answer in terms of $\pi$.

a)   A box with length = 3 ft, width = 2 ft, and height = 1.5 ft

b)   A sphere with radius = 3 in.

**Example 7**

***Application***

A large truck has a fuel tank in the shape of a right circular cylinder. Its radius is 1 ft, and it is 4 ft long.

a)   How many cubic feet of diesel fuel will the tank hold? (Use 3.14 for $\pi$.)

b)   How many gallons will it hold? Round to the nearest gallon. (1 ft$^3$ ≈ 7.48 gal)

c)   If diesel fuel costs \$1.75 per gallon, how much will it cost to fill the tank?

***Solution***

a)   We're asked to determine how much fuel the tank will hold. We must find the *volume* of the tank.

$$\text{Volume of a cylinder} = \pi r^2 h$$
$$\approx (3.14)(1\text{ ft})^2(4\text{ ft})$$
$$= 12.56\text{ ft}^3$$

The tank will hold 12.56 ft$^3$ of diesel fuel.

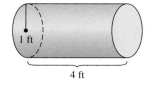

1 ft

4 ft

b)   We must convert 12.56 ft$^3$ to gallons. Since 1 ft$^3$ ≈ 7.48 gal, we can change units by multiplying:

$$12.56\ \cancel{\text{ft}^3} \cdot \left(\dfrac{7.48\text{ gal}}{1\ \cancel{\text{ft}^3}}\right) = 93.9488\text{ gal}$$

$$\approx 94\text{ gal}$$

We can divide out units in fractions the same way we can divide out common factors.

The tank will hold approximately 94 gal.

c)   Diesel fuel costs \$1.75 per gallon. We can figure out the total cost of the fuel the same way we did in (b).

$1.75 *per* gallon

↓

$$94\ \cancel{\text{gal}} \cdot \left(\dfrac{\$1.75}{\cancel{\text{gal}}}\right) = \$164.50$$        Divide out the units of gallons.

It will cost about \$164.50 to fill the tank.

**You Try 7**

A large truck has a fuel tank in the shape of a right circular cylinder. Its radius is 1 ft, and it is 3 ft long.

a) How many cubic feet of diesel fuel will the tank hold? (Use 3.14 for $\pi$).

b) How many gallons of fuel will it hold? Round to the nearest gallon. ($1 \text{ ft}^3 \approx 7.48$ gal)

c) If diesel fuel costs $1.75 per gallon, how much will it cost to fill the tank?

---

### Answers to You Try Exercises

1) $118°$    2) $m\angle A = 130°$; $m\angle B = 25°$    3) $P = 38$ cm; $A = 88 \text{ cm}^2$    4) a) $C = 10\pi$ in.;
$C \approx 31.4$ in.    b) $A = 25\pi \text{ in}^2$; $A \approx 78.5 \text{ in}^2$    5) $P = 70$ in.; $A = 300 \text{ in}^2$    6) a) $9 \text{ ft}^3$    b) $36\pi \text{ in}^3$
7) a) $9.42 \text{ ft}^3$    b) 70 gal    c) $122.50

---

# 1.3 Exercises

**Objective 1: Identify Angles and Parallel and Perpendicular Lines**

1) An angle whose measure is between $0°$ and $90°$ is a(n) _____ angle.

2) An angle whose measure is $90°$ is a(n) _____ angle.

3) An angle whose measure is $180°$ is a(n) _____ angle.

4) An angle whose measure is between $90°$ and $180°$ is a(n) _____ angle.

5) If the sum of two angles is $180°$, the angles are _____.
If the sum of two angles is $90°$, the angles are _____.

6) If two angles are supplementary, can both of them be obtuse? Explain.

Find the complement of each angle.

7) $59°$                          8) $84°$

9) $12°$                          10) $40°$

Find the supplement of each angle.

11) $143°$                        12) $62°$

13) $38°$                         14) $155°$

Find the measure of the missing angles.

15)

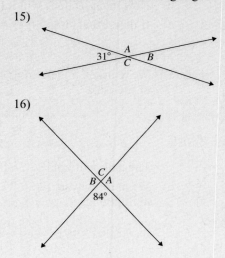

16)

**Objective 2: Identify Triangles**

17) The sum of the angles in a triangle is _____ degrees.

Find the missing angle and classify each triangle as acute, obtuse, or right.

18)

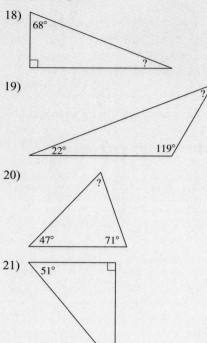

19)

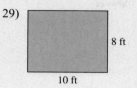

20)

21)

22) Can a triangle contain more than one obtuse angle? Explain.

Classify each triangle as equilateral, isosceles, or scalene.

23)

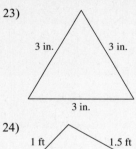

24)

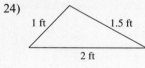

25)

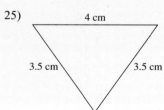

26) What can you say about the measures of the angles in an equilateral triangle?

27) True or False: A right triangle can also be isosceles.

28) True or False: If a triangle has two sides of equal length, then the angles opposite these sides are equal.

**Objective 3: Use Area, Perimeter, and Circumference Formulas**

Find the area and perimeter of each figure. Include the correct units.

29)

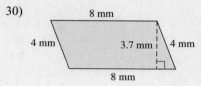

30)

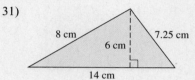

31)

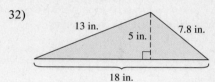

32)

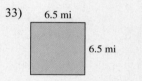

33)

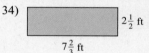

34)

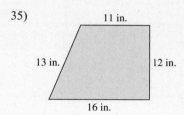

35)

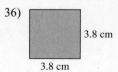

36)

For 37–40, find (a) the area and (b) the circumference of the circle. Give an exact answer for each and give an approximation using 3.14 for $\pi$. Include the correct units.

37) 38)

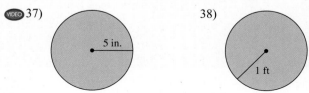

39)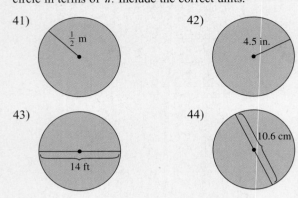

2.5 m

40)

7 cm

For 41–44, find the exact area and circumference of the circle in terms of $\pi$. Include the correct units.

41)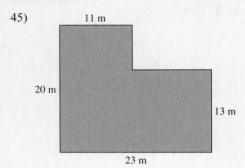

$\frac{1}{2}$ m

42)

4.5 in.

43)

14 ft

44)

10.6 cm

Find the area and perimeter of each figure. Include the correct units.

45)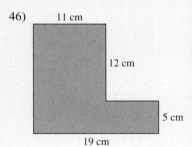

11 m

20 m

13 m

23 m

46)

11 cm

12 cm

5 cm

19 cm

47)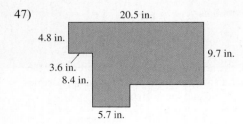

20.5 in.

4.8 in.

9.7 in.

3.6 in.

8.4 in.

5.7 in.

48)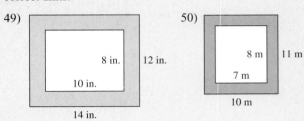

7 ft

5 ft

4 ft

6 ft

5 ft

7 ft

Find the area of the shaded region. Use 3.14 for $\pi$. Include the correct units.

49)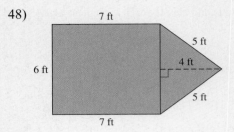

8 in.    12 in.

10 in.

14 in.

50)

8 m    11 m

7 m

10 m

51)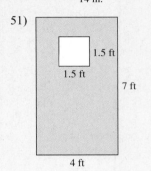

1.5 ft

1.5 ft    7 ft

4 ft

52)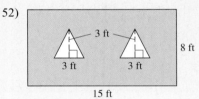

3 ft

8 ft

3 ft    3 ft

15 ft

53)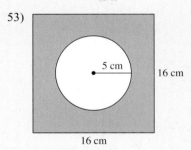

5 cm    16 cm

16 cm

54)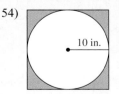

10 in.

## Objective 4: Use Volume Formulas

Find the volume of each figure. Where appropriate, give the answer in terms of $\pi$. Include the correct units.

**VIDEO** 55)

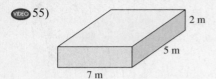

56)

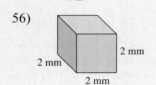

57)

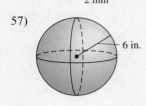

58)

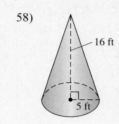

59)

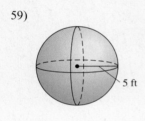

60)

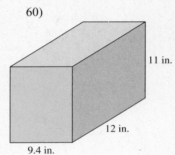

61)

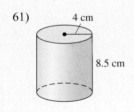

62)

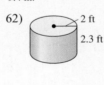

## Mixed Exercises: Objectives 3 and 4

*Applications of Perimeter, Area, and Volume:* Use 3.14 for $\pi$ and include the correct units.

63) To lower her energy costs, Yun would like to replace her rectangular storefront window with low-emissivity (low-e) glass that costs $20.00/ft². The window measures 9 ft by 6.5 ft, and she can spend at most $900.

   a) How much glass does she need?

   b) Can she afford the low-e glass for this window?

64) An insulated rectangular cooler is 15" long, 10" wide, and 13.6" high. What is the volume of the cooler?

65) A fermentation tank at a winery is in the shape of a right circular cylinder. The diameter of the tank is 6 ft, and it is 8 ft tall.

   a) How many cubic feet of wine will the tank hold?

   b) How many gallons of wine will the tank hold? Round to the nearest gallon. (1 ft³ ≈ 7.48 gallons)

66) Yessenia wants a custom-made area rug measuring 5 ft by 8 ft. She has budgeted $500. She likes the Alhambra carpet sample that costs $9.80/ft² and the Sahara pattern that costs $12.20/ft². Can she afford either of these fabrics to make the area rug, or does she have to choose the cheaper one to remain within her budget? Support your answer by determining how much it would cost to have the rug made in each pattern.

67) The lazy Susan on a table in a Chinese restaurant has a 10-inch radius. (A lazy Susan is a rotating tray used to serve food.)

   a) What is the perimeter of the lazy Susan?

   b) What is its area?

68) Find the perimeter of home plate given the dimensions below.

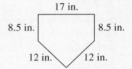

69) A rectangular reflecting pool is 30 ft long, 19 ft wide and 1.5 ft deep. How many gallons of water will this pool hold? (1 ft³ ≈ 7.48 gallons)

70) Ralph wants to childproof his house now that his daughter has learned to walk. The round, glass-top coffee table in his living room has a diameter of 36 inches. How much soft padding does Ralph need to cover the edges around the table?

71) Nadia is remodeling her kitchen and her favorite granite countertop costs $80.00/ft², including installation. The layout of the countertop is shown below, where the counter has a uniform width of $2\frac{1}{4}$ ft. If she can spend at most $2500.00, can she afford her first-choice granite?

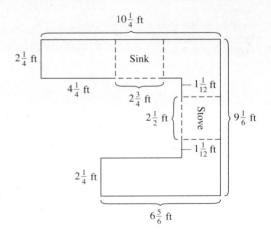

72) A container of lip balm is in the shape of a right circular cylinder with a radius of 1.5 cm and a height of 2 cm. How much lip balm will the container hold?

73) The radius of a women's basketball is approximately 4.6 in. Find its circumference to the nearest tenth of an inch.

74) The chamber of a rectangular laboratory water bath measures $6'' \times 11\frac{3}{4}'' \times 5\frac{1}{2}''$.

a) How many cubic inches of water will the water bath hold?

b) How many liters of water will the water bath hold? ($1 \text{ in}^3 \approx 0.016$ liter)

75) A town's public works department will install a flower garden in the shape of a trapezoid. It will be enclosed by decorative fencing that costs $23.50/ft.

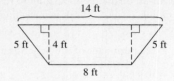

a) Find the area of the garden.

b) Find the cost of the fence.

76) Jaden is making decorations for the bulletin board in his fifth-grade classroom. Each equilateral triangle has a height of 15.6 inches and sides of length 18 inches.

a) Find the area of each triangle.

b) Find the perimeter of each triangle.

77) The top of a counter-height pub table is in the shape of an equilateral triangle. Each side has a length of 18 inches, and the height of the triangle is 15.6 inches. What is the area of the table top?

78) The dimensions of Riyad's home office are $10' \times 12'$. He plans to install laminated hardwood flooring that costs $2.69/ft². How much will the flooring cost?

79) Salt used to melt road ice in winter is piled in the shape of a right circular cone. The radius of the base is 12 ft, and the pile is 8 ft high. Find the volume of salt in the pile.

80) Find the volume of the ice cream pictured below. Assume that the right circular cone is completely filled and that the scoop on top is half of a sphere.

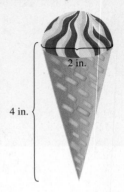

---

## Section 1.4 Sets of Numbers and Absolute Value

### Objectives

1. Identify and Graph Numbers on a Number Line
2. Compare Numbers Using Inequality Symbols
3. Find the Additive Inverse and Absolute Value of a Number

### 1. Identify and Graph Numbers on a Number Line

In Section 1.1, we defined the following sets of numbers:

Natural numbers: $\{1, 2, 3, 4, \ldots\}$

Whole numbers: $\{0, 1, 2, 3, 4, \ldots\}$

We will begin this section by discussing other sets of numbers.

On a **number line**, positive numbers are to the right of zero and negative numbers are to the left of zero.

> **Definition**
>
> The set of **integers** includes the set of natural numbers, their negatives, and zero. The set of *integers* is $\{\ldots, -3, -2, -1, 0, 1, 2, 3, \ldots\}$.

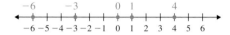

**Example 1**

Graph each number on a number line.

$$4, 1, -6, 0, -3$$

**Solution**

4 and 1 are to the right of zero since they are positive.

$-3$ is three units to the left of zero, and $-6$ is six units to the left of zero. ∎

**You Try 1**

Graph each number on a number line. 2, $-4$, 5, $-1$, $-2$

Positive and negative numbers are also called **signed numbers**.

**Example 2**

Given the set of numbers $\left\{ 4, -7, 0, \dfrac{3}{4}, -6, 10, -3 \right\}$, list the

a)  whole numbers        b)  natural numbers        c)  integers

**Solution**

a)  whole numbers: 0, 4, 10

b)  natural numbers: 4, 10

c)  integers: $-7, -6, -3, 0, 4, 10$ ∎

**You Try 2**

Given the set of numbers $\left\{ -1, 5, \dfrac{2}{7}, 8, -\dfrac{4}{5}, 0, -12 \right\}$, list the

a)  whole numbers        b)  natural numbers        c)  integers

Notice in Example 2 that $\dfrac{3}{4}$ did not belong to any of these sets. That is because the whole numbers, natural numbers, and integers do not contain any fractional parts. $\dfrac{3}{4}$ is a *rational number*.

---

**Definition**

A **rational number** is any number of the form $\dfrac{p}{q}$, where $p$ and $q$ are integers and $q \neq 0$.

That is, a rational number is any number that can be written as a fraction where the numerator and denominator are integers and the denominator does not equal zero.

---

Rational numbers include much more than numbers like $\dfrac{3}{4}$, which are already in fractional form.

## Example 3

Explain why each of the following numbers is rational.

a)  7          b)  0.8          c)  −5

d)  $6\dfrac{1}{4}$          e)  $0.\overline{3}$          f)  $\sqrt{4}$

### Solution

| Rational Number | Reason |
|---|---|
| 7 | 7 can be written as $\dfrac{7}{1}$. |
| 0.8 | 0.8 can be written as $\dfrac{8}{10}$. |
| −5 | −5 can be written as $\dfrac{-5}{1}$. |
| $6\dfrac{1}{4}$ | $6\dfrac{1}{4}$ can be written as $\dfrac{25}{4}$. |
| $0.\overline{3}$ | $0.\overline{3}$ can be written as $\dfrac{1}{3}$. |
| $\sqrt{4}$ | $\sqrt{4} = 2$ and $2 = \dfrac{2}{1}$. |

$\sqrt{4}$ is read as "the square root of 4." This means, "What number times itself equals 4?" That number is 2.  ■

## You Try 3

Explain why each of the following numbers is rational.

a)  12          b)  0.7          c)  −8          d)  $2\dfrac{3}{4}$          e)  $0.\overline{6}$          f)  $\sqrt{100}$

To summarize, the set of rational numbers includes

1)  Integers, whole numbers, and natural numbers.

2)  Repeating decimals.

3)  Terminating decimals.

4)  Fractions and mixed numbers.

The set of rational numbers does *not* include nonrepeating, nonterminating decimals. These decimals cannot be written as the quotient of two integers. Numbers such as these are called *irrational numbers*.

### Definition

The set of numbers that cannot be written as the quotient of two integers is called the set of **irrational numbers**. Written in decimal form, an *irrational number* is a nonrepeating, nonterminating decimal.

**Example 4**

Explain why each of the following numbers is irrational.

a)  0.8271316…          b)  $\pi$          c)  $\sqrt{3}$

### Solution

| Irrational Number | Reason |
|---|---|
| 0.827136… | It is a nonrepeating, nonterminating decimal. |
| $\pi$ | $\pi \approx 3.14159265\ldots$ It is a nonrepeating, nonterminating decimal. |
| $\sqrt{3}$ | 3 is not a perfect square, and the decimal equivalent of the square root of a nonperfect square is a nonrepeating, nonterminating decimal. Here, $\sqrt{3} \approx 1.73205\ldots$. |

**You Try 4**

Explain why each of the following numbers is irrational.

a)  2.41895…          b)  $\sqrt{2}$

If we put together the sets of numbers we have discussed up to this point, we get the *real numbers.*

### Definition

The set of **real numbers** consists of the rational and irrational numbers.

We summarize the information next with examples of the different sets of numbers:

**Real Numbers**

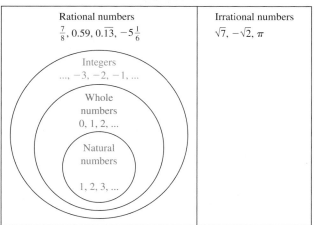

From the figure we can see, for example, that all whole numbers $\{0, 1, 2, 3, \ldots\}$ are integers, but not all integers are whole numbers ($-3$, for example).

## Example 5

Given the set of numbers $\left\{ -16, 3.82, 0, 29, 0.\overline{7}, -\dfrac{11}{12}, \sqrt{10}, 5.302981\ldots \right\}$, list the

a)  integers            b)  natural numbers         c)  whole numbers
d)  rational numbers    e)  irrational numbers      f)  real numbers

### Solution

a)  integers: $-16, 0, 29$
b)  natural numbers: $29$
c)  whole numbers: $0, 29$
d)  rational numbers: $-16, 3.82, 0, 29, 0.\overline{7}, -\dfrac{11}{12}$ Each of these numbers can be written as the quotient of two integers.
e)  irrational numbers: $\sqrt{10}, 5.302981\ldots$
f)  real numbers: All of the numbers in this set are real.
$$\left\{ -16, 3.82, 0, 29, 0.\overline{7}, -\dfrac{11}{12}, \sqrt{10}, 5.302981\ldots \right\}$$

## You Try 5

Given the set of numbers $\left\{ \dfrac{9}{8}, \sqrt{14}, 34, -41, 6.5, 0.\overline{21}, 0, 7.412835\ldots \right\}$, list the

a)  whole numbers       b)  integers
c)  rational numbers    d)  irrational numbers

## 2. Compare Numbers Using Inequality Symbols

Let's review the inequality symbols.

$<$ less than            $\leq$ less than or equal to
$>$ greater than         $\geq$ greater than or equal to
$\neq$ not equal to      $\approx$ approximately equal to

We use these symbols to compare numbers as in $5 > 2$, $6 \leq 17$, $4 \neq 9$, and so on. How do we compare negative numbers?

### Note

As we move to the *left* on the number line, the numbers get smaller. As we move to the *right* on the number line, the numbers get larger.

**Example 6**

Insert $>$ or $<$ to make the statement true. Look at the number line, if necessary.

a)  4 __ 2          b)  −3 __ 1          c)  −2 __ −5          d)  −4 __ −1

**Solution**

a)  $4 \geq 2$      4 is to the right of 2.
b)  $-3 \leq 1$    −3 is to the left of 1.
c)  $-2 \geq -5$  −2 is to the right of −5.
d)  $-4 \leq -1$  −4 is to the left of −1.

**You Try 6**

Insert $>$ or $<$ to make the statement true.

a)  7 __ 3          b)  −5 __ −1          c)  −6 __ −14

**Application of Signed Numbers**

**Example 7**

Use a signed number to represent the change in each situation.

a)  During the recession, the number of employees at a manufacturing company decreased by 850.

b)  From October 2008 to March 2009, the number of Facebook users increased by over 23,000,000. (www.insidefacebook.com)

**Solution**

a)  −850          The negative number represents a decrease in the number of employees.

b)  23,000,000   The positive number represents an increase in the number of Facebook users.

**You Try 7**

Use a signed number to represent the change.
After getting off the highway, Huda decreased his car's speed by 25 mph.

### 3. Find the Additive Inverse and Absolute Value of a Number

Notice that both −3 and 3 are a distance of 3 units from 0 but are on opposite sides of 0. We say that 3 and −3 are *additive inverses*.

**Definition**

Two numbers are **additive inverses** if they are the same distance from 0 on the number line but on the opposite side of 0. Therefore, if $a$ is any real number, then $-a$ is its additive inverse.

Furthermore, $-(-a) = a$. We can see this on the number line.

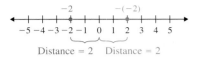

Find $-(-2)$.

**Solution**

So, beginning with $-2$, the number on the opposite side of zero and 2 units away from zero is 2.

$-(-2) = 2$

**You Try 8**

Find $-(-13)$.

We can explain "distance from zero" in another way: *absolute value.* The absolute value of a number is the distance between that number and 0 on the number line. It just describes the distance, *not* what side of zero the number is on. Therefore, the absolute value of a number is always positive or zero.

**Definition**

If $a$ is any real number, then the **absolute value of $a$**, denoted by $|a|$, is

i)  $a$ if $a \geq 0$

ii)  $-a$ if $a < 0$

Remember, $|a|$ is never negative.

**Example 9**

Evaluate each.

a)  $|6|$        b)  $|-5|$        c)  $|0|$        d)  $-|12|$        e)  $|14 - 5|$

**Solution**

a)  $|6| = 6$          6 is 6 units from 0.

b)  $|-5| = 5$          $-5$ is 5 units from 0.

c)  $|0| = 0$

d)  $-|12| = -12$          First, evaluate $|12|$: $|12| = 12$. Then, apply the negative symbol to get $-12$.

e)  $|14 - 5| = |9|$          The absolute value symbols work like parentheses. First, evaluate what is inside: $14 - 5 = 9$.

$= 9$          Find the absolute value.

**You Try 9**

Evaluate each.

a)  $|19|$        b)  $|-8|$        c)  $-|7|$        d)  $|20 - 9|$

**Answers to You Try Exercises**

1)

$-4 \quad -2 -1 \quad\quad 2 \quad\quad 5$

$-6\ -5\ -4\ -3\ -2\ -1\ \ 0\ \ 1\ \ 2\ \ 3\ \ 4\ \ 5\ \ 6$

2) a) $0, 5, 8$   b) $5, 8$   c) $-12, -1, 0, 5, 8$

3) a) $12 = \dfrac{12}{1}$   b) $0.7 = \dfrac{7}{10}$   c) $-8 = \dfrac{-8}{1}$   d) $2\dfrac{3}{4} = \dfrac{11}{4}$   e) $0.\overline{6} = \dfrac{2}{3}$   f) $\sqrt{100} = 10$ and $10 = \dfrac{10}{1}$

4) a) It is a nonrepeating, nonterminating decimal.   b) 2 is not a perfect square, so the decimal equivalent of $\sqrt{2}$ is a nonrepeating, nonterminating decimal.   5) a) $34, 0$   b) $34, -41, 0$

c) $\dfrac{9}{8}, 34, -41, 6.5, 0.\overline{21}, 0$   d) $\sqrt{14}, 7.412835\ldots$   6) a) $>$   b) $<$   c) $>$   7) $-25$

8) $13$   9) a) $19$   b) $8$   c) $-7$   d) $11$

# 1.4 Exercises

## Objective 1: Identify and Graph Numbers on a Number Line

1) In your own words, explain the difference between the set of rational numbers and the set of irrational numbers. Give two examples of each type of number.

2) In your own words, explain the difference between the set of whole numbers and the set of natural numbers. Give two examples of each type of number.

In Exercises 3 and 4, given each set of numbers, list the

a) natural numbers        b) whole numbers

c) integers        d) rational numbers

e) irrational numbers        f) real numbers

3) $\left\{ 17, 3.8, \dfrac{4}{5}, 0, \sqrt{10}, -25, 6.\overline{7}, -2\dfrac{1}{8}, 9.721983\ldots \right\}$

4) $\left\{ -6, \sqrt{23}, 21, 5.\overline{62}, 0.4, 3\dfrac{2}{9}, 0, -\dfrac{7}{8}, 2.074816\ldots \right\}$

Determine whether each statement is true or false.

5) Every whole number is a real number.

6) Every real number is an integer.

7) Every rational number is a whole number.

8) Every whole number is an integer.

9) Every natural number is a whole number.

10) Every integer is a rational number.

Graph the numbers on a number line. Label each.

11) $5, -2, \dfrac{3}{2}, -3\dfrac{1}{2}, 0$

12) $-4, 3, \dfrac{7}{8}, 4\dfrac{1}{3}, -2\dfrac{1}{4}$

13) $-6.8, -\dfrac{3}{8}, 0.2, 1\dfrac{8}{9}, -4\dfrac{1}{3}$

14) $-3.25, \dfrac{2}{3}, 2, -1\dfrac{3}{8}, 4.1$

## Objective 3: Find the Additive Inverse and Absolute Value of a Number

15) What does the absolute value of a number represent?

16) If $a$ is a real number and if $|a|$ is not a positive number, then what is the value of $a$?

Find the additive inverse of each.

17) $8$

18) $6$

19) $-15$

20) $-1$

21) $-\dfrac{3}{4}$

22) $4.7$

Evaluate.

23) $|-10|$

24) $|9|$

25) $\left| \dfrac{9}{4} \right|$

26) $\left| -\dfrac{5}{6} \right|$

27)  $-|-14|$

28)  $-|27|$

29)  $|17 - 4|$

30)  $-|10 - 6|$

31)  $-\left|-4\frac{1}{7}\right|$

32)  $|-9.6|$

Write each group of numbers from smallest to largest.

**VIDEO** 33)  $7, -2, 3.8, -10, 0, \dfrac{9}{10}$

34)  $2.6, 2.06, -1, -5\dfrac{3}{8}, 3, \dfrac{7}{4}$

35)  $7\dfrac{5}{6}, -5, -6.5, -6.51, 7\dfrac{1}{3}, 2$

36)  $-\dfrac{3}{4}, 0, -0.5, 4, -1, \dfrac{15}{2}$

**Mixed Exercises: Objectives 2 and 3**

Decide whether each statement is true or false.

37)  $16 \geq -11$

38)  $-19 < -18$

39)  $\dfrac{7}{11} \leq \dfrac{5}{9}$

40)  $-1.7 \geq -1.6$

41)  $-|-28| = 28$

42)  $-|13| = -13$

**VIDEO** 43)  $-5\dfrac{3}{10} < -5\dfrac{3}{4}$

44)  $\dfrac{3}{2} \leq \dfrac{3}{4}$

Use a signed number to represent the change in each situation.

45)  In 2007, Alex Rodriguez of the New York Yankees had 156 RBIs (runs batted in) while in 2008 he had 103 RBIs. That was a decrease of 53 RBIs. (http://newyork.yankees.mlb.com)

46)  In 2006, Madonna's *Confessions* tour grossed about $194 million. Her *Sticky and Sweet* tour grossed about $230 million in 2008, an increase of $36 million compared to the *Confessions* tour. (www.billboard.com)

47)  In January 2009, an estimated 2.6 million people visited the Twitter website. In February 2009, there were about 4 million visitors to the site. This is an increase of 1.4 million people. (www.techcrunch.com)

48)  According to the *Statistical Abstract of the United States,* the population of Louisiana decreased by about 58,000 from April 1, 2000 to July 1, 2008. (www.census.gov)

49)  From 2006 to 2007, the number of new housing starts decreased by about 419,000. (www.census.gov)

50)  Research done by the U.S. Department of Agriculture has found that the per capita consumption of bottled water increased by 2.1 gallons from 2005 to 2006. (www.census.gov)

## Section 1.5 Addition and Subtraction of Real Numbers

**Objectives**

1. Add Integers Using a Number Line
2. Add Real Numbers with the Same Sign
3. Add Real Numbers with Different Signs
4. Subtract Real Numbers
5. Solve Applied Problems
6. Apply the Order of Operations to Real Numbers
7. Translate English Expressions to Mathematical Expressions

In Section 1.4, we defined real numbers. In this section, we will discuss adding and subtracting real numbers.

### 1. Add Integers Using a Number Line

Let's use a number line to add numbers.

**Example 1**

Use a number line to add each pair of numbers.

a)  $2 + 5$          b)  $-1 + (-4)$          c)  $2 + (-5)$          d)  $-8 + 12$

**Solution**

a)  $2 + 5$: Start at 2 and move 5 units to the right.

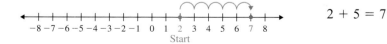

$2 + 5 = 7$

Start

b)  $-1 + (-4)$: Start at $-1$ and move 4 units to the left.
(Move to the left when adding a negative.)

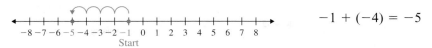

$-1 + (-4) = -5$

c)  $2 + (-5)$: Start at 2 and move 5 units to the left.

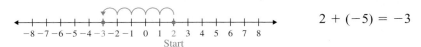

$2 + (-5) = -3$

d)  $-8 + 12$: Start at $-8$ and move 12 units to the right.

$-8 + 12 = 4$

---

**You Try 1**

Use a number line to add each pair of numbers.

a)  $1 + 3$          b)  $-3 + (-2)$          c)  $8 + (-6)$          d)  $-10 + 7$

## 2. Add Real Numbers with the Same Sign

We found that

$$2 + 5 = 7, \qquad -1 + (-4) = -5, \qquad 2 + (-5) = -3, \qquad -8 + 12 = 4.$$

Notice that when we add two numbers with the same sign, the result has the same sign as the numbers being added.

> **Procedure** Adding Numbers with the Same Sign
>
> To add numbers with the same sign, find the absolute value of each number and add them. The sum will have the same sign as the numbers being added.

Apply this rule to $-1 + (-4)$.

The result will be negative
↓
$$-1 + (-4) = \underbrace{-(|-1| + |-4|)}_{\substack{\text{Add the} \\ \text{absolute} \\ \text{value of} \\ \text{each number.}}} = -(1 + 4) = -5$$

---

**Example 2**

Add.

a)  $-5 + (-4)$          b)  $-23 + (-41)$

**Solution**

a)  $-5 + (-4) = -(|-5| + |-4|) = -(5 + 4) = -9$

b)  $-23 + (-41) = -(|-23| + |-41|) = -(23 + 41) = -64$

## You Try 2

Add.

a)   $-6 + (-10)$          b)   $-38 + (-56)$

### 3. Add Real Numbers with Different Signs

In Example 1, we found that $2 + (-5) = -3$ and $-8 + 12 = 4$.

> **Procedure**   Adding Numbers with Different Signs
>
> To add two numbers with different signs, find the absolute value of each number. Subtract the smaller absolute value from the larger. The sum will have the sign of the number with the larger absolute value.

Let's apply this to $2 + (-5)$ and $-8 + 12$.

$2 + (-5)$:      $|2| = 2$     $|-5| = 5$
Since $2 < 5$, subtract $5 - 2$ to get 3. Since $|-5| > |2|$, the sum will be negative.
$2 + (-5) = -3$

$-8 + 12$:      $|-8| = 8$     $|12| = 12$
Subtract $12 - 8$ to get 4. Since $|12| > |-8|$, the sum will be positive.
$-8 + 12 = 4$

## Example 3

Add.

a)   $-17 + 5$          b)   $9.8 + (-6.3)$          c)   $\dfrac{1}{5} + \left(-\dfrac{2}{3}\right)$          d)   $-8 + 8$

**Solution**

a)   $-17 + 5 = -12$          The sum will be negative since the number with the larger absolute value, $|-17|$, is negative.

b)   $9.8 + (-6.3) = 3.5$          The sum will be positive since the number with the larger absolute value, $|9.8|$, is positive.

c)   $\dfrac{1}{5} + \left(-\dfrac{2}{3}\right) = \dfrac{3}{15} + \left(-\dfrac{10}{15}\right)$          Get a common denominator.
The sum will be negative since the number with the

$\qquad\qquad = -\dfrac{7}{15}$          larger absolute value, $\left|-\dfrac{10}{15}\right|$, is negative.

d)   $-8 + 8 = 0$          ∎

>
>
> **Note**
>
> *The sum of a number and its additive inverse is always 0. That is, if $a$ is a real number, then* $a + (-a) = 0$. *Notice in part d) of Example 3 that* $-8$ *and 8 are additive inverses.*

## You Try 3

Add.

a)   $20 + (-19)$          b)   $-14 + (-2)$          c)   $-\dfrac{3}{7} + \dfrac{1}{4}$          d)   $7.2 + (-7.2)$

## 4. Subtract Real Numbers

We can use the additive inverse to subtract numbers. Let's start with a basic subtraction problem and use a number line to find $8 - 5$.

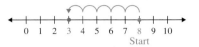

Start at 8. Then to subtract 5, move 5 units to the left to get 3.

$$8 - 5 = 3$$

We use the same procedure to find $8 + (-5)$. This leads us to a definition of subtraction:

> **Definition**
>
> If $a$ and $b$ are real numbers, then $a - b = a + (-b)$.

The definition tells us that to subtract $a - b$,

1) Change subtraction to addition.
2) Find the additive inverse of $b$.
3) Add $a$ and the additive inverse of $b$.

### Example 4

Subtract.

a)  $4 - 9$          b)  $-10 - 8$          c)  $24 - 11$          d)  $6 - (-25)$

**Solution**

a)  $4 - 9 = 4 + (-9) = -5$

        Change to      Additive inverse
        addition             of 9

b)  $-10 - 8 = -10 + (-8) = -18$

        Change to      Additive inverse
        addition             of 8

c)  $24 - 11 = 24 + (-11)$
             $= 13$

d)  $6 - (-25) = 6 + 25 = 31$

        Change to      Additive inverse
        addition             of $-25$

### You Try 4

Subtract.

a)  $2 - 14$          b)  $-9 - 13$          c)  $31 - 14$          d)  $23 - (-34)$

In part d) of Example 4, $6 - (-25)$ changed to $6 + 25$. This illustrates that *subtracting a negative number is equivalent to adding a positive number.* Therefore, $-7 - (-15) = -7 + 15 = 8$.

### 5. Solve Applied Problems

We can use signed numbers to solve real-life problems.

**Example 5**

According to the National Weather Service, the coldest temperature ever recorded in Wyoming was −66°F on February 9, 1944. The record high was 115°F on August 8, 1983. What is the difference between these two temperatures? (www.ncdc.noaa.gov)

**Solution**

$$\text{Difference} = \text{Highest temperature} - \text{Lowest temperature}$$
$$= \qquad 115 \qquad - \qquad (-66)$$
$$= 115 + 66$$
$$= 181$$

The difference between the temperatures is 181°F.

**You Try 5**

The best score in a golf tournament was −16, and the worst score was +9. What is the difference between these two scores?

### 6. Apply the Order of Operations to Real Numbers

We discussed the order of operations in Section 1.2. Let's explore it further with the real numbers.

**Example 6**

Simplify.

a)   $(10 - 18) + (-4 + 6)$        b)   $-13 - (-21 + 5)$

c)   $|-31 - 4| - 7|9 - 4|$

**Solution**

a)   $(10 - 18) + (-4 + 6) = -8 + 2$     First, perform the operations in parentheses.
$$= -6 \qquad\qquad \text{Add.}$$

b)   $-13 - (-21 + 5) = -13 - (-16)$     First, perform the operations in parentheses.
$$= -13 + 16 \qquad \text{Change to addition.}$$
$$= 3 \qquad\qquad \text{Add.}$$

c)   $|-31 - 4| - 7|9 - 4| = |-31 + (-4)| - 7|9 - 4|$
$$= |-35| - 7|5| \qquad \text{Perform the operations in the absolute values.}$$
$$= 35 - 7(5) \qquad \text{Evaluate the absolute values.}$$
$$= 35 - 35$$
$$= 0$$

**You Try 6**

Simplify.

a)   $[12 + (-5)] - [-16 + (-8)]$        b)   $-\dfrac{4}{9} + \left(\dfrac{1}{6} - \dfrac{2}{3}\right)$        c)   $-|7 - 15| - |4 - 2|$

## 7. Translate English Expressions to Mathematical Expressions

Knowing how to translate from English expressions to mathematical expressions is a skill students need to learn algebra. Here, we will discuss how to "translate" from English to mathematics.

Let's look at some key words and phrases you may encounter.

| **English Expression** | **Mathematical Operation** |
|---|---|
| sum, more than, increased by | addition |
| difference between, less than, decreased by | subtraction |

Here are some examples:

**Example 7**

Write a mathematical expression for each and simplify.

a)   9 more than $-2$          b)   10 less than 41          c)   $-8$ decreased by 17

d)   the sum of 13 and $-4$          e)   8 less than the sum of $-11$ and $-3$

**Solution**

a)   9 more than $-2$

*9 more than* a quantity means we *add 9* to the quantity, in this case, $-2$.

$$-2 + 9 = 7$$

b)   10 less than 41

*10 less than* a quantity means we *subtract 10 from* that quantity, in this case, 41.

$$41 - 10 = 31$$

c)   $-8$ decreased by 17

If $-8$ is being *decreased by 17*, then we subtract 17 *from* $-8$.

$$-8 - 17 = -8 + (-17)$$
$$= -25$$

d)   the sum of 13 and $-4$

*Sum* means add.   $13 + (-4) = 9$

e)   8 less than the sum of $-11$ and $-3$.

*8 less than* means we are subtracting 8 *from* something. From what? From the *sum of $-11$ and $-3$.*

*Sum* means add, so we must find the sum of $-11$ and $-3$ and subtract 8 from it.

$$[-11 + (-3)] - 8 = -14 - 8 \qquad \text{First, perform the operation in the brackets.}$$
$$= -14 + (-8) \qquad \text{Change to addition.}$$
$$= -22 \qquad \text{Add.}$$ ■

  **You Try 7**

Write a mathematical expression for each and simplify.

a)   $-14$ increased by 6          b)   27 less than 15

c)   The sum of 23 and $-7$ decreased by 5

**Answers to You Try Exercises**

1) a) 4   b) −5   c) 2   d) −3    2) a) −16   b) −94    3) a) 1   b) −16   c) $-\dfrac{5}{28}$   d) 0

4) a) −12   b) −22   c) 17   d) 57    5) 25    6) a) 31   b) $-\dfrac{17}{18}$   c) −10

7) a) −14 + 6; −8   b) 15 − 27; −12   c) [23 + (−7)] − 5; 11

## 1.5 Exercises

**Mixed Exercises: Objectives 1–4 and 6**

1) Explain, in your own words, how to subtract two negative numbers.

2) Explain, in your own words, how to add two negative numbers.

3) Explain, in your own words, how to add a positive and a negative number.

Use a number line to represent each sum or difference.

4) −8 + 5

5) 6 − 11

6) −1 − 5

7) −2 + (−7)

8) 10 + (−6)

Add or subtract as indicated.

9) 8 + (−15)

10) −12 + (−6)

11) −3 − 11

12) −7 + 13

13) −31 + 54

14) 19 − (−14)

15) −26 − (−15)

16) −20 − (−30)

17) −352 − 498

18) 217 + (−521)

19) $-\dfrac{7}{12} + \dfrac{3}{4}$

20) $\dfrac{3}{10} - \dfrac{11}{15}$

21) $-\dfrac{1}{6} - \dfrac{7}{8}$

22) $\dfrac{2}{9} - \left(-\dfrac{2}{5}\right)$

23) $-\dfrac{4}{9} - \left(-\dfrac{4}{15}\right)$

24) $-\dfrac{1}{8} + \left(-\dfrac{3}{4}\right)$

25) 19.4 + (−16.7)

26) −31.3 − (−19.82)

27) −25.8 − (−16.57)

28) 7.3 − 21.9

29) 9 − (5 − 11)

30) −2 + (3 − 8)

31) −1 + (−6 − 4)

32) 14 − (−10 − 2)

33) (−3 − 1) − (−8 + 6)

34) [14 + (−9)] + (1 − 8)

35) −16 + 4 + 3 − 10

36) 8 − 28 + 3 − 7

37) 5 − (−30) − 14 + 2

38) −17 − (−9) + 1 − 10

39) $\dfrac{4}{9} - \left(\dfrac{2}{3} + \dfrac{5}{6}\right)$

40) $-\dfrac{1}{2} + \left(\dfrac{3}{5} - \dfrac{3}{10}\right)$

41) $\left(\dfrac{1}{8} - \dfrac{1}{2}\right) + \left(\dfrac{3}{4} - \dfrac{1}{6}\right)$

42) $\dfrac{11}{12} - \left(\dfrac{3}{8} - \dfrac{2}{3}\right)$

43) (2.7 + 3.8) − (1.4 − 6.9)

44) −9.7 − (−5.5 + 1.1)

45) $|7 - 11| + |6 + (-13)|$

46) $|8 - (-1)| - |3 + 12|$

47) $-|2 - (-3)| - 2|-5 + 8|$

48) $|-6 + 7| + 5|-20 - (-11)|$

Determine whether each statement is true or false. For any real numbers $a$ and $b$,

49) $|a + b| = |a| + |b|$

50) $|a - b| = |b - a|$

51) $|a + b| = a + b$

52) $|a| + |b| = a + b$

53) $-b - (-b) = 0$

54) $a + (-a) = 0$

**Objective 5: Solve Applied Problems**

*Applications of Signed Numbers:* Write an expression for each and simplify. Answer the question with a complete sentence.

55) Tiger Woods won his first Masters championship in 1997 at age 21 with a score of −18. When he won the championship in 2005, his score was 6 strokes higher. What was Tiger's score when he won the Masters in 2005? (www.masters.com)

56) In 1999, the U.S. National Park System recorded 287.1 million visits while in 2007 there were 275.6 million visits. What was the difference in the number of visits from 1999 to 2007? (www.nationalparkstraveler.com)

57) In 2006, China's carbon emissions were 6,110,000 thousand metric tons and the carbon emissions of the United States totaled 5,790,000 thousand metric tons. By how much did China's carbon emissions exceed those of the United States? (www.pbl.nl)

58) The budget of the Cincinnati Public Schools was $428,554,470 in the 2006–2007 school year. This was $22,430 less than the previous school year. What was the budget in the 2005–2006 school year? (www.cps-k12.org)

59) From 2007 to 2008, the number of flights going through O'Hare Airport in Chicago decreased by 45,407. There were 881,566 flights in 2008. How many flights went through O'Hare in 2007? (www.ohare.com)

60) The lowest temperature ever recorded in Minneapolis was −41°F while the highest temperature on record was 149° greater than that. What was the warmest temperature ever recorded in Minneapolis? (www.weather.com)

61) The bar graph shows the total number of daily newspapers in the United States in various years. Use a signed number to represent the change in the number of dailies over the given years. (www.naa.org)

Number of Daily Newspapers in the U.S.

a) 1970–1980    b) 1980–1990

c) 1990–2000    d) 1960–2000

62) The bar graph shows the TV ratings for the World Series over a 5-year period. Each ratings number represents the percentage of people watching TV at the time of the World Series who were tuned into the games. Use a signed number to represent the change in ratings over the given years. (www.baseball-almanac.com)

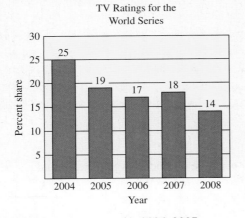

TV Ratings for the World Series

a) 2004–2005    b) 2006–2007

c) 2007–2008    d) 2004–2008

63) The bar graph shows the average number of days a woman was in the hospital for childbirth. Use a signed number to represent the change in hospitalization time over the given years. (www.cdc.gov)

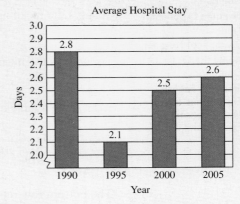

Average Hospital Stay

a) 1990–1995    b) 1995–2000

c) 2000–2005    d) 1990–2005

64) The bar graph shows snowfall totals for different seasons in Syracuse, NY. Use a signed number to represent the difference in snowfall totals over different years. (www.erh.noaa.gov)

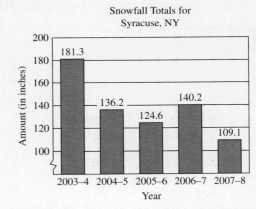

Snowfall Totals for Syracuse, NY

a) 2003–4 to 2004–5    b) 2005–6 to 2006–7

c) 2006–7 to 2007–8    d) 2003–4 to 2007–8

**Objective 7: Translate English Expressions to Mathematical Expressions**

Write a mathematical expression for each and simplify.

65) 7 more than 5    66) 3 more than 11

67) 16 less than 10    68) 15 less than 4

69) −8 less than 9    70) −25 less than −19

71) The sum of −21 and 13    72) The sum of −7 and 20

73) −20 increased by 30    74) −37 increased by 22

75) 23 decreased by 19    76) 8 decreased by 18

77) 18 less than the sum of −5 and 11

78) 35 less than the sum of −17 and 3

## Section 1.6 Multiplication and Division of Real Numbers

**Objectives**

1. **Multiply Real Numbers**
2. **Evaluate Exponential Expressions**
3. **Divide Real Numbers**
4. **Apply the Order of Operations**
5. **Translate English Expressions to Mathematical Expressions**

### 1. Multiply Real Numbers

What is the meaning of $4 \cdot 5$? It is repeated addition.

$$4 \cdot 5 = 5 + 5 + 5 + 5 = 20$$

So, what is the meaning of $4 \cdot (-5)$? It, too, represents repeated addition.

$$4 \cdot (-5) = -5 + (-5) + (-5) + (-5) = -20$$

Let's make a table of some products:

| × | 5 | 4 | ③ | 2 | 1 | 0 | −1 | −2 | −3 | −4 | −5 |
|---|---|---|---|---|---|---|----|----|----|----|----|
| ④ | 20 | 16 | 12 | 8 | 4 | 0 | −4 | −8 | −12 | −16 | −20 |

$$4 \cdot 3 = 12$$

The bottom row represents the product of 4 and the number above it ($4 \cdot 3 = 12$). Notice that as the numbers in the first row decrease by 1, the numbers in the bottom row decrease by 4. Therefore, once we get to $4 \cdot (-1)$, the product is negative. From the table we can see that,

**Note**

The product of a positive number and a negative number is negative.

---

**Example 1**

Multiply.

a)  $-6 \cdot 9$        b)  $\dfrac{3}{8} \cdot (-12)$        c)  $-5 \cdot 0$

**Solution**

a)  $-6 \cdot 9 = -54$

b)  $\dfrac{3}{8} \cdot (-12) = \dfrac{3}{\overset{}{\underset{2}{8}}} \cdot \left( -\dfrac{\overset{3}{\cancel{12}}}{1} \right) = -\dfrac{9}{2}$

c)  $-5 \cdot 0 = 0$        The product of zero and any real number is zero.    ■

---

 **You Try 1**

Multiply.

a)  $-7 \cdot 3$        b)  $\dfrac{8}{15} \cdot (-10)$

---

What is the sign of the product of two negative numbers? Again, we'll make a table.

| × | 3 | 2 | 1 | 0 | −1 | −2 | −3 |
|---|---|---|---|---|----|----|----|
| −4 | −12 | −8 | −4 | 0 | 4 | 8 | 12 |

As we decrease the numbers in the top row by 1, the numbers in the bottom row *increase* by 4. When we reach $-4 \cdot (-1)$, our product is a positive number, 4. The table illustrates that,

**Note**

The product of two negative numbers is positive.

We can summarize our findings this way:

> **Procedure**   Multiplying Real Numbers
>
> 1) The product of two positive numbers is positive.
> 2) The product of two negative numbers is positive.
> 3) The product of a positive number and a negative number is negative.
> 4) The product of any real number and zero is zero.

---

**Example 2**

Multiply.

a)  $-8 \cdot (-5)$          b)  $-1.5 \cdot 6$

c)  $-\dfrac{3}{8} \cdot \left(-\dfrac{4}{5}\right)$          d)  $-5 \cdot (-2) \cdot (-3)$

**Solution**

a)  $-8 \cdot (-5) = 40$          The product of two negative numbers is positive.

b)  $-1.5 \cdot 6 = -9$          The product of a negative number and a positive number is negative.

c)  $-\dfrac{3}{8} \cdot \left(-\dfrac{4}{5}\right) = -\dfrac{3}{\overset{}{\underset{2}{8}}} \cdot \left(-\dfrac{\overset{1}{4}}{5}\right)$

   $= \dfrac{3}{10}$          The product of two negatives is positive.

d)  $\underbrace{-5 \cdot (-2)}_{10} \cdot (-3) = 10 \cdot (-3)$          Order of operations—multiply from left to right.

   $= -30$          ∎

---

**You Try 2**

Multiply.

a)  $-6 \cdot 7$          b)  $-\dfrac{8}{9} \cdot \dfrac{3}{4}$          c)  $-4 \cdot (-1) \cdot (-5) \cdot (-2)$

---

**Note**

It is helpful to know that

1) An **even number** of negative factors in a product gives a positive result.

   $$-3 \cdot 1 \cdot (-2) \cdot (-1) \cdot (-4) = 24 \qquad \text{Four negative factors}$$

2) An **odd number** of negative factors in a product gives a negative result.

   $$5 \cdot (-3) \cdot (-1) \cdot (-2) \cdot (3) = -90 \qquad \text{Three negative factors}$$

## 2. Evaluate Exponential Expressions

In Section 1.2 we discussed exponential expressions. Recall that exponential notation is a shorthand way to represent repeated multiplication:

$$2^4 = 2 \cdot 2 \cdot 2 \cdot 2 = 16$$

Now we will discuss exponents and negative numbers. Consider a base of $-2$ raised to different powers. (The $-2$ is in parentheses to indicate that it is the base.)

$$(-2)^1 = -2$$
$$(-2)^2 = -2 \cdot (-2) = 4$$
$$(-2)^3 = -2 \cdot (-2) \cdot (-2) = -8$$
$$(-2)^4 = -2 \cdot (-2) \cdot (-2) \cdot (-2) = 16$$
$$(-2)^5 = -2 \cdot (-2) \cdot (-2) \cdot (-2) \cdot (-2) = -32$$
$$(-2)^6 = -2 \cdot (-2) \cdot (-2) \cdot (-2) \cdot (-2) \cdot (-2) = 64$$

Do you notice that

1)    $-2$ raised to an *odd* power gives a negative result?

and

2)    $-2$ raised to an *even* power gives a positive result?

This will always be true.

**Note**

A negative number raised to an *odd* power will give a *negative* result. A negative number raised to an *even* power will give a *positive* result.

### Example 3

Evaluate.

a)    $(-6)^2$          b)    $(-10)^3$

**Solution**

a)    $(-6)^2 = 36$          b)    $(-10)^3 = -1000$                          ■

### You Try 3

Evaluate.

a)    $(-9)^2$          b)    $(-5)^3$

How do $(-2)^4$ and $-2^4$ differ? Let's identify their bases and evaluate each.

$(-2)^4$:    Base $= -2$          $(-2)^4 = 16$

$-2^4$:    Since there are no parentheses,

$-2^4$ is equivalent to $-1 \cdot 2^4$. Therefore, the base is 2.

$$-2^4 = -1 \cdot 2^4$$
$$= -1 \cdot 2 \cdot 2 \cdot 2 \cdot 2$$
$$= -1 \cdot 16$$
$$= -16$$

So, $(-2)^4 = 16$ and $-2^4 = -16$.

When working with exponential expressions, be able to identify the base.

## Example 4

Evaluate.

a)  $(-5)^3$        b)  $-9^2$        c)  $\left(-\dfrac{1}{7}\right)^2$

**Solution**

a)  $(-5)^3$: Base $= -5$        $(-5)^3 = -5 \cdot (-5) \cdot (-5) = -125$

b)  $-9^2$: Base $= 9$        $-9^2 = -1 \cdot 9^2$
$$= -1 \cdot 9 \cdot 9$$
$$= -81$$

c)  $\left(-\dfrac{1}{7}\right)^2$: Base $= -\dfrac{1}{7}$     $\left(-\dfrac{1}{7}\right)^2 = -\dfrac{1}{7} \cdot \left(-\dfrac{1}{7}\right) = \dfrac{1}{49}$

### You Try 4

Evaluate.

a)  $-3^4$        b)  $(-11)^2$        c)  $-8^2$        d)  $-\left(-\dfrac{2}{3}\right)^3$

## 3. Divide Real Numbers

Here are the rules for dividing signed numbers:

| **Procedure**   Dividing Signed Numbers |
| --- |
| 1)   The quotient of two positive numbers is a positive number. |
| 2)   The quotient of two negative numbers is a positive number. |
| 3)   The quotient of a positive and a negative number is a negative number. |

## Example 5

Divide.

a)  $-36 \div 9$        b)  $-\dfrac{1}{10} \div \left(-\dfrac{3}{5}\right)$        c)  $\dfrac{-8}{-1}$        d)  $\dfrac{-24}{42}$

**Solution**

a)  $-36 \div 9 = -4$

b)  $-\dfrac{1}{10} \div \left(-\dfrac{3}{5}\right) = -\dfrac{1}{10} \cdot \left(-\dfrac{5}{3}\right)$        When dividing by a fraction, multiply by the reciprocal.

$$= -\dfrac{1}{\overset{}{\underset{2}{10}}} \cdot \left(-\dfrac{\overset{1}{5}}{3}\right) = \dfrac{1}{6}$$

c)  $\dfrac{-8}{-1} = 8$     The quotient of two negative numbers is positive, and $\dfrac{8}{1}$ simplifies to 8.

d)  $\dfrac{-24}{42} = -\dfrac{24}{42}$     The quotient of a negative number and a positive number is negative, so reduce $\dfrac{24}{42}$.

$= -\dfrac{4}{7}$     24 and 42 each divide by 6.

It is important to note here in part d) that there are three ways to write the answer: $-\dfrac{4}{7}, \dfrac{-4}{7},$ or $\dfrac{4}{-7}$. These are equivalent. However, we usually write the negative sign in front of the entire fraction as in $-\dfrac{4}{7}$.  ∎

**You Try 5**

Divide.

a)  $-\dfrac{8}{5} \div \left(-\dfrac{6}{5}\right)$     b)  $\dfrac{-30}{-10}$     c)  $\dfrac{21}{-56}$

## 4. Apply the Order of Operations

**Example 6**

Simplify.

a)  $-24 \div 12 - 2^2$     b)  $-5(-3) - 4(2 - 3)$

**Solution**

a)  $-24 \div 12 - 2^2 = -24 \div 12 - 4$     Simplify exponent first.
$= -2 - 4$     Perform division before subtraction.
$= -6$

b)  $-5(-3) - 4(2 - 3) = -5(-3) - 4(-1)$     Simplify the difference in parentheses.
$= 15 - (-4)$     Find the products.
$= 15 + 4$
$= 19$     ∎

**You Try 6**

Simplify.

a)  $-13 - 4(-5 + 2)$     b)  $(-10)^2 + 2[8 - 5(4)]$

## 5. Translate English Expressions to Mathematical Expressions

Here are some words and phrases you may encounter and how they would translate to mathematical expressions:

| English Expression | Mathematical Operation |
| --- | --- |
| times, product of | multiplication |
| divided by, quotient of | division |

**Example 7**

Write a mathematical expression for each and simplify.

a) The quotient of $-56$ and 7

b) The product of 4 and the sum of 15 and $-6$

c) Twice the difference of $-10$ and $-3$

d) Half of the sum of $-8$ and 3

**Solution**

a) The quotient of $-56$ and 7:
Quotient means division with $-56$ in the numerator and 7 in the denominator.
The expression is $\dfrac{-56}{7} = -8$.

b) The product of 4 and the sum of 15 and $-6$:
The *sum of 15 and $-6$* means we must add the two numbers. *Product* means multiply.

$$\overbrace{4[15 + (-6)]}^{\substack{\text{Sum of 15} \\ \text{and } -6}} = 4(9) = 36$$
$$\underbrace{\phantom{4[15 + (-6)]}}_{\substack{\text{Product of 4} \\ \text{and the sum}}}$$

c) Twice the difference of $-10$ and $-3$:
The *difference of $-10$ and $-3$* will be in parentheses with $-3$ being subtracted from $-10$. *Twice* means "two times."

$$2[-10 - (-3)] = 2(-10 + 3)$$
$$= 2(-7)$$
$$= -14$$

d) Half of the sum of $-8$ and 3:
The *sum of $-8$ and 3* means that we will add the two numbers. They will be in parentheses. *Half of* means multiply by $\dfrac{1}{2}$.

$$\frac{1}{2}(-8 + 3) = \frac{1}{2}(-5) = -\frac{5}{2}$$

**You Try 7**

Write a mathematical expression for each and simplify.

a) 12 less than the product of $-7$ and 4

b) Twice the sum of 19 and $-11$

c) The sum of $-41$ and $-23$, divided by the square of $-2$

---

**Answers to You Try Exercises**

1) a) $-21$  b) $-\dfrac{16}{3}$     2) a) $-42$  b) $-\dfrac{2}{3}$  c) 40     3) a) 81  b) $-125$

4) a) $-81$  b) 121  c) $-64$  d) $\dfrac{8}{27}$     5) a) $\dfrac{4}{3}$  b) 3  c) $-\dfrac{3}{8}$     6) a) $-1$  b) 76

7) a) $(-7) \cdot 4 - 12$; $-40$  b) $2[19 + (-11)]$; 16  c) $\dfrac{-41 + (-23)}{(-2)^2}$; $-16$

## 1.6 Exercises

**Objective 1: Multiply Real Numbers**

Fill in the blank with *positive* or *negative*.

1) The product of a positive number and a negative number is _____.

2) The product of two negative numbers is _____.

Multiply.

3) $-8 \cdot 7$

4) $4 \cdot (-9)$

5) $-15 \cdot (-3)$

6) $-23 \cdot (-48)$

7) $-4 \cdot 3 \cdot (-7)$

8) $-5 \cdot (-1) \cdot (-11)$

9) $\dfrac{4}{33} \cdot \left(-\dfrac{11}{10}\right)$

10) $-\dfrac{14}{27} \cdot \left(-\dfrac{15}{28}\right)$

11) $(-0.5)(-2.8)$

12) $(-6.1)(5.7)$

13) $-9 \cdot (-5) \cdot (-1) \cdot (-3)$

14) $-1 \cdot (-6) \cdot (4) \cdot (-2) \cdot (3)$

15) $\dfrac{3}{10} \cdot (-7) \cdot (8) \cdot (-1) \cdot (-5)$

16) $-\dfrac{5}{6} \cdot (-4) \cdot 0 \cdot 3$

**Objective 2: Evaluate Exponential Expressions**

17) For what values of $k$ is $k^5$ a negative quantity?

18) For what values of $k$ is $k^5$ a positive quantity?

19) For what values of $k$ is $-k^2$ a negative quantity?

20) Explain the difference between how you would evaluate $(-8)^2$ and $-8^2$. Then, evaluate each.

Evaluate.

21) $(-6)^2$

22) $-6^2$

23) $-5^3$

24) $(-2)^4$

25) $(-3)^2$

26) $(-1)^5$

27) $-7^2$

28) $-4^3$

29) $-2^5$

30) $(-12)^2$

**Objective 3: Divide Real Numbers**

Fill in the blank with *positive* or *negative*.

31) The quotient of two negative numbers is _____.

32) The quotient of a negative number and a positive number is _____.

Divide.

33) $-50 \div (-5)$

34) $-84 \div 12$

35) $\dfrac{64}{-16}$

36) $\dfrac{-54}{-9}$

37) $\dfrac{-2.4}{0.3}$

38) $\dfrac{16}{-0.5}$

39) $-\dfrac{12}{13} \div \left(-\dfrac{6}{5}\right)$

40) $20 \div \left(-\dfrac{15}{7}\right)$

41) $-\dfrac{0}{7}$

42) $\dfrac{0}{-6}$

43) $\dfrac{270}{-180}$

44) $\dfrac{-64}{-320}$

**Objective 4: Apply the Order of Operations**

Use the order of operations to simplify.

45) $7 + 8(-5)$

46) $-40 \div 2 - 10$

47) $(9 - 14)^2 - (-3)(6)$

48) $-23 - 6^2 \div 4$

49) $10 - 2(1 - 4)^3 \div 9$

50) $-7(4) + (-8 + 6)^4 + 5$

51) $\left(-\dfrac{3}{4}\right)(8) - 2[7 - (-3)(-6)]$

52) $-2^5 - (-3)(4) + 5[(-9 + 30) \div 7]$

53) $\dfrac{-46 - 3(-12)}{(-5)(-2)(-4)}$

54) $\dfrac{(8)(-6) + 10 - 7}{(-5 + 1)^2 - 12 + 5}$

**Objective 5: Translate English Expressions to Mathematical Expressions**

Write a mathematical expression for each and simplify.

55) The product of $-12$ and 6

56) The quotient of $-80$ and $-4$

57) 9 more than the product of $-7$ and $-5$

58) The product of $-10$ and 2 increased by 11

59) The quotient of 63 and $-9$ increased by 7

60) 8 more than the quotient of 54 and $-6$

61) 19 less than the product of $-4$ and $-8$

62) The product of $-16$ and $-3$ decreased by 20

63) The quotient of $-100$ and 4 decreased by the sum of $-7$ and 2

64) The quotient of $-35$ and 5 increased by the product of $-11$ and $-2$

65) Twice the sum of 18 and $-31$

66) Twice the difference of $-5$ and $-15$

67) Two-thirds of $-27$

68) Half of $-30$

69) The product of 12 and $-5$ increased by half of 36

70) One third of $-18$ decreased by half the sum of $-21$ and $-5$

## Section 1.7 Algebraic Expressions and Properties of Real Numbers

### Objectives

1. Identify the Terms and Coefficients in an Expression
2. Evaluate Algebraic Expressions
3. Identify Like Terms
4. Use the Commutative Properties
5. Use the Associative Properties
6. Use the Identity and Inverse Properties
7. Use the Distributive Property
8. Combine Like Terms
9. Translate English Expressions to Mathematical Expressions

### 1. Identify the Terms and Coefficients in an Expression

Here is an algebraic expression:

$$8x^3 - 5x^2 + \frac{2}{7}x + 4$$

$x$ is the *variable*. A **variable** is a symbol, usually a letter, used to represent an unknown number. The *terms* of this algebraic expression are $8x^3$, $-5x^2$, $\frac{2}{7}x$, and 4. A **term** is a number or a variable or a product or quotient of numbers and variables. *4* is the **constant** or **constant term**. The value of a constant does not change. Each term has a **coefficient**.

| Term | Coefficient |
|------|-------------|
| $8x^3$ | 8 |
| $-5x^2$ | $-5$ |
| $\frac{2}{7}x$ | $\frac{2}{7}$ |
| 4 | 4 |

### Definition

An **algebraic expression** is a collection of numbers, variables, and grouping symbols connected by operation symbols such as $+$, $-$, $\times$, and $\div$.

Examples of expressions:

$$3c + 4, \qquad 9(p^2 - 7p - 2), \qquad -4a^2b^2 + 5ab - 8a + 1.$$

---

**Example 1**

List the terms and coefficients of $4x^2y + 7xy - x + \dfrac{y}{9} - 12$.

### Solution

| Term | Coefficient |
|------|-------------|
| $4x^2y$ | 4 |
| $7xy$ | 7 |
| $-x$ | $-1$ |
| $\dfrac{y}{9}$ | $\dfrac{1}{9}$ |
| $-12$ | $-12$ |

The minus sign indicates a negative coefficient.

$\dfrac{y}{9}$ can be rewritten as $\dfrac{1}{9}y$.

$-12$ is also called the "constant."

**You Try 1**

List the terms and coefficients of $-15r^3 + r^2 - 4r + 8$.

Next, we will use our knowledge of operations with real numbers to evaluate algebraic expressions.

### 2. Evaluate Algebraic Expressions

We can **evaluate** an algebraic expression by substituting a value for a variable and simplifying. The value of an algebraic expression changes depending on the value that is substituted.

**Example 2**

Evaluate $3x - 8$ when    a) $x = 5$    and    b) $x = -4$.

**Solution**

a)   $3x - 8$   when $x = 5$          Substitute 5 for $x$.
   $= 3(5) - 8$                        Use parentheses when substituting a value for a variable.
   $= 15 - 8$                          Multiply.
   $= 7$                               Subtract.

b)   $3x - 8$   when $x = -4$         Substitute $-4$ for $x$.
   $= 3(-4) - 8$                       Use parentheses when substituting a value for a variable.
   $= -12 - 8$                         Multiply.
   $= -20$                             ■

**You Try 2**

Evaluate $6x + 5$ when $x = -2$.

**Example 3**

Evaluate $2a^2 - 7ab + 9$ when $a = -2$ and $b = 3$.

**Solution**

   $2a^2 - 7ab + 9$   when $a = -2$ and $b = 3$   Substitute $-2$ for $a$ and 3 for $b$.
   $= 2(-2)^2 - 7(-2)(3) + 9$                      Use parentheses when substituting a value for a variable.

   $= 2(4) - 7(-6) + 9$                            Evaluate exponent; multiply.
   $= 8 - (-42) + 9$                               Multiply.
   $= 8 + 42 + 9$
   $= 59$                                          ■

**You Try 3**

Evaluate $d^2 + 3cd - 10c - 1$ when $c = \dfrac{1}{2}$ and $d = -4$.

In algebra, it is important to be able to identify *like terms*.

### 3. Identify Like Terms

In the expression $15a + 11a - 8a + 3a$, there are four **terms**: $15a, 11a, -8a, 3a$. In fact, they are **like terms**. *Like terms contain the same variables with the same exponents.*

**Example 4**

Determine whether the following groups of terms are like terms.

a)   $4y^2, -9y^2, \dfrac{2}{3}y^2$                 b)   $-5x^6, 0.8x^9, 3x^4$

c)   $6a^2b^3, a^2b^3, -\dfrac{5}{8}a^2b^3$         d)   $9c, 4d$

**Solution**

a)   $4y^2, -9y^2, \dfrac{2}{3}y^2$

Yes. Each contains the variable $y$ with an exponent of 2. They are $y^2$-terms.

b)   $-5x^6, 0.8x^9, 3x^4$

No. Although each contains the variable $x$, the exponents are not the same.

c)   $6a^2b^3, a^2b^3, -\dfrac{5}{8}a^2b^3$

Yes. Each contains $a^2$ and $b^3$.

d)   $9c, 4d$

No. The terms contain different variables.                                   ■

**You Try 4**

Determine whether the following groups of terms are like terms.

a)   $2k^2, -9k^2, \dfrac{1}{5}k^2$        b)   $-xy^2, 8xy^2, 7xy^2$        c)   $3r^3s^2, -10r^2s^3$

After we discuss the properties of real numbers, we will use them to help us combine like terms.

### Properties of Real Numbers

Like the order of operations, the properties of real numbers guide us in our work with numbers and variables. We begin with the commutative properties of real numbers.

True or false?

1)   $7 + 3 = 3 + 7$        *True*: $7 + 3 = 10$ and $3 + 7 = 10$

2)   $8 - 2 = 2 - 8$        *False*: $8 - 2 = 6$ but $2 - 8 = -6$

3)   $(-6)(5) = (5)(-6)$        *True*: $(-6)(5) = -30$ and $(5)(-6) = -30$

### 4. Use the Commutative Properties

In 1) we see that adding 7 and 3 in any order still equals 10. The third equation shows that multiplying $(-6)(5)$ and $(5)(-6)$ both equal $-30$. But, 2) illustrates that changing the order in which numbers are subtracted does *not* necessarily give the same result: $8 - 2 \neq 2 - 8$. Therefore, subtraction is **not commutative**, while the addition and multiplication of real numbers **is commutative**. This gives us our first property of real numbers:

> **Property**   Commutative Properties
>
> If *a* and *b* are real numbers, then
>
> 1)  $a + b = b + a$      Commutative property of addition
>
> 2)  $ab = ba$      Commutative property of multiplication

We have already shown that subtraction is not commutative. Is division commutative? No. For example,

$$20 \div 4 \overset{?}{=} 4 \div 20$$
$$5 \neq \frac{1}{5}$$

**Example 5**

Use the commutative property to rewrite each expression.

a)   $12 + 5$        b)   $k \cdot 3$

**Solution**

a)   $12 + 5 = 5 + 12$        b)   $k \cdot 3 = 3 \cdot k$ or $3k$      ∎

**You Try 5**

Use the commutative property to rewrite each expression.

a)   $1 + 16$        b)   $n \cdot 6$

## 5. Use the Associative Properties

Another important property involves the use of grouping symbols. Let's determine whether these two statements are true:

$$(9 + 4) + 2 \overset{?}{=} 9 + (4 + 2)$$
$$13 + 2 \overset{?}{=} 9 + 6$$
$$15 = 15$$
$$\text{TRUE}$$

and

$$(2 \cdot 3)4 \overset{?}{=} 2(3 \cdot 4)$$
$$(6)4 \overset{?}{=} 2(12)$$
$$24 = 24$$
$$\text{TRUE}$$

We can generalize and say that when adding or multiplying real numbers, the way in which we group them to evaluate them will not affect the result. Notice that the *order* in which the numbers are written does not change.

> **Property** Associative Properties
>
> If *a*, *b*, and *c* are real numbers, then
>
> 1)  $(a + b) + c = a + (b + c)$      Associative property of addition
>
> 2)  $(ab)c = a(bc)$      Associative property of multiplication

Sometimes, applying the associative property can simplify calculations.

## Example 6

Apply the associative property to simplify $\left(7 \cdot \dfrac{2}{5}\right)5$.

### Solution

By the associative property, $\left(7 \cdot \dfrac{2}{5}\right)5 = 7 \cdot \left(\dfrac{2}{\cancel{5}} \cdot \cancel{5}^{1}\right)$

$$= 7 \cdot 2$$
$$= 14 \qquad \blacksquare$$

**You Try 6**

Apply the associative property to simplify $\left(9 \cdot \dfrac{4}{3}\right)3$.

## Example 7

Use the associative property to simplify each expression.

a)  $-6 + (10 + y)$     b)  $\left(-\dfrac{3}{11} \cdot \dfrac{8}{5}\right)\dfrac{5}{8}$

### Solution

a)  $-6 + (10 + y) = (-6 + 10) + y$
$$= 4 + y$$

b)  $\left(-\dfrac{3}{11} \cdot \dfrac{8}{5}\right)\dfrac{5}{8} = -\dfrac{3}{11}\left(\dfrac{8}{5} \cdot \dfrac{5}{8}\right)$

$$= -\dfrac{3}{11}(1) \qquad \text{A number times its reciprocal equals 1.}$$

$$= -\dfrac{3}{11} \qquad \qquad \blacksquare$$

**You Try 7**

Use the associative property to simplify each expression.

a)  $(k + 3) + 9$     b)  $\left(-\dfrac{9}{7} \cdot \dfrac{8}{5}\right)\dfrac{5}{8}$

The identity properties of addition and multiplication are also ones we need to know.

### 6. Use the Identity and Inverse Properties

For addition we know that, for example,

$$5 + 0 = 5, \qquad 0 + \dfrac{2}{3} = \dfrac{2}{3}, \qquad -14 + 0 = -14.$$

When zero is added to a number, the value of the number is unchanged. *Zero* is the **identity element for addition** (also called the **additive identity**).

What is the identity element for multiplication?

$$-4(1) = -4 \qquad 1(3.82) = 3.82 \qquad \frac{9}{2}(1) = \frac{9}{2}$$

When a number is multiplied by 1, the value of the number is unchanged. *One* is the **identity element for multiplication** (also called the **multiplicative identity**).

---

**Property**    Identity Properties

If $a$ is a real number, then

1)  $a + 0 = 0 + a = a$    Identity property of addition

2)  $a \cdot 1 = 1 \cdot a = a$    Identity property of multiplication

---

The next properties we will discuss give us the additive and multiplicative identities as results. In Section 1.4, we introduced an **additive inverse**.

| Number | Additive Inverse |
|:------:|:----------------:|
| 3 | $-3$ |
| $-11$ | 11 |
| $-\dfrac{7}{9}$ | $\dfrac{7}{9}$ |

Let's add each number and its additive inverse:

$$3 + (-3) = 0, \qquad -11 + 11 = 0, \qquad -\frac{7}{9} + \frac{7}{9} = 0.$$

**Note**

The sum of a number and its additive inverse is zero (the identity element for addition).

Given a number such as $\dfrac{3}{5}$, we know that its **reciprocal** (or **multiplicative inverse**) is $\dfrac{5}{3}$. We have also established the fact that the product of a number and its reciprocal is 1 as in

$$\frac{3}{5} \cdot \frac{5}{3} = 1$$

Therefore, multiplying a number $b$ by its reciprocal (multiplicative inverse) $\dfrac{1}{b}$ gives us the identity element for multiplication, 1. That is,

$$b \cdot \frac{1}{b} = \frac{1}{b} \cdot b = 1$$

---

**Property**    Inverse Properties

If $a$ is any real number and $b$ is a real number not equal to 0, then

1)  $a + (-a) = -a + a = 0$    Inverse property of addition

2)  $b \cdot \dfrac{1}{b} = \dfrac{1}{b} \cdot b = 1$    Inverse property of multiplication

---

## Example 8

Which property is illustrated by each statement?

a)  $0 + 12 = 12$

b)  $-9.4 + 9.4 = 0$

c)  $\dfrac{1}{7} \cdot 7 = 1$

d)  $2(1) = 2$

### Solution

a)  $0 + 12 = 12$    Identity property of addition

b)  $-9.4 + 9.4 = 0$    Inverse property of addition

c)  $\dfrac{1}{7} \cdot 7 = 1$    Inverse property of multiplication

d)  $2(1) = 2$    Identity property of multiplication    ∎

### You Try 8

Which property is illustrated by each statement?

a)  $5 \cdot \dfrac{1}{5} = 1$

b)  $-26 + 26 = 0$

c)  $2.7(1) = 2.7$

d)  $-4 + 0 = -4$

## 7. Use the Distributive Property

The last property we will discuss is the **distributive property**. It involves both multiplication and addition or multiplication and subtraction.

### Property   Distributive Properties

If $a$, $b$, and $c$ are real numbers, then

1)  $a(b + c) = ab + ac$    and    $(b + c)a = ba + ca$

2)  $a(b - c) = ab - ac$    and    $(b - c)a = ba - ca$

## Example 9

Evaluate using the distributive property.

a)  $3(2 + 8)$

b)  $-6(7 - 8)$

c)  $-(6 + 3)$

### Solution

a)  $3(2 + 8) = 3 \cdot 2 + 3 \cdot 8$    Apply distributive property.

  $= 6 + 24$

  $= 30$

*Note*: We would get the same result if we would apply the order of operations:

$$3(2 + 8) = 3(10)$$
$$= 30$$

b) $-6(7 - 8) = -6 \cdot 7 - (-6)(8)$    Apply distributive property.
$\qquad\qquad\quad = -42 - (-48)$
$\qquad\qquad\quad = -42 + 48$
$\qquad\qquad\quad = 6$

c) $-(6 + 3) = -1(6 + 3)$
$\qquad\qquad\quad = -1 \cdot 6 + (-1)(3)$    Apply distributive property.
$\qquad\qquad\quad = -6 + (-3)$
$\qquad\qquad\quad = -9$

A negative sign in front of parentheses is the same as multiplying by $-1$.    ■

**You Try 9**

**Evaluate using the distributive property.**

a) $2(11 - 5)$    b) $-5(3 - 7)$    c) $-(4 + 9)$

The distributive property can be applied when there are more than two terms in parentheses and when there are variables.

**Example 10**

Use the distributive property to rewrite each expression. Simplify if possible.

a) $-2(3 + 8 - 5)$    b) $7(x + 4)$    c) $-(-5c + 4d - 6)$

***Solution***

a) $-2(3 + 8 - 5) = -2 \cdot 3 + (-2)(8) - (-2)(5)$    Apply distributive property.
$\qquad\qquad\qquad\quad = -6 + (-16) - (-10)$    Multiply.
$\qquad\qquad\qquad\quad = -6 + (-16) + 10$
$\qquad\qquad\qquad\quad = -12$

b) $7(x + 4) = 7x + 7 \cdot 4$    Apply distributive property.
$\qquad\qquad\quad = 7x + 28$

c) $-(-5c + 4d - 6) = -1(-5c + 4d - 6)$
$\qquad\qquad\qquad\qquad = -1(-5c) + (-1)(4d) - (-1)(6)$    Apply distributive property.
$\qquad\qquad\qquad\qquad = 5c + (-4d) - (-6)$    Multiply.
$\qquad\qquad\qquad\qquad = 5c - 4d + 6$    ■

**You Try 10**

**Use the distributive property to rewrite each expression. Simplify if possible.**

a) $6(a + 2)$    b) $5(2x - 7y - 4z)$    c) $-(-r + 4s - 9)$

The properties stated previously are summarized next.

**Summary**   Properties of Real Numbers

If $a$, $b$, and $c$ are real numbers, then

| | |
|---|---|
| **Commutative Properties:** | $a + b = b + a$ and $ab = ba$ |
| **Associative Properties:** | $(a + b) + c = a + (b + c)$ and $(ab)c = a(bc)$ |
| **Identity Properties:** | $a + 0 = 0 + a = a$ |
| | $a \cdot 1 = 1 \cdot a = a$ |
| **Inverse Properties:** | $a + (-a) = -a + a = 0$ |
| | $b \cdot \dfrac{1}{b} = \dfrac{1}{b} \cdot b = 1$ $(b \neq 0)$ |
| **Distributive Properties:** | $a(b + c) = ab + ac$ and $(b + c)a = ba + ca$ |
| | $a(b - c) = ab - ac$ and $(b - c)a = ba - ca$ |

## 8. Combine Like Terms

To simplify an expression like $15a + 11a - 8a + 3a$, we combine like terms using the distributive property.

$$
\begin{aligned}
15a + 11a - 8a + 3a &= (15 + 11 - 8 + 3)a && \text{Distributive property} \\
&= (26 - 8 + 3)a && \text{Order of operations} \\
&= (18 + 3)a && \text{Order of operations} \\
&= 21a
\end{aligned}
$$

We can add and subtract only those terms that are like terms.

### Example 11

Combine like terms.

a)  $-9k + 2k$      b)  $n + 8 - 4n + 3$      c)  $\dfrac{3}{5}t^2 + \dfrac{1}{4}t^2$

d)  $10x^2 + 6x - 2x^2 + 5x$

**Solution**

a)  We can use the distributive property to combine like terms.

$$-9k + 2k = (-9 + 2)k = -7k$$

Notice that using the distributive property to combine like terms is the same as combining the coefficients of the terms and leaving the variable and its exponent the same.

b)  $\begin{aligned}[t] n + 8 - 4n + 3 &= n - 4n + 8 + 3 \\ &= -3n + 11 \end{aligned}$      Rewrite like terms together.
         Remember, $n$ is the same as $1n$.

c)  $\begin{aligned}[t] \dfrac{3}{5}t^2 + \dfrac{1}{4}t^2 &= \dfrac{12}{20}t^2 + \dfrac{5}{20}t^2 \\ &= \dfrac{17}{20}t^2 \end{aligned}$      Get a common denominator.

d)  $\begin{aligned}[t] 10x^2 + 6x - 2x^2 + 5x &= 10x^2 - 2x^2 + 6x + 5x \\ &= 8x^2 + 11x \end{aligned}$      Rewrite like terms together.

$8x^2 + 11x$ cannot be simplified more because the terms are *not* like terms.      ■

**You Try 11**

Combine like terms.

a)  $6z + 5z$    b)  $q - 9 - 4q + 11$    c)  $\dfrac{5}{6}c^2 - \dfrac{2}{3}c^2$

d)  $2y^2 + 8y + y^2 - 3y$

If an expression contains parentheses, we use the distributive property to clear the parentheses, and then combine like terms.

**Example 12**

Combine like terms.

a)  $5(2c + 3) - 3c + 4$    b)  $3(2n + 1) - (6n - 11)$

c)  $\dfrac{3}{8}(8 - 4p) + \dfrac{5}{6}(2p - 6)$

**Solution**

a)  $5(2c + 3) - 3c + 4 = 10c + 15 - 3c + 4$    Distributive property
$$= 10c - 3c + 15 + 4 \qquad \text{Rewrite like terms together.}$$
$$= 7c + 19$$

b)  $3(2n + 1) - (6n - 11) = 3(2n + 1) - 1(6n - 11)$    Remember, $-(6n - 11)$ is the same as $-1(6n - 11)$.
$$= 6n + 3 - 6n + 11 \qquad \text{Distributive property}$$
$$= 6n - 6n + 3 + 11 \qquad \text{Rewrite like terms together.}$$
$$= 0n + 14 \qquad\qquad 0n = 0$$
$$= 14$$

c)  $\dfrac{3}{8}(8 - 4p) + \dfrac{5}{6}(2p - 6) = \dfrac{3}{8}(8) - \dfrac{3}{8}(4p) + \dfrac{5}{6}(2p) - \dfrac{5}{6}(6)$    Distributive property

$$= 3 - \dfrac{3}{2}p + \dfrac{5}{3}p - 5 \qquad \text{Multiply.}$$

$$= -\dfrac{3}{2}p + \dfrac{5}{3}p + 3 - 5 \qquad \text{Rewrite like terms together.}$$

$$= -\dfrac{9}{6}p + \dfrac{10}{6}p + 3 - 5 \qquad \text{Get a common denominator.}$$

$$= \dfrac{1}{6}p - 2 \qquad \text{Combine like terms.}$$

**You Try 12**

Combine like terms.

a)  $9d^2 - 7 + 2d^2 + 3$    b)  $10 - 3(2k + 5) + k - 6$

## 9. Translate English Expressions to Mathematical Expressions

Translating from English to a mathematical expression is a skill that is necessary to solve applied problems. We will practice writing mathematical expressions.

Read the phrase carefully, choose a variable to represent the unknown quantity, then translate the phrase to a mathematical expression.

### Example 13

Write a mathematical expression for each and simplify. Define the unknown with a variable.

a)   Seven more than twice a number
b)   The sum of a number and four times the same number

**Solution**

a)   Seven more than twice a number
   i)   **Define the unknown.** This means that you should clearly state on your paper what the variable represents.

$$\text{Let } x = \text{the number.}$$

   ii)  **Slowly, break down the phrase.** How do you write an expression for "seven more than" something?

$$+ \; 7$$

   iii) **What does "twice a number" mean?** It means two times the number. Since our number is represented by $x$, "twice a number" is $2x$.
   iv)  **Put the information together:**

Seven more than twice a number

$$2x \qquad + \qquad 7$$

The expression is $2x + 7$.

b)   The sum of a number and four times the same number
   i)   **Define the unknown.**

$$\text{Let } y = \text{the number.}$$

   ii)  **Slowly, break down the phrase.** What does *sum* mean? **Add.** So, we have to add a number and four times the same number:

Number + 4(Number)

   iii) Since $y$ represents the number, *four times the number* is $4y$.
   iv)  Therefore, to translate from English to a mathematical expression, we know that we must add the number, $y$, to four times the number, $4y$. Our expression is $y + 4y$. It simplifies to $5y$.    ∎

### You Try 13

Write a mathematical expression for each and simplify. Let *x* equal the unknown number.

a)   Five less than twice a number
b)   The sum of a number and two times the same number

## Using Technology

A graphing calculator can be used to evaluate an algebraic expression. This is especially valuable when evaluating expressions for several values of the given variables.

We will evaluate the expression $\dfrac{x^2 - 2xy}{3x + y}$ when $x = -3$ and $y = 8$.

### Method 1

Substitute the values for the variables and evaluate the arithmetic expression on the home screen. Each value substituted for a variable should be enclosed in parentheses to guarantee a correct answer. For example $(-3)^2$ gives the result 9, whereas $-3^2$ gives the result $-9$. Be careful to press the negative key $\boxed{(-)}$ when entering a negative sign and the minus key $\boxed{-}$ when entering the minus operator.

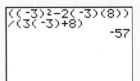

### Method 2

Store the given values in the variables and evaluate the algebraic expression on the home screen.

To store $-3$ in the variable $x$, press $\boxed{(-)}$ $\boxed{3}$ $\boxed{\text{STO>}}$ $\boxed{\text{X,T,o,n}}$ $\boxed{\text{ENTER}}$.

To store $8$ in the variable $y$, press $\boxed{8}$ $\boxed{\text{STO>}}$ $\boxed{\text{ALPHA}}$ $\boxed{1}$ $\boxed{\text{ENTER}}$.

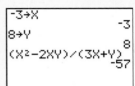

Enter $\dfrac{x^2 - 2xy}{3x + y}$ on the home screen.

The advantage of Method 2 is that we can easily store two different values in $x$ and $y$. For example, store 5 in $x$ and $-2$ in $y$. It is not necessary to enter the expression again because the calculator can recall previous entries.

Press $\boxed{\text{2nd}}$ $\boxed{\text{ENTER}}$ three times; then press $\boxed{\text{ENTER}}$.

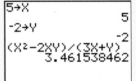

To convert this decimal to a fraction, press $\boxed{\text{MATH}}$ $\boxed{\text{ENTER}}$ $\boxed{\text{ENTER}}$.

Evaluate each expression when $x = -5$ and $y = 2$.

1. $3y - 4x$    2. $2xy - 5y$    3. $y^3 - 2x^2$
4. $\dfrac{x - y}{4x}$    5. $\dfrac{2x + 5y}{x - y}$    6. $\dfrac{x - y^2}{2x}$

---

### Answers to You Try Exercises

1)

| Term | Coeff. |
|------|--------|
| $-15r^3$ | $-15$ |
| $r^2$ | 1 |
| $-4r$ | $-4$ |
| 8 | 8 |

2) $-7$    3) 4    4) a) yes   b) yes   c) no

5) a) $16 + 1$   b) $6n$    6) 36    7) a) $k + 12$

b) $-\dfrac{9}{7}$ or $-1\dfrac{2}{7}$    8) a) inverse property of multiplication

b) inverse property of addition   c) identity property of multiplication   d) identity property of addition

9) a) 12   b) 20   c) $-13$    10) a) $6a + 12$

b) $10x - 35y - 20z$   c) $r - 4s + 9$    11) a) $11z$

b) $-3q + 2$   c) $\dfrac{1}{6}c^2$    d) $3y^2 + 5y$

12) a) $11d^2 - 4$   b) $-5k - 11$

13) a) $2x - 5$   b) $x + 2x$; $3x$

---

**Answers to Technology Exercises**

1. 26    2. $-30$    3. $-42$    4. $\dfrac{7}{20}$    5. 0    6. $\dfrac{9}{10}$

---

# 1.7 Exercises

**Objective 1: Identify the Terms and Coefficients in an Expression**

For each expression, list the terms and their coefficients. Also, identify the constant.

1) $7p^2 - 6p + 4$

2) $-8z + \dfrac{5}{6}$

3) $x^2 y^2 + 2xy - y + 11$

4) $w^3 - w^2 + 9w - 5$

5) $-2g^5 + \dfrac{g^4}{5} + 3.8g^2 + g - 1$

6) $121c^2 - d^2$

**Objective 2: Evaluate Algebraic Expressions**

7) Evaluate $4c + 3$ when

a) $c = 2$    b) $c = -5$

8) Evaluate $8m - 5$ when

a) $m = 3$    b) $m = -1$

Evaluate each expression when $x = 3$, $y = -5$, and $z = -2$.

9) $x + 4y$

10) $3z - y$

11) $z^2 - xy - 19$

12) $x^2 + 4yz$

13) $\dfrac{x^3}{2y + 1}$

14) $\dfrac{z^3}{x^2 - 1}$

15) $\dfrac{z^2 - y^2}{2y - 4(x + z)}$

16) $\dfrac{10 + 3(y + 2z)}{x^3 - z^4}$

**Objective 3: Identify Like Terms**

17) Are $9k$ and $9k^2$ *like* terms? Why or why not?

18) Are $\dfrac{3}{4}n$ and $8n$ *like* terms? Why or why not?

19) Are $a^3b$ and $-7a^3b$ *like* terms? Why or why not?

20) Write three *like* terms that are $x^2$-terms.

**Mixed Exercises: Objectives 4–7**

21) What is the identity element for multiplication?

22) What is the identity element for addition?

23) What is the additive inverse of 5?

24) What is the multiplicative inverse of 8?

Which property of real numbers is illustrated by each example? Choose from the commutative, associative, identity, inverse, or distributive property.

25) $9(2 + 8) = 9 \cdot 2 + 9 \cdot 8$

26) $(-16 + 7) + 3 = -16 + (7 + 3)$

27) $14 \cdot 1 = 14$

28) $\left(\dfrac{9}{2}\right)\left(\dfrac{2}{9}\right) = 1$

29) $-10 + 18 = 18 + (-10)$

30) $4 \cdot 6 - 4 \cdot 1 = 4(6 - 1)$

31) $5(2 \cdot 3) = (5 \cdot 2) \cdot 3$

32) $11 \cdot 7 = 7 \cdot 11$

Rewrite each expression using the indicated property.

33) $p + 19$; commutative

34) $5(m + n)$; distributive

35) $8 + (1 + 9)$; associative

36) $-2c + 0$; identity

37) $3(k - 7)$; distributive

38) $10 + 9x$; commutative

39) $y + 0$; identity

40) $\left(4 \cdot \dfrac{2}{7}\right) \cdot 7$; associative

41) Is $2a - 7$ equivalent to $7 - 2a$? Why or why not?

42) Is $6 + t$ equivalent to $t + 6$? Why or why not?

Rewrite each expression using the distributive property. Simplify if possible.

43) $2(1 + 9)$

44) $3(9 + 4)$

VIDEO 45) $-2(5 + 7)$

46) $-5(3 + 7)$

47) $4(8 - 3)$

48) $-6(5 - 11)$

49) $-(10 - 4)$

50) $-(3 + 9)$

51) $8(y + 3)$

52) $4(k + 11)$

53) $-10(z + 6)$

54) $-7(m + 5)$

55) $-3(x - 4y - 6)$

56) $6(2a - 5b + 1)$

VIDEO 57) $-(-8c + 9d - 14)$

58) $-(x - 10y - 4z)$

**Objective 8: Combine Like Terms**

Combine like terms and simplify.

VIDEO 59) $10p + 9 + 14p - 2$

60) $11 - k^2 + 12k^2 - 3 + 6k^2$

61) $-18y^2 - 2y^2 + 19 + y^2 - 2 + 13$

62) $-7x - 3x - 1 + 9x + 6 - 2x$

63) $\dfrac{4}{9} + 3r - \dfrac{2}{3} + \dfrac{1}{5}r$

64) $6a - \dfrac{3}{8}a + 2 + \dfrac{1}{4} - \dfrac{3}{4}a$

65) $2(3w + 5) + w$

66) $-8d^2 + 6(d^2 - 3) + 7$

VIDEO 67) $9 - 4(3 - x) - 4x + 3$

68) $m + 11 + 3(2m - 5) + 1$

69) $3g - (8g + 3) + 5$

70) $-6 + 4(10b - 11) - 8(5b + 2)$

71) $-5(t - 2) - (10 - 2t)$

72) $11 + 8(3u - 4) - 2(u + 6) + 9$

73) $3[2(5x + 7) - 11] + 4(7 - x)$

74) $22 - [6 + 5(2w - 3)] - (7w + 16)$

VIDEO 75) $\dfrac{4}{5}(2z + 10) - \dfrac{1}{2}(z + 3)$

76) $\dfrac{2}{3}(6c - 7) + \dfrac{5}{12}(2c + 5)$

77) $1 + \dfrac{3}{4}(10t - 3) + \dfrac{5}{8}\left(t + \dfrac{1}{10}\right)$

78) $\dfrac{7}{15} - \dfrac{9}{10}(2y + 1) - \dfrac{2}{5}(4y - 3)$

79) $2.5(x - 4) - 1.2(3x + 8)$

80) $9.4 - 3.8(2a + 5) + 0.6 + 1.9a$

**Objective 9: Translate English Expressions to Mathematical Expressions**

Write a mathematical expression for each phrase, and combine like terms if possible. Let $x$ represent the unknown quantity.

81) Eighteen more than a number

82) Eleven more than a number

83) Six subtracted from a number

84) Eight subtracted from a number

85) Three less than a number

86) Fourteen less than a number

87) The sum of twelve and twice a number

88) Five added to the sum of a number and six

VIDEO 89) Seven less than the sum of three and twice a number

90) Two more than the sum of a number and nine

91) The sum of a number and fifteen decreased by five

92) The sum of $-8$ and twice a number increased by three

# Chapter 1: Summary

| Definition/Procedure | Example |
|---|---|

## 1.1 Review of Fractions

### Reducing Fractions
A fraction is in **lowest terms** when the numerator and denominator have no common factors other than 1. **(p. 2)**

Write $\dfrac{36}{48}$ in lowest terms. Divide 36 and 48 by a common

factor, 12. Since $36 \div 12 = 3$ and $48 \div 12 = 4$, $\dfrac{36}{48} = \dfrac{3}{4}$.

### Multiplying Fractions
To multiply fractions, multiply the numerators and multiply the denominators. Common factors can be divided out either before or after multiplying. **(p. 6)**

Multiply $\dfrac{21}{45} \cdot \dfrac{9}{14}$.

$\dfrac{\overset{3}{\cancel{21}}}{\underset{5}{\cancel{45}}} \cdot \dfrac{\overset{1}{\cancel{9}}}{\underset{2}{\cancel{14}}}$  ← 9 and 45 each divide by 9.

← 21 and 14 each divide by 7.

$= \dfrac{3}{5} \cdot \dfrac{1}{2} = \dfrac{3}{10}$

### Dividing Fractions
To divide fractions, multiply the first fraction by the reciprocal of the second. **(p. 7)**

Divide $\dfrac{7}{5} \div \dfrac{4}{3}$.

$\dfrac{7}{5} \div \dfrac{4}{3} = \dfrac{7}{5} \cdot \dfrac{3}{4} = \dfrac{21}{20}$ or $1\dfrac{1}{20}$

### Adding and Subtracting Fractions
To add or subtract fractions,
1) Identify the least common denominator (LCD).
2) Write each fraction as an equivalent fraction using the LCD.
3) Add or subtract.
4) Express the answer in lowest terms. **(p. 12)**

Add $\dfrac{5}{11} + \dfrac{2}{11}$.  $\dfrac{5}{11} + \dfrac{2}{11} = \dfrac{7}{11}$

Subtract $\dfrac{8}{9} - \dfrac{3}{4}$.  $\dfrac{8}{9} - \dfrac{3}{4} = \dfrac{32}{36} - \dfrac{27}{36} = \dfrac{5}{36}$

## 1.2 Exponents and Order of Operations

### Exponents
An **exponent** represents repeated multiplication. **(p. 17)**

Write $9 \cdot 9 \cdot 9 \cdot 9 \cdot 9$ in exponential form.
$9 \cdot 9 \cdot 9 \cdot 9 \cdot 9 = 9^5$

Evaluate $2^4$.  $2^4 = 2 \cdot 2 \cdot 2 \cdot 2 = 16$

### Order of Operations
**P**arentheses, **E**xponents, **M**ultiplication, **D**ivision, **A**ddition, **S**ubtraction **(p. 18)**

Evaluate $8 + (5 - 1)^2 - 6 \cdot 3$.
$8 + (5 - 1)^2 - 6 \cdot 3$

| | |
|---|---|
| $= 8 + 4^2 - 6 \cdot 3$ | Parentheses |
| $= 8 + 16 - 6 \cdot 3$ | Exponents |
| $= 8 + 16 - 18$ | Multiply. |
| $= 24 - 18$ | Add. |
| $= 6$ | Subtract. |

## 1.3 Geometry Review

### Important Angles
The definitions for an acute angle, an obtuse angle, and a right angle can be found on **p. 22.**

Two angles are **complementary** if the sum of their angles is 90°.

Two angles are **supplementary** if the sum of their angles is 180°. **(p. 22)**

The measure of an angle is 73°. Find the measure of its complement and its supplement.

The measure of its complement is 17° since $90° - 73° = 17°$.

The measure of its supplement is 107° since $180° - 73° = 107°$.

| Definition/Procedure | Example |
|---|---|

**Triangle Properties**

The sum of the measures of the angles of any triangle is 180°.

An **equilateral triangle** has three sides of equal length. Each angle measures 60°.

An **isosceles triangle** has two sides of equal length. The angles opposite the sides have the same measure.

A **scalene triangle** has no sides of equal length. No angles have the same measure. **(p. 23)**

Find the measure of ∠C.

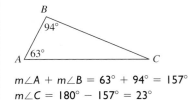

$m\angle A + m\angle B = 63° + 94° = 157°$
$m\angle C = 180° - 157° = 23°$

**Perimeter and Area**

The formulas for the perimeter and area of a rectangle, square, triangle, parallelogram, and trapezoid can be found on **p. 24.**

Find the area and perimeter of this rectangle.

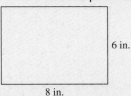

6 in.

8 in.

Area = (Length)(Width)   Perimeter = 2(Length) + 2(Width)
   = (8 in.)(6 in.)      = 2(8 in.) + 2(6 in.)
   = 48 in$^2$      = 16 in. + 12 in. = 28 in.

**Volume**

The formulas for the volume of a rectangular solid, cube, right circular cylinder, sphere, and right circular cone can be found on **p. 28.**

Find the volume of the cylinder pictured here.

9 cm

4 cm

Give an exact answer and give an approximation using 3.14 for $\pi$.

$V = \pi r^2 h$             $V = 144\pi \text{ cm}^3$
  $= \pi(4 \text{ cm})^2(9 \text{ cm})$     $\approx 144(3.14) \text{ cm}^3$
  $= \pi(16 \text{ cm}^2)(9 \text{ cm})$     $= 452.16 \text{ cm}^3$
  $= 144\pi \text{ cm}^3$

## 1.4 Sets of Numbers and Absolute Value

**Natural numbers:** $\{1, 2, 3, 4, \ldots\}$
**Whole numbers:** $\{0, 1, 2, 3, 4, \ldots\}$
**Integers:** $\{\ldots, -3, -2, -1, 0, 1, 2, 3, \ldots\}$

A **rational number** is any number of the form $\frac{p}{q}$, where $p$ and $q$ are integers and $q \neq 0$. **(p. 35)**

The following numbers are rational:   $-3, 10, \frac{5}{8}, 7.4, 2.\bar{3}$

An **irrational number** cannot be written as the quotient of two integers. **(p. 36)**

The following numbers are irrational:   $\sqrt{6}, 9.2731\ldots$

The set of **real numbers** includes the rational and irrational numbers. **(p. 37)**

Any number that can be represented on the number line is a real number.

The **additive inverse** of $a$ is $-a$. **(p. 39)**

The additive inverse of 4 is $-4$.

**Absolute Value**

$|a|$ is the distance of $a$ from zero. **(p. 40)**

$|-6| = 6$

| Definition/Procedure | Example |
|---|---|

## 1.5 Addition and Subtraction of Real Numbers

**Adding Real Numbers**

To add numbers with the **same sign**, add the absolute value of each number. The sum will have the same sign as the numbers being added. **(p. 43)**

$-3 + (-9) = -12$

To add two numbers with **different signs**, subtract the smaller absolute value from the larger. The sum will have the sign of the number with the larger absolute value. **(p. 44)**

$-20 + 15 = -5$

**Subtracting Real Numbers**

To subtract $a - b$, change subtraction to addition and add the additive inverse of $b$: $a - b = a + (-b)$. **(p. 45)**

$2 - 11 = 2 + (-11) = -9$

$-17 - (-7) = -17 + 7 = -10$

## 1.6 Multiplication and Division of Real Numbers

**Multiplying Real Numbers**

The product of two real numbers with the *same* sign is positive.

$8 \cdot 3 = 24 \qquad -7 \cdot (-8) = 56$

The product of a positive number and a negative number is *negative*.

$-2 \cdot 5 = -10 \qquad 9 \cdot (-1) = -9$

An *even number* of negative factors in a product gives a *positive* result.

$\underbrace{(-1)(-6)(-3)(2)(-4)}_{\text{4 negative factors}} = 144$

An *odd number* of negative factors in a product gives a *negative* result. **(p. 51)**

$\underbrace{(5)(-2)(-3)(1)(-1)}_{\text{3 negative factors}} = -30$

**Evaluating Exponential Expressions (p. 52)**

Evaluate $(-3)^4$. The base is $-3$.
$(-3)^4 = (-3)(-3)(-3)(-3) = 81$

Evaluate $-3^4$. The base is 3.
$-3^4 = -1 \cdot 3^4 = -1 \cdot 3 \cdot 3 \cdot 3 \cdot 3 = -81$

**Dividing real numbers**

The quotient of two numbers with the *same* sign is positive.

$\dfrac{40}{2} = 20 \qquad -18 \div (-3) = 6$

The quotient of two numbers with *different* signs is negative. **(p. 53)**

$\dfrac{-56}{8} = -7 \qquad 48 \div (-4) = -12$

## 1.7 Algebraic Expressions and Properties of Real Numbers

An **algebraic expression** is a collection of numbers, variables, and grouping symbols connected by operation symbols such as $+, -, \times,$ and $\div$. **(p. 57)**

$4y^2 - 7y + \dfrac{3}{5}$

**Important terms**

Variable        Constant

Term        Coefficient

We can evaluate expressions for different values of the variables. **(p. 58)**

Evaluate $2xy - 5y + 1$ when $x = -3$ and $y = 4$.

Substitute $-3$ for $x$ and 4 for $y$ and simplify.

$$\begin{aligned}
2xy - 5y + 1 &= 2(-3)(4) - 5(4) + 1 \\
&= -24 - 20 + 1 \\
&= -24 + (-20) + 1 \\
&= -43
\end{aligned}$$

| Definition/Procedure | Example |
|---|---|
| **Like Terms**<br>**Like terms** contain the same variables with the same exponents.<br>**(p. 59)** | In the group of terms $5k^2$, $-8k$, $-4k^2$, $\frac{1}{3}k$,<br><br>$5k^2$ and $-4k^2$ are like terms and $-8k$ and $\frac{1}{3}k$ are like terms. |
| **Properties of Real Numbers**<br>If $a$, $b$, and $c$ are real numbers, then the following properties hold. | |
| **Commutative Properties:**<br>$a + b = b + a$<br>$ab = ba$ | $10 + 3 = 3 + 10$<br>$(-6)(5) = (5)(-6)$ |
| **Associative Properties:**<br>$(a + b) + c = a + (b + c)$<br>$(ab)c = a(bc)$ | $(9 + 4) + 2 = 9 + (4 + 2)$<br>$(5 \cdot 2)8 = 5 \cdot (2 \cdot 8)$ |
| **Identity Properties:**<br>$a + 0 = 0 + a = a$<br>$a \cdot 1 = 1 \cdot a = a$ | $7 + 0 = 7 \qquad \frac{2}{3} \cdot 1 = \frac{2}{3}$ |
| **Inverse Properties:**<br>$a + (-a) = -a + a = 0$<br>$b \cdot \dfrac{1}{b} = \dfrac{1}{b} \cdot b = 1$ | $11 + (-11) = 0 \qquad 5 \cdot \dfrac{1}{5} = 1$ |
| **Distributive Properties:**<br>$a(b + c) = ab + ac$ and $(b + c)a = ba + ca$<br>$a(b - c) = ab - ac$ and $(b - c)a = ba - ca$ **(p. 65)** | $6(5 + 8) = 6 \cdot 5 + 6 \cdot 8$<br>$\qquad\quad = 30 + 48$<br>$\qquad\quad = 78$<br>$9(w - 2) = 9w - 9 \cdot 2$<br>$\qquad\qquad = 9w - 18$ |
| **Combining Like Terms**<br>We can simplify expressions by combining like terms. **(p. 65)** | Combine like terms and simplify.<br>$4n^2 - 3n + 1 - 2(6n^2 - 5n + 7)$<br>$= 4n^2 - 3n + 1 - 12n^2 + 10n - 14$    Distributive property<br>$= -8n^2 + 7n - 13$    Combine like terms. |
| **Writing Mathematical Expressions (p. 67)** | Write a mathematical expression for the following:<br>*Sixteen more than twice a number*<br><br>Let $x =$ the number.<br><br><u>Sixteen more than</u>    <u>twice a number</u><br>$\qquad +16 \qquad\qquad\qquad 2x$<br><br>$\qquad\qquad\quad 2x + 16$ |

**(1.1)**

1) **Find all factors of each number.**

   a) 16  b) 37

2) **Find the prime factorization of each number.**

   a) 28  b) 66

3) **Write each fraction in lowest terms.**

   a) $\dfrac{12}{30}$  b) $\dfrac{414}{702}$

**Perform the indicated operation. Write the answer in lowest terms.**

4) $\dfrac{4}{11} \cdot \dfrac{3}{5}$  5) $\dfrac{45}{64} \cdot \dfrac{32}{75}$

6) $\dfrac{5}{8} \div \dfrac{3}{10}$  7) $35 \div \dfrac{7}{8}$

8) $4\dfrac{2}{3} \cdot 1\dfrac{1}{8}$  9) $\dfrac{30}{49} \div 2\dfrac{6}{7}$

10) $\dfrac{2}{9} + \dfrac{4}{9}$  11) $\dfrac{2}{3} + \dfrac{1}{4}$

12) $\dfrac{9}{40} + \dfrac{7}{16}$  13) $\dfrac{1}{5} + \dfrac{1}{3} + \dfrac{1}{6}$

14) $\dfrac{21}{25} - \dfrac{11}{25}$  15) $\dfrac{5}{8} - \dfrac{2}{7}$

16) $3\dfrac{2}{9} + 5\dfrac{3}{8}$  17) $9\dfrac{3}{8} - 2\dfrac{5}{6}$

18) A pattern for a skirt calls for $1\dfrac{7}{8}$ yd of fabric. If Mary Kate wants to make one skirt for herself and one for her twin, how much fabric will she need?

**(1.2) Evaluate.**

19) $3^4$  20) $2^6$

21) $\left(\dfrac{3}{4}\right)^3$  22) $(0.6)^2$

23) $13 - 7 + 4$  24) $8 \cdot 3 + 20 \div 4$

25) $\dfrac{12 - 56 \div 8}{(1 + 5)^2 - 2^4}$

**(1.3)**

26) The complement of $51°$ is _____.

27) The supplement of $78°$ is _____.

28) Is this triangle acute, obtuse, or right? Find the missing angle.

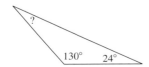

**Find the area and perimeter of each figure. Include the correct units.**

29) 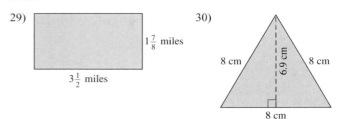  30)

31) 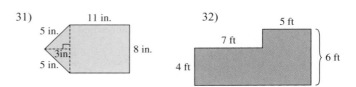  32)

**Find a) the area and b) the circumference of each circle. Give an exact answer for each and give an approximation using 3.14 for $\pi$. Include the correct units.**

33) 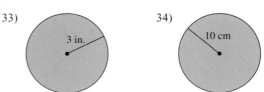  34)

**Find the area of the shaded region. Use 3.14 for $\pi$. Include the correct units.**

35)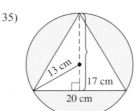

**Find the volume of each figure. Where appropriate, give the answer in terms of $\pi$. Include the correct units.**

36) 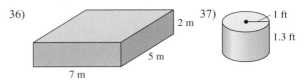  37)

38) 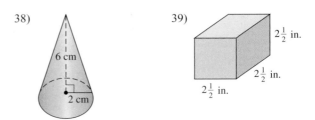  39)

40) The radius of a basketball is approximately 4.7 inches. Find its circumference to the nearest tenth of an inch.

**(1.4)**

41) Given this set of numbers,

$$\left\{ \frac{7}{15}, -16, 0, 3.\overline{2}, 8.5, \sqrt{31}, 4, 6.01832\ldots \right\}$$

list the

a) integers

b) rational numbers

c) natural numbers

d) whole numbers

e) irrational numbers

42) Graph and label these numbers on a number line.

$$-3.5, 4, \frac{9}{10}, 2\frac{1}{3}, -\frac{3}{4}, -5$$

43) Evaluate.

a) $|-18|$        b) $-|7|$

**(1.5) Add or subtract as indicated.**

44) $-38 + 13$

45) $-21 - (-40)$

46) $-1.9 + 2.3$

47) $\frac{5}{12} - \frac{5}{8}$

48) The lowest temperature on record in the country of Greenland is $-87°$F. The coldest temperature ever reached on the African continent was in Morocco and is $76°$ higher than Greenland's record low. What is the lowest temperature ever recorded in Africa?
(www.ncdc.noaa.gov)

**(1.6) Multiply or divide as indicated.**

49) $\left(-\frac{3}{2}\right)(8)$

50) $(-4.9)(-3.6)$

51) $(-4)(3)(-2)(-1)(-3)$

52) $\left(-\frac{2}{3}\right)(-5)(2)(-6)$

53) $-108 \div 9$

54) $\frac{56}{-84}$

55) $-3\frac{1}{8} \div \left(-\frac{5}{6}\right)$

56) $-\frac{9}{10} \div 12$

**Evaluate.**

57) $-6^2$

58) $(-6)^2$

59) $(-2)^6$

60) $-1^{10}$

61) $3^3$

62) $(-5)^3$

**Use the order of operations to simplify.**

63) $56 \div (-7) - 1$

64) $15 - (2 - 5)^3$

65) $-11 + 4 \cdot 3 + (-8 + 6)^5$

66) $\dfrac{1 + 6(7 - 3)}{2[3 - 2(8 - 1)] - 3}$

**Write a mathematical expression for each and simplify.**

67) The quotient of $-120$ and $-3$

68) Twice the sum of 22 and $-10$

69) 15 less than the product of $-4$ and 7

70) 11 more than half of $-18$

**(1.7)**

71) List the terms and coefficients of

$$5z^4 - 8z^3 + \frac{3}{5}z^2 - z + 14.$$

72) Evaluate $9x - 4y$ when $x = -3$ and $y = 7$.

73) Evaluate $\dfrac{2a + b}{a^3 - b^2}$ when $a = -3$ and $b = 5$.

**Which property of real numbers is illustrated by each example? Choose from the commutative, associative, identity, inverse, or distributive property.**

74) $12 + (5 + 3) = (12 + 5) + 3$

75) $\left(\dfrac{2}{5}\right)\left(\dfrac{5}{2}\right) = 1$

76) $0 + 19 = 19$

77) $-4(7 + 2) = -4(7) + (-4)(2)$

78) $8 \cdot 3 = 3 \cdot 8$

**Rewrite each expression using the distributive property. Simplify if possible.**

79) $7(3 - 9)$

80) $(10 + 4)5$

81) $-(15 - 3)$

82) $-6(9p - 4q + 1)$

**Combine like terms and simplify.**

83) $9m - 14 + 3m + 4$

84) $-5c + d - 2c + 8d$

85) $15y^2 + 8y - 4 + 2y^2 - 11y + 1$

86) $7t + 10 - 3(2t + 3)$

87) $\dfrac{3}{2}(5n - 4) + \dfrac{1}{4}(n + 6)$

88) $1.4(a + 5) - (a + 2)$

1) Find the prime factorization of 210.

2) Write in lowest terms:

   a) $\dfrac{45}{72}$

   b) $\dfrac{420}{560}$

**Perform the indicated operations. Write all answers in lowest terms.**

3) $\dfrac{7}{16} \cdot \dfrac{10}{21}$

4) $\dfrac{5}{12} + \dfrac{2}{9}$

5) $10\dfrac{2}{3} - 3\dfrac{1}{4}$

6) $\dfrac{4}{9} \div 12$

7) $\dfrac{3}{5} - \dfrac{17}{20}$

8) $-31 - (-14)$

9) $16 + 8 \div 2$

10) $\dfrac{1}{8} \cdot \left(-\dfrac{2}{3}\right)$

11) $-15 \cdot (-4)$

12) $-9.5 + 5.8$

13) $23 - 6[-4 + (9 - 11)^4]$

14) $\dfrac{7 \cdot 2 - 4}{48 \div 3 - 8^0}$

15) An extreme sports athlete has reached an altitude of 14,693 ft while ice climbing and has dived to a depth of 518 ft below sea level. What is the difference between these two elevations?

16) Evaluate.

   a) $5^3$

   b) $-2^4$

   c) $|-43|$

   d) $-|18 - 40| - 3|9 - 4|$

17) The supplement of 31° is _____.

18) Find the missing angle, and classify the triangle as acute, obtuse, or right.

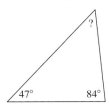

19) Find the area and perimeter of each figure. Include the correct units.

   a)

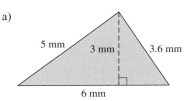

   b)

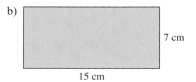

   c)

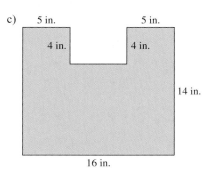

20) Find the volume of this figure:

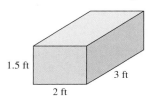

21) The radius of the pitcher's mound on a major-league baseball diamond is 9 ft.

   a) Find the exact area of the pitcher's mound.

   b) Find the approximate area of the pitcher's mound using 3.14 for $\pi$.

22) Given this set of numbers,

   $\{3\dfrac{1}{5}, 22, -7, \sqrt{43}, 0, 6.2, 1.\overline{5}, 8.0934\ldots\}$ list the

   a) whole numbers

   b) natural numbers

   c) irrational numbers

   d) integers

   e) rational numbers

23) Graph the numbers on a number line. Label each.

   $4, -5, \dfrac{2}{3}, -3\dfrac{1}{2}, -\dfrac{5}{6}, 2.2$

24) Write a mathematical expression for each and simplify.

   a) The sum of $-4$ and 27

   b) The product of 5 and $-6$ subtracted from 17

25) List the terms and coefficients of

$$4p^3 - p^2 + \frac{1}{3}p - 10.$$

26) Evaluate $\dfrac{x^2 - y^2}{6y + x}$ when $x = 3$ and $y = -4$.

27) Which property of real numbers is illustrated by each example? Choose from the commutative, associative, identity, inverse, or distributive property.

a) $9 \cdot 5 = 5 \cdot 9$

b) $16 + (4 + 7) = (16 + 4) + 7$

c) $\left(\dfrac{10}{3}\right)\left(\dfrac{3}{10}\right) = 1$

d) $8(1 - 4) = 8 \cdot 1 - 8 \cdot 4$

28) Rewrite each expression using the distributive property. Simplify if possible.

a) $-4(2 + 7)$

b) $3(8m - 3n + 11)$

29) Combine like terms and simplify.

a) $-8k^2 + 3k - 5 + 2k^2 + k - 9$

b) $\dfrac{4}{3}(6c - 5) - \dfrac{1}{2}(4c + 3)$

30) Write a mathematical expression for "nine less than twice a number." Let $x$ represent the number.

# CHAPTER 2

# The Rules of Exponents

## Algebra at Work: Custom Motorcycle Shop

The people who build custom motorcycles use a lot of mathematics to do their jobs. Mark is building a chopper frame and needs to make the supports for the axle. He has to punch holes in the plates that will be welded to the frame.

Mark has to punch holes with a diameter of 1 in. in mild steel that is $\frac{3}{8}$ in. thick. The press punches two holes at a time. To determine how much power is needed to do this job, he uses a formula containing an exponent, $P = \dfrac{t^2 dN}{3.78}$. After substituting the numbers into the expression, he calculates that the power needed to punch these holes is 0.07 hp.

In this chapter, we will learn more about working with expressions containing exponents.

## Section 2.1A The Product Rule and Power Rules

### Objectives

1. Evaluate Exponential Expressions
2. Use the Product Rule for Exponents
3. Use the Power Rule $(a^m)^n = a^{mn}$
4. Use the Power Rule $(ab)^n = a^n b^n$
5. Use the Power Rule $\left(\dfrac{a}{b}\right)^n = \dfrac{a^n}{b^{n'}}$

Where $b \neq 0$

### 1. Evaluate Exponential Expressions

Recall from Chapter 1 that exponential notation is used as a shorthand way to represent a multiplication problem.

For example, $3 \cdot 3 \cdot 3 \cdot 3 \cdot 3$ can be written as $3^5$.

#### Definition

An **exponential expression** of the form $a^n$, where $a$ is any real number and $n$ is a positive integer, is equivalent to $\underbrace{a \cdot a \cdot a \cdot \cdots \cdot a}_{n \text{ factors of } a}$. We say that $a$ is the **base** and $n$ is the **exponent.**

We can also evaluate an exponential expression.

 **Example 1**

Identify the base and the exponent in each expression and evaluate.

a)  $2^4$      b)  $(-2)^4$       c)  $-2^4$

**Solution**

a)  $2^4$      *2 is the base, 4 is the exponent. Therefore,* $2^4 = 2 \cdot 2 \cdot 2 \cdot 2 = 16$.

b)  $(-2)^4$      $-2$ *is the base, 4 is the exponent. Therefore,*
$(-2)^4 = (-2) \cdot (-2) \cdot (-2) \cdot (-2) = 16$.

c)  $-2^4$      It may be very tempting to say that the base is $-2$. However, there are no parentheses in this expression. Therefore, *2 is the base, and 4 is the exponent.* To evaluate,

$$-2^4 = -1 \cdot 2^4 = -1 \cdot 2 \cdot 2 \cdot 2 \cdot 2$$
$$= -16$$

**BE CAREFUL**

The expressions $(-a)^n$ and $-a^n$ are not always equivalent:

$$(-a)^n = \underbrace{(-a) \cdot (-a) \cdot (-a) \cdot \ldots \cdot (-a)}_{n \text{ factors of } -a}$$

$$-a^n = -1 \cdot \underbrace{a \cdot a \cdot a \cdot \ldots \cdot a}_{n \text{ factors of } a}$$

 **You Try 1**

Identify the base and exponent in each expression and evaluate.

a)  $5^3$      b)  $-8^2$      c)  $\left(-\dfrac{2}{3}\right)^3$

## 2. Use the Product Rule for Exponents

Is there a rule to help us *multiply* exponential expressions? Let's rewrite each of the following products as a single power of the base using what we already know:

1) $\overbrace{\quad}^{\substack{\text{3 factors} \\ \text{of 2}}} \overbrace{\quad}^{\substack{2 \\ \text{factors} \\ \text{of 2}}}$
$2^3 \cdot 2^2 = 2 \cdot 2 \cdot 2 \cdot 2 \cdot 2$
$= \underbrace{2 \cdot 2 \cdot 2 \cdot 2 \cdot 2}_{\text{5 factors of 2}}$
$= 2^5$

2) $\overbrace{\quad}^{\substack{\text{4 factors} \\ \text{of 5}}} \overbrace{\quad}^{\substack{3 \\ \text{factors} \\ \text{of 5}}}$
$5^4 \cdot 5^3 = 5 \cdot 5 \cdot 5 \cdot 5 \cdot 5 \cdot 5 \cdot 5$
$= \underbrace{5 \cdot 5 \cdot 5 \cdot 5 \cdot 5 \cdot 5 \cdot 5}_{\text{7 factors of 5}}$
$= 5^7$

Let's summarize: $2^3 \cdot 2^2 = 2^5,\qquad 5^4 \cdot 5^3 = 5^7$

Do you notice a pattern? *When you multiply expressions with the same base, keep the same base and add the exponents.* This is called the **product rule** for exponents.

> **Property**   Product Rule
>
> Let $a$ be any real number and let $m$ and $n$ be positive integers. Then,
> $$a^m \cdot a^n = a^{m+n}$$

### Example 2

Find each product.

a)  $2^2 \cdot 2^4$     b)  $x^9 \cdot x^6$     c)  $5c^3 \cdot 7c^9$     d)  $(-k)^8 \cdot (-k) \cdot (-k)^{11}$

**Solution**

a)  $2^2 \cdot 2^4 = 2^{2+4} = 2^6 = 64$      Since the bases are the same, add the exponents.

b)  $x^9 \cdot x^6 = x^{9+6} = x^{15}$

c)  $5c^3 \cdot 7c^9 = (5 \cdot 7)(c^3 \cdot c^9)$      Associative and commutative properties
$= 35c^{12}$

d)  $(-k)^8 \cdot (-k) \cdot (-k)^{11} = (-k)^{8+1+11} = (-k)^{20}$      Product rule   ∎

### You Try 2

Find each product.

a)  $3 \cdot 3^2$     b)  $y^{10} \cdot y^4$     c)  $-6m^5 \cdot 9m^{11}$     d)  $h^4 \cdot h^6 \cdot h^4$     e)  $(-3)^2 \cdot (-3)^2$

**BE CAREFUL**

Can the product rule be applied to $4^3 \cdot 5^2$? **No!** The bases are not the same, so we cannot add the exponents. To evaluate $4^3 \cdot 5^2$, we would evaluate $4^3 = 64$ and $5^2 = 25$, then multiply:

$$4^3 \cdot 5^2 = 64 \cdot 25 = 1600$$

## 3. Use the Power Rule $(a^m)^n = a^{mn}$

What does $(2^2)^3$ mean?
We can rewrite $(2^2)^3$ first as $2^2 \cdot 2^2 \cdot 2^2$.

$2^2 \cdot 2^2 \cdot 2^2 = 2^{2+2+2}$      Use the product rule for exponents.
$= 2^6$      Add the exponents.
$= 64$      Simplify.

Notice that $(2^2)^3 = 2^{2+2+2}$, or $2^{2 \cdot 3}$. This leads us to the basic power rule for exponents: *When you raise a power to another power, keep the base and multiply the exponents.*

**Property   Basic Power Rule**

Let $a$ be any real number and let $m$ and $n$ be positive integers. Then,

$$(a^m)^n = a^{mn}$$

## Example 3

Simplify using the power rule.

a)   $(3^8)^4$          b)   $(n^3)^7$          c)   $((-f)^4)^3$

**Solution**

a)   $(3^8)^4 = 3^{8 \cdot 4} = 3^{32}$      b)   $(n^3)^7 = n^{3 \cdot 7} = n^{21}$      c)   $((-f)^4)^3 = (-f)^{4 \cdot 3} = (-f)^{12}$ ∎

**You Try 3**

Simplify using the power rule.

a)   $(5^4)^3$          b)   $(j^6)^5$          c)   $((-2)^3)^2$

## 4. Use the Power Rule $(ab)^n = a^n b^n$

We can use another power rule to simplify an expression such as $(5c)^3$. We can rewrite and simplify $(5c)^3$ as $5c \cdot 5c \cdot 5c = 5 \cdot 5 \cdot 5 \cdot c \cdot c \cdot c = 5^3 c^3 = 125c^3$. *To raise a product to a power, raise each factor to that power.*

**Property   Power Rule for a Product**

Let $a$ and $b$ be real numbers and let $n$ be a positive integer. Then,

$$(ab)^n = a^n b^n$$

**BE CAREFUL**

Notice that $(ab)^n = a^n b^n$ is different from $(a + b)^n$. $(a + b)^n \neq a^n + b^n$. We will study this in Chapter 6.

## Example 4

Simplify each expression.

a)   $(9y)^2$          b)   $\left(\dfrac{1}{4}t\right)^3$          c)   $(5c^2)^3$          d)   $3(6ab)^2$

**Solution**

a)   $(9y)^2 = 9^2 y^2 = 81y^2$                              b)   $\left(\dfrac{1}{4}t\right)^3 = \left(\dfrac{1}{4}\right)^3 \cdot t^3 = \dfrac{1}{64}t^3$

c)   $(5c^2)^3 = 5^3 \cdot (c^2)^3 = 125c^{2 \cdot 3} = 125c^6$

d)   $3(6ab)^2 = 3[6^2 \cdot (a)^2 \cdot (b)^2]$      The 3 is not in parentheses; therefore, it will not be squared.

$\qquad\qquad = 3(36a^2 b^2)$

$\qquad\qquad = 108a^2 b^2$ ∎

**You Try 4**

Simplify.

a) $(k^4)^7$      b) $(2k^{10}m^3)^6$      c) $(-r^2s^8)^3$      d) $-4(3tu)^2$

## 5. Use the Power Rule $\left(\dfrac{a}{b}\right)^n = \dfrac{a^n}{b^n}$, Where $b \neq 0$

Another power rule allows us to simplify an expression like $\left(\dfrac{2}{x}\right)^4$. We can rewrite and simplify $\left(\dfrac{2}{x}\right)^4$ as $\dfrac{2}{x} \cdot \dfrac{2}{x} \cdot \dfrac{2}{x} \cdot \dfrac{2}{x} = \dfrac{2 \cdot 2 \cdot 2 \cdot 2}{x \cdot x \cdot x \cdot x} = \dfrac{2^4}{x^4} = \dfrac{16}{x^4}$. *To raise a quotient to a power, raise both the numerator and denominator to that power.*

> **Property**  Power Rule for a Quotient
>
> Let $a$ and $b$ be real numbers and let $n$ be a positive integer. Then,
>
> $$\left(\frac{a}{b}\right)^n = \frac{a^n}{b^n}, \text{ where } b \neq 0$$

**Example 5**

Simplify using the power rule for quotients.

a) $\left(\dfrac{3}{8}\right)^2$      b) $\left(\dfrac{5}{x}\right)^3$      c) $\left(\dfrac{t}{u}\right)^9$

**Solution**

a) $\left(\dfrac{3}{8}\right)^2 = \dfrac{3^2}{8^2} = \dfrac{9}{64}$      b) $\left(\dfrac{5}{x}\right)^3 = \dfrac{5^3}{x^3} = \dfrac{125}{x^3}$      c) $\left(\dfrac{t}{u}\right)^9 = \dfrac{t^9}{u^9}$ ■

**You Try 5**

Simplify using the power rule for quotients.

a) $\left(\dfrac{5}{12}\right)^2$      b) $\left(\dfrac{2}{d}\right)^5$      c) $\left(\dfrac{u}{v}\right)^6$

Let's summarize the rules of exponents we have learned in this section:

> **Summary**   The Product and Power Rules of Exponents
>
> In the rules below, $a$ and $b$ are any real numbers and $m$ and $n$ are positive integers.

| Rule | | Example |
|---|---|---|
| Product rule | $a^m \cdot a^n = a^{m+n}$ | $p^4 \cdot p^{11} = p^{4+11} = p^{15}$ |
| Basic power rule | $(a^m)^n = a^{mn}$ | $(c^8)^3 = c^{8 \cdot 3} = c^{24}$ |
| Power rule for a product | $(ab)^n = a^n b^n$ | $(3z)^4 = 3^4 \cdot z^4 = 81z^4$ |
| Power rule for a quotient | $\left(\dfrac{a}{b}\right)^n = \dfrac{a^n}{b^n}, (b \neq 0)$ | $\left(\dfrac{w}{2}\right)^4 = \dfrac{w^4}{2^4} = \dfrac{w^4}{16}$ |

---

**Answers to You Try Exercises**

1) a) base: 5; exponent: 3; $5^3 = 125$    b) base: 8; exponent: 2; $-8^2 = -64$    c) base: $-\dfrac{2}{3}$; exponent: 3;

$\left(-\dfrac{2}{3}\right)^3 = -\dfrac{8}{27}$    2) a) 27    b) $y^{14}$    c) $-54m^{16}$    d) $h^{14}$    e) 81    3) a) $5^{12}$    b) $j^{30}$    c) 64    4) a) $k^{28}$

b) $64k^{60}m^{18}$    c) $-r^6s^{24}$    d) $-36t^2u^2$    5) a) $\dfrac{25}{144}$    b) $\dfrac{32}{d^5}$    c) $\dfrac{u^6}{v^6}$

---

## 2.1A Exercises

**Objective 1: Evaluate Exponential Expressions**

Rewrite each expression using exponents.

1) $9 \cdot 9 \cdot 9 \cdot 9 \cdot 9 \cdot 9$

2) $4 \cdot 4 \cdot 4 \cdot 4 \cdot 4 \cdot 4 \cdot 4$

3) $\left(\dfrac{1}{7}\right)\left(\dfrac{1}{7}\right)\left(\dfrac{1}{7}\right)\left(\dfrac{1}{7}\right)$

4) $(0.8)(0.8)(0.8)$

5) $(-5)(-5)(-5)(-5)(-5)(-5)(-5)$

6) $(-c)(-c)(-c)(-c)(-c)$

7) $(-3y)(-3y)(-3y)(-3y)(-3y)(-3y)(-3y)(-3y)$

8) $\left(-\dfrac{5}{4}t\right)\left(-\dfrac{5}{4}t\right)\left(-\dfrac{5}{4}t\right)\left(-\dfrac{5}{4}t\right)$

Identify the base and the exponent in each.

9) $6^8$

10) $9^4$

11) $(0.05)^7$

12) $(0.3)^{10}$

13) $(-8)^5$

14) $(-7)^6$

15) $(9x)^8$

16) $(13k)^3$

17) $(-11a)^2$

18) $(-2w)^9$

19) $5p^4$

20) $-3m^5$

21) $-\dfrac{3}{8}y^2$

22) $\dfrac{5}{9}t^7$

23) Evaluate $(3 + 4)^2$ and $3^2 + 4^2$. Are they equivalent? Why or why not?

24) Evaluate $(7 - 3)^2$ and $7^2 - 3^2$. Are they equivalent? Why or why not?

25) For any values of $a$ and $b$, does $(a + b)^2 = a^2 + b^2$? Why or why not?

26) Does $-2^4 = (-2)^4$? Why or why not?

27) Are $3t^4$ and $(3t)^4$ equivalent? Why or why not?

28) Is there any value of $a$ for which $(-a)^2 = -a^2$? Support your answer with an example.

Evaluate.

29) $2^5$

30) $9^2$

31) $(11)^2$

32) $4^3$

33) $(-2)^4$

34) $(-5)^3$

35) $-3^4$

36) $-6^2$

37) $-2^3$

38) $-8^2$

39) $\left(\dfrac{1}{5}\right)^3$

40) $\left(\dfrac{3}{2}\right)^4$

**Objective 2: Use the Product Rule for Exponents**

Evaluate the expression using the product rule, where applicable.

41) $2^2 \cdot 2^3$

42) $5^2 \cdot 5$

43) $3^2 \cdot 3^2$

44) $2^3 \cdot 2^3$

45) $5^2 \cdot 2^3$

46) $4^3 \cdot 3^2$

47) $\left(\dfrac{1}{2}\right)^4 \cdot \left(\dfrac{1}{2}\right)^2$

48) $\left(\dfrac{4}{3}\right) \cdot \left(\dfrac{4}{3}\right)^2$

Simplify the expression using the product rule. Leave your answer in exponential form.

49) $8^3 \cdot 8^9$

50) $6^4 \cdot 6^3$

51) $5^2 \cdot 5^4 \cdot 5^5$

52) $12^4 \cdot 12 \cdot 12^2$

53) $(-7)^2 \cdot (-7)^3 \cdot (-7)^3$

54) $(-3)^5 \cdot (-3) \cdot (-3)^6$

55) $b^2 \cdot b^4$

56) $x^4 \cdot x^3$

57) $k \cdot k^2 \cdot k^3$

58) $n^6 \cdot n^5 \cdot n^2$

59) $8y^3 \cdot y^2$

60) $10c^8 \cdot c^2 \cdot c$

61) $(9m^4)(6m^{11})$

62) $(-10p^8)(-3p)$

63) $(-6r)(7r^4)$

64) $(8h^5)(-5h^2)$

65) $(-7t^6)(t^3)(-4t^7)$

66) $(3k^2)(-4k^5)(2k^4)$

67) $\left(\frac{5}{3}x^2\right)(12x)(-2x^3)$

68) $\left(\frac{7}{10}y^9\right)(-2y^4)(3y^2)$

69) $\left(\frac{8}{21}b\right)(-6b^8)\left(-\frac{7}{2}b^6\right)$

70) $(12c^3)\left(\frac{14}{15}c^2\right)\left(\frac{5}{7}c^6\right)$

**Mixed Exercises: Objectives 3–5**

Simplify the expression using one of the power rules.

71) $(y^3)^4$

72) $(x^5)^8$

73) $(w^{11})^7$

74) $(a^3)^2$

75) $(3^3)^2$

76) $(2^2)^2$

77) $((-5)^3)^2$

78) $((-4)^5)^3$

79) $\left(\frac{1}{3}\right)^4$

80) $\left(\frac{5}{2}\right)^3$

81) $\left(\frac{6}{a}\right)^2$

82) $\left(\frac{v}{4}\right)^3$

83) $\left(\frac{m}{n}\right)^5$

84) $\left(\frac{t}{u}\right)^{12}$

85) $(10y)^4$

86) $(7w)^2$

87) $(-3p)^4$

88) $(2m)^5$

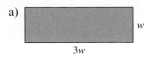 89) $(-4ab)^3$

90) $(-2cd)^4$

91) $6(xy)^3$

92) $-8(mn)^5$

93) $-9(tu)^4$

94) $2(ab)^6$

**Mixed Exercises: Objectives 2–5**

95) Find the area and perimeter of each rectangle.

a)

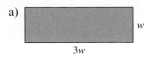

$w$

$3w$

b)

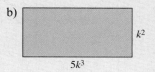

$k^2$

$5k^3$

96) Find the area.

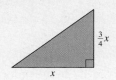

$x$

$\frac{5}{2}x$

[VIDEO] 97) Find the area.

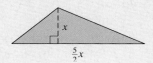

$\frac{3}{4}x$

$x$

98) The shape and dimensions of the Millers' family room are given below. They will have wall-to-wall carpeting installed, and the carpet they have chosen costs \$2.50/ft$^2$.

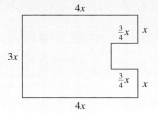

$4x$

$\frac{3}{4}x$   $x$

$3x$

$\frac{3}{4}x$   $x$

$4x$

a) Write an expression for the amount of carpet they will need. (Include the correct units.)

b) Write an expression for the cost of carpeting the family room. (Include the correct units.)

## Section 2.1B  Combining the Rules

**Objective**

1. **Combine the Product Rule and Power Rules of Exponents**

## 1. Combine the Product Rule and Power Rules of Exponents

Now that we have learned the product rule and the power rules for exponents, let's think about how to combine the rules.

If we were asked to evaluate $2^3 \cdot 3^2$, we would follow the order of operations. What would be the first step?

$$2^3 \cdot 3^2 = 8 \cdot 9 \qquad \text{Evaluate exponents.}$$
$$= 72 \qquad \text{Multiply.}$$

When we combine the rules of exponents, we follow the order of operations.

## Example 1

Simplify.

a) $(2c)^3(3c^8)^2$    b) $2(5k^4m^3)^3$    c) $\dfrac{(6t^5)^2}{(2u^4)^3}$

**Solution**

a) $(2c)^3(3c^8)^2$

Because evaluating exponents comes before multiplying in the order of operations, *evaluate the exponents first.*

$$(2c)^3(3c^8)^2 = (2^3c^3)(3^2)(c^8)^2 \qquad \text{Use the power rule.}$$
$$= (8c^3)(9c^{16}) \qquad \text{Use the power rule and evaluate exponents.}$$
$$= 72c^{19} \qquad \text{Product rule}$$

b) $2(5k^4m^3)^3$

Which operation should be performed first, multiplying $2 \cdot 5$ or simplifying $(5k^4m^3)^3$? In the order of operations, we evaluate exponents before multiplying, so *we will begin by simplifying* $(5k^4m^3)^3$.

$$2(5k^4m^3)^3 = 2 \cdot (5)^3(k^4)^3(m^3)^3 \qquad \text{Order of operations and power rule}$$
$$= 2 \cdot 125k^{12}m^9 \qquad \text{Power rule}$$
$$= 250k^{12}m^9 \qquad \text{Multiply.}$$

c) $\dfrac{(6t^5)^2}{(2u^4)^3}$

What comes first in the order of operations, dividing or evaluating exponents? *Evaluating exponents.*

$$\frac{(6t^5)^2}{(2u^4)^3} = \frac{36t^{10}}{8u^{12}} \qquad \text{Power rule}$$
$$= \frac{\overset{9}{\cancel{36}}t^{10}}{\underset{2}{\cancel{8}}u^{12}} \qquad \text{Divide out the common factor of 4.}$$
$$= \frac{9t^{10}}{2u^{12}}$$

**BE CAREFUL**

When simplifying the expression in Example 1c, $\dfrac{(6t^5)^2}{(2u^4)^3}$, it may be tempting to simplify before applying the product rule, like this:

$$\frac{(\overset{3}{\cancel{6}}t^5)^2}{(2\underset{1}{\cancel{u^4}})^3} \neq \frac{(3t^5)^2}{(u^4)^3} = \frac{9t^{10}}{u^{12}} \quad \textbf{Wrong!}$$

You can see, however, that because we did not follow the rules for the order of operations, we did **not** get the correct answer.

## You Try 1

Simplify.

a) $-4(2a^9b^6)^4$    b) $(7x^{10}y)^2(-x^4y^5)^4$    c) $\dfrac{10(m^2n^3)^5}{(5p^4)^2}$    d) $\left(\dfrac{1}{6}w^7\right)^2(3w^{11})^3$

**Answers to You Try Exercises**

1) a) $-64a^{36}b^{24}$   b) $49x^{36}y^{22}$   c) $\dfrac{2m^{10}n^{15}}{5p^8}$   d) $\dfrac{3}{4}w^{47}$

## 2.1B Exercises

**Objective 1: Combine the Product Rule and Power Rules of Exponents**

1) When evaluating expressions involving exponents, always keep in mind the order of _____.

2) The first step in evaluating $(9-3)^2$ is _____.

Simplify.

3) $(k^9)^2(k^3)^2$

4) $(d^5)^3(d^2)^4$

5) $(5z^4)^2(2z^6)^3$

6) $(3r)^2(6r^8)^2$

7) $6ab(-a^{10}b^2)^3$

8) $-5pq^4(-p^4q)^4$

9) $(9+2)^2$

10) $(8-5)^3$

11) $(-4t^6u^2)^3(u^4)^5$

12) $(-m^2)^6(-2m^9)^4$

13) $8(6k^7l^2)^2$

14) $5(-7c^4d)^2$

15) $\left(\dfrac{3}{g^5}\right)^3\left(\dfrac{1}{6}\right)^2$

16) $\left(-\dfrac{2}{5}z^5\right)^3(10z)^2$

17) $\left(\dfrac{7}{8}n^2\right)^2(-4n^9)^2$

18) $\left(\dfrac{2}{3}d^8\right)^4\left(\dfrac{9}{2}d^3\right)^2$

19) $h^4(10h^3)^2(-3h^9)^2$

20) $-v^6(-2v^5)^5(-v^4)^3$

21) $3w^{11}(7w^2)^2(-w^6)^5$

22) $5z^3(-4z)^2(2z^3)^2$

23) $\dfrac{(12x^3)^2}{(10y^5)^2}$

24) $\dfrac{(-3a^4)^3}{(6b)^2}$

25) $\dfrac{(4d^9)^2}{(-2c^5)^6}$

26) $\dfrac{(-5m^7)^3}{(5n^{12})^2}$

27) $\dfrac{8(a^4b^7)^9}{(6c)^2}$

28) $\dfrac{(3x^5)^3}{21(yz^2)^6}$

29) $\dfrac{r^4(r^5)^7}{2t(11t^2)^2}$

30) $\dfrac{k^5(k^2)^3}{7m^{10}(2m^3)^2}$

31) $\left(\dfrac{4}{9}x^3y\right)^2\left(\dfrac{3}{2}x^6y^4\right)^3$

32) $(6s^8t^3)^2\left(-\dfrac{10}{3}st^4\right)^2$

33) $\left(-\dfrac{2}{5}c^9d^2\right)^3\left(\dfrac{5}{4}cd^6\right)^2$

34) $-\dfrac{11}{12}\left(\dfrac{3}{2}m^3n^{10}\right)^2$

35) $\left(\dfrac{5x^5y^2}{z^4}\right)^3$

36) $\left(-\dfrac{7a^4b}{8c^6}\right)^2$

37) $\left(-\dfrac{3t^4u^9}{2v^7}\right)^4$

38) $\left(\dfrac{2pr^8}{q^{11}}\right)^5$

39) $\left(\dfrac{12w^5}{4x^3y^6}\right)^2$

40) $\left(\dfrac{10b^3c^5}{15a}\right)^2$

41) The length of a side of a square is $5l^2$ units.

   a) Write an expression for its perimeter.

   b) Write an expression for its area.

42) The width of a rectangle is $2w$ units, and the length of the rectangle is $7w$ units.

   a) Write an expression for its area.

   b) Write an expression for its perimeter.

43) The length of a rectangle is $x$ units, and the width of the rectangle is $\dfrac{3}{8}x$ units.

   a) Write an expression for its area.

   b) Write an expression for its perimeter.

44) The width of a rectangle is $4y^3$ units, and the length of the rectangle is $\dfrac{13}{2}y^3$ units.

   a) Write an expression for its perimeter.

   b) Write an expression for its area.

## Section 2.2A Real-Number Bases

**Objectives**

1. Use 0 as an Exponent
2. Use Negative Integers as Exponents

Thus far, we have defined an exponential expression such as $2^3$. The exponent of 3 indicates that $2^3 = 2 \cdot 2 \cdot 2$ (3 factors of 2) so that $2^3 = 2 \cdot 2 \cdot 2 = 8$. Is it possible to have an exponent of zero or a negative exponent? If so, what do they mean?

### 1. Use 0 as an Exponent

> **Definition**
>
> **Zero as an Exponent:** If $a \neq 0$, then $a^0 = 1$.

How can this be possible? Let's look at an example involving the product rule to help us understand why $a^0 = 1$.

Let's evaluate $2^0 \cdot 2^3$. Using the product rule, we get:

$$2^0 \cdot 2^3 = 2^{0+3} = 2^3 = 8$$

But we know that $2^3 = 8$. Therefore, if $2^0 \cdot 2^3 = 8$, then $2^0 = 1$. This is one way to understand that $a^0 = 1$.

### Example 1

Evaluate each expression.

a)  $5^0$    b)  $-8^0$    c)  $(-7)^0$    d)  $-3(2^0)$

**Solution**

a)  $5^0 = 1$    b)  $-8^0 = -1 \cdot 8^0 = -1 \cdot 1 = -1$

c)  $(-7)^0 = 1$    d)  $-3(2^0) = -3(1) = -3$    ∎

### You Try 1

Evaluate.

a)  $9^0$    b)  $-2^0$    c)  $(-5)^0$    d)  $3^0(-2)$

### 2. Use Negative Integers as Exponents

So far we have worked with exponents that are zero or positive. What does a negative exponent mean?

Let's use the product rule to find $2^3 \cdot 2^{-3}$.

$$2^3 \cdot 2^{-3} = 2^{3+(-3)} = 2^0 = 1.$$

Remember that a number multiplied by its reciprocal is 1, and here we have that a quantity, $2^3$, times another quantity, $2^{-3}$, is 1. Therefore, $2^3$ and $2^{-3}$ are reciprocals!

This leads to the definition of a negative exponent.

> **Definition**
>
> **Negative Exponent:** If $n$ is any integer and $a$ and $b$ are not equal to zero, then
>
> $$a^{-n} = \left(\frac{1}{a}\right)^n = \frac{1}{a^n} \quad \text{and} \quad \left(\frac{a}{b}\right)^{-n} = \left(\frac{b}{a}\right)^n.$$

Therefore, to rewrite an expression of the form $a^{-n}$ with a positive exponent, *take the reciprocal of the base and make the exponent positive.*

**Example 2**

Evaluate each expression.

a)  $2^{-3}$       b)  $\left(\dfrac{3}{2}\right)^{-4}$       c)  $\left(\dfrac{1}{5}\right)^{-3}$       d)  $(-7)^{-2}$

**Solution**

a)  $2^{-3}$: The reciprocal of 2 is $\dfrac{1}{2}$, so $2^{-3} = \left(\dfrac{1}{2}\right)^{3} = \dfrac{1^3}{2^3} = \dfrac{1}{8}$.

Above we found that $2^{3} \cdot 2^{-3} = 1$ using the product rule, but now we can evaluate the product using the definition of a negative exponent.

$$2^{3} \cdot 2^{-3} = 8 \cdot \left(\dfrac{1}{2}\right)^{3} = 8 \cdot \dfrac{1}{8} = 1$$

b)  $\left(\dfrac{3}{2}\right)^{-4}$:  The reciprocal of $\dfrac{3}{2}$ is $\dfrac{2}{3}$, so $\left(\dfrac{3}{2}\right)^{-4} = \left(\dfrac{2}{3}\right)^{4} = \dfrac{2^4}{3^4} = \dfrac{16}{81}$.

 **BE CAREFUL**   Notice that a negative exponent does not make the answer negative!

c)  $\left(\dfrac{1}{5}\right)^{-3}$:  The reciprocal of $\dfrac{1}{5}$ is 5, so $\left(\dfrac{1}{5}\right)^{-3} = 5^{3} = 125$.

d)  $(-7)^{-2}$: The reciprocal of $-7$ is $-\dfrac{1}{7}$, so

$$(-7)^{-2} = \left(-\dfrac{1}{7}\right)^{2} = \left(-1 \cdot \dfrac{1}{7}\right)^{2} = (-1)^{2}\left(\dfrac{1}{7}\right)^{2} = 1 \cdot \dfrac{1^2}{7^2} = \dfrac{1}{49}$$

 **You Try 2**

Evaluate.

a)  $(10)^{-2}$       b)  $\left(\dfrac{1}{4}\right)^{-2}$       c)  $\left(\dfrac{2}{3}\right)^{-3}$       d)  $-5^{-3}$

**Answers to You Try Exercises**

1) a) 1  b) $-1$  c) 1  d) $-2$     2) a) $\dfrac{1}{100}$  b) 16  c) $\dfrac{27}{8}$  d) $-\dfrac{1}{125}$

## 2.2A Exercises

**Mixed Exercises: Objectives 1 and 2**

1) True or False: Raising a positive base to a negative exponent will give a negative result. (Example: $2^{-4}$)

2) True or False: $8^0 = 1$.

3) True or False: The reciprocal of 4 is $\frac{1}{4}$.

4) True or False: $3^{-2} - 2^{-2} = 1^{-2}$.

Evaluate.

5) $2^0$                    6) $(-4)^0$

7) $-5^0$                   8) $-1^0$

9) $0^8$                    10) $-(-9)^0$

11) $(5)^0 + (-5)^0$        12) $\left(\frac{4}{7}\right)^0 - \left(\frac{7}{4}\right)^0$

13) $6^{-2}$         14) $9^{-2}$

15) $2^{-4}$               16) $11^{-2}$

17) $5^{-3}$               18) $2^{-5}$

19) $\left(\frac{1}{8}\right)^{-2}$        20) $\left(\frac{1}{10}\right)^{-3}$

21) $\left(\frac{1}{2}\right)^{-5}$        22) $\left(\frac{1}{4}\right)^{-2}$

23) $\left(\frac{4}{3}\right)^{-3}$        24) $\left(\frac{2}{5}\right)^{-3}$

25) $\left(\frac{9}{7}\right)^{-2}$        26) $\left(\frac{10}{3}\right)^{-2}$

27) $\left(-\frac{1}{4}\right)^{-3}$       28) $\left(-\frac{1}{12}\right)^{-2}$

29) $\left(-\frac{3}{8}\right)^{-2}$       30) $\left(-\frac{5}{2}\right)^{-3}$

31) $-2^{-6}$              32) $-4^{-3}$

33) $-1^{-5}$              34) $-9^{-2}$

35) $2^{-3} - 4^{-2}$       36) $5^{-2} + 2^{-2}$

37) $2^{-2} + 3^{-2}$       38) $4^{-1} - 6^{-2}$

39) $-9^{-2} + 3^{-3} + (-7)^0$    40) $6^0 - 9^{-1} + 4^0 + 3^{-2}$

---

## Section 2.2B  Variable Bases

### Objectives
1. Use 0 as an Exponent
2. Rewrite an Exponential Expression with Positive Exponents

### 1. Use 0 as an Exponent

We can apply 0 as an exponent to bases containing variables.

**Example 1**

Evaluate each expression. Assume the variable does not equal zero.

a) $t^0$          b) $(-k)^0$          c) $-(11p)^0$

**Solution**

a) $t^0 = 1$          b) $(-k)^0 = 1$

c) $-(11p)^0 = -1 \cdot (11p)^0 = -1 \cdot 1 = -1$    ∎

**You Try 1**

Evaluate. Assume the variable does not equal zero.

a) $p^0$          b) $(10x)^0$          c) $-(7s)^0$

## 2. Rewrite an Exponential Expression with Positive Exponents

Next, let's apply the definition of a negative exponent to bases containing variables. As in Example 1, we will assume the variable does not equal zero since having zero in the denominator of a fraction will make the fraction undefined.

Recall that $2^{-4} = \left(\dfrac{1}{2}\right)^4 = \dfrac{1}{16}$. That is, to rewrite the expression with a positive exponent, we take the reciprocal of the base.

What is the reciprocal of $x$? The reciprocal is $\dfrac{1}{x}$.

### Example 2

Rewrite the expression with positive exponents. Assume the variable does not equal zero.

a)  $x^{-6}$           b)  $\left(\dfrac{2}{n}\right)^{-6}$          c)  $3a^{-2}$

**Solution**

a)  $x^{-6} = \left(\dfrac{1}{x}\right)^6 = \dfrac{1^6}{x^6} = \dfrac{1}{x^6}$          b)  $\left(\dfrac{2}{n}\right)^{-6} = \left(\dfrac{n}{2}\right)^6$     The reciprocal of $\dfrac{2}{n}$ is $\dfrac{n}{2}$.

$$= \dfrac{n^6}{2^6} = \dfrac{n^6}{64}$$

c)  $3a^{-2} = 3 \cdot \left(\dfrac{1}{a}\right)^2$     Remember, the base is $a$, *not* $3a$, since there are no parentheses. Therefore, the exponent of $-2$ applies only to $a$.

$$= 3 \cdot \dfrac{1}{a^2} = \dfrac{3}{a^2}$$

### You Try 2

Rewrite the expression with positive exponents. Assume the variable does not equal zero.

a)  $m^{-4}$        b)  $\left(\dfrac{1}{z}\right)^{-7}$        c)  $-2y^{-3}$

How could we rewrite $\dfrac{x^{-2}}{y^{-2}}$ with only positive exponents? One way would be to apply the

power rule for exponents: $\dfrac{x^{-2}}{y^{-2}} = \left(\dfrac{x}{y}\right)^{-2} = \left(\dfrac{y}{x}\right)^2 = \dfrac{y^2}{x^2}$

Let's do the same for $\dfrac{a^{-5}}{b^{-5}}$: $\dfrac{a^{-5}}{b^{-5}} = \left(\dfrac{a}{b}\right)^{-5} = \left(\dfrac{b}{a}\right)^5 = \dfrac{b^5}{a^5}$

Notice that to rewrite the original expression with only positive exponents, the terms with the negative exponents "switch" their positions in the fraction. We can generalize this way:

### Definition

If $m$ and $n$ are any integers and $a$ and $b$ are real numbers not equal to zero, then

$$\dfrac{a^{-m}}{b^{-n}} = \dfrac{b^n}{a^m}$$

## Example 3

Rewrite the expression with positive exponents. Assume the variables do not equal zero.

a) $\dfrac{c^{-8}}{d^{-3}}$    b) $\dfrac{5p^{-6}}{q^{7}}$    c) $t^{-2}u^{-1}$

d) $\dfrac{2xy^{-3}}{3z^{-2}}$    e) $\left(\dfrac{ab}{4c}\right)^{-3}$

**Solution**

a) $\dfrac{c^{-8}}{d^{-3}} = \dfrac{d^{3}}{c^{8}}$    To make the exponents positive, "switch" the positions of the terms in the fraction.

b) $\dfrac{5p^{-6}}{q^{7}} = \dfrac{5}{p^{6}q^{7}}$    Since the exponent on $q$ is positive, we do not change its position in the expression.

c) $t^{-2}u^{-1} = \dfrac{t^{-2}u^{-1}}{1}$

$= \dfrac{1}{t^{2}u^{1}} = \dfrac{1}{t^{2}u}$    Move $t^{-2}u^{-1}$ to the denominator to write with positive exponents.

d) $\dfrac{2xy^{-3}}{3z^{-2}} = \dfrac{2xz^{2}}{3y^{3}}$    To make the exponents positive, "switch" the positions of the factors with negative exponents in the fraction.

e) $\left(\dfrac{ab}{4c}\right)^{-3} = \left(\dfrac{4c}{ab}\right)^{3}$    To make the exponent positive, use the reciprocal of the base.

$= \dfrac{4^{3}c^{3}}{a^{3}b^{3}}$    Power rule

$= \dfrac{64c^{3}}{a^{3}b^{3}}$    Simplify.

## You Try 3

Rewrite the expression with positive exponents. Assume the variables do not equal zero.

a) $\dfrac{n^{-6}}{y^{-2}}$    b) $\dfrac{z^{-9}}{3k^{-4}}$    c) $8x^{-5}y$    d) $\dfrac{8d^{-4}}{6m^{2}n^{-1}}$    e) $\left(\dfrac{3x^{2}}{y}\right)^{-2}$

**Answers to You Try Exercises**

1) a) 1    b) 1    c) $-1$    2) a) $\dfrac{1}{m^{4}}$    b) $z^{7}$    c) $-\dfrac{2}{y^{3}}$    3) a) $\dfrac{y^{2}}{n^{6}}$    b) $\dfrac{k^{4}}{3z^{9}}$    c) $\dfrac{8y}{x^{5}}$

d) $\dfrac{4n}{3m^{2}d^{4}}$    e) $\dfrac{y^{2}}{9x^{4}}$

## 2.2B Exercises

**Objective 1:  Use 0 as an Exponent**

1) Identify the base in each expression.

   a) $w^0$                 b) $-3n^{-5}$

   c) $(2p)^{-3}$          d) $4c^0$

2) True or False: $6^0 - 4^0 = (6 - 4)^0$

Evaluate. Assume the variables do not equal zero.

3) $r^0$                      4) $(5m)^0$

5) $-2k^0$               6) $-z^0$

7) $x^0 + (2x)^0$        8) $\left(\dfrac{7}{8}\right)^0 - \left(\dfrac{3}{5}\right)^0$

**Objective 2:  Rewrite an Exponential Expression with Positive Exponents**

Rewrite each expression with only positive exponents. Assume the variables do not equal zero.

9) $d^{-3}$               10) $y^{-7}$

11) $p^{-1}$             12) $a^{-5}$

(VIDEO) 13) $\dfrac{a^{-10}}{b^{-3}}$         14) $\dfrac{h^{-2}}{k^{-1}}$

15) $\dfrac{y^{-8}}{x^{-5}}$          16) $\dfrac{v^{-2}}{w^{-7}}$

17) $\dfrac{t^5}{8u^{-3}}$          18) $\dfrac{9x^{-4}}{y^5}$

19) $5m^6n^{-2}$        20) $\dfrac{1}{9}a^{-4}b^3$

21) $\dfrac{2}{t^{-11}u^{-5}}$         22) $\dfrac{7r}{2t^{-9}u^2}$

(VIDEO) 23) $\dfrac{8a^6b^{-1}}{5c^{-10}d}$      24) $\dfrac{17k^{-8}h^5}{20m^{-7}n^{-2}}$

25) $\dfrac{2z^4}{x^{-7}y^{-6}}$        26) $\dfrac{1}{a^{-2}b^{-2}c^{-1}}$

(VIDEO) 27) $\left(\dfrac{a}{6}\right)^{-2}$        28) $\left(\dfrac{3}{y}\right)^{-4}$

29) $\left(\dfrac{2n}{q}\right)^{-5}$       30) $\left(\dfrac{w}{5v}\right)^{-3}$

(VIDEO) 31) $\left(\dfrac{12b}{cd}\right)^{-2}$     32) $\left(\dfrac{2tu}{v}\right)^{-6}$

33) $-9k^{-2}$         34) $3g^{-5}$

35) $3t^{-3}$           36) $8h^{-4}$

37) $-m^{-9}$        38) $-d^{-5}$

39) $\left(\dfrac{1}{z}\right)^{-10}$       40) $\left(\dfrac{1}{k}\right)^{-6}$

41) $\left(\dfrac{1}{j}\right)^{-1}$        42) $\left(\dfrac{1}{c}\right)^{-7}$

43) $5\left(\dfrac{1}{n}\right)^{-2}$     44) $7\left(\dfrac{1}{t}\right)^{-8}$

(VIDEO) 45) $c\left(\dfrac{1}{d}\right)^{-3}$    46) $x^2\left(\dfrac{1}{y}\right)^{-2}$

## Section 2.3  The Quotient Rule

**Objective**

1. Use the Quotient Rule for Exponents

### 1. Use the Quotient Rule for Exponents

In this section, we will discuss how to simplify the quotient of two exponential expressions with the same base. Let's begin by simplifying $\dfrac{8^6}{8^4}$. One way to simplify this expression is to write the numerator and denominator without exponents:

$$\frac{8^6}{8^4} = \frac{8 \cdot 8 \cdot 8 \cdot 8 \cdot 8 \cdot 8}{8 \cdot 8 \cdot 8 \cdot 8} \qquad \text{Divide out common factors.}$$
$$= 8 \cdot 8 = 8^2 = 64$$

Therefore,

$$\frac{8^6}{8^4} = 8^2 = 64$$

Do you notice a relationship between the exponents in the original expression and the exponent we get when we simplify?

$$\frac{8^6}{8^4} = 8^{6-4} = 8^2 = 64$$

That's right. We *subtracted* the exponents.

---

**Property**    Quotient Rule for Exponents

If $m$ and $n$ are any integers and $a \neq 0$, then

$$\frac{a^m}{a^n} = a^{m-n}$$

---

Notice that the base in the numerator and denominator is $a$. *To divide expressions with the same base, keep the base and subtract the denominator's exponent from the numerator's exponent.*

## Example 1

Simplify. Assume the variables do not equal zero.

a) $\dfrac{2^9}{2^3}$     b) $\dfrac{t^{10}}{t^4}$     c) $\dfrac{3}{3^{-2}}$     d) $\dfrac{n^5}{n^7}$     e) $\dfrac{3^2}{2^4}$

### Solution

a) $\dfrac{2^9}{2^3} = 2^{9-3} = 2^6 = 64$     Since the bases are the same, subtract the exponents.

b) $\dfrac{t^{10}}{t^4} = t^{10-4} = t^6$     Since the bases are the same, subtract the exponents.

c) $\dfrac{3}{3^{-2}} = \dfrac{3^1}{3^{-2}} = 3^{1-(-2)}$     Since the bases are the same, subtract the exponents.

$\qquad\qquad = 3^3 = 27$     Be careful when subtracting the negative exponent!

d) $\dfrac{n^5}{n^7} = n^{5-7} = n^{-2}$     Same base; subtract the exponents.

$\qquad = \left(\dfrac{1}{n}\right)^2 = \dfrac{1}{n^2}$     Write with a positive exponent.

e) $\dfrac{3^2}{2^4} = \dfrac{9}{16}$     Since the bases are not the same, we cannot apply the quotient rule. Evaluate the numerator and denominator separately. ∎

---

 **You Try 1**

Simplify. Assume the variables do not equal zero.

a) $\dfrac{5^7}{5^4}$     b) $\dfrac{c^4}{c^{-1}}$     c) $\dfrac{k^2}{k^{10}}$     d) $\dfrac{2^3}{2^7}$

We can apply the quotient rule to expressions containing more than one variable. Here are more examples:

## Example 2

Simplify. Assume the variables do not equal zero.

a) $\dfrac{x^8 y^7}{x^3 y^4}$ 
b) $\dfrac{12a^{-5}b^{10}}{8a^{-3}b^2}$

### Solution

a) $\dfrac{x^8 y^7}{x^3 y^4} = x^{8-3}y^{7-4}$      Subtract the exponents.

$= x^5 y^3$

b) $\dfrac{12a^{-5}b^{10}}{8a^{-3}b^2}$      We will reduce $\dfrac{12}{8}$ in addition to applying the quotient rule.

$\dfrac{\overset{3}{\cancel{12}}a^{-5}b^{10}}{\underset{2}{\cancel{8}}a^{-3}b^2} = \dfrac{3}{2}a^{-5-(-3)}b^{10-2}$      Subtract the exponents.

$= \dfrac{3}{2}a^{-5+3}b^8 = \dfrac{3}{2}a^{-2}b^8 = \dfrac{3b^8}{2a^2}$ ∎

## You Try 2

Simplify. Assume the variables do not equal zero.

a) $\dfrac{r^4 s^{10}}{rs^3}$ 
b) $\dfrac{30m^6 n^{-8}}{42m^4 n^{-3}}$

---

### Answers to You Try Exercises

1) a) 125   b) $c^5$   c) $\dfrac{1}{k^8}$   d) $\dfrac{1}{16}$    2) a) $r^3 s^7$   b) $\dfrac{5m^2}{7n^5}$

# 2.3 Exercises

## Objective 1: Use the Quotient Rule for Exponents

State what is wrong with the following steps and then simplify correctly.

1) $\dfrac{a^5}{a^3} = a^{3-5} = a^{-2} = \dfrac{1}{a^2}$

2) $\dfrac{4^3}{2^6} = \left(\dfrac{4}{2}\right)^{3-6} = 2^{-3} = \dfrac{1}{2^3} = \dfrac{1}{8}$

Simplify using the quotient rule. Assume the variables do not equal zero.

3) $\dfrac{d^{10}}{d^5}$

4) $\dfrac{z^{11}}{z^7}$

5) $\dfrac{m^9}{m^5}$

6) $\dfrac{a^6}{a}$

7) $\dfrac{8t^{15}}{t^8}$

8) $\dfrac{4k^4}{k^2}$

9) $\dfrac{6^{12}}{6^{10}}$

10) $\dfrac{4^4}{4}$

11) $\dfrac{3^{12}}{3^8}$

12) $\dfrac{2^7}{2^4}$

13) $\dfrac{2^5}{2^9}$

14) $\dfrac{9^5}{9^7}$

31) $\dfrac{15w^2}{w^{10}}$

32) $\dfrac{-7p^3}{p^{12}}$

VIDEO 15) $\dfrac{5^6}{5^9}$

16) $\dfrac{8^4}{8^6}$

VIDEO 33) $\dfrac{-6k}{k^4}$

34) $\dfrac{21h^3}{h^7}$

17) $\dfrac{10d^4}{d^2}$

18) $\dfrac{3x^6}{x^2}$

35) $\dfrac{a^4b^9}{ab^2}$

36) $\dfrac{p^5q^7}{p^2q^3}$

19) $\dfrac{20c^{11}}{30c^6}$

20) $\dfrac{35t^7}{56t^2}$

37) $\dfrac{10k^{-2}l^{-6}}{15k^{-5}l^2}$

38) $\dfrac{28tu^{-2}}{14t^5u^{-9}}$

21) $\dfrac{y^3}{y^8}$

22) $\dfrac{m^4}{m^{10}}$

39) $\dfrac{300x^7y^3}{30x^{12}y^8}$

40) $\dfrac{63a^{-3}b^2}{7a^7b^8}$

VIDEO 23) $\dfrac{x^{-3}}{x^6}$

24) $\dfrac{u^{-20}}{u^{-9}}$

VIDEO 41) $\dfrac{6v^{-1}w}{54v^2w^{-5}}$

42) $\dfrac{3a^2b^{-11}}{18a^{-10}b^6}$

25) $\dfrac{t^{-6}}{t^{-3}}$

26) $\dfrac{y^8}{y^{15}}$

43) $\dfrac{3c^5d^{-2}}{8cd^{-3}}$

44) $\dfrac{9x^{-5}y^2}{4x^{-2}y^6}$

27) $\dfrac{a^{-1}}{a^9}$

28) $\dfrac{m^{-9}}{m^{-3}}$

45) $\dfrac{(x+y)^9}{(x+y)^2}$

46) $\dfrac{(a+b)^9}{(a+b)^4}$

29) $\dfrac{t^4}{t}$

30) $\dfrac{c^7}{c^{-1}}$

47) $\dfrac{(c+d)^{-5}}{(c+d)^{-11}}$

48) $\dfrac{(a+2b)^{-3}}{(a+2b)^{-4}}$

## Putting It All Together

### Objective

1. Combine the Rules of Exponents

### 1. Combine the Rules of Exponents

Let's review all the rules for simplifying exponential expressions and then see how we can combine the rules to simplify expressions.

---

**Summary**   Rules of Exponents

In the rules stated here, $a$ and $b$ are any real numbers and $m$ and $n$ are positive integers.

| | |
|---|---|
| Product rule | $a^m \cdot a^n = a^{m+n}$ |
| Basic power rule | $(a^m)^n = a^{mn}$ |
| Power rule for a product | $(ab)^n = a^n b^n$ |
| Power rule for a quotient | $\left(\dfrac{a}{b}\right)^n = \dfrac{a^n}{b^n}, \quad (b \neq 0)$ |
| Quotient rule | $\dfrac{a^m}{a^n} = a^{m-n}, \quad (a \neq 0)$ |

Changing from negative to positive exponents, where $a \neq 0, b \neq 0$, and $m$ and $n$ are any integers:

$$\dfrac{a^{-m}}{b^{-n}} = \dfrac{b^n}{a^m} \qquad \left(\dfrac{a}{b}\right)^{-m} = \left(\dfrac{b}{a}\right)^m$$

In the following definitions, $a \neq 0$, and $n$ is any integer.

| | |
|---|---|
| Zero as an exponent | $a^0 = 1$ |
| Negative number as an exponent | $a^{-n} = \dfrac{1}{a^n}$ |

---

| Example 1 |

Simplify using the rules of exponents. Assume all variables represent nonzero real numbers.

a)  $(2t^{-6})^3(3t^2)^2$     b)  $\left(\dfrac{7c^{10}d^7}{c^4d^2}\right)^2$     c)  $\dfrac{w^{-3} \cdot w^4}{w^6}$     d)  $\left(\dfrac{12a^{-2}b^9}{30ab^{-2}}\right)^{-3}$

### Solution

a)  $(2t^{-6})^3 (3t^2)^2$ — We must follow the order of operations. Therefore, evaluate the exponents first.

$$(2t^{-6})^3 \cdot (3t^2)^2 = 2^3 t^{(-6)(3)} \cdot 3^2 t^{(2)(2)}$$ — Apply the power rule.

$$= 8t^{-18} \cdot 9t^4$$ — Simplify.

$$= 72t^{-18+4}$$ — Multiply $8 \cdot 9$ and add the exponents.

$$= 72t^{-14}$$

$$= \frac{72}{t^{14}}$$ — Write the answer using a positive exponent.

b)  $\left(\dfrac{7c^{10}d^7}{c^4d^2}\right)^2$ — How can we begin this problem? We can use the quotient rule to simplify the expression before squaring it.

$$\left(\frac{7c^{10}d^7}{c^4d^2}\right)^2 = (7c^{10-4}d^{7-2})^2$$ — Apply the quotient rule in the parentheses.

$$= (7c^6d^5)^2$$ — Simplify.

$$= 7^2 c^{(6)(2)}d^{(5)(2)}$$ — Apply the power rule.

$$= 49c^{12}d^{10}$$

c)  $\dfrac{w^{-3} \cdot w^4}{w^6}$ — Let's begin by simplifying the numerator:

$$\frac{w^{-3} \cdot w^4}{w^6} = \frac{w^{-3+4}}{w^6}$$ — Add the exponents in the numerator.

$$= \frac{w^1}{w^6}$$

Now, we can apply the quotient rule:

$$= w^{1-6} = w^{-5}$$ — Subtract the exponents.

$$= \frac{1}{w^5}$$ — Write the answer using a positive exponent.

d)  $\left(\dfrac{12a^{-2}b^9}{30ab^{-2}}\right)^{-3}$ — Eliminate the negative exponent *outside* the parentheses by taking the reciprocal of the base. Notice that we have *not* eliminated the negatives on the exponents *inside* the parentheses.

$$\left(\frac{12a^{-2}b^9}{30ab^{-2}}\right)^{-3} = \left(\frac{30ab^{-2}}{12a^{-2}b^9}\right)^3$$

We could apply the exponent of 3 to the quantity inside the parentheses, but we could also reduce $\dfrac{30}{12}$ first and apply the quotient rule before cubing the quantity.

$$\left(\frac{30\,ab^{-2}}{12a^{-2}\,b^9}\right)^3 = \left(\frac{5}{2}a^{1-(-2)}b^{-2-9}\right)^3 \qquad \text{Reduce } \frac{30}{12} \text{ and subtract the exponents.}$$

$$= \left(\frac{5}{2}a^3 b^{-11}\right)^3$$

$$= \frac{125}{8}a^9 b^{-33} \qquad \text{Apply the power rule.}$$

$$= \frac{125a^9}{8b^{33}} \qquad \text{Write the answer using positive exponents.}$$

**You Try 1**

Simplify using the rules of exponents.

a) $\left(\dfrac{m^{12}n^3}{m^4 n}\right)^4$     b) $(-p^{-5})^4(6p^7)^2$     c) $\left(\dfrac{9x^4 y^{-5}}{54x^3 y}\right)^{-2}$

It is possible for variables to appear in exponents. The same rules apply.

**Example 2**

Simplify using the rules of exponents. Assume that the variables represent nonzero integers. Write your final answer so that the exponents have positive coefficients.

a) $c^{4x} \cdot c^{2x}$     b) $\dfrac{x^{5y}}{x^{9y}}$

**Solution**

a) $c^{4x} \cdot c^{2x} = c^{4x+2x} = c^{6x}$     The bases are the same, so apply the product rule. Add the exponents.

b) $\dfrac{x^{5y}}{x^{9y}} = x^{5y-9y}$     The bases are the same, so apply the quotient rule. Subtract the exponents.

$= x^{-4y}$

$= \dfrac{1}{x^{4y}}$     Write the answer with a positive coefficient in the exponent.

**You Try 2**

Simplify using the rules of exponents. Assume that the variables represent nonzero integers. Write your final answer so that the exponents have positive coefficients.

a) $8^{2k} \cdot 8^k \cdot 8^{10k}$     b) $(w^3)^{-2p}$

**Answers to You Try Exercises**

1) a) $m^{32}n^8$  b) $\dfrac{36}{p^6}$  c) $\dfrac{36y^{12}}{x^2}$     2) a) $8^{13k}$  b) $\dfrac{1}{w^{6p}}$

## Putting It All Together
## Summary Exercises

### Objective 1: Combine the Rules of Exponents

Use the rules of exponents to evaluate.

1) $\left(\dfrac{2}{3}\right)^4$

2) $(2^2)^3$

3) $\dfrac{3^9}{3^5 \cdot 3^4}$

4) $\dfrac{(-5)^6 \cdot (-5)^2}{(-5)^5}$

5) $\left(\dfrac{10}{3}\right)^{-2}$

6) $\left(\dfrac{3}{7}\right)^{-2}$

7) $(9-6)^2$

8) $(3-8)^3$

9) $10^{-2}$

10) $2^{-3}$

11) $\dfrac{2^7}{2^{12}}$

12) $\dfrac{3^{19}}{3^{15}}$

13) $\left(-\dfrac{5}{3}\right)^{-7} \cdot \left(-\dfrac{5}{3}\right)^{4}$

14) $\left(\dfrac{1}{8}\right)^{-2}$

15) $3^{-2} - 12^{-1}$

16) $2^{-2} + 3^{-2}$

**Simplify.** Assume all variables represent nonzero numbers. The final answer should not contain negative exponents.

17) $-10(-3g^4)^3$

18) $7(2d^3)^3$

19) $\dfrac{33s}{s^{12}}$

20) $\dfrac{c^{-7}}{c^{-2}}$

21) $\left(\dfrac{2xy^4}{3x^{-9}y^{-2}}\right)^4$

22) $\left(\dfrac{a^6b^5}{10a^3}\right)^3$

23) $\left(\dfrac{9m^8}{n^3}\right)^{-2}$

24) $\left(\dfrac{3s^{-6}}{r^2}\right)^{-4}$

25) $(-b^5)^3$

26) $(h^{11})^8$

27) $(-3m^5n^2)^3$

28) $(13a^6b)^2$

29) $\left(-\dfrac{9}{4}z^5\right)\left(\dfrac{8}{3}z^{-2}\right)$

30) $(15w^3)\left(-\dfrac{3}{5}w^6\right)$

31) $\left(\dfrac{s^7}{t^3}\right)^{-6}$

32) $\dfrac{m^{-3}}{n^{14}}$

33) $(-ab^3c^5)^2\left(\dfrac{a^4}{bc}\right)^3$

34) $\dfrac{(4v^3)^2}{(6v^8)^2}$

35) $\left(\dfrac{48u^{-7}v^2}{36u^3v^{-5}}\right)^{-3}$

36) $\left(\dfrac{xy^5}{9x^{-2}y}\right)^{-2}$

37) $\left(\dfrac{-3t^4u}{t^2u^{-4}}\right)^3$

38) $\left(\dfrac{k^7m^7}{12k^{-1}m^6}\right)^2$

39) $(h^{-3})^6$

40) $(-d^4)^{-5}$

41) $\left(\dfrac{h}{2}\right)^4$

42) $13f^{-2}$

43) $-7c^4(-2c^2)^3$

44) $5p^3(4p^6)^2$

45) $(12a^7)^{-1}(6a)^2$

46) $(9r^2s^2)^{-1}$

47) $\left(\dfrac{9}{20}r^4\right)(4r^{-3})\left(\dfrac{2}{33}r^9\right)$

48) $\left(\dfrac{f^8 \cdot f^{-3}}{f^2 \cdot f^9}\right)^6$

49) $\dfrac{(a^2b^{-5}c)^{-3}}{(a^4b^{-3}c)^{-2}}$

50) $\dfrac{(x^{-1}y^7z^4)^3}{(x^4yz^{-5})^{-3}}$

51) $\dfrac{(2mn^{-2})^3(5m^2n^{-3})^{-1}}{(3m^{-3}n^3)^{-2}}$

52) $\dfrac{(4s^3t^{-1})^2(5s^2t^{-3})^{-2}}{(4s^3t^{-1})^3}$

53) $\left(\dfrac{4n^{-3}m}{n^8m^2}\right)^0$

54) $\left(\dfrac{7qr^4}{37r^{-19}}\right)^0$

55) $\left(\dfrac{49c^4d^8}{21c^4d^5}\right)^{-2}$

56) $\dfrac{(2x^4y)^{-2}}{(5xy^3)^2}$

**Simplify.** Assume that the variables represent nonzero integers. Write your final answer so that the exponents have positive coefficients.

57) $(p^{2c})^6$

58) $(5d^{4t})^2$

59) $y^m \cdot y^{3m}$

60) $x^{-5c} \cdot x^{9c}$

61) $t^{5b} \cdot t^{-8b}$

62) $a^{-4y} \cdot a^{-3y}$

63) $\dfrac{25c^{2x}}{40c^{9x}}$

64) $-\dfrac{3y^{-10a}}{8y^{-2a}}$

## Section 2.4 Scientific Notation

### Objectives

1. **Multiply a Number by a Power of Ten**
2. **Understand Scientific Notation**
3. **Write a Number in Scientific Notation**
4. **Perform Operations with Numbers in Scientific Notation**

The distance from the Earth to the Sun is approximately 150,000,000 km. A single rhinovirus (cause of the common cold) measures 0.00002 mm across. Performing operations on very large or very small numbers like these can be difficult. This is why scientists and economists, for example, often work with such numbers in a shorthand form called *scientific notation*. Writing numbers in scientific notation together with applying rules of exponents can simplify calculations with very large and very small numbers.

### 1. Multiply a Number by a Power of Ten

Before discussing scientific notation further, we need to understand some principles behind the notation. Let's look at multiplying numbers by positive powers of 10.

**Example 1**

Multiply.

a)  $3.4 \times 10^1$       b)  $0.0857 \times 10^3$       c)  $97 \times 10^2$

**Solution**

a)  $3.4 \times 10^1 = 3.4 \times 10 = 34$

b)  $0.0857 \times 10^3 = 0.0857 \times 1000 = 85.7$

c)  $97 \times 10^2 = 97 \times 100 = 9700$

Notice that when we multiply each of these numbers by a positive power of 10, the result is *larger* than the original number. In fact, the exponent determines how many places to the *right* the decimal point is moved.

$$3.40 \times 10^1 = 3.4 \times 10^1 = 34 \qquad 0.0857 \times 10^3 = 85.7$$
$$\text{1 place to the right} \qquad\qquad \text{3 places to the right}$$

$$97 \times 10^2 = 97.00 \times 10^2 = 9700$$
$$\text{2 places to the right}$$

**You Try 1**

Multiply by moving the decimal point the appropriate number of places.

a)  $6.2 \times 10^2$       b)  $5.31 \times 10^5$       c)  $0.000122 \times 10^4$

What happens to a number when we multiply by a *negative* power of 10?

**Example 2**

Multiply.

a)  $41 \times 10^{-2}$       b)  $367 \times 10^{-4}$       c)  $5.9 \times 10^{-1}$

**Solution**

a)  $41 \times 10^{-2} = 41 \times \dfrac{1}{100} = \dfrac{41}{100} = 0.41$

b)  $367 \times 10^{-4} = 367 \times \dfrac{1}{10,000} = \dfrac{367}{10,000} = 0.0367$

c)  $5.9 \times 10^{-1} = 5.9 \times \dfrac{1}{10} = \dfrac{5.9}{10} = 0.59$

Is there a pattern? When we multiply each of these numbers by a negative power of 10, the result is *smaller* than the original number. The exponent determines how many places to the *left* the decimal point is moved:

$$41 \times 10^{-2} = 41. \times 10^{-2} = 0.41 \qquad 367 \times 10^{-4} = 0367. \times 10^{-4} = 0.0367$$
$$\text{2 places to the left} \qquad\qquad\qquad \text{4 places to the left}$$

$$5.9 \times 10^{-1} = 5.9 \times 10^{-1} = 0.59$$
$$\text{1 place to the left}$$

**You Try 2**

Multiply.

a)  $83 \times 10^{-2}$          b)  $45 \times 10^{-3}$

It is important to understand the previous concepts to understand how to use scientific notation.

## 2. Understand Scientific Notation

**Definition**

A number is in **scientific notation** if it is written in the form $a \times 10^n$ where $1 \leq |a| < 10$ and $n$ is an integer.

Multiplying $|a|$ by a *positive* power of 10 will result in a number that is *larger* than $|a|$. Multiplying $|a|$ by a *negative* power of 10 will result in a number that is *smaller* than $|a|$. The double inequality $1 \leq |a| < 10$ means that $a$ is a number that has *one* nonzero digit to the left of the decimal point.

Here are some examples of numbers written in scientific notation: $3.82 \times 10^{-5}$, $1.2 \times 10^{3}$, and $7 \times 10^{-2}$.

The following numbers are *not* in scientific notation:

$$51.94 \times 10^{4} \qquad\qquad 0.61 \times 10^{-3} \qquad\qquad 300 \times 10^{6}$$
$$\uparrow \qquad\qquad\qquad \uparrow \qquad\qquad\qquad \uparrow$$

| 2 digits to left of decimal point | Zero is to left of decimal point | 3 digits to left of decimal point |

Now let's convert a number written in scientific notation to a number without exponents.

**Example 3**

Rewrite without exponents.

a)  $5.923 \times 10^{4}$          b)  $7.4 \times 10^{-3}$          c)  $1.8875 \times 10^{3}$

**Solution**

a)  $5.923 \times 10^{4} \rightarrow 5.9230 = 59,230$      Remember, multiplying by a positive power of
    4 places to the right                                10 will make the result *larger* than 5.923.

b)  $7.4 \times 10^{-3} \rightarrow 007.4 = 0.0074$      Multiplying by a negative power of 10 will
    3 places to the left                                make the result *smaller* than 7.4.

c)  $1.8875 \times 10^{3} \rightarrow 1.8875 = 1887.5$
    3 places to the right

**You Try 3**

Rewrite without exponents.

a)  $3.05 \times 10^{4}$          b)  $8.3 \times 10^{-5}$          c)  $6.91853 \times 10^{3}$

### 3. Write a Number in Scientific Notation

We will write the number 48,000 in scientific notation.

To write the number 48,000 in scientific notation, first locate its decimal point.

$$48,000.$$

Next, determine where the decimal point will be when the number is in scientific notation:

$$48,000.$$
$$\wedge$$
Decimal point
will be here.

Therefore, $48,000 = 4.8 \times 10^n$, where $n$ is an integer. Will $n$ be positive or negative? We can see that 4.8 must be multiplied by a *positive* power of 10 to make it larger, 48,000.

$$48000.$$
$$\wedge \qquad \nwarrow$$
Decimal point     Decimal point
will be here.     starts here.

Now we count four places between the original and the final decimal place locations.

$$48000.$$
$$\underset{1\,2\,3\,4}{}$$

We use the number of spaces, 4, as the exponent of 10.

$$48,000 = 4.8 \times 10^4$$

---

**Example 4**

Write each number in scientific notation.

**Solution**

a)  The distance from the Earth to the Sun is approximately 150,000,000 km.

$$150,000,000.$$
$$\wedge \qquad \nwarrow$$
Decimal point    Decimal point
will be here.     is here.

$$150,000,000. \qquad \text{Move decimal point eight places.}$$

$$150,000,000 \text{ km} = 1.5 \times 10^8 \text{ km}$$

b)  A single rhinovirus measures 0.00002 mm across.

$$0.00002 \text{ mm} \qquad\qquad\qquad 0.00002 \text{ mm} = 2 \times 10^{-5} \text{ mm}$$
$$\qquad \wedge$$
Decimal point
will be here.

■

---

**Summary   How to Write a Number in Scientific Notation**

1)  Locate the decimal point in the original number.

2)  Determine where the decimal point will be when converting to scientific notation. Remember, there will be *one* nonzero digit to the left of the decimal point.

3)  Count how many places you must move the decimal point to take it from its original place to its position for scientific notation.

4)  If the absolute value of the resulting number is *smaller* than the absolute value of the original number, you will multiply the result by a *positive* power of 10. Example: $350.9 = 3.509 \times 10^2$.

If the absolute value of the resulting number is *larger* than the absolute value of the original number, you will multiply the result by a *negative* power of 10. Example: $0.0000068 = 6.8 \times 10^{-6}$.

**You Try 4**

Write each number in scientific notation.

a)   The gross domestic product of the United States in 2008 was approximately $14,264,600,000,000.

b)   The diameter of a human hair is approximately 0.001 in.

## 4. Perform Operations with Numbers in Scientific Notation

We use the rules of exponents to perform operations with numbers in scientific notation.

**Example 5**

Perform the operations and simplify.

a)   $(-2 \times 10^3)(4 \times 10^2)$          b)   $\dfrac{3 \times 10^3}{4 \times 10^5}$

**Solution**

a)   $(-2 \times 10^3)(4 \times 10^2) = (-2 \times 4)(10^3 \times 10^2)$      Commutative property
$= -8 \times 10^5$      Add the exponents.
$= -800,000$

b)   $\dfrac{3 \times 10^3}{4 \times 10^5} = \dfrac{3}{4} \times \dfrac{10^3}{10^5}$
$= 0.75 \times 10^{-2}$      Write $\dfrac{3}{4}$ in decimal form.
$= 7.5 \times 10^{-3}$      Use scientific notation.
or $0.0075$

**You Try 5**

Perform the operations and simplify.

a)   $(2.6 \times 10^2)(5 \times 10^4)$          b)   $\dfrac{7.2 \times 10^{-9}}{6 \times 10^{-5}}$

**Using Technology**

We can use a graphing calculator to convert a very large or very small number to scientific notation, or to convert a number in scientific notation to a number written without an exponent. Suppose we are given a very large number such as 35,000,000,000. If you enter any number with more than 10 digits on the home screen on your calculator and press ENTER, the number will automatically be displayed in scientific notation as shown on the screen below. A small number with more than two zeros to the right of the decimal point (such as .000123) will automatically be displayed in scientific notation as shown below.

The E shown in the screen refers to a power of 10, so 3.5 E 10 is the number $3.5 \times 10^{10}$ in scientific notation. 1.23 E-4 is the number $1.23 \times 10^{-4}$ in scientific notation.

```
35000000000
            3.5E10
.000123
         1.23E-4
```

If a large number has 10 or fewer digits, or if a small number has fewer than three zeros to the right of the decimal point, then the number will not automatically be displayed in scientific notation. To display the number using scientific notation, press **MODE**, select SCI, and press ⌐ENTER⌐. When you return to the home screen, all numbers will be displayed in scientific notation as shown below.

A number written in scientific notation can be entered directly into your calculator. For example, the number $2.38 \times 10^7$ can be entered directly on the home screen by typing 2.38 followed by **2nd** ⌐,⌐ ⌐7⌐ ⌐ENTER⌐ as shown here. If you wish to display this number without an exponent, change the mode back to NORMAL and enter the number on the home screen as shown.

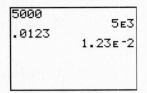

Write each number without an exponent, using a graphing calculator.

1.  $3.4 \times 10^5$          2.  $9.3 \times 10^7$          3.  $1.38 \times 10^{-3}$

Write each number in scientific notation, using a graphing calculator.

4.  186,000          5.  5280          6.  0.0469

---

### Answers to You Try Exercises

1) a) 620   b) 531,000   c) 1.22     2) a) 0.83   b) 0.045     3) a) 30,500   b) 0.000083
c) 6918.53     4) a) $1.42646 \times 10^{13}$ dollars   b) $1.0 \times 10^{-3}$ in.     5) a) 13,000,000   b) 0.00012

---

### Answers to Technology Exercises

1) 314,000     2) 93,000,000     3) .00138     4) $1.86 \times 10^5$     5) $5.28 \times 10^3$
6) $4.69 \times 10^{-2}$

---

## 2.4 Exercises

**Mixed Exercises: Objectives 1 and 2**

Determine whether each number is in scientific notation.

1)  $7.23 \times 10^5$

2)  $24.0 \times 10^{-3}$

3)  $0.16 \times 10^{-4}$

4)  $-2.8 \times 10^4$

5)  $-37 \times 10^{-2}$

6)  $0.9 \times 10^{-1}$

7)  $-5 \times 10^6$

8)  $7.5 \times 2^{-10}$

9)  Explain, in your own words, how to determine whether a number is expressed in scientific notation.

10) Explain, in your own words, how to write $4.1 \times 10^{-3}$ without an exponent.

11) Explain, in your own words, how to write $-7.26 \times 10^4$ without an exponent.

Multiply.

12) $980.2 \times 10^4$

13) $71.765 \times 10^2$

14) $0.1502 \times 10^8$

15) $40.6 \times 10^{-3}$

16) $0.0674 \times 10^{-1}$

17) $1,200,006 \times 10^{-7}$

**Objective 1: Multiply a Number by a Power of Ten**

Write each number without an exponent.

18) $1.92 \times 10^6$

19) $-6.8 \times 10^{-5}$

20) $2.03449 \times 10^3$

21) $-5.26 \times 10^4$

22) $-7 \times 10^{-4}$

23) $8 \times 10^{-6}$

24) $-9.5 \times 10^{-3}$

25) $6.021967 \times 10^5$

26) $6 \times 10^4$

VIDEO 27) $3 \times 10^6$

28) $-9.815 \times 10^{-2}$

29) $-7.44 \times 10^{-4}$

30) $4.1 \times 10^{-6}$

Write the following quantities without an exponent.

31) About $2.4428 \times 10^7$ Americans played golf at least 2 times in a year. Write this number using scientific notation. (Statistical Abstract of the U.S., www.census.gov)

32) In 2009, the social network website Facebook claimed that over $2.20 \times 10^8$ photos were uploaded each week. (www.facebook.com)

33) The radius of one hydrogen atom is about $2.5 \times 10^{-11}$ meters.

34) The length of a household ant is $2.54 \times 10^{-3}$ meters.

### Objective 3: Write a Number in Scientific Notation

Write each number in scientific notation.

VIDEO 35) 2110.5

36) 38.25

37) 0.000096

38) 0.00418

39) $-7,000,000$

40) 62,000

41) 3400

42) $-145,000$

VIDEO 43) 0.0008

44) $-0.00000022$

45) $-0.076$

46) 990

47) 6000

48) $-500,000$

Write each number in scientific notation.

49) The total weight of the Golden Gate Bridge is 380,800,000 kg. (www.goldengatebridge.com)

50) A typical hard drive may hold approximately 160,000,000,000 bytes of data.

51) The diameter of an atom is about 0.00000001 cm.

52) The oxygen-hydrogen bond length in a water molecule is 0.000000001 mm.

### Objective 4: Perform Operations with Numbers in Scientific Notation

Perform the operation as indicated. Write the final answer without an exponent.

53) $\dfrac{6 \times 10^9}{2 \times 10^5}$

54) $(7 \times 10^2)(2 \times 10^4)$

VIDEO 55) $(2.3 \times 10^3)(3 \times 10^2)$

56) $\dfrac{8 \times 10^7}{4 \times 10^4}$

57) $\dfrac{8.4 \times 10^{12}}{-7 \times 10^9}$

58) $\dfrac{-4.5 \times 10^{-6}}{-1.5 \times 10^{-8}}$

59) $(-1.5 \times 10^{-8})(4 \times 10^6)$

60) $(-3 \times 10^{-2})(-2.6 \times 10^{-3})$

61) $\dfrac{-3 \times 10^5}{6 \times 10^8}$

62) $\dfrac{2 \times 10^1}{5 \times 10^4}$

63) $(9.75 \times 10^4) + (6.25 \times 10^4)$

64) $(4.7 \times 10^{-3}) + (8.8 \times 10^{-3})$

65) $(3.19 \times 10^{-5}) + (9.2 \times 10^{-5})$

66) $(2 \times 10^2) + (9.7 \times 10^2)$

For each problem, express each number in scientific notation, then solve the problem.

67) Humans shed about $1.44 \times 10^7$ particles of skin every day. How many particles would be shed in a year? (Assume 365 days in a year.)

68) Scientists send a lunar probe to land on the moon and send back data. How long will it take for pictures to reach the Earth if the distance between the Earth and the moon is 360,000 km and if the speed of light is $3 \times 10^5$ km/sec?

69) In Wisconsin in 2001, approximately 1,300,000 cows produced $2.21 \times 10^{10}$ lb of milk. On average, how much milk did each cow produce? (www.nass.usda.gov)

70) The average snail can move $1.81 \times 10^{-3}$ mi in 5 hours. What is its rate of speed in miles per hour?

71) A photo printer delivers approximately $1.1 \times 10^6$ droplets of ink per square inch. How many droplets of ink would a 4 in. × 6 in. photo contain?

72) In 2007, 3,500,000,000,000 prescription drug orders were filled in the United States. If the average price of each prescription was roughly $65.00, how much did the United States pay for prescription drugs last year? (National Conference of State Legislatures, www.ncsl.org)

73) In 2006, American households spent a total of about $7.3 \times 10^{11}$ dollars on food. If there were roughly 120,000,000 households in 2006, how much money did the average household spend on food? (Round to the closest dollar.) (www.census.gov)

74) Find the population density of Australia if the estimated population in 2009 was about 22,000,000 people and the country encompasses about 2,900,000 sq mi. (Australian Bureau of Statistics, www.abs.gov.au)

75) When one of the U.S. space shuttles enters orbit, it travels at about 7800 m/sec. How far does it travel in 2 days? (Hint: Change days to seconds, and write all numbers in scientific notation before doing the computations.) (hypertextbook.com)

76) According to Nielsen Media Research, over 92,000,000 people watched Super Bowl XLIII in 2009 between the Pittsburgh Steelers and the Arizona Cardinals. The California Avocado Commission estimates that about 736,000,000 ounces of avocados were eaten during that Super Bowl, mostly in the form of guacamole. On average, how many ounces of avocados did each viewer eat during the Super Bowl?

77) In 2007, the United States produced about $6 \times 10^9$ metric tons of carbon emissions. The U.S. population that year was about 300 million. Find the amount of carbon emissions produced per person that year. (www.eia.doe.gov, U.S. Census Bureau)

| Definition/Procedure | Example |
|---|---|

### 2.1A  The Product Rule and Power Rules

**Exponential Expression:**

$a^n = \underbrace{a \cdot a \cdot a \cdot \cdots \cdot a}_{n \text{ factors of } a}$

$a$ is the **base**, $n$ is the exponent. **(p. 80)**

$5^4 = 5 \cdot 5 \cdot 5 \cdot 5$

5 is the **base**, 4 is the exponent.

---

**Product Rule:** $a^m \cdot a^n = a^{m+n}$ **(p. 81)**

$x^8 \cdot x^2 = x^{10}$

---

**Basic Power Rule:** $(a^m)^n = a^{mn}$ **(p. 82)**

$(t^3)^5 = t^{15}$

---

**Power Rule for a Product:**

$(ab)^n = a^n b^n$ **(p. 82)**

$(2c)^4 = 2^4 c^4 = 16c^4$

---

**Power Rule for a Quotient:**

$\left(\dfrac{a}{b}\right)^n = \dfrac{a^n}{b^n}$, where $b \neq 0$. **(p. 83)**

$\left(\dfrac{w}{5}\right)^3 = \dfrac{w^3}{5^3} = \dfrac{w^3}{125}$

### 2.1B  Combining the Rules

Remember to follow the order of operations. **(p. 85)**

Simplify $(3y^4)^2 (2y^9)^3$.

$= 9y^8 \cdot 8y^{27}$    Exponents come before multiplication.

$= 72y^{35}$    Use the product rule and multiply coefficients.

### 2.2A  Real-Number Bases

**Zero Exponent:** If $a \neq 0$, then $a^0 = 1$. **(p. 88)**

$(-9)^0 = 1$

---

**Negative Exponent:**

For $a \neq 0$, $a^{-n} = \left(\dfrac{1}{a}\right)^n = \dfrac{1}{a^n}$. **(p. 88)**

Evaluate. $\left(\dfrac{5}{2}\right)^{-3} = \left(\dfrac{2}{5}\right)^3 = \dfrac{2^3}{5^3} = \dfrac{8}{125}$

### 2.2B  Variable Bases

If $a \neq 0$ and $b \neq 0$, then $\left(\dfrac{a}{b}\right)^{-m} = \left(\dfrac{b}{a}\right)^m$. **(p. 91)**

Rewrite $p^{-10}$ with a positive exponent (assume $p \neq 0$).

$p^{-10} = \left(\dfrac{1}{p}\right)^{10} = \dfrac{1}{p^{10}}$

---

If $a \neq 0$ and $b \neq 0$, then $\dfrac{a^{-m}}{b^{-n}} = \dfrac{b^n}{a^m}$. **(p. 92)**

Rewrite each expression with positive exponents. Assume the variables represent nonzero real numbers.

a) $\dfrac{x^{-3}}{y^{-7}} = \dfrac{y^7}{x^3}$    b) $\dfrac{14m^{-6}}{n^{-1}} = \dfrac{14n}{m^6}$

| Definition/Procedure | Example |
|---|---|

## 2.3 The Quotient Rule

**Quotient Rule:** If $a \neq 0$, then $\dfrac{a^m}{a^n} = a^{m-n}$. **(p. 94)**

Simplify.

$$\frac{4^9}{4^6} = 4^{9-6} = 4^3 = 64$$

## Putting It All Together

**Combine the Rules of Exponents (p. 96)**

Simplify.

$$\left(\frac{a^4}{2a^7}\right)^{-5} = \left(\frac{2a^7}{a^4}\right)^5 = (2a^3)^5 = 32a^{15}$$

## 2.4 Scientific Notation

**Scientific Notation**

A number is in **scientific notation** if it is written in the form $a \times 10^n$, where $1 \leq |a| < 10$ and $n$ is an integer. That is, $a$ is a number that has one nonzero digit to the left of the decimal point. **(p. 101)**

Write in scientific notation.

a) $78{,}000 \rightarrow 78{,}000 \rightarrow 7.8 \times 10^4$

b) $0.00293 \rightarrow 0.00293 \rightarrow 2.93 \times 10^{-3}$

**Converting from Scientific Notation (p. 101)**

Write without exponents.

a) $5 \times 10^{-4} \rightarrow 0005. \rightarrow 0.0005$

b) $1.7 \times 10^6 = 1.700000 \rightarrow 1{,}700{,}000$

**Performing Operations (p. 103)**

Multiply. $(4 \times 10^2)(2 \times 10^4)$

$\quad = (4 \times 2)(10^2 \times 10^4)$

$\quad = 8 \times 10^6$

$\quad = 8{,}000{,}000$

**(2.1 A)**

1) Write in exponential form.

   a) $8 \cdot 8 \cdot 8 \cdot 8 \cdot 8 \cdot 8$

   b) $(-7)(-7)(-7)(-7)$

2) Identify the base and the exponent.

   a) $-6^5$

   b) $(4t)^3$

   c) $4t^3$

   d) $-4t^3$

3) Use the rules of exponents to simplify.

   a) $2^3 \cdot 2^2$

   b) $\left(\dfrac{1}{3}\right)^2 \cdot \left(\dfrac{1}{3}\right)$

   c) $(7^3)^4$

   d) $(k^5)^6$

4) Use the rules of exponents to simplify.

   a) $(3^2)^2$

   b) $8^3 \cdot 8^7$

   c) $(m^4)^9$

   d) $p^9 \cdot p^7$

5) Simplify using the rules of exponents.

   a) $(5y)^3$

   b) $(-7m^4)(2m^{12})$

   c) $\left(\dfrac{a}{b}\right)^6$

   d) $6(xy)^2$

   e) $\left(\dfrac{10}{9}c^4\right)(2c)\left(\dfrac{15}{4}c^3\right)$

6) Simplify using the rules of exponents.

   a) $\left(\dfrac{x}{y}\right)^{10}$

   b) $(-2z)^5$

   c) $(6t^7)\left(-\dfrac{5}{8}t^5\right)\left(\dfrac{2}{3}t^2\right)$

   d) $-3(ab)^4$

   e) $(10j^6)(4j)$

**(2.1 B)**

7) Simplify using the rules of exponents.

   a) $(z^5)^2(z^3)^4$

   b) $-2(3c^5d^8)^2$

   c) $(9-4)^3$

   d) $\dfrac{(10t^3)^2}{(2u^7)^3}$

8) Simplify using the rules of exponents.

   a) $\left(\dfrac{-20d^4c}{5b^3}\right)^3$

   b) $(-2y^8z)^3(3yz^2)^2$

   c) $\dfrac{x^7 \cdot (x^2)^5}{(2y^3)^4}$

   d) $(6-8)^2$

**(2.2 A)**

9) Evaluate.

   a) $8^0$

   b) $-3^0$

   c) $9^{-1}$

   d) $3^{-2} - 2^{-2}$

   e) $\left(\dfrac{4}{5}\right)^{-3}$

10) Evaluate.

   a) $(-12)^0$

   b) $5^0 + 4^0$

   c) $-6^{-2}$

   d) $2^{-4}$

   e) $\left(\dfrac{10}{3}\right)^{-2}$

**(2.2 B)**

11) Rewrite the expression with positive exponents. Assume the variables do not equal zero.

   a) $v^{-9}$

   b) $\left(\dfrac{9}{c}\right)^{-2}$

   c) $\left(\dfrac{1}{y}\right)^{-8}$

   d) $-7k^{-9}$

   e) $\dfrac{19z^{-4}}{a^{-1}}$

   f) $20m^{-6}n^5$

   g) $\left(\dfrac{2j}{k}\right)^{-5}$

12) Rewrite the expression with positive exponents. Assume the variables do not equal zero.

   a) $\left(\dfrac{1}{x}\right)^{-5}$

   b) $3p^{-4}$

   c) $a^{-8}b^{-3}$

   d) $\dfrac{12k^{-3}r^5}{16mn^{-6}}$

   e) $\dfrac{c^{-1}d^{-1}}{15}$

   f) $\left(-\dfrac{m}{4n}\right)^{-3}$

   g) $\dfrac{10b^4}{a^{-9}}$

**(2.3)**

In Exercises 13–16, assume the variables represent nonzero real numbers. The answers should not contain negative exponents.

13) Simplify using the rules of exponents.

   a) $\dfrac{3^8}{3^6}$

   b) $\dfrac{r^{11}}{r^3}$

   c) $\dfrac{48t^{-2}}{32t^3}$

   d) $\dfrac{21xy^2}{35x^{-6}y^3}$

14) Simplify using the rules of exponents.

a) $\dfrac{2^9}{2^{15}}$  b) $\dfrac{d^4}{d^{-10}}$

c) $\dfrac{m^{-5}n^3}{mn^8}$  d) $\dfrac{100a^8b^{-1}}{25a^7b^{-4}}$

15) Simplify by applying one or more of the rules of exponents.

a) $(-3s^4t^5)^4$  b) $\dfrac{(2a^6)^5}{(4a^7)^2}$

c) $\left(\dfrac{z^4}{y^3}\right)^{-6}$  d) $(-x^3y)^5(6x^{-2}y^3)^2$

e) $\left(\dfrac{cd^{-4}}{c^8d^{-9}}\right)^5$  f) $\left(\dfrac{14m^5n^5}{7m^4n}\right)^3$

g) $\left(\dfrac{3k^{-1}t}{5k^{-7}t^4}\right)^{-3}$  h) $\left(\dfrac{40}{21}x^{10}\right)(3x^{-12})\left(\dfrac{49}{20}x^2\right)$

16) Simplify by applying one or more of the rules of exponents.

a) $\left(\dfrac{4}{3}\right)^8\left(\dfrac{4}{3}\right)^{-2}\left(\dfrac{4}{3}\right)^{-3}$  b) $\left(\dfrac{k^{10}}{k^4}\right)^3$

c) $\left(\dfrac{x^{-4}y^{11}}{xy^2}\right)^{-2}$  d) $(-9z^5)^{-2}$

e) $\left(\dfrac{g^2\cdot g^{-1}}{g^{-7}}\right)^{-4}$  f) $(12p^{-3})\left(\dfrac{10}{3}p^5\right)\left(\dfrac{1}{4}p^2\right)^2$

g) $\left(\dfrac{30u^2v^{-3}}{40u^7v^{-7}}\right)^{-2}$  h) $-5(3h^4k^9)^2$

17) Simplify. Assume that the variables represent nonzero integers. Write your final answer so that the exponents have positive coefficients.

a) $y^{3k}\cdot y^{7k}$  b) $(x^{5p})^2$

c) $\dfrac{z^{12c}}{z^{5c}}$  d) $\dfrac{t^{6d}}{t^{11d}}$

**(2.4)**
**Write each number without an exponent.**

18) $9.38\times 10^5$  19) $-4.185\times 10^2$

20) $9\times 10^3$  21) $6.7\times 10^{-4}$

22) $1.05\times 10^{-6}$  23) $2\times 10^4$

24) $8.8\times 10^{-2}$

**Write each number in scientific notation.**

25) 0.0000575  26) 36,940

27) 32,000,000  28) 0.0000004

29) 178,000  30) 66

31) 0.0009315

**Write the number without exponents.**

32) Before 2010, golfer Tiger Woods earned over $7\times 10^7$ dollars per year in product endorsements. (www.forbes.com)

**Perform the operation as indicated. Write the final answer without an exponent.**

33) $\dfrac{8\times 10^6}{2\times 10^{13}}$  34) $\dfrac{-1\times 10^9}{5\times 10^{12}}$

35) $(9\times 10^{-8})(4\times 10^7)$  36) $(5\times 10^3)(3.8\times 10^{-8})$

37) $\dfrac{-3\times 10^{10}}{-4\times 10^6}$  38) $(-4.2\times 10^2)(3.1\times 10^3)$

**For each problem, write each of the numbers in scientific notation, then solve the problem. Write the answer without exponents.**

39) Eight porcupines have a total of about $2.4\times 10^5$ quills on their bodies. How many quills would one porcupine have?

40) In 2002, Nebraska had approximately $4.6\times 10^7$ acres of farmland and about 50,000 farms. What was the average size of a Nebraska farm in 2002? (www.nass.usda.gov)

41) One molecule of water has a mass of $2.99\times 10^{-23}$ g. Find the mass of 100,000,000 molecules.

42) At the end of 2008, the number of SMS text messages sent in one month in the United States was 110.4 billion. If 270.3 million people used SMS text messaging, about how many did each person send that month? (Round to the nearest whole number.) (www.ctia.org/advocacy/research/index.cfm/AID/10323)

43) When the polls closed on the west coast on November 4, 2008, and Barack Obama was declared the new president, there were about 143,000 visits per second to news websites. If the visits continued at that rate for 3 minutes, how many visits did the news websites receive during that time? (www.xconomy.com)

**Write in exponential form.**

1) $(-3)(-3)(-3)$

2) $x \cdot x \cdot x \cdot x \cdot x$

**Use the rules of exponents to simplify.**

3) $5^2 \cdot 5$

4) $\left(\dfrac{1}{x}\right)^5 \cdot \left(\dfrac{1}{x}\right)^2$

5) $(8^3)^{12}$

6) $p^7 \cdot p^{-2}$

**Evaluate.**

7) $3^4$

8) $8^0$

9) $2^{-5}$

10) $4^{-2} + 2^{-3}$

11) $\left(-\dfrac{3}{4}\right)^3$

12) $\left(\dfrac{10}{7}\right)^{-2}$

**Simplify using the rules of exponents. Assume all variables represent nonzero real numbers. The final answer should not contain negative exponents.**

13) $(5n^6)^3$

14) $(-3p^4)(10p^8)$

15) $\dfrac{m^{10}}{m^4}$

16) $\dfrac{a^9 b}{a^5 b^7}$

17) $\left(\dfrac{-12t^{-6}u^8}{4t^5 u^{-1}}\right)^{-3}$

18) $(2y^{-4})^6 \left(\dfrac{1}{2}y^5\right)^3$

19) $\left(\dfrac{(9x^2 y^{-2})^3}{4xy}\right)^0$

20) $\dfrac{(2m + n)^3}{(2m + n)^2}$

21) $\dfrac{12a^4 b^{-3}}{20c^{-2}d^3}$

22) $\left(\dfrac{y^{-7} \cdot y^3}{y^5}\right)^{-2}$

23) Simplify $t^{10k} \cdot t^{3k}$. Assume that the variables represent nonzero integers.

24) Rewrite $7.283 \times 10^5$ without exponents.

25) Write 0.000165 in scientific notation.

26) Divide. Write the answer without exponents. $\dfrac{-7.5 \times 10^{12}}{1.5 \times 10^8}$

27) Write the number without an exponent: In 2002, the population of Texas was about $2.18 \times 10^7$. (U.S. Census Bureau)

28) An electron is a subatomic particle with a mass of $9.1 \times 10^{-28}$ g. What is the mass of 2,000,000,000 electrons? Write the answer without exponents.

1) Write $\dfrac{90}{150}$ in lowest terms.

**Perform the indicated operations. Write the answer in lowest terms.**

2) $\dfrac{2}{15} + \dfrac{1}{10} + \dfrac{7}{20}$

3) $\dfrac{4}{15} \div \dfrac{20}{21}$

4) $-144 \div (-12)$

5) $-26 + 5 - 7$

6) $-9^2$

7) $(-1)^5$

8) $(5 + 1)^2 - 2[17 + 5(10 - 14)]$

9) Glen Crest High School is building a new football field. The dimensions of a regulation-size field are $53\dfrac{1}{3}$ yd by 120 yd. (There are 10 yd of end zone on each end.) The sod for the field will cost $1.80/yd^2$.

   a)  Find the perimeter of the field.

   b)  How much will it cost to sod the field?

10) Evaluate $2p^2 - 11q$ when $p = 3$ and $q = -4$.

11) State the formula for the volume of a sphere.

12) Given this set of numbers $\left\{ 3, -4, -2.1\overline{3}, \sqrt{11}, 2\dfrac{2}{3} \right\}$, list the

   a)  integers

   b)  irrational numbers

   c)  natural numbers

   d)  rational numbers

   e)  whole numbers

13) Evaluate $4x^3 + 2x - 3$ when $x = 4$.

14) Rewrite $\dfrac{3}{4}(6m - 20n + 7)$ using the distributive property.

15) Combine like terms and simplify:
    $5(t^2 + 7t - 3) - 2(4t^2 - t + 5)$

16) Let $x$ represent the unknown quantity, and write a mathematical expression for "thirteen less than half of a number."

**Simplify using the rules of exponents. The answer should not contain negative exponents. Assume the variables represent nonzero real numbers.**

17) $4^3 \cdot 4^7$

18) $\left(\dfrac{x}{y}\right)^{-3}$

19) $\left(\dfrac{32x^3}{8x^{-2}}\right)^{-1}$

20) $-(3rt^{-3})^4$

21) $(4z^3)(-7z^5)$

22) $\dfrac{n^2}{n^9}$

23) $(-2a^{-6}b)^5$

24) Write 0.000729 in scientific notation.

25) Perform the indicated operation. Write the final answer without an exponent. $(6.2 \times 10^5)(9.4 \times 10^{-2})$

# CHAPTER 3

# Linear Equations and Inequalities

## Algebra at Work: Landscape Architecture

A landscape architect must have excellent problem-solving skills.

Matthew is designing the driveway, patio, and walkway for this new home. The village has a building code which states that, at most, 70% of the lot can be covered with an impervious surface such as the house, driveway, patio, and walkway leading up to the front door. So, he cannot design just anything.

To begin, Matthew must determine the area of the land and find 70% of that number to determine how much land can be covered with hard surfaces. He must subtract the area covered by the house to determine how much land he has left for the driveway, patio, and walkway. Using his design experience and problem-solving skills, he must come up with a plan for building the driveway, patio, and walkway that will not only please his client but will meet building codes as well.

In this chapter, we will learn different strategies for solving many different types of problems.

## Section 3.1 Solving Linear Equations Part I

### Objectives

1. Define a Linear Equation in One Variable
2. Use the Addition and Subtraction Properties of Equality
3. Use the Multiplication and Division Properties of Equality
4. Solve Equations of the Form $ax + b = c$

### 1. Define a Linear Equation in One Variable

What is an equation? It is a mathematical statement that two expressions are equal. For example, $4 + 3 = 7$ is an equation.

**Note**

An equation contains an "=" sign and an expression does not.

$3x + 5 = 17$ is an *equation*.

$3x + 5x$ is an *expression*.

We can **solve** equations, and we can **simplify** expressions.

There are many different types of algebraic equations, and in Sections 3.1–3.3 we will learn how to solve *linear* equations. Here are some examples of linear equations in one variable:

$$p - 1 = 4, \qquad 3x + 5 = 17, \qquad 8(n + 1) - 7 = 2n + 3, \qquad -\frac{5}{6}y + \frac{1}{3} = y - 2$$

**Definition**

A **linear equation in one variable** is an equation that can be written in the form $ax + b = 0$, where $a$ and $b$ are real numbers and $a \neq 0$.

The exponent of the variable, $x$, in a linear equation is 1. For this reason, linear equations are also known as first-degree equations. Equations like $k^2 - 13k + 36 = 0$ and $\sqrt{w - 3} = 2$ are not linear equations and are presented later in the text.

To **solve an equation** means to find the value or values of the variable that make the equation true. For example, the solution of the equation $p - 1 = 4$ is $p = 5$ since substituting 5 for the variable makes the equation true.

$$p - 1 = 4$$
$$5 - 1 = 4 \quad \text{True}$$

Usually, we use set notation to list all the solutions of an equation. The **solution set** of an equation is the set of all numbers that make the equation true. Therefore, $\{5\}$ is the solution set of the equation $p - 1 = 4$. We also say that 5 *satisfies* the equation $p - 1 = 4$.

### 2. Use the Addition and Subtraction Properties of Equality

Begin with the true statement $8 = 8$. What happens if we add the same number, say 2, to each side? Is the statement still true? Yes.

$$8 = 8$$
$$8 + 2 = 8 + 2$$
$$10 = 10 \quad \text{True}$$

Will a statement remain true if we *subtract* the same number from each side? Let's begin with the true statement $5 = 5$ and subtract 9 from each side:

$$5 = 5$$
$$5 - 9 = 5 - 9$$
$$-4 = -4 \quad \text{True}$$

When we subtracted 9 from each side of the equation, the new statement was true.

$$8 = 8 \text{ and } 8 + 2 = 8 + 2 \text{ are } \textit{equivalent equations.}$$
$$5 = 5 \text{ and } 5 - 9 = 5 - 9 \text{ are } \textit{equivalent equations} \text{ as well.}$$

Adding the same number to both sides of an equation or subtracting the same number from both sides of an equation will produce equivalent equations. We can use these principles to solve an algebraic equation because doing so will not change the equation's solution.

> **Property** Addition and Subtraction Properties of Equality
>
> Let $a, b$, and $c$ be expressions representing real numbers. Then,
>
> 1) If $a = b$, then $a + c = b + c$    Addition property of equality
>
> 2) If $a = b$, then $a - c = b - c$    Subtraction property of equality

**Example 1**

Solve $x - 8 = 3$. Check the solution.

**Solution**

Remember, to solve the equation means to find the value of the variable that makes the statement true. To do this, we want to get the variable on a side by itself. We call this **isolating the variable.**

On the left side of the equal sign, the 8 is being **subtracted from** the $x$. To isolate $x$, we perform the "opposite" operation—that is, we **add 8** to each side.

$$x - 8 = 3$$
$$x - 8 + 8 = 3 + 8 \qquad \text{Add 8 to each side.}$$
$$x = 11 \qquad \text{Simplify.}$$

Check: Substitute 11 for $x$ in the original equation.

$$x - 8 = 3$$
$$11 - 8 = 3$$
$$3 = 3 \quad \checkmark$$

The solution set is $\{11\}$.

 **You Try 1**

Solve $b - 5 = 9$. Check the solution.

**Example 2**

Solve $t + 3.4 = 8.6$. Check the solution.

**Solution**

Here, 3.4 is being added to $t$. To get the $t$ by itself, subtract 3.4 from each side.

$$t + 3.4 = 8.6$$
$$t + 3.4 - 3.4 = 8.6 - 3.4 \qquad \text{Subtract 3.4 from each side.}$$
$$t = 5.2 \qquad \text{Simplify.}$$

Check: Substitute 5.2 for $t$ in the original equation.

$$t + 3.4 = 8.6$$
$$5.2 + 3.4 = 8.6$$
$$8.6 = 8.6 \quad \checkmark$$

The solution set is $\{5.2\}$.

**You Try 2**

Solve $r + 7.3 = 4.9$. Check the solution.

Equations can contain variables on either side of the equal sign.

**Example 3**

Solve $-5 = m + 4$. Check the solution.

**Solution**

Notice that the 4 is being added to the variable, $m$. We will subtract 4 from each side to isolate the variable.

$$-5 = m + 4$$
$$-5 - 4 = m + 4 - 4 \qquad \text{Subtract 4 from each side.}$$
$$-9 = m \qquad \text{Simplify.}$$

Check: Substitute $-9$ for $m$ in the original equation.

$$-5 = m + 4$$
$$-5 = -9 + 4$$
$$-5 = -5 \quad \checkmark$$

The solution set is $\{-9\}$.    ■

**You Try 3**

Solve $-2 = y + 6$. Check the solution.

## 3. Use the Multiplication and Division Properties of Equality

We have just seen that we can add a quantity to each side of an equation or subtract a quantity from each side of an equation, and we will obtain an equivalent equation. It is also true that if we multiply both sides of an equation by the same nonzero number or divide both sides of an equation by the same nonzero number, then we will obtain an equivalent equation.

---

**Property**    Multiplication and Division Properties of Equality

Let $a, b,$ and $c$ be expressions representing real numbers where $c \neq 0$. Then,

1)    If $a = b$, then $ac = bc$       Multiplication property of equality

2)    If $a = b$, then $\dfrac{a}{c} = \dfrac{b}{c}$       Division property of equality

---

**Example 4**

Solve $3k = -9.6$. Check the solution.

**Solution**

On the left-hand side of the equation, the $k$ is being **multiplied** by 3. So, we will perform the "opposite" operation and **divide** each side by 3.

$$3k = -9.6$$

$$\frac{3k}{3} = \frac{-9.6}{3} \qquad \text{Divide each side by 3.}$$

$$k = -3.2 \qquad \text{Simplify.}$$

Check: Substitute $-3.2$ for $k$ in the original equation.

$$3k = -9.6$$
$$3(-3.2) = -9.6$$
$$-9.6 = -9.6 \;\checkmark$$

The solution set is $\{-3.2\}$.

### You Try 4

Solve $-8w = 42.4$. Check the solution.

### Example 5

Solve $-m = 19$.

**Solution**

The negative sign in front of the $m$ tells us that the coefficient of $m$ is $-1$. Since $m$ is being **multiplied** by $-1$, we will **divide** each side by $-1$.

$$-m = 19$$
$$\frac{-1m}{-1} = \frac{19}{-1} \qquad \text{Rewrite } -m \text{ as } -1m; \text{ divide each side by } -1.$$
$$m = -19 \qquad \text{Simplify.}$$

The check is left to the student. The solution set is $\{-19\}$.

### You Try 5

Solve $-a = -25$. Check the solution.

### Example 6

Solve $\dfrac{x}{4} = 5$.

**Solution**

The $x$ is being **divided** by 4. Therefore, we will **multiply** each side by 4 to get the $x$ on a side by itself.

$$\frac{x}{4} = 5$$
$$4 \cdot \frac{x}{4} = 4 \cdot 5 \qquad \text{Multiply each side by 4.}$$
$$1x = 20 \qquad \text{Simplify.}$$
$$x = 20$$

The check is left to the student. The solution set is $\{20\}$.

### You Try 6

Solve $\dfrac{h}{7} = 8$. Check the solution.

## Example 7

Solve $\dfrac{3}{8}y = 12$.

### Solution

On the left-hand side, the $y$ is being **multiplied** by $\dfrac{3}{8}$. So, we could divide each side by $\dfrac{3}{8}$. However, recall that dividing a quantity by a fraction is the same as **multiplying by the reciprocal** of the fraction. Therefore, we will multiply each side by the reciprocal of $\dfrac{3}{8}$.

$$\frac{3}{8}y = 12$$

$$\frac{8}{3} \cdot \frac{3}{8}y = \frac{8}{3} \cdot 12 \qquad \text{The reciprocal of } \frac{3}{8} \text{ is } \frac{8}{3}. \text{ Multiply each side by } \frac{8}{3}.$$

$$1y = \frac{8}{\overset{1}{\cancel{3}}} \cdot \overset{4}{\cancel{12}} \qquad \text{Perform the multiplication.}$$

$$y = 32 \qquad \text{Simplify.}$$

The check is left to the student. The solution set is $\{32\}$.    ■

## You Try 7

Solve $-\dfrac{5}{9}c = 20$. Check the solution.

### 4. Solve Equations of the Form $ax + b = c$

So far, we have not combined the properties of addition, subtraction, multiplication, and division to solve an equation. But that is exactly what we must do to solve equations like

$$3p + 7 = 31 \qquad \text{and} \qquad 4x + 9 - 6x + 2 = 17$$

## Example 8

Solve $3p + 7 = 31$.

### Solution

In this equation, there is a number, 7, being **added** to the term containing the variable, and the variable is being multiplied by a number, 3. **In general, we first eliminate the number being added to or subtracted from the variable.** Then we eliminate the coefficient.

$$3p + 7 = 31$$
$$3p + 7 - 7 = 31 - 7 \qquad \text{Subtract 7 from each side.}$$
$$3p = 24 \qquad \text{Combine like terms.}$$
$$\frac{3p}{3} = \frac{24}{3} \qquad \text{Divide by 3.}$$
$$p = 8 \qquad \text{Simplify.}$$

$$\text{Check:} \quad 3p + 7 = 31$$
$$3(8) + 7 = 31$$
$$24 + 7 = 31$$
$$31 = 31 \quad ✓$$

The solution set is $\{8\}$.    ■

**You Try 8**

Solve $2n + 9 = 15$.

**Example 9**

Solve $-\dfrac{6}{5}c - 1 = 13$.

**Solution**

On the left-hand side, the $c$ is being multiplied by $-\dfrac{6}{5}$, and 1 is being subtracted from the $c$-term. To solve the equation, begin by eliminating the number being subtracted from the $c$-term.

$$-\frac{6}{5}c - 1 = 13$$

$$-\frac{6}{5}c - 1 + 1 = 13 + 1 \qquad \text{Add 1 to each side.}$$

$$-\frac{6}{5}c = 14 \qquad \text{Combine like terms.}$$

$$-\frac{5}{6} \cdot \left(-\frac{6}{5}c\right) = -\frac{5}{6} \cdot 14 \qquad \text{Multiply each side by the reciprocal of } -\frac{6}{5}.$$

$$1c = -\frac{5}{\overset{}{\underset{3}{6}}} \cdot \overset{7}{14} \qquad \text{Simplify.}$$

$$c = -\frac{35}{3}$$

The check is left to the student. The solution set is $\left\{-\dfrac{35}{3}\right\}$.

**You Try 9**

Solve $-\dfrac{4}{9}z + 3 = -7$.

**Example 10**

Solve $-8.85 = 2.1y - 5.49$.

**Solution**

The variable is on the right-hand side of the equation. First, we will add 5.49 to each side, then we will divide by 2.1.

$$-8.85 = 2.1y - 5.49$$

$$-8.85 + 5.49 = 2.1y - 5.49 + 5.49 \qquad \text{Add 5.49 to each side.}$$

$$-3.36 = 2.1y \qquad \text{Combine like terms.}$$

$$\frac{-3.36}{2.1} = \frac{2.1y}{2.1} \qquad \text{Divide each side by 2.1.}$$

$$-1.6 = y \qquad \text{Simplify.}$$

Verify that $-1.6$ is the solution. The solution set is $\{-1.6\}$.

**You Try 10**

Solve $6.7 = -0.4t - 5.3$.

---

### Answers to You Try Exercises

1) $\{14\}$   2) $\{-2.4\}$   3) $\{-8\}$   4) $\{-5.3\}$   5) $\{25\}$   6) $\{56\}$

7) $\{-36\}$   8) $\{3\}$   9) $\left\{\dfrac{45}{2}\right\}$   10) $\{-30\}$

---

## 3.1 Exercises

### Objective 1: Define a Linear Equation in One Variable

Identify each as an expression or an equation.

1) $9c + 4 - 2c$

2) $5n - 3 = 6$

3) $y + 10(8y + 1) = -13$

4) $4 + 2(5p - 3)$

5) Can we solve $-6x + 10x$? Why or why not?

6) Can we solve $-6x + 10x = 28$? Why or why not?

7) Which of the following are linear equations in one variable?

   a) $2k + 9 - 7k + 1$   b) $-8 = \dfrac{3}{2}n - 1$

   c) $4(3t - 1) + 9t = 12$   d) $w^2 + 13w + 36 = 0$

8) Explain how to check the solution of an equation.

Determine whether the given value is a solution to the equation.

9) $a - 4 = -9$;   $a = 5$

10) $-5c = -10$;   $c = 2$

11) $-12y = 8$;   $y = -\dfrac{2}{3}$

12) $5 = a + 14$;   $a = -9$

13) $1.3 = 2p - 1.7$;   $p = 1.5$

14) $20m + 3 = 16$;   $m = \dfrac{4}{5}$

### Objective 2: Use the Addition and Subtraction Properties of Equality

Solve each equation and check the solution.

15) $n - 5 = 12$

16) $z + 8 = -2$

17) $b + 10 = 4$

18) $x - 3 = 9$

19) $-16 = k - 12$

20) $23 = r + 14$

21) $6 = 6 + y$

22) $5 = -1 + k$

23) $a - 2.9 = -3.6$

24) $w + 4.7 = 9.1$

25) $12 = x + 7.2$

26) $-8.3 = y - 5.6$

27) $h + \dfrac{5}{6} = \dfrac{1}{3}$

28) $b - \dfrac{3}{8} = \dfrac{3}{10}$

29) $-\dfrac{2}{5} = -\dfrac{5}{4} + c$

30) $-\dfrac{8}{9} = d + 2$

31) Write an equation that can be solved with the subtraction property of equality and that has a solution set of $\{-7\}$.

32) Write an equation that can be solved with the addition property of equality and that has a solution set of $\{10\}$.

### Objective 3: Use the Multiplication and Division Properties of Equality

Solve each equation and check the solution.

33) $2n = 8$

34) $9k = 72$

35) $-5z = 35$

36) $-8b = -24$

37) $-48 = -4r$

38) $-54 = 6m$

39) $63 = -28y$

40) $75 = 50c$

41) $10n = 2.3$

42) $-4x = -28.4$

43) $-7 = -0.5d$

44) $-3.9 = 1.3p$

45) $-x = 1$

46) $-h = -3$

47) $-6.5 = -v$

48) $\dfrac{4}{9} = -t$

49) $\dfrac{a}{4} = 12$

50) $\dfrac{w}{5} = 4$

51) $-\dfrac{m}{3} = 13$

52) $7 = -\dfrac{n}{11}$

53) $\dfrac{w}{6} = -\dfrac{3}{4}$

54) $-\dfrac{a}{21} = -\dfrac{2}{3}$

55) $\dfrac{1}{5}q = -9$

56) $\dfrac{1}{8}y = 3$

57) $-\dfrac{1}{6}m = -14$

58) $-\dfrac{1}{3}z = 9$

59) $\dfrac{5}{12} = \dfrac{1}{4}c$

60) $-\dfrac{5}{9} = -\dfrac{1}{6}r$

61) $\dfrac{1}{3}y = -\dfrac{11}{15}$

62) $-\dfrac{1}{7}a = \dfrac{3}{2}$

63) $-\dfrac{5}{3}d = -30$

64) $35 = \dfrac{5}{3}k$

65) $-21 = \dfrac{3}{2}d$

66) $-\dfrac{4}{7}w = -36$

85) $2 - \dfrac{5}{6}t = -2$

86) $5 + \dfrac{3}{4}h = -1$

87) $\dfrac{3}{4} = \dfrac{1}{2} - \dfrac{1}{6}z$

88) $1 = \dfrac{3}{5} - \dfrac{2}{3}k$

89) $0.2p + 9.3 = 5.7$

90) $0.5x - 2.6 = 4.9$

91) $3.8c - 7.62 = 2.64$

92) $4.3a + 1.98 = -14.36$

93) $14.74 = -20.6 - 5.7u$

94) $10.5 - 9.2m = -36.42$

**Objective 4: Solve Equations of the Form $ax + b = c$**

Solve each equation and check the solution.

67) $5z + 8 = 43$

68) $2y - 5 = 3$

69) $4n - 15 = -19$

70) $7c + 4 = 18$

71) $8d - 15 = -15$

72) $10m + 7 = 7$

73) $-11 = 5t - 9$

74) $7 = 4k + 13$

75) $-6h + 19 = 3$

76) $-8q - 11 = 9$

77) $10 = 3 - 7y$

78) $-6 = 9 - 3p$

79) $\dfrac{1}{2}d + 7 = 12$

80) $\dfrac{1}{3}x + 4 = 11$

81) $\dfrac{4}{5}b - 9 = -13$

82) $-\dfrac{12}{7}r + 5 = 3$

83) $-1 = \dfrac{10}{11}c + 5$

84) $2 = -\dfrac{9}{4}a - 10$

**Mixed Exercises: Objectives 1–4**

Solve and check each equation.

95) $-9z = 6$

96) $u - 23 = 52$

97) $3a - 11 = 16$

98) $20 = -\dfrac{1}{4}x$

99) $-\dfrac{c}{6} = -9$

100) $6w - 1 = -19$

101) $-34 = n + 15$

102) $67.9 = 7y$

103) $-\dfrac{1}{7}p = -8$

104) $-6 = \dfrac{m}{8}$

105) $8.33 - 6.35d = 17.22$

106) $\dfrac{10}{7}r + 3 = 18$

107) $-15 = 9 + \dfrac{4}{5}c$

108) $-7.92q + 41.95 = 22.15$

109) $-\dfrac{3}{4}k + \dfrac{2}{5} = -2$

110) $\dfrac{2}{3} = \dfrac{1}{2} - \dfrac{3}{8}t$

## Section 3.2  Solving Linear Equations Part II

### Objectives

1. Summarize the Steps for Solving a Linear Equation
2. Solve Equations Containing Variables on One Side of the Equal Sign
3. Solve Equations Containing Variables on Both Sides of the Equal Sign

### 1. Summarize the Steps for Solving a Linear Equation

In Section 3.1, we learned how to solve equations such as

$$x - 8 = 3 \qquad 3k = -9.6 \qquad \dfrac{3}{8}y = 12 \qquad -8.85 = 2.1y - 5.49$$

Each of these equations contains only one variable term. In this section, we will discuss how to solve equations in which more than one term contains a variable and where variables appear on both sides of the equal sign. We can use the following steps.

**Procedure   How to Solve a Linear Equation**

**Step 1:** **Clear parentheses** and **combine like terms** on each side of the equation.

**Step 2:** **Get the variable on one side of the equal sign and the constant on the other side of the equal sign** (isolate the variable) using the addition or subtraction property of equality.

**Step 3:** **Solve for the variable** using the multiplication or division property of equality.

**Step 4:** **Check the solution** in the original equation.

## 2. Solve Equations Containing Variables on One Side of the Equal Sign

Let's start by solving equations containing variables on only one side of the equal sign.

### Example 1

Solve $4x + 9 - 6x + 2 = 17$.

**Solution**

**Step 1:** Since there are two $x$-terms on the left side of the equal sign, begin by combining like terms.

$$4x + 9 - 6x + 2 = 17$$
$$-2x + 11 = 17 \qquad \text{Combine like terms.}$$

**Step 2:** Isolate the variable.

$$-2x + 11 - 11 = 17 - 11 \qquad \text{Subtract 11 from each side.}$$
$$-2x = 6 \qquad \text{Combine like terms.}$$

**Step 3:** Solve for $x$ using the division property of equality.

$$\frac{-2x}{-2} = \frac{6}{-2} \qquad \text{Divide each side by } -2.$$
$$y = -3 \qquad \text{Simplify.}$$

**Step 4:** Check:

$$4x + 9 - 6x + 2 = 17$$
$$4(-3) + 9 - 6(-3) + 2 = 17$$
$$-12 + 9 + 18 + 2 = 17$$
$$17 = 17 ✓$$

The solution set is $\{-3\}$.  ∎

### You Try 1

Solve $15 - 7u - 6 + 2u = -1$.

### Example 2

Solve $2(1 - 3h) - 5(2h + 3) = -21$.

**Solution**

**Step 1:** Clear the parentheses and combine like terms.

$$2(1 - 3h) - 5(2h + 3) = -21$$
$$2 - 6h - 10h - 15 = -21 \qquad \text{Distribute.}$$
$$-16h - 13 = -21 \qquad \text{Combine like terms.}$$

**Step 2:** Isolate the variable.

$$-16h - 13 + 13 = -21 + 13 \qquad \text{Add 13 to each side.}$$
$$-16h = -8 \qquad \text{Combine like terms.}$$

**Step 3:** Solve for $h$ using the division property of equality.

$$\frac{-16h}{-16} = \frac{-8}{-16} \qquad \text{Divide each side by } -16.$$
$$h = \frac{1}{2} \qquad \text{Simplify.}$$

**Step 4:** The check is left to the student. The solution set is $\left\{\dfrac{1}{2}\right\}$.  ∎

**You Try 2**

Solve $-3(4y - 3) + 4(y + 1) = 15$.

**Example 3**

Solve $\dfrac{1}{2}(3b + 8) + \dfrac{3}{4} = -\dfrac{1}{2}$.

**Solution**

**Step 1:** Clear the parentheses and combine like terms.

$$\frac{1}{2}(3b + 8) + \frac{3}{4} = -\frac{1}{2}$$

$$\frac{3}{2}b + 4 + \frac{3}{4} = -\frac{1}{2} \qquad \text{Distribute.}$$

$$\frac{3}{2}b + \frac{16}{4} + \frac{3}{4} = -\frac{1}{2} \qquad \text{Get a common denominator for the like terms.}$$

$$\frac{3}{2}b + \frac{19}{4} = -\frac{1}{2} \qquad \text{Combine like terms.}$$

**Step 2:** Isolate the variable.

$$\frac{3}{2}b + \frac{19}{4} - \frac{19}{4} = -\frac{1}{2} - \frac{19}{4} \qquad \text{Subtract } \frac{19}{4} \text{ from each side.}$$

$$\frac{3}{2}b = -\frac{2}{4} - \frac{19}{4} \qquad \text{Get a common denominator.}$$

$$\frac{3}{2}b = -\frac{21}{4} \qquad \text{Simplify.}$$

**Step 3:** Solve for $b$ using the multiplication property of equality.

$$\frac{2}{3} \cdot \frac{3}{2}b = \frac{2}{3} \cdot \left(-\frac{21}{4}\right) \qquad \text{Multiply both sides by the reciprocal of } \frac{3}{2}.$$

$$b = \frac{\overset{1}{2}}{\underset{1}{3}} \cdot \left(-\frac{\overset{-7}{21}}{\underset{2}{4}}\right) \qquad \text{Perform the multiplication.}$$

$$b = -\frac{7}{2} \qquad \text{Simplify.}$$

**Step 4:** The check is left to the student. The solution set is $\left\{ -\dfrac{7}{2} \right\}$. ■

**You Try 3**

Solve $\dfrac{1}{3}(2m - 1) + \dfrac{5}{9} = \dfrac{4}{3}$.

In the next section, we will learn another way to solve an equation containing several fractions like the one in Example 3.

### 3. Solve Equations Containing Variables on Both Sides of the Equal Sign

Now let's see how we use the steps to solve equations containing variables on both sides of the equal sign. Remember that we want to get the variables on one side of the equal sign and the constants on the other side so that we can combine like terms and isolate the variable.

**Example 4**

Solve $3y - 11 + 7y = 6y + 9$

**Solution**

**Step 1:**    Combine like terms on the left side of the equal sign.

$$3y - 11 + 7y = 6y + 9$$
$$10y - 11 = 6y + 9 \qquad \text{Combine like terms.}$$

**Step 2:**    Isolate the variable using the addition and subtraction properties of equality. Combine like terms so that there is a single variable term on one side of the equation and a constant on the other side.

$$10y - 6y - 11 = 6y - 6y + 9 \qquad \text{Subtract } 6y \text{ from each side.}$$
$$4y - 11 = 9 \qquad \text{Combine like terms.}$$
$$4y - 11 + 11 = 9 + 11 \qquad \text{Add 11 to each side.}$$
$$4y = 20 \qquad \text{Combine like terms.}$$

**Step 3:**    Solve for $y$ using the division property of equality.

$$\frac{4y}{4} = \frac{20}{4} \qquad \text{Divide each side by 4.}$$
$$y = 5 \qquad \text{Simplify.}$$

**Step 4:**    Check:

$$10y - 11 = 6y + 9$$
$$10(5) - 11 = 6(5) + 9$$
$$50 - 11 = 30 + 9$$
$$39 = 39 \checkmark$$

The solution set is $\{5\}$.    ■

 **You Try 4**

Solve $-3k + 4 = 8k - 15 - 6k - 11$.

**Example 5**

Solve $9t + 4 - (7t - 2) = t + 6(t + 1)$.

**Solution**

**Step 1:**    Clear the parentheses and combine like terms.

$$9t + 4 - (7t - 2) = t + 6(t + 1)$$
$$9t + 4 - 7t + 2 = t + 6t + 6 \qquad \text{Distribute.}$$
$$2t + 6 = 7t + 6 \qquad \text{Combine like terms.}$$

**Step 2:**    Isolate the variable.

$$2t - 7t + 6 = 7t - 7t + 6 \qquad \text{Subtract } 7t \text{ from each side.}$$
$$-5t + 6 = 6 \qquad \text{Combine like terms.}$$
$$-5t + 6 - 6 = 6 - 6 \qquad \text{Subtract 6 from each side.}$$
$$-5t = 0 \qquad \text{Combine like terms.}$$

**Step 3:**    Solve for $t$ using the division property of equality.

$$\frac{-5t}{-5} = \frac{0}{-5} \qquad \text{Divide each side by } -5.$$
$$t = 0 \qquad \text{Simplify.}$$

**Step 4:**    Check:

$$9t + 4 - (7t - 2) = t + 6(t + 1)$$
$$9(0) + 4 - [7(0) - 2] = 0 + 6[(0) + 1]$$
$$0 + 4 - (0 - 2) = 0 + 6(1)$$
$$4 - (-2) = 0 + 6$$
$$6 = 6 \checkmark$$

The solution set is $\{0\}$.

## You Try 5

Solve $5 + 3(a + 4) = 7a - (9 - 10a) + 4$.

---

**Answers to You Try Exercises**

1) $\{2\}$    2) $\left\{-\dfrac{1}{4}\right\}$    3) $\left\{\dfrac{5}{3}\right\}$    4) $\{6\}$    5) $\left\{\dfrac{11}{7}\right\}$

# 3.2 Exercises

**Objective 1: Summarize the Steps for Solving a Linear Equation**

1) Explain, in your own words, the steps for solving a linear equation.

2) What is the first step for solving $8n + 3 + 2n - 9 = 13$? Do not solve the equation.

**Objective 2: Solve Equations Containing Variables on One Side of the Equal Sign**

Solve each equation.

**Fill It In**

Fill in the blanks with either the missing mathematical step or reason for the given step.

3) $3x + 7 + 5x + 4 = 27$
   $8x + 11 = 27$    _____

   _____    Subtraction property of equality

   $8x = 16$    _____

   _____    Division property of equality

   _____    Simplify.

   The solution set is _____.

4) $5 - 2(3k + 1) + 2k = 23$

   _____    Distribute.

   _____    Combine like terms.

   $-4k + 3 - 3 = 23 - 3$    _____

   $-4k = 20$

   $\dfrac{-4k}{-4} = \dfrac{20}{-4}$    _____

   _____    Simplify.

   The solution set is _____.

Solve each equation and check the solution.

5) $6a - 10 + 4a + 9 = 39$

6) $7m + 11 + 2m - 5 = 33$

7) $-15 + 8y - 10y + 1 = 8$

8) $3 - 3p - 2p + 9 = 2$

9) $30 = 5c + 14 - 11c + 1$

10) $-42 = 4x - 17 + 5x + 8$

11) $5 - 3m + 9m + 10 - 7m = -4$

12) $1 + 10z - 14 - 2z - 9z = -5$

13) $5 = -12p + 7 + 4p - 12$

14) $-40 = 13t + 2 - 4t - 11 - 5t + 3$

15) $\frac{1}{4}n + 2 + \frac{1}{2}n - \frac{3}{2} = \frac{11}{4}$

16) $\frac{1}{6} + \frac{1}{2}w - \frac{4}{3} + \frac{1}{3}w = \frac{1}{2}$

17) $4.2d - 1.7 - 2.2d + 4.3 = -1.4$

18) $5.9h + 2.8 - 3.7 - 3.9 = 1.1$

19) $2(5x + 3) - 3x + 4 = -11$

20) $6(2c - 1) + 3 - 7c = 42$

21) $7(b - 5) + 5(b + 4) = 45$

22) $4(z - 2) + 3(z + 8) = -12$

23) $8 - 3(2k + 9) + 2(7 + k) = 2$

24) $1 - 5(3y + 2) + 3(4 + y) = -12$

25) $-23 = 4(3x - 7) - (8x - 5)$

26) $38 = 9(2a + 3) - (10a - 7)$

27) $8 = 5(4n + 3) - 3(2n - 7) - 20$

28) $-4 = 3(2z - 5) - 2(5z - 1) + 9$

29) $2(7u - 3) - (u + 9) - 3(2u + 1) = 24$

30) $6(4h + 7) + 2(h - 5) - (h - 11) = 18$

31) $\frac{1}{3}(3w + 4) - \frac{2}{3} = -\frac{1}{3}$

32) $\frac{3}{4}(2r - 5) + \frac{1}{2} = \frac{5}{4}$

33) $\frac{1}{2}(c - 2) + \frac{1}{4}(2c + 1) = \frac{5}{4}$

34) $\frac{2}{3}(m + 3) - \frac{4}{15}(3m + 7) = \frac{4}{5}$

35) $\frac{4}{3}(t + 1) - \frac{1}{6}(4t - 3) = 2$

36) $\frac{1}{4}(3x - 2) - \frac{1}{2}(x - 1) = -\frac{1}{7}$

**Objective 3: Solve Equations Containing Variables on Both Sides of the Equal Sign**

Solve each equation and check the solution.

37) $9y + 8 = 4y - 17$

38) $12b - 5 = 8b + 11$

39) $5k - 6 = 7k - 8$

40) $3v + 14 = 9v - 22$

41) $-15w + 4 = 24 - 7w$

42) $-7x + 13 = 3 - 13x$

43) $1.8z - 1.1 = 1.4z + 1.7$

44) $2.2q + 1.9 = 2.8q + 7.3$

45) $18 - h + 5h - 11 = 9h + 19 - 3h$

46) $4m - 1 - 6m + 7 = 11m + 3 - 10m$

47) $2t + 7 - 6t + 12 = 4t + 5 - 7t - 1$

48) $4 + 8a - 17 + 3a = 7a + 1 + 5a - 10$

49) $6.1r + 1.6 - 3.7r - 0.3 = r - 1.7 + 0.2r - 0.6$

50) $-7.5k + 3.2 + 3.8k + 0.9 = 0.1k - 2.1 - 3.4k + 7$

51) $1 + 5(4n - 7) = 4(7n - 3) - 30$

52) $10 + 2(z - 9) = 3(z + 1) - 6$

53) $2(1 - 8c) = 5 - 3(6c + 1) + 4c$

54) $13u + 6 - 5(2u - 3) = 1 + 4(u + 5)$

55) $9 - (8p - 5) + 4p = 6(2p + 1)$

56) $2(6d + 5) = 16 - (7d - 4) + 11d$

57) $-3(4r + 9) + 2(3r + 8) = r - (9r - 5)$

58) $2(3t - 4) - 6(t + 1) = -t + 4(t + 10)$

59) $m + \frac{1}{2}(3m + 4) - 5 = \frac{2}{3}(2m - 1) + \frac{5}{6}$

60) $2x - \frac{2}{3}(5x - 6) - 10 = \frac{1}{4}(3x + 1) + \frac{1}{2}$

## Section 3.3  Solving Linear Equations Part III

### Objectives

1. **Solve Equations Containing Fractions**
2. **Solve Equations Containing Decimals**
3. **Solve Equations with No Solution or an Infinite Number of Solutions**
4. **Use the Five Steps for Solving Applied Problems**

When equations contain fractions or decimals, they may appear difficult to solve. We begin this section by learning how to eliminate the fractions or decimals so that it will be easier to solve such equations.

### 1. Solve Equations Containing Fractions

To solve $\frac{1}{2}(3b + 8) + \frac{3}{4} = -\frac{1}{2}$ in Section 3.2, we began by using the distributive property to clear the parentheses, and we worked with the fractions throughout the solving process. But, there is another way we can solve equations containing several fractions. Before applying the steps for solving a linear equation, we can eliminate the fractions from the equation.

**Procedure**   Eliminating Fractions from an Equation

To eliminate the fractions, determine the least common denominator for all the fractions in the equation. Then multiply both sides of the equation by the least common denominator (LCD).

Let's solve the equation we solved in Section 3.2 using this new approach.

### Example 1

Solve $\dfrac{1}{2}(3b + 8) + \dfrac{3}{4} = -\dfrac{1}{2}$.

**Solution**

The least common denominator of all the fractions in the equation is 4. Multiply both sides of the equation by 4 to eliminate the fractions.

$$4\left[\frac{1}{2}(3b + 8) + \frac{3}{4}\right] = 4\left(-\frac{1}{2}\right)$$

**Step 1:** Distribute the 4, clear the parentheses, and combine like terms.

$$4 \cdot \frac{1}{2}(3b + 8) + 4 \cdot \frac{3}{4} = -2 \qquad \text{Distribute.}$$
$$2(3b + 8) + 3 = -2 \qquad \text{Multiply.}$$
$$6b + 16 + 3 = -2 \qquad \text{Distribute.}$$
$$6b + 19 = -2 \qquad \text{Combine like terms.}$$

**Step 2:** Isolate the variable.

$$6b + 19 - 19 = -2 - 19 \qquad \text{Subtract 19 from each side.}$$
$$6b = -21 \qquad \text{Combine like terms.}$$

**Step 3:** Solve for $b$ using the division property of equality.

$$\frac{6b}{6} = \frac{-21}{6} \qquad \text{Divide each side by 6.}$$
$$b = -\frac{7}{2} \qquad \text{Simplify.}$$

**Step 4:** The check is left to the student. The solution set is $\left\{-\dfrac{7}{2}\right\}$. This is the same as the result we obtained in Section 3.2, Example 3. ■

### You Try 1

Solve $\dfrac{1}{6}x + \dfrac{5}{4} = \dfrac{1}{2}x - \dfrac{5}{12}$.

## 2. Solve Equations Containing Decimals

Just as we can eliminate the fractions from an equation to make it easier to solve, we can eliminate decimals from an equation before applying the four-step equation-solving process.

**Procedure**   Eliminating Decimals from an Equation

To eliminate the decimals from an equation, multiply both sides of the equation by the smallest power of 10 that will eliminate all decimals from the problem.

### Example 2

Solve $0.05a + 0.2(a + 3) = 0.1$.

**Solution**

We want to eliminate the decimals. The number containing a decimal place farthest to the right is 0.05. The 5 is in the hundredths place. Therefore, multiply both sides of the equation by 100 to eliminate all decimals in the equation.

$$100[0.05a + 0.2(a + 3)] = 100(0.1)$$

**Step 1:** Distribute the 100, clear the parentheses, and combine like terms.

$$
\begin{aligned}
100 \cdot (0.05a) + 100[0.2(a + 3)] &= 10 && \text{Distribute.} \\
5a + 20(a + 3) &= 10 && \text{Multiply.} \\
5a + 20a + 60 &= 10 && \text{Distribute.} \\
25a + 60 &= 10 && \text{Combine like terms.}
\end{aligned}
$$

**Step 2:** Isolate the variable.

$$
\begin{aligned}
25a + 60 - 60 &= 10 - 60 && \text{Subtract 60 from each side.} \\
25a &= -50 && \text{Combine like terms.}
\end{aligned}
$$

**Step 3:** Solve for $a$ using the division property of equality.

$$
\begin{aligned}
\frac{25a}{25} &= \frac{-50}{25} && \text{Divide each side by 25.} \\
a &= -2 && \text{Simplify.}
\end{aligned}
$$

**Step 4:** The check is left to the student. The solution set is $\{-2\}$.

### You Try 2

Solve $0.08k - 0.2(k + 5) = -0.1$.

## 3. Solve Equations with No Solution or an Infinite Number of Solutions

Does every equation have a solution? Consider the next example.

### Example 3

Solve $11w - 9 = 5w + 2(3w + 1)$.

**Solution**

$$
\begin{aligned}
11w - 9 &= 5w + 2(3w + 1) \\
11w - 9 &= 5w + 6w + 2 && \text{Distribute.} \\
11w - 9 &= 11w + 2 && \text{Combine like terms.} \\
11w - 11w - 9 &= 11w - 11w + 2 && \text{Subtract } 11w \text{ from each side.} \\
-9 &= 2 && \text{False}
\end{aligned}
$$

Notice that the variable has "dropped out." Is $-9 = 2$ a true statement? No! This means that there is no value for $w$ that will make the statement true. This means that the equation has *no solution*. We say that the solution set is the **empty set,** or **null set,** and it is denoted by $\varnothing$.

We have seen that a linear equation may have one solution or no solution. There is a third possibility—a linear equation may have an infinite number of solutions.

**Example 4**

Solve $10r - 3r + 15 = 7r + 15$.

**Solution**

$$
\begin{aligned}
10r - 3r + 15 &= 7r + 15 \\
7r + 15 &= 7r + 15 &&\text{Combine like terms.} \\
7r - 7r + 15 &= 7r - 7r + 15 &&\text{Subtract } 7r \text{ from each side.} \\
15 &= 15 &&\text{True}
\end{aligned}
$$

Here, the variable has "dropped out," and we are left with an equation, $15 = 15$, that is true. This means that any real number we substitute for $r$ will make the original equation true. Therefore, this equation has an *infinite number of solutions*. The solution set is **{all real numbers}**. ■

**You Try 3**

Solve.

a) $9 - 4(3c + 1) = 15 - 12c - 10$       b) $12z + 7 - 10z = 2z + 5$

---

**Summary**   Outcomes When Solving Linear Equations

There are three possible outcomes when solving a linear equation. The equation may have

1) **one solution.** Solution set: {a real number}. An equation that is true for some values and not for others is called a **conditional equation.**

or

2) **no solution.** In this case, the variable will drop out, and there will be a false statement such as $-9 = 2$. Solution set: $\varnothing$. An equation that has no solution is called a **contradiction.**

or

3) **an infinite number of solutions.** In this case, the variable will drop out, and there will be a true statement such as $15 = 15$. Solution set: {all real numbers}. An equation that has all real numbers as its solution set is called an **identity.**

---

## 4. Use the Five Steps for Solving Applied Problems

Mathematical equations can be used to describe many situations in the real world. To do this, we must learn how to translate information presented in English into an algebraic equation. We will begin slowly and work our way up to more challenging problems. Yes, it may be difficult at first, but with patience and persistence, you can do it!

Although no single method will work for solving all applied problems, the following approach is suggested to help in the problem-solving process.

---

**Procedure**   Steps for Solving Applied Problems

**Step 1:** **Read** the problem carefully, more than once if necessary, until you understand it. Draw a picture, if applicable. Identify what you are being asked to find.

**Step 2:** **Choose a variable** to represent an unknown quantity. If there are any other unknowns, define them in terms of the variable.

**Step 3:** **Translate** the problem from English into an equation using the chosen variable. Here are some suggestions for doing so:
  • Restate the problem in your own words.
  • Read and think of the problem in "small parts."
  • Make a chart to separate these "small parts" of the problem to help you translate into mathematical terms.
  • Write an equation in English, then translate it into an algebraic equation.

**Step 4:** **Solve** the equation.

**Step 5:** **Check** the answer in the original problem, and **interpret** the solution as it relates to the problem. Be sure your answer makes sense in the context of the problem.

## Example 5

Write the following statement as an equation, and find the number.

*Nine more than twice a number is fifteen. Find the number.*

### Solution

**Step 1:**   **Read** the problem carefully. We must find an unknown number.

**Step 2:**   **Choose a variable** to represent the unknown.

Let $x$ = the number.

**Step 3:**   **Translate** the information that appears in English into an algebraic equation by rereading the problem slowly and "in parts."

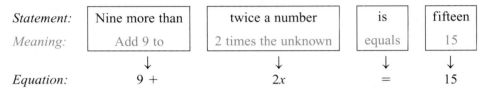

| *Statement:* | Nine more than | twice a number | is | fifteen |
|---|---|---|---|---|
| *Meaning:* | Add 9 to | 2 times the unknown | equals | 15 |
| | ↓ | ↓ | ↓ | ↓ |
| *Equation:* | 9 + | $2x$ | = | 15 |

The equation is $9 + 2x = 15$.

**Step 4:**   **Solve** the equation.

$$9 + 2x = 15$$
$$9 - 9 + 2x = 15 - 9 \qquad \text{Subtract 9 from each side.}$$
$$2x = 6 \qquad \text{Combine like terms.}$$
$$x = 3 \qquad \text{Divide each side by 2.}$$

**Step 5:**   **Check** the answer. Does the answer make sense? Nine more than twice three is $9 + 2(3) = 15$. The answer is correct. The number is 3.   ■

## You Try 4

Write the following statement as an equation, and find the number. *Three more than twice a number is twenty-nine.*

Sometimes, dealing with subtraction in an application can be confusing. So let's look at an arithmetic problem first.

## Example 6

What is two less than seven?

### Solution

To solve this problem, do we subtract $7 - 2$ or $2 - 7$? "Two less than seven" is written as $7 - 2$, and $7 - 2 = 5$. Five is two less than seven. To get the correct answer, the 2 is *subtracted from* the 7.   ■

Keep this in mind as you read the next problem.

**Example 7**

Write the following statement as an equation, and find the number.

*Five less than three times a number is the same as the number increased by seven. Find the number.*

**Solution**

***Step 1:*** **Read** the problem carefully. We must find an unknown number.

***Step 2:*** **Choose a variable** to represent the unknown.

$$\text{Let } x = \text{the number.}$$

***Step 3:*** **Translate** the information that appears in English into an algebraic equation by rereading the problem slowly and "in parts."

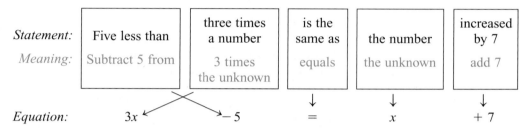

The equation is $3x - 5 = x + 7$.

***Step 4:*** **Solve** the equation.

$$
\begin{array}{ll}
3x - 5 = x + 7 & \\
3x - x - 5 = x - x + 7 & \text{Subtract } x \text{ from each side.} \\
2x - 5 = 7 & \text{Combine like terms.} \\
2x - 5 + 5 = 7 + 5 & \text{Add 5 to each side.} \\
2x = 12 & \text{Combine like terms.} \\
x = 6 & \text{Divide each side by 2.}
\end{array}
$$

***Step 5:*** **Check** the answer. Does it make sense? Five less than three times 6 is $3(6) - 5 = 13$. The number increased by seven is $6 + 7 = 13$. The answer is correct. The number is 6. ▪

 **You Try 5**

Write the following statement as an equation, and find the number.
*Three less than five times a number is the same as the number increased by thirteen.*

## Using Technology

We can use a graphing calculator to solve a linear equation in one variable. First, enter the left side of the equation in $Y_1$ and the right side of the equation in $Y_2$. Then graph the equations. The $x$-coordinate of the point of intersection is the solution to the equation.

We will solve $x + 2 = -3x + 7$ algebraically and by using a graphing calculator, and then compare the results. First, use algebra to solve $x + 2 = -3x + 7$. You should get $\dfrac{5}{4}$.

Next, use a graphing calculator to solve $x + 2 = -3x + 7$.

1) Enter $x + 2$ in $Y_1$ by pressing $\boxed{Y=}$ and entering $x + 2$ to the right of $\backslash Y_1 =$. Then press $\boxed{\text{ENTER}}$.

2) Enter $-3x + 7$ in $Y_2$ by pressing the $\boxed{Y=}$ key and entering $-3x + 7$ to the right of $\backslash Y_2 =$. Press $\boxed{\text{ENTER}}$.

3) Press $\boxed{\text{ZOOM}}$ and select 6:ZStandard to graph the equations.

4) To find the intersection point, press $\boxed{\text{2nd}}$ $\boxed{\text{TRACE}}$ and select 5:intersect. Press $\boxed{\text{ENTER}}$ three times. The $x$-coordinate of the intersection point is shown on the left side of the screen, and is stored in the variable $x$.

5) Return to the home screen by pressing $\boxed{\text{2nd}}$ $\boxed{\text{MODE}}$. Press $\boxed{\text{X,T,}\Theta\text{,n}}$ $\boxed{\text{ENTER}}$ to display the solution. Since the result in this case is a decimal value, we can convert it to a fraction by pressing $\boxed{\text{X,T,}\Theta\text{,n}}$ $\boxed{\text{MATH}}$, selecting Frac, then pressing $\boxed{\text{ENTER}}$.

The calculator then gives us the solution set $\left\{\dfrac{5}{4}\right\}$.

Solve each equation algebraically, then verify your answer using a graphing calculator.

1) $x + 6 = -2x - 3$

2) $2x + 3 = -x - 4$

3) $\dfrac{5}{6}x + \dfrac{1}{2} = \dfrac{1}{6}x - \dfrac{3}{4}$

4) $0.3x - 1 = -0.2x - 5$

5) $3x - 7 = -x + 5$

6) $6x - 7 = 5$

---

### Answers to You Try Exercises

1) $\{5\}$    2) $\{-7.5\}$    3) a) {all real numbers}   b) $\varnothing$    4) $2x + 3 = 29; 13$

5) $5x - 3 = x + 13; 4$

---

### Answers to Technology Exercises

1) $\{-3\}$    2) $\left\{-\dfrac{7}{3}\right\}$    3) $\left\{-\dfrac{15}{8}\right\}$    4) $\{-8\}$    5) $\{3\}$    6) $\{2\}$

## 3.3 Exercises

**Mixed Exercises: Objectives 1 and 2**

1) If an equation contains fractions, what is the first step you can perform to make it easier to solve?

2) If an equation contains decimals, what is the first step you can perform to make it easier to solve?

3) How can you eliminate the fractions from the equation $\frac{3}{8}x - \frac{1}{2} = \frac{1}{8}x + \frac{3}{4}$?

4) How can you eliminate the decimals from the equation $0.02n + 0.1(n - 3) = 0.06$?

Solve each equation by first clearing the fractions or decimals.

5) $\frac{3}{8}x - \frac{1}{2} = \frac{1}{8}x + \frac{3}{4}$

6) $\frac{1}{2}c + \frac{7}{4} = \frac{5}{4}c - \frac{1}{2}$

7) $\frac{4}{7}t + \frac{1}{14} = \frac{3}{14}t + \frac{3}{2}$

8) $\frac{1}{2} - \frac{7}{12}k = \frac{1}{6}k + \frac{11}{4}$

9) $\frac{1}{3} - \frac{1}{2}m = \frac{1}{6}m + \frac{7}{9}$

10) $\frac{1}{15}p - \frac{1}{2} = \frac{1}{5}p - \frac{3}{10}$

11) $\frac{1}{3} + \frac{1}{9}(k + 5) - \frac{k}{4} = 2$

12) $\frac{5}{8}(2w + 3) + \frac{5}{4}w = \frac{3}{4}(4w + 1)$

13) $\frac{3}{4}(y + 7) + \frac{1}{2}(3y - 5) = \frac{9}{4}(2y - 1)$

14) $\frac{2}{3}(5z - 2) - \frac{4}{9}(3z - 2) = \frac{1}{3}(2z + 1)$

15) $\frac{1}{2}(4r + 1) - r = \frac{2}{5}(2r - 3) + \frac{3}{2}$

16) $\frac{2}{3}(3h - 5) + 1 = \frac{3}{2}(h - 2) + \frac{1}{6}h$

17) $0.06d + 0.13 = 0.31$

18) $0.09x - 0.14 = 0.4$

19) $0.04n - 0.05(n + 2) = 0.1$

20) $0.07t + 0.02(3t + 8) = -0.1$

21) $0.2(c - 4) + 1 = 0.15(c + 2)$

22) $0.12(5q - 1) - q = 0.15(7 - 2q)$

23) $0.35a - a = 0.03(5a + 4)$

24) $0.3(x - 2) + 1 = 0.25(x + 9)$

25) $0.06w + 0.1(20 - w) = 0.08(20)$

26) $0.17m + 0.05(16 - m) = 0.11(16)$

27) $0.07k + 0.15(200) = 0.09(k + 200)$

28) $0.2p + 0.08(120) = 0.16(p + 120)$

**Objective 3: Solve Equations with No Solution or an Infinite Number of Solutions**

29) How do you know that an equation has no solution?

30) How do you know that the solution set of an equation is {all real numbers}?

Determine whether each of the following equations has a solution set of {all real numbers} or has no solution, $\varnothing$.

31) $9(c + 6) - 2c = 4c + 1 + 3c$

32) $-21n + 22 = 3(4 - 7n) + 10$

33) $5t + 2(t + 3) - 4t = 4(t + 1) - (t - 2)$

34) $8z + 11 + 5z - 9 = 16 - 6(3 - 2z) + z$

35) $\frac{5}{6}k - \frac{2}{3} = \frac{1}{6}(5k - 4) + \frac{1}{2}$

36) $0.4y + 0.3(20 - y) = 0.1y + 6$

**Mixed Exercises: Objectives 1–3**

The following set of exercises contains equations from Sections 3.1–3.3. Solve each equation.

37) $\frac{n}{5} = 20$

38) $z + 18 = -5$

39) $-19 = 6 - p$

40) $-a = 34$

41) $-5.4 = -0.9m$

42) $\frac{15}{7}h = 25$

43) $51 = 4y - 13$

44) $3c + 8 = 5c + 11$

45) $9 - (7k - 2) + 2k = 4(k + 3) + 5$

46) $0.3t + 0.18(5000 - t) = 0.21(5000)$

47) $-\dfrac{5}{4}r + 17 = 7$

48) $-2.3 = 2.4z + 1.3$

VIDEO 49) $8(3t + 4) = 10t - 3 + 7(2t + 5)$

50) $-6 - (a + 9) + 7 = 3a + 2(4a - 1)$

51) $\dfrac{5}{3}w + \dfrac{2}{5} = w - \dfrac{7}{3}$

52) $2d + 7 = -4d + 3(2d - 5)$

53) $7(2q + 3) - 3(q + 5) = 6$

54) $-11 = \dfrac{4}{5}k - 17$

55) $0.16h + 0.4(2000) = 0.22(2000 + h)$

56) $\dfrac{4}{9} + \dfrac{2}{3}(c - 1) + \dfrac{5}{9}c = \dfrac{2}{9}(5c + 3)$

VIDEO 57) $-9r + 4r - 11 + 2 = 3r + 7 - 8r + 9$

58) $2u - 4.6 = -4.6$

59) $\dfrac{1}{2}(2r + 9) - \dfrac{1}{3}(r + 12) = 1$

60) $t + 18 = 3(5 - t) + 4t + 3$

**Objective 4: Use the Five Steps for Solving Applied Problems**

61) What are the five steps for solving applied problems?

62) If you are solving an applied problem in which you have to find the length of a side of a rectangle, would a solution of $-18$ be reasonable? Explain your answer.

Write each statement as an equation, and find the number.

63) Twelve more than a number is five.

64) Fifteen more than a number is nineteen.

65) Nine less than a number is twelve.

66) Fourteen less than a number is three.

67) Five more than twice a number is seventeen.

68) Seven more than twice a number is twenty-three.

69) Eleven more than twice a number is thirteen.

70) Eighteen more than twice a number is eight.

VIDEO 71) Three times a number decreased by eight is forty.

72) Five less than four times a number is forty-three.

73) Three-fourths of a number is thirty-three.

74) Two-thirds of a number is twenty-six.

75) Nine less than half a number is three.

76) Two less than one-fourth of a number is three.

77) Three less than twice a number is the same as the number increased by eight.

78) Twelve less than five times a number is the same as the number increased by sixteen.

79) Ten more than one-third of a number is the same as the number decreased by two.

80) A number decreased by nine is the same as seven more than half the number.

81) If twenty-four is subtracted from a number, the result is the same as the number divided by nine.

82) If forty-five is subtracted from a number, the result is the same as the number divided by four.

83) If two-thirds of a number is added to the number, the result is twenty-five.

84) If three-eighths of a number is added to twice the number, the result is thirty-eight.

85) When a number is decreased by twice the number, the result is thirteen.

86) When three times a number is subtracted from the number, the result is ten.

## Section 3.4 Applications of Linear Equations

### Objectives

1. **Solve Problems Involving General Quantities**
2. **Solve Problems Involving Lengths**
3. **Solve Consecutive Integer Problems**

In the previous section, we learned the five steps for solving applied problems and used this procedure to solve problems involving unknown numbers. Now we will apply this problem-solving technique to other types of applications.

## 1. Solve Problems Involving General Quantities

**Example 1**

Write an equation and solve.

Swimmers Michael Phelps and Natalie Coughlin both competed in the 2004 Olympics in Athens and in the 2008 Olympics in Beijing, where they won a total of 27 medals. Phelps won five more medals than Coughlin. How many Olympic medals has each athlete won? (http://swimming.teamusa.org)

### Solution

**Step 1:** **Read** the problem carefully, and identify what we are being asked to find.

We must find the number of medals each Olympian won.

**Step 2:** **Choose a variable** to represent an unknown, and define the other unknown in terms of this variable.

In the statement "Phelps won five more medals than Coughlin," the number of medals that Michael Phelps won is expressed *in terms of* the number of medals won by Natalie Coughlin. Therefore, let

$x$ = the number of medals Coughlin won

Define the other unknown (the number of medals that Michael Phelps won) in terms of $x$. The statement "Phelps won five more medals than Coughlin" means

number of Coughlin's medals $+ 5$ = number of Phelps' medals

$x + 5$ = number of Phelps' medals

**Step 3:** **Translate** the information that appears in English into an algebraic equation. One approach is to restate the problem in your own words.

Since these two athletes won a total of 27 medals, we can think of the situation in this problem as:

The number of medals Coughlin won plus the number of medals Phelps won is 27.

Let's write this as an equation.

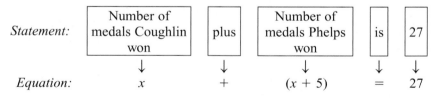

The equation is $x + (x + 5) = 27$.

**Step 4:** **Solve** the equation.

$$x + (x + 5) = 27$$
$$2x + 5 = 27$$
$$2x + 5 - 5 = 27 - 5 \qquad \text{Subtract 5 from each side.}$$
$$2x = 22 \qquad \text{Combine like terms.}$$
$$\frac{2x}{2} = \frac{22}{2} \qquad \text{Divide each side by 2.}$$
$$x = 11 \qquad \text{Simplify.}$$

***Step 5:*** **Check** the answer and **interpret** the solution as it relates to the problem.

Since $x$ represents the number of medals that Natalie Coughlin won, she won 11 medals.

The expression $x + 5$ represents the number of medals Michael Phelps won, so he won $x + 5 = 11 + 5 = 16$ medals.

The answer makes sense because the total number of medals they won was $11 + 16 = 27$.  ■

## You Try 1

Write an equation and solve.

An employee at a cellular phone store is doing inventory. The store has 23 more conventional cell phones in stock than smart phones. If the store has a total of 73 phones, how many of each type of phone is in stock?

## Example 2

Write an equation and solve.

Nick has half as many songs on his iPod as Mariah. Together they have a total of 4887 songs. How many songs does each of them have?

### Solution

***Step 1:*** **Read** the problem carefully, and identify what we are being asked to find.

We must find the number of songs on Nick's iPod and the number on Mariah's iPod.

***Step 2:*** **Choose a variable** to represent an unknown, and define the other unknown in terms of this variable.

In the sentence "Nick has half as many songs on his iPod as Mariah," the number of songs Nick has is expressed *in terms of* the number of songs Mariah has. Therefore, let

$x =$ the number of songs on Mariah's iPod

Define the other unknown in terms of $x$.

$\frac{1}{2}x =$ the number of songs on Nick's iPod

***Step 3:*** **Translate** the information that appears in English into an algebraic equation. One approach is to restate the problem in your own words.

Since Mariah and Nick have a total of 4887 songs, we can think of the situation in this problem as:

The number of Mariah's songs plus the number of Nick's songs equals 4887.

Let's write this as an equation.

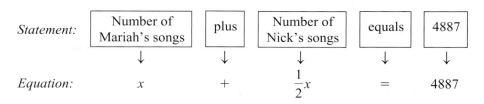

| *Statement:* | Number of Mariah's songs | plus | Number of Nick's songs | equals | 4887 |
|---|---|---|---|---|---|
| | ↓ | ↓ | ↓ | ↓ | ↓ |
| *Equation:* | $x$ | $+$ | $\frac{1}{2}x$ | $=$ | 4887 |

The equation is $x + \frac{1}{2}x = 4887$.

**Step 4:** **Solve** the equation.

$$x + \frac{1}{2}x = 4887$$

$$\frac{3}{2}x = 4887 \qquad \text{Combine like terms.}$$

$$\frac{2}{3} \cdot \frac{3}{2}x = \frac{2}{3} \cdot 4887 \qquad \text{Multiply by the reciprocal of } \frac{3}{2}.$$

$$x = 3258 \qquad \text{Multiply.}$$

**Step 5:** **Check** the answer and **interpret** the solution as it relates to the problem.

Mariah has 3258 songs on her iPod.

The expression $\frac{1}{2}x$ represents the number of songs on Nick's iPod, so there are

$\frac{1}{2}(3258) = 1629$ songs on Nick's iPod.

The answer makes sense because the total number of songs on their iPods is 3258 + 1629 = 4887 songs.

 **You Try 2**

Write an equation and solve.
    Terrance and Janay are in college. Terrance has earned twice as many credits as Janay. How many credits does each student have if together they have earned 51 semester hours?

## 2. Solve Problems Involving Lengths

**Example 3**

Write an equation and solve.
    A plumber has a section of PVC pipe that is 12 ft long. He needs to cut it into two pieces so that one piece is 2 ft shorter than the other. How long will each piece be?

### Solution

**Step 1:**   **Read** the problem carefully, and identify what we are being asked to find.

We must find the length of each of two pieces of pipe.

A picture will be very helpful in this problem.

**Step 2:**   **Choose a variable** to represent an unknown, and define the other unknown in terms of this variable.

One piece of pipe must be 2 ft shorter than the other piece. Therefore, let

$$x = \text{the length of one piece}$$

Define the other unknown in terms of $x$.

$$x - 2 = \text{the length of the second piece}$$

**Step 3:**   **Translate** the information that appears in English into an algebraic equation. Let's label the picture with the expressions representing the unknowns and then restate the problem in our own words.

From the picture we can see that the

length of one piece plus the length of the second piece equals 12 ft.

Let's write this as an equation.

| *Statement:* | Length of one piece | plus | Length of second piece | equals | 12 ft |
|---|---|---|---|---|---|
| | ↓ | ↓ | ↓ | ↓ | ↓ |
| *Equation:* | $x$ | $+$ | $(x - 2)$ | $=$ | $12$ |

The equation is $x + (x - 2) = 12$.

**Step 4:**   **Solve** the equation.

$$
\begin{aligned}
x + (x - 2) &= 12 \\
2x - 2 &= 12 \\
2x - 2 + 2 &= 12 + 2 &&\text{Add 2 to each side.} \\
2x &= 14 &&\text{Combine like terms.} \\
\frac{2x}{2} &= \frac{14}{2} &&\text{Divide each side by 2.} \\
x &= 7 &&\text{Simplify.}
\end{aligned}
$$

**Step 5:**   **Check** the answer and **interpret** the solution as it relates to the problem.

One piece of pipe is 7 ft long.

The expression $x - 2$ represents the length of the other piece of pipe, so the length of the other piece is $x - 2 = 7 - 2 = 5$ ft.

The answer makes sense because the length of the original pipe was $7\ \text{ft} + 5\ \text{ft} = 12\ \text{ft}$.  ■

### You Try 3

Write the following statement as an equation, and find the number.

An electrician has a 20-ft wire. He needs to cut the wire so that one piece is 4 ft shorter than the other. What will be the length of each piece?

### 3. Solve Consecutive Integer Problems

*Consecutive* means one after the other, in order. In this section, we will look at consecutive integers, consecutive even integers, and consecutive odd integers.

**Consecutive integers** differ by 1. Look at the consecutive integers 5, 6, 7, and 8. If $x = 5$, then $x + 1 = 6$, $x + 2 = 7$, and $x + 3 = 8$. Therefore, to define the unknowns for consecutive integers, let

$$x = \text{first integer}$$
$$x + 1 = \text{second integer}$$
$$x + 2 = \text{third integer}$$
$$x + 3 = \text{fourth integer}$$

and so on.

---

**Example 4**

The sum of three consecutive integers is 87. Find the integers.

**Solution**

**Step 1:** **Read** the problem carefully, and identify what we are being asked to find.

We must find three consecutive integers with a sum of 87.

**Step 2:** **Choose a variable** to represent an unknown, and define the other unknowns in terms of this variable.

There are three unknowns. We will let $x$ represent the first consecutive integer and then define the other unknowns in terms of $x$.

$$x = \text{the first integer}$$

Define the other unknowns in terms of $x$.

$$x + 1 = \text{the second integer} \qquad x + 2 = \text{the third integer}$$

**Step 3:** **Translate** the information that appears in English into an algebraic equation. What does the original statement mean?

"The sum of three consecutive integers is 87" means that when the three numbers are *added,* the sum is 87.

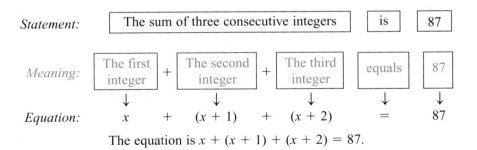

The equation is $x + (x + 1) + (x + 2) = 87$.

**Step 4:**    **Solve** the equation.

$$x + (x + 1) + (x + 2) = 87$$
$$3x + 3 = 87$$
$$3x + 3 - 3 = 87 - 3 \qquad \text{Subtract 3 from each side.}$$
$$3x = 84 \qquad \text{Combine like terms.}$$
$$\frac{3x}{3} = \frac{84}{3} \qquad \text{Divide each side by 3.}$$
$$x = 28 \qquad \text{Simplify.}$$

**Step 5:**    **Check** the answer and **interpret** the solution as it relates to the problem.

The first integer is 28. The second integer is 29 since $x + 1 = 28 + 1 = 29$, and the third integer is 30 since $x + 2 = 28 + 2 = 30$.

The answer makes sense because their sum is $28 + 29 + 30 = 87$.  ∎

**You Try 4**

The sum of three consecutive integers is 162. Find the integers.

Next, let's look at **consecutive even integers,** which are even numbers that differ by 2, such as $-10$, $-8$, $-6$, and $-4$. If $x$ is the first even integer, we have

| $-10$ | $-8$ | $-6$ | $-4$ |
|---|---|---|---|
| $x$ | $x + 2$ | $x + 4$ | $x + 6$ |

Therefore, to define the unknowns for consecutive even integers, let

$$x = \text{the first even integer}$$
$$x + 2 = \text{the second even integer}$$
$$x + 4 = \text{the third even integer}$$
$$x + 6 = \text{the fourth even integer}$$

and so on.

Will the expressions for **consecutive odd integers** be any different? No! When we count by consecutive odds, we are still counting by 2's. Look at the numbers 9, 11, 13, and 15 for example. If $x$ is the first odd integer, we have

| 9 | 11 | 13 | 15 |
|---|---|---|---|
| $x$ | $x + 2$ | $x + 4$ | $x + 6$ |

To define the unknowns for consecutive odd integers, let

$$x = \text{the first odd integer}$$
$$x + 2 = \text{the second odd integer}$$
$$x + 4 = \text{the third odd integer}$$
$$x + 6 = \text{the fourth odd integer}$$

**Example 5**

The sum of two consecutive odd integers is 19 more than five times the larger integer. Find the integers.

**Solution**

**Step 1:**    **Read** the problem carefully, and identify what we are being asked to find.

We must find two consecutive odd integers.

**Step 2:** **Choose a variable** to represent an unknown, and define the other unknown in terms of this variable.

There are two unknowns. We will let $x$ represent the first consecutive odd integer and then define the other unknown in terms of $x$.

$$x = \text{the first odd integer}$$
$$x + 2 = \text{the second odd integer}$$

**Step 3:** **Translate** the information that appears in English into an algebraic equation. Read the problem slowly and carefully, breaking it into small parts.

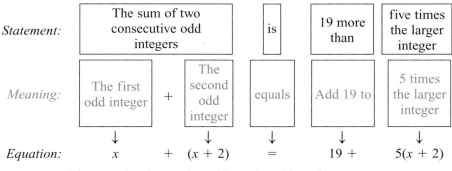

The equation is $x + (x + 2) = 19 + 5(x + 2)$.

**Step 4:** **Solve** the equation.

| | |
|---|---|
| $x + (x + 2) = 19 + 5(x + 2)$ | |
| $2x + 2 = 19 + 5x + 10$ | Combine like terms; distribute. |
| $2x + 2 = 5x + 29$ | Combine like terms. |
| $2x + 2 - 2 = 5x + 29 - 2$ | Subtract 2 from each side. |
| $2x = 5x + 27$ | Combine like terms. |
| $2x - 5x = 5x - 5x + 27$ | Subtract $5x$ from each side. |
| $-3x = 27$ | Combine like terms. |
| $\dfrac{-3x}{-3} = \dfrac{27}{-3}$ | Divide each side by $-3$. |
| $x = -9$ | Simplify. |

**Step 5:** **Check** the answer and **interpret** the solution as it relates to the problem.

The first odd integer is $-9$. The second integer is $-7$ since $x + 2 = -9 + 2 = -7$.

Check these numbers in the original statement of the problem. The sum of $-9$ and $-7$ is $-16$. Then, 19 more than five times the larger integer is $19 + 5(-7) = 19 + (-35) = -16$. The numbers are $-9$ and $-7$.  ■

**You Try 5**

The sum of two consecutive even integers is 16 less than three times the larger number. Find the integers.

**Answers to You Try Exercises**

1) smart phones: 25; conventional phones: 48   2) Janay: 17 hours; Terrance: 34 hours
3) 8 ft and 12 ft   4) 53, 54, 55   5) 12 and 14

## 3.4 Exercises

### Objective 1: Solve Problems Involving General Quantities

1) During the month of June, a car dealership sold 14 more compact cars than SUVs. Write an expression for the number of compact cars sold if $c$ SUVs were sold.

2) During a Little League game, the Tigers scored 3 more runs than the Eagles. Write an expression for the number of runs the Tigers scored if the Eagles scored $r$ runs.

3) A restaurant had 37 fewer customers on a Wednesday night than on a Thursday night. If there were $c$ customers on Thursday, write an expression for the number of customers on Wednesday.

4) After a storm rolled through Omaha, the temperature dropped 15 degrees. If the temperature before the storm was $t$ degrees, write an expression for the temperature after the storm.

5) Due to the increased use of e-mail to send documents, the shipping expenses of a small business in 2010 were half of what they were in 2000. Write an expression for the cost of shipping in 2010 if the cost in 2000 was $s$ dollars.

6) A coffee shop serves three times as many cups of regular coffee as decaffeinated coffee. If the shop serves $d$ cups of decaffeinated coffee, write an expression for the number of cups of regular coffee it sells.

7) An electrician cuts a 14-foot wire into two pieces. If one is $x$ feet long, how long is the other piece?

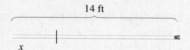

14 ft

$x$

8) Ralph worked for a total of 8.5 hours one day, some at his office and some at home. If he worked $h$ hours in his office, write an expression for the number of hours he worked at home.

9) If you are asked to find the number of children in a class, why would 26.5 not be a reasonable answer?

10) If you are asked to find the length of a piece of wire, why would $-7$ not be a reasonable answer?

11) If you are asked to find consecutive odd integers, why would $-10$ not be a reasonable answer?

12) If you are asked to find the number of workers at an ice cream shop, why would $5\frac{1}{4}$ not be a reasonable answer?

Solve using the five-step method. See Examples 1 and 2.

13) The wettest April (greatest rainfall amount) for Albuquerque, NM, was recorded in 1905. The amount was 1.2 inches more than the amount recorded for the second-wettest April, in 2004. If the total rainfall for these two months was 7.2 inches, how much rain fell in April of each year? (www.srh.noaa.gov)

14) Bo-Lin applied to three more colleges than his sister Liling. Together they applied to 13 schools. To how many colleges did each apply?

15) Miguel Indurain of Spain won the Tour de France two fewer times than Lance Armstrong. They won a total of 12 titles. How many times did each cyclist win this race? (www.letour.fr)

16) In 2009, an Apple MacBook weighed 11 lb less than the Apple Macintosh Portable did in 1989. Find the weight of each computer if they weighed a total of 21 lb. (http://oldcomputers.net; www.apple.com)

17) A 12-oz cup of regular coffee at Starbucks has 13 times the amount of caffeine found in the same-sized serving of decaffeinated coffee. Together they contain 280 mg of caffeine. How much caffeine is in each type of coffee? (www.starbucks.com)

18) A farmer plants soybeans and corn on his 540 acres of land. He plants twice as many acres with soybeans as with corn. How many acres are planted with each crop?

19) In the sophomore class at Dixon High School, the number of students taking French is two-thirds of the number

taking Spanish. How many students are studying each language if the total number of students in French and Spanish is 310?

20) A serving of salsa contains one-sixth of the number of calories of the same-sized serving of guacamole. Find the number of calories in each snack if they contain a total of 175 calories.

### Objective 2: Solve Problems Involving Lengths

Solve using the five-step method. See Example 3.

VIDEO 21) A plumber has a 36-in. pipe. He must cut it into two pieces so that one piece is 14 inches longer than the other. How long is each piece?

22) A 40-in. board is to be cut into two pieces so that one piece is 8 inches shorter than the other. Find the length of each piece.

23) Trisha has a 28.5-inch piece of wire to make a necklace and a bracelet. She has to cut the wire so that the piece for the necklace will be twice as long as the piece for the bracelet. Find the length of each piece.

24) Ethan has a 20-ft piece of rope that he will cut into two pieces. One piece will be one-fourth the length of the other piece. Find the length of each piece of rope.

25) Derek orders a 6-ft sub sandwich for himself and two friends. Cory wants his piece to be 2 feet longer than Tamara's piece, and Tamara wants half as much as Derek. Find the length of each person's sub.

26) A 24-ft pipe must be cut into three pieces. The longest piece will be twice as long as the shortest piece, and the medium-sized piece will be 4 feet longer than the shortest piece. Find the length of each piece of pipe.

### Objective 3: Solve Consecutive Integer Problems

Solve using the five-step method. See Examples 4 and 5.

27) The sum of three consecutive integers is 126. Find the integers.

28) The sum of two consecutive integers is 171. Find the integers.

29) Find two consecutive even integers such that twice the smaller is 16 more than the larger.

30) Find two consecutive odd integers such that the smaller one is 12 more than one-third the larger.

VIDEO 31) Find three consecutive odd integers such that their sum is five more than four times the largest integer.

32) Find three consecutive even integers such that their sum is 12 less than twice the smallest.

33) Two consecutive page numbers in a book add up to 215. Find the page numbers.

34) The addresses on the west side of Hampton Street are consecutive even numbers. Two consecutive house numbers add up to 7446. Find the addresses of these two houses.

### Mixed Exercises: Objectives 1–3

Solve using the five-step method.

35) In a fishing derby, Jimmy caught six more trout than his sister Kelly. How many fish did each person catch if they caught a total of 20 fish?

36) Five times the sum of two consecutive integers is two more than three times the larger integer. Find the integers.

37) A 16-ft steel beam is to be cut into two pieces so that one piece is 1 foot longer than twice the other. Find the length of each piece.

38) A plumber has a 9-ft piece of copper pipe that has to be cut into three pieces. The longest piece will be 4 ft longer than the shortest piece. The medium-sized piece will be three times the length of the shortest. Find the length of each piece of pipe.

39) The attendance at the 2008 Lollapalooza Festival was about 15,000 more than three times the attendance at Bonnaroo that year. The total number of people attending those festivals was about 295,000. How many people went to each event? (www.chicagotribune.com, www.ilmc.com)

40) A cookie recipe uses twice as much flour as sugar. If the total amount of these ingredients is $2\frac{1}{4}$ cups, how much sugar and how much flour are in these cookies?

41) The sum of three consecutive page numbers in a book is 174. What are the page numbers?

42) At a ribbon-cutting ceremony, the mayor cuts a 12-ft ribbon into two pieces so that the length of one piece is 2 ft shorter than the other. Find the length of each piece.

43) During season 7 of *The Biggest Loser,* Tara lost 15 lb more than Helen. The amount of weight that Mike lost was 73 lb less than twice Helen's weight loss. They lost a combined 502 lb. How much weight did each contestant lose? (www.msnbc.msn.com)

44) Find three consecutive odd integers such that three times the middle number is 23 more than the sum of the other two.

45) A builder is installing hardwood floors. He has to cut a 72-in piece into three separate pieces so that the smallest piece is one-third the length of the longest piece, and the third piece is 12 inches shorter than the longest. How long is each piece?

46) In 2008, there were 395 fewer cases of tuberculosis in the United States than in 2007. If the total number of TB cases in those two years was 26,191 how many

people tested positive for TB in 2007 and in 2008? (www.cdc.gov)

47) In 2008, Lil Wayne's "The Carter III" sold 0.73 million more copies than Coldplay's "Viva La Vida. . . ." Taylor Swift came in third place with her "Fearless" album, selling 0.04 million fewer copies than Coldplay. The three artists sold a total of 7.14 million albums. How many albums did each artist sell? (www.billboard.com)

48) Workers cutting down a large tree have a rope that is 33 ft long. They need to cut it into two pieces so that one piece is half the length of the other piece. How long is each piece of rope?

49) One-sixth of the smallest of three consecutive even integers is three less than one-tenth the sum of the other even integers. Find the integers.

50) Caedon's mom is a math teacher, and when he asks her on which pages he can find the magazine article on LeBron James, she says, "The article is on three consecutive pages so that 62 less than four times the last page number is the same as the sum of all the page numbers." On what page does the LeBron James article begin?

## Section 3.5 Applications Involving Percentages

### Objectives

1. Solve Problems Involving Percent Change
2. Solve Problems Involving Simple Interest
3. Solve Mixture Problems

Problems involving percents are everywhere—at the mall, in a bank, in a laboratory, just to name a few places. In this section, we begin learning how to solve different types of applications involving percents.

Before trying to solve a percent problem using algebra, let's look at an arithmetic problem we might see in a store. Relating an algebra problem to an arithmetic problem can make it easier to solve an application that requires the use of algebra.

### 1. Solve Problems Involving Percent Change

### Example 1

A hat that normally sells for $60.00 is marked down 40%. What is the sale price?

**Solution**

Concentrate on the **procedure** used to obtain the answer. This is the same procedure we will use to solve algebra problems with percent increase and percent decrease.

$$\text{Sale price} = \text{Original price} - \text{Amount of discount}$$

How much is the discount? It is 40% of $60.00.

Change the percent to a decimal. The amount of the discount is calculated by multiplying:

$$\text{Amount of discount} = (\text{Rate of discount})(\text{Original price})$$
$$\text{Amount of discount} = (0.40) \cdot (\$60.00) = \$24.00$$

$$\begin{aligned}\text{Sale price} &= \text{Original price} - \text{Amount of discount} \\ &= \$60.00 - (0.40)(\$60.00) \\ &= \$60.00 - \$24.00 \\ &= \$36.00\end{aligned}$$

The sale price is $36.00.

**You Try 1**

A pair of running shoes that normally sells for $80.00 is marked down 30%. What is the sale price?

Next, let's solve an algebra problem involving percent change.

**Example 2**

The sale price of a video game is $48.75 after a 25% discount. What was the original price of the game?

**Solution**

**Step 1:**   **Read** the problem carefully, and identify what we are being asked to find.

We must find the original price of the video game.

**Step 2:**   **Choose a variable** to represent the unknown.

$x$ = the original price of the video game

**Step 3:**   **Translate** the information that appears in English into an algebraic equation. One way to figure out how to write an algebraic equation is to relate this problem to the arithmetic problem in Example 1. To find the sale price of the hat in Example 1, we found that

Sale price = Original price − Amount of discount

where we found the amount of the discount by multiplying the rate of the discount by the original price. We will write an algebraic equation using the same procedure.

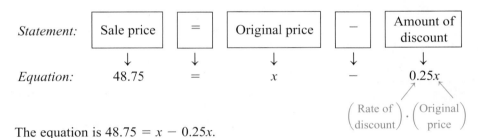

The equation is $48.75 = x - 0.25x$.

**Step 4:**   **Solve** the equation.

$$48.75 = x - 0.25x$$
$$48.75 = 0.75x \qquad \text{Combine like terms.}$$
$$\frac{48.75}{0.75} = \frac{0.75x}{0.75} \qquad \text{Divide each side by 0.75.}$$
$$x = 65 \qquad \text{Simplify.}$$

**Step 5:**   **Check** the answer and **interpret** the solution as it relates to the problem.

The original price of the video game was $65.00.

The answer makes sense because the amount of the discount is $(0.25)(\$65.00) = \$16.25$, which makes the sale price $65.00 − $16.25 = $48.75.

**You Try 2**

A circular saw is on sale for $120.00 after a 20% discount. What was the original price of the saw?

## 2. Solve Problems Involving Simple Interest

When customers invest their money in bank accounts, their accounts earn interest. There are different ways to calculate the amount of interest earned from an investment, and in this section we will discuss *simple interest*. **Simple interest** calculations are based on the initial amount of money deposited in an account. This is known as the **principal.**

The formula used to calculate simple interest is $I = PRT$, where

$$I = \text{interest (simple) earned}$$
$$P = \text{principal (initial amount invested)}$$
$$R = \text{annual interest rate (expressed as a decimal)}$$
$$T = \text{amount of time the money is invested (in years)}$$

We will begin with two arithmetic problems. The procedures used will help you understand more clearly how we arrive at the algebraic equation in Example 5.

### Example 3

If $600 is invested for 1 year in an account earning 4% simple interest, how much interest will be earned?

#### Solution

We are given that $P = \$600$, $R = 0.04$, $T = 1$. We need to find $I$.

$$I = PRT$$
$$I = (600)(0.04)(1)$$
$$I = 24$$

The interest earned will be $24.    ∎

### You Try 3

If $1400 is invested for 1 year in an account earning 3% simple interest, how much interest will be earned?

### Example 4

Gavin invests $1000 in an account earning 6% interest and $7000 in an account earning 3% interest. After 1 year, how much interest will he have earned?

#### Solution

Gavin will earn interest from two accounts. Therefore,

Total interest earned = Interest from 6% account + Interest from 3% account

$$\underset{\substack{P \quad R \quad T}}{\text{Total interest earned} = (1000)(0.06)(1)} + \underset{\substack{P \quad R \quad T}}{(7000)(0.03)(1)}$$
$$= \quad\quad 60 \quad\quad + \quad\quad 210$$
$$= \$270$$

Gavin will earn a total of $270 in interest from the two accounts.    ∎

### You Try 4

Taryn invests $2500 in an account earning 4% interest and $6000 in an account earning 5.5% interest. After 1 year, how much interest will she have earned?

**Note**

When money is invested for 1 year, $T = 1$. Therefore, the formula $I = PRT$ can be written as $I = PR$.

In the next example, we will use the same procedure for solving an algebraic problem that we used for solving the arithmetic problems in Examples 3 and 4.

**Example 5**

Samira had $8000 to invest. She invested some of it in a savings account that paid 4% simple interest and the rest in a certificate of deposit that paid 6% simple interest. In 1 year, she earned a total of $360 in interest. How much did Samira invest in each account?

**Solution**

**Step 1:**   **Read** the problem carefully, and identify what we are being asked to find.

We must find the amounts Samira invested in the 4% account and in the 6% account.

**Step 2:**   **Choose a variable** to represent an unknown, and define the other unknown in terms of this variable.

Let $x =$ amount Samira invested in the 4% account.

How do we write an expression, in terms of $x$, for the amount invested in the 6% account?

Total invested      Amount invested in 4% account

$\downarrow$                                          $\downarrow$

8000        $-$              $x$               $=$ Amount invested in the 6% account

We define the unknowns as:

$x =$ amount Samira invested in the 4% account
$8000 - x =$ amount Samira invested in the 6% account

**Step 3:**   **Translate** the information that appears in English into an algebraic equation. Use the "English equation" we used in Example 4. Remember, since $T = 1$, we can compute the interest using $I = PR$.

Total interest earned $=$ Interest from 4% account $+$ Interest from 6% account

|         |     | $P$  $R$      |     | $P$        $R$          |
|         |     |               |     |                         |
|   360   | $=$ | $(x)(0.04)$   | $+$ | $(8000 - x)(0.06)$      |

The equation is $360 = 0.04x + 0.06(8000 - x)$.

We can also get the equation by organizing the information in a table:

| Amount Invested, in Dollars $P$ | Interest Rate $R$ | Interest Earned After 1 Year $I$ |
|---|---|---|
| $x$ | 0.04 | $0.04x$ |
| $8000 - x$ | 0.06 | $0.06(8000 - x)$ |

Total interest earned $=$ Interest from 4% account $+$ Interest from 6% account
360                  $=$        $0.04x$        $+$     $0.06(8000 - x)$

The equation is $360 = 0.04x + 0.06(8000 - x)$.

Either way of organizing the information will lead us to the correct equation.

**Step 4:** **Solve** the equation. Begin by multiplying both sides of the equation by 100 to eliminate the decimals.

$$360 = 0.04x + 0.06(8000 - x)$$
$$100(360) = 100[0.04x + 0.06(8000 - x)]$$

| | |
|---|---|
| $36,000 = 4x + 6(8000 - x)$ | Multiply by 100. |
| $36,000 = 4x + 48,000 - 6x$ | Distribute. |
| $36,000 = -2x + 48,000$ | Combine like terms. |
| $-12,000 = -2x$ | Subtract 48,000. |
| $6000 = x$ | Divide by $-2$. |

**Step 5:** **Check** the answer and **interpret** the solution as it relates to the problem.

Samira invested $6000 at 4% interest. The amount invested at 6% is $8000 − x or $8000 − $6000 = $2000.

Check:

Total interest earned = Interest from 4% account + Interest from 6% account
360       =       6000(0.04)     +     2000(0.06)
            =           240         +        120
            = 360 ∎

**You Try 5**

Jeff inherited $10,000 from his grandfather. He invested part of it at 3% simple interest and the rest at 5% simple interest. Jeff earned a total of $440 in interest after 1 year. How much did he deposit in each account?

## 3. Solve Mixture Problems

Percents can also be used to solve mixture problems. Let's look at an arithmetic example before solving a problem using algebra.

**Example 6**

The state of Illinois mixes ethanol (made from corn) in its gasoline to reduce pollution. If a customer purchases 12 gallons of gasoline and it has a 10% ethanol content, how many gallons of ethanol are in the 12 gallons of gasoline?

**Solution**

Write an equation in English first:

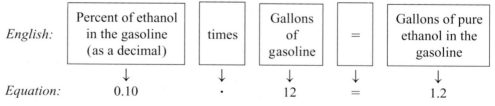

The equation is $0.10(12) = 1.2$.

We can also organize the information in a table:

| Percent of Ethanol in the Gasoline (as a decimal) | Gallons of Gasoline | Gallons of Pure Ethanol in the Gasoline |
|---|---|---|
| 0.10 | 12 | $0.10(12) = 1.2$ |

Either way, we find that there are 1.2 gallons of ethanol in 12 gallons of gasoline. ■

We will use this same idea to help us solve the next mixture problem.

### Example 7

A chemist needs to make 24 liters (L) of an 8% acid solution. She will make it from some 6% acid solution and some 12% acid solution that is in the storeroom. How much of the 6% solution and the 12% solution should she use?

*Solution*

**Step 1:** **Read** the problem carefully, and identify what we are being asked to find.

We must find the amount of 6% acid solution and the amount of 12% acid solution she should use.

**Step 2:** **Choose a variable** to represent an unknown, and define the other unknown in terms of this variable. Let

$x$ = the number of liters of 6% acid solution needed

Define the other unknown (the amount of 12% acid solution needed) in terms of $x$. Since she wants to make a total of 24 L of acid solution,

$24 - x$ = the number of liters of 12% acid solution needed

**Step 3:** **Translate** the information that appears in English into an algebraic equation.

Let's begin by arranging the information in a table. *Remember, to obtain the expression in the last column, multiply the percent of acid in the solution by the number of liters of solution to get the number of liters of acid in the solution.*

| | Percent of Acid in Solution (as a decimal) | Liters of Solution | Liters of Acid in Solution |
|---|---|---|---|
| Mix these | 0.06 | $x$ | $0.06x$ |
| | 0.12 | $24 - x$ | $0.12(24 - x)$ |
| to make → | 0.08 | 24 | $0.08(24)$ |

Now, write an equation in English. Since we make the 8% solution by mixing the 6% and 12% solutions,

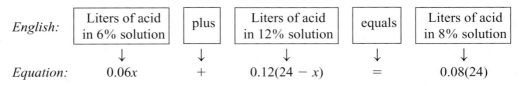

*English:*  | Liters of acid in 6% solution | plus | Liters of acid in 12% solution | equals | Liters of acid in 8% solution |

*Equation:*    $0.06x$    $+$    $0.12(24 - x)$    $=$    $0.08(24)$

The equation is $0.06x + 0.12(24 - x) = 0.08(24)$.

***Step 4:***    **Solve** the equation.

$$0.06x + 0.12(24 - x) = 0.08(24)$$
$$100[0.06x + 0.12(24 - x)] = 100[0.08(24)] \qquad \text{Multiply by 100 to eliminate decimals.}$$

$$6x + 12(24 - x) = 8(24)$$
$$6x + 288 - 12x = 192 \qquad \text{Distribute.}$$
$$-6x + 288 = 192 \qquad \text{Combine like terms.}$$
$$-6x = -96 \qquad \text{Subtract 288 from each side.}$$
$$x = 16 \qquad \text{Divide by } -6.$$

***Step 5:***    **Check** the answer and **interpret** the solution as it relates to the problem.

The chemist needs 16 L of the 6% solution.

Find the other unknown, the amount of 12% solution needed.

$$24 - x = 24 - 16 = 8\,\text{L of 12\% solution.}$$

Check:

Acid in 6% solution + Acid in 12% solution = Acid in 8% solution

$$0.06(16) \quad + \quad 0.12(8) \quad = \quad 0.08(24)$$
$$0.96 \quad + \quad 0.96 \quad = \quad 1.92$$
$$1.92 = 1.92 \qquad ■$$

**You Try 6**

Write an equation and solve.

How many milliliters (mL) of a 10% alcohol solution and how many milliliters of a 20% alcohol solution must be mixed to obtain 30 mL of a 16% alcohol solution?

**Answers to You Try Exercises**

1)   $56.00    2)   $150.00    3)   $42    4)   $430    5)   $3000 at 3% and $7000 at 5%

6)   12 mL of the 10% solution and 18 mL of the 20% solution

# 3.5 Exercises

**Objective 1: Solve Problems Involving Percent Change**

Find the sale price of each item.

1) A USB thumb drive that regularly sells for $50.00 is marked down 15%.

2) A surfboard that retails for $525.00 is on sale at 20% off.

3) A sign reads, "Take 30% off the original price of all Blu-ray Disc movies." The original price on the movie you want to buy is $29.50.

4) The $100.00 basketball shoes Frank wants are now on sale at 20% off.

5) At the end of the summer, the bathing suit that sold for $49.00 is marked down 60%.

6) An advertisement states that a flat-screen TV that regularly sells for $899.00 is being discounted 25%.

Solve using the five-step method. See Example 2.

7) A digital camera is on sale for $119 after a 15% discount. What was the original price of the camera?

8) Candace paid $21.76 for a hardcover book that was marked down 15%. What was the original selling price of the book?

9) In March, a store discounted all of its calendars by 75%. If Bruno paid $4.40 for a calendar, what was its original price?

10) An appliance store advertises 20% off all of its dishwashers. Mr. Petrenko paid $479.20 for the dishwasher. Find its original price.

11) The sale price of a coffeemaker is $40.08. This is 40% off the original price. What was the original price of the coffeemaker?

12) Katrina paid $25.50 for a box fan that was marked down 15%. Find the original retail price of the box fan.

13) In 2009, there were about 1224 acres of farmland in Crane County. This is 32% less than the number of acres of farmland in 2000. Calculate the number of acres of farmland in Crane County in 2000.

14) One hundred forty countries participated in the 1984 Summer Olympics in Los Angeles. This was 75% more than the number of countries that took part in the Summer Olympics in Moscow 4 years earlier. How many countries participated in the 1980 Olympics in Moscow? (www.mapsofworld.com)

15) In 2006, there were 12,440 Starbucks stores worldwide. This is approximately 1126% more stores than 10 years earlier. How many Starbucks stores were there in 1996? (Round to the nearest whole number.) (www.starbucks.com)

16) McDonald's total revenue in 2003 was $17.1 billion. This is a 28.5% increase over the 1999 revenue. What was McDonald's revenue in 1999? (Round to the tenths place.) (www.mcdonalds.com)

17) From 2001 to 2003, the number of employees at Kmart's corporate headquarters decreased by approximately 34%. If 2900 people worked at the headquarters in 2003, how many worked there in 2001? (Round to the hundreds place.) (www.detnews.com)

18) Jet Fi's salary this year is 14% higher than it was 3 years ago. If he earns $37,050 this year, what did he earn 3 years ago?

**Objective 2: Solve Problems Involving Simple Interest**

Solve.

19) Kristi invests $300 in an account for 1 year earning 3% simple interest. How much interest was earned from this account?

20) Last year, Mr. Doubtfire deposited $14,000 into an account earning 8.5% simple interest for 1 year. How much interest was earned?

21) Jake Thurmstrom invested $6500 in an account earning 7% simple interest. How much money will be in the account 1 year later?

22) If $4000 is deposited into an account for 1 year earning 5.5% simple interest, how much money will be in the account after 1 year?

23) Rachel Rays has a total of $4500 to invest for 1 year. She deposits $3000 into an account earning 6.5% annual simple interest and the rest into an account earning 8% annual simple interest. How much interest did Rachel earn?

24) Bob Farker plans to invest a total of $11,000 for 1 year. Into the account earning 5.2% simple interest he will deposit $6000, and into an account earning 7% simple interest he will deposit the rest. How much interest will Bob earn?

Solve using the five-step method. See Example 5.

25) Amir Sadat receives a $15,000 signing bonus upon accepting his new job. He plans to invest some of it at 6% annual simple interest and the rest at 7% annual simple interest. If he will earn $960 in interest after 1 year, how much will Amir invest in each account?

26) Angelica invested part of her $15,000 inheritance in an account earning 5% simple interest and the rest in an account earning 4% simple interest. How much did Angelica invest in each account if she earned $680 in total interest after 1 year?

27) Barney's money earned $204 in interest after 1 year. He invested some of his money in an account earning 6% simple interest and $450 more than that amount in an account earning 5% simple interest. Find the amount Barney invested in each account.

28) Saori Yamachi invested some money in an account earning 7.4% simple interest and three times that amount in an account earning 12% simple interest. She earned $1085 in interest after 1 year. How much did Saori invest in each account?

29) Last year, Taz invested a total of $7500 in two accounts earning simple interest. Some of it he invested at 9.5%, and the rest he invested at 6.5%. How much did he invest in each account if he earned a total of $577.50 in interest last year?

30) Luke has $3000 to invest. He deposits a portion of it into an account earning 4% simple interest and the rest at 6.5% simple interest. After 1 year, he has earned $170 in interest. How much did Luke deposit into each account?

## Objective 3: Solve Mixture Problems

Solve.

31) How many ounces of alcohol are in 50 oz of a 6% alcohol solution?

32) How many milliliters of acid are in 50 mL of a 5% acid solution?

33) Seventy-five milliliters of a 10% acid solution are mixed with 30 mL of a 2.5% acid solution. How much acid is in the mixture?

34) Fifty ounces of a 9% alcohol solution are mixed with 60 ounces of a 7% alcohol solution. How much alcohol is in the mixture?

Solve using the five-step method. See Example 7.

35) How many ounces of a 4% acid solution and how many ounces of a 10% acid solution must be mixed to make 24 oz of a 6% acid solution?

36) How many milliliters of a 17% alcohol solution must be added to 40 mL of a 3% alcohol solution to make a 12% alcohol solution?

37) How many liters of a 25% antifreeze solution must be mixed with 4 liters of a 60% antifreeze solution to make a mixture that is 45% antifreeze?

38) How many milliliters of an 8% hydrogen peroxide solution and how many milliliters of a 2% hydrogen peroxide solution should be mixed to get 300 mL of a 4% hydrogen peroxide solution?

39) All-Mixed-Up Nut Shop sells a mix consisting of cashews and pistachios. How many pounds of cashews, which sell for $7.00 per pound, should be mixed with 4 pounds of pistachios, which sell for $4.00 per pound, to get a mix worth $5.00 per pound?

40) Creative Coffees blends its coffees for customers. How much of the Aromatic coffee, which sells for $8.00 per pound, and how much of the Hazelnut coffee, which sells for $9.00 per pound, should be mixed to make 3 pounds of the Smooth blend to be sold at $8.75 per pound?

41) An alloy that is 50% silver is mixed with 500 g of a 5% silver alloy. How much of the 50% alloy must be used to obtain an alloy that is 20% silver?

42) A pharmacist needs to make 20 cubic centimeters (cc) of a 0.05% steroid solution to treat allergic rhinitis. How much of a 0.08% solution and a 0.03% solution should she use?

43) How much pure acid must be added to 6 gallons of a 4% acid solution to make a 20% acid solution?

44) How many milliliters of pure alcohol and how many milliliters of a 4% alcohol solution must be combined to make 480 milliliters of an 8% alcohol solution?

### Mixed Exercises: Objectives 1–3

Solve using the five-step method.

45) In her gift shop, Cheryl sells all stuffed animals for 60% more than what she paid her supplier. If one of these toys sells for $14.00 in her shop, what did it cost Cheryl?

46) Aaron has $7500 to invest. He will invest some of it in a long-term IRA paying 4% simple interest and the rest in a short-term CD earning 2.5% simple interest. After 1 year, Aaron's investments have earned $225 in interest. How much did Aaron invest in each account?

47) In Johnson County, 8330 people were collecting unemployment benefits in September 2010. This is 2% less than the number collecting the benefits in September 2009. How many people in Johnson County were getting unemployment benefits in September 2009?

48) Andre bought a new car for $15,225. This is 13% less than the car's sticker price. What was the sticker price of the car?

49) Erica invests some money in three different accounts. She puts some of it in a CD earning 3% simple interest and twice as much in an IRA paying 4% simple interest. She also decides to invest $1000 more than what she's invested in the CD into a mutual fund earning 5% simple interest. Determine how much money Erica invested in each account if she earned $370 in interest after 1 year.

50) Gil marks up the prices of his fishing poles by 55%. Determine what Gil paid his supplier for his best-selling fishing pole if Gil charges his customers $124.

51) Find the original price of a desk lamp if it costs $25.60 after a 20% discount.

52) It is estimated that in 2003 the number of Internet users in Slovakia was 40% more than the number of users in Kenya. If Slovakia had 700,000 Internet users in 2003, how many people used the Internet in Kenya that year? (*2003 CIA World Factbook*, www.theodora.com)

53) Zoe's current salary is $40,144. This is 4% higher than last year's salary. What was Zoe's salary last year?

54) Jackson earns $284 in interest from 1-year investments. He invested some money in an account earning 6% simple interest, and he deposited $1500 more than that amount into an account paying 5% simple interest. How much did Jackson invest in each account?

55) How many ounces of a 9% alcohol solution and how many ounces of a 17% alcohol solution must be mixed to get 12 ounces of a 15% alcohol solution?

56) How many milliliters of a 4% acid solution and how many milliliters of a 10% acid solution must be mixed to obtain 54 mL of a 6% acid solution?

57) How many pounds of peanuts that sell for $1.80 per pound should be mixed with cashews that sell for $4.50 per pound so that a 10-pound mixture is obtained that will sell for $2.61 per pound?

58) Sally invested $4000 in two accounts, some of it at 3% simple interest and the rest in an account earning 5% simple interest. How much did she invest in each account if she earned $144 in interest after 1 year?

59) Diego inherited $20,000 and put some of it into an account earning 4% simple interest and the rest into an

account earning 7% simple interest. He earned a total of $1130 in interest after a year. How much did he deposit into each account?

60) How much pure acid and how many liters of a 10% acid solution should be mixed to get 12 liters of a 40% acid solution?

61) How many ounces of pure orange juice and how many ounces of a citrus fruit drink containing 5% fruit juice should be mixed to get 76 ounces of a fruit drink that is 25% fruit juice?

62) A store owner plans to make 10 pounds of a candy mix worth $1.92/lb. How many pounds of gummi bears worth $2.40/lb and how many pounds of jelly beans worth $1.60/lb must be combined to make the candy mix?

63) The number of plastic surgery procedures performed in the United States in 2003 was 293% more than the number performed in 1997. If approximately 8,253,000 cosmetic procedures were performed in 2003, how many took place in 1997?

(American Society for Aesthetic Plastic Surgery)

## Section 3.6 Geometry Applications and Solving Formulas

### Objectives
1. Substitute Values into a Formula, and Find the Unknown Variable
2. Solve Problems Using Formulas from Geometry
3. Solve Problems Involving Angle Measures
4. Solve a Formula for a Specific Variable

A **formula** is a rule containing variables and mathematical symbols to state relationships between certain quantities.

Some examples of formulas we have used already are

$$P = 2l + 2w \qquad A = \frac{1}{2}bh \qquad C = 2\pi r \qquad I = PRT$$

In this section we will solve problems using *formulas,* and then we will learn how to solve a formula for a specific variable.

### 1. Substitute Values into a Formula, and Find the Unknown Variable

**Example 1**

The formula for the area of a triangle is $A = \frac{1}{2}bh$. If $A = 30$ when $b = 8$, find $h$.

**Solution**

The only unknown variable is $h$ since we are given the values of $A$ and $b$. Substitute $A = 30$ and $b = 8$ into the formula, and solve for $h$.

$$A = \frac{1}{2}bh$$

$$30 = \frac{1}{2}(8)h \qquad \text{Substitute the given values.}$$

Since $h$ is the only remaining variable in the equation, we can solve for it.

$$30 = 4h \qquad \text{Multiply.}$$
$$\frac{30}{4} = \frac{4h}{4} \qquad \text{Divide by 4.}$$
$$\frac{15}{2} = h \qquad \text{Simplify.}$$

**You Try 1**

The area of a trapezoid is $A = \dfrac{1}{2}h(b_1 + b_2)$. If $A = 21$ when $b_1 = 10$ and $b_2 = 4$, find $h$.

## 2. Solve Problems Using Formulas from Geometry

Next we will solve applied problems using concepts and formulas from geometry. Unlike in Example 1, you will not be given a formula. You will need to know the geometry formulas that we reviewed in Section 1.3. They are also found at the end of the book.

**Example 2**

A soccer field is in the shape of a rectangle and has an area of 9000 yd². Its length is 120 yd. What is the width of the field?

### Solution

**Step 1:** **Read** the problem carefully, and identify what we are being asked to find.

We must find the length of the soccer field.

A picture will be very helpful in this problem.

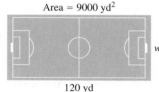

Area = 9000 yd²

120 yd

**Step 2:** **Choose a variable** to represent the unknown.

$w$ = the width of the soccer field

Label the picture with the length, 120 yd, and the width, $w$.

**Step 3:** **Translate** the information that appears in English into an algebraic equation. We will use a known geometry formula. How do we know which formula to use? List the information we are given and what we want to find:

The field is in the shape of a rectangle; its area = 9000 yd² and its length = 120 yd. We must find the width. Which formula involves the area, length, and width of a rectangle?

$$A = lw$$

Substitute the known values into the formula for the area of a rectangle, and solve for $w$.

$$A = lw$$
$$9000 = 120w \qquad \text{Substitute the known values.}$$

**Step 4:** **Solve** the equation.

$$9000 = 120w$$
$$\frac{9000}{120} = \frac{120w}{120} \qquad \text{Divide by 120.}$$
$$75 = w \qquad \text{Simplify.}$$

**Step 5:**   **Check** the answer and **interpret** the solution as it relates to the problem.

If $w = 75$ yd, then $l \cdot w = 120$ yd $\cdot$ 75 yd $= 9000$ yd$^2$. Therefore, the width of the soccer field is 75 yd.   ◼

**Note**

Remember to include the correct units in your answer!

**You Try 2**

Write an equation and solve.
The area of a rectangular room is 270 ft$^2$. Find the length of the room if the width is 15 ft.

**Example 3**

Stewart wants to put a rectangular safety fence around his backyard pool. He calculates that he will need 120 feet of fencing and that the length will be 4 feet longer than the width. Find the dimensions of the safety fence.

**Solution**

**Step 1:**   **Read** the problem carefully, and identify what we are being asked to find.

We must find the length and width of the safety fence.

Draw a picture.

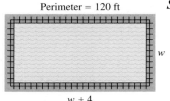

Perimeter = 120 ft

$w$

$w + 4$

**Step 2:**   **Choose a variable** to represent an unknown, and define the other unknown in terms of this variable.

The length is 4 feet longer than the width. Therefore, let

$w =$ the width of the safety fence

Define the other unknown in terms of $w$.

$w + 4 =$ the length of the safety fence

Label the picture with the expressions for the width and length.

**Step 3:**   **Translate** the information that appears in English into an algebraic equation.

Use a known geometry formula. What does the 120 ft of fencing represent? *Since the fencing will go around the pool, the 120 ft represents the perimeter of the rectangular safety fence.* We need to use a formula that involves the length, width, and perimeter of a rectangle. The formula we will use is

$$P = 2l + 2w$$

Substitute the known values and expressions into the formula.

$$P = 2l + 2w$$
$$120 = 2(w + 4) + 2w \qquad \text{Substitute.}$$

***Step 4:***   **Solve** the equation.

$$120 = 2(w + 4) + 2w$$
$$120 = 2w + 8 + 2w \qquad \text{Distribute.}$$
$$120 = 4w + 8 \qquad \text{Combine like terms.}$$
$$120 - 8 = 4w + 8 - 8 \qquad \text{Subtract 8 from each side.}$$
$$112 = 4w \qquad \text{Combine like terms.}$$
$$\frac{112}{4} = \frac{4w}{4} \qquad \text{Divide each side by 4.}$$
$$28 = w \qquad \text{Simplify.}$$

***Step 5:***   **Check** the answer and **interpret** the solution as it relates to the problem.

The width of the safety fence is 28 ft. The length is $w + 4 = 28 + 4 = 32$ ft.

The answer makes sense because the perimeter of the fence is 2(32 ft) + 2(28 ft) = 64 ft + 56 ft = 120 ft.   ∎

**You Try 3**

Write an equation and solve.
  Marina wants to make a rectangular dog run in her backyard. It will take 46 feet of fencing to enclose it, and the length will be 1 foot less than three times the width. Find the dimensions of the dog run.

## 3. Solve Problems Involving Angle Measures

Recall from Section 1.3 that the sum of the angle measures in a triangle is 180°. We will use this fact in our next example.

**Example 4**

Find the missing angle measures.

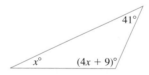

**Solution**

***Step 1:***   **Read** the problem carefully, and identify what we are being asked to find.

Find the missing angle measures.

***Step 2:***   The unknowns are already defined. We must find $x$, the measure of one angle, and then $4x + 9$, the measure of the other angle.

***Step 3:***   **Translate** the information into an algebraic equation. Since the sum of the angles in a triangle is 180°, we can write

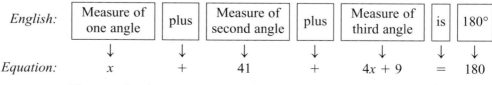

| *English:* | Measure of one angle | plus | Measure of second angle | plus | Measure of third angle | is | 180° |
|---|---|---|---|---|---|---|---|
| | ↓ | ↓ | ↓ | ↓ | ↓ | ↓ | ↓ |
| *Equation:* | $x$ | + | 41 | + | $4x + 9$ | = | 180 |

The equation is $x + 41 + (4x + 9) = 180$.

**Step 4:**   **Solve** the equation.

$$x + 41 + (4x + 9) = 180$$
$$5x + 50 = 180 \qquad \text{Combine like terms.}$$
$$5x + 50 - 50 = 180 - 50 \qquad \text{Subtract 50 from each side.}$$
$$5x = 130 \qquad \text{Combine like terms.}$$
$$\frac{5x}{5} = \frac{130}{5} \qquad \text{Divide each side by 5.}$$
$$x = 26 \qquad \text{Simplify.}$$

**Step 5:**   **Check** the answer and **interpret** the solution as it relates to the problem.

One angle, $x$, has a measure of 26°. The other unknown angle measure is $4x + 9 = 4(26) + 9 = 113°$.

The answer makes sense because the sum of the angle measure is $26° + 41° + 113° = 180°$.

 **You Try 4**

Find the missing angle measures.

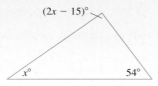

Let's look at another type of problem involving angle measures.

**Example 5**

Find the measure of each indicated angle.

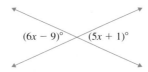

**Solution**

The indicated angles are *vertical angles*, and **vertical angles** have the same measure. (See Section 1.3.) Since their measures are the same, set $6x - 9 = 5x + 1$ and solve for $x$.

$$6x - 9 = 5x + 1$$
$$6x - 9 + 9 = 5x + 1 + 9 \qquad \text{Add 9 to each side.}$$
$$6x = 5x + 10 \qquad \text{Combine like terms.}$$
$$6x - 5x = 5x - 5x + 10 \qquad \text{Subtract } 5x \text{ from each side.}$$
$$x = 10 \qquad \text{Combine like terms.}$$

Be careful! Although $x = 10$, the angle measure is *not* 10. To find the angle measures, substitute $x = 10$ into the expressions for the angles.

The measure of the angle on the left is $6x - 9 = 6(10) - 9 = 51°$. The other angle measure is also 51° since these are vertical angles. We can verify this by substituting 10 into the expression for the other angle, $5x + 1$: $5x + 1 = 5(10) + 1 = 51°$.

**You Try 5**

Find the measure of each indicated angle.

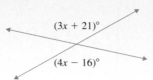

$(3x + 21)°$

$(4x - 16)°$

In Section 1.3, we learned that two angles are **complementary** if the sum of their angles is 90°, and two angles are **supplementary** if the sum of their angles is 180°.

For example, if the measure of $\angle A$ is 71°, then

a)  the measure of its complement is $90° - 71° = 19°$.

b)  the measure of its supplement is $180° - 71° = 109°$.

Now let's say the measure of an angle is $x$. Using the same reasoning as above,

a)  the measure of its complement is $90 - x$.

b)  the measure of its supplement is $180 - x$.

We will use these ideas to solve the problem in Example 6.

**Example 6**

The supplement of an angle is 34° more than twice the complement of the angle. Find the measure of the angle.

**Solution**

*Step 1:*   **Read** the problem carefully, and identify what we are being asked to find.

We must find the measure of the angle.

*Step 2:*   **Choose a variable** to represent an unknown, and define the other unknowns in terms of this variable.

This problem has three unknowns: the measures of the angle, its complement, and its supplement. Choose a variable to represent the original angle, then define the other unknowns in terms of this variable.

$$x = \text{the measure of the angle}$$

Define the other unknowns in terms of $x$.

$$90 - x = \text{the measure of the complement}$$
$$180 - x = \text{the measure of the supplement}$$

*Step 3:*   **Translate** the information that appears in English into an algebraic equation.

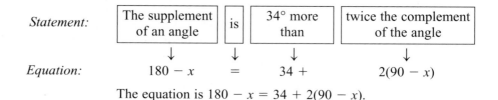

| *Statement:* | The supplement of an angle | is | 34° more than | twice the complement of the angle |
|---|---|---|---|---|
| | ↓ | ↓ | ↓ | ↓ |
| *Equation:* | $180 - x$ | $=$ | $34 +$ | $2(90 - x)$ |

The equation is $180 - x = 34 + 2(90 - x)$.

***Step 4:*** **Solve** the equation.

$$180 - x = 34 + 2(90 - x)$$
$$180 - x = 34 + 180 - 2x \qquad \text{Distribute.}$$
$$180 - x = 214 - 2x \qquad \text{Combine like terms.}$$
$$180 - 180 - x = 214 - 180 - 2x \qquad \text{Subtract 180 from each side.}$$
$$-x = 34 - 2x \qquad \text{Combine like terms.}$$
$$-x + 2x = 34 - 2x + 2x \qquad \text{Add } 2x \text{ to each side.}$$
$$x = 34 \qquad \text{Simplify.}$$

***Step 5:*** **Check** the answer and **interpret** the solution as it relates to the problem.

The measure of the angle is 34°.

To check the answer, we first need to find its complement and supplement. The complement is $90° - 34° = 56°$, and its supplement is $180° - 34° = 146°$. Now we can check these values in the original statement: The supplement is 146°. Thirty-four degrees more than twice the complement is $34° + 2(56°) = 34° + 112° = 146°$. ∎

**You Try 6**

Write an equation and solve.
 Twice the complement of an angle is 18° less than the supplement of the angle. Find the measure of the angle.

## 4. Solve a Formula for a Specific Variable

The formula $P = 2l + 2w$ allows us to find the perimeter of a rectangle when we know its length ($l$) and width ($w$). But what if we were solving problems where we repeatedly needed to find the value of $w$? Then, we could rewrite $P = 2l + 2w$ so that it is solved for $w$:

$$w = \frac{P - 2l}{2}$$

Doing this means that we have *solved the formula $P = 2l + 2w$ for the specific variable w.*
 Solving a formula for a specific variable may seem confusing at first because the formula contains more than one letter. Keep in mind that we will solve for a specific variable the same way we have been solving equations up to this point.
 We'll start by solving $3x + 4 = 19$ step-by-step for $x$ and then applying the same procedure to solving $ax + b = c$ for $x$.

**Example 7**

Solve $3x + 4 = 19$ and $ax + b = c$ for $x$.

### Solution

Look at these equations carefully, and notice that they have the same form. Read the parts of the solution in numerical order.

**Part 1**   Solve $3x + 4 = 19$.

Don't quickly run through the solution of this equation. **The emphasis here is on the steps used to solve the equation and why we use those steps!**

$$3\boxed{x} + 4 = 19$$

We are solving for $x$. We'll put a box around it. What is the first step? "Get rid of" what is being added to the $3x$; that is, "get rid of" the 4 on the left. Subtract 4 from each side.

$$3\boxed{x} + 4 - 4 = 19 - 4$$

Combine like terms.

$$3\boxed{x} = 15$$

**Part 3**   We left off needing to solve $3\boxed{x} = 15$ for $x$. We need to eliminate the 3 on the left. Since $x$ is being multiplied by 3, we will **divide** each side by 3.

$$\frac{3\boxed{x}}{3} = \frac{15}{3}$$

Simplify.

$$x = 5$$

**Part 2**   Solve $ax + b = c$ for $x$.

Since we are solving for $x$, we'll put a box around it.

$$a\boxed{x} + b = c$$

The goal is to get the $x$ on a side by itself. What do we do first? As in part 1, "get rid of" what is being added to the $ax$ term; that is, "get rid of" the $b$ on the left. Since $b$ is being added to $ax$, we will subtract it from each side. (We are performing the same steps as in part 1!)

$$a\boxed{x} + b - b = c - b$$

Combine like terms.

$$a\boxed{x} = c - b$$

We cannot combine the terms on the right, so it remains $c - b$.

**Part 4**   Now, we have to solve $a\boxed{x} = c - b$ for $x$. We need to eliminate the $a$ on the left. Since $x$ is being multiplied by $a$, we will **divide** each side by $a$.

$$\frac{a\boxed{x}}{a} = \frac{c - b}{a}$$

These are the same steps used in part 3!

Simplify.

$$\frac{a\boxed{x}}{a} = \frac{c - b}{a}$$

$$x = \frac{c - b}{a} \text{ or } \frac{c}{a} - \frac{b}{a}$$

**Note**

To obtain the result $x = \dfrac{c}{a} - \dfrac{b}{a}$, we distributed the $a$ in the denominator to each term in the numerator. Either form of the answer is correct.

When you are solving a formula for a specific variable, think about the steps you use to solve an equation in one variable.

**You Try 7**

Solve $rt - n = k$ for $t$.

**Example 8**

$U = \dfrac{1}{2}LI^2$ is a formula used in physics. Solve this equation for $L$.

**Solution**

$$U = \dfrac{1}{2}\boxed{L}I^2 \qquad \text{Solve for } L. \text{ Put it in a box.}$$

$$2U = 2 \cdot \dfrac{1}{2}\boxed{L}I^2 \qquad \text{Multiply by 2 to eliminate the fraction.}$$

$$\dfrac{2U}{I^2} = \dfrac{\boxed{L}I^2}{I^2} \qquad \text{Divide each side by } I^2.$$

$$\dfrac{2U}{I^2} = L \qquad \text{Simplify.}$$

**Example 9**

$A = \dfrac{1}{2}h(b_1 + b_2)$ is the formula for the area of a trapezoid. Solve it for $b_1$.

**Solution**

There are two ways to solve this for $b_1$.

**Method 1:**   We will put $b_1$ in a box to remind us that this is what we must solve for. In Method 1, we will start by eliminating the fraction.

$$2A = 2 \cdot \dfrac{1}{2}h(\boxed{b_1} + b_2) \qquad \text{Multiply each side by 2.}$$

$$2A = h(\boxed{b_1} + b_2) \qquad \text{Simplify.}$$

$$\dfrac{2A}{h} = \dfrac{h(\boxed{b_1} + b_2)}{h} \qquad \text{Divide each side by } h.$$

$$\dfrac{2A}{h} = \boxed{b_1} + b_2$$

$$\dfrac{2A}{h} - b_2 = \boxed{b_1} + b_2 - b_2 \qquad \text{Subtract } b_2 \text{ from each side.}$$

$$\dfrac{2A}{h} - b_2 = b_1 \qquad \text{Simplify.}$$

**Method 2:**   Another way to solve $A = \dfrac{1}{2}h(b_1 + b_2)$ for $b_1$ is to begin by distributing $\dfrac{1}{2}h$ on the right.

$$A = \dfrac{1}{2}h\boxed{b_1} + \dfrac{1}{2}hb_2 \qquad \text{Distribute.}$$

$$2A = 2\left(\dfrac{1}{2}h\boxed{b_1} + \dfrac{1}{2}hb_2\right) \qquad \text{Multiply by 2 to eliminate the fractions.}$$

$$2A = h\boxed{b_1} + hb_2 \qquad \text{Distribute.}$$

$$2A - hb_2 = h\boxed{b_1} + hb_2 - hb_2 \qquad \text{Subtract } hb_2 \text{ from each side.}$$

$$2A - hb_2 = h\boxed{b_1} \qquad \text{Simplify.}$$

$$\dfrac{2A - hb_2}{h} = \dfrac{h\boxed{b_1}}{h} \qquad \text{Divide by } h.$$

$$\dfrac{2A - hb_2}{h} = b_1 \qquad \text{Simplify.}$$

Therefore, $b_1$ can be written as $b_1 = \dfrac{2A}{h} - \dfrac{hb_2}{h}$   or   $b_1 = \dfrac{2A}{h} - b_2$. These two forms are equivalent.  ■

### You Try 8

Solve for the indicated variable.

a)   $t = \dfrac{qr}{s}$ for $q$          b)   $R = t(k - c)$ for $c$

---

### Answers to You Try Exercises

1)  3     2)  18 ft     3)  6 ft × 17 ft     4)  47°, 79°     5)  132°, 132°     6)  18°

7)  $t = \dfrac{k + n}{r}$     8)  a) $q = \dfrac{st}{r}$   b) $c = \dfrac{kt - R}{t}$ or $c = k - \dfrac{R}{t}$

---

# 3.6 Exercises

### Objective 1: Substitute Values into a Formula, and Find the Unknown Variable

1) If you are using the formula $A = \dfrac{1}{2}bh$, is it reasonable to get an answer of $h = -6$? Explain your answer.

2) If you are finding the area of a rectangle and the lengths of the sides are given in inches, the area of the rectangle would be expressed in which unit?

3) If you are asked to find the volume of a sphere and the radius is given in centimeters, the volume would be expressed in which unit?

4) If you are asked to find the perimeter of a football field and the length and width are given in yards, the perimeter of the field would be expressed in which unit?

Substitute the given values into the formula and solve for the remaining variable.

5) $A = lw$; If $A = 44$ when $l = 16$, find $w$.

6) $A = \dfrac{1}{2}bh$; If $A = 21$ when $h = 14$, find $b$.

7) $I = PRT$; If $I = 240$ when $R = 0.04$ and $T = 2$, find $P$.

8) $I = PRT$; If $I = 600$ when $P = 2500$ and $T = 4$, find $R$.

9) $d = rt$
   (Distance formula: $distance = rate \cdot time$);
   If $d = 150$ when $r = 60$, find $t$.

10) $d = rt$
   (Distance formula: $distance = rate \cdot time$);
   If $r = 36$ and $t = 0.75$, find $d$.

11) $C = 2\pi r$; If $r = 4.6$, find $C$.

12) $C = 2\pi r$; If $C = 15\pi$, find $r$.

13) $P = 2l + 2w$; If $P = 11$ when $w = \dfrac{3}{2}$, find $l$.

14) $P = s_1 + s_2 + s_3$ (Perimeter of a triangle); If $P = 11.6$ when $s_2 = 2.7$ and $s_3 = 3.8$, find $s_1$.

15) $V = lwh$; If $V = 52$ when $l = 6.5$ and $h = 2$, find $w$.

16) $V = \dfrac{1}{3}Ah$ (Volume of a pyramid); If $V = 16$ when $A = 24$, find $h$.

17) $V = \dfrac{1}{3}\pi r^2 h$; If $V = 48\pi$ when $r = 4$, find $h$.

18) $V = \dfrac{1}{3}\pi r^2 h$; If $V = 50\pi$ when $r = 5$, find $h$.

19) $S = 2\pi r^2 + 2\pi rh$ (Surface area of a right circular cylinder); If $S = 154\pi$ when $r = 7$, find $h$.

20) $S = 2\pi r^2 + 2\pi rh$; If $S = 132\pi$ when $r = 6$, find $h$.

21) $A = \dfrac{1}{2}h(b_1 + b_2)$; If $A = 136$ when $b_1 = 7$ and $h = 16$, find $b_2$.

22) $A = \dfrac{1}{2}h(b_1 + b_2)$; If $A = 1.5$ when $b_1 = 3$ and $b_2 = 1$, find $h$.

**Objective 2: Solve Problems Using Formulas from Geometry**

Use a known formula to solve. See Example 2.

23) The area of a tennis court is 2808 ft². Find the length of the court if it is 36 ft wide.

24) A rectangular tabletop has an area of 13.5 ft². What is the width of the table if it is 4.5 ft long?

25) A rectangular flower box holds 1232 in³ of soil. Find the height of the box if it is 22 in. long and 7 in. wide.

26) A rectangular storage box is 2.5 ft wide, 4 ft long, and 1.5 ft high. What is the storage capacity of the box?

27) The center circle on a soccer field has a radius of 10 yd. What is the area of the center circle? Use 3.14 for $\pi$.

28) The face of the clock on Big Ben in London has a radius of 11.5 feet. What is the area of this circular clock face? Use 3.14 for $\pi$. (www.bigben.freeservers.com)

29) Abbas drove 134 miles on the highway in 2 hours. What was his average speed?

30) If Reza drove 108 miles at 72 mph, without stopping, for how long did she drive?

31) A stainless steel garbage can is in the shape of a right circular cylinder. If its radius is 6 inches and its volume is $864\pi$ in³, what is the height of the can?

32) A coffee can in the shape of a right circular cylinder has a volume of $50\pi$ in³. Find the height of the can if its diameter is 5 inches.

33) A flag is in the shape of a triangle and has an area of 6 ft². Find the length of the base if its height is 4 ft.

34) A championship banner hanging from the rafters of a stadium is in the shape of a triangle and has an area of 20 ft². How long is the banner if its base is 5 ft?

35) Leilani invested $1500 in a bank account for 2 years and earned $75 in interest. What interest rate did she receive?

36) The backyard of a house is in the shape of a trapezoid as pictured here. If the area of the yard is 6750 ft²:

a) Find the length of the missing side, x.

b) How much fencing would be needed to completely enclose the plot?

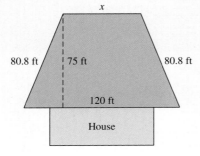

Use a known formula to solve. See Example 3.

37) Vivian is making a rectangular wooden picture frame that will have a width that is 10 in. shorter than its length. If she will use 92 in. of wood, what are the dimensions of the frame?

38) A construction crew is making repairs next to a school, so they have to enclose the rectangular area with a fence. They determine that they will need 176 ft of fencing for the work area, which is 22 ft longer than it is wide. Find the dimensions of the fenced area.

39) The "lane" on a basketball court is a rectangle that has a perimeter of 62 ft. Find the dimensions of the "lane" given that its length is 5 ft less than twice the width.

40) A rectangular whiteboard in a classroom is twice as long as it is high. Its perimeter is 24 ft. What are the dimensions of the whiteboard?

41) One base of a trapezoid is 2 in. longer than three times the other base. Find the lengths of the bases if the trapezoid is 5 in. high and has an area of 25 in².

42) A caution flag on the side of a road is shaped like a trapezoid. One base of the trapezoid is 1 ft shorter than the other base. Find the lengths of the bases if the trapezoid is 4 ft high and has an area of 10 ft².

43) A triangular sign in a store window has a perimeter of 5.5 ft. Two of the sides of the triangle are the same length while the third side is 1 foot longer than those sides. Find the lengths of the sides of the sign.

44) A triangle has a perimeter of 31 in. The longest side is 1 in. less than twice the shortest side, and the third side is 4 in. longer than the shortest side. Find the lengths of the sides.

**Objective 3: Solve Problems Involving Angle Measures**

Find the missing angle measures.

45)

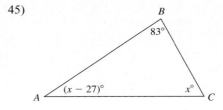

46)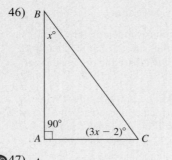

VIDEO 47)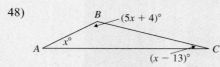

48)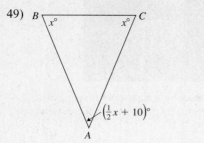

49)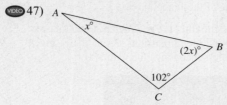

50)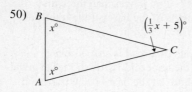

Find the measure of each indicated angle.

51)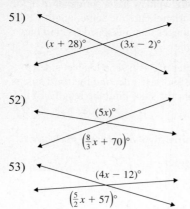

52)

$(5x)°$

$\left(\frac{8}{3}x + 70\right)°$

53)

$(4x − 12)°$

$\left(\frac{5}{2}x + 57\right)°$

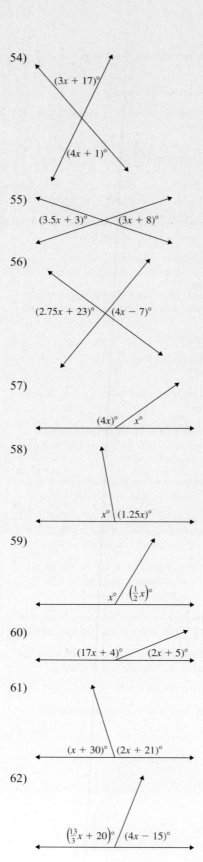

54)

$(3x + 17)°$

$(4x + 1)°$

55)

$(3.5x + 3)°$    $(3x + 8)°$

56)

$(2.75x + 23)°$    $(4x − 7)°$

57)

$(4x)°$    $x°$

58)

$x°$    $(1.25x)°$

59)

$x°$    $\left(\frac{1}{2}x\right)°$

60)

$(17x + 4)°$    $(2x + 5)°$

61)

$(x + 30)°$    $(2x + 21)°$

62)

$\left(\frac{13}{3}x + 20\right)°$    $(4x − 15)°$

63) If $x$ = the measure of an angle, write an expression for its supplement.

64) If $x$ = the measure of an angle, write an expression for its complement.

Write an equation and solve.

65) The supplement of an angle is 63° more than twice the measure of its complement. Find the measure of the angle.

66) Twice the complement of an angle is 49° less than its supplement. Find the measure of the angle.

67) Six times an angle is 12° less than its supplement. Find the measure of the angle.

68) An angle is 1° less than 12 times its complement. Find the measure of the angle.

69) Four times the complement of an angle is 40° less than twice the angle's supplement. Find the angle, its complement, and its supplement.

70) Twice the supplement of an angle is 30° more than eight times its complement. Find the angle, its complement, and its supplement.

71) The sum of an angle and half its supplement is seven times its complement. Find the measure of the angle.

72) The sum of an angle and three times its complement is 62° more than its supplement. Find the measure of the angle.

73) The sum of four times an angle and twice its complement is 270°. Find the angle.

74) The sum of twice an angle and half its supplement is 192°. Find the angle.

**Objective 4: Solve a Formula for a Specific Variable**

75) Solve for $x$.

a) $x + 16 = 37$    b) $x + h = y$

c) $x + r = c$

76) Solve for $t$.

a) $t - 8 = 17$    b) $t - p = z$

c) $t - k = n$

77) Solve for $c$.

a) $8c = 56$    b) $ac = d$

c) $mc = v$

78) Solve for $k$.

a) $9k = 54$    b) $nk = t$

c) $wk = h$

79) Solve for $a$.

a) $\dfrac{a}{4} = 11$    b) $\dfrac{a}{y} = r$

c) $\dfrac{a}{w} = d$

80) Solve for $d$.

a) $\dfrac{d}{6} = 3$    b) $\dfrac{d}{t} = q$

c) $\dfrac{d}{x} = a$

81) Solve for $d$.

a) $8d - 7 = 17$    b) $kd - a = z$

82) Solve for $w$.

a) $5w + 18 = 3$    b) $pw + r = \pi$

83) Solve for $h$.

a) $9h + 23 = 17$    b) $qh + v = n$

84) Solve for $b$.

a) $12b - 5 = 17$    b) $mb - c = a$

Solve each formula for the indicated variable.

85) $F = ma$ for $m$ (Physics)

86) $C = 2\pi r$ for $r$

87) $n = \dfrac{c}{v}$ for $c$ (Physics)

88) $f = \dfrac{R}{2}$ for $R$ (Physics)

89) $E = \sigma T^4$ for $\sigma$ (Meteorology)

90) $p = \rho g y$ for $\rho$ (Geology)

91) $V = \dfrac{1}{3}\pi r^2 h$ for $h$

92) $d = rt$ for $r$

93) $R = \dfrac{E}{I}$ for $E$ (Electricity)

94) $A = \dfrac{1}{2}bh$ for $b$

95) $I = PRT$ for $R$

96) $I = PRT$ for $P$

97) $P = 2l + 2w$ for $l$

98) $A = P + PRT$ for $T$ (Finance)

99) $H = \dfrac{D^2 N}{2.5}$ for $N$ (Auto mechanics)

100) $V = \dfrac{AH}{3}$ for $A$ (Geometry)

101) $A = \dfrac{1}{2}h(b_1 + b_2)$ for $b_2$

102) $A = \pi(R^2 - r^2)$ for $r^2$ (Geometry)

For Exercises 103 and 104, refer to the figure below.

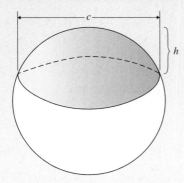

The surface area, $S$, of the spherical segment shown in the figure is given by $S = \dfrac{\pi}{4}(4h^2 + c^2)$, where $h$ is the height of the segment and $c$ is the diameter of the segment's base.

103) Solve the formula for $h^2$.

104) Solve the formula for $c^2$.

105) The perimeter, $P$, of a rectangle is $P = 2l + 2w$, where $l$ = length and $w$ = width.

   a) Solve $P = 2l + 2w$ for $w$.

   b) Find the width of the rectangle with perimeter 28 cm and length 11 cm.

106) The area, $A$, of a triangle is $A = \dfrac{1}{2}bh$, where $b$ = length of the base and $h$ = height.

   a) Solve $A = \dfrac{1}{2}bh$ for $h$.

   b) Find the height of the triangle that has an area of 39 cm$^2$ and a base of length 13 cm.

107) The formula $C = \dfrac{5}{9}(F - 32)$ can be used to convert from degrees Fahrenheit, $F$, to degrees Celsius, $C$.

   a) Solve this formula for $F$.

   b) The average high temperature in Paris, France, in May is 20°C. Use the result in part a) to find the equivalent temperature in degrees Fahrenheit. (www.bbc.co.uk)

108) The average low temperature in Buenos Aires, Argentina, in June is 5°C. Use the result in Exercise 107 a) to find the equivalent temperature in degrees Fahrenheit. (www.bbc.co.uk)

---

## Section 3.7 Applications of Linear Equations to Proportions, Money Problems, and $d = rt$

### Objectives

1. **Use Ratios**
2. **Solve a Proportion**
3. **Solve Problems Involving Denominations of Money**
4. **Solve Problems Involving Distance, Rate, and Time**

### 1. Use Ratios

We hear about *ratios* and use them in many ways in everyday life. For example, if a survey on cell phone use revealed that 80 teenagers prefer texting their friends while 25 prefer calling their friends, we could write the ratio of teens who prefer texting to teens who prefer calling as

$$\dfrac{\text{Number who prefer texting}}{\text{Number who prefer calling}} = \dfrac{80}{25} = \dfrac{16}{5}$$

Here is a formal definition of a ratio:

> **Definition**
>
> A **ratio** is a quotient of two quantities. The ratio of the number $x$ to the number $y$, where $y \neq 0$, can be written as $\dfrac{x}{y}$, $x$ to $y$, or $x:y$.

A percent is actually a ratio. For example, we can think of 39% as $\dfrac{39}{100}$ or as the ratio of 39 to 100.

**Example 1**

Write the ratio of 4 feet to 2 yards.

**Solution**

Write each quantity with the same units. Let's change yards to feet. Since there are 3 feet in 1 yard,

$$2 \text{ yards} = 2 \cdot 3 \text{ feet} = 6 \text{ feet}$$

Then the ratio of 4 feet to 2 yards is

$$\frac{4 \text{ feet}}{2 \text{ yds}} = \frac{4 \text{ feet}}{6 \text{ feet}} = \frac{4}{6} = \frac{2}{3}$$ ∎

**You Try 1**

Write the ratio of 3 feet to 24 inches.

We can use ratios to help us figure out which item in a store gives us the most value for our money. To do this, we will determine the *unit price* of each item. The **unit price** is the ratio of the price of the item to the amount of the item.

**Example 2**

A store sells Haagen-Dazs vanilla ice cream in three different sizes. The sizes and prices are listed here. Which size is the best buy?

| Size | Price |
|------|-------|
| 4 oz | $1.00 |
| 14 oz | $3.49 |
| 28 oz | $7.39 |

**Solution**

For each carton of ice cream, we must find the unit price, or how much the ice cream costs per ounce. We will find the unit price by dividing.

$$\text{Unit price} = \frac{\text{Price of ice cream}}{\text{Number of ounces in the container}} = \text{Cost per ounce}$$

| Size | Unit Price |
|------|-----------|
| 4 oz | $\dfrac{\$1.00}{4 \text{ oz}} = \$0.250 \text{ per oz}$ |
| 14 oz | $\dfrac{\$3.49}{14 \text{ oz}} = \$0.249 \text{ per oz}$ |
| 28 oz | $\dfrac{\$7.39}{28 \text{ oz}} = \$0.264 \text{ per oz}$ |

We round the answers to the thousandths place because, as you can see, there is not much difference in the unit price. Since the 14-oz carton of ice cream has the smallest unit price, it is the best buy. ∎

    **You Try 2**

> A store sells Gatorade fruit punch in three different sizes. A 20-oz bottle costs $1.00, a 32-oz bottle sells for $1.89, and the price of a 128-oz bottle is $5.49. Which size is the best buy, and what is its unit price?

## 2. Solve a Proportion

We have learned that a ratio is a way to compare two quantities. If two ratios are equivalent, like $\dfrac{4}{6}$ and $\dfrac{2}{3}$, we can set them equal to make a *proportion*.

> **Definition**
>
> A **proportion** is a statement that two ratios are equal.

How can we be certain that a proportion is true? We can find the **cross products.** If the cross products are equal, then the proportion is true. If the cross products are not equal, then the proportion is false.

> **Property**
>
> **Cross Products**    If $\dfrac{a}{b} = \dfrac{c}{d}$, then $ad = bc$ provided that $b \neq 0$ and $d \neq 0$.

We will see later in the book that finding the cross products is the same as multiplying both sides of the equation by the least common denominator of the fractions.

**Example 3**

Determine whether each proportion is true or false.

a)  $\dfrac{5}{7} = \dfrac{15}{21}$        b)  $\dfrac{2}{9} = \dfrac{7}{36}$

### *Solution*
a)    Find the cross products.

Multiply.

$$\dfrac{5}{7} \bowtie \dfrac{15}{21}$$

Multiply.

$$5 \cdot 21 = 7 \cdot 15$$
$$105 = 105 \qquad \text{True}$$

The cross products are equal, so the proportion is true.

b)    Find the cross products.

Multiply.

$$\dfrac{2}{9} \bowtie \dfrac{7}{36}$$

Multiply.

$$2 \cdot 36 = 9 \cdot 7$$
$$72 = 63 \qquad \text{False}$$

The cross products are not equal, so the proportion is false.    ■

**You Try 3**

Determine whether each proportion is true or false.

a) $\dfrac{4}{9} = \dfrac{24}{56}$          b) $\dfrac{3}{8} = \dfrac{12}{32}$

We can use cross products to solve equations.

**Example 4**

Solve each proportion.

a) $\dfrac{16}{24} = \dfrac{x}{3}$          b) $\dfrac{k + 2}{2} = \dfrac{k - 4}{5}$

**Solution**

Find the cross products.

a) $\dfrac{16}{24} = \dfrac{x}{3}$   Multiply.   Multiply.          b) $\dfrac{k + 2}{2} = \dfrac{k - 4}{5}$   Multiply.   Multiply.

| | |
|---|---|
| $16 \cdot 3 = 24 \cdot x$   Set the cross products equal. | $5(k + 2) = 2(k - 4)$   Set the cross products equal. |
| $48 = 24x$   Multiply. | $5k + 10 = 2k - 8$   Distribute. |
| $2 = x$   Divide by 24. | $3k + 10 = -8$   Subtract $2k$. |
| | $3k = -18$   Subtract 10. |
| | $k = -6$   Divide by 3. |

The solution set is $\{2\}$.          The solution set is $\{-6\}$.   ■

**You Try 4**

Solve each proportion.

a) $\dfrac{2}{3} = \dfrac{w}{27}$          b) $\dfrac{b - 6}{12} = \dfrac{b + 2}{20}$

Proportions are often used to solve real-world problems. When we solve problems by setting up a proportion, we must be sure that the numerators contain the same quantities and the denominators contain the same quantities.

**Example 5**

Write an equation and solve.

Cailen is an artist, and she wants to make turquoise paint by mixing the green and blue paints that she already has. To make turquoise, she will have to mix 4 parts of green with 3 parts of blue. If she uses 6 oz of green paint, how much blue paint should she use?

**Solution**

*Step 1:*   **Read** the problem carefully, and identify what we are being asked to find.

We must find the amount of blue paint needed.

*Step 2:*   **Choose a variable** to represent the unknown.

$x$ = the number of ounces of blue paint

**Step 3:** **Translate** the information that appears in English into an algebraic equation. Write a proportion. We will write our ratios in the form of

$$\frac{\text{Amount of green paint}}{\text{Amount of blue paint}}$$ so that the numerators contain the same quantities and the denominators contain the same quantities.

$$\text{Amount of green paint} \rightarrow \frac{4}{3} = \frac{6}{x} \leftarrow \text{Amount of green paint}$$
$$\text{Amount of blue paint} \rightarrow \qquad\qquad \leftarrow \text{Amount of blue paint}$$

The equation is $\dfrac{4}{3} = \dfrac{6}{x}$.

**Step 4:** **Solve** the equation.

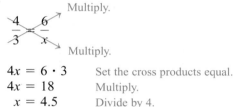

$$4x = 6 \cdot 3 \qquad \text{Set the cross products equal.}$$
$$4x = 18 \qquad \text{Multiply.}$$
$$x = 4.5 \qquad \text{Divide by 4.}$$

**Step 5:** **Check** the answer and **interpret** the solution as it relates to the problem.

Cailen should mix 4.5 oz of blue paint with the 6 oz of green paint to make the turquoise paint she needs. The check is left to the student. ■

**You Try 5**

Write an equation and solve.
If 3 lb of coffee costs $21.60, how much would 5 lb of the same coffee cost?

Another application of proportions is for solving similar triangles.

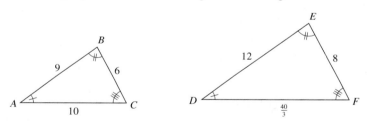

$$m\angle A = m\angle D, \quad m\angle B = m\angle E, \quad \text{and} \quad m\angle C = m\angle F$$

We say that $\triangle ABC$ and $\triangle DEF$ are *similar triangles*. Two triangles are **similar** if they have the same shape, the corresponding angles have the same measure, and the corresponding sides are proportional.

The ratio of each of the corresponding sides is $\dfrac{3}{4}$:

$$\frac{9}{12} = \frac{3}{4}; \quad \frac{6}{8} = \frac{3}{4}; \quad \frac{10}{\frac{40}{3}} = 10 \cdot \frac{3}{40} = \frac{3}{4}.$$

We can use a proportion to find the length of an unknown side in two similar triangles.

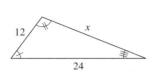

**Example 6**

Given the following similar triangles, find *x*.

**Solution**

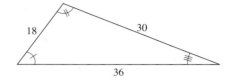

$$\frac{12}{18} = \frac{x}{30}$$    Set the ratios of two corresponding sides equal to each other. (Set up a proportion.)

$$12 \cdot 30 = 18 \cdot x$$    Solve the proportion.
$$360 = 18x$$    Multiply.
$$20 = x$$    Divide by 18.

**You Try 6**

Given the following similar triangles, find *x*.

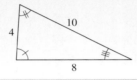 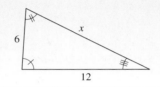

## 3. Solve Problems Involving Denominations of Money

Many applications in algebra involve the number of coins or bills and their values. We will begin this topic with two arithmetic examples; then we will solve an algebraic problem.

**Example 7**

Determine the amount of money you have in cents *and* in dollars if you have

a)   9 nickels          b)   2 quarters          c)   9 nickels and 2 quarters

**Solution**

You may be able to figure out these answers quickly and easily, but what is important here is to understand the *procedure* that is used to do this arithmetic problem so that you can apply the same procedure to algebra. So, read this carefully!

a) and b):    Let's begin with part a), finding the value of 9 nickels.

Value in Cents
$$5 \cdot 9 = 45¢$$
Value of a nickel ↗   Number ↑ of nickels   Value ↖ of 9 nickels

Value in Dollars
$$0.05 \cdot 9 = \$0.45$$
Value of a nickel ↗   Number ↑ of nickels   Value ↖ of 9 nickels

Here's how we find the value of 2 quarters:

Value in Cents
$$25 \cdot 2 = 50¢$$
Value of a quarter ↗   Number ↑ of quarters   Value ↖ of 2 quarters

Value in Dollars
$$0.25 \cdot 2 = \$0.50$$
Value of a quarter ↗   Number ↑ of quarters   Value ↖ of 2 quarters

A table can help us organize the information, so let's put both part a) and part b) in a table so that we can see a pattern.

Value of the Coins **(in cents)**                    Value of the Coins **(in dollars)**

|  | Value of the Coin | Number of Coins | Total Value of the Coins |
|---|---|---|---|
| Nickels | 5 | 9 | $5 \cdot 9 = 45$ |
| Quarters | 25 | 2 | $25 \cdot 2 = 50$ |

|  | Value of the Coin | Number of Coins | Total Value of the Coins |
|---|---|---|---|
| Nickels | 0.05 | 9 | $0.05 \cdot 9 = 0.45$ |
| Quarters | 0.25 | 2 | $0.25 \cdot 2 = 0.50$ |

In each case, notice that we find the total value of the coins by multiplying:

$$\begin{array}{ccc} \text{Value of} & & \text{Number of} & & \text{Value of all} \\ \text{the coin} & \cdot & \text{coins} & = & \text{of the coins} \end{array}$$

c)   Now let's write an equation in English to find the total value of the 9 nickels and 2 quarters.

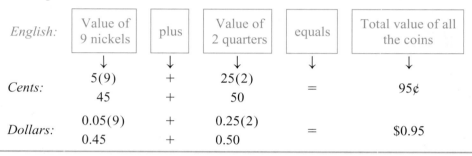

We will use the same procedure that we just used to solve these arithmetic problems to write algebraic expressions to represent the value of a collection of coins.

## Example 8

Write expressions for the amount of money you have in cents *and* in dollars if you have

a)   *n* nickels          b)   *q* quarters          c)   *n* nickels and *q* quarters

**Solution**

a) and b):   Let's use tables just like we did in Example 7. We will put parts a) and b) in the same table.

Value of the Coins **(in cents)**                    Value of the Coins **(in dollars)**

|  | Value of the Coin | Number of Coins | Total Value of the Coins |
|---|---|---|---|
| Nickels | 5 | $n$ | $5 \cdot n = 5n$ |
| Quarters | 25 | $q$ | $25 \cdot q = 25q$ |

|  | Value of the Coin | Number of Coins | Total Value of the Coins |
|---|---|---|---|
| Nickels | 0.05 | $n$ | $0.05 \cdot n = 0.05n$ |
| Quarters | 0.25 | $q$ | $0.25 \cdot q = 0.25q$ |

If you have *n* nickels, then the expression for the amount of money in cents is $5n$. The amount of money in dollars is $0.05n$.

If you have *q* quarters, then the expression for the amount of money in cents is $25q$. The amount of money in dollars is $0.25q$.

c)  Write an equation in English to find the total value of $n$ nickels and $q$ quarters. It is based on the same idea that we used in Example 7.

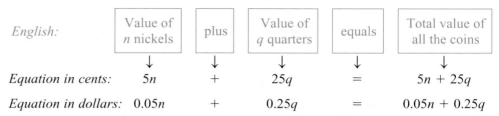

| English: | Value of $n$ nickels | plus | Value of $q$ quarters | equals | Total value of all the coins |
|---|---|---|---|---|---|
| | ↓ | ↓ | ↓ | ↓ | ↓ |
| Equation in cents: | $5n$ | $+$ | $25q$ | $=$ | $5n + 25q$ |
| Equation in dollars: | $0.05n$ | $+$ | $0.25q$ | $=$ | $0.05n + 0.25q$ |

The expression in cents is $5n + 25q$. The expression in dollars is $0.05n + 0.25q$. ■

**You Try 7**

Determine the amount of money you have in cents and in dollars if you have

a)  8 dimes          b)  67 pennies          c)  8 dimes and 67 pennies

d)  $d$ dimes          e)  $p$ pennies          f)  $d$ dimes and $p$ pennies

Now we are ready to solve an algebraic application involving denominations of money.

**Example 9**

Write an equation and solve.

At the end of the day, Annah counts the money in the cash register at the bakery where she works. There are twice as many dimes as nickels, and they are worth a total of $5.25. How many dimes and nickels are in the cash register?

**Solution**

**Step 1:**  **Read** the problem carefully, and identify what we are being asked to find.

We must find the number of dimes and nickels in the cash register.

**Step 2:**  **Choose a variable** to represent an unknown, and define the other unknown in terms of this variable.

In the statement "there are twice as many dimes as nickels," the number of dimes is expressed *in terms of* the number of nickels. Therefore, let

$n$ = the number of nickels

Define the other unknown (the number of dimes) in terms of $n$:

$2n$ = the number of dimes

**Step 3:**  **Translate** the information that appears in English into an algebraic equation.

Let's begin by making a table to write an expression for the value of the nickels and the value of the dimes. We will write the expression in terms of dollars because the total value of the coins, $5.25, is given in dollars.

| | Value of the Coin | Number of Coins | Total Value of the Coins |
|---|---|---|---|
| Nickels | 0.05 | $n$ | $0.05n$ |
| Dimes | 0.10 | $2n$ | $0.10 \cdot (2n)$ |

Write an equation in English and substitute the expressions we found in the table and the total value of the coins to get an algebraic equation.

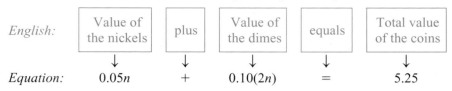

| English: | Value of the nickels | plus | Value of the dimes | equals | Total value of the coins |
|---|---|---|---|---|---|
| | ↓ | ↓ | ↓ | ↓ | ↓ |
| Equation: | $0.05n$ | $+$ | $0.10(2n)$ | $=$ | $5.25$ |

**Step 4:** **Solve** the equation.

$$0.05n + 0.10(2n) = 5.25$$
$$100[0.05n + 0.10(2n)] = 100(5.25) \qquad \text{Multiply by 100 to eliminate the decimals.}$$
$$5n + 10(2n) = 525 \qquad \text{Distribute.}$$
$$5n + 20n = 525 \qquad \text{Multiply.}$$
$$25n = 525 \qquad \text{Combine like terms.}$$
$$\frac{25n}{25} = \frac{525}{25} \qquad \text{Divide each side by 25.}$$
$$n = 21 \qquad \text{Simplify.}$$

**Step 5:** **Check** the answer and **interpret** the solution as it relates to the problem.

There were 21 nickels in the cash register and $2(21) = 42$ dimes in the register.

**Check:** The value of the nickels is $\$0.05(21) = \$1.05$, and the value of the dimes is $\$0.10(42) = \$4.20$. Their total value is $\$1.05 + \$4.20 = \$5.25$. ∎

**You Try 8**

Write an equation and solve.
   A collection of coins consists of pennies and quarters. There are three times as many pennies as quarters, and the coins are worth $8.40. How many of each type of coin is in the collection?

## 4. Solve Problems Involving Distance, Rate, and Time

An important mathematical relationship is one involving distance, rate, and time. These quantities are related by the formula

### Distance = Rate × Time

and is also written as $d = rt$. We use this formula often in mathematics and in everyday life. Let's use the formula $d = rt$ to answer the following question: If you drive on a highway at a rate of 65 mph for 3 hours, how far will you drive? Using $d = rt$, we get

$$d = rt$$
$$d = (65 \text{ mph}) \cdot (3 \text{ hr}) \qquad \text{Substitute the values.}$$
$$d = 195 \text{ mi}$$

Notice that the rate is in miles per *hour*, and the time is in *hours*. That is, the units are consistent, and they must always be consistent to correctly solve a problem like this. If the time had been expressed in minutes, we would have had to convert minutes to hours.

Next we will use the relationship $d = rt$ to solve two algebraic applications.

## Example 10

Write an equation and solve.

Two planes leave St. Louis, one flying east and the other flying west. The westbound plane travels 100 mph faster than the eastbound plane, and after 1.5 hours they are 750 miles apart. Find the speed of each plane.

**Solution**

**Step 1:** **Read** the problem carefully, and identify what we are being asked to find.

We must find the speed of the eastbound and westbound planes.

We will draw a picture to help us see what is happening in this problem.

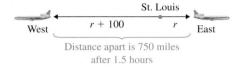

Distance apart is 750 miles
after 1.5 hours

**Step 2:** **Choose a variable** to represent an unknown, and define the other unknown in terms of this variable.

The westbound plane is traveling 100 mph faster than the eastbound plane, so let

$$r = \text{the rate of the eastbound plane}$$
$$r + 100 = \text{the rate of the westbound plane}$$

Label the picture.

**Step 3:** **Translate** the information that appears in English into an algebraic equation.

Let's make a table using the equation $d = rt$. Fill in the time, 1.5 hr, and the rates first, then multiply those together to fill in the values for the distance.

|  | $d$ | $r$ | $t$ |
|---|---|---|---|
| Eastbound | $1.5r$ | $r$ | $1.5$ |
| Westbound | $1.5(r + 100)$ | $r + 100$ | $1.5$ |

We will write an equation in English to help us write an algebraic equation. The picture shows that

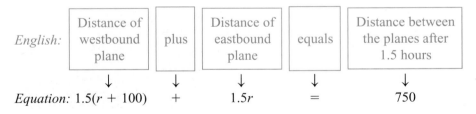

The expressions for the distances in the equation come from the table.

The equation is $1.5(r + 100) + 1.5r = 750$.

**Step 4:**   **Solve** the equation.

$$1.5(r + 100) + 1.5r = 750$$
$$10[1.5(r + 100) + 1.5r] = 10(750) \qquad \text{Multiply by 10 to eliminate the decimals.}$$
$$15(r + 100) + 15r = 7500 \qquad \text{Distribute.}$$
$$15r + 1500 + 15r = 7500 \qquad \text{Distribute.}$$
$$30r + 1500 = 7500 \qquad \text{Combine like terms.}$$
$$30r = 6000 \qquad \text{Subtract 1500.}$$
$$\frac{30r}{30} = \frac{6000}{30} \qquad \text{Divide each side by 30.}$$
$$x = 200 \qquad \text{Simplify.}$$

**Step 5:**   **Check** the answer and **interpret** the solution as it relates to the problem.

The speed of the eastbound plane is 200 mph, and the speed of the westbound plane is 200 + 100 = 300 mph.

**Check** to see that 1.5(200) + 1.5(300) = 300 + 450 = 750 miles.  ∎

**You Try 9**

Write an equation and solve.
   Two drivers leave Albany, Oregon, on Interstate 5. Dhaval heads south traveling 4 mph faster than Pradeep, who is driving north. After $\frac{1}{2}$ hr, they are 62 miles apart. How fast is each man driving?

**Example 11**

Write an equation and solve.
   Alex and Jenny are taking a cross-country road trip on their motorcycles. Jenny leaves a rest area first traveling at 60 mph. Alex leaves 30 minutes later, traveling on the same highway, at 70 mph. How long will it take Alex to catch Jenny?

**Solution**

**Step 1:**   **Read** the problem carefully, and identify what we are being asked to find.

We must determine how long it takes Alex to catch Jenny.

We will use a picture to help us see what is happening in this problem.

Since both girls leave the same rest area and travel on the same highway, when Alex catches Jenny they have driven the *same* distance.

**Step 2:**   **Choose a variable** to represent an unknown, and define the other unknown in terms of this variable.

Alex's time is in terms of Jenny's time, so let

$t$ = the number of hours Jenny has been riding when Alex catches her

Alex leaves 30 minutes ($\frac{1}{2}$ hour) after Jenny, so Alex travels $\frac{1}{2}$ hour *less than* Jenny.

$$t - \frac{1}{2} = \text{the number of hours it takes Alex to catch Jenny}$$

Label the picture.

**Step 3:**    **Translate** the information that appears in English into an algebraic equation.

Let's make a table using the equation $d = rt$. Fill in the time and the rates first; then multiply those together to fill in the values for the distance.

|  | $d$ | $r$ | $t$ |
|---|---|---|---|
| Jenny | $60t$ | $60$ | $t$ |
| Alex | $70(t - \frac{1}{2})$ | $70$ | $t - \frac{1}{2}$ |

We will write an equation in English to help us write an algebraic equation. The picture shows that

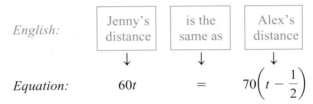

*English:*    Jenny's distance    is the same as    Alex's distance

*Equation:*    $60t$    $=$    $70\left(t - \dfrac{1}{2}\right)$

The expressions for the distances come from the table.

The equation is $60t = 70\left(t - \dfrac{1}{2}\right)$.

**Step 4:**    **Solve** the equation.

$$60t = 70\left(t - \frac{1}{2}\right)$$

| | |
|---|---|
| $60t = 70t - 35$ | Distribute. |
| $-10t = -35$ | Subtract $70t$. |
| $\dfrac{-10t}{-10} = \dfrac{-35}{-10}$ | Divide each side by $-10$. |
| $t = 3.5$ | Simplify. |

**Step 5:**    **Check** the answer and **interpret** the solution as it relates to the problem.

Remember, Jenny's time is $t$. Alex's time is $t - \dfrac{1}{2} = 3\dfrac{1}{2} - \dfrac{1}{2} = 3$ hr.

It took Alex 3 hr to catch Jenny.

**Check** to see that Jenny travels 60 mph $\cdot$ (3.5 hr) = 210 miles, and Alex travels 70 mph $\cdot$ (3 hr) = 210 miles. The girls traveled the same distance.    ■

**You try 10**

Write an equation and solve.
    Brad leaves home driving 40 mph. Angelina leaves the house 30 minutes later driving the same route at 50 mph. How long will it take Angelina to catch Brad?

**Answers to You Try Exercises**

1) $\dfrac{3}{2}$     2)   128-oz bottle; $0.043/oz     3)   a) false   b) true     4)   a) {18}  b) {18}

5)   $36.00     6)   15     7)   a) 80¢; $0.80   b) 67¢; $0.67   c) 147¢; $1.47   d) 10d cents; 0.10d dollars

e)   p cents; 0.01p dollars   f)   10d + p cents; 0.10d + 0.01p dollars     8)   30 quarters, 90 pennies

9)   Dhaval: 64 mph; Pradeep: 60 mph     10)   2 hr

# 3.7 Exercises

## Objective 1: Use Ratios

1)  Write three ratios that are equivalent to $\dfrac{3}{4}$.

2)  Is 0.65 equivalent to the ratio 13 to 20? Explain.

3)  Is a percent a type of ratio? Explain.

4)  Write 57% as a ratio.

Write as a ratio in lowest terms.

5)  16 girls to 12 boys

6)  9 managers to 90 employees

7)  4 coaches to 50 team members

8)  30 blue marbles to 18 red marbles

9)  20 feet to 80 feet          10)  7 minutes to 4 minutes

11)  2 feet to 36 inches        12)  30 minutes to 3 hours

13)  18 hours to 2 days         14)  20 inches to 3 yards

A store sells the same product in different sizes. Determine which size is the best buy based on the unit price of each item.

15)  Batteries                  16)  Cat litter

| Number | Price   |
|--------|---------|
| 8      | $ 6.29  |
| 16     | $12.99  |

| Size  | Price   |
|-------|---------|
| 30 lb | $ 8.48  |
| 50 lb | $12.98  |

17)  Mayonnaise                 18)  Applesauce

| Size  | Price  |
|-------|--------|
| 8 oz  | $2.69  |
| 15 oz | $3.59  |
| 48 oz | $8.49  |

| Size  | Price  |
|-------|--------|
| 16 oz | $1.69  |
| 24 oz | $2.29  |
| 48 oz | $3.39  |

19)  Cereal                     20)  Shampoo

| Size  | Price  |
|-------|--------|
| 11 oz | $4.49  |
| 16 oz | $5.15  |
| 24 oz | $6.29  |

| Size  | Price  |
|-------|--------|
| 14 oz | $3.19  |
| 25 oz | $5.29  |
| 32 oz | $6.99  |

## Objective 2: Solve a Proportion

21)  What is the difference between a ratio and a proportion?

22)  In the proportion $\dfrac{a}{b} = \dfrac{c}{d}$, can $b = 0$? Explain.

Determine whether each proportion is true or false.

23)  $\dfrac{4}{7} = \dfrac{20}{35}$     24)  $\dfrac{54}{64} = \dfrac{7}{8}$

25)  $\dfrac{72}{54} = \dfrac{8}{7}$     26)  $\dfrac{120}{140} = \dfrac{30}{35}$

27)  $\dfrac{8}{10} = \dfrac{2}{\frac{5}{2}}$     28)  $\dfrac{3}{4} = \dfrac{\frac{1}{2}}{\frac{2}{3}}$

Solve each proportion.

29)  $\dfrac{8}{36} = \dfrac{c}{9}$     30)  $\dfrac{n}{3} = \dfrac{20}{15}$

31)  $\dfrac{w}{15} = \dfrac{32}{12}$     32)  $\dfrac{8}{14} = \dfrac{d}{21}$

33)  $\dfrac{40}{24} = \dfrac{30}{a}$     34)  $\dfrac{10}{x} = \dfrac{12}{54}$

35)  $\dfrac{2}{k} = \dfrac{9}{12}$     36)  $\dfrac{15}{27} = \dfrac{m}{6}$

37)  $\dfrac{3z + 10}{14} = \dfrac{2}{7}$     38)  $\dfrac{8t - 9}{20} = \dfrac{3}{4}$

39) $\dfrac{r + 7}{9} = \dfrac{r - 5}{3}$    40) $\dfrac{b + 6}{5} = \dfrac{b + 10}{15}$

41) $\dfrac{3h + 15}{16} = \dfrac{2h + 5}{4}$    42) $\dfrac{a + 7}{8} = \dfrac{4a - 11}{6}$

43) $\dfrac{4m - 1}{6} = \dfrac{6m}{10}$    44) $\dfrac{9w + 8}{10} = \dfrac{5 - 3w}{12}$

Set up a proportion and solve.

45) If 4 containers of yogurt cost $2.36, find the cost of 6 containers of yogurt.

46) Find the cost of 3 scarves if 2 scarves cost $29.00.

47) A marinade for chicken uses 2 parts of lime juice for every 3 parts of orange juice. If the marinade uses $\dfrac{1}{3}$ cup of lime juice, how much orange juice should be used?

48) The ratio of salt to baking soda in a cookie recipe is 0.75 to 1. If a recipe calls for $1\dfrac{1}{2}$ teaspoons of salt, how much baking soda is in the cookie dough?

49) A 12-oz serving of Mountain Dew contains 55 mg of caffeine. How much caffeine is in an 18-oz serving of Mountain Dew?
(www.energyfiend.com)

50) An 8-oz serving of Red Bull energy drink contains about 80 mg of caffeine. Approximately how much caffeine is in 12 oz of Red Bull?
(www.energyfiend.com)

51) Approximately 9 out of 10 smokers began smoking before the age of 21. In a group of 400 smokers, about how many of them started before they reached their 21st birthday?
(www.lungusa.org)

52) Ridgemont High School administrators estimate that 2 out of 3 members of its student body attended the homecoming football game. If there are 1941 students in the school, how many went to the game?

53) At the end of a week, Ernest put 20 lb of yard waste and some kitchen scraps on the compost pile. If the ratio of yard waste to kitchen scraps was 5 to 2, how many pounds of kitchen scraps did he put on the pile?

54) On a map of the United States, 1 inch represents 120 miles. If two cities are 3.5 inches apart on the map, what is the actual distance between the two cities?

55) On July 4, 2009, the exchange rate was such that $20.00 (American) was worth 14.30 Euros. How many Euros could you get for $50.00?
(www.xe.com)

56) On July 4, 2009, the exchange rate was such that 100 British pounds were worth $163.29 (American). How many dollars could you get for 280 British pounds?
(www.xe.com)

Given the following similar triangles, find $x$.

57)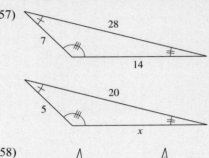

58)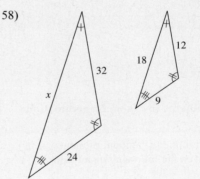

59)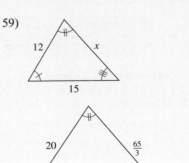

60)

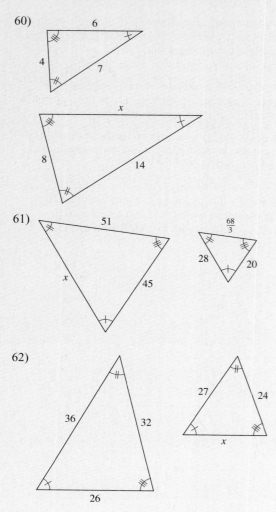

61)

62)

## Objective 3: Solve Problems Involving Denominations of Money

For Exercises 63–68, determine the amount of money a) in dollars and b) in cents given the following quantities.

63) 7 dimes

64) 17 nickels

65) 422 pennies

66) 14 quarters

67) 9 nickels and 7 quarters

68) 73 pennies and 14 dimes

For Exercises 69–74, write an expression that represents the amount of money in a) dollars and b) cents given the following quantities.

69) $q$ quarters

70) $p$ pennies

71) $d$ dimes

72) $n$ nickels

73) $p$ pennies and $n$ nickels

74) $q$ quarters and $d$ dimes

Solve using the five-step method.

75) Turtle and Vince combine their coins to find they have all dimes and quarters. They have 8 more quarters than dimes, and the coins are worth a total of $5.15. How many dimes and quarters do they have?

76) Johnny saves all of his nickels and dimes in a jar. One day he counted them and found that there were 131 coins worth $9.20. How many pennies and how many nickels were in the jar?

77) Eric has been saving his paper route money. He has $73.00 consisting of $5 bills and $1 bills. If he has a total of 29 bills, how many $5 bills and how many $1 bills does he have?

78) A bank employee is servicing the automated teller machine after a busy Friday night. She finds the machine contains only $20 bills and $10 bills and that there are twice as many $20 bills remaining as there are $10 bills. If there is a total of $600.00 left in the machine, how many of the bills are twenties, and how many are tens?

79) The community pool charges $9.00 for adults and $7.00 for children. The total revenue for a particular cloudy day is $437.00. Determine the number of adults and the number of children who went to the pool that day if twice as many children paid for admission as adults.

80) At the convenience store, Sandeep buys 12 more 44¢ stamps than 28¢ stamps. If he spends $11.04 on the stamps, how many of each type did he buy?

81) Carlos attended two concerts with his friends at the American Airlines Arena in Miami. He bought five tickets to see Marc Anthony and two tickets to the Santana concert for $563. If the Santana ticket cost $19.50 less than the Marc Anthony ticket, find the cost of a ticket to each concert. (www.pollstaronline.com)

82) Both of the pop groups Train and Maroon 5 played at the House of Blues in North Myrtle Beach, South Carolina, in 2003. If Train tickets cost $14.50 more than Maroon 5 tickets and four Maroon 5 tickets and four Train tickets would have cost $114, find the cost of a ticket to each concert. (www.pollstaronline.com)

## Objective 4: Solve Problems Involving Distance, Rate, and Time

83) If you use the formula $d = rt$ to find the distance traveled by a car when its rate is given in miles per hour and its time traveled is given in hours, what would be the units of its distance?

84) If you use the formula $d = rt$ to find the distance traveled by a car when its rate is given in miles per hour and its time traveled is given in minutes, what must you do before you substitute the rate and time into the formula?

Solve using the five-step method.

85) Two planes leave San Francisco, one flying north and the other flying south. The southbound plane travels 50 mph faster than the northbound plane, and after 2 hours they are 900 miles apart. Find the speed of each plane.

86) Two cars leave Indianapolis, one driving east and the other driving west. The eastbound car travels 8 mph slower than the westbound car, and after 3 hours they are 414 miles apart. Find the speed of each car.

87) When Lance and Danica pass each other on their bikes going in opposite directions, Lance is riding at 22 mph, and Danica is pedaling at 18 mph. If they continue at those speeds, after how long will they be 200 miles apart?

88) A car and a truck leave the same location, the car headed east and the truck headed west. The truck's speed is 10 mph less than the speed of the car. After 3 hours, the car and truck are 330 miles apart. Find the speed of each vehicle.

89) Ahmad and Davood leave the same location traveling the same route, but Davood leaves 20 minutes after Ahmad. If Ahmad drives 30 mph and Davood drives 36 mph, how long will it take Davood to catch Ahmad?

90) Nayeli and Elena leave the gym to go to work traveling the same route, but Nayeli leaves 10 minutes after Elena. If Elena drives 60 mph and Nayeli drives 72 mph, how long will it take Nayeli to catch Elena?

91) A truck and a car leave the same intersection traveling in the same direction. The truck is traveling at 35 mph, and the car is traveling at 45 mph. In how many minutes will they be 6 miles apart?

92) Greg is traveling north on a road while Peter is traveling south on the same road. They pass by each other at noon, Greg driving 30 mph and Peter driving 40 mph. At what time will they be 105 miles apart?

93) Nick and Scott leave opposite ends of a bike trail 13 miles apart and travel toward each other. Scott is traveling 2 mph slower than Nick. Find each of their speeds if they meet after 30 minutes.

94) At 3:00 P.M., a truck and a car leave the same intersection traveling in the same direction. The truck is traveling at 30 mph, and the car is traveling at 42 mph. At what time will they be 9 miles apart?

95) A passenger train and a freight train leave cities 400 miles apart and travel toward each other. The passenger train is traveling 20 mph faster than the freight train. Find the speed of each train if they pass each other after 5 hours.

96) A freight train passes the Old Towne train station at 11:00 A.M. going 30 mph. Ten minutes later, a passenger train, headed in the same direction on an adjacent track, passes the same station at 45 mph. At what time will the passenger train catch the freight train?

## Mixed Exercises: Objectives 2–4

Solve using the five-step method.

97) If the exchange rate between the American dollar and the Japanese yen is such that $4.00 = 442 yen, how many yen could be exchanged for $70.00? (moneycentral.msn.com)

98) A collection of coins contains 73 coins, all nickels and quarters. If the value of the coins is $14.05, determine the number of each type of coin in the collection.

99) Sherri is riding her bike at 10 mph when Bill passes her going in the opposite direction at 14 mph. How long will it take before the distance between them is 6 miles?

100) The ratio of sugar to flour in a brownie recipe is 1 to 2. If the recipe used 3 cups of flour, how much sugar is used?

101) At the end of her shift, a cashier has a total of $6.70 in dimes and quarters. There are 11 more dimes than quarters. How many of each of these coins does she have?

102) Paloma leaves Mateo's house traveling 30 mph. Mateo leaves 15 minutes later, trying to catch up to Paloma, going 40 mph. If they drive along the same route, how long will it take Mateo to catch Paloma?

103) A jet flying at an altitude of 30,000 ft passes over a small plane flying at 15,000 ft headed in the same direction. The jet is flying twice as fast as the small plane, and 45 minutes later they are 150 miles apart. Find the speed of each plane.

104) Tickets for a high school play cost $3.00 each for children and $5.00 each for adults. The revenue from one performance was $663, and 145 tickets were sold. How many adult tickets and how many children's tickets were sold?

105) A survey of teenage girls found that 3 out of 5 of them earned money by babysitting. If 400 girls were surveyed, how many of them were babysitters?

106) A car and a tour bus leave the same location and travel in opposite directions. The car's speed is 12 mph more than the speed of the bus. If they are 270 miles apart after $2\frac{1}{2}$ hours, how fast is each vehicle traveling?

## Section 3.8 Solving Linear Inequalities in One Variable

### Objectives

1. **Use Graphs and Set and Interval Notations**
2. **Solve Inequalities Using the Addition and Subtraction Properties of Inequality**
3. **Solve Inequalities Using the Multiplication Property of Inequality**
4. **Solve Inequalities Using a Combination of the Properties**
5. **Solve Three-Part Inequalities**
6. **Solve Applications Involving Linear Inequalities**

Recall the inequality symbols

$<$ "is less than"         $\leq$ "is less than or equal to"

$>$ "is greater than"      $\geq$ "is greater than or equal to"

We will use the symbols to form *linear inequalities in one variable.*
Some examples of linear inequalities in one variable are $2x + 11 \leq 19$ and $y > -4$.

---

**Definition**

A **linear inequality in one variable** can be written in the form $ax + b < c$, $ax + b \leq c$, $ax + b > c$, or $ax + b \geq c$, where $a, b,$ and $c$ are real numbers and $a \neq 0$.

---

The solution to a linear inequality is a set of numbers that can be represented in one of three ways:

1) On a graph

2) In *set notation*

3) In *interval notation*

In this section, we will learn how to solve linear inequalities in one variable and how to represent the solution in each of those three ways.

### Graphing an Inequality and Using the Notations

### 1. Use Graphs and Set and Interval Notations

**Example 1**

Graph each inequality and express the solution in set notation and interval notation.

a)  $x \leq 2$          b)  $k > -3$

**Solution**

a)  $x \leq 2$

When we graph $x \leq 2$, we are finding the solution set of $x \leq 2$. What value(s) of $x$ will make the inequality true? The largest solution is 2. Also, any number *less than* 2 will make $x \leq 2$ true. We represent this **on the number line** as follows:

The graph illustrates that the solution is the set of all numbers less than and including 2.
Notice that the dot on 2 is shaded. This tells us that 2 is included in the solution set. The shading to the left of 2 indicates that *any* real number (not just integers) in this region is a solution.
We can write the solution set in **set notation** this way: $\{x \mid x \leq 2\}$. This means

In **interval notation** we write

$$(-\infty, 2]$$

↗           ↖

$-\infty$ is not a
number. $x$
gets infinitely
more negative
without bound.
Use a "(" instead
of a bracket.

The bracket
indicates the
2 is included
in the
interval.

**Note**

The variable does not appear anywhere in interval notation.

b)   $k > -3$

We will plot $-3$ as an *open circle* on the number line because the symbol is ">" and *not* "$\geq$." The inequality $k > -3$ means that we must find the set of all numbers, $k$, *greater than* (but *not* equal to) $-3$. Shade to the right of $-3$.

$$\xleftarrow{\hspace{1em}} \overset{\diamondsuit}{\underset{-4\,-3\,-2\,-1\ \ 0\ \ 1\ \ 2\ \ 3\ \ 4\ \ 5\ \ 6}{\rule{0pt}{0pt}}} \xrightarrow{\hspace{1em}}$$

The graph illustrates that the solution is the set of all numbers greater than $-3$ but not including $-3$.

We can write the solution set in *set notation* this way: $\{k|\ k > -3\}$

In *interval notation* we write

$$(-3, \infty\,)$$

↗           ↖

The "(" indicates
that $-3$ is the
lower bound of
the interval but
that it is not
included.

$\infty$ is not a number.
$k$ gets increasingly
bigger without bound.
Use a "(" instead of
a bracket.

Hints for using interval notation:

1)   The variable never appears in interval notation.
2)   A number *included* in the solution set gets a bracket: $x \leq -2 \rightarrow (-\infty, -2]$
3)   A number *not included* in the solution set gets a parenthesis: $k > -3 \rightarrow (-3, \infty)$
4)   The symbols $-\infty$ and $\infty$ *always* get parentheses.
5)   The smaller number is always placed to the left. The larger number is placed to the right.
6)   Even if we are not asked to graph the solution set, the graph may be helpful in writing the interval notation correctly.   ∎

**You Try 1**

Graph each inequality and express the solution in interval notation.

a)   $z \geq -1$          b)   $n < 4$

## 2. Solve Inequalities Using the Addition and Subtraction Properties of Inequality

The addition and subtraction properties of equality help us to solve equations. Similar properties hold for inequalities as well.

---

**Property**    Addition and Subtraction Properties of Inequality

Let $a, b,$ and $c$ be real numbers. Then,

1) $a < b$ and $a + c < b + c$ are equivalent

and

2) $a < b$ and $a - c < b - c$ are equivalent.

Adding the same number to both sides of an inequality or subtracting the same number from both sides of an inequality will not change the solution.

---

**Note**

The above properties hold for any of the inequality symbols.

---

**Example 2**

Solve $n - 9 \geq -8$. Graph the solution set and write the answer in interval and set notations.

### Solution

$$n - 9 \geq -8$$
$$n - 9 + 9 \geq -8 + 9 \qquad \text{Add 9 to each side.}$$
$$n \geq 1$$

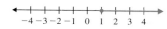

The solution set in interval notation is $[1, \infty)$.
In set notation, we write $\{n \mid n \geq 1\}$.    ■

---

**You Try 2**

Solve $q - 5 \geq -3$. Graph the solution set and write the answer in interval and set notations.

---

## 3. Solve Inequalities Using the Multiplication Property of Inequality

Let's see how multiplication works in inequalities.

Begin with an inequality we know is true and multiply both sides by a *positive* number.

$$2 < 5 \qquad \text{True}$$
$$3(2) < 3(5) \qquad \text{Multiply by 3.}$$
$$6 < 15 \qquad \text{True}$$

Begin again with $2 < 5$ and multiply both sides by a *negative* number.

$$2 < 5 \qquad \text{True}$$
$$-3(2) < -3(5) \qquad \text{Multiply by } -3.$$
$$-6 < -15 \qquad \text{False}$$

To make $-6 < -15$ into a *true* statement, we must *reverse the direction of the inequality symbol.*

$$-6 > -15 \qquad \text{True}$$

If you begin with a true inequality and *divide* by a positive number or by a negative number, the results will be the same as above since division can be defined in terms of multiplication. This leads us to the multiplication property of inequality.

---

**Property**   Multiplication Property of Inequality

Let $a$, $b$, and $c$ be real numbers.

1) If $c$ is a *positive* number, then $a < b$ and $ac < bc$ are equivalent inequalities and have the same solutions.

2) If $c$ is a *negative* number, then $a < b$ and $ac > bc$ are equivalent inequalities and have the same solutions.

---

It is also true that if $c > 0$ and $a < b$, then $\dfrac{a}{c} < \dfrac{b}{c}$. If $c < 0$ and $a < b$, then $\dfrac{a}{c} > \dfrac{b}{c}$.

For the most part, the procedures used to solve linear inequalities are the same as those for solving linear equations **except** *when you multiply or divide an inequality by a negative number, you must reverse the direction of the inequality symbol.*

### Example 3

Solve each inequality. Graph the solution set and write the answer in interval and set notations.

a)   $-6t \leq 12$          b)   $6t \leq -12$

**Solution**

a)   $-6t \leq 12$

First, divide each side by $-6$. *Since we are dividing by a negative number, we must remember to reverse the direction of the inequality symbol.*

$$-6t \leq 12$$
$$\frac{-6t}{-6} \geq \frac{12}{-6} \qquad \text{Divide by } -6, \text{ so reverse the inequality symbol.}$$
$$t \geq -2$$

Interval notation: $[-2, \infty)$          Set notation: $\{t \mid t \geq -2\}$

b)   $6t \leq -12$

First, divide by 6. Since we are dividing by a *positive* number, the inequality symbol remains the same.

$$6t \leq -12$$
$$\frac{6t}{6} \leq \frac{-12}{6} \qquad \text{Divide by 6. Do } not \text{ reverse the inequality symbol.}$$
$$t \leq -2$$

Interval notation: $(-\infty, -2]$          Set notation: $\{t \mid t \leq -2\}$   ∎

### You Try 3

Solve $-\dfrac{1}{2}t < 3$. Graph the solution set and write the answer in interval and set notations.

### 4. Solve Inequalities Using a Combination of the Properties

Often it is necessary to combine the properties to solve an inequality.

    **Example 4**

Solve $3(1 - 4a) + 15 < 2(2a + 5)$. Graph the solution set and write the answer in interval and set notations.

**Solution**

$$
\begin{array}{ll}
3(1 - 4a) + 15 < 2(2a + 5) & \\
3 - 12a + 15 < 4a + 10 & \text{Distribute.} \\
18 - 12a < 4a + 10 & \text{Combine like terms.} \\
18 - 12a - 4a < 4a - 4a + 10 & \text{Subtract } 4a \text{ from each side.} \\
18 - 16a < 10 & \\
18 - 18 - 16a < 10 - 18 & \text{Subtract 18 from each side.} \\
-16a < -8 & \\
\dfrac{-16a}{-16} > \dfrac{-8}{-16} & \text{Divide both sides by } -16. \\
& \text{Reverse the inequality symbol.} \\
a > \dfrac{1}{2} & \text{Simplify.}
\end{array}
$$

The solution set in interval notation is $\left(\dfrac{1}{2}, \infty\right)$.

In set notation, we write $\left\{ a \,\middle|\, a > \dfrac{1}{2} \right\}$.

**You Try 4**

Solve $5(b + 2) - 3 > 4(2b - 1)$. Graph the solution set and write the answer in interval and set notations.

### 5. Solve Three-Part Inequalities

A **three-part inequality** states that one number is between two other numbers. Some examples are $5 < 8 < 12$, $-4 \le x \le 1$, and $0 < r + 2 < 5$. They are also called **compound inequalities** because they contain more than one inequality symbol.

The inequality $-4 \le x \le 1$ means that $x$ is *between* $-4$ and $1$, and $-4$ and $1$ are included in the interval.

On a number line, the inequality would be represented as

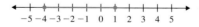

Notice that the **lower bound** of the interval on the number line is $-4$ (including $-4$), and the **upper bound** is $1$ (including $1$). Therefore, we can write the interval notation as

$$[-4, 1]$$

The endpoint, $-4$,          The endpoint, $1$,
is included in the          is included in the
interval, so use            interval, so use
a bracket.                  a bracket.

The set notation to represent $-4 \le x \le 1$ is $\{x \mid -4 \le x \le 1\}$.

Next, we will solve the inequality $0 < r + 2 < 5$. To solve this type of compound inequality, you must remember that *whatever operation you perform on one part of the inequality must be performed on all three parts.* All properties of inequalities apply.

### Example 5

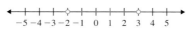

Solve $0 < r + 2 < 5$. Graph the solution set, and write the answer in interval notation.

**Solution**

$$0 < r + 2 < 5$$
$$0 - 2 < r + 2 - 2 < 5 - 2 \qquad \text{To get the } r \text{ by itself, subtract 2}$$
$$-2 < r < 3 \qquad\qquad\quad \text{from each part of the inequality.}$$

The solution set is $(-2, 3)$.  ■

> **Note**
>
> Use parentheses here since $-2$ and 3 are not included in the solution set.

### You Try 5

Solve $-1 \le 5w + 4 \le 14$. Graph the solution set, and write the answer in interval notation.

We can eliminate fractions in an inequality by multiplying by the LCD of all of the fractions.

### Example 6

Solve $-\dfrac{5}{6} < \dfrac{1}{4}p + \dfrac{5}{12} \le \dfrac{4}{3}$. Graph the solution set, and write the answer in interval notation.

**Solution**

The LCD of the fractions is 12. Multiply by 12 to eliminate the fractions.

$$-\frac{5}{6} < \frac{1}{4}p + \frac{5}{12} \le \frac{4}{3}$$
$$12\left(-\frac{5}{6}\right) < 12\left(\frac{1}{4}p + \frac{5}{12}\right) \le 12\left(\frac{4}{3}\right) \qquad \text{Multiply all parts of the inequality by 12.}$$
$$-10 < 3p + 5 \le 16$$
$$-10 - 5 < 3p + 5 - 5 \le 16 - 5 \qquad \text{Subtract 5 from each part.}$$
$$-15 < 3p \le 11 \qquad \text{Combine like terms.}$$
$$-\frac{15}{3} < \frac{3p}{3} \le \frac{11}{3} \qquad \text{Divide each part by 3.}$$
$$-5 < p \le \frac{11}{3} \qquad \text{Simplify.}$$

The solution set is $\left(-5, \dfrac{11}{3}\right]$.  ■

**You Try 6**

Solve $-\dfrac{3}{2} \le \dfrac{1}{4}x - \dfrac{3}{2} < \dfrac{1}{8}$. Graph the solution set, and write the answer in interval notation.

Remember, if we multiply or divide an inequality by a negative number, we reverse the direction of the inequality symbol. When solving a compound inequality like these, reverse *both* symbols.

**Example 7**

Solve $3 < -2m + 7 < 13$. Graph the solution set, and write the answer in interval notation.

**Solution**

$$3 < -2m + 7 < 13$$

$$3 - 7 < -2m + 7 - 7 < 13 - 7 \qquad \text{Subtract 7 from each part.}$$

$$-4 < -2m < 6$$

$$\frac{-4}{-2} > \frac{-2m}{-2} > \frac{6}{-2} \qquad \begin{array}{l}\text{Divide by } -2 \text{ and reverse the direction} \\ \text{of the inequality symbols.}\end{array}$$

$$2 > m > -3 \qquad \text{Simplify.}$$

Think carefully about what $2 > m > -3$ means. It means "$m$ is less than 2 *and* m is greater than $-3$." This is especially important to understand when writing the correct interval notation.

The graph of the solution set is

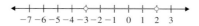

Even though our result is $2 > m > -3$, $-3$ is actually the lower bound of the solution set and 2 is the upper bound. The inequality $2 > m > -3$ can also be written as $-3 < m < 2$.

Interval notation: $(-3, 2)$.

Lower bound on the left    Upper bound on the right

**You Try 7**

Solve $-1 < -4p + 11 < 15$. Graph the solution set, and write the answer in interval notation.

## 6. Solve Applications Involving Linear Inequalities

Certain phrases in applied problems indicate the use of inequality symbols:

|  |  |
|---|---|
| at least: $\ge$ | no less than: $\ge$ |
| at most: $\le$ | no more than: $\le$ |

There are others. Next, we will look at an example of a problem involving the use of an inequality symbol. We will use the same steps that were used to solve applications involving equations.

**Example 8**

Keisha is planning a baby shower for her sister. The restaurant charges $450 for the first 25 people plus $15 for each additional guest. If Keisha can spend at most $700, find the greatest number of people who can attend the shower.

**Solution**

**Step 1:**   **Read** the problem carefully. We must find the greatest number of people who can attend the shower.

**Step 2:**   **Choose a variable** to represent the unknown quantity. We know that the first 25 people will cost $450, but we do not know how many *additional* guests Keisha can afford to invite.

$$x = \text{number of people } \textbf{over} \text{ the first 25 who attend the shower}$$

**Step 3:**   **Translate** from English to an algebraic inequality.

| *English:* | Cost of first 25 people | $+$ | Cost of additional guests | is at most | $700 |
|---|---|---|---|---|---|
| *Inequality:* | 450 | $+$ | $15x$ | $\leq$ | 700 |

The inequality is $450 + 15x \leq 700$.

**Step 4:**   **Solve** the inequality.

$$450 + 15x \leq 700$$
$$15x \leq 250 \qquad \text{Subtract 450.}$$
$$x \leq 16.\overline{6} \qquad \text{Divide by 15.}$$

**Step 5:**   **Check** the answer and **interpret** the solution as it relates to the problem.

The result was $x \leq 16.\overline{6}$, where $x$ represents the number of additional people who can attend the baby shower. Since it is not possible to have $16.\overline{6}$ people, and $x \leq 16.\overline{6}$, in order to stay within budget, Keisha can afford to pay for at most 16 additional guests *over* the initial 25.

Therefore, the greatest number of people who can attend the shower is

$$\begin{array}{ccc} \text{The first 25} & + & \text{Additional} & = & \text{Total} \\ \downarrow & & \downarrow & & \\ 25 & + & 16 & = & 41 \end{array}$$

At most, 41 people can attend the baby shower.

Does the answer make sense?

$$\text{Total cost of shower} = \$450 + \$15(16) = \$450 + \$240 = \$690$$

We can see that one more guest (at a cost of $15) would put Keisha over budget.

**You Try 8**

Tristan's basic mobile phone plan gives him 500 minutes of calling per month for $40.00. Each additional minute costs $0.25. If he can spend at most $55.00 per month on his phone bill, find the greatest number of minutes Tristan can talk each month.

## Answers to You Try Exercises

1)  a) $[-1, \infty)$

b) $(-\infty, 4)$

2)  interval: $[2, \infty)$, set: $\{q \mid q \geq 2\}$

3)  interval: $(-6, \infty)$, set: $\{t \mid t > -6\}$

4)  interval: $\left(-\infty, \dfrac{11}{3}\right)$, set: $\left\{b \mid b < \dfrac{11}{3}\right\}$

5)  $[-1, 2]$

6)  $\left[0, \dfrac{13}{2}\right)$

7)  $(-1, 3)$

8)  560

# 3.8 Exercises

**Objective 1: Use Graphs and Set and Interval Notations**

1)  When do you use brackets when writing a solution set in interval notation?

2)  When do you use parentheses when writing a solution set in interval notation?

Write each set of numbers in interval notation.

3)

4)

5)

6)

Graph the inequality. Express the solution in a) set notation and b) interval notation.

7)  $k \leq 2$

8)  $y \geq 3$

9)  $c < \dfrac{5}{2}$

10)  $n > -\dfrac{11}{3}$

11)  $a \geq -4$

12)  $x < -1$

**Mixed Exercises:  Objectives 2 and 3**

13)  When solving an inequality, when do you change the direction of the inequality symbol?

14)  What is the solution set of $-4x \leq 12$?

a) $(-\infty, -3]$    b) $[-3, \infty)$

c) $(-\infty, 3]$    d) $[3, \infty)$

Solve each inequality. Graph the solution set and write the answer in a) set notation and b) interval notation.

15)  $k + 9 \geq 7$

16)  $t - 3 \leq 2$

17)  $c - 10 \leq -6$

18)  $x + 12 \geq 8$

19)  $-3 + d < -4$

20)  $1 + k > 1$

21)  $16 < z + 11$

22)  $-5 > p - 7$

23)  $5m > 15$

24)  $10r > 40$

25)  $12x < -21$

26)  $6y < -22$

27)  $-4b \leq 32$

28)  $-7b \geq 21$

29)  $-24a < -40$

30)  $-12n > -36$

31)  $\dfrac{1}{3}k \geq -5$

32)  $\dfrac{1}{2}w < -3$

33)  $-\dfrac{3}{8}c < -3$

34)  $-\dfrac{7}{2}d \geq 35$

**Objective 4: Solve Inequalities Using a Combination of the Properties**

Solve each inequality. Graph the solution set and write the answer in interval notation.

35) $4p - 11 \le 17$

36) $6y + 5 > -13$

37) $9 - 2w \le 11$

38) $17 - 7x \ge 20$

39) $-\frac{3}{4}m + 10 > 1$

40) $\frac{1}{2}k - 3 < -2$

41) $3c + 10 > 5c + 13$

42) $a + 2 < 2a + 3$

43) $3(n + 1) - 16 \le 2(6 - n)$

44) $-6 - (t + 8) \le 2(11 - 3t) + 4$

VIDEO 45) $\frac{8}{3}(2k + 1) > \frac{1}{6}k + \frac{8}{3}$

46) $\frac{11}{6} + \frac{3}{2}(d - 2) > \frac{2}{3}(d + 5) + \frac{1}{2}d$

47) $0.05x + 0.09(40 - x) > 0.07(40)$

48) $0.02c + 0.1(30) < 0.08(30 + c)$

**Objective 5: Solve Three-Part Inequalities**

Write each set of numbers in interval notation.

49) 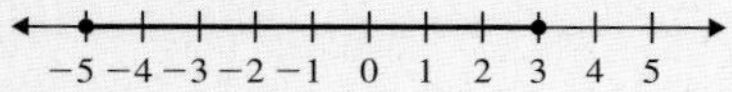


50) 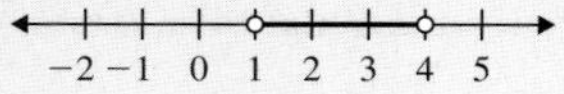


51) 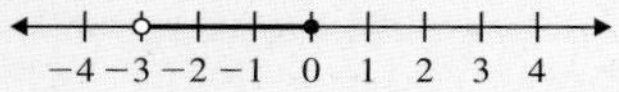


52) 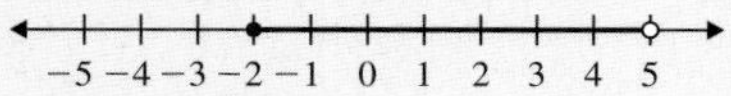


Graph the inequality. Express the solution in a) set notation and b) interval notation.

53) $-4 < y < 0$

54) $1 \le t \le 4$

55) $-3 \le k \le 2$

56) $-2 < p < 1$

VIDEO 57) $\frac{1}{2} < n \le 3$

58) $-2 \le a < 3$

Solve each inequality. Graph the solution set and write the answer in interval notation.

59) $-11 \le b - 8 \le -7$

60) $4 < k + 9 < 10$

61) $-10 < 2a < 7$

62) $-5 \le 5m \le -2$

63) $-5 \le 4x - 13 \le 7$

64) $-4 < 2y - 7 < -1$

65) $-17 < \frac{3}{2}c - 5 < 1$

66) $2 \le \frac{1}{2}n + 3 \le 5$

VIDEO 67) $-6 \le 4c - 13 < -1$

68) $-4 \le 3w - 1 \le 3$

69) $4 \le \frac{k + 11}{4} \le 5$

70) $0 < \frac{5t + 2}{3} < \frac{7}{3}$

71) $-7 \le 8 - 5y < 3$

72) $-9 < 7 - 4m < 9$

73) $2 < 10 - p \le 5$

74) $6 \le 4 - 3b < 10$

**Objective 6: Solve Applications Involving Linear Inequalities**

Write an inequality for each problem and solve.

75) Leslie is planning a party for her daughter at Princess Party Palace. The cost of a party is \$180 for the first 10 children plus \$16.00 for each additional child. If Leslie can spend at most \$300, find the greatest number of children who can attend the party.

76) Big-City Parking Garage charges \$36.00 for the first 4 hours plus \$3.00 for each additional half-hour. Eduardo has \$50.00 for parking. For how long can Eduardo park his car in this garage?

77) Heinrich is planning an Oktoberfest party at the House of Bratwurst. It costs \$150.00 to rent a tent plus \$11.50 per person for food. If Heinrich can spend at most \$450.00, find the greatest number of people he can invite to the party.

78) A marketing company plans to hold a meeting in a large conference room at a hotel. The cost of renting the room is \$500, and the hotel will provide snacks and beverages for an additional \$8.00 per person. If the company has budgeted \$1000.00 for the room and refreshments, find the greatest number of people who can attend the meeting.

79) A taxi in a large city charges \$2.50 plus \$0.40 for every $\frac{1}{5}$ of a mile. How many miles can you go if you have \$14.50?

80) A taxi in a small city charges \$2.00 plus \$0.30 for every $\frac{1}{4}$ of a mile. How many miles can you go if you have \$14.00?

VIDEO 81) Melinda's first two test grades in Psychology were 87 and 94. What does she need to make on the third test to maintain an average of at least 90?

82) Eliana's first three test scores in Algebra were 92, 85, and 96. What does she need to make on the fourth test to maintain an average of at least 90?

| Definition/Procedure | Example |
|---|---|

## 3.1 Solving Linear Equations Part I

**The Addition and Subtraction Properties of Equality**
1) If $a = b$, then $a + c = b + c$.
2) If $a = b$, then $a - c = b - c$. **(p. 114)**

Solve $\quad 3 + b = 20$
$\qquad 3 - 3 + b = 20 - 3 \qquad$ Subtract 3 from each side.
$\qquad\qquad b = 17$

The solution set is $\{17\}$.

---

**The Multiplication and Division Properties of Equality**
1) If $a = b$, then $ac = bc$.
2) If $a = b$, then $\dfrac{a}{c} = \dfrac{b}{c}$ $(c \neq 0)$. **(p. 116)**

Solve $\quad \dfrac{3}{5}m = 6$

$\qquad \dfrac{5}{3} \cdot \dfrac{3}{5}m = \dfrac{5}{3} \cdot 6 \qquad$ Multiply each side by $\dfrac{5}{3}$.

$\qquad\qquad m = 10$

The solution set is $\{10\}$.

## 3.2 Solving Linear Equations Part II

**How to Solve a Linear Equation**
*Step 1:* **Clear parentheses** and **combine like terms** on each side of the equation.
*Step 2:* **Get the variable on one side of the equal sign and the constant on the other side of the equal sign** (isolate the variable) using the addition or subtraction property of equality.
*Step 3:* **Solve for the variable** using the multiplication or division property of equality.
*Step 4:* **Check the solution** in the original equation. **(p. 121)**

Solve $2(c + 2) + 11 = 5c + 9$.
$\quad 2c + 4 + 11 = 5c + 9 \qquad$ Distribute.
$\qquad 2c + 15 = 5c + 9 \qquad$ Combine like terms.
$2c - 5c + 15 = 5c - 5c + 9 \qquad$ Get variable terms on one side.
$\qquad -3c + 15 = 9$
$\qquad\qquad -3c = -6 \qquad$ Get constants on one side.
$\qquad\quad \dfrac{-3c}{-3} = \dfrac{-6}{-3} \qquad$ Division property of equality
$\qquad\qquad c = 2$

The solution set is $\{2\}$.

## 3.3 Solving Linear Equations Part III

**Solve Equations Containing Fractions or Decimals**
To eliminate the fractions, determine the least common denominator (LCD) for all of the fractions in the equation. Then, multiply both sides of the equation by the LCD. **(p. 127)**

To eliminate the decimals from an equation, multiply both sides of the equation by the smallest power of 10 that will eliminate all decimals from the problem. **(p. 127)**

$\dfrac{3}{4}y - 3 = \dfrac{1}{4}y - \dfrac{2}{3} \qquad$ LCD = 12

$12\left(\dfrac{3}{4}y - 3\right) = 12\left(\dfrac{1}{4}y - \dfrac{2}{3}\right) \qquad$ Multiply each side of the equation by 12.

$\qquad 9y - 36 = 3y - 8 \qquad$ Distribute.
$9y - 3y - 36 = 3y - 3y - 8 \qquad$ Get the $y$ terms on one side.
$\qquad 6y - 36 = -8$
$6y - 36 + 36 = -8 + 36 \qquad$ Get the constants on the other side.

$\qquad\qquad 6y = 28$
$\qquad\qquad \dfrac{6y}{6} = \dfrac{28}{6} \qquad$ Divide each side by 6.

$\qquad\qquad y = \dfrac{28}{6} = \dfrac{14}{3} \qquad$ Reduce.

| Definition/Procedure | Example |
|---|---|

**Steps for Solving Applied Problems**
1) **Read** and reread the problem. Draw a picture, if applicable.
2) **Choose a variable** to represent an unknown. Define other unknown quantities in terms of the variable.
3) **Translate** from English to math.
4) **Solve** the equation.
5) **Check** the answer in the original problem, and **interpret** the solution as it relates to the problem. **(p. 129)**

Nine less than twice a number is the same as the number plus thirteen. Find the number.
1) Read the problem carefully, then read it again.
2) Choose a variable to represent the unknown.

$$x = \text{the number}$$

3) "Nine less than twice a number is the same as the number plus thirteen" means $2x - 9 = x + 13$.
4) Solve the equation.

$$2x - 9 = x + 13$$
$$2x - 9 + 9 = x + 13 + 9$$
$$2x = x + 22$$
$$x = 22$$

The number is 22.

### 3.4 Applications of Linear Equations

The "**Steps for Solving Applied Problems**" can be used to solve problems involving general quantities, lengths, and consecutive integers. **(p. 135)**

The sum of three consecutive even integers is 72. Find the integers.
1) Read the problem carefully, then read it again.
2) Define the unknowns.

$$x = \text{the first even integer}$$
$$x + 2 = \text{the second even integer}$$
$$x + 4 = \text{the third even integer}$$

3) "The sum of three consecutive even integers is 72" means

| First even | + | Second even | + | Third even | = | 72 |
|---|---|---|---|---|---|---|
| $x$ | + | $(x + 2)$ | + | $(x + 4)$ | = | 72 |

*Equation:* $x + (x + 2) + (x + 4) = 72$

4) Solve $x + (x + 2) + (x + 4) = 72$

$$3x + 6 = 72$$
$$3x + 6 - 6 = 72 - 6$$
$$3x = 66$$
$$\frac{3x}{3} = \frac{66}{3}$$
$$x = 22$$

5) Find the values of all the unknowns.

$$x = 22, \quad x + 2 = 24, \quad x + 4 = 26$$

The numbers are 22, 24, and 26.

| Definition/Procedure | Example |
|---|---|

## 3.5 Applications Involving Percentages

The **"Steps for Solving Applied Problems"** can be used to solve applications involving percent change, interest earned on investments, and mixture problems. **(p. 144)**

A Lady Gaga poster is on sale for $7.65 after a 15% discount. Find the original price.
1) Read the problem carefully, then read it again.
2) Choose a variable to represent the unknown.
$x$ = the original price of the poster
3) Write an equation in English, then translate it to math.

| Original price of the poster | $-$ | Amount of the discount | $=$ | Sale price of the poster |
|---|---|---|---|---|
| $x$ | $-$ | $0.15x$ | $=$ | $7.65$ |

*Equation:* $x - 0.15x = 7.65$
4) Solve $x - 0.15x = 7.65$
$$0.85x = 7.65$$
$$\frac{0.85x}{0.85} = \frac{7.65}{0.85}$$
$$x = 9.00$$
5) The original price of the Lady Gaga poster was $9.00.

## 3.6 Geometry Applications and Solving Formulas

Formulas from geometry can be used to solve applications. **(p. 153)**

A rectangular bulletin board has an area of 180 in². It is 12 in. wide. Find its length.

Use $A = lw$.    Formula for the area of a rectangle

$A = 180$ in², $w = 12$ in.    Find $l$.

$$A = lw$$
$$180 = l(12) \qquad \text{Substitute values into } A = lw.$$
$$\frac{180}{12} = \frac{l(12)}{12}$$
$$15 = l$$

The length is 15 inches.

To solve a formula for a specific variable, think about the steps involved in solving a linear equation in one variable. **(p. 159)**

Solve $C = kr - w$ for $k$.
$$C + w = \boxed{k}r - w + w \qquad \text{Add } w \text{ to each side.}$$
$$C + w = \boxed{k}r$$
$$\frac{C + w}{r} = \frac{\boxed{k}r}{r} \qquad \text{Divide each side by } r.$$
$$\frac{C + w}{r} = k$$

| Definition/Procedure | Example |
|---|---|

## 3.7 Applications of Linear Equations to Proportions, Money Problems, and $d = rt$

A **proportion** is a statement that two ratios are equivalent.

We can use the same principles for solving equations involving similar triangles, money, and distance = rate · time. **(p. 168)**

If Geri can watch 4 movies in 3 weeks, how long will it take her to watch 7 movies?

1) Read the problem carefully twice.
2) Choose a variable to represent the unknown.

   $x$ = number of weeks to watch 7 movies

3) Set up a proportion. $\dfrac{4 \text{ movies}}{3 \text{ weeks}} = \dfrac{7 \text{ movies}}{x \text{ weeks}}$

   *Equation:* $\dfrac{4}{3} = \dfrac{7}{x}$

4) Solve $\dfrac{4}{3} = \dfrac{7}{x}$.

   $4x = 3(7)$      Set cross products equal.

   $\dfrac{4x}{4} = \dfrac{21}{4}$

   $x = \dfrac{21}{4} = 5\dfrac{1}{4}$

5) It will take Geri $5\dfrac{1}{4}$ weeks to watch 7 movies.

## 3.8 Solving Linear Inequalities in One Variable

We solve linear inequalities in very much the same way we solve linear equations *except that when we multiply or divide by a negative number, we must reverse the direction of the inequality symbol.*

We can graph the solution set, write the solution in set notation, or write the solution in interval notation. **(p. 184)**

Solve $x - 9 \leq -7$. Graph the solution set and write the answer in both set notation and interval notation.

$$x - 9 \leq -7$$
$$x - 9 + 9 \leq -7 + 9$$
$$x \leq 2$$

$\{x \mid x \leq 2\}$      Set notation

$(-\infty, 2]$      Interval notation

**Sections 3.1–3.3**
**Determine whether the given value is a solution to the equation.**

1) $\frac{3}{2}k - 5 = 1; \quad k = -4$

2) $5 - 2(3p + 1) = 9p - 2; \quad p = \frac{1}{3}$

3) How do you know that an equation has no solution?

4) What can you do to make it easier to solve an equation with fractions?

**Solve each equation.**

5) $h + 14 = -5$

6) $w - 9 = 16$

7) $-7g = 56$

8) $-0.78 = -0.6t$

9) $4 = \frac{c}{9}$

10) $-\frac{10}{3}y = 16$

11) $23 = 4m - 7$

12) $\frac{1}{6}v - 7 = -3$

13) $4c + 9 + 2(c - 12) = 15$

14) $\frac{5}{9}x + \frac{1}{6} = -\frac{3}{2}$

15) $2z + 11 = 8z + 15$

16) $8 - 5(2y - 3) = 14 - 9y$

17) $k + 3(2k - 5) = 4(k - 2) - 7$

18) $10 - 7b = 4 - 5(2b + 9) + 3b$

19) $0.18a + 0.1(20 - a) = 0.14(20)$

20) $16 = -\frac{12}{5}d$

21) $3(r + 4) - r = 2(r + 6)$

22) $\frac{1}{2}(n - 5) - 1 = \frac{2}{3}(n - 6)$

**Write each statement as an equation, and find the number.**

23) Nine less than twice a number is twenty-five.

24) One more than two-thirds of a number is the same as the number decreased by three.

**Section 3.4**
**Solve using the five-step method.**

25) Kendrick received 24 fewer e-mails on Friday than he did on Thursday. If he received a total of 126 e-mails on those two days, how many did he get on each day?

26) The number of Michael Jackson solo albums sold the week after his death was 42.2 times the number sold the previous week. If total of 432,000 albums were sold during those two weeks, how many albums were sold the week after his death? (http://abcnews.go.com)

27) A plumber has a 36-inch pipe that he has to cut into two pieces so that one piece is 8 in. longer than the other. Find the length of each piece.

28) The sum of three consecutive integers is 249. Find the integers.

**Section 3.5**
**Solve using the five-step method.**

29) Today's typical hip implant weighs about 50% less than it did 20 years ago. If an implant weighs about 3 lb today, how much did it weigh 20 years ago?

30) By mid-February of 2009, the number of out-of-state applicants to the University of Colorado had decreased by about 19% compared to the same time the previous year. If the school received about 11,500 out-of-state applications in 2009, how many did it receive in 2008? Round the answer to the nearest hundred. (www.dailycamera.com)

31) Jose had $6000 to invest. He put some of it into a savings account earning 2% simple interest and the rest into an account earning 4% simple interest. If he earned $210 of interest in 1 year, how much did he invest in each account?

32) How many milliliters of a 10% hydrogen peroxide solution and how many milliliters of a 2% hydrogen peroxide solution should be mixed to obtain 500 mL of a 4% hydrogen peroxide solution?

**Section 3.6**
**Substitute the given values into the formula and solve for the remaining variable.**

33) $P = 2l + 2w$; If $P = 32$ when $l = 9$, find $w$.

34) $V = \frac{1}{3}\pi r^2 h$; If $V = 60\pi$ when $r = 6$, find $h$.

**Use a known formula to solve.**

35) The base of a triangle measures 12 in. If the area of the triangle is 42 in², find the height.

36) The Statue of Liberty holds a tablet in her left hand that is inscribed with the date, in Roman numerals, that the Declaration of Independence was signed. The length of this rectangular tablet is 120 in. more than the width, and the perimeter of the tablet is 892 in. What are the dimensions of the tablet? (www.nps.gov)

37) Find the missing angle measures.

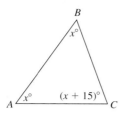

**Find the measure of each indicated angle.**

38)

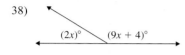

$(2x)°$ $(9x + 4)°$

39)

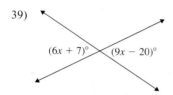

$(6x + 7)°$ $(9x - 20)°$

**Solve using the five-step method.**

40) The sum of the supplement of an angle and twice the angle is 10° more than four times the measure of its complement. Find the measure of the angle.

**Solve for the indicated variable.**

41) $p - n = z$   for $p$          42) $r = ct + a$   for $t$

43) $A = \dfrac{1}{2}bh$   for $b$          44) $M = \dfrac{1}{4}k(d + D)$   for $D$

**Section 3.7**

45) Can 15% be written as a ratio? Explain.

46) What is the difference between a ratio and a proportion?

47) Write the ratio of 12 girls to 15 boys in lowest terms.

48) A store sells olive oil in three different sizes. Which size is the best buy, and what is its unit price?

| Size | Price |
|------|-------|
| 17 oz | $ 8.69 |
| 25 oz | $11.79 |
| 101 oz | $46.99 |

**Solve each proportion.**

49) $\dfrac{x}{15} = \dfrac{8}{10}$          50) $\dfrac{2c + 3}{6} = \dfrac{c - 4}{2}$

**Set up a proportion and solve.**

51) The 2007 Youth Risk Behavior Survey found that about 9 out of 20 high school students drank some amount of alcohol in the 30 days preceding the survey. If a high school has 2500 students, how many would be expected to have used alcohol within a 30-day period? (www.cdc.gov)

52) Given these two similar triangles, find $x$.

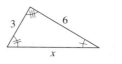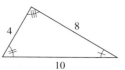

**Solve using the five-step method.**

53) At the end of his shift, Bruno had $340 worth of tips, all in $10 and $20 bills. If he had two more $20 bills than $10 bills, how many of each bill did Bruno have?

54) At Ralph's grocery store, green peppers cost $0.88 each and red peppers cost $0.95 each. Chung-Hee buys twice as many green peppers as red peppers and spends $5.42. How many green peppers and how many red peppers did he buy?

55) Jared and Meg leave opposite ends of a hiking trail 11 miles apart and travel toward each other. Jared is jogging 1 mph slower than Meg. Find each of their speeds if they meet after an hour.

56) Ceyda jogs past the library at 9:00 A.M. going 4 mph. Twenty minutes later, Turgut runs past the library at 6 mph following the same trail. At what time will Turgut catch up to Ceyda?

**Section 3.8**
**Solve each inequality. Graph the solution set, and write the answer in interval notation.**

57) $w + 8 > 5$

58) $-6k \le 15$

59) $5x - 2 \le 18$

60) $3(3c + 8) - 7 > 2(7c + 1) - 5$

61) $-19 \le 7p + 9 \le 2$

62) $-3 < \dfrac{3}{4}a - 6 \le 0$

63) $\dfrac{1}{2} < \dfrac{1 - 4t}{6} < \dfrac{3}{2}$

64) **Write an inequality and solve.** Gia's scores on her first three History tests were 94, 88, and 91. What does she need to make on her fourth test to have an average of at least 90?

**Mixed Exercises: Solving Equations and Applications**
**Solve each equation.**

65) $-8k + 13 = -7$

66) $-7 - 4(3w - 2) = 1 - 9w$

67) $29 = -\dfrac{4}{7}m + 5$

68) $\dfrac{c}{20} = \dfrac{18}{12}$

69) $10p + 11 = 5(2p + 3) - 1$

70) $0.14a + 0.06(36 - a) = 0.12(36)$

71) $\dfrac{2x + 9}{5} = \dfrac{x + 1}{2}$

72) $14 = 8 - h$

73) $\dfrac{5}{6} - \dfrac{3}{4}(r + 2) = \dfrac{1}{2}r + \dfrac{7}{12}$

74) $\dfrac{1}{4}d + \dfrac{9}{4} = 1 + \dfrac{1}{4}(d + 5)$

**Solve using the five-step method.**

75) How many ounces of a 5% alcohol solution must be mixed with 60 oz of a 17% alcohol solution to obtain a 9% alcohol solution?

76) A library offers free tutoring after school for children in grades 1–5. The number of students who attended on Friday was half the number who attended on Thursday. How many students came for tutoring each day if the total number of students served on both days was 42?

77) The sum of two consecutive odd integers is 21 less than three times the larger integer. Find the numbers.

78) Blair and Serena emptied their piggy banks before heading to the new candy store. They have 45 coins in nickels and quarters, for a total of $8.65. How many nickels and quarters did they have?

79) The perimeter of a triangle is 35 cm. One side is 3 cm longer than the shortest side, and the longest side is twice as long as the shortest. How long is each side of the triangle?

80) Yvette and Celeste leave the same location on their bikes and head in opposite directions. Yvette travels at 10 mph, and Celeste travels at 12 mph. How long will it take before they are 33 miles apart?

81) A 2008 poll revealed that 9 out of 25 residents of Quebec, Canada, wanted to secede from the rest of the country. If 1000 people were surveyed, how many said they would like to see Quebec separate from the rest of Canada? (www.bloomberg.com)

82) If a certain environmental bill is passed by Congress, the United States would have to reduce greenhouse gas emissions by 17% from 2005 levels by the year 2020. The University of New Hampshire has been at the forefront of reducing emissions, and if this bill is passed they would be required to have a greenhouse gas emission level of about 56,440 MTCDE (metric tons carbon dioxide equivalents) by 2020. Find their approximate emission level in 2005. (www.sustainableunh.unh.edu, www.knoxnews.com)

**Solve each equation.**

1) $-18y = 14$

2) $16 = 7 - a$

3) $\frac{8}{3}n - 11 = 5$

4) $3c - 2 = 8c + 13$

5) $\frac{1}{2} - \frac{1}{6}(x - 5) = \frac{1}{3}(x + 1) + \frac{2}{3}$

6) $7(3k + 4) = 11k + 8 + 10k - 20$

7) $\frac{9 - w}{4} = \frac{3w + 1}{2}$

8) What is the difference between a ratio and a proportion?

**Solve using the five-step method.**

9) The sum of three consecutive even integers is 114. Find the numbers.

10) How many milliliters of a 20% acid solution should be mixed with 50 mL of an 8% acid solution to obtain an 18% acid solution?

11) Ray buys 14 gallons of gas and pays $40.60. His wife, Debra, goes to the same gas station later that day and buys 11 gallons of the same gasoline. How much did she spend?

12) The tray table on the back of an airplane seat is in the shape of a rectangle. It is 5 in. longer than it is wide and has a perimeter of 50 in. Find the dimensions.

13) Two cars leave the same location at the same time, one traveling east and the other going west. The westbound car is driving 6 mph faster than the eastbound car. After 2.5 hr they are 345 miles apart. What is the speed of each car?

**Solve for the indicated variable.**

14) $B = \frac{an}{4}$ for $a$

15) $S = 2\pi r^2 + 2\pi rh$ for $h$

16) Find the measure of each indicated angle.

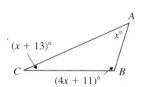

**Solve. Graph the solution set, and write the answer in interval notation.**

17) $6m + 19 \le 7$

18) $1 - 2(3x - 5) < 2x + 5$

19) $-\frac{5}{6} < \frac{4c - 1}{6} \le \frac{3}{2}$

20) Write an inequality and solve.

Anton has grades of 87 and 76 on his first two Biology tests. What does he need on the third test to keep an average of at least 80?

**Perform the operations and simplify.**

1) $\dfrac{3}{8} - \dfrac{5}{6}$

2) $\dfrac{5}{8} \cdot 12$

3) $26 - 14 \div 2 + 5 \cdot 7$

4) $-82 + 15 + 10(1 - 3)$

5) $-39 - |7 - 15|$

6) Find the area of a triangle with a base of length 9 cm and height of 6 cm.

**Given the set of numbers** $\left\{\dfrac{3}{4}, -5, \sqrt{11}, 2.5, 0, 9, 0.\overline{4}\right\}$ **identify**

7) the integers.

8) the rational numbers.

9) the whole numbers.

10) Which property is illustrated by $6(5 + 2) = 6 \cdot 5 + 6 \cdot 2$?

11) Does the commutative property apply to the subtraction of real numbers? Explain.

12) Combine like terms.

$11y^2 - 14y + 6y^2 + y - 5y$

**Simplify. The answer should not contain any negative exponents.**

13) $\dfrac{35r^{16}}{28r^4}$

14) $(-2m^5)^3 (3m^9)^2$

15) $(-12z^{10})\left(\dfrac{3}{8}z^{-16}\right)$

16) $\left(\dfrac{10c^{12}d^2}{5c^9d^{-3}}\right)^{-2}$

17) Write $0.00000895$ in scientific notation.

**Solve.**

18) $8t - 17 = 10t + 6$

19) $\dfrac{3}{2}n + 14 = 20$

20) $3(7w - 5) - w = -7 + 4(5w - 2)$

21) $\dfrac{x + 3}{10} = \dfrac{2x - 1}{4}$

22) $-\dfrac{1}{2}c + \dfrac{1}{5}(2c - 3) = \dfrac{3}{10}(2c + 1) - \dfrac{3}{4}c$

**Solve using the five-step method.**

23) Stu and Phil were racing from Las Vegas back to Napa Valley. Stu can travel 140 miles by train in the time it takes Phil to travel 120 miles by car. What are the speeds of the train and the car if the train is traveling 10 mph faster than the car?

**Solve. Write the answer in interval notation.**

24) $7k + 4 \geq 9k + 16$

25) $-17 < 6b - 11 < 4$

# CHAPTER 4

# Linear Equations in Two Variables

## Algebra at Work: Landscape Architecture

We will take a final look at how mathematics is used in land-scape architecture.

A landscape architect uses slope in many different ways. David explains that one important application of slope is in designing driveways after a new house has been built. Towns often have building codes that restrict the slope or steepness of a driveway. In this case, the rise of the land is the difference in height between the top and the bottom of the driveway. The run is the linear horizontal distance between those two points. By finding $\dfrac{rise}{run}$, a landscape architect knows how to design the driveway so that it meets the town's building code. This is especially important in cold weather climates, where if a driveway is too steep, a car will too easily slide into the street. If it doesn't meet the code, the driveway may have to be removed and rebuilt, or coils that radiate heat might have to be installed under the driveway to melt the snow in the wintertime. Either way, a mistake in calculating slope could cost the landscape company or the client a lot of extra money.

In Chapter 4, we will learn about slope, its meaning, and different ways to use it.

## Section 4.1 Introduction to Linear Equations in Two Variables

### Objectives

1. **Define a Linear Equation in Two Variables**
2. **Decide Whether an Ordered Pair Is a Solution of a Given Equation**
3. **Complete Ordered Pairs for a Given Equation**
4. **Plot Ordered Pairs**
5. **Solve Applied Problems Involving Ordered Pairs**

Graphs are everywhere—online, in newspapers, and in books. The accompanying graph shows how many billions of dollars consumers spent shopping online for consumer electronics during the years 2001–2007.

We can get different types of information from this graph. For example, in the year 2001, consumers spent about $1.5 billion on electronics, while in 2007, they spent about $8.4 billion on electronics. The graph also illustrates a general trend in online shopping: More and more people are buying their consumer electronics online.

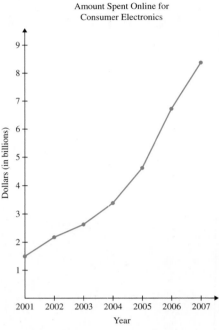

Source: www.census.gov

### 1. Define a Linear Equation in Two Variables

Later in this section, we will see that graphs like this one are based on the *Cartesian coordinate system*, also known as the *rectangular coordinate system*, which gives us a way to graphically represent the relationship between two quantities. We will also learn about different ways to represent relationships between two quantities, like year and online spending, when we learn about *linear equations in two variables*. Let's begin with a definition.

### Definition

A **linear equation in two variables** can be written in the form $Ax + By = C$, where $A$, $B$, and $C$ are real numbers and where both $A$ and $B$ do not equal zero.

Some examples of linear equations in two variables are

$$5x - 2y = 11 \qquad y = \frac{3}{4}x + 1 \qquad -3a + b = 2 \qquad y = x \qquad x = -3$$

(We can write $x = -3$ as $x + 0y = -3$; therefore it is a linear equation in two variables.)

### 2. Decide Whether an Ordered Pair Is a Solution of a Given Equation

A solution to a linear equation in two variables is written as an *ordered pair* so that when the values are substituted for the appropriate variables, we obtain a true statement.

---

**Example 1**

Determine whether each ordered pair is a solution of $5x - 2y = 11$.

a)  $(1, -3)$    b)  $\left(\dfrac{3}{5}, 4\right)$

#### Solution

a)  Solutions to the equation $5x - 2y = 11$ are written in the form $(x, y)$ where $(x, y)$ is called an *ordered pair*. Therefore, the ordered pair $(1, -3)$ means that $x = 1$ and $y = -3$.

$$\underset{\substack{\nearrow \\ x\text{-coordinate}}}{(1,} \underset{\substack{\nwarrow \\ y\text{-coordinate}}}{-3)}$$

To determine whether $(1, -3)$ is a solution of $5x - 2y = 11$, we substitute $1$ for $x$ and $-3$ for $y$. Remember to put these values in parentheses.

$$5x - 2y = 11$$
$$5(1) - 2(-3) = 11 \qquad \text{Substitute } x = 1 \text{ and } y = -3.$$
$$5 + 6 = 11 \qquad \text{Multiply.}$$
$$11 = 11 \qquad \text{True}$$

Since substituting $x = 1$ and $y = -3$ into the equation gives the true statement $11 = 11$, $(1, -3)$ *is a solution* of $5x - 2y = 11$. We say that $(1, -3)$ *satisfies* $5x - 2y = 11$.

b)   The ordered pair $\left(\dfrac{3}{5}, 4\right)$ tells us that $x = \dfrac{3}{5}$ and $y = 4$.

$$5x - 2y = 11$$
$$5\left(\dfrac{3}{5}\right) - 2(4) = 11 \qquad \text{Substitute } \dfrac{3}{5} \text{ for } x \text{ and } 4 \text{ for } y.$$
$$3 - 8 = 11 \qquad \text{Multiply.}$$
$$-5 = 11 \qquad \text{False}$$

Since substituting $\left(\dfrac{3}{5}, 4\right)$ into the equation gives the false statement $-5 = 11$, the ordered pair is *not* a solution to the equation. ∎

### You Try 1

Determine whether each ordered pair is a solution of the equation $y = -\dfrac{3}{4}x + 5$.

a)   $(12, -4)$        b)   $(0, 7)$        c)   $(-8, 11)$

If the variables in the equation are not $x$ and $y$, then the variables in the ordered pairs are written in alphabetical order. For example, solutions to $-3a + b = 2$ are ordered pairs of the form $(a, b)$.

## 3. Complete Ordered Pairs for a Given Equation

Often, we are given the value of one variable in an equation and we can find the value of the other variable that makes the equation true.

### Example 2

Complete the ordered pair $(-3, \quad)$ for $y = 2x + 10$.

**Solution**

To complete the ordered pair $(-3, \quad)$, we must find the value of $y$ from $y = 2x + 10$ when $x = -3$.

$$y = 2x + 10$$
$$y = 2(-3) + 10 \qquad \text{Substitute } -3 \text{ for } x.$$
$$y = -6 + 10$$
$$y = 4$$

When $x = -3$, $y = 4$. The ordered pair is $(-3, 4)$. ∎

### You Try 2

Complete the ordered pair $(5, \quad)$ for $y = 3x - 7$.

If we want to complete more than one ordered pair for a particular equation, we can organize the information in a **table of values.**

## Example 3

Complete the table of values for each equation and write the information as ordered pairs.

a)  $-x + 3y = 8$

| x | y |
|---|---|
| 1 |   |
|   | $-4$ |
|   | $\frac{2}{3}$ |

b)  $y = 2$

| x | y |
|---|---|
| 7 |   |
| $-5$ |   |
| 0 |   |

### Solution

a)  $-x + 3y = 8$

| x | y |
|---|---|
| 1 |   |
|   | $-4$ |
|   | $\frac{2}{3}$ |

The first ordered pair is $(1, \ \ )$, and we must find $y$.

$$-x + 3y = 8$$
$$-(1) + 3y = 8 \qquad \text{Substitute 1 for } x.$$
$$-1 + 3y = 8$$
$$3y = 9 \qquad \text{Add 1 to each side.}$$
$$y = 3 \qquad \text{Divide by 3.}$$

The ordered pair is $(1, 3)$.

The second ordered pair is $( \ \ , -4)$, and we must find $x$.

$$-x + 3y = 8$$
$$-x + 3(-4) = 8 \qquad \text{Substitute } -4 \text{ for } y.$$
$$-x + (-12) = 8 \qquad \text{Multiply.}$$
$$-x = 20 \qquad \text{Add 12 to each side.}$$
$$x = -20 \qquad \text{Divide by } -1.$$

The ordered pair is $(-20, -4)$.

The third ordered pair is $\left( \ \ , \dfrac{2}{3} \right)$, and we must find $x$.

$$-x + 3y = 8$$
$$-x + 3\left(\frac{2}{3}\right) = 8 \qquad \text{Substitute } \frac{2}{3} \text{ for } y.$$
$$-x + 2 = 8 \qquad \text{Multiply.}$$
$$-x = 6 \qquad \text{Subtract 2 from each side.}$$
$$x = -6 \qquad \text{Divide by } -1.$$

The ordered pair is $\left( -6, \dfrac{2}{3} \right)$.

As you complete each ordered pair, fill in the table of values. The completed table will look like this:

| x | y |
|---|---|
| 1 | 3 |
| $-20$ | $-4$ |
| $-6$ | $\frac{2}{3}$ |

b)  $y = 2$

| x | y |
|---|---|
| 7 |   |
| -5 |  |
| 0 |   |

The first ordered pair is (7,   ), and we must find $y$. The equation $y = 2$ means that *no matter the value of x, y always equals 2*. Therefore, when $x = 7$, $y = 2$.

The ordered pair is (7, 2).

Since $y = 2$ for every value of $x$, we can complete the table of values as follows:

The ordered pairs are (7, 2), (-5, 2), and (0, 2).

| x | y |
|---|---|
| 7 | 2 |
| -5 | 2 |
| 0 | 2 |

■

## You Try 3

Complete the table of values for each equation and write the information as ordered pairs.

a) $x - 2y = 9$

| x | y |
|---|---|
| 5 |   |
| 12 |  |
|   | -7 |
|   | $\frac{5}{2}$ |

b) $x = -3$

| x | y |
|---|---|
|   | 1 |
|   | 3 |
|   | -8 |

## 4. Plot Ordered Pairs

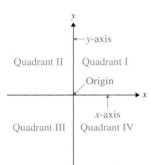

When we completed the table of values for the last two equations, we were finding solutions to each linear equation in two variables.

How can we represent the solutions graphically? We will use the **Cartesian coordinate system**, also known as the **rectangular coordinate system**, to graph the ordered pairs, $(x, y)$.

In the Cartesian coordinate system, we have a horizontal number line, called the $x$-axis, and a vertical number line, called the $y$-axis.

The $x$-axis and $y$-axis in the Cartesian coordinate system determine a flat surface called a **plane.** The axes divide this plane into four **quadrants**, as shown in the figure. The point at which the $x$-axis and $y$-axis intersect is called the **origin.** The arrow at one end of the $x$-axis and one end of the $y$-axis indicates the positive direction on each axis.

Ordered pairs can be represented by **points** in the plane. Therefore, to graph the ordered pair (4, 2) we *plot the point* (4, 2). We will do this in Example 4.

## Example 4

Plot the point (4, 2).

### Solution

Since $x = 4$, we say that the *x-coordinate* of the point is 4. Likewise, the *y-coordinate* is 2.

The *origin* has coordinates (0, 0). The **coordinates** of a point tell us how far from the origin, in the *x*-direction and *y*-direction, the point is located. So, the coordinates of the point (4, 2) tell us that to locate the point we do the following:

(4, 2)

First, from the origin, move 4 units to the right along the *x*-axis.

Then, from the current position, move 2 units up, parallel to the *y*-axis.

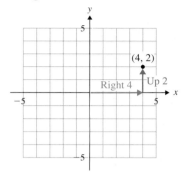

---

## Example 5

Plot the points.

a)  (−2, 5)       b)  (1, −4)       c)  $\left(\dfrac{5}{2}, 3\right)$

d)  (−5, −2)      e)  (0, 1)        f)  (−4, 0)

**Solution**

The points are plotted on the following graph.

a)                 (−2, 5)                           b)                 (1, −4)

First            Then                               First            Then
From the origin,   From the                         From the origin,   From the
move left 2 units  current position,                move right 1 unit  current position,
on the *x*-axis.   move 5 units                     on the *x*-axis.   move 4 units
                   up, parallel to                                     down, parallel
                   the *y*-axis.                                       to the *y*-axis.

c)  $\left(\dfrac{5}{2}, 3\right)$

Think of $\dfrac{5}{2}$ as $2\dfrac{1}{2}$. From the origin, move right $2\dfrac{1}{2}$ units, then up 3 units.

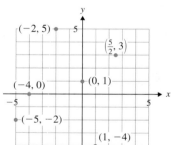

d)  (−5, −2)   From the origin, move left 5 units, then down 2 units.

e)  (0, 1)     The *x*-coordinate of 0 means that we don't move in the *x*-direction (horizontally). From the origin, move up 1 on the *y*-axis.

f)  (−4, 0)    From the origin, move left 4 units. Since the *y*-coordinate is zero, we do not move in the *y*-direction (vertically).

---

## You Try 4

Plot the points.

a)  (3, 1)     b)  (−2, 4)     c)  (0, −5)     d)  (2, 0)     e)  (−4, −3)     f)  $\left(1, \dfrac{7}{2}\right)$

**Note**
The coordinate system should always be labeled to indicate how many units each mark represents.

We can graph sets of ordered pairs for a linear equation in two variables.

**Example 6**

Complete the table of values for $2x - y = 5$, then plot the points.

| x | y |
|---|---|
| 0 | |
| 1 | |
| | 3 |

**Solution**

The first ordered pair is (0,  ), and we must find $y$.

$$2x - y = 5$$
$$2(0) - y = 5 \quad \text{Substitute 0 for } x.$$
$$0 - y = 5$$
$$-y = 5$$
$$y = -5 \quad \text{Divide by } -1.$$

The ordered pair is (0, −5).

The third ordered pair is (  , 3), and we must find $x$.

$$2x - y = 5$$
$$2x - (3) = 5 \quad \text{Substitute 3 for } y.$$
$$2x = 8 \quad \text{Add 3 to each side.}$$
$$x = 4 \quad \text{Divide by 2.}$$

The ordered pair is (4, 3).

The second ordered pair is (1,  ), and we must find $y$.

$$2x - y = 5$$
$$2(1) - y = 5 \quad \text{Substitute 1 for } x.$$
$$2 - y = 5$$
$$-y = 3 \quad \text{Subtract 2 from each side.}$$
$$y = -3 \quad \text{Divide by } -1.$$

The ordered pair is (1, −3).

Each of the points (0, −5), (1, −3), and (4, 3) satisfies the equation $2x - y = 5$.

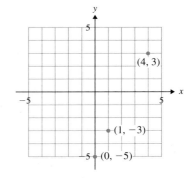

**You Try 5**

Complete the table of values for $3x + y = 1$, then plot the points.

| x | y |
|---|---|
| 0 | |
| −1 | |
| | −5 |

## 5. Solve Applied Problems Involving Ordered Pairs

Next, we will look at an application of ordered pairs.

**Example 7**

The length of an 18-year-old female's hair is measured to be 250 millimeters (mm) (almost 10 in.). The length of her hair after $x$ days can be approximated by

$$y = 0.30x + 250$$

where $y$ is the length of her hair in millimeters.

a)  Find the length of her hair (i) 10 days, (ii) 60 days, and (iii) 90 days after the initial measurement and write the results as ordered pairs.

b)  Graph the ordered pairs.

c)  How long would it take for her hair to reach a length of 274 mm (almost 11 in.)?

**Solution**

a)  The problem states that in the equation $y = 0.30x + 250$,

$x$ = number of days after the hair was measured
$y$ = length of the hair (in millimeters)

| x | y |
|---|---|
| 10 | |
| 60 | |
| 90 | |

We must determine the length of her hair after 10 days, 60 days, and 90 days. We can organize the information in a table of values.

i) $x = 10$:     $y = 0.30x + 250$
   $y = 0.30(10) + 250$     Substitute 10 for $x$.
   $y = 3 + 250$     Multiply.
   $y = 253$

After 10 days, her hair is 253 mm long. We can write this as the ordered pair (10, 253).

ii) $x = 60$:     $y = 0.30x + 250$
   $y = 0.30(60) + 250$     Substitute 60 for $x$.
   $y = 18 + 250$     Multiply.
   $y = 268$

After 60 days, her hair is 268 mm long. We can write this as the ordered pair (60, 268).

iii) $x = 90$:     $y = 0.30x + 250$
   $y = 0.30(90) + 250$     Substitute 90 for $x$.
   $y = 27 + 250$     Multiply.
   $y = 277$

After 90 days, her hair is 277 mm long. We can write this as the ordered pair (90, 277).

We can complete the table of values:

| x | y |
|---|---|
| 10 | 253 |
| 60 | 268 |
| 90 | 277 |

The ordered pairs are (10, 253), (60, 268), and (90, 277).

b) Graph the ordered pairs.

The x-axis represents the number of days after the hair was measured. Since it does not make sense to talk about a negative number of days, we will not continue the x-axis in the negative direction.

The y-axis represents the length of the female's hair. Likewise, a negative number does not make sense in this situation, so we will not continue the y-axis in the negative direction.

The scales on the x-axis and y-axis are different. This is because the size of the numbers they represent are quite different.

Here are the ordered pairs we must graph: (10, 253), (60, 268), and (90, 277).

The x-values are 10, 60, and 90, so we will let each mark in the x-direction represent 10 units.

The y-values are 253, 268, and 277. While the numbers are rather large, they do not actually differ by much. We will begin labeling the y-axis at 250, but each mark in the y-direction will represent 3 units. Because there is a large jump in values from 0 to 250 on the y-axis, we indicate this with "⸜" on the axis between the 0 and 250.

Notice also that we have labeled both axes. The ordered pairs are plotted on the following graph.

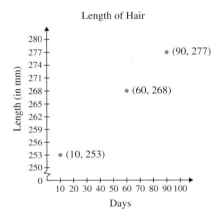

c) We must determine how many days it would take for the hair to grow to a length of 274 mm.

The length, 274 mm, is the y-value. We must find the value of x that corresponds to y = 274 since x represents the number of days.

The equation relating x and y is $y = 0.30x + 250$. We will substitute 274 for y and solve for x.

$$y = 0.30x + 250$$
$$274 = 0.30x + 250$$
$$24 = 0.30x$$
$$80 = x$$

It will take 80 days for her hair to grow to a length of 274 mm.

**Answers to You Try Exercises**

1) a) yes  b) no  c) yes    2) $(5, 8)$    3) a) $(5, -2), (12, \frac{3}{2}), (-5, -7), (14, \frac{5}{2})$
b) $(-3, 1), (-3, 3), (-3, -8)$    4) a) b) c) d) e) f)    5) $(0, 1), (-1, 4), (2, -5)$

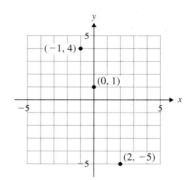

# 4.1 Exercises

**Mixed Exercises: Objectives 1 and 2**

The graph shows the number of gallons of diet soda consumed per person for the years 2002–2006. (U.S. Dept of Agriculture)

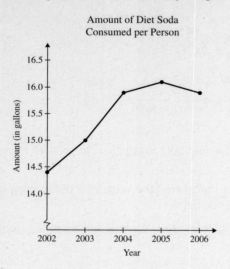

1) How many gallons of diet soda were consumed per person in 2005?

2) During which year was the consumption level about 15.0 gallons per person?

3) During which two years was consumption the same, and how much diet soda was consumed each of these years?

4) During which year did people drink the least amount of diet soda?

5) What was the general consumption trend from 2002 to 2005?

6) Compare the consumption level in 2002 with that in 2006.

The bar graph shows the public high school graduation rate in certain states in 2006. (www.higheredinfo.org)

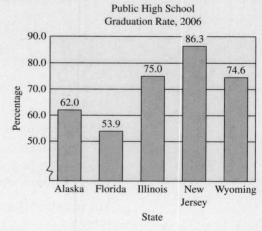

7) Which state had the highest graduation rate, and what percentage of its public high school students graduated?

8) Which states graduated between 70% and 80% of its students?

9) How does the graduation rate of Florida compare with that of New Jersey?

10) Which state had a graduation rate of about 62%?

11) Explain the difference between a linear equation in one variable and a linear equation in two variables. Give an example of each.

12) True or False: $3x + 6y^2 = -1$ is a linear equation in two variables.

Determine whether each ordered pair is a solution of the given equation.

13) $2x + 5y = 1$; $(-2, 1)$    14) $2x + 7y = -4$; $(2, -5)$

15) $-3x - 2y = -15$; $(7, -3)$  16) $y = 5x - 6$; $(3, 9)$

17) $y = -\dfrac{3}{2}x - 7$; $(8, 5)$    18) $5y = \dfrac{2}{3}x + 1$; $(6, 1)$

19) $y = -7$; $(9, -7)$    20) $x = 8$; $(-10, 8)$

### Objective 3: Complete Ordered Pairs for a Given Equation

Complete the ordered pair for each equation.

21) $y = 3x - 7$; $(4, \ )$    22) $y = -2x + 3$; $(6, \ )$

VIDEO 23) $2x - 15y = 13$; $\left( \ , -\dfrac{4}{3} \right)$    24) $-x + 10y = 8$; $\left( \ , \dfrac{2}{5} \right)$

25) $x = 5$; $( \ , -200)$    26) $y = -10$; $(12, \ )$

Complete the table of values for each equation.

27) $y = 2x - 4$

| x | y |
|---|---|
| 0 |   |
| 1 |   |
| -1 |  |
| -2 |  |

28) $y = -5x + 1$

| x | y |
|---|---|
| 0 |   |
| 1 |   |
| 2 |   |
| -1 |  |

29) $y = 4x$

| x | y |
|---|---|
| 0 |   |
| $\dfrac{1}{2}$ |   |
|   | 12 |
|   | -20 |

30) $y = 9x - 8$

| x | y |
|---|---|
| 0 |   |
| $-\dfrac{1}{3}$ |   |
|   | -17 |
|   | 1 |

31) $5x + 4y = -8$

| x | y |
|---|---|
| 0 |   |
|   | 0 |
| 1 |   |
| $-\dfrac{12}{5}$ |   |

32) $2x - y = 12$

| x | y |
|---|---|
|   | 0 |
| 0 |   |
|   | -2 |
| $\dfrac{5}{2}$ |   |

33) $y = -2$

| x | y |
|---|---|
| 0 |   |
| -3 |  |
| 8 |   |
| 17 |  |

34) $x = 20$

| x | y |
|---|---|
|   | 0 |
|   | 3 |
|   | -4 |
|   | -9 |

35) Explain, in words, how to complete the table of values for $x = -13$.

| x | y |
|---|---|
|   | 0 |
|   | 2 |
|   | -1 |

36) Explain, in words, how to complete the ordered pair $( \ , -3)$ for $y = -x - 2$.

### Objective 4: Plot Ordered Pairs

Name each point with an ordered pair, and identify the quadrant in which each point lies.

37)

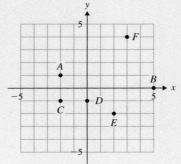

38)

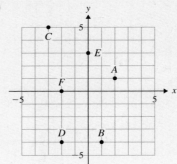

Graph each ordered pair and explain how you plotted the points.

39) $(2, 4)$    40) $(4, 1)$

41) $(-3, -5)$    42) $(-2, 1)$

Graph each ordered pair.

43) $(-6, 1)$    44) $(-2, -3)$

45) $(0, -1)$    46) $(4, -5)$

47) $(0, 4)$    48) $(-5, 0)$

49) $(-2, 0)$    50) $(0, -1)$

51) $\left( -2, \dfrac{3}{2} \right)$    52) $\left( \dfrac{4}{3}, 3 \right)$

53) $\left( 3, -\dfrac{1}{4} \right)$    54) $\left( -2, -\dfrac{9}{4} \right)$

55) $\left( 0, -\dfrac{11}{5} \right)$    56) $\left( -\dfrac{9}{2}, -\dfrac{2}{3} \right)$

## Mixed Exercises: Objectives 3 and 4

Complete the table of values for each equation, and plot the points.

57) $y = -4x + 3$

| x | y |
|---|---|
| 0 |   |
|   | 0 |
| 2 |   |
|   | 7 |

58) $y = -3x + 4$

| x | y |
|---|---|
| 0 |   |
|   | 0 |
|   | -2 |
| -1 |   |

59) $y = x$

| x | y |
|---|---|
| 0 |   |
| -1 |   |
|   | 3 |
|   | -5 |

60) $y = -2x$

| x | y |
|---|---|
|   | 0 |
|   | 4 |
| 2 |   |
| -3 |   |

61) $3x + 4y = 12$

| x | y |
|---|---|
| 0 |   |
|   | 0 |
| 1 |   |
|   | 6 |

62) $2x - 3y = 6$

| x | y |
|---|---|
| 0 |   |
|   | 0 |
|   | 2 |
| 2 |   |

63) $y + 1 = 0$

| x | y |
|---|---|
| 0 |   |
| 1 |   |
| -3 |   |
| -1 |   |

64) $x = 3$

| x | y |
|---|---|
|   | 0 |
|   | -2 |
|   | 3 |
|   | 1 |

65) $y = \dfrac{1}{4}x + 2$

| x | y |
|---|---|
| 0 |   |
| -2 |   |
| 4 |   |
| -1 |   |

66) $y = -\dfrac{5}{2}x + 3$

| x | y |
|---|---|
| 0 |   |
| 4 |   |
| 2 |   |
| 1 |   |

VIDEO 67) For $y = \dfrac{2}{3}x - 7$,

a) find $y$ when $x = 3$, $x = 6$, and $x = -3$. Write the results as ordered pairs.

b) find $y$ when $x = 1$, $x = 5$, and $x = -2$. Write the results as ordered pairs.

c) why is it easier to find the $y$-values in part a) than in part b)?

68) Which ordered pair is a solution to every linear equation of the form $y = mx$, where $m$ is a real number?

Fill in the blank with *positive, negative,* or *zero.*

69) The $x$-coordinate of every point in quadrant III is _____.

70) The $y$-coordinate of every point in quadrant I is _____.

71) The $x$-coordinate of every point in quadrant II is _____.

72) The $y$-coordinate of every point in quadrant II is _____.

73) The $x$-coordinate of every point in quadrant I is _____.

74) The $y$-coordinate of every point in quadrant IV is _____.

75) The $x$-coordinate of every point on the $y$-axis is _____.

76) The $y$-coordinate of every point on the $x$-axis is _____.

## Objective 5: Solve Applied Problems Involving Ordered Pairs

77) The graph shows the number of people who visited Las Vegas from 2003 to 2008. (www.lvcva.com)

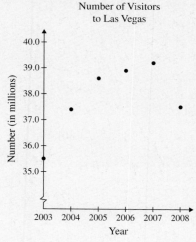

Number of Visitors to Las Vegas

a) If a point on the graph is represented by the ordered pair $(x, y)$, then what do $x$ and $y$ represent?

b) What does the ordered pair (2004, 37.4) represent in the context of this problem?

c) Approximately how many people went to Las Vegas in 2006?

d) In which year were there approximately 38.6 million visitors?

e) Approximately how many more people visited Las Vegas in 2008 than in 2003?

f) Represent the following with an ordered pair: During which year did Las Vegas have the most visitors, and how many visitors were there?

78) The graph shows the average amount of time people spent commuting to work in the Los Angeles metropolitan area from 2003 to 2007. (American Community Survey, U.S. Census)

Average Commute Time in
Los Angeles Area

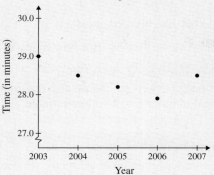

a) If a point on the graph is represented by the ordered pair $(x, y)$, then what do $x$ and $y$ represent?

b) What does the ordered pair (2004, 28.5) represent in the context of this problem?

c) Which year during this time period had the shortest commute? What was the approximate commute time?

d) When was the average commute 28.5 minutes?

e) Write an ordered pair to represent when the average commute time was 28.2 minutes.

79) The percentage of deadly highway crashes involving alcohol is given in the table. (www.bts.gov)

| Year | Percentage |
|------|-----------|
| 1985 | 52.9 |
| 1990 | 50.6 |
| 1995 | 42.4 |
| 2000 | 41.4 |
| 2005 | 40.0 |

a) Write the information as ordered pairs $(x, y)$ where $x$ represents the year and $y$ represents the percentage of accidents involving alcohol.

b) Label a coordinate system, choose an appropriate scale, and graph the ordered pairs.

c) Explain the meaning of the ordered pair (2000, 41.4) in the context of the problem.

80) The average annual salary of a social worker is given in the table. (www.bts.gov)

| Year | Salary |
|------|--------|
| 2005 | $42,720 |
| 2006 | $44,950 |
| 2007 | $47,170 |
| 2008 | $48,180 |

a) Write the information as ordered pairs $(x, y)$ where $x$ represents the year and $y$ represents the average annual salary.

b) Label a coordinate system, choose an appropriate scale, and graph the ordered pairs.

c) Explain the meaning of the ordered pair (2007, 47,170) in the context of the problem.

81) The amount of sales tax paid by consumers in Seattle in 2009 is given by $y = 0.095x$, where $x$ is the price of an item in dollars and $y$ is the amount of tax to be paid.

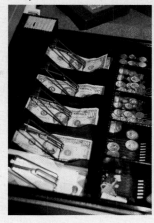

a) Complete the table of values, and write the information as ordered pairs.

b) Label a coordinate system, choose an appropriate scale, and graph the ordered pairs.

| x | y |
|------|---|
| 100.00 | |
| 140.00 | |
| 210.72 | |
| 250.00 | |

c) Explain the meaning of the ordered pair (140.00, 13.30) in the context of the problem.

d) How much tax would a customer pay if the cost of an item was $210.72?

e) Look at the graph. Is there a pattern indicated by the points?

f) If a customer paid $19.00 in sales tax, what was the cost of the item purchased?

82) Kyle is driving from Atlanta to Oklahoma City. His distance from Atlanta, $y$ (in miles), is given by $y = 66x$, where $x$ represents the number of hours driven.

a) Complete the table of values, and write the information as ordered pairs.

| x | y |
|-----|---|
| 1 | |
| 1.5 | |
| 2 | |
| 4.5 | |

b) Label a coordinate system, choose an appropriate scale, and graph the ordered pairs.

c) Explain the meaning of the ordered pair (4.5, 297) in the context of the problem.

d) Look at the graph. Is there a pattern indicated by the points?

e) What does the 66 in $y = 66x$ represent?

f) How many hours of driving time will it take for Kyle to get to Oklahoma City if the distance between Atlanta and Oklahoma City is about 860 miles?

## Section 4.2 Graphing by Plotting Points and Finding Intercepts

### Objectives

1. Graph a Linear Equation by Plotting Points
2. Graph a Linear Equation in Two Variables by Finding the Intercepts
3. Graph a Linear Equation of the Form $Ax + By = 0$
4. Graph Linear Equations of the Forms $x = c$ and $y = d$
5. Model Data with a Linear Equation

In Example 3 of Section 4.1 we found that the ordered pairs $(1, 3)$, $(-20, -4)$, and $\left(-6, \dfrac{2}{3}\right)$ are three solutions to the equation $-x + 3y = 8$. But how many solutions does the equation have? It has an infinite number of solutions. Every linear equation in two variables has an infinite number of solutions because we can choose any real number for one of the variables and we will get another real number for the other variable.

> **Property** Solutions of Linear Equations in Two Variables
>
> Every linear equation in two variables has an infinite number of solutions, and the solutions are ordered pairs.

How can we represent all of the solutions to a linear equation in two variables? We can represent them with a graph, and that graph is a line.

> **Property** The Graph of a Linear Equation in Two Variables
>
> The graph of a linear equation in two variables, $Ax + By = C$, is a straight line. Each point on the line is a solution to the equation.

### 1. Graph a Linear Equation by Plotting Points

**Example 1**

Complete the table of values and graph $4x - y = 5$.

| x | y |
|---|---|
| 2 | 3 |
| 0 | -5 |
| 1 | |
| | 0 |

**Solution**

When $x = 1$, we get

$$4x - y = 5$$
$$4(1) - y = 5 \qquad \text{Substitute 1 for } x.$$
$$4 - y = 5$$
$$-y = 1$$
$$y = -1 \qquad \text{Solve for } y.$$

When $y = 0$, we get

$$4x - y = 5$$
$$4x - (0) = 5 \qquad \text{Substitute 0 for } y.$$
$$4x = 5$$
$$x = \frac{5}{4} \qquad \text{Solve for } x.$$

The completed table of values is

| x | y |
|---|---|
| 2 | 3 |
| 0 | -5 |
| 1 | -1 |
| $\dfrac{5}{4}$ | 0 |

This gives us the ordered pairs $(2, 3)$, $(0, -5)$, $(1, -1)$, and $\left(\dfrac{5}{4}, 0\right)$. Each is a solution to the equation $4x - y = 5$.

Plot the points. They lie on a straight line. We draw the line through these points to get the graph.

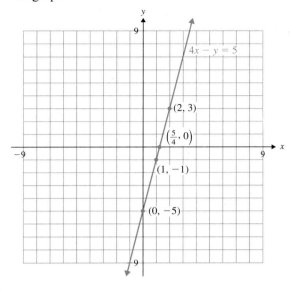

The line represents all solutions to the equation $4x - y = 5$. Every point on the line is a solution to the equation. The arrows on the ends of the line indicate that the line extends indefinitely in each direction. Although it is true that we need to find only two points to graph a line, it is best to plot at least three as a check.

 **You Try 1**

Complete the table of values and graph $x - 2y = 3$.

| x | y |
|---|---|
| 1 | −1 |
| 3 | 0 |
| 0 |   |
|   | 2 |
| 5 |   |

**Example 2**

Graph $-x + 2y = 4$.

**Solution**

We will find three ordered pairs that satisfy the equation. Let's complete a table of values for $x = 0$, $x = 2$, and $x = -4$.

$x = 0$:
$$-x + 2y = 4$$
$$-(0) + 2y = 4$$
$$2y = 4$$
$$y = 2$$

$x = 2$:
$$-x + 2y = 4$$
$$-(2) + 2y = 4$$
$$-2 + 2y = 4$$
$$2y = 6$$
$$y = 3$$

$x = -4$:
$$-x + 2y = 4$$
$$-(-4) + 2y = 4$$
$$4 + 2y = 4$$
$$2y = 0$$
$$y = 0$$

We get the table of values

| x | y |
|---|---|
| 0 | 2 |
| 2 | 3 |
| −4 | 0 |

Plot the points $(0, 2)$, $(-4, 0)$, and $(2, 3)$, and draw the line through them.

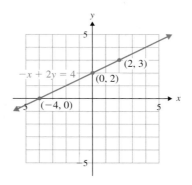

 **You Try 2**

Graph each line.

a)   $3x + 2y = 6$          b)   $y = 4x - 8$

## 2. Graph a Linear Equation in Two Variables by Finding the Intercepts

In Example 2, the line crosses the $x$-axis at $-4$ and crosses the $y$-axis at 2. These points are called **intercepts**.

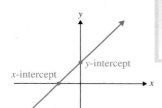

### Definitions
The **x-intercept** of the graph of an equation is the point where the graph intersects the $x$-axis.
The **y-intercept** of the graph of an equation is the point where the graph intersects the $y$-axis.

What is the $y$-coordinate of any point on the $x$-axis? It is zero. Likewise, the $x$-coordinate of any point on the $y$-axis is zero.

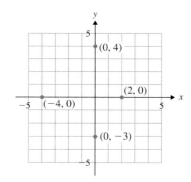

We use these facts to find the intercepts of the graph of an equation.

### Procedure   Finding Intercepts
To find the *x-intercept* of the graph of an equation, let $y = 0$ and solve for $x$.
To find the *y-intercept* of the graph of an equation, let $x = 0$ and solve for $y$.

Finding intercepts is very helpful for graphing linear equations in two variables.

**Example 3**

Graph $y = -\dfrac{1}{3}x + 1$ by finding the intercepts and one other point.

**Solution**

We will begin by finding the intercepts.

*x-intercept*: Let $y = 0$, and solve for $x$.
$$0 = -\frac{1}{3}x + 1$$
$$-1 = -\frac{1}{3}x$$
$$3 = x \qquad \text{Multiply both sides by } -3 \text{ to solve for } x.$$

The *x*-intercept is (3, 0).

*y-intercept*: Let $x = 0$, and solve for $y$.
$$y = -\frac{1}{3}(0) + 1$$
$$y = 0 + 1$$
$$y = 1$$

The *y*-intercept is (0, 1).

We must find another point. Let's look closely at the equation $y = -\dfrac{1}{3}x + 1$. The coefficient of $x$ is $-\dfrac{1}{3}$. If we choose a value for $x$ that is a multiple of 3 (the denominator of the fraction), then $-\dfrac{1}{3}x$ will not be a fraction.

Let $x = -3$.
$$y = -\frac{1}{3}x + 1$$
$$y = -\frac{1}{3}(-3) + 1$$
$$y = 1 + 1$$
$$y = 2$$

The third point is $(-3, 2)$.

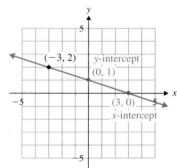

Plot the points, and draw the line through them. See the graph above.  ■

  **You Try 3**

Graph $y = \dfrac{4}{3}x - 2$ by finding the intercepts and one other point.

### 3. Graph a Linear Equation of the Form $Ax + By = 0$

Sometimes the $x$- and $y$-intercepts are the same point.

**Example 4**

Graph $-2x + y = 0$.

**Solution**

If we begin by finding the $x$-intercept, let $y = 0$ and solve for $x$.

$$-2x + y = 0$$
$$-2x + (0) = 0$$
$$-2x = 0$$
$$x = 0$$

The $x$-intercept is $(0, 0)$. But this is the same as the $y$-intercept since we find the $y$-intercept by substituting 0 for $x$ and solving for $y$. Therefore, *the $x$- and $y$-intercepts are the same point.*

Instead of the intercepts giving us two points on the graph of $-2x + y = 0$, we have only one. We will find two other points on the line.

$x = 2$:    $-2x + y = 0$          $x = -2$:    $-2x + y = 0$
              $-2(2) + y = 0$                        $-2(-2) + y = 0$
              $-4 + y = 0$                            $4 + y = 0$
              $y = 4$                                $y = -4$

The ordered pairs $(2, 4)$ and $(-2, -4)$ are also solutions to the equation. Plot all three points on the graph and draw the line through them.

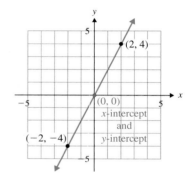

**Property    The Graph of $Ax + By = 0$**

If $A$ and $B$ are nonzero real numbers, then the graph of $Ax + By = 0$ is a line passing through the origin, $(0, 0)$.

**You Try 4**

Graph $x - y = 0$.

### 4. Graph Linear Equations of the Forms $x = c$ and $y = d$

In Section 4.1 we said that an equation like $x = -2$ is a linear equation in two variables since it can be written in the form $x + 0y = -2$. The same is true for $y = 3$. It can be written as $0x + y = 3$. Let's see how we can graph these equations.

**Example 5**

Graph $x = -2$.

**Solution**

The equation $x = -2$ means that *no matter the value of y, x always equals* $-2$. We can make a table of values where we choose any value for $y$, but $x$ is always $-2$.

Plot the points, and draw a line through them. The graph of $x = -2$ is a *vertical line.*

| x | y |
|----|----|
| -2 | 0 |
| -2 | 1 |
| -2 | -2 |

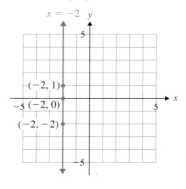

We can generalize the result as follows:

**Property   The Graph of $x = c$**

If $c$ is a constant, then the graph of $x = c$ is a *vertical line* going through the point $(c, 0)$.

 **You Try 5**

Graph $x = 2$.

**Example 6**

Graph $y = 3$.

**Solution**

The equation $y = 3$ means that *no matter the value of x, y always equals* 3. Make a table of values where we choose any value for $x$, but $y$ is always 3.

| x | y |
|----|----|
| 0 | 3 |
| 2 | 3 |
| -2 | 3 |

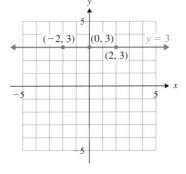

Plot the points, and draw a line through them. The graph of $y = 3$ is a **horizontal line.**

We can generalize the result as follows:

**Property   The Graph of $y = d$**

If $d$ is a constant, then the graph of $y = d$ is a *horizontal line* going through the point $(0, d)$.

 **You Try 6**

Graph $y = -4$.

### 5. Model Data with a Linear Equation

Linear equations are often used to model (or describe mathematically) real-world data. We can use these equations to learn what has happened in the past or predict what will happen in the future.

**Example 7**

The average annual cost of college tuition and fees at private, 4-year institutions can be modeled by

$$y = 907x + 12,803$$

where $x$ is the number of years after 1996 and $y$ is the average tuition and fees, in dollars. (The College Board)

a)   Find the $y$-intercept of the graph of this equation and explain its meaning.

b)   Find the approximate cost of tuition and fees in 2000 and 2006. Write the information as ordered pairs.

c)   Graph $y = 907x + 12,803$.

d)   Use the graph to approximate the average cost of tuition and fees in 2005. Is this the same result as when you use the equation to estimate the average cost?

**Solution**

a)   To find the $y$-intercept, let $x = 0$.

$$y = 907(0) + 12,803$$
$$y = 12,803$$

The $y$-intercept is (0, 12,803). What does this represent?

The problem states that $x$ is the number of years *after* 1996. Therefore, $x = 0$ represents zero years after 1996, which is the year 1996.

The $y$-intercept (0, 12,803) tells us that in 1996 the average cost of tuition and fees at a private 4-year institution was $12,803.

b)   The approximate cost of tuition and fees in

2000:   First, realize that $x \neq 2000$. $x$ is the number of years *after* 1996. Since 2000 is 4 years after 1996, $x = 4$. Let $x = 4$ in $y = 907x + 12,803$ and find $y$.

$$y = 907(4) + 12,803$$
$$y = 3628 + 12,803$$
$$y = 16,431$$

In 2000, the approximate cost of college tuition and fees at these schools was $16,431. We can write this information as the ordered pair (4, 16,431).

2006:   Begin by finding $x$. 2006 is 10 years after 1996, so $x = 10$.

$$y = 907(10) + 12,803$$
$$y = 9070 + 12,803$$
$$y = 21,873$$

In 2006, the approximate cost of college tuition and fees at private 4-year schools was $21,873.

The ordered pair (10, 21,873) can be written from this information.

c)   We will plot the points (0, 12,803), (4, 16,431), and (10, 21,873). Label the axes, and choose an appropriate scale for each.

The $x$-coordinates of the ordered pairs range from 0 to 10, so we will let each mark in the $x$-direction represent 2 units.

The $y$-coordinates of the ordered pairs range from 12,803 to 21,873. We will let each mark in the $y$-direction represent 2000 units.

d)  Using the graph to estimate the cost of tuition and fees in 2005, we locate $x = 9$ on the $x$-axis since 2005 is 9 years after 1996. When $x = 9$, we move straight up the graph to $y \approx 21,000$. Our approximation from the graph is $21,000.
    If we use the equation and let $x = 9$, we get

$$y = 907x + 12,803$$
$$y = 907(9) + 12,803$$
$$y = 8163 + 12,803$$
$$y = 20,966$$

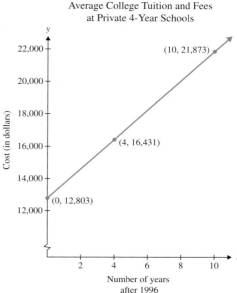

Average College Tuition and Fees at Private 4-Year Schools

From the equation we find that the cost of college tuition and fees at private 4-year schools was about $20,966. The numbers are not exactly the same, but they are close.   ■

## Using Technology

A graphing calculator can be used to graph an equation and to verify information that we find using algebra. We will graph the equation $y = -\frac{1}{2}x + 2$ and then find the intercepts both algebraically and using the calculator.
First, enter the equation into the calculator. Press ZOOM and select 6:Zstandard to graph the equation.

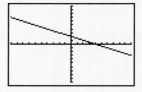

1)  Find, algebraically, the $y$-intercept of the graph of $y = -\frac{1}{2}x + 2$. Is it consistent with the graph of the equation?

2)  Find, algebraically, the $x$-intercept of the graph of $y = -\frac{1}{2}x + 2$. Is it consistent with the graph of the equation?

Now let's verify the intercepts using the graphing calculator. To find the $y$-intercept, press TRACE after displaying the graph. The cursor is automatically placed at the center $x$-value on the screen, which is at the point $(0, 2)$ as shown next on the left. To find the $x$-intercept, press TRACE, type 4, and press ENTER. The calculator displays $(4, 0)$ as shown next on the right. This is consistent with the intercepts found in 1 and 2, using algebra.

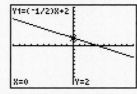   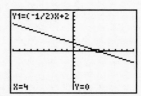

Use algebra to find the $x$- and $y$-intercepts of the graph of each equation. Then, use the graphing calculator to verify your results.

1)  $y = 2x - 4$          2)  $y = x + 3$          3)  $y = -x + 5$
4)  $2x - 5y = 10$        5)  $3x + 4y = 24$       6)  $3x - 7y = 21$

## Answers to You Try Exercises

1)    $\left(0, -\frac{3}{2}\right), (7, 2), (5, 1)$

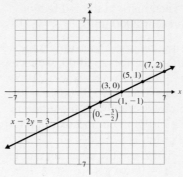

2)    a)

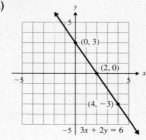

b)

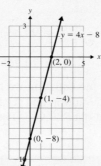

3)

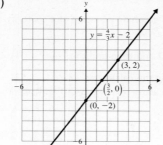

4)

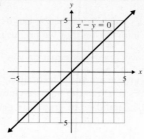

5)

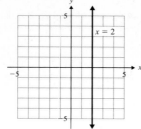

6)

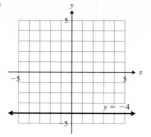

**Answers to Technology Exercises**

1)  $(2, 0), (0, -4)$         2)  $(-3, 0), (0, 3)$         3)  $(5, 0), (0, 5)$
4)  $(5, 0), (0, -2)$         5)  $(8, 0), (0, 6)$         6)  $(7, 0), (0, -3)$

# 4.2 Exercises

### Objective 1: Graph a Linear Equation by Plotting Points

1)  The graph of a linear equation in two variables is a
_____.

2)  Every linear equation in two variables has how many
solutions?

Complete the table of values and graph each equation.

3)  $y = -2x + 4$

| x | y |
|---|---|
| 0 | |
| -1 | |
| 2 | |
| 3 | |

4)  $y = 3x - 2$

| x | y |
|---|---|
| 0 | |
| 1 | |
| 2 | |
| -1 | |

5)  $y = \dfrac{3}{2}x + 7$

| x | y |
|---|---|
| 0 | |
| 2 | |
| -2 | |
| -4 | |

6)  $y = -\dfrac{5}{3}x + 3$

| x | y |
|---|---|
| 0 | |
| -3 | |
| 3 | |
| 6 | |

7)  $2x = 3 - y$

| x | y |
|---|---|
| | 0 |
| 0 | |
| $\frac{1}{2}$ | |
| | 5 |

8)  $-x + 5y = 10$

| x | y |
|---|---|
| 0 | |
| | 0 |
| | 4 |
| -3 | |

9)  $x = -\dfrac{4}{9}$

| x | y |
|---|---|
| | 5 |
| | 0 |
| | -1 |
| | -2 |

10)  $y + 5 = 0$

| x | y |
|---|---|
| 0 | |
| -3 | |
| -1 | |
| 2 | |

### Mixed Exercises: Objectives 1–4

11)  What is the y-intercept of the graph of an equation? How
do you find it?

12)  What is the x-intercept of the graph of an equation? How
do you find it?

Graph each equation by finding the intercepts and at least
one other point.

13)  $y = x - 1$                     14)  $y = -x + 3$

15)  $3x - 4y = 12$                16)  $2x - 7y = 14$

17)  $x = -\dfrac{4}{3}y - 2$         18)  $x = \dfrac{5}{4}y - 5$

19)  $2x - y = 8$                     20)  $3x + y = -6$

21)  $y = -x$                         22)  $y = 3x$

23)  $4x - 3y = 0$                   24)  $6y - 5x = 0$

25)  $x = 5$                           26)  $y = -4$

27)  $y = 0$                           28)  $x = 0$

29)  $x - \dfrac{4}{3} = 0$           30)  $y + 1 = 0$

31)  $4x - y = 9$                     32)  $x + 3y = -5$

33)  Which ordered pair is a solution to every linear equation
of the form $Ax + By = 0$?

34)  True or False: The graph of $Ax + By = 0$ will always
pass through the origin.

## Objective 5: Model Data with a Linear Equation

35) The cost of downloading popular songs from iTunes is given by $y = 1.29x$, where $x$ represents the number of songs downloaded and $y$ represents the cost, in dollars.

   a) Make a table of values using $x = 0, 4, 7,$ and 12, and write the information as ordered pairs.

   b) Explain the meaning of each ordered pair in the context of the problem.

   c) Graph the equation. Use an appropriate scale.

   d) How many songs could you download for $11.61?

36) The force, $y$, measured in newtons (N), required to stretch a particular spring $x$ meters is given by $y = 100x$.

   a) Make a table of values using $x = 0, 0.5, 1.0,$ and 1.5, and write the information as ordered pairs.

   b) Explain the meaning of each ordered pair in the context of the problem.

   c) Graph the equation. Use an appropriate scale.

   d) If the spring was pulled with a force of 80 N, how far did it stretch?

37) The number of doctorate degrees awarded in science and engineering in the United States from 2003 to 2007 can be modeled by $y = 1662x + 24,916$, where $x$ represents the number of years after 2003, and $y$ represents the number of doctorate degrees awarded. The actual data are graphed here. (www.nsf.gov)

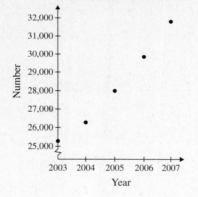

Number of Science and Engineering Doctorates Awarded in the U.S.

a) From the graph, estimate the number of science and engineering doctorates awarded in 2004 and 2007.

b) Determine the number of degrees awarded during the same years using the equation. Are the numbers close?

c) Graph the line that models the data given on the original graph.

d) What is the $y$-intercept of the graph of this equation, and what does it represent? How close is it to the actual point plotted on the given graph?

e) If the trend continues, how many science and engineering doctorates will be awarded in 2012? Use the equation.

38) The amount of money Americans spent on skin and scuba diving equipment from 2004 to 2007 can be modeled by $y = 8.6x + 350.6$, where $x$ represents the number of years after 2004, and $y$ represents the amount spent on equipment in millions of dollars. The actual data are graphed here. (www.census.gov)

Amount Spent on Skin and Scuba Equipment

a) From the graph, estimate the amount spent in 2005 and 2006.

b) Determine the amount of money spent during the same years using the equation. Are the numbers close?

c) Graph the line that models the data given on the original graph.

d) What is the $y$-intercept of the graph of this equation, and what does it represent? How close is it to the actual point plotted on the given graph?

e) If the trend continues, how much will be spent on skin and scuba gear in 2014? Use the equation.

# Section 4.3 The Slope of a Line

**Objectives**

1. Understand the Concept of Slope
2. Find the Slope of a Line Given Two Points on the Line
3. Use Slope to Solve Applied Problems
4. Find the Slope of Horizontal and Vertical Lines
5. Use Slope and One Point on a Line to Graph the Line

## 1. Understand the Concept of Slope

In Section 4.2, we learned to graph lines by plotting points. You may have noticed that some lines are steeper than others. Their "slants" are different, too.

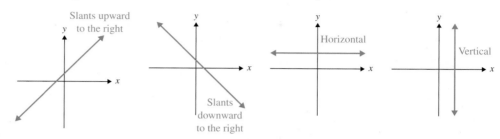

We can describe the steepness of a line with its *slope*.

> **Property   Slope of a Line**
>
> The **slope** of a line measures its steepness. It is the ratio of the vertical change in y to the horizontal change in x. Slope is denoted by m.

We can also think of slope as a rate of change. *Slope* is the rate of change between two points. More specifically, it describes the rate of change in *y* to the change in *x*.

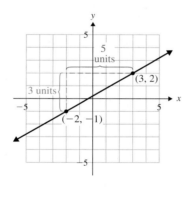

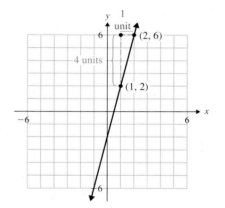

$$\text{Slope} = \frac{3}{5} \begin{array}{l} \leftarrow \text{vertical change} \\ \leftarrow \text{horizontal change} \end{array} \qquad \text{Slope} = 4 \text{ or } \frac{4}{1} \begin{array}{l} \leftarrow \text{vertical change} \\ \leftarrow \text{horizontal change} \end{array}$$

For example, in the graph on the left, the line changes 3 units vertically for every 5 units it changes horizontally. Its slope is $\frac{3}{5}$. The line on the right changes 4 units vertically for every 1 unit of horizontal change. It has a slope of $\frac{4}{1}$ or 4.

Notice that the line with slope 4 is steeper than the line that has a slope of $\frac{3}{5}$.

**Note**

As the magnitude of the slope gets larger, the line gets steeper.

Here is an application of slope.

**Example 1**

A sign along a highway through the Rocky Mountains is shown on the left. What does it mean?

### Solution

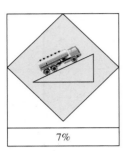

7%

Percent means "out of 100." Therefore, we can write 7% as $\dfrac{7}{100}$. We can interpret $\dfrac{7}{100}$ as the ratio of the vertical change in the road to horizontal change in the road.

$$\text{The slope of the road is } \dfrac{7}{100}. \quad \begin{array}{l} \longleftarrow \text{ Vertical change} \\ \longleftarrow \text{ Horizontal change} \end{array}$$

The highway rises 7 ft for every 100 horizontal feet.    ■

**You Try 1**

The slope of a conveyer belt is $\dfrac{5}{12}$, where the dimensions of the ramp are in inches. What does this mean?

## 2. Find the Slope of a Line Given Two Points on the Line

Here is line $L$. The points $(x_1, y_1)$ and $(x_2, y_2)$ are two points on line $L$. *We will find the ratio of the vertical change in y to the horizontal change in x between the points $(x_1, y_1)$ and $(x_2, y_2)$.*

To get from $(x_1, y_1)$ to $(x_2, y_2)$, we move *vertically* to point $P$ then *horizontally* to $(x_2, y_2)$. The $x$-coordinate of point $P$ is $x_1$, and the $y$-coordinate of $P$ is $y_2$.

When we moved *vertically* from $(x_1, y_1)$ to point $P(x_1, y_2)$, how far did we go? We moved a vertical distance $y_2 - y_1$.

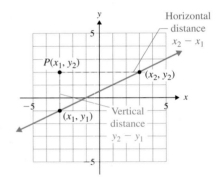

**Note**

The vertical change is $y_2 - y_1$ and is called the **rise**.

Then we moved *horizontally* from point $P(x_1, y_2)$ to $(x_2, y_2)$. How far did we go? We moved a horizontal distance $x_2 - x_1$.

**Note**

The horizontal change is $x_2 - x_1$ and is called the **run**.

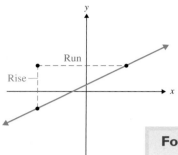

We said that the slope of a line is the ratio of the vertical change (rise) to the horizontal change (run). Therefore,

---

**Formula   The Slope of a Line**

The **slope**, $m$, of a line containing the points $(x_1, y_1)$ and $(x_2, y_2)$ is given by

$$m = \frac{\text{Vertical change}}{\text{Horizontal change}} = \frac{y_2 - y_1}{x_2 - x_1}.$$

---

We can also think of slope as:

$$\frac{\text{Rise}}{\text{Run}} \quad \text{or} \quad \frac{\text{Change in } y}{\text{Change in } x}.$$

Let's look at some different ways to determine the slope of a line.

**Example 2**

Determine the slope of each line.

a)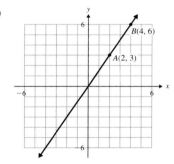

b)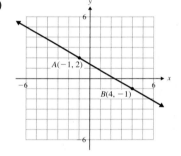

**Solution**

a)  We will find the slope in two ways.

   i)  First, we will find the vertical change and the horizontal change by counting these changes as we go from $A$ to $B$.

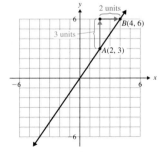

Vertical change (change in $y$) from $A$ to $B$: 3 units

Horizontal change (change in $x$) from $A$ to $B$: 2 units

$$\text{Slope} = \frac{\text{Change in } y}{\text{Change in } x} = \frac{3}{2} \quad \text{or} \quad m = \frac{3}{2}$$

ii) We can also find the slope using the formula.

Let $(x_1, y_1) = (2, 3)$ and $(x_2, y_2) = (4, 6)$.

$$m = \frac{y_2 - y_1}{x_2 - x_1} = \frac{6 - 3}{4 - 2} = \frac{3}{2}.$$

You can see that we get the same result either way we find the slope.

b)  i)  First, find the slope by counting the vertical change and horizontal change as we go from $A$ to $B$.

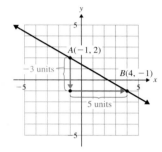

Vertical change (change in $y$) from $A$ to $B$: $-3$ units

Horizontal change (change in $x$) from $A$ to $B$: $5$ units

$$\text{Slope} = \frac{\text{Change in } y}{\text{Change in } x} = \frac{-3}{5} = -\frac{3}{5}$$

$$\text{or} \quad m = -\frac{3}{5}$$

ii) We can also find the slope using the formula.

Let $(x_1, y_1) = (-1, 2)$ and $(x_2, y_2) = (4, -1)$.

$$m = \frac{y_2 - y_1}{x_2 - x_1} = \frac{-1 - 2}{4 - (-1)} = \frac{-3}{5} = -\frac{3}{5}.$$

Again, we obtain the same result using either method for finding the slope.  ∎

**Note**

The slope of $-\dfrac{3}{5}$ can be thought of as $\dfrac{-3}{5}$, $\dfrac{3}{-5}$, or $-\dfrac{3}{5}$.

**You Try 2**

Determine the slope of each line by

a)  counting the vertical change and horizontal change.        b)  using the formula.

a)

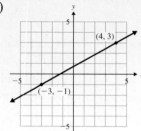

b)

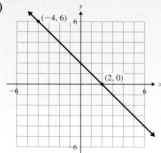

Notice that in Example 2a, the line has a positive slope and slants upward from left to right. As the value of $x$ increases, the value of $y$ increases as well. The line in Example 2b has a negative slope and slants downward from left to right. Notice, in this case, that as the line goes from left to right, the value of $x$ increases while the value of $y$ decreases. We can summarize these results with the following general statements.

---

**Property**   Positive and Negative Slopes

A line with a **positive slope** slants upward from left to right. As the value of $x$ increases, the value of $y$ increases as well.

    A line with a **negative slope** slants downward from left to right. As the value of $x$ increases, the value of $y$ decreases.

---

### 3. Use Slope to Solve Applied Problems

**Example 3**

The graph models the number of members of a certain health club from 2006 to 2010.

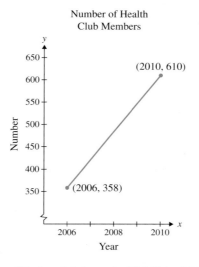

Number of Health
Club Members

a)   How many members did the club have in 2006? in 2010?

b)   What does the sign of the slope of the line segment mean in the context of the problem?

c)   Find the slope of the line segment, and explain what it means in the context of the problem.

**Solution**

a)   We can determine the number of members by reading the graph. In 2006, there were 358 members, and in 2010 there were 610 members.

b)   The positive slope tells us that from 2006 to 2010 the number of members was increasing.

c)   Let $(x_1, y_1) = (2006, 358)$ and $(x_2, y_2) = (2010, 610)$.

$$\text{Slope} = \frac{y_2 - y_1}{x_2 - x_1} = \frac{610 - 358}{2010 - 2006} = \frac{252}{4} = 63$$

The slope of the line is 63. Therefore, the number of members of the health club between 2006 and 2010 increased by 63 per year.   ∎

## 4. Find the Slope of Horizontal and Vertical Lines

**Example 4**

Find the slope of the line containing each pair of points.

a)   $(-4, 1)$ and $(2, 1)$          b)   $(2, 4)$ and $(2, -3)$

**Solution**

a)   Let $(x_1, y_1) = (-4, 1)$ and $(x_2, y_2) = (2, 1)$.

$$m = \frac{y_2 - y_1}{x_2 - x_1} = \frac{1 - 1}{2 - (-4)} = \frac{0}{6} = 0$$

If we plot the points, we see that they lie on a horizontal line. Each point on the line has a $y$-coordinate of 1, so $y_2 - y_1$ *always* equals zero.
   *The slope of every horizontal line is zero.*

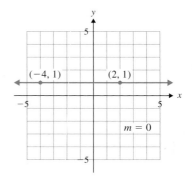

b)   Let $(x_1, y_1) = (2, 4)$ and $(x_2, y_2) = (2, -3)$.

$$m = \frac{y_2 - y_1}{x_2 - x_1} = \frac{-3 - 4}{2 - 2} = \frac{-7}{0} \quad \text{undefined}$$

We say that the slope is undefined. Plotting these points gives us a vertical line. Each point on the line has an $x$-coordinate of 2, so $x_2 - x_1$ *always* equals zero.
   *The slope of every vertical line is undefined.*

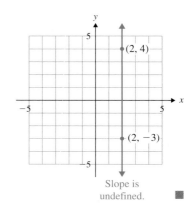

Slope is undefined. ■

 **You Try 3**

Find the slope of the line containing each pair of points.
a)   $(4, 9)$ and $(-3, 9)$          b)   $(-7, 2)$ and $(-7, 0)$

---

**Property**   Slopes of Horizontal and Vertical Lines

The slope of a horizontal line, $y = d$, is **zero**. The slope of a vertical line, $x = c$, is **undefined**. ($c$ and $d$ are constants.)

---

## 5. Use Slope and One Point on a Line to Graph the Line

We have seen how we can find the slope of a line given two points on the line. Now, we will see how we can use the slope and *one* point on the line to graph the line.

## Example 5

Graph the line containing the point

a)   $(-1, -2)$ with a slope of $\dfrac{3}{2}$.

b)   $(0, 1)$ with a slope of $-3$.

### Solution

a)   Plot the point.

Use the slope to find another point on the line.

Plot this point, and draw a line through the two points.

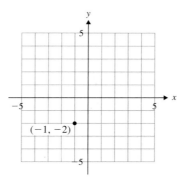

$$m = \frac{3}{2} = \frac{\text{Change in } y}{\text{Change in } x}$$

To get from the point $(-1, -2)$ to another point on the line, move up 3 units in the $y$-direction and right 2 units in the $x$-direction.

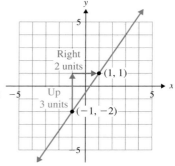

b)   Plot the point $(0, 1)$.

What does the slope, $m = -3$, mean?

$$m = -3 = \frac{-3}{1} = \frac{\text{Change in } y}{\text{Change in } x}$$

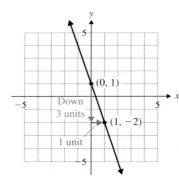

To get from $(0, 1)$ to another point on the line, we will move *down* 3 units in the $y$-direction and *right* 1 unit in the $x$-direction. We end up at $(1, -2)$.

Plot this point, and draw a line through $(0, 1)$ and $(1, -2)$.

In part b), we could have written $m = -3$ as $m = \dfrac{3}{-1}$. This would have given us a different point on the same line.   ■

## You Try 4

Graph the line containing the point

a)   $(-2, 1)$ with a slope of $-\dfrac{3}{2}$.

b)   $(0, -3)$ with a slope of 2.

c)   $(3, 2)$ with an undefined slope.

## Using Technology

When we look at the graph of a linear equation, we should be able to estimate its slope. Use the equation $y = x$ as a guideline.

**Step 1:**  Graph the equation $y = x$.

We can make the graph a thick line (so we can tell it apart from the others) by moving the arrow all the way to the left and pressing $\boxed{\text{ENTER}}$ :

     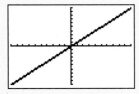

**Step 2:**  Keeping this equation, graph the equation $y = 2x$:

     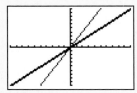

   a.  Is the new graph steeper or flatter than the graph of $y = x$?

   b.  Make a guess as to whether $y = 3x$ will be steeper or flatter than $y = x$. Test your guess by graphing $y = 3x$.

**Step 3:**  Clear the equation $y = 2x$ and graph the equation $y = 0.5x$:

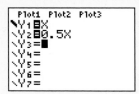     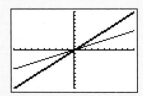

   a.  Is the new graph steeper or flatter than the graph of $y = x$?

   b.  Make a guess as to whether $y = 0.65x$ will be steeper or flatter than $y = x$. Test your guess by graphing $y = 0.65x$.

**Step 4:**  Test similar situations, except with negative slopes: $y = -x$

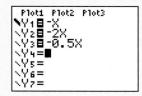     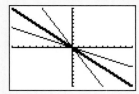

Did you notice that we have the same relationship, except in the opposite direction? That is, $y = 2x$ is steeper than $y = x$ in the positive direction, and $y = -2x$ is steeper than $y = -x$, but in the negative direction. And $y = 0.5x$ is flatter than $y = x$ in the positive direction, and $y = -0.5x$ is flatter than $y = -x$, but in the negative direction.

**Answers to You Try Exercises**

1) The belt rises 5 in. for every 12 horizontal inches.  2) a) $m = \dfrac{4}{7}$  b) $m = -1$

3) a) $m = 0$  b) undefined

4) a)  b)  c)

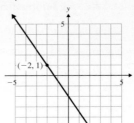  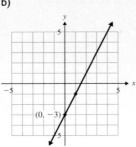  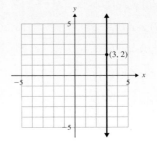

## 4.3 Exercises

**Objective 1: Understand the Concept of Slope**

1) Explain the meaning of slope.

2) Describe the slant of a line with a negative slope.

3) Describe the slant of a line with a positive slope.

4) The slope of a horizontal line is _____.

5) The slope of a vertical line is _____.

6) If a line contains the points $(x_1, y_1)$ and $(x_2, y_2)$, write the formula for the slope of the line.

**Mixed Exercises: Objectives 2 and 4**

Determine the slope of each line by

    a) counting the vertical change and the horizontal change as you move from one point to the other on the line;

    and

    b) using the slope formula. (See Example 2.)

 7)

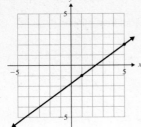

8)

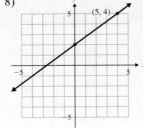

9)

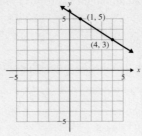

10)

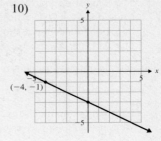

11)

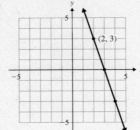

12)

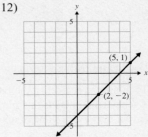

13)

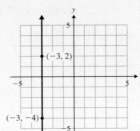

14)

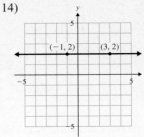

15) Graph a line with a positive slope and a negative y-intercept.

16) Graph a line with a negative slope and a positive x-intercept.

Use the slope formula to find the slope of the line containing each pair of points.

17) (2, 1) and (0, −3)

18) (0, 3) and (9, 6)

19) (2, −6) and (−1, 6)

20) (−3, 9) and (2, 4)

21) (−4, 3) and (1, −8)

22) (2, 0) and (−5, 4)

23) (−2, −2) and (−2, 7)

24) (0, −6) and (−9, −6)

25) (3, 5) and (−1, 5)

26) (1, 3) and (1, −1)

27) $\left(\dfrac{2}{3}, \dfrac{5}{2}\right)$ and $\left(-\dfrac{1}{2}, 2\right)$

28) $\left(-\dfrac{1}{5}, \dfrac{3}{4}\right)$ and $\left(\dfrac{1}{3}, -\dfrac{3}{5}\right)$

29) (3.5, −1.4) and (7.5, 1.6)

30) (−1.7, 10.2) and (0.8, −0.8)

**Objective 3: Use Slope to Solve Applied Problems**

31) The longest run at Ski Dubai, an indoor ski resort in the Middle East, has a vertical drop of about 60 m with a horizontal distance of about 395 m. What is the slope of this ski run? (www.skidxb.com)

32) The federal government requires that all wheelchair ramps in new buildings have a maximum slope of $\dfrac{1}{12}$. Does the following ramp meet this requirement? Give a reason for your answer. (www.access-board.gov)

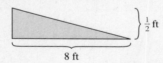
$\frac{1}{2}$ ft

8 ft

Use the following information for Exercises 33 and 34.

To minimize accidents, the Park District Risk Management Agency recommends that playground slides and sledding hills have a maximum slope of about 0.577.
(Illinois Parks and Recreation)

33) Does this slide meet the agency's recommendations?

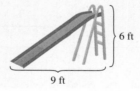
6 ft

9 ft

34) Does this sledding hill meet the agency's recommendations?

75 ft

140 ft

35) In Granby, Colorado, the first 50 ft of a driveway cannot have a slope of more than 5%. If the first 50 ft of a driveway rises 0.75 ft for every 20 ft of horizontal distance, does this driveway meet the building code?
(http://co.grand.co.us)

Use the following information for Exercises 36–38.

The steepness (slope) of a roof on a house in a certain town cannot exceed $\dfrac{7}{12}$, also known as a 7:12 *pitch*. The first number refers to the rise of the roof. The second number refers to how far over you must go (the run) to attain that rise.

36) Find the slope of a roof with a 12:20 pitch.

37) Find the slope of a roof with a 12:26 pitch.

38) Does the slope in Exercise 36 meet the town's building code? Give a reason for your answer.

39) The graph shows the approximate number of people in the United States injured in a motor vehicle accident from 2003 to 2007. (www.bts.gov)

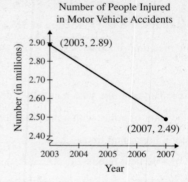

a) Approximately how many people were injured in 2003? in 2005?

b) Without computing the slope, determine whether it is positive or negative.

c) What does the sign of the slope mean in the context of the problem?

d) Find the slope of the line segment, and explain what it means in the context of the problem.

40) The graph shows the approximate number of prescriptions filled by mail order from 2004 to 2007. (www.census.gov)

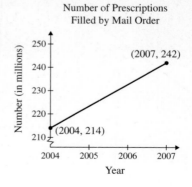

a) Approximately how many prescriptions were filled by mail order in 2004? in 2007?

b) Without computing the slope, determine whether it is positive or negative.

c) What does the sign of the slope mean in the context of the problem?

d) Find the slope of the line segment, and explain what it means in the context of the problem.

**Objective 5: Use Slope and One Point on a Line to Graph the Line**

Graph the line containing the given point and with the given slope.

41) $(2, 1)$; $m = \dfrac{3}{4}$

42) $(1, 2)$; $m = \dfrac{1}{3}$

43) $(-2, -3)$; $m = \dfrac{1}{4}$

44) $(-5, 0)$; $m = \dfrac{2}{5}$

45) $(1, 2)$; $m = -\dfrac{3}{4}$

46) $(1, -3)$; $m = -\dfrac{2}{5}$

47) $(-1, -3)$; $m = 3$

48) $(0, -2)$; $m = -2$

VIDEO 49) $(6, 2)$; $m = -4$

50) $(4, 3)$; $m = -5$

51) $(3, -4)$; $m = -1$

52) $(-1, -2)$; $m = 0$

53) $(-2, 3)$; $m = 0$

54) $(2, 0)$; slope is undefined.

55) $(-1, -4)$; slope is undefined.

56) $(0, 0)$; $m = 1$

57) $(0, 0)$; $m = -1$

## Section 4.4 The Slope-Intercept Form of a Line

**Objectives**

1. Define the Slope-Intercept Form of a Line
2. Graph a Line Expressed in Slope-Intercept Form
3. Rewrite an Equation in Slope-Intercept Form and Graph the Line
4. Use Slope to Determine Whether Two Lines Are Parallel or Perpendicular

In Section 4.1, we learned that a linear equation in two variables can be written in the form $Ax + By = C$ (this is called **standard form**), where $A$, $B$, and $C$ are real numbers and where both $A$ and $B$ do not equal zero. Equations of lines can take other forms, too, and we will look at one of those forms in this section.

### 1. Define the Slope-Intercept Form of a Line

We know that if $(x_1, y_1)$ and $(x_2, y_2)$ are points on a line, then the slope of the line is

$$m = \frac{y_2 - y_1}{x_2 - x_1}$$

Recall that to find the $y$-intercept of a line, we let $x = 0$ and solve for $y$. Let one of the points on a line be the $y$-intercept $(0, b)$, where $b$ is a number. Let another point on the line be $(x, y)$.

See the graph on the left.

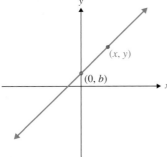

Substitute the points $(0, b)$ and $(x, y)$ into the slope formula:

Subtract $y$-coordinates.
$$\downarrow$$
$$m = \frac{y_2 - y_1}{x_2 - x_1} = \frac{y - b}{x - 0} = \frac{y - b}{x}$$
$$\uparrow$$
Subtract $x$-coordinates.

Solve $m = \dfrac{y - b}{x}$ for $y$.

$$mx = \dfrac{y - b}{x} \cdot x \qquad \text{Multiply by } x \text{ to eliminate the fraction.}$$

$$mx = y - b$$

$$mx + b = y - b + b \qquad \text{Add } b \text{ to each side to solve for } y.$$

$$mx + b = y$$

OR

$$y = mx + b \qquad \text{Slope-intercept form}$$

---

**Definition**

The **slope-intercept form of a line** is $y = mx + b$, where $m$ is the slope and $(0, b)$ is the $y$-intercept.

---

When an equation is in the form $y = mx + b$, we can quickly recognize the $y$-intercept and slope to graph the line.

## 2. Graph a Line Expressed in Slope-Intercept Form

**Example 1**

Graph each equation.

a) $y = -\dfrac{4}{3}x + 2$     b) $y = 4x - 3$     c) $y = \dfrac{1}{2}x$

**Solution**

Notice that each equation is in slope-intercept form, $y = mx + b$, where $m$ is the slope and $(0, b)$ is the $y$-intercept.

a)  Graph $y = -\dfrac{4}{3}x + 2$.

$$\text{Slope} = -\dfrac{4}{3}, \quad y\text{-intercept is } (0, 2).$$

Graph the line by first plotting the $y$-intercept and then using the slope to locate another point on the line.

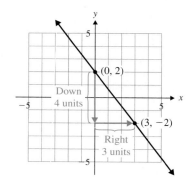

b)  Graph $y = 4x - 3$.

$$\text{Slope} = 4, \quad y\text{-intercept is } (0, -3).$$

Plot the $y$-intercept first, then use the slope to locate another point on the line. Since the slope is 4, think of it as $\dfrac{4}{1}.$  $\leftarrow$ Change in $y$
$\leftarrow$ Change in $x$

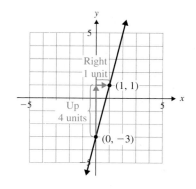

c) The equation $y = \dfrac{1}{2}x$ is the same as $y = \dfrac{1}{2}x + 0$.

Identify the slope and $y$-intercept.

$$\text{Slope} = \frac{1}{2}, \quad y\text{-intercept is } (0, 0).$$

Plot the $y$-intercept, then use the slope to locate another point on the line.

$\dfrac{1}{2}$ is equivalent to $\dfrac{-1}{-2}$, so we can use $\dfrac{-1}{-2}$ as the slope to locate yet another point on the line.

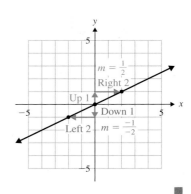

 **You Try 1**

Graph each line using the slope and $y$-intercept.

a) $y = \dfrac{1}{4}x + 1$      b) $y = x - 3$      c) $y = -2x$

## 3. Rewrite an Equation in Slope-Intercept Form and Graph the Line

Lines are not always written in slope-intercept form. They may be written in *standard form* (like $7x + 4y = 12$) or in another form such as $2x = 2y + 10$. We can put equations like these into slope-intercept form by solving the equation for $y$.

**Example 2**

Put each equation into slope-intercept form, and graph.

a) $7x + 4y = 12$      b) $2x = 2y + 10$

**Solution**

a) The slope-intercept form of a line is
$y = mx + b$. We must solve the equation for $y$.

$$
\begin{aligned}
7x + 4y &= 12 \\
4y &= -7x + 12 && \text{Add } -7x \text{ to each side.} \\
y &= \frac{-7}{4}x + 3 && \text{Divide each side by 4.}
\end{aligned}
$$

$$\text{Slope} = -\frac{7}{4} \text{ or } \frac{-7}{4}; \quad y\text{-intercept is } (0, 3).$$

$$\left(\text{We could also have thought of the slope as } \frac{7}{-4}.\right)$$

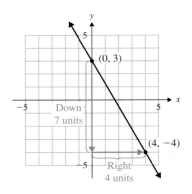

b)   We must solve $2x = 2y + 10$ for $y$.

$$2x = 2y + 10$$
$$2x - 10 = 2y \qquad \text{Subtract 10 from each side.}$$
$$x - 5 = y \qquad \text{Divide each side by 2.}$$

The slope-intercept form is $y = x - 5$. We can also think of this as $y = 1x - 5$.

$$\text{slope} = 1, \quad y\text{-intercept is } (0, -5).$$

We will think of the slope as $\dfrac{1}{1}$. $\begin{array}{l} \leftarrow \text{Change in } y \\ \leftarrow \text{Change in } x \end{array}$

$\left(\text{We could also think of it as } \dfrac{-1}{-1}.\right)$

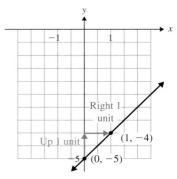

 **You Try 2**

Put each equation into slope-intercept form, and graph.

a)   $10x - 5y = -5$          b)   $2x = -3 - 3y$

---

**Summary**    Different Methods for Graphing a Line Given Its Equation

We have learned that we can use different methods for graphing lines. Given the equation of a line we can:

1)   Make a table of values, plot the points, and draw the line through the points.
2)   Find the $x$-intercept by letting $y = 0$ and solving for $x$, and find the $y$-intercept by letting $x = 0$ and solving for $y$. Plot the points, then draw the line through the points.
3)   Put the equation into slope-intercept form, $y = mx + b$, identify the slope and $y$-intercept, then graph the line.

---

### 4. Use Slope to Determine Whether Two Lines Are Parallel or Perpendicular

Recall that two lines in a plane are **parallel** if they do not intersect. If we are given the equations of two lines, how can we determine whether they are parallel?

Here are the equations of two lines:

$$2x - 3y = -3 \qquad\qquad y = \frac{2}{3}x - 5$$

We will graph each line. To graph the first line, we write it in slope-intercept form.

$$-3y = -2x - 3 \qquad \text{Add } -2x \text{ to each side.}$$
$$y = \frac{-2}{-3}x - \frac{3}{-3} \qquad \text{Divide by } -3.$$
$$y = \frac{2}{3}x + 1 \qquad \text{Simplify.}$$

The slope-intercept form of the first line is $y = \dfrac{2}{3}x + 1$, and the second line is already in slope-intercept form, $y = \dfrac{2}{3}x - 5$. Now, graph each line.

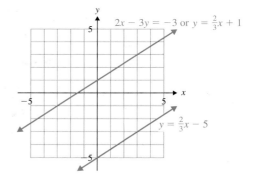

These lines are parallel. Their slopes are the same, but they have different $y$-intercepts. (If the $y$-intercepts were the same, they would be the same line.) This is how we determine whether two (nonvertical) lines are parallel. They have the same slope, but different $y$-intercepts.

---

**Property** Parallel Lines

Parallel lines have the same slope. (If two lines are vertical, they are parallel. However, their slopes are undefined.)

---

### Example 3

Determine whether each pair of lines is parallel.

a)   $2x + 8y = 12$
      $x + 4y = -20$

b)   $y = -5x + 2$
      $5x - y = 7$

**Solution**

a)   To determine whether the lines are parallel, we must find the slope of each line. If the slopes are the same, but the $y$-intercepts are different, the lines are parallel.

Write each equation in slope-intercept form.

$$2x + 8y = 12 \qquad\qquad x + 4y = -20$$
$$8y = -2x + 12 \qquad\qquad 4y = -x - 20$$
$$y = -\frac{2}{8}x + \frac{12}{8} \qquad\qquad y = -\frac{x}{4} - \frac{20}{4}$$
$$y = -\frac{1}{4}x + \frac{3}{2} \qquad\qquad y = -\frac{1}{4}x - 5$$

Each line has a slope of $-\dfrac{1}{4}$. Their $y$-intercepts are different. Therefore, $2x + 8y = 12$ and $x + 4y = -20$ are parallel lines.

b)   Again, we must find the slope of each line. $y = -5x + 2$ is already in slope-intercept form. Its slope is $-5$.

Write $5x - y = 7$ in slope-intercept form.

$$-y = -5x + 7 \qquad \text{Add } -5x \text{ to each side.}$$
$$y = 5x - 7 \qquad \text{Divide each side by } -1.$$

The slope of $y = -5x + 2$ is $-5$. The slope of $5x - y = 7$ is $5$. The slopes are different; therefore, the lines are not parallel.

The slopes of two lines can tell us about another relationship between the lines. The slopes can tell us whether two lines are *perpendicular*.

Recall that two lines are **perpendicular** if they intersect at 90° angles.

The graphs of two perpendicular lines and their equations are on the left. We will see how their slopes are related.

Find the slope of each line by writing them in slope-intercept form.

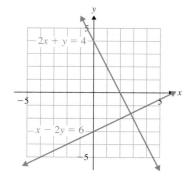

Line 1:   $2x + y = 4$

$$y = -2x + 4$$

$$m = -2$$

Line 2:   $x - 2y = 6$

$$-2y = -x + 6$$

$$y = \frac{-x}{-2} + \frac{6}{-2}$$

$$y = \frac{1}{2}x - 3$$

$$m = \frac{1}{2}$$

How are the slopes related? They are **negative reciprocals**. That is, if the slope of one line is $a$, then the slope of a line perpendicular to it is $-\dfrac{1}{a}$. This is how we determine whether two lines are perpendicular (where neither one is vertical).

---

**Property**   **Perpendicular Lines**

Perpendicular lines have slopes that are negative reciprocals of each other.

---

### Example 4

Determine whether each pair of lines is perpendicular.

a)   $15x - 12y = -4$
        $4x - 5y = 10$

b)   $2x - 9y = -9$
        $9x + 2y = 8$

**Solution**

a)   To determine whether the lines are perpendicular, we must find the slope of each line. If the slopes are negative reciprocals, then the lines are perpendicular.

Write each equation in slope-intercept form.

$$15x - 12y = -4$$
$$-12y = -15x - 4$$
$$y = \frac{-15}{-12}x - \frac{4}{-12}$$
$$y = \frac{5}{4}x + \frac{1}{3}$$
$$m = \frac{5}{4}$$

$$4x - 5y = 10$$
$$-5y = -4x + 10$$
$$y = \frac{-4}{-5}x + \frac{10}{-5}$$
$$y = \frac{4}{5}x - 2$$
$$m = \frac{4}{5}$$

The slopes are reciprocals, but they are not *negative* reciprocals. Therefore, the lines are *not* perpendicular.

b)   Begin by writing each equation in slope-intercept form so that we can find their slopes.

$$2x - 9y = -9 \qquad\qquad 9x + 2y = 8$$
$$-9y = -2x - 9 \qquad\qquad 2y = -9x + 8$$
$$y = \frac{-2}{-9}x - \frac{9}{-9} \qquad\qquad y = -\frac{9}{2}x + \frac{8}{2}$$
$$y = \frac{2}{9}x + 1 \qquad\qquad y = -\frac{9}{2}x + 4$$
$$m = \frac{2}{9} \qquad\qquad m = -\frac{9}{2}$$

The slopes are negative reciprocals; therefore, the lines are perpendicular.

 **You Try 3**

Determine whether each pair of lines is parallel, perpendicular, or neither.

a)   $5x - y = -2$          b)   $y = \dfrac{8}{3}x + 9$          c)   $x + 2y = 8$          d)   $x = 7$

$\phantom{a)}$ $3x + 15y = -20$          $-32x + 12y = 15$          $2x = 4y + 3$          $y = -4$

**Answers to You Try Exercises**

1)  a)

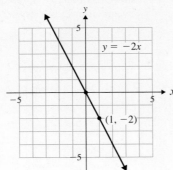

b)

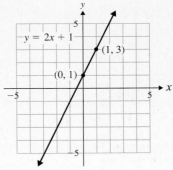

c)

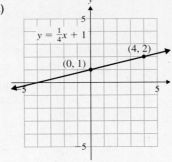

2)  a)

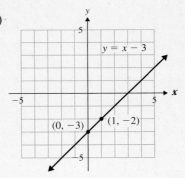

b)

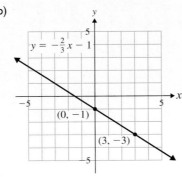

3)  a)  perpendicular
    b)  parallel
    c)  neither
    d)  perpendicular

## 4.4 Exercises

**Mixed Exercises: Objectives 1 and 2**

1) The slope-intercept form of a line is $y = mx + b$. What is the slope? What is the $y$-intercept?

2) How do you put an equation that is in standard form, $Ax + By = C$, into slope-intercept form?

Each of the following equations is in slope-intercept form. Identify the slope and the $y$-intercept, then graph each line using this information.

VIDEO 3) $y = \frac{2}{5}x - 6$

4) $y = \frac{7}{5}x - 1$

5) $y = -\frac{3}{2}x + 3$

6) $y = -\frac{1}{3}x + 2$

7) $y = \frac{3}{4}x + 2$

8) $y = \frac{2}{3}x + 5$

9) $y = -2x - 3$

10) $y = 3x - 1$

11) $y = 5x$

12) $y = -2x + 5$

13) $y = -\frac{3}{2}x - \frac{7}{2}$

14) $y = \frac{3}{5}x + \frac{3}{4}$

15) $y = 6$

16) $y = -4$

**Objective 3: Rewrite an Equation in Slope-Intercept Form and Graph the Line**

Put each equation into slope-intercept form, if possible, and graph.

VIDEO 17) $x + 3y = -6$         18) $x + 2y = -8$

19) $4x + 3y = 21$       20) $2x - 5y = 5$

21) $2 = x + 3$           22) $x + 12 = 4$

23) $2x = 18 - 3y$       24) $98 = 49y - 28x$

25) $y + 2 = -3$          26) $y + 3 = 3$

27) Kolya has a part-time job, and his gross pay can be described by the equation $P = 8.50h$, where $P$ is his gross pay, in dollars, and $h$ is the number of hours worked.

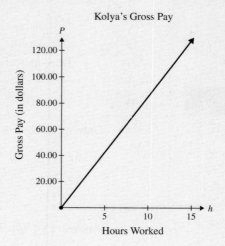

Kolya's Gross Pay

a) What is the $P$-intercept? What does it mean in the context of the problem?

b) What is the slope? What does it mean in the context of the problem?

c) Use the graph to find Kolya's gross pay when he works 12 hours. Confirm your answer using the equation.

28) The number of people, $y$, leaving on cruises from Florida from 2002 to 2006 can be approximated by $y = 137,000x + 4,459,000$, where $x$ is the number of years after 2002. (www.census.gov)

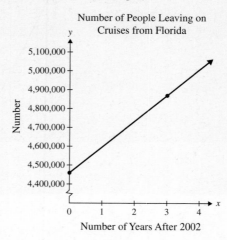

Number of People Leaving on Cruises from Florida

a) What is the $y$-intercept? What does it mean in the context of the problem?

b) What is the slope? What does it mean in the context of the problem?

c) Use the graph to determine how many people left on cruises from Florida in 2005. Confirm your answer using the equation.

29) A Tasmanian devil is a marsupial that lives in Australia. Once a joey leaves its mother's pouch, its weight for the first 8 weeks can be approximated by $y = 2x + 18$, where $x$ represents the number of weeks it has been out of the pouch and $y$ represents its weight, in ounces.
(Wikipedia and Animal Planet)

a) What is the $y$-intercept, and what does it represent?

b) How much does a Tasmanian devil weigh 3 weeks after emerging from the pouch?

c) Explain the meaning of the slope in the context of this problem.

d) How long would it take for a joey to weigh 32 oz?

30) The number of active physicians in Idaho, $y$, from 2002 to 2006 can be approximated by $y = 74.7x + 2198.8$, where $x$ represents the number of years after 2002.
(www.census.gov)

a) What is the $y$-intercept and what does it represent?

b) How many doctors were practicing in 2006?

c) Explain the meaning of the slope in the context of this problem.

d) If the current trend continues, how many practicing doctors would Idaho have in 2015?

31) On a certain day in 2009, the exchange rate between the American dollar and the Indian rupee was given by $r = 48.2d$, where $d$ represents the number of dollars and $r$ represents the number of rupees.

a) What is the $r$-intercept and what does it represent?

b) What is the slope? What does it mean in the context of the problem?

c) If Madhura is going to India to visit her family, how many rupees could she get for $80.00?

d) How many dollars could be exchanged for 2410 rupees?

32) The value of a car, $v$, in dollars, $t$ years after it is purchased is given by $v = -1800t + 20,000$.

a) What is the $v$-intercept and what does it represent?

b) What is the slope? What does it mean in the context of the problem?

c) What is the car worth after 3 years?

d) When will the car be worth $11,000?

Write the slope-intercept form for the equation of a line with the given slope and $y$-intercept.

VIDEO 33) $m = -4$; $y$-int: $(0, 7)$

34) $m = -7$; $y$-int: $(0, 4)$

35) $m = \dfrac{9}{5}$; $y$-int: $(0, -3)$

36) $m = \dfrac{7}{4}$; $y$-int: $(0, -2)$

37) $m = -\dfrac{5}{2}$; $y$-int: $(0, -1)$

38) $m = \dfrac{1}{4}$; $y$-int: $(0, 7)$

39) $m = 1$; $y$-int: $(0, 2)$

40) $m = -1$; $y$-int: $(0, 0)$

41) $m = 0$; $y$-int: $(0, 0)$

42) $m = 0$; $y$-int: $(0, -8)$

## Objective 4: Use Slope to Determine Whether Two Lines Are Parallel or Perpendicular

43) How do you know whether two lines are perpendicular?

44) How do you know whether two lines are parallel?

Determine whether each pair of lines is parallel, perpendicular, or neither.

45) $y = -x - 5$
    $y = x + 8$

46) $y = \dfrac{3}{4}x + 2$
    $y = \dfrac{3}{4}x - 1$

VIDEO 47) $y = \dfrac{2}{9}x + 4$
    $4x - 18y = 9$

48) $y = \dfrac{4}{5}x + 2$
    $5x + 4y = 12$

49) $3x - y = 4$
    $2x - 5y = -9$

50) $-4x + 3y = -5$
    $4x - 6y = -3$

51) $-x + y = -21$
    $y = 2x + 5$

52) $x + 3y = 7$
    $y = 3x$

53) $x + 7y = 4$
    $y - 7x = 4$

54) $5y - 3x = 1$
    $3x - 5y = -8$

55) $y = -\dfrac{1}{2}x$

$x + 2y = 4$

56) $-4x + 6y = 5$

$2x - 3y = -12$

57) $x = -1$

$y = 6$

58) $y = 12$

$y = 4$

59) $x = -4.3$

$x = 0$

60) $x = 7$

$y = 0$

Lines $L_1$ and $L_2$ contain the given points. Determine whether lines $L_1$ and $L_2$ are parallel, perpendicular, or neither.

VIDEO 61) $L_1$:  $(-1, -7), (2, 8)$
   $L_2$:  $(10, 2), (0, 4)$

62) $L_1$:  $(0, -3), (-4, -11)$
   $L_2$:  $(-2, 0), (3, 10)$

63) $L_1$:  $(1, 10), (3, 8)$
   $L_2$:  $(2, 4), (-5, -17)$

64) $L_1$:  $(-1, 4), (2, -8)$
   $L_2$:  $(8, 5), (0, 3)$

65) $L_1$:  $(-3, 6), (4, -1)$
   $L_2$:  $(-6, -5), (-10, -1)$

66) $L_1$:  $(5, -5), (7, 11)$
   $L_2$:  $(-3, 0), (6, 3)$

67) $L_1$:  $(-6, 2), (-6, 1)$
   $L_2$:  $(4, 0), (4, -5)$

68) $L_1$:  $(8, 1), (7, 1)$
   $L_2$:  $(12, -1), (-2, -1)$

69) $L_1$:  $(7, 2), (7, 5)$
   $L_2$:  $(-2, 0), (1, 0)$

70) $L_1$:  $(-6, 4), (-6, -1)$
   $L_2$:  $(-1, 10), (-3, 10)$

## Section 4.5 Writing an Equation of a Line

### Objectives

1. **Rewrite an Equation in Standard Form**
2. **Write an Equation of a Line Given Its Slope and y-Intercept**
3. **Use the Point-Slope Formula to Write an Equation of a Line Given Its Slope and a Point on the Line**
4. **Use the Point-Slope Formula to Write an Equation of a Line Given Two Points on the Line**
5. **Write Equations of Horizontal and Vertical Lines**
6. **Write an Equation of a Line That is Parallel or Perpendicular to a Given Line**
7. **Write a Linear Equation to Model Real-World Data**

So far in this chapter, we have been graphing lines given their equations. Now we will write an equation of a line when we are given information about it.

### 1. Rewrite an Equation in Standard Form

In Section 4.4, we practiced writing equations of lines in slope-intercept form. Here we will discuss how to write a line in **standard form, $Ax + By = C$, with the additional conditions that $A$, $B$, and $C$ are integers and $A$ is positive.**

| Example 1 |

Rewrite each linear equation in standard form.

a)  $3x + 8 = -2y$

b)  $y = -\dfrac{3}{4}x + \dfrac{1}{6}$

**Solution**

a)  In standard form, the $x$- and $y$-terms are on the same side of the equation.

$$3x + 8 = -2y$$
$$3x = -2y - 8 \quad \text{Subtract 8 from each side.}$$
$$3x + 2y = -8 \quad \text{Add } 2y \text{ to each side; the equation is now in standard form.}$$

b)  Since an equation $Ax + By = C$ is considered to be in standard form when $A$, $B$, and $C$ are integers, the first step in writing $y = -\dfrac{3}{4}x + \dfrac{1}{6}$ in standard form is to eliminate the fractions.

$$y = -\dfrac{3}{4}x + \dfrac{1}{6}$$
$$12 \cdot y = 12\left(-\dfrac{3}{4}x + \dfrac{1}{6}\right) \quad \text{Multiply both sides of the equation by 12.}$$
$$12y = -9x + 2$$
$$9x + 12y = 2 \quad \text{Add } 9x \text{ to each side.}$$

The standard form is $9x + 12y = 2$. ■

**You Try 1**

Rewrite each equation in standard form.

a)   $5x = 3 + 11y$          b)   $y = \dfrac{1}{3}x - 7$

In the rest of this section, we will learn how to write equations of lines given information about their graphs.

## 2. Write an Equation of a Line Given Its Slope and *y*-Intercept

**Procedure   Write an Equation of a Line Given Its Slope and *y*-Intercept**

**If we are given the slope and *y*-intercept of a line, use $y = mx + b$ and substitute those values into the equation.**

**Example 2**

Find an equation of the line with slope $= -6$ and *y*-intercept $(0, 5)$.

***Solution***

Since we are told the slope and *y*-intercept, use $y = mx + b$.

$$m = -6 \qquad \text{and} \qquad b = 5$$

Substitute these values into $y = mx + b$ to get $y = -6x + 5$.  ■

**You Try 2**

Find an equation of the line with slope $= \dfrac{5}{8}$ and *y*-intercept $(0, -9)$.

## 3. Use the Point-Slope Formula to Write an Equation of a Line Given Its Slope and a Point on the Line

When we are given the slope of a line and a point on that line, we can use another method to find its equation. This method comes from the formula for the slope of a line.

Let $(x_1, y_1)$ be a given point on a line, and let $(x, y)$ be any other point on the same line, as shown in the figure. The slope of that line is

$$m = \frac{y - y_1}{x - x_1} \qquad \text{Definition of slope}$$
$$m(x - x_1) = y - y_1 \qquad \text{Multiply each side by } x - x_1.$$
$$y - y_1 = m(x - x_1) \qquad \text{Rewrite the equation.}$$

We have found the *point-slope form* of the equation of a line.

*y*

$(x, y)$ Another point

$(x_1, y_1)$ Given point

*x*

**Formula   Point-Slope Form of a Line**

The **point-slope form of a line** is

$$y - y_1 = m(x - x_1)$$

where $(x_1, y_1)$ is a point on the line and $m$ is its slope.

**Note**

**If we are given the slope of the line and a point on the line, we can use the point-slope formula to find an equation of the line.**

The point-slope formula will help us write an equation of a line. We will not express our final answer in this form. We will write our answer in either slope-intercept form or in standard form.

## Example 3

Find an equation of the line containing the point $(-4, 3)$ with slope $= \dfrac{1}{2}$. Express the answer in slope-intercept form.

### Solution

First, ask yourself, *"What kind of information am I given?"* Since the problem tells us the slope of the line and a point on the line, we will use the point-slope formula: $y - y_1 = m(x - x_1)$.

Substitute $\dfrac{1}{2}$ for $m$. Substitute $(-4, 3)$ for $(x_1, y_1)$.

(Notice we do *not* substitute anything for $x$ and $y$.)

$$y - y_1 = m(x - x_1)$$
$$y - 3 = \frac{1}{2}(x - (-4)) \qquad \text{Substitute } -4 \text{ for } x_1 \text{ and } 3 \text{ for } y_1; \text{ let } m = \frac{1}{2}.$$
$$y - 3 = \frac{1}{2}(x + 4)$$
$$y - 3 = \frac{1}{2}x + 2 \qquad \text{Distribute.}$$

We must write our answer in slope-intercept form, $y = mx + b$, so solve the equation for $y$.

$$y = \frac{1}{2}x + 5 \qquad \text{Add 3 to each side.}$$

The equation is $y = \dfrac{1}{2}x + 5$.

## You Try 3

Find an equation of the line containing the point $(5, 3)$ with slope $= 2$. Express the answer in slope-intercept form.

## Example 4

A line has slope $-4$ and contains the point $(1, 5)$. Find the standard form for the equation of the line.

### Solution

Although we are told to find the *standard form* for the equation of the line, we do not try to immediately "jump" to standard form. First, ask yourself, *"What information am I given?"*

We are given the *slope* and a *point on the line*. Therefore, we will begin by using the point-slope formula. Our *last* step will be to put the equation in standard form.

Use $y - y_1 = m(x - x_1)$. Substitute $-4$ for $m$. Substitute $(1, 5)$ for $(x_1, y_1)$.

$$y - y_1 = m(x - x_1)$$
$$y - 5 = -4(x - 1) \qquad \text{Substitute 1 for } x_1 \text{ and 5 for } y_1; \text{ let } m = -4.$$
$$y - 5 = -4x + 4 \qquad \text{Distribute.}$$

To write the answer in standard form, we must get the $x$- and $y$-terms on the same side of the equation so that the coefficient of $x$ is positive.

$$4x + y - 5 = 4 \qquad \text{Add } 4x \text{ to each side.}$$
$$4x + y = 9 \qquad \text{Add 5 to each side.}$$

The standard form of the equation is $4x + y = 9$. ■

**You Try 4**

A line has slope $-8$ and contains the point $(-4, 5)$. Find the standard form for the equation of the line.

### 4. Use the Point-Slope Formula to Write an Equation of a Line Given Two Points on the Line

We are now ready to discuss how to write an equation of a line when we are given two points on a line.

**To write an equation of a line given two points on the line,**

**a) use the points to find the slope of line**
   *then*
**b) use the slope and *either one* of the points in the point-slope formula.**

**Example 5**

Write an equation of the line containing the points $(4, 9)$ and $(2, 6)$. Express the answer in slope-intercept form.

**Solution**

We are given two points on the line, so first, we will find the slope.

$$m = \frac{6 - 9}{2 - 4} = \frac{-3}{-2} = \frac{3}{2}$$

We will use the slope and *either one* of the points in the point-slope formula. (Each point will give the same result.) We will use $(4, 9)$.

Substitute $\frac{3}{2}$ for $m$. Substitute $(4, 9)$ for $(x_1, y_1)$.

$$y - y_1 = m(x - x_1)$$
$$y - 9 = \frac{3}{2}(x - 4) \qquad \text{Substitute 4 for } x_1 \text{ and 9 for } y_1; \text{ let } m = \frac{3}{2}.$$

$$y - 9 = \frac{3}{2}x - 6 \qquad \text{Distribute.}$$

$$y = \frac{3}{2}x + 3 \qquad \text{Add 9 to each side to solve for } y.$$

The equation is $y = \frac{3}{2}x + 3$. ■

**You Try 5**

Find the slope-intercept form for the equation of the line containing the points $(4, 2)$ and $(1, -5)$.

### 5. Write Equations of Horizontal and Vertical Lines

Earlier we learned that the slope of a horizontal line is zero and that it has equation $y = d$, where $d$ is a constant. The slope of a vertical line is undefined, and its equation is $x = c$, where $c$ is a constant.

> **Formula**    Equations of Horizontal and Vertical Lines
>
> **Equation of a Horizontal Line:** The equation of a horizontal line containing the point $(c, d)$ is $y = d$.
>
> **Equation of a Vertical Line:** The equation of a vertical line containing the point $(c, d)$ is $x = c$.

**Example 6**

Write an equation of the horizontal line containing the point $(7, -1)$.

**Solution**

The equation of a horizontal line has the form $y = d$, where $d$ is the $y$-coordinate of the point. The equation of the line is $y = -1$.

**You Try 6**

Write an equation of the horizontal line containing the point $(3, -8)$.

> **Summary**    Writing Equations of Lines
>
> If you are given
>
> 1) **the slope and y-intercept of the line,** use $y = mx + b$ and substitute those values into the equation.
>
> 2) **the slope of the line and a point on the line,** use the point-slope formula:
>
> $$y - y_1 = m(x - x_1)$$
>
> Substitute the slope for $m$ and the point you are given for $(x_1, y_1)$. Write your answer in slope-intercept or standard form.
>
> 3) **two points on the line,** find the slope of the line and then use the slope and *either one* of the points in the point-slope formula. Write your answer in slope-intercept or standard form.
>
> The equation of a **horizontal line** containing the point $(c, d)$ is **$y = d$.**
>
> The equation of a **vertical line** containing the point $(c, d)$ is **$x = c$.**

### 6. Write an Equation of a Line that is Parallel or Perpendicular to a Given Line

In Section 4.4, we learned that parallel lines have the same slope, and perpendicular lines have slopes that are negative reciprocals of each other. We will use this information to write the equation of a line that is parallel or perpendicular to a given line.

## Example 7

A line contains the point $(2, -2)$ and is parallel to the $y = \dfrac{1}{2}x + 1$. Write the equation of the line in slope-intercept form.

### Solution

Let's look at the graph on the left to help us understand what is happening in this example. We must find the equation of the line in red. It is the line containing the point $(2, -2)$ that is parallel to $y = \dfrac{1}{2}x + 1$.

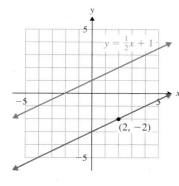

The line $y = \dfrac{1}{2}x + 1$ has $m = \dfrac{1}{2}$. Therefore, the red line will have $m = \dfrac{1}{2}$ as well.

We know the slope, $\dfrac{1}{2}$, and a point on the line, $(2, -2)$, so we use the point-slope formula to find its equation.

Substitute $\dfrac{1}{2}$ for $m$. Substitute $(2, -2)$ for $(x_1, y_1)$.

$$y - y_1 = m(x - x_1)$$

$$y - (-2) = \dfrac{1}{2}(x - 2) \qquad \text{Substitute 2 for } x_1 \text{ and } -2 \text{ for } y_1; \text{ let } m = \dfrac{1}{2}.$$

$$y + 2 = \dfrac{1}{2}x - 1 \qquad \text{Distribute.}$$

$$y = \dfrac{1}{2}x - 3 \qquad \text{Subtract 2 from each side.}$$

The equation is $y = \dfrac{1}{2}x - 3$.

## You Try 7

A line contains the point $(-6, 2)$ and is parallel to the line $y = -\dfrac{3}{2}x + \dfrac{1}{4}$. Write the equation of the line in slope-intercept form.

## Example 8

Find the standard form for the equation of the line that contains the point $(-4, 3)$ and that is perpendicular to $3x - 4y = 8$.

### Solution

Begin by finding the slope of $3x - 4y = 8$ by putting it into slope-intercept form.

$$3x - 4y = 8$$

$$-4y = -3x + 8 \qquad \text{Add } -3x \text{ to each side.}$$

$$y = \dfrac{-3}{-4}x + \dfrac{8}{-4} \qquad \text{Divide by } -4.$$

$$y = \dfrac{3}{4}x - 2 \qquad \text{Simplify.}$$

$$m = \dfrac{3}{4}$$

*Then,* determine the slope of the line containing $(-4, 3)$ by finding the *negative reciprocal* of the slope of the given line.

$$m_{\text{perpendicular}} = -\frac{4}{3}$$

The line we want has $m = -\frac{4}{3}$ and contains the point $(-4, 3)$. Use the point-slope formula to find an equation of the line.

Substitute $-\frac{4}{3}$ for $m$. Substitute $(-4, 3)$ for $(x_1, y_1)$.

$$y - y_1 = m(x - x_1)$$

$$y - 3 = -\frac{4}{3}(x - (-4)) \qquad \text{Substitute } -4 \text{ for } x_1 \text{ and } 3 \text{ for } y_1; \text{ let } m = -\frac{4}{3}.$$

$$y - 3 = -\frac{4}{3}(x + 4)$$

$$y - 3 = -\frac{4}{3}x - \frac{16}{3} \qquad \text{Distribute.}$$

Since we are asked to write the equation in standard form, eliminate the fractions by multiplying each side by 3.

$$3(y - 3) = 3\left(-\frac{4}{3}x - \frac{16}{3}\right)$$

$$\begin{aligned} 3y - 9 &= -4x - 16 \qquad &\text{Distribute.} \\ 3y &= -4x - 7 \qquad &\text{Add 9 to each side.} \\ 4x + 3y &= -7 \qquad &\text{Add } 4x \text{ to each side.} \end{aligned}$$

The equation is $4x + 3y = -7$. ∎

**You Try 8**

Find the equation of the line perpendicular to $5x - y = -6$ containing the point $(10, 0)$. Write the equation in standard form.

## 7. Write a Linear Equation to Model Real-World Data

Equations of lines are often used to describe real-world situations. We will look at an example in which we must find the equation of a line when we are given some data.

**Example 9**

Since 2003, vehicle consumption of E85 fuel (ethanol, 85%) has increased by about 8262.4 thousand gallons per year. In 2006, approximately 61,029.4 thousand gallons were used. (Statistical Abstract of the United States)

a)   Write a linear equation to model these data. Let $x$ represent the number of years after 2003, and let $y$ represent the amount of E85 fuel (in thousands of gallons) consumed.

b)   How much E85 fuel did vehicles use in 2003? in 2005?

### Solution

a)  The statement "vehicle consumption of E85 fuel … has increased by about 8262.4 thousand gallons per year" tells us the rate of change of fuel use with respect to time. Therefore, this is the *slope*. It will be *positive* since consumption is increasing.

$$m = 8262.4$$

The statement "In 2006, approximately 61,029.4 thousand gallons were used" gives us a point on the line.

If $x$ = the number of years after 2003, then the year 2006 corresponds to $x = 3$.

If $y$ = number of gallons (in thousands) of E85 consumed, then 61,029.4 thousand gallons corresponds to $y = 61{,}029.4$.

A point on the line is **(3, 61,029.4).**

Now that we know the slope and a point on the line, we can write an equation of the line using the point-slope formula.

Substitute $8262.4$ for $m$. Substitute $(3, 61{,}029.4)$ for $(x_1, y_1)$.

$$
\begin{aligned}
y - y_1 &= m(x - x_1) \\
y - 61{,}029.4 &= 8262.4(x - 3) && \text{Substitute 3 for } x_1 \text{ and } 61{,}029.4 \text{ for } y_1. \\
y - 61{,}029.4 &= 8262.4x - 24{,}787.2 && \text{Distribute.} \\
y &= 8262.4x + 36{,}242.2 && \text{Add } 61{,}029.4 \text{ to each side.}
\end{aligned}
$$

The equation is $y = 8262.4x + 36{,}242.2$.

b)  To determine the amount of E85 used in 2003, let $x = 0$ since $x$ = the number of years *after* 2003.

$$
\begin{aligned}
y &= 8262.4(0) + 36{,}242.2 && \text{Substitute } x = 0. \\
y &= 36{,}242.2
\end{aligned}
$$

In 2003, vehicles used about 36,242.2 thousand gallons of E85 fuel. Notice that the equation is in slope-intercept form, $y = mx + b$, and our result is $b$. That is because when we find the $y$-intercept we let $x = 0$.

To determine how much E85 fuel was used in 2005, let $x = 2$ since 2005 is 2 years after 2003.

$$
\begin{aligned}
y &= 8262.4(2) + 36{,}242.2 && \text{Substitute } x = 2. \\
y &= 16{,}524.8 + 36{,}242.2 && \text{Multiply.} \\
y &= 52{,}767.0
\end{aligned}
$$

In 2003, vehicles used approximately 52,767.0 thousand gallons of E85.    ■

### Using Technology

We can use a graphing calculator to explore what we have learned about perpendicular lines.

1.  Graph the line $y = -2x + 4$. What is its slope?

2.  Find the slope of the line perpendicular to the graph of $y = -2x + 4$.

3.  Find the equation of the line perpendicular to $y = -2x + 4$ that passes through the point $(6, 0)$. Express the equation in slope-intercept form.

4.  Graph both the original equation and the equation of the perpendicular line:

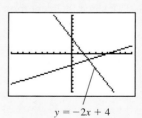

$y = -2x + 4$

5. Do the lines above appear to be perpendicular?

6. Press ZOOM and choose 5:Zsquare.

7. Do the graphs look perpendicular now? Because the viewing window on a graphing calculator is a rectangle, *squaring* the window will give a more accurate picture of the graphs of the equations.

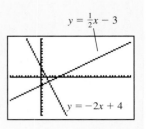

---

### Answers to You Try Exercises

1) a) $5x - 11y = 3$    b) $x - 3y = 21$    2) $y = \dfrac{5}{8}x - 9$    3) $y = 2x - 7$    4) $8x + y = -27$

5) $y = \dfrac{7}{3}x - \dfrac{22}{3}$    6) $y = -8$    7) $y = -\dfrac{3}{2}x - 7$    8) $x + 5y = 10$

---

### Answers to Technology Exercises

1) $-2$    2) $\dfrac{1}{2}$    3) $y = \dfrac{1}{2}x - 3$    5) No, because they do not meet at 90° angles.

7) Yes, because they meet at 90° angles.

---

## 4.5 Exercises

**Objective 1: Rewrite an Equation in Standard Form**

Rewrite each equation in standard form.

1) $y = -2x - 4$

2) $y = 3x + 5$

3) $x = y + 1$

4) $x = -4y - 9$

5) $y = \dfrac{4}{5}x + 1$

6) $y = \dfrac{2}{3}x - 6$

7) $y = -\dfrac{1}{3}x - \dfrac{5}{4}$

8) $y = -\dfrac{1}{4}x + \dfrac{2}{5}$

14) $m = -\dfrac{2}{5}$, $y$-int: $(0, -4)$; standard form

15) $m = \dfrac{2}{7}$, $y$-int: $(0, -3)$; standard form

16) $m = 1$, $y$-int: $(0, 0)$; slope-intercept form

17) $m = -1$, $y$-int: $(0, 0)$; slope-intercept form

18) $m = \dfrac{5}{9}$, $y$-int: $\left(0, -\dfrac{1}{3}\right)$; standard form

**Objective 2: Write an Equation of a Line Given Its Slope and y-Intercept**

9) Explain how to find an equation of a line when you are given the slope and $y$-intercept of the line.

Find an equation of the line with the given slope and $y$-intercept. Express your answer in the indicated form.

10) $m = -3$, $y$-int: $(0, 3)$; slope-intercept form

11) $m = -7$, $y$-int: $(0, 2)$; slope-intercept form

12) $m = 1$, $y$-int: $(0, -4)$; standard form

13) $m = -4$, $y$-int: $(0, 6)$; standard form

**Objective 3: Use the Point-Slope Formula to Write an Equation of a Line Given Its Slope and a Point on the Line**

19) a) If $(x_1, y_1)$ is a point on a line with slope $m$, then the point-slope formula is _____.

   b) Explain how to find an equation of a line when you are given the slope and a point on the line.

Find an equation of the line containing the given point with the given slope. Express your answer in the indicated form.

20) $(2, 3)$, $m = 4$; slope-intercept form

21) $(5, 7)$, $m = 1$; slope-intercept form

22) $(-2, 5)$, $m = -3$; slope-intercept form

23) $(4, -1)$, $m = -5$; slope-intercept form

24) $(-1, -2)$, $m = 2$; standard form

VIDEO 25) $(-2, -1)$, $m = 4$; standard form

26) $(9, 3)$, $m = -\dfrac{1}{3}$; standard form

27) $(-5, 8)$, $m = \dfrac{2}{5}$; standard form

28) $(-2, -3)$, $m = \dfrac{1}{8}$; slope-intercept form

29) $(5, 1)$, $m = -\dfrac{5}{4}$; slope-intercept form

30) $(4, 0)$, $m = -\dfrac{3}{16}$; standard form

31) $(-3, 0)$, $m = \dfrac{5}{6}$; standard form

32) $\left(\dfrac{1}{4}, -1\right)$, $m = 3$; slope-intercept form

### Objective 4: Use the Point-Slope Formula to Write an Equation of a Line Given Two Points on the Line

33) Explain how to find an equation of a line when you are given two points on the line.

Find an equation of the line containing the two given points. Express your answer in the indicated form.

34) $(-2, 1)$ and $(8, 11)$; slope-intercept form

35) $(-1, 7)$ and $(3, -5)$; slope-intercept form

36) $(6, 8)$ and $(-1, -4)$; slope-intercept form

37) $(4, 5)$ and $(7, 11)$; slope-intercept form

38) $(2, -1)$ and $(5, 1)$; standard form

VIDEO 39) $(-2, 4)$ and $(1, 3)$; slope-intercept form

40) $(-1, 10)$ and $(3, -2)$; standard form

41) $(-5, 1)$ and $(4, -2)$; standard form

42) $(4.2, 1.3)$ and $(-3.4, -17.7)$; slope-intercept form

43) $(-3, -11)$ and $(3, -1)$; standard form

44) $(-6, 0)$ and $(3, -1)$; standard form

45) $(-2.3, 8.3)$ and $(5.1, -13.9)$; slope-intercept form

46) $(-7, -4)$ and $(14, 2)$; standard form

Write the slope-intercept form of the equation of each line, if possible.

47)

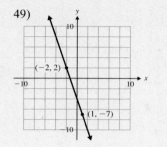

48)

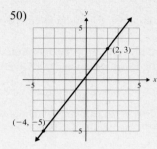

49)

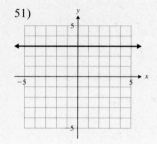

50)

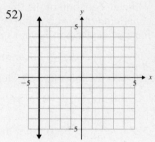

51)

52)

### Mixed Exercises: Objectives 2–5

Write the slope-intercept form of the equation of the line, if possible, given the following information.

53) contains $(-4, 7)$ and $(2, -1)$

54) $m = 2$ and contains $(-3, -2)$

55) $m = 1$ and contains $(3, 5)$

56) $m = \dfrac{7}{5}$ and y-intercept $(0, -4)$

57) y-intercept $(0, 6)$ and $m = 7$

58) contains $(-3, -3)$ and $(1, -7)$

VIDEO 59) vertical line containing $(3, 5)$

60) vertical line containing $\left(\dfrac{1}{2}, 6\right)$

61) horizontal line containing $(2, 3)$

62) horizontal line containing $(5, -4)$

63) $m = -4$ and $y$-intercept $(0, -4)$

64) $m = -\dfrac{2}{3}$ and contains $(3, -1)$

65) $m = -3$ and contains $(10, -10)$

66) contains $(0, 3)$ and $(5, 0)$

67) contains $(-4, -4)$ and $(2, -1)$

68) $m = -1$ and $y$-intercept $(0, 0)$

## Objective 6: Write an Equation of a Line That Is Parallel or Perpendicular to a Given Line

69) What can you say about the equations of two parallel lines?

70) What can you say about the equations of two perpendicular lines?

Write an equation of the line *parallel* to the given line and containing the given point. Write the answer in slope-intercept form or in standard form, as indicated.

71) $y = 4x + 9$; $(0, 2)$; slope-intercept form

72) $y = 8x + 3$; $(0, -3)$; slope-intercept form

73) $y = 4x + 2$; $(-1, -4)$; standard form

74) $y = \dfrac{2}{3}x - 6$; $(6, 6)$; standard form

75) $x + 2y = 22$; $(-4, 7)$; standard form

76) $3x + 5y = -6$; $(-5, 8)$; standard form

77) $15x - 3y = 1$; $(-2, -12)$; slope-intercept form

78) $x + 6y = 12$; $(-6, 8)$; slope-intercept form

Write an equation of the line *perpendicular* to the given line and containing the given point. Write the answer in slope-intercept form or in standard form, as indicated.

79) $y = -\dfrac{2}{3}x + 4$; $(4, 2)$; slope-intercept form

80) $y = -\dfrac{5}{3}x + 10$; $(10, 5)$; slope-intercept form

81) $y = -5x + 1$; $(10, 0)$; standard form

82) $y = \dfrac{1}{4}x - 9$; $(-1, 7)$; standard form

83) $y = x$; $(4, -9)$; slope-intercept form

84) $x + y = 9$; $(4, 4)$; slope-intercept form

85) $x + 3y = 18$; $(4, 2)$; standard form

86) $12x - 15y = 10$; $(16, -25)$; standard form

Write the slope-intercept form (if possible) of the equation of the line meeting the given conditions.

87) parallel to $3x + y = 8$ containing $(-4, 0)$

88) perpendicular to $x - 5y = -4$ containing $(3, 5)$

89) perpendicular to $y = x - 2$ containing $(2, 9)$

90) parallel to $y = 4x - 1$ containing $(-3, -8)$

91) parallel to $y = 1$ containing $(-3, 4)$

92) parallel to $x = -3$ containing $(-7, -5)$

93) perpendicular to $x = 0$ containing $(9, 2)$

94) perpendicular to $y = 4$ containing $(-4, -5)$

95) perpendicular to $21x - 6y = 2$ containing $(4, -1)$

96) parallel to $-3x + 4y = 8$ containing $(9, 4)$

97) parallel to $y = 0$ containing $\left(4, -\dfrac{3}{2}\right)$

98) perpendicular to $y = \dfrac{7}{3}$ containing $(-7, 9)$

## Objective 7: Write a Linear Equation to Model Real-World Data

99) The graph shows the average annual wage of a mathematician in the United States from 2005 to 2008. $x$ represents the number of years after 2005 so that $x = 0$ represents 2005, $x = 1$ represents 2006, and so on. Let $y$ represent the average annual wage of a mathematician. (www.bls.gov)

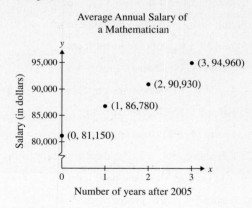

Average Annual Salary of a Mathematician

a) Write a linear equation to model these data. Use the data points for 2005 and 2008, and round the slope to the nearest tenth.

b) Explain the meaning of the slope in the context of this problem.

c) If the current trend continues, find the average salary of a mathematician in 2014.

100) The graph shows a dieter's weight over a 12-week period. Let $y$ represent his weight $x$ weeks after beginning his diet.

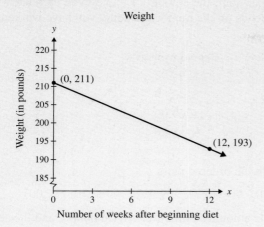

Weight

Weight (in pounds)

(0, 211)

(12, 193)

Number of weeks after beginning diet

a) Write a linear equation to model these data. Use the data points for week 0 and week 12.

b) What is the meaning of the slope in the context of this problem?

c) If he keeps losing weight at the same rate, what will he weigh 13 weeks after he started his diet?

101) In 2007, a grocery store chain had an advertising budget of $500,000 per year. Every year since then its budget has been cut by $15,000 per year. Let $y$ represent the advertising budget, in dollars, $x$ years after 2007.

a) Write a linear equation to model these data.

b) Explain the meaning of the slope in the context of the problem.

c) What was the advertising budget in 2010?

d) If the current trend continues, in what year will the advertising budget be $365,000?

102) A temperature of $-10°C$ is equivalent to $14°F$, while $15°C$ is the same as $59°F$. Let $F$ represent the temperature on the Fahrenheit scale and $C$ represent the temperature on the Celsius scale.

a) Write a linear equation to convert from degrees Celsius to degrees Fahrenheit. That is, write an equation for $F$ in terms of $C$.

b) Explain the meaning of the slope in the context of the problem.

c) Convert $24°C$ to degrees Fahrenheit.

d) Change $95°F$ to degrees Celsius.

103) A kitten weighs an average of 100 g at birth and should gain about 8 g per day for the first few weeks of life. Let $y$ represent the weight of a kitten, in grams, $x$ days after birth. (http://veterinarymedicine.dvm360.com)

a) Write a linear equation to model these data.

b) Explain the meaning of the slope in the context of the problem.

c) How much would an average kitten weigh 5 days after birth? 2 weeks after birth?

d) How long would it take for a kitten to reach a weight of 284 g?

104) In 2000, Red Delicious apples cost an average of $0.82 per lb, and in 2007 they cost $1.12 per lb. Let $y$ represent the cost of a pound of Red Delicious apples $x$ years after 2000. (www.census.gov)

a) Write a linear equation to model these data. Round the slope to the nearest hundredth.

b) Explain the meaning of the slope in the context of the problem.

c) Find the cost of a pound of apples in 2003.

d) When was the average cost about $1.06/lb?

105) If a woman wears a size 6 on the U.S. shoe size scale, her European size is 38. A U.S. women's size 8.5 corresponds to a European size 42. Let $A$ represent the U.S. women's shoe size, and let $E$ represent that size on the European scale.

a) Write a liner equation that models the European shoe size in terms of the U.S. shoe size.

b) If a woman's U.S. shoe size is 7.5, what is her European shoe size? (Round to the nearest unit.)

106) If a man's foot is 11.5 inches long, his U.S. shoe size is 12.5. A man wears a size 8 if his foot is 10 inches long. Let $L$ represent the length of a man's foot, and let $S$ represent his shoe size.

a) Write a linear equation that describes the relationship between shoe size in terms of the length of a man's foot.

b) If a man's foot is 10.5 inches long, what is his shoe size?

# Section 4.6 Introduction to Functions

## Objectives

1. **Define and Identify Relation, Function, Domain, and Range**
2. **Use the Vertical Line Test**
3. **Find the Domain and Range of a Relation from Its Graph**
4. **Find the Domain of a Function Using Its Equation**
5. **Use Function Notation**
6. **Define and Graph a Linear Function**
7. **Solve Problems Using Linear Functions**

If you have a job and you earn $9.50 per hour, the amount of money you earn each week before deductions (gross earnings) depends on the number of hours you have worked.

| Hours Worked | Gross Earnings |
|---|---|
| 10 hours | $ 95.00 |
| 15 hours | $142.50 |
| 22 hours | $209.00 |
| 30 hours | $285.00 |

We can express these relationships with the ordered pairs

$(10, 95.00)$        $(15, 142.50)$        $(22, 209.00)$        $(30, 285.00)$

where the first coordinate represents the amount of time worked (in hours), and the second coordinate represents the gross earnings (in dollars).

We can also describe this relationship with the equation

$$y = 9.50x$$

where $y$ is the gross earnings, in dollars, and $x$ is the number of hours worked.

## 1. Define and Identify Relation, Function, Domain, and Range

### Relations and Functions

If we form a set of ordered pairs from the ones listed above, we get a *relation:*

$\{(10, 95.00), (15, 142.50), (22, 209.00), (30, 285.00)\}$

### Definition

A **relation** is any set of ordered pairs.

### Definition

The **domain** of a relation is the set of all values of the first coordinates in the set of ordered pairs. The **range** of a relation is the set of all values of the second coordinates in the set of ordered pairs.

The domain of the given relation is $\{10, 15, 22, 30\}$. The range of the relation is $\{95.00, 142.50, 209.00, 285.00\}$.

The relation $\{(10, 95.00), (15, 142.50), (22, 209.00), (30, 285.00)\}$ is also a *function* because every first coordinate corresponds to *exactly one* second coordinate. A function is a very important concept in mathematics.

### Definition

A **function** is a special type of relation. If each element of the domain corresponds to *exactly one* element of the range, then the relation is a function.

Relations and functions can be represented in another way—as a *correspondence* or a *mapping* from one set, the domain, to another, the range. In this representation, the domain is the set of all values in the first set, and the range is the set of all values in the second set.

Identify the domain and range of each relation, and determine whether each relation is a function.

a)   $\{(-3, -2), (4, 0), (5, 3), (5, -3)\}$

b)   $\left\{(-4, -1), (-2, 0), (0, 1), \left(3, \dfrac{5}{2}\right), (4, 3)\right\}$

c)

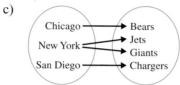

### Solution

a)   The *domain* is the set of first coordinates, $\{-3, 4, 5\}$. (We write the 5 in the set only once even though it appears in two ordered pairs.) The *range* is the set of second coordinates, $\{-2, 0, 3, -3\}$.

   To determine whether or not this relation is a function, ask yourself, *"Does every first coordinate correspond to exactly one second coordinate?"* *No.* In the ordered pairs (5, 3) and (5, −3), the same first coordinate, 5, corresponds to two different second coordinates, 3 and −3. Therefore, this relation is *not* a function.

b)   The *domain* is $\{-4, -2, 0, 3, 4\}$. The *range* is $\left\{-1, 0, 1, \dfrac{5}{2}, 3\right\}$.

   Ask yourself, "Does every first coordinate correspond to *exactly one* second coordinate?" *Yes.* This relation *is* a function.

c)   The *domain* is {Chicago, New York, San Diego}. The *range* is {Bears, Jets, Giants, Chargers}.

   One of the elements in the domain, New York, corresponds to *two* elements in the range, Jets and Giants. Therefore, this relation is *not* a function.   ∎

Identify the domain and range of each relation, and determine whether each relation is a function.

a)  $\{(-5, -3), (-4, 0), (2, 7), (8, 8), (10, 15)\}$          b)   $\{(-2, -13), (-2, 13), (0, -3), (1, -2)\}$

c)

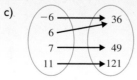

## 2. Use the Vertical Line Test

We know that a relation is a function if each element of the domain corresponds to *exactly one* element of the range.

   If the ordered pairs of a relation are such that the first coordinates represent *x*-values and the second coordinates represent *y*-values (the ordered pairs are in the form $(x, y)$), then we can think of the definition of a function in this way:

## Definition

A relation is a **function** if each *x*-value corresponds to exactly one *y*-value.

What does a function look like when it is graphed? Let's look at the graphs of the ordered pairs in the relations of Example 1a) and 1b).

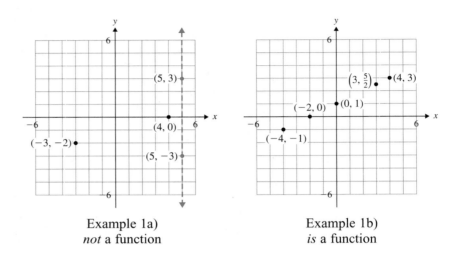

Example 1a)
*not* a function

Example 1b)
*is* a function

The relation in Example 1a) is *not* a function since the *x*-value of 5 corresponds to *two different y*-values, 3 and −3. Notice that we can draw a vertical line that intersects the graph in more than one point—the line through (5, 3) and (5, −3).

The relation in Example 1b), however, *is* a function—each *x*-value corresponds to only one *y*-value. Here we cannot draw a vertical line through more than one point on this graph.

This leads us to the **vertical line test** for a function.

## Procedure   The Vertical Line Test

If there is no vertical line that can be drawn through a graph so that it intersects the graph more than once, then the graph represents a function.

If a vertical line *can* be drawn through a graph so that it intersects the graph more than once, then the graph does *not* represent a function.

## Example 2

Use the vertical line test to determine whether each graph, in blue, represents a function.

a)

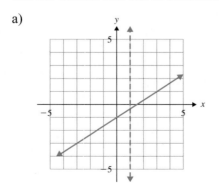

b)

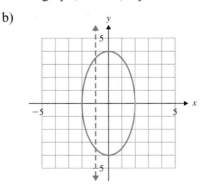

### Solution

a)  Anywhere a vertical line is drawn through the graph, the line will intersect the graph only once. *This graph represents a function.*

b)  This graph fails the vertical line test because we can draw a vertical line through the graph that intersects it more than once. *This graph does* not *represent a function.*    ∎

We can identify the domain and range of a relation or function from its graph.

### 3. Find the Domain and Range of a Relation from Its Graph

**Example 3**

Identify the domain and range of each relation in Example 2.

### Solution

a)

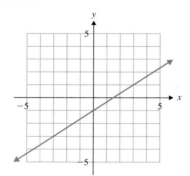

The arrows on the graph indicate that the graph continues without bound.

The domain of this function is the set of $x$-values on the graph. Since the graph continues indefinitely in the $x$-direction, the domain is the set of all real numbers. *The domain is* $(-\infty, \infty)$.

The range is the set of $y$-values on the graph. Since the arrows show that the graph continues indefinitely in the $y$-direction, *the range is* the set of all real numbers or $(-\infty, \infty)$.

b)

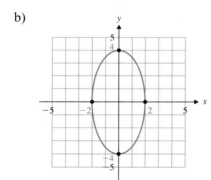

The set of $x$-values on the graph includes all real numbers from $-2$ to $2$. *The domain is* $[-2, 2]$.

The set of $y$-values on the graph includes all real numbers from $-4$ to $4$. *The range is* $[-4, 4]$.    ∎

Part (a) in Example 3 suggests the following point:

**Note**

A linear equation of the form $y = mx + b$ is a function.

Equations of the form $x = c$ are vertical lines with undefined slope and are not functions.

**You Try 2**

Use the vertical line test to determine whether each relation is also a function. Then, identify the domain and range.

a)

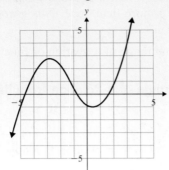

b)

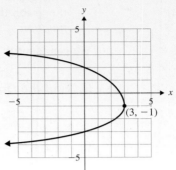

## 4. Find the Domain of a Function Using Its Equation

We have seen how to determine the domain of a relation written as a set of ordered pairs, as a correspondence (or mapping), and as a graph. Next, we will discuss how to determine the domain of a relation written as an equation.

Sometimes, it is helpful to ask yourself, "Is there any number that *cannot* be substituted for $x$?"

**Example 4**

Determine the domain of each function.

a)  $y = 3x - 1$        b)  $y = x^2$        c)  $y = \dfrac{1}{x}$

**Solution**

a)  To determine the domain of $y = 3x - 1$, ask yourself, "Is there any number that *cannot* be substituted for $x$?" *No.* Any real number can be substituted for $x$, and $y = 3x - 1$ will be defined. The domain consists of all real numbers and can be written as $(-\infty, \infty)$.

b)  Ask yourself, "Is there any number that *cannot* be substituted for $x$ in $y = x^2$?" *No.* Any real number can be substituted for $x$. The domain is all real numbers, $(-\infty, \infty)$.

c)  To determine the domain of $y = \dfrac{1}{x}$, ask yourself, "Is there any number that *cannot* be substituted for $x$?" *Yes. $x$ cannot equal zero because a fraction is undefined if its denominator equals zero.* Any other number can be substituted for $x$ and $y = \dfrac{1}{x}$ will be defined.

The domain consists of all real numbers *except* 0. We can write the domain in interval notation as $(-\infty, 0) \cup (0, \infty)$.  ∎

---

**Procedure**    Finding the Domain of a Function

To find the domain of a function written as an equation in terms of $x$:

1)  Ask yourself, "Is there any number that *cannot* be substituted for $x$?"

2)  If $x$ is in the denominator of a fraction, determine what value of $x$ will make the denominator equal 0 by setting the expression equal to zero. Solve for $x$. This $x$-value is *not* in the domain.

The domain consists of all real numbers that can be substituted for $x$.

**You Try 3**

Determine the domain of each function.

a)  $y = x + 8$          b)  $y = x^2 + 3$          c)  $y = \dfrac{5}{x - 2}$

## 5. Use Function Notation

We can use *function notation* to name functions. If a relation is a function, then $f(x)$ can be used in place of $y$. $f(x)$ *is the same as* $y$.

For example, $y = x + 2$ is a function. We can also write $y = x + 2$ as $f(x) = x + 2$. *They mean the same thing.*

> **Definition**
>
> If $y$ represents a function, then the **function notation** $f(x)$ is the same as $y$. We read $f(x)$ as "$f$ of $x$," and we can say $y = f(x)$.

**Example 5**

a)  Evaluate $y = x + 2$ for $x = 5$.

b)  If $f(x) = x + 2$, find $f(5)$.

**Solution**

a)  To evaluate $y = x + 2$ for $x = 5$ means to substitute 5 for $x$ and find the corresponding value of $y$.

$$y = x + 2$$
$$y = 5 + 2 \qquad \text{Substitute 5 for } x.$$
$$y = 7$$

When $x = 5$, $y = 7$. We can also say that the ordered pair $(5, 7)$ satisfies $y = x + 2$.

b)  To find $f(5)$ (read as "$f$ of 5") means to find the value of the function when $x = 5$.

$$f(x) = x + 2$$
$$f(5) = 5 + 2 \qquad \text{Substitute 5 for } x.$$
$$f(5) = 7$$

We can also say that the ordered pair $(5, 7)$ satisfies $f(x) = x + 2$ where the ordered pair represents $(x, f(x))$.  ■

> **Note**
>
> Example 5 illustrates that evaluating $y = x + 2$ for $x = 5$ and finding $f(5)$ when $f(x) = x + 2$ are *exactly* the same thing. Remember, $f(x)$ is another name for $y$.

**You Try 4**

a)  Evaluate $y = 3x - 7$ for $x = -2$.          b)  If $f(x) = 3x - 7$, find $f(-2)$.

Different letters can be used to name functions. $g(x)$ is read as "$g$ of $x$," $h(x)$ is read as "$h$ of $x$," and so on. Also, the function notation does *not* indicate multiplication; $f(x)$ does *not* mean $f$ times $x$.

**BE CAREFUL**

$f(x)$ does *not* mean $f$ times $x$.

We can also think of a function as a machine—we put values into the machine, and the equation determines the values that come out. We can think of the function in Example 5b) like this:

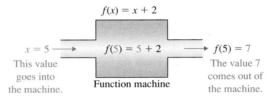

$f(x) = x + 2$

$x = 5 \longrightarrow$     $f(5) = 5 + 2$     $\longrightarrow f(5) = 7$

This value          goes into                                The value 7
the machine.    **Function machine**    comes out of
                                                            the machine.

Sometimes, we call evaluating a function for a certain value *finding a function value*.

## Example 6

Let $f(x) = 5x - 8$ and $g(x) = x^2 - 3x + 1$. Find the following function values.

a)  $f(4)$          b)  $g(-2)$

**Solution**

a)  "Find $f(4)$" means to find the value of the function when $x = 4$. Substitute 4 for $x$.

$$f(x) = 5x - 8$$
$$f(4) = 5(4) - 8 = 20 - 8 = 12$$
$$f(4) = 12$$

We can also say that the ordered pair $(4, 12)$ satisfies $f(x) = 5x - 8$.

b)  To find $g(-2)$, substitute $-2$ for every $x$ in the function $g(x)$.

$$g(x) = x^2 - 3x + 1$$
$$g(-2) = (-2)^2 - 3(-2) + 1 = 4 + 6 + 1 = 11$$
$$g(-2) = 11$$

The ordered pair $(-2, 11)$ satisfies $g(x) = x^2 - 3x + 1$. ■

## You Try 5

Let $f(x) = -6x + 5$ and $h(x) = x^2 + 9x - 7$. Find the following function values.

a)  $f(1)$          b)  $f(-3)$          c)  $h(-2)$          d)  $h(0)$

## 6. Define and Graph a Linear Function

Earlier in this chapter, we learned that a linear equation can have the form $y = mx + b$. Similarly, a **linear function** has the form $f(x) = mx + b$. Its domain is all real numbers.

**Example 7**

Graph $f(x) = -\dfrac{1}{2}x + 2$ using the slope and $y$-intercept.

**Solution**

$$f(x) = -\dfrac{1}{2}x + 2$$

$$m = -\dfrac{1}{2} \qquad y\text{-int: } (0, 2)$$

To graph this function, first plot the $y$-intercept, $(0, 2)$, then use the slope to locate another point on the line.

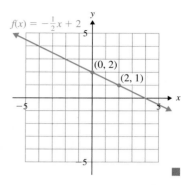

**You Try 6**

Graph $f(x) = 3x - 4$ using the slope and $y$-intercept.

## 7. Solve Problems Using Linear Functions

The domain of a function does not have to be represented by $x$. When using functions to solve problems, we often choose a more "meaningful" letter to represent a quantity. The same is true for naming the function.

**Example 8**

A compact disk is read at 44.1 kHz (kilohertz). This means that a CD player scans 44,100 samples of sound per second on a CD to produce the sound that we hear. The function

$$S(t) = 44,100t$$

tells us how many samples of sound, $S(t)$, are read after $t$ seconds. (www.mediatechnics.com)

a)   How many samples of sound are read in 20 sec?
b)   How many samples of sound are read in 1.5 min?
c)   How long would it take the CD player to scan 1,764,000 samples of sound?
d)   What is the smallest value $t$ could equal?

**Solution**

a)   To determine how much sound is read in 20 sec, let $t = 20$ and find $S(20)$.

$$S(t) = 44,100t$$
$$S(20) = 44,100(20) \qquad \text{Substitute 20 for } t.$$
$$S(20) = 882,000 \qquad \text{Multiply.}$$

The CD player has read 882,000 samples of sound.

b)   To determine how much sound is read in 1.5 min, do we let $t = 1.5$ and find $S(1.5)$? *No.* Recall that $t$ is in *seconds*. Change 1.5 min to seconds before substituting for $t$. We must use the correct units in the function.

$$1.5 \text{ min} = 90 \text{ sec}$$

Let $t = 90$ and find $S(90)$.

$$S(t) = 44{,}100t$$
$$S(90) = 44{,}100(90)$$
$$S(90) = 3{,}969{,}000$$

It has read 3,969,000 samples of sound.

c)  Since we are asked to determine *how long* it would take a CD player to scan 1,764,000 samples of sound, we will be solving for $t$. What do we substitute for $S(t)$? We substitute 1,764,000 for $S(t)$ and find $t$. That is, find $t$ when $S(t) = 1{,}764{,}000$.

$$S(t) = 44{,}100t$$
$$1{,}764{,}000 = 44{,}100t \qquad \text{Substitute } 1{,}764{,}000 \text{ for } S(t).$$
$$40 = t \qquad \text{Divide by } 44{,}100.$$

It will take 40 sec for the CD player to scan 1,764,000 samples of sound.

d)  Since $t$ represents the number of seconds a CD has been playing, the smallest value that makes sense for $t$ is 0.  ∎

## Using Technology

A graphing calculator can be used to represent a function as a graph and also as a table of values. Consider the function $f(x) = 2x - 5$. To graph the function, press [Y=], then type $2x - 5$ to the right of \Y1=. Press [ZOOM] and select 6:ZStandard to graph the equation as shown on the left below. We can select a point on the graph. For example, press [TRACE], type 4, and press [ENTER]. The point $(4, 3)$ is displayed on the screen as shown below on the right.

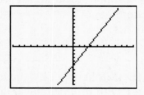
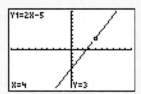

The function can also be represented as a table on a graphing calculator. To set up the table, press [2nd] [WINDOW], move the cursor after TblStart =, and enter a number such as 0 to set the starting $x$-value for the table. Enter 1 after ΔTbl = to set the increment between $x$-values as shown on the left below. Then press [2nd] [GRAPH] to display the table as shown on the right below.

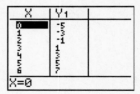

The point $(4, 3)$ is represented in the table above as well as on the graph.

Given the function, find the function value on a graph and a table using a graphing calculator.

1)  $f(x) = 3x - 4; f(2)$    2)  $f(x) = 4x - 1; f(1)$    3)  $f(x) = -3x + 7; f(1)$
4)  $f(x) = 2x + 5; f(-1)$    5)  $f(x) = 2x - 7; f(-1)$    6)  $f(x) = -x + 5; f(4)$

...

**Answers to You Try Exercises**

1)   a) domain: $\{-5, -4, 2, 8, 10\}$; range: $\{-3, 0, 7, 8, 15\}$; function   b) domain: $\{-2, 0, 1\}$; range: $\{-13, 13, -3, -2\}$; not a function   c) domain: $\{-6, 6, 7, 11\}$; range: $\{36, 49, 121\}$; function

2)   a) function; domain: $(-\infty, \infty)$; range: $(-\infty, \infty)$   b) not a function; domain: $(-\infty, 3]$; range: $(-\infty, \infty)$

3)   a) $(-\infty, \infty)$   b) $(-\infty, \infty)$   c) $(-\infty, 2) \cup (2, \infty)$       4)   a) $y = -13$   b) $f(-2) = -13$

5)   a) $f(1) = -1$   b) $f(-3) = 23$   c) $h(-2) = -21$   d) $h(0) = -7$

6)

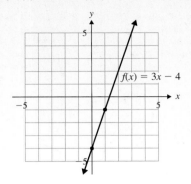

$f(x) = 3x - 4$

**Answers to Technology Exercises**

1) 2          2) 3          3) 4          4) 3          5) −9          6) 1

---

# 4.6 Exercises

**Mixed Exercises: Objectives 1–3**

1) a) What is a relation?

   b) What is a function?

   c) Give an example of a relation that is also a function.

2) Give an example of a relation that is *not* a function.

Identify the domain and range of each relation, and determine whether each relation is a function.

3) $\{(4, 10), (-4, 2), (1, -4), (8, -3)\}$

4) $\{(1, 3), (3, 4), (3, 2), (4, 5), (9, 1)\}$

5) $\{(9, -1), (25, -3), (1, 1), (9, 5), (25, 7)\}$

6) $\{(-5, 1), (3, 3), (-2, 4), (6, -7), (5, -1)\}$

7) $\left\{(-4, -2), \left(-3, -\dfrac{1}{2}\right), \left(-1, -\dfrac{1}{2}\right), (0, -2)\right\}$

8) $\{(-4, 1), (4, 7), (2, 7), (-2, 8)\}$

9) $\{(-2.3, 6.2), (3.0, 7.8), (3.0, 3.1), (-4.1, -5.7)\}$

10) $\{(5, 2), (6, 4), (1, 3), (0, 3)\}$

11)

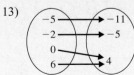

12)

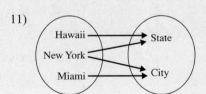

13)

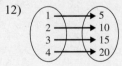

14)

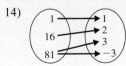

15)

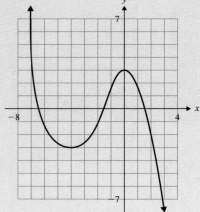

16)

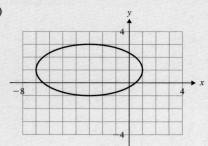

17)

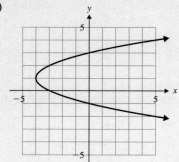

18)

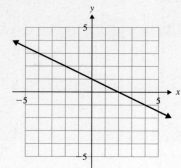

 19)

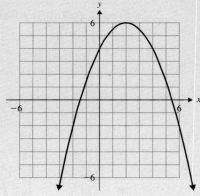

20)

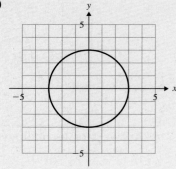

## Objective 4: Find the Domain of a Function Using Its Equation

Determine the domain of each function.

21) $y = 3x + 1$                  22) $y = -7x - 2$

23) $y = -\dfrac{2}{5}x - 8$      24) $y = \dfrac{1}{4}x + 9$

25) $y = x^2 - 4$                 26) $y = -x^2 + 2$

27) $y = x^3$                     28) $y = 2x^4$

29) $y = \dfrac{2}{x}$            30) $y = \dfrac{9}{x}$

31) $y = \dfrac{12}{x - 5}$       32) $y = \dfrac{4}{x + 9}$

33) $y = \dfrac{1}{x + 1}$        34) $y = \dfrac{2}{x - 3}$

35) $y = \dfrac{6}{x - 20}$       36) $y = \dfrac{1}{x + 11}$

## Objective 5: Use Function Notation

37) What is the meaning of the notation $y = f(x)$?

38) Does $y = f(x)$ mean "$y = f$ times $x$"? Explain.

39) a) Evaluate $y = 2x - 1$ for $x = 6$.

   b) If $f(x) = 2x - 1$, find $f(6)$.

40) a) Evaluate $y = -x - 5$ for $x = -3$.

   b) If $f(x) = -x - 5$, find $f(-3)$.

41) a) Evaluate $y = -4x + 3$ for $x = 3$.

   b) If $f(x) = -4x + 3$, find $f(3)$.

42) a) Evaluate $y = -5x - 6$ for $x = -1$.

   b) If $f(x) = -5x - 6$, find $f(-1)$.

For Exercises 43–52, let $f(x) = 2x - 11$ and $g(x) = x^2 - 4x + 2$, and find the following function values.

43) $f(4)$          44) $f(1)$          45) $f(0)$

46) $f\left(\dfrac{11}{2}\right)$       47) $g(2)$          48) $g(-1)$

49) $g(1)$          50) $g(0)$          51) $g\left(\dfrac{1}{2}\right)$

52) $g\left(-\dfrac{1}{3}\right)$

For each function $f$ in Exercises 53–60, find $f(-2)$ and $f(5)$.

53) $f = \{(-1, 18), (-2, 12), (0, 4), (5, 4), (4, -3)\}$

54) $f = \{(5, 10), (0, 7), (6, 4), (4, -5), (-2, -4)\}$

55) $f = \left\{\left(-\dfrac{3}{4}, -1\right), (-2, 4), (5, 2), (8, 10)\right\}$

56) $f = \left\{(-8, -1), \left(-2, \dfrac{3}{4}\right), (10, 8), (5, -1)\right\}$

57)

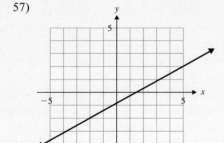

58)

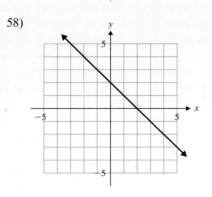

59)

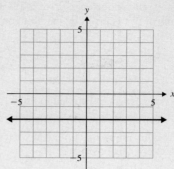

60)

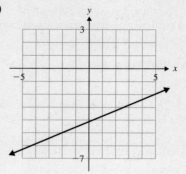

61) $f(x) = -3x - 2$. Find $x$ so that $f(x) = 10$.

62) $f(x) = -x + 7$. Find $x$ so that $f(x) = -2$.

63) $h(x) = -\dfrac{3}{2}x - 5$. Find $x$ so that $h(x) = 1$.

64) $g(x) = \dfrac{3}{4}x + 6$. Find $x$ so that $g(x) = 12$.

**Objective 6: Define and Graph a Linear Function**

Graph each function using the slope and $y$-intercept.

65) $f(x) = -4x - 1$          66) $f(x) = -3x - 5$

67) $g(x) = -\dfrac{1}{3}x - 4$          68) $h(x) = \dfrac{3}{5}x - 6$

69) $h(x) = 5x - 1$          70) $g(x) = 7x - 2$

71) $g(x) = \dfrac{3}{2}x - \dfrac{5}{2}$          72) $h(x) = -\dfrac{7}{3}x + 1$

**Objective 7: Solve Problems Using Linear Functions**

73) The amount Fiona pays a babysitter, $A$ (in dollars), for babysitting $t$ hours can be described by the function $A(t) = 9t$.

   a) How much does she owe the babysitter for working 4 hr?

   b) How much does she owe the babysitter for working 6.5 hr?

   c) For how long could Fiona hire the sitter for $76.50?

   d) What is the slope of this linear function, and what does it mean in the context of this problem?

74) The number of miles, $N$, that Ahmed can drive in his hybrid car on $g$ gallons of gas is given by the function $N(g) = 46g$.

   a) How far can Ahmed's car go on 5 gallons of gas?

   b) How far can he drive on 11 gallons of gas?

   c) How many gallons of gas would he need to drive 437 miles?

   d) What is the slope of this linear function, and what does it mean in the context of this problem?

75) A website sells concert tickets and charges a one-time $12.00 service fee. The total cost, $C$ (in dollars), of buying $t$ tickets to a particular concert from this website is given by the function $C(t) = 59.50t + 12.00$.

   a) Find the cost of 2 tickets.

   b) Find the cost of 4 tickets.

   c) If Yoshiko paid $428.50 for tickets from this website, how many did she buy?

76) When Ezra works an 8-hour day, he earns $90.40 plus a 5% commission on everything he sells. Ezra's daily gross pay, $P$ (in dollars), can be described by the function $P(s) = 0.05s + 90.40$, where $s$ is the value of his sales, in dollars.

   a) How much would he earn if he sold $400.00 worth of merchandise?

   b) Find Ezra's gross pay if his sales totaled $290.00.

   c) If Ezra's gross pay on a Saturday was $143.40, find the value of the merchandise he sold that day.

When skeletal remains are found at a crime scene, forensic scientists use formulas based on the person's sex, ethnicity, and bone length to estimate the person's height. Use the following formulas to answer the questions. (The *tibia* is a bone in the lower leg, and the *humerus* is a bone in the upper arm.) (http://forensics.rice.edu)

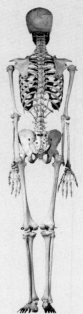

77) The height, $h$ (in cm), of an Asian male in terms of the length of his tibia, $t$ (in cm), is $h(t) = 2.39t + 81.45$. Find the height of an Asian man if his tibia is 41.6 cm long. Round to the nearest tenth of a cm.

78) The height, $h$ (in cm), of a Caucasian male in terms of the length of his tibia, $t$ (in cm), is $h(t) = 2.42t + 81.93$. Find the height of a Caucasian man if his tibia is 41.6 cm long. Round to the nearest tenth of a cm.

79) The height, $h$ (in cm), of an African-American female in terms of the length of her humerus, $L$ (in cm) is $h(L) = 3.08L + 64.67$. Find the height of an African-American woman if her humerus is 33.5 cm long. Round to the nearest tenth of a cm.

80) The height, $h$ (in cm), of a Caucasian female in terms of the length of her humerus, $L$ (in cm) is $h(L) = 3.36L + 57.97$. Find the height of a Caucasian woman if her humerus is 33.5 cm long. Round to the nearest tenth of a cm.

| Definition/Procedure | Example |
|---|---|

## 4.1 Introduction to Linear Equations in Two Variables

A **linear equation in two variables** can be written in the form $Ax + By = C$, where $A$, $B$, and $C$ are real numbers and where both $A$ and $B$ do not equal zero.

To determine whether an ordered pair is a solution of an equation, substitute the values for the variables. **(p. 202)**

Is $(4, -1)$ a solution of $3x - 5y = 17$?
Substitute 4 for $x$ and $-1$ for $y$.

$$3x - 5y = 17$$
$$3(4) - 5(-1) = 17$$
$$12 + 5 = 17$$
$$17 = 17 \checkmark$$

Yes, $(4, -1)$ is a solution.

## 4.2 Graphing by Plotting Points and Finding Intercepts

The graph of a linear equation in two variables, $Ax + By = C$, is a straight line. Each point on the line is a solution to the equation.

We can graph the line by plotting the points and drawing the line through them. **(p. 214)**

Graph $y = \frac{1}{3}x + 2$ by plotting points.

Make a table of values. Plot the points, and draw a line through them.

| x | y |
|---|---|
| 0 | 2 |
| 3 | 3 |
| -3 | 1 |

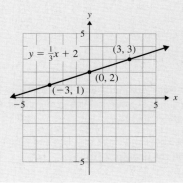

The **x-intercept** of an equation is the point where the graph intersects the x-axis. To find the *x-intercept* of the graph of an equation, let $y = 0$ and solve for $x$.

The **y-intercept** of an equation is the point where the graph intersects the y-axis. To find the *y-intercept* of the graph of an equation, let $x = 0$ and solve for $y$. **(p. 216)**

Graph $2x + 5y = -10$ by finding the intercepts and another point on the line.
*x-intercept*: Let $y = 0$, and solve for $x$.

$$2x + 5(0) = -10$$
$$2x = -10$$
$$x = -5$$

The *x-intercept* is $(-5, 0)$.

*y-intercept*: Let $x = 0$, and solve for $y$.

$$2(0) + 5y = -10$$
$$5y = -10$$
$$y = -2$$

The *y-intercept* is $(0, -2)$.
Another point on the line is $(5, -4)$.
Plot the points, and draw the line through them.

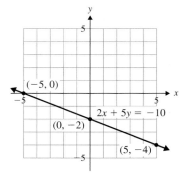

| Definition/Procedure | Example |
|---|---|
| If $c$ is a constant, then the graph of $x = c$ is a *vertical line* going through the point $(c, 0)$.<br><br>If $d$ is a constant, then the graph of $y = d$ is a *horizontal line* going through the point $(0, d)$. **(p. 219)** | Graph $x = -2$.　　　　Graph $y = 4$.<br><br>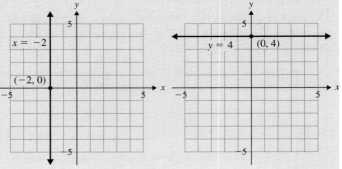 |

### 4.3 The Slope of a Line

| | |
|---|---|
| The **slope** of a line is the ratio of the vertical change in $y$ to the horizontal change in $x$. Slope is denoted by $m$.<br><br>The slope of a line containing the points $(x_1, y_1)$ and $(x_2, y_2)$ is<br><br>$$m = \frac{y_2 - y_1}{x_2 - x_1}.$$<br><br>The slope of a horizontal line is zero.<br>The slope of a vertical line is undefined. **(p. 225)** | Find the slope of the line containing the points $(4, -3)$ and $(-1, 5)$.<br><br>$$m = \frac{y_2 - y_1}{x_2 - x_1}$$<br>$$= \frac{5 - (-3)}{-1 - 4} = \frac{8}{-5} = -\frac{8}{5}$$<br><br>The slope of the line is $-\dfrac{8}{5}$. |
| If we know the slope of a line and a point on the line, we can graph the line. **(p. 230)** | Graph the line containing the point $(-2, 3)$ with a slope of $-\dfrac{5}{6}$.<br><br>Start with the point $(-2, 3)$, and use the slope to plot another point on the line.<br><br>$$m = \frac{-5}{6} = \frac{\text{Change in } y}{\text{Change in } x}$$<br><br>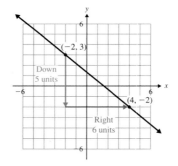 |

| Definition/Procedure | Example |
|---|---|

## 4.4 The Slope-Intercept Form of a Line

The **slope-intercept form of a line** is $y = mx + b$, where $m$ is the slope and $(0, b)$ is the y-intercept.

If a line is written in slope-intercept form, we can use the y-intercept and the slope to graph the line. **(p. 236)**

Write the equation in slope-intercept form and graph it.

$$8x - 3y = 6$$
$$-3y = -8x + 6$$
$$y = \frac{-8}{-3}x + \frac{6}{-3}$$
$$y = \frac{8}{3}x - 2 \qquad \text{Slope-intercept form}$$

$m = \frac{8}{3}$, y-intercept $(0, -2)$

Plot $(0, -2)$, then use the slope to locate another point on the line. We will think of the slope as

$$m = \frac{8}{3} = \frac{\text{Change in } y}{\text{Change in } x}.$$

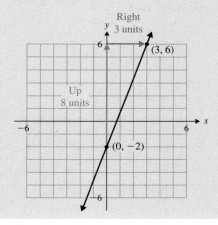

**Parallel lines** have the same slope.

**Perpendicular lines** have slopes that are negative reciprocals of each other. **(p. 238)**

Determine whether the lines $2x + y = 18$ and $x - 2y = 7$ are parallel, perpendicular, or neither.

Put each line into slope-intercept form to find their slopes.

$$2x + y = 18 \qquad\qquad x - 2y = 7$$
$$y = -2x + 18 \qquad\quad -2y = -x + 7$$
$$y = \frac{1}{2}x - \frac{7}{2}$$

$$m = -2 \qquad\qquad\qquad m = \frac{1}{2}$$

The lines are *perpendicular* since their slopes are negative reciprocals of each other.

## 4.5 Writing an Equation of a Line

To write the equation of a line given its slope and y-intercept, use $y = mx + b$ and substitute those values into the equation. **(p. 245)**

Find an equation of the line with slope = 5 and y-intercept $(0, -3)$.

Use $y = mx + b$.    Substitute 5 for $m$ and $-3$ for $b$.
$$y = 5x - 3$$

| Definition/Procedure | Example |
|---|---|

If $(x_1, y_1)$ is a point on a line and $m$ is the slope of the line, then the equation of the line is given by $y - y_1 = m(x - x_1)$. This is the **point-slope formula.**

**If we are given the slope of the line and a point on the line, we can use the point-slope formula to find an equation of the line. (p. 245)**

Find an equation of the line containing the point $(7, -2)$ with slope $= 3$. Express the answer in standard form.

$$\text{Use } y - y_1 = m(x - x_1).$$

Substitute 3 for $m$. Substitute $(7, -2)$ for $(x_1, y_1)$.

$$y - (-2) = 3(x - 7)$$
$$y + 2 = 3x - 21$$
$$-3x + y = -23$$
$$3x - y = 23 \qquad \text{Standard form}$$

---

**To write an equation of a line given two points on the line,**

**a) use the points to find the slope of the line**

**then**

**b) use the slope and *either one* of the points in the point-slope formula. (p. 247)**

Find an equation of the line containing the points $(4, 1)$ and $(-4, 5)$. Express the answer in slope-intercept form.

$$m = \frac{5 - 1}{-4 - 4} = \frac{4}{-8} = -\frac{1}{2}$$

We will use $m = -\frac{1}{2}$ and the point $(4, 1)$ in the point-slope formula.

$$y - y_1 = m(x - x_1)$$

Substitute $-\frac{1}{2}$ for $m$. Substitute $(4, 1)$ for $(x_1, y_1)$.

$$y - 1 = -\frac{1}{2}(x - 4) \qquad \text{Substitute.}$$

$$y - 1 = -\frac{1}{2}x + 2 \qquad \text{Distribute.}$$

$$y = -\frac{1}{2}x + 3 \qquad \text{Slope-intercept form}$$

---

The equation of a **horizontal line** containing the point $(c, d)$ is $y = d$.

The equation of a **vertical line** containing the point $(c, d)$ is $x = c$. **(p. 248)**

The equation of a horizontal line containing the point $(3, -2)$ is $y = -2$.

The equation of a vertical line containing the point $(6, 4)$ is $x = 6$.

---

To write an equation of the line parallel or perpendicular to a given line, we must first find the slope of the given line. **(p. 248)**

Write an equation of the line parallel to $4x - 5y = 20$ containing the point $(4, -3)$. Express the answer in slope-intercept form.

Find the slope of $4x - 5y = 20$.
$$-5y = -4x + 20$$
$$y = \frac{4}{5}x - 4 \qquad\qquad m = \frac{4}{5}$$

The slope of the parallel line is also $\frac{4}{5}$. Since this line contains $(4, -3)$, use the point-slope formula to write its equation.

$$y - y_1 = m(x - x_1)$$
$$y - (-3) = \frac{4}{5}(x - 4) \qquad \text{Substitute values.}$$
$$y + 3 = \frac{4}{5}x - \frac{16}{5} \qquad \text{Distribute.}$$
$$y = \frac{4}{5}x - \frac{31}{5} \qquad \text{Slope-intercept form}$$

| Definition/Procedure | Example |
|---|---|

## 4.6 Introduction to Functions

A **relation** is any set of ordered pairs. A relation can also be represented as a correspondence or mapping from one set to another. **(p. 256)**

The **domain** of a relation is the set of values of the first coordinates in the set of ordered pairs.

The **range** of a relation is the set of all values of the second coordinates in the set of ordered pairs. **(p. 256)**

A **function** is a relation in which each element of the domain corresponds to exactly one element of the range.

Alternative definition: A relation is a **function** if each *x*-value corresponds to one *y*-value. **(p. 256)**

**Relations:**

a) $\{(-4, -12), (-1, -3), (3, 9), (5, 15)\}$

b)
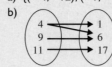

In a), the domain is $\{-4, -1, 3, 5\}$, and the range is $\{-12, -3, 9, 15\}$.

In b), the domain is $\{4, 9, 11\}$, and the range is $\{1, 6, 17\}$.

The relation in a) *is* a function.

The relation in b) *is not* a function.

### The Vertical Line Test

If no vertical line can be drawn through a graph that intersects the graph more than once, then the graph represents a function.

If a vertical line *can* be drawn that intersects the graph more than once, then the graph does not represent a function. **(p. 258)**

This graph represents a function. Anywhere a vertical line is drawn, it will intersect the graph only once.

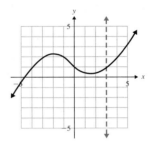

The graph below is not the graph of a function. A vertical line can be drawn so that it intersects the graph more than once.

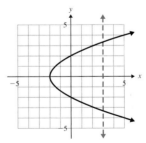

| Definition/Procedure | Example |
|---|---|

When determining the domain of a relation, it can be helpful to keep these tips in mind.

1) Ask yourself, "Is there any number that *cannot* be substituted for $x$?"

2) If $x$ is in the denominator of a fraction, determine what value of $x$ will make the denominator equal 0 by setting the expression equal to zero. Solve for $x$. This $x$-value is *not* in the domain. **(p. 260)**

Determine the domain of $f(x) = \dfrac{6}{x+3}$.

$$x + 3 = 0 \qquad \text{Set the denominator} = 0.$$
$$x = -3 \qquad \text{Solve.}$$

When $x = -3$, the denominator of $f(x) = \dfrac{6}{x+3}$ equals zero.

The domain contains all real numbers *except* $-3$. The domain of the function is $(-\infty, -3) \cup (-3, \infty)$.

---

$y = f(x)$ is called **function notation** and it is read as "$y$ equals $f$ of $x$."

Finding a function value means evaluating the function for the given value of the variable. **(p. 261)**

If $f(x) = 2x - 9$, find $f(4)$.

Substitute 2 for $x$ and evaluate.

$$f(4) = 2(4) - 9 = 8 - 9 = -1$$

Therefore, $f(4) = -1$.

---

A **linear function** has the form

$$f(x) = mx + b$$

where $m$ and $b$ are real numbers, $m$ is the *slope* of the line, and $(0, b)$ is the *y-intercept*.

The domain of a linear function is all real numbers. **(p. 262)**

Graph $f(x) = -\dfrac{1}{2}x + 2$ using the slope and $y$-intercept.

The slope is $-\dfrac{1}{2}$ and the $y$-intercept is $(0, 2)$. Plot the $y$-intercept and use the slope to locate another point on the line.

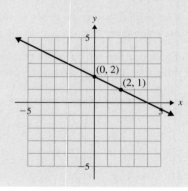

**(4.1) Determine whether each ordered pair is a solution of the given equation.**

1) $5x - y = 13$; $(2, -3)$   2) $2x + 3y = 8$; $(-1, 5)$

3) $y = -\dfrac{4}{3}x + \dfrac{7}{3}$; $(4, -3)$   4) $x = 6$; $(6, 2)$

**Complete the ordered pair for each equation.**

5) $y = -2x + 4$; $(-5, \quad)$   6) $y = \dfrac{5}{2}x - 3$; $(6, \quad)$

7) $y = -9$; $(7, \quad)$   8) $8x - 7y = -10$; $(\quad, 4)$

**Complete the table of values for each equation.**

9) $y = x - 14$

| x | y |
|---|---|
| 0 |   |
| 6 |   |
| −3 |   |
| −8 |   |

10) $3x - 2y = 9$

| x | y |
|---|---|
|   | 0 |
| 0 |   |
| 2 |   |
|   | −1 |

**Plot the ordered pairs on the same coordinate system.**

11) a) $(4, 0)$          b) $(-2, 3)$

    c) $(5, 1)$          d) $(-1, -4)$

12) a) $(0, -3)$          b) $(-4, 4)$

    c) $(1, \frac{3}{2})$          d) $(-\frac{1}{3}, -2)$

13) The cost of renting a pick-up for one day is given by $y = 0.5x + 45.00$, where $x$ represents the number of miles driven and $y$ represents the cost, in dollars.

    a) Complete the table of values, and write the information as ordered pairs.

| x | y |
|---|---|
| 10 |   |
| 18 |   |
| 29 |   |
| 36 |   |

    b) Label a coordinate system, choose an appropriate scale, and graph the ordered pairs.

    c) Explain the meaning of the ordered pair $(58, 74)$ in the context of the problem.

14) Fill in the blank with positive, negative, or zero.

    a) The $x$-coordinate of every point in quadrant III is _____.

    b) The $y$-coordinate of every point in quadrant II is _____.

**(4.2) Complete the table of values and graph each equation.**

15) $y = -2x + 4$

| x | y |
|---|---|
| 0 |   |
| 1 |   |
| 2 |   |
| 3 |   |

16) $2x + 3y = 6$

| x | y |
|---|---|
| 0 |   |
| 3 |   |
| −2 |   |
| −3 |   |

**Graph each equation by finding the intercepts and at least one other point.**

17) $x - 2y = 2$          18) $3x - y = -3$

19) $y = -\dfrac{1}{2}x + 1$          20) $2x + y = 0$

21) $y = 4$          22) $x = -1$

**(4.3) Determine the slope of each line.**

23)

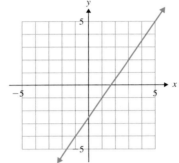

24)

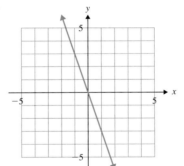

**Use the slope formula to find the slope of the line containing each pair of points.**

25) $(5, 8)$ and $(1, -12)$          26) $(-3, 4)$ and $(1, -1)$

27) $(-7, -2)$ and $(2, 4)$          28) $(7, 3)$ and $(15, 1)$

29) $\left(-\dfrac{1}{4}, 1\right)$ and $\left(\dfrac{3}{4}, -6\right)$          30) $(3.7, 2.3)$ and $(5.8, 6.5)$

31) $(-2, 5)$ and $(4, 5)$          32) $(-9, 3)$ and $(-9, 2)$

33) Christine collects old record albums. The graph shows the value of an original, autographed copy of one of her albums from 1975.

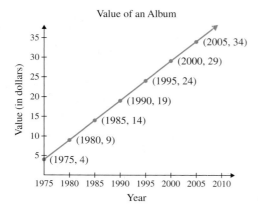

Value of an Album

a) How much did she pay for the album in 1975?

b) Is the slope of the line positive or negative? What does the sign of the slope mean in the context of the problem?

c) Find the slope. What does it mean in the context of the problem?

**Graph the line containing the given point and with the given slope.**

34) $(3, -4)$; $m = 2$

35) $(-2, 2)$; $m = -3$

36) $(1, 3)$; $m = -\dfrac{1}{2}$

37) $(-4, 1)$; slope undefined

38) $(-2, -3)$; $m = 0$

**(4.4) Identify the slope and y-intercept, then graph the line.**

39) $y = -x + 5$

40) $y = 4x - 2$

41) $y = \dfrac{2}{5}x - 6$

42) $y = -\dfrac{1}{2}x + 5$

43) $x + 3y = -6$

44) $18 = 6y - 15x$

45) $x + y = 0$

46) $y + 6 = 1$

47) The value of the squash crop in the United States since 2003 can be modeled by $y = 7.9x + 197.6$, where $x$ represents the number of years after 2003, and $y$ represents the value of the crop in millions of dollars. (U.S. Dept. of Agriculture)

Value of the Squash Crop in the U.S.

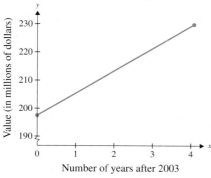

Number of years after 2003

a) What is the $y$-intercept? What does it mean in the context of the problem?

b) Has the value of the squash crop been increasing or decreasing since 2003? By how much per year?

c) Use the graph to estimate the value of the squash crop in the United States in 2005. Then use the equation to determine this number.

**Determine whether each pair of lines is parallel, perpendicular, or neither.**

48) $y = \dfrac{3}{5}x - 8$
    $5x + 3y = 3$

49) $x - 4y = 20$
    $-x + 4y = 6$

50) $5x + y = 4$
    $2x + 10y = 1$

51) $x = 7$
    $y = -3$

52) Write the point-slope formula for the equation of a line with slope $m$ and which contains the point $(x_1, y_1)$.

**Write the *slope-intercept form* of the equation of the line, if possible, given the following information.**

53) $m = 6$ and contains $(-1, 4)$

54) $m = -5$ and $y$-intercept $(0, -3)$

55) $m = -\dfrac{3}{4}$ and $y$-intercept $(0, 7)$

56) contains $(-4, 2)$ and $(-2, 5)$

57) contains $(4, 1)$ and $(6, -3)$

58) $m = \dfrac{2}{3}$ and contains $(5, -2)$

59) horizontal line containing $(3, 7)$

60) vertical line containing $(-5, 1)$

**Write the *standard form* of the equation of the line given the following information.**

61) contains $(4, 5)$ and $(-1, -10)$

62) $m = -\dfrac{1}{2}$ and contains $(3, 0)$

63) $m = \dfrac{5}{2}$ and contains $\left(1, -\dfrac{3}{2}\right)$

64) contains $(-4, 1)$ and $(4, 3)$

65) $m = -4$ and $y$-intercept $(0, 0)$

66) $m = -\dfrac{3}{7}$ and $y$-intercept $(0, 1)$

67) contains $(6, 1)$ and $(2, 5)$

68) $m = \dfrac{3}{4}$ and contains $\left(-2, \dfrac{7}{2}\right)$

69) Mr. Romanski works as an advertising consultant, and his salary has been growing linearly. In 2005 he earned

$62,000, and in 2010 he earned $79,500. Let $y$ represent Mr. Romanski's salary, in dollars, $x$ years after 2005.

a) Write a linear equation to model these data.

b) Explain the meaning of the slope in the context of the problem.

c) How much did he earn in 2008?

d) If the trend continues, in what year could he expect to earn $93,500?

**Write an equation of the line _parallel_ to the given line and containing the given point. Write the answer in slope-intercept form or in standard form, as indicated.**

70) $y = 2x + 10$; $(2, -5)$; slope-intercept form

71) $y = -8x + 8$; $(-1, 14)$; slope-intercept form

72) $3x + y = 5$; $(-3, 5)$; standard form

73) $x - 2y = 6$; $(4, 11)$; standard form

74) $3x + 4y = 1$; $(-1, 2)$; slope-intercept form

75) $x + 5y = 10$; $(15, 7)$; slope-intercept form

**Write an equation of the line _perpendicular_ to the given line and containing the given point. Write the answer in slope-intercept form or in standard form, as indicated.**

76) $y = -\dfrac{1}{5}x + 7$; $(1, 7)$; slope-intercept form

77) $y = -x + 9$; $(3, -9)$; slope-intercept form

78) $4x - 3y = 6$; $(8, -5)$; slope-intercept form

79) $2x + 3y = -3$; $(-4, -4)$; slope-intercept form

80) $x + 8y = 8$; $(-2, -7)$; standard form

81) Write an equation of the line parallel to $y = 5$ containing $(8, 4)$

82) Write an equation of the line perpendicular to $x = -2$ containing $(4, -3)$.

**(4.6) Identify the domain and range of each relation, and determine whether each relation is a function.**

83) $\{(-6, 0), (5, 1), (8, 1), (10, 4)\}$

84) $\{(-2, 4), (2, 4), (3, 9), (3, 16)\}$

85)

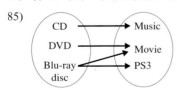

86)

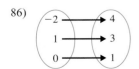

87)

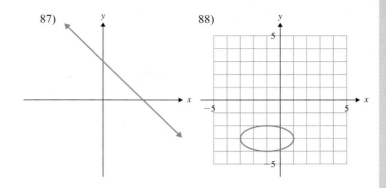

88)

**Determine the domain of each function.**

89) $y = 2x - 5$

90) $y = \dfrac{1}{5}x + 6$

91) $y = \dfrac{6}{x}$

92) $y = \dfrac{3}{x + 4}$

93) $y = x^2 + 3$

94) $y = x^3 - 9$

**For each function, $f$, find $f(2)$ and $f(-1)$.**

95) $f = \{(0, 9), (-2, 5), (-1, -8), (2, 7)\}$

96)

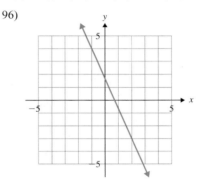

97) Let $f(x) = -3x + 8$, $g(x) = x^2 - 9x + 4$. Find each of the following and simplify.

a) $f(3)$

b) $f(-5)$

c) $g(4)$

d) $g(0)$

98) $h(x) = 2x + 11$. Find $x$ so that $h(x) = -7$.

**Graph each function using the slope and y-intercept.**

99) $f(x) = \dfrac{1}{4}x - 2$

100) $g(x) = -3x + 1$

101) $h(x) = 2$

102) Between August 2008 and January 2009, approximately 374,000 people were signing up for Facebook every day. This can be modeled by the function $N(x) = 374,000x$ where $N$ is the total number of people who have signed up for Facebook in $x$ days. (www.slate.com)

a) How many people signed up for Facebook in 4 days?

b) How many people signed up for Facebook after 1 week?

c) How long would it take to sign up 4,488,000 new users?

1)  Is $(-3, -2)$ a solution of $2x - 7y = 8$?

2)  Complete the table of values and graph $y = \dfrac{3}{2}x - 2$.

| x | y |
|---|---|
| 0 |   |
| -2 |   |
| 4 |   |
|   | 1 |

3)  Fill in the blanks with *positive* or *negative*. In quadrant IV, the $x$-coordinate of every point is _____ and the $y$-coordinate is _____.

4)  For $3x - 4y = 6$,

   a)  find the $x$-intercept.

   b)  find the $y$-intercept.

   c)  find one other point on the line.

   d)  graph the line.

5)  Graph $y = -3$.

6)  Graph $x + y = 0$.

7)  Find the slope of the line containing the points

   a)  $(3, -1)$ and $(-5, 9)$.

   b)  $(8, 6)$ and $(11, 6)$.

8)  Graph the line containing the point $(-1, 4)$ with slope $-\dfrac{3}{2}$.

9)  Graph the line containing the point $(2, 3)$ with an undefined slope.

10)  Put $3x - 2y = 10$ into slope-intercept form. Then, graph the line.

11)  Write the slope-intercept form for the equation of the line with slope 7 and $y$-intercept $(0, -10)$.

12)  Write the standard form for the equation of a line with slope $-\dfrac{1}{3}$ containing the point $(-3, 5)$.

13)  Determine whether $4x + 18y = 9$ and $9x - 2y = -6$ are parallel, perpendicular, or neither.

14)  Find the slope-intercept form of the equation of the line

   a)  perpendicular to $y = 2x - 9$ containing $(-6, 10)$.

   b)  parallel to $3x - 4y = -4$ containing $(11, 8)$.

15)  The graph shows the number of children attending a neighborhood school from 2005 to 2010. Let $y$ represent the number of children attending the school $x$ years after 2005.

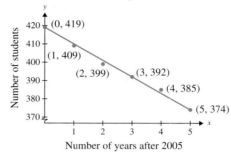

a)  According to the graph, how many children attended this school in 2007?

b)  Write a linear equation (in slope-intercept form) to model these data. Use the data points for 2005 and 2010.

c)  Use the equation in part b) to determine the number of students attending the school in 2007. How does your answer in part a) compare to the number predicted by the equation?

d)  Explain the meaning of the slope in the context of the problem.

e)  What is the $y$-intercept? What does it mean in the context of the problem?

f)  If the current trend continues, how many children can be expected to attend this school in 2013?

16)  What is a function?

**Identify the domain and range of each relation, and determine whether each relation is a function.**

17)  $\{(0, 0), (1, -1), (16, 2), (16, -2)\}$

18)

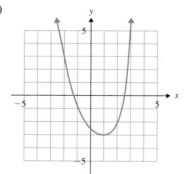

**Determine the domain of each function.**

19) $y = 9x + 2$

20) $y = \dfrac{5}{x + 8}$

**For each function, $f$, find $f(-3)$.**

21) $f = \{(-5, 9), (-3, 4), (1, 0), (6, -3)\}$

22)

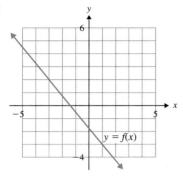

**Let $f(x) = -6x + 11$ and $g(x) = x^2 - 5x - 3$. Find each of the following and simplify.**

23) $f(5)$

24) $f\left(-\dfrac{3}{2}\right)$

25) $g(-2)$

26) $g(0)$

27) Graph the function $f(x) = 2x + 1$.

28) A cell phone company has a plan that charges $9.99 per month plus $0.06 per minute. The cost of this plan, $C$ (in dollars), can be described by the function $C(x) = 0.06x + 9.99$, where $x$ is the number of minutes the phone is used.

   a) Find the amount of a customer's bill if she used the phone for 82 minutes during a month.

   b) For how many minutes did she use the phone if her bill was $16.83?

1) Write $\dfrac{336}{792}$ in lowest terms.

2) A rectangular picture frame measures 7 in. by 12.5 in. Find the perimeter of the frame.

**Evaluate.**

3) $-3^4$

4) $\dfrac{24}{35} \cdot \dfrac{49}{60}$

5) $\dfrac{3}{8} - 2$

6) $4 + 2^6 \div |5 - 13|$

7) Write an expression for "9 less than twice 17" and simplify.

**Simplify. The answer should not contain any negative exponents.**

8) $(5k^6)(-4k^9)$

9) $\left(\dfrac{30w^5}{15w^{-3}}\right)^{-4}$

**Solve.**

10) $-\dfrac{2}{5}y + 9 = 15$

11) $\dfrac{3}{2}(7c - 5) - 1 = \dfrac{2}{3}(2c + 1)$

12) $7 + 2(p - 6) = 8(p + 3) - 6p + 1$

13) Solve. Write the solution in interval notation.
$3x + 14 \le 7x + 4$

14) The Chase family put their house on the market for $306,000. This is 10% less than what they paid for it 3 years ago. What did they pay for the house?

15) Find the missing angle measures.

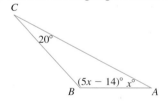

16) A 24-ft rope is cut into two pieces so that one piece is 4 ft longer than the other. Find the length of each piece.

17) Find the slope of the line containing the points $(4, -11)$ and $(10, 5)$.

18) Graph $3x + y = 4$.

19) Write an equation of the line with slope $\dfrac{3}{8}$ containing the point $(16, 5)$. Express the answer in standard form.

20) Write an equation of the line perpendicular to $4x + 3y = -6$ containing the point $(-8, -6)$. Express the answer in slope-intercept form.

21) Determine the domain of $y = \dfrac{5}{6}x - 9$.

**Let $f(x) = 2x + 8$ and $g(x) = x^2 + 10x - 3$. Find each of the following.**

22) $f(9)$

23) $f(0)$

24) $g(-5)$

25) Graph $f(x) = -\dfrac{1}{3}x + 2$.

# CHAPTER 5

# Solving Systems of Linear Equations

## Algebra at Work: Custom Motorcycles

We will take another look at how algebra is used in a custom motorcycle shop.

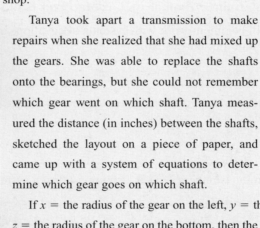

Tanya took apart a transmission to make repairs when she realized that she had mixed up the gears. She was able to replace the shafts onto the bearings, but she could not remember which gear went on which shaft. Tanya measured the distance (in inches) between the shafts, sketched the layout on a piece of paper, and came up with a system of equations to determine which gear goes on which shaft.

If $x$ = the radius of the gear on the left, $y$ = the radius of the gear on the right, and $z$ = the radius of the gear on the bottom, then the system of equations Tanya must solve to determine where to put each gear is

$$x + y = 2.650$$
$$x + z = 2.275$$
$$y + z = 1.530$$

Solving this system, Tanya determines that $x = 1.698$ in., $y = 0.952$ in., and $z = 0.578$ in. Now she knows on which shaft to place each gear.

In this chapter, we will learn how to write and solve systems of two and three equations.

## Section 5.1 Solving Systems by Graphing

### Objectives

1. Determine Whether an Ordered Pair Is a Solution of a System
2. Solve a Linear System by Graphing
3. Solve a Linear System by Graphing: Special Cases
4. Determine the Number of Solutions of a System Without Graphing

What is a system of linear equations? A **system of linear equations** consists of two or more linear equations with the same variables. In Sections 5.1–5.3, we will learn how to solve systems of two equations in two variables. Some examples of such systems are

$$2x + 5y = 5 \qquad\qquad y = \frac{1}{3}x - 8 \qquad\qquad -3x + y = 1$$
$$x + 4y = -1 \qquad\qquad 5x - 6y = 10 \qquad\qquad x = -2$$

In the third system, we see that $x = -2$ is written with only one variable. However, we can think of it as an equation in two variables by writing it as $x + 0y = -2$.

It is also possible to solve systems of inequalities. In Section 5.5, we will learn how to solve linear inequalities in two variables.

### 1. Determine Whether an Ordered Pair Is a Solution of a System

We will begin our work with systems of equations by determining whether an ordered pair is a solution of the system.

### Definition

A **solution of a system** of two equations in two variables is an ordered pair that is a solution of each equation in the system.

---

### Example 1

Determine whether $(2, 3)$ is a solution of each system of equations.

a) $y = x + 1$
   $x + 2y = 8$

b) $4x - 5y = -7$
   $3x + y = 4$

### Solution

a) If $(2, 3)$ is a solution of $\begin{array}{l} y = x + 1 \\ x + 2y = 8 \end{array}$ then when we substitute 2 for $x$ and 3 for $y$, the ordered pair will make each equation true.

$$
\begin{array}{ll}
y = x + 1 & \\
3 \stackrel{?}{=} 2 + 1 & \text{Substitute.} \\
& \\
3 = 3 & \text{True}
\end{array}
\qquad
\begin{array}{ll}
x + 2y = 8 & \\
2 + 2(3) \stackrel{?}{=} 8 & \text{Substitute.} \\
2 + 6 \stackrel{?}{=} 8 & \\
8 = 8 & \text{True}
\end{array}
$$

Since $(2, 3)$ is a solution of each equation, it is a solution of the system.

b) We will substitute 2 for $x$ and 3 for $y$ to see whether $(2, 3)$ satisfies (is a solution of) each equation.

$$
\begin{array}{ll}
4x - 5y = -7 & \\
4(2) - 5(3) \stackrel{?}{=} -7 & \text{Substitute.} \\
8 - 15 \stackrel{?}{=} -7 & \\
-7 = -7 & \text{True}
\end{array}
\qquad
\begin{array}{ll}
3x + y = 4 & \\
3(2) + 3 \stackrel{?}{=} 4 & \text{Substitute.} \\
6 + 3 \stackrel{?}{=} 4 & \\
9 = 4 & \text{False}
\end{array}
$$

Although $(2, 3)$ is a solution of the first equation, it does *not* satisfy $3x + y = 4$. Therefore, $(2, 3)$ is *not* a solution of the system.    ∎

**You Try 1**

Determine whether $(-4, 3)$ is a solution of each system of equations.

a)  $3x + 5y = 3$  
     $-2x - y = -5$

b)  $y = \dfrac{1}{2}x + 5$  
     $-x + 3y = 13$

If we are given a system and no solution is given, how do we *find* the solution to the system of equations? In this chapter, we will discuss three methods for solving systems of equations:

1) Graphing (this section)
2) Substitution (Section 5.2)
3) Elimination (Section 5.3)

Let's begin with the graphing method.

## 2. Solve a Linear System by Graphing

To **solve a system of equations in two variables** means to find the ordered pair (or pairs) that satisfies each equation in the system.

Recall from Chapter 4 that the graph of a linear equation is a line. This line represents all solutions of the equation.

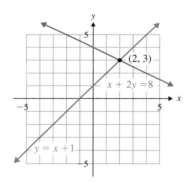

If two lines intersect at one point, that point of intersection is a solution of each equation.

For example, the graph shows the lines representing the two equations in Example 1(a). The solution to that system is their point of intersection, $(2, 3)$.

---

**Definition**

*When solving a system of equations by graphing,* the point of intersection is the solution of the system. If a system has at least one solution, we say that the system is **consistent**. The equations are **independent** if the system has one solution.

---

**Example 2**

Solve the system by graphing.

$$y = \frac{1}{3}x - 2$$

$$2x + 3y = 3$$

### Solution

Graph each line on the same axes. The first equation is in slope-intercept form, and we see that $m = \dfrac{1}{3}$ and $b = -2$. Its graph is in blue.

Let's graph $2x + 3y = 3$ by plotting points.

| x | y |
|---|---|
| 0 | 1 |
| −3 | 3 |
| 3 | −1 |

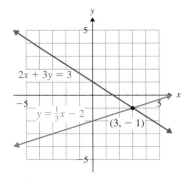

The point of intersection is $(3, -1)$. Therefore, the solution to the system is $(3, -1)$.

This is a consistent system.

**Note**

It is important that you use a straightedge to graph the lines. If the graph is not precise, it will be difficult to correctly locate the point of intersection. Furthermore, if the solution of a system contains numbers that are not integers, it may be impossible to accurately read the point of intersection. This is one reason why solving a system by graphing is not always the best way to find the solution. But it can be a useful method, and it is one that is used to solve problems not only in mathematics, but in areas such as business, economics, and chemistry as well.

**You Try 2**

Solve the system by graphing.

$$3x + 2y = 2$$
$$y = \frac{1}{2}x - 3$$

### 3. Solve a Linear System by Graphing: Special Cases

Do two lines *always* intersect? No! Then if we are trying to solve a system of two linear equations by graphing and the graphs do not intersect, what does this tell us about the solution to the system?

**Example 3**

Solve the system by graphing.

$$-2x - y = 1$$
$$2x + y = 3$$

### Solution

Graph each line on the same axes.

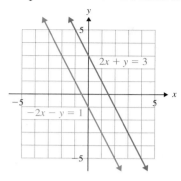

The lines are parallel; they will never intersect. Therefore, there is *no solution* to the system. We write the solution set as $\varnothing$.

---

**Definition**

When solving a system of equations by graphing, if the lines are parallel, then the system has **no solution.** We write this as $\varnothing$. Furthermore, a system that has no solution is **inconsistent,** and the equations are **independent.**

---

What if the graphs of the equations in a system are the same line?

**Example 4**

---

Solve the system by graphing.

$$y = \frac{2}{3}x + 2$$
$$12y - 8x = 24$$

### Solution

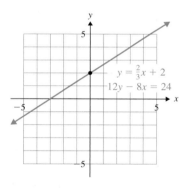

If we write the second equation in slope-intercept form, we see that it is the same as the first equation. This means that the graph of each equation is the same line. Therefore, each point on the line satisfies each equation. The system has an *infinite number of solutions* of the form $y = \dfrac{2}{3}x + 2$.

The solution set is $\left\{ (x, y) \,\middle|\, y = \dfrac{2}{3}x + 2 \right\}$.

We read this as "the set of all ordered pairs $(x, y)$ such that $y = \dfrac{2}{3}x + 2$."

---

We could have used either equation to write the solution set in Example 4. However, we will use either the equation that is written in slope-intercept form or the equation written in standard form with integer coefficients that have no common factor other than 1.

---

**Definition**

When solving a system of equations by graphing, if the graph of each equation is the same line, then the system has an **infinite number of solutions.** The system is **consistent,** and the equations are **dependent.**

---

We will summarize what we have learned so far about solving a system of linear equations by graphing:

---

**Procedure**    Solving a System by Graphing

To solve a system by graphing, graph each line on the same axes.

1) If the lines intersect at a single point, then the point of intersection is the solution of the system. The system is *consistent*, and the equations are *independent*. (See Figure 5.1a.)

2) If the lines are parallel, then the system has *no solution*. We write the solution set as ∅. The system is *inconsistent*. The equations are *independent*. (See Figure 5.1b.)

3) If the graphs are the same line, then the system has an *infinite number of solutions*. We say that the system is *consistent*, and the equations are *dependent*. (See Figure 5.1c.)

---

Figure 5.1

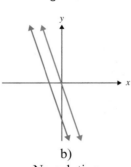    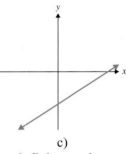

| a) | b) | c) |
|---|---|---|
| One solution—the point of intersection | No solution | Infinite number of solutions |
| Consistent system | Inconsistent system | Consistent system |
| Independent equations | Independent equations | Dependent equations |

**You Try 3**

Solve each system by graphing.

a)    $-x = y + 4$         b)    $2x - 6y = 9$
       $x + y = 1$                $-4x + 12y = -18$

### 4. Determine the Number of Solutions of a System Without Graphing

The graphs of lines can lead us to the solution of a system. We can also determine the number of solutions a system has without graphing.

We saw in Example 4 that if a system has lines with the same slope and the same $y$-intercept (they are the same line), then the system has an *infinite number of solutions*.

Example 3 shows that if a system contains lines with the same slope and different $y$-intercepts, then the lines are parallel and the system has *no solution*.

Finally, we learned in Example 2 that if the lines in a system have different slopes, then they will intersect and the system has *one solution*.

**Example 5**

Without graphing, determine whether each system has no solution, one solution, or an infinite number of solutions.

a) $y = \dfrac{3}{4}x + 7$

$5x + 8y = -8$

b) $4x - 8y = 10$

$-6x + 12y = -15$

c) $9x + 6y = -13$

$3x + 2y = 4$

**Solution**

a) The first equation is already in slope-intercept form, so write the second equation in slope-intercept form.

$$5x + 8y = -8$$
$$8y = -5x - 8$$
$$y = -\dfrac{5}{8}x - 1$$

The slopes, $\dfrac{3}{4}$ and $-\dfrac{5}{8}$, are different; therefore, this system has *one solution*.

b) Write each equation in slope-intercept form.

$$4x - 8y = 10$$
$$-8y = -4x + 10$$
$$y = \dfrac{-4}{-8}x + \dfrac{10}{-8}$$
$$y = \dfrac{1}{2}x - \dfrac{5}{4}$$

$$-6x + 12y = -15$$
$$12y = 6x - 15$$
$$y = \dfrac{6}{12}x - \dfrac{15}{12}$$
$$y = \dfrac{1}{2}x - \dfrac{5}{4}$$

The equations are the same: they have the same slope and $y$-intercept. Therefore, this system has an *infinite number of solutions.*

c) Write each equation in slope-intercept form.

$$9x + 6y = -13$$
$$6y = -9x - 13$$
$$y = \dfrac{-9}{6}x - \dfrac{13}{6}$$
$$y = -\dfrac{3}{2}x - \dfrac{13}{6}$$

$$3x + 2y = 4$$
$$2y = -3x + 4$$
$$y = \dfrac{-3}{2}x + \dfrac{4}{2}$$
$$y = -\dfrac{3}{2}x + 2$$

The equations have the same slope but different $y$-intercepts. If we graphed them, the lines would be parallel. Therefore, this system has *no solution.*  ∎

**You Try 4**

Without graphing, determine whether each system has no solution, one solution, or an infinite number of solutions.

a)  $-2x = 4y - 8$

$x + 2y = -6$

b)  $y = -\dfrac{5}{6}x + 1$

$10x + 12y = 12$

c)  $-5x + 3y = 12$

$3x - y = 2$

## Using Technology

In this section, we have learned that the solution of a system of equations is the point at which their graphs intersect. We can solve a system by graphing using a graphing calculator. On the calculator, we will solve the following system by graphing:

$$x + y = 5$$
$$y = 2x - 3$$

Begin by entering each equation using the [Y=] key. Before entering the first equation, we must solve for y.

$$x + y = 5$$
$$y = -x + 5$$

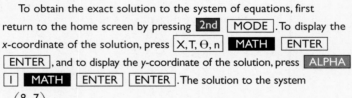

Enter $-x + 5$ in Y1 and $2x - 3$ in Y2, press [ZOOM], and select 6: ZStandard to graph the equations.

Since the lines intersect, the system has a solution. How can we find that solution? Once you see from the graph that the lines intersect, press [2nd] [TRACE]. Select 5: intersect and then press [ENTER] three times. The screen will move the cursor to the point of intersection and display the solution to the system of equations on the bottom of the screen.

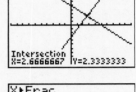

To obtain the exact solution to the system of equations, first return to the home screen by pressing [2nd] [MODE]. To display the x-coordinate of the solution, press [X,T,Θ,n] [MATH] [ENTER] [ENTER], and to display the y-coordinate of the solution, press [ALPHA] [1] [MATH] [ENTER] [ENTER]. The solution to the system is $\left(\dfrac{8}{3}, \dfrac{7}{3}\right)$.

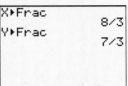

Use a graphing calculator to solve each system.

1) $y = x + 4$
   $y = -x + 2$

2) $y = -3x + 7$
   $y = x - 5$

3) $y = -4x - 2$
   $y = x + 5$

4) $5x + y = -1$
   $4x - y = 2$

5) $5x + 2y = 7$
   $2x + 4y = 3$

6) $3x + 2y = -2$
   $-x - 3y = -5$

---

## Answers to You Try Exercises

1)  a) no    b) yes

2)  $(2, -2)$

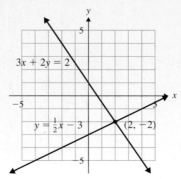

3)  a) $\varnothing$

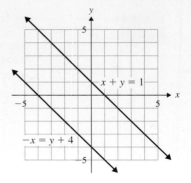

b)    infinite number of solutions of the form $\{(x, y) \mid 2x - 6y = 9\}$

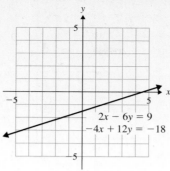

$2x - 6y = 9$
$-4x + 12y = -18$

4)   a) no solution   b) infinite number of solutions   c) one solution

---

**Answers to Technology Exercises**

1)  $(-1, 3)$          2)  $(3, -2)$          3)  $\left(-\dfrac{7}{5}, \dfrac{18}{5}\right)$

4)  $\left(\dfrac{1}{9}, -\dfrac{14}{9}\right)$      5)  $\left(\dfrac{11}{8}, \dfrac{1}{16}\right)$      6)  $\left(-\dfrac{16}{7}, \dfrac{17}{7}\right)$

---

# 5.1 Exercises

**Objective 1: Determine Whether an Ordered Pair Is a Solution of a System**

Determine whether the ordered pair is a solution of the system of equations.

1)   $x + 2y = -6$
     $-x - 3y = 13$
     $(8, -7)$

2)   $y - x = 4$
     $x + 3y = 8$
     $(-1, 3)$

3)   $5x + y = 21$
     $2x - 3y = 11$
     $(4, 1)$

4)   $7x + 2y = 14$
     $-5x + 6y = -12$
     $(2, 0)$

5)   $5y - 4x = -5$
     $6x + 2y = -21$
     $\left(-\dfrac{5}{2}, -3\right)$

6)      $x = 9y - 7$
     $18y = 7x + 4$
     $\left(-1, \dfrac{2}{3}\right)$

7)   $y = -x + 11$
     $x = 5y - 2$
     $(0, 9)$

8)   $x = -y$
     $y = \dfrac{5}{8}x - 13$
     $(8, -8)$

**Mixed Exercises: Objectives 2 and 3**

9)   If you are solving a system of equations by graphing, how do you know whether the system has no solution?

10)  If you are solving a system of equations by graphing, how do you know whether the system has an infinite number of solutions?

Solve each system of equations by graphing. If the system is inconsistent or the equations are dependent, identify this.

VIDEO 11)  $y = -\dfrac{2}{3}x + 3$
     $y = x - 2$

12)  $y = \dfrac{1}{2}x + 2$
     $y = 2x - 1$

13)  $y = x + 1$
     $y = -\dfrac{1}{2}x + 4$

14)  $y = -2x + 3$
     $y = x - 3$

15)  $x + y = -1$
     $x - 2y = 14$

16)  $2x - 3y = 6$
     $x + y = -7$

17)  $x - 2y = 7$
     $-3x + y = -1$

18)  $-x + 2y = 4$
     $3x + 4y = -12$

VIDEO 19)  $\dfrac{3}{4}x - y = 0$
     $3x - 4y = 20$

20)  $y = -x$
     $4x + 4y = 2$

21)  $y = \dfrac{1}{3}x - 2$
     $4x - 12y = 24$

22)  $5x + 5y = 5$
     $x + y = 1$

23)  $x = 8 - 4y$
     $3x + 2y = 4$

24)  $x - y = 0$
     $7x - 3y = 12$

VIDEO 25)  $y = -3x + 1$
     $12x + 4y = 4$

26)  $2x - y = 1$
     $-2x + y = -3$

27) $x + y = 0$

   $y = \dfrac{1}{2}x + 3$

28) $x = -2$

   $y = -\dfrac{5}{2}x - 1$

29) $-3x + y = -4$

       $y = -1$

30) $5x + 2y = 6$

   $-15x - 6y = -18$

31) $y = \dfrac{3}{5}x - 6$

   $-3x + 5y = 10$

32) $y - x = -2$

   $2x + y = -5$

Write a system of equations so that the given ordered pair is a solution of the system.

33) $(2, 5)$

34) $(3, 1)$

35) $(-4, -3)$

36) $(6, -1)$

37) $\left(-\dfrac{1}{3}, 4\right)$

38) $\left(0, \dfrac{3}{2}\right)$

For Exercises 39–42, determine which ordered pair could be a solution to the system of equations that is graphed. Explain why you chose that ordered pair.

39)

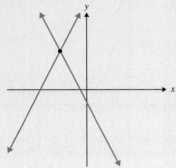

A. $(2, -6)$   B. $(3, 4)$
C. $(-3, 4)$   D. $(-2, -3)$

40)

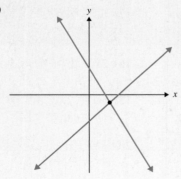

A. $\left(\dfrac{7}{2}, -\dfrac{1}{2}\right)$   B. $(-4, -1)$

C. $\left(\dfrac{9}{4}, \dfrac{3}{4}\right)$   D. $\left(-\dfrac{10}{3}, -\dfrac{2}{3}\right)$

41)

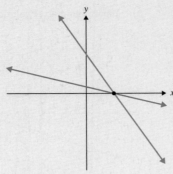

A. $(0, 3.8)$   B. $(4.1, 0)$
C. $(-2.1, 0)$   D. $(0, 5)$

42)

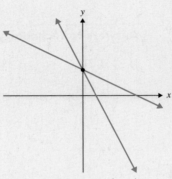

A. $(4, 0)$   B. $\left(\dfrac{1}{3}, 0\right)$

C. $(0, -3)$   D. $(0, 2)$

**Objective 4: Determine the Number of Solutions of a System Without Graphing**

43) How do you determine, *without graphing*, that a system of equations has exactly one solution?

44) How do you determine, *without graphing*, that a system of equations has no solution?

Without graphing, determine whether each system has no solution, one solution, or an infinite number of solutions.

45) $y = 5x - 4$

   $y = -3x + 7$

46) $y = \dfrac{2}{3}x + 9$

   $y = \dfrac{2}{3}x + 1$

47) $y = -\dfrac{3}{8}x + 1$

   $6x + 16y = -9$

48) $y = -\dfrac{1}{4}x + 3$

   $2x + 8y = 24$

49) $-15x + 9y = 27$
$10x - 6y = -18$

50) $7x - y = 6$
$x + y = 13$

51) $3x + 12y = 9$
$x - 4y = 3$

52) $6x - 4y = -10$
$-21x + 14y = 35$

53) $x = 5$
$x = -1$

54) $y = x$
$y = 2$

55) The graph shows the percentage of foreign students in U.S. institutions of higher learning from Hong Kong and Malaysia from 1980 to 2005.   (http://nces.ed.gov)

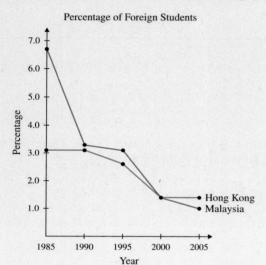

Percentage of Foreign Students

a) When was there a greater percentage of students from Malaysia?

b) Write the point of intersection of the graphs as an ordered pair in the form (year, percentage) and explain its meaning.

c) During which years did the percentage of students from Hong Kong remain the same?

d) During which years did the percentage of students from Malaysia decrease the most? How can this be related to the slope of this line segment?

56) The graph shows the approximate number of veterans living in Connecticut and Iowa from 2003 to 2007. (www.census.gov)

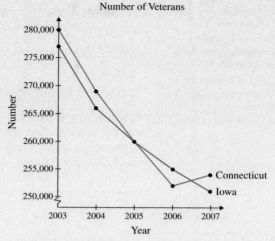

Number of Veterans

a) In which year were there fewer veterans living in Connecticut than in Iowa? Approximately how many were living in each state?

b) Write the data point for Iowa in 2003 as an ordered pair of the form (year, number) and explain its meaning.

c) Write the point of intersection of the graphs for the year 2005 as an ordered pair in the form (year, number) and explain its meaning.

d) Which line segment on the Connecticut graph has a positive slope? How can this be explained in the context of this problem?

Solve each system using a graphing calculator.

57) $y = -2x + 2$
$y = x - 7$

58) $y = x + 1$
$y = 3x + 3$

59) $x - y = 3$
$x + 4y = 8$

60) $2x + 3y = 3$
$y - x = -4$

61) $4x + 5y = -17$
$3x - 7y = 4.45$

62) $-5x + 6y = 22.8$
$3x - 2y = -5.2$

## Section 5.2 Solving Systems by the Substitution Method

### Objectives

1. **Solve a Linear System by Substitution**
2. **Solve a System Containing Fractions or Decimals**
3. **Solve a System by Substitution: Special Cases**

In Section 5.1, we learned to solve a system of equations by graphing. This method, however, is not always the *best* way to solve a system. If your graphs are not precise, you may read the solution incorrectly. And, if a solution consists of numbers that are not integers, like $\left(\dfrac{2}{3}, -\dfrac{1}{4}\right)$, it may not be possible to accurately identify the point of intersection of the graphs.

### 1. Solve a Linear System by Substitution

Another way to solve a system of equations is to use the *substitution method*. When we use the **substitution method**, we solve one of the equations for one of the variables in terms of the other. Then we substitute that expression into the other equation. We can do this because solving a system means finding the ordered pair, or pairs, that satisfy *both* equations. *The substitution method is especially good when one of the variables has a coefficient of 1 or −1.*

---

| **Example 1** |
| --- |

Solve the system using substitution.

$$2x + 3y = -1$$
$$y = 2x - 3$$

**Solution**

The second equation is already solved for $y$; it tells us that $y$ *equals* $2x - 3$. Therefore, we can substitute $2x - 3$ for $y$ in the first equation, then solve for $x$.

$$\begin{aligned}
2x + 3y &= -1 & &\text{First equation}\\
2x + 3(2x - 3) &= -1 & &\text{Substitute.}\\
2x + 6x - 9 &= -1 & &\text{Distribute.}\\
8x - 9 &= -1 &\\
8x &= 8 &\\
x &= 1 &
\end{aligned}$$

We have found that $x = 1$, but we still need to find $y$. Substitute $x = 1$ into *either* equation, and solve for $y$. In this case, we will substitute $x = 1$ into the second equation since it is already solved for $y$.

$$\begin{aligned}
y &= 2x - 3 & &\text{Second equation}\\
y &= 2(1) - 3 & &\text{Substitute.}\\
y &= 2 - 3 &\\
y &= -1 &
\end{aligned}$$

Check $x = 1$, $y = -1$ in *both* equations.

$$\begin{aligned}
2x + 3y &= -1 \\
2(1) + 3(-1) &\overset{?}{=} -1 & &\text{Substitute.}\\
2 - 3 &\overset{?}{=} -1 &\\
-1 &= -1 & &\text{True}
\end{aligned}$$

$$\begin{aligned}
y &= 2x - 3 \\
-1 &\overset{?}{=} 2(1) - 3 & &\text{Substitute.}\\
-1 &\overset{?}{=} 2 - 3 &\\
-1 &= -1 & &\text{True}
\end{aligned}$$

We write the solution as an ordered pair, $(1, -1)$.

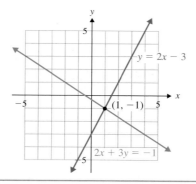

If we solve the system in Example 1 by graphing, we can see that the lines intersect at $(1, -1)$, giving us the same solution we obtained using the substitution method.

Let's summarize the steps we use to solve a system by the substitution method:

---

**Procedure    Solving a System by the Substitution Method**

**Step 1:** Solve one of the equations for one of the variables. If possible, solve for a variable that has a coefficient of 1 or $-1$.

**Step 2:** Substitute the expression found in *Step 1* into the *other* equation. The equation you obtain should contain only one variable.

**Step 3:** Solve the equation you obtained in *Step 2*.

**Step 4:** Substitute the value found in *Step 3* into either of the equations to obtain the value of the other variable.

**Step 5:** Check the values in each of the original equations, and write the solution as an ordered pair.

---

**Example 2**

Solve the system by the substitution method.

$$x - 2y = 7 \qquad (1)$$
$$2x + 3y = -21 \qquad (2)$$

**Solution**

We will follow the steps listed above.

1)  For which variable should we solve? The $x$ in the first equation is the only variable with a coefficient of 1 or $-1$. Therefore, we will solve the first equation for $x$.

$$x - 2y = 7 \qquad \text{First equation (1)}$$
$$x = 2y + 7 \qquad \text{Add } 2y.$$

2)  Substitute $2y + 7$ for the $x$ in equation (2).

$$2x + 3y = -21 \qquad \text{Second equation (2)}$$
$$2(2y + 7) + 3y = -21 \qquad \text{Substitute.}$$

3)  Solve this new equation for $y$.

$$2(2y + 7) + 3y = -21$$
$$4y + 14 + 3y = -21 \qquad \text{Distribute.}$$
$$7y + 14 = -21$$
$$7y = -35$$
$$y = -5$$

4) To determine the value of $x$, we can substitute $-5$ for $y$ in either equation. We will use equation (1).

$$\begin{aligned} x - 2y &= 7 & \text{Equation (1)} \\ x - 2(-5) &= 7 & \text{Substitute.} \\ x + 10 &= 7 \\ x &= -3 \end{aligned}$$

5) The check is left to the reader. The solution of the system is $(-3, -5)$. ∎

**You Try 1**

Solve the system by the substitution method.

$$\begin{aligned} -3x + 4y &= -2 \\ 6x - y &= -3 \end{aligned}$$

If no variable in the system has a coefficient of 1 or $-1$, solve for any variable.

## 2. Solve a System Containing Fractions or Decimals

If a system contains an equation with fractions, first multiply the equation by the least common denominator to eliminate the fractions. Likewise, if an equation in the system contains decimals, begin by multiplying the equation by the lowest power of 10 that will eliminate the decimals.

**Example 3**

Solve the system by the substitution method.

$$\frac{3}{10}x - \frac{1}{5}y = 1 \qquad (1)$$

$$-\frac{1}{12}x + \frac{1}{3}y = \frac{5}{6} \qquad (2)$$

**Solution**

Before applying the steps for solving the system, eliminate the fractions in each equation.

$$\frac{3}{10}x - \frac{1}{5}y = 1$$

$$10\left(\frac{3}{10}x - \frac{1}{5}y = 1\right) = 10 \cdot 1 \qquad \text{Multiply by the LCD: 10.}$$

$$3x - 2y = 10 \qquad (3) \qquad \text{Distribute.}$$

$$-\frac{1}{12}x + \frac{1}{3}y = \frac{5}{6}$$

$$12\left(-\frac{1}{12}x + \frac{1}{3}y = \frac{5}{6}\right) = 12 \cdot \frac{5}{6} \qquad \text{Multiply by the LCD: 12.}$$

$$-x + 4y = 10 \qquad (4) \qquad \text{Distribute.}$$

From the original equations, we obtain an equivalent system of equations.

$$\begin{aligned} 3x - 2y &= 10 & (3) \\ -x + 4y &= 10 & (4) \end{aligned}$$

Now, we will work with equations (3) and (4).
Apply the steps:

1) The $x$ in equation (4) has a coefficient of $-1$. Solve this equation for $x$.

$$\begin{aligned} -x + 4y &= 10 & \text{Equation (4)} \\ -x &= 10 - 4y & \text{Subtract } 4y. \\ x &= -10 + 4y & \text{Divide by } -1. \end{aligned}$$

2)  Substitute $-10 + 4y$ for $x$ in equation (3).

$$3x - 2y = 10 \qquad (3)$$
$$3(-10 + 4y) - 2y = 10 \qquad \text{Substitute.}$$

3)  Solve the equation above for $y$.

$$3(-10 + 4y) - 2y = 10$$
$$-30 + 12y - 2y = 10 \qquad \text{Distribute.}$$
$$-30 + 10y = 10$$
$$10y = 40$$
$$y = 4 \qquad \text{Divide by 10.}$$

4)  Find $x$ by substituting 4 for $y$ in either equation (3) or (4). Let's use equation (4) since it has smaller coefficients.

$$-x + 4y = 10 \qquad (4)$$
$$-x + 4(4) = 10 \qquad \text{Substitute.}$$
$$-x + 16 = 10$$
$$-x = -6$$
$$x = 6 \qquad \text{Divide by } -1.$$

5)  Check $x = 6$ and $y = 4$ in the original equations. The solution of the system is (6, 4). ■

**You Try 2**

Solve each system by the substitution method.

a)  $-\dfrac{1}{6}x + \dfrac{1}{3}y = \dfrac{2}{3}$

$\dfrac{3}{2}x - \dfrac{5}{2}y = -7$

b)  $0.1x + 0.03y = -0.05$

$0.1x - 0.1y = 0.6$

## 3. Solve a System by Substitution: Special Cases

We saw in Section 5.1 that a system may have no solution or an infinite number of solutions. If we are solving a system by graphing, we know that a system has no solution if the lines are parallel, and a system has an infinite number of solutions if the graphs are the same line.

When we solve a system by *substitution*, how do we know whether the system is inconsistent or dependent? Read Examples 4 and 5 to find out.

**Example 4**

Solve the system by substitution.

$$3x + y = 5 \qquad (1)$$
$$12x + 4y = -7 \qquad (2)$$

**Solution**

1)  $\qquad y = -3x + 5 \qquad$ Solve equation (1) for $y$.

2)  $\qquad 12x + 4y = -7 \qquad$ Substitute $-3x + 5$ for $y$ in equation (2).
$12x + 4(-3x + 5) = -7$

3)  $12x + 4(-3x + 5) = -7 \qquad$ Solve the resulting equation for $x$.
$12x - 12x + 20 = -7 \qquad$ Distribute.
$20 = -7 \qquad$ False

*Since the variables drop out, and we get a false statement, there is no solution to the system. The system is inconsistent, so the solution set is $\varnothing$.*

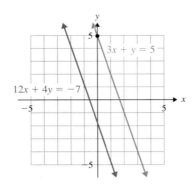

The graph of the equations in the system supports our work. The lines are parallel; therefore, the system has no solution. ■

## Example 5

Solve the system by substitution.

$$2x - 6y = 10 \qquad (1)$$
$$x = 3y + 5 \qquad (2)$$

**Solution**

1) Equation (2) is already solved for $x$.

2) $\qquad\qquad 2x - 6y = 10 \qquad$ Substitute $3y + 5$ for $x$ in equation (1).
$$2(3y + 5) - 6y = 10$$

3) $\qquad\qquad 2(3y + 5) - 6y = 10 \qquad$ Solve the equation for $y$.
$$6y + 10 - 6y = 10$$
$$10 = 10 \qquad \text{True}$$

*Since the variables drop out and we get a true statement, the system has an infinite number of solutions. The equations are dependent, and the solution set is $\{(x, y) \mid x - 3y = 5\}$.*

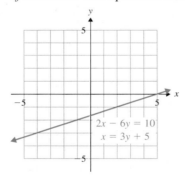

The graph shows that the equations in the system are the same line, therefore the system has an infinite number of solutions. ■

### Note

When you are solving a system of equations and the variables drop out:

1) If you get a *false statement*, like $3 = 5$, then the system has *no solution* and is *inconsistent*.

2) If you get a *true statement*, like $-4 = -4$, then the system has an *infinite number of solutions*. The equations are *dependent*.

## You Try 3

Solve each system by substitution.

a) $\qquad -20x + 5y = 3$
$\qquad\qquad 4x - y = -1$

b) $\qquad x - 3y = 5$
$\qquad\ 4x - 12y = 20$

## Answers to You Try Exercises

1) $\left(-\dfrac{2}{3}, -1\right)$ 　 2) a) $(-8, -2)$　b) $(1, -5)$ 　 3) a) $\varnothing$　b) $\{(x, y) \mid x - 3y = 5\}$

## 5.2 Exercises

**Mixed Exercises: Objectives 1 and 3**

1) If you were asked to solve this system by substitution, why would it be easiest to begin by solving for $y$ in the first equation?

$$7x + y = 1$$
$$-2x + 5y = 9$$

2) When is the best time to use substitution to solve a system?

3) When solving a system of linear equations, how do you know whether the system has no solution?

4) When solving a system of linear equations, how do you know whether the system has an infinite number of solutions?

Solve each system by substitution.

5) $y = 4x - 3$
$5x + y = 15$

6) $y = 3x + 10$
$-5x + 2y = 14$

7) $x = 7y + 11$
$4x - 5y = -2$

8) $x = 9 - y$
$-3x + 4y = 8$

9) $x + 2y = -3$
$4x + 5y = -6$

10) $x + 4y = 1$
$5x + 3y = 5$

11) $2y - 7x = -14$
$4x - y = 7$

12) $-2x - y = 3$
$3x + 2y = -3$

13) $9y - 18x = 5$
$2x - y = 3$

14) $2x + 30y = 9$
$x = 6 - 15y$

15) $x - 2y = 10$
$3x - 6y = 30$

16) $6x + y = -6$
$-12x - 2y = 12$

17) $10x + y = -5$
$-5x + 2y = 10$

18) $2y - x = 4$
$x + 6y = 8$

19) $x = -\dfrac{3}{5}y + 7$
$x + 4y = 24$

20) $y = \dfrac{3}{2}x - 5$
$2x - y = 5$

21) $4y = 2x + 4$
$2y - x = 2$

22) $3x + y = -12$
$6x = 10 - 2y$

23) $2x + 3y = 6$
$5x + 2y = -7$

24) $2x - 5y = -4$
$8x - 9y = 6$

25) $6x - 7y = -4$
$9x - 2y = 11$

26) $4x + 6y = -13$
$7x - 4y = -1$

27) $18x + 6y = -66$
$12x + 4y = -19$

28) $6y - 15x = -12$
$5x - 2y = 4$

**Objective 2: Solve a System Containing Fractions or Decimals**

29) If an equation in a system contains fractions, what should you do first to make the system easier to solve?

30) If an equation in a system contains decimals, what should you do first to make the system easier to solve?

Solve each system by substitution.

31) $\dfrac{1}{4}x - \dfrac{1}{2}y = 1$
$\dfrac{2}{3}x + \dfrac{1}{6}y = \dfrac{25}{6}$

32) $\dfrac{2}{9}x + \dfrac{2}{9}y = 2$
$\dfrac{7}{4}x - \dfrac{1}{8}y = \dfrac{3}{4}$

33) $\dfrac{1}{6}x + \dfrac{4}{3}y = \dfrac{13}{3}$
$\dfrac{2}{5}x + \dfrac{3}{2}y = \dfrac{18}{5}$

34) $\dfrac{1}{10}x + \dfrac{1}{2}y = \dfrac{1}{5}$
$-\dfrac{1}{3}x + \dfrac{1}{2}y = \dfrac{3}{2}$

35) $\dfrac{x}{10} - \dfrac{y}{2} = \dfrac{13}{10}$
$\dfrac{x}{3} + \dfrac{5}{4}y = -\dfrac{3}{2}$

36) $-\dfrac{x}{3} + \dfrac{y}{2} = \dfrac{5}{3}$
$\dfrac{x}{5} - \dfrac{4}{5}y = -1$

37) $y - \dfrac{5}{2}x = -2$
$\dfrac{3}{4}x - \dfrac{3}{10}y = \dfrac{3}{5}$

38) $-\dfrac{2}{15}x - \dfrac{1}{3}y = \dfrac{2}{3}$
$\dfrac{2}{3}x + \dfrac{5}{3}y = \dfrac{1}{2}$

39) $\dfrac{3}{4}x + \dfrac{1}{2}y = 6$
$x = 3y + 8$

40) $\dfrac{5}{3}x - \dfrac{4}{3}y = -\dfrac{4}{3}$
$y = 2x + 4$

41) $0.2x - 0.1y = 0.1$
$0.01x + 0.04y = 0.23$

42) $0.01x - 0.09y = -0.5$
$0.02x + 0.05y = 0.38$

43) $0.6x - 0.1y = 1$
$-0.4x + 0.5y = -1.1$

44) $0.8x + 0.7y = -1.7$
$0.6x - 0.1y = 0.6$

45) $0.02x + 0.01y = -0.44$
$-0.1x - 0.2y = 4$

46) $0.3x + 0.1y = 3$
$0.01x - 0.05y = -0.06$

47) $2.8x + 0.7y = 0.1$
$0.04x + 0.01y = -0.06$

48) $0.1x - 0.3y = -1.2$
$1.5y - 0.5x = 6$

Solve by substitution. Begin by combining like terms.

49) $8 + 2(3x - 5) - 7x + 6y = 16$
$9(y - 2) + 5x - 13y = -4$

50) $3 + 4(2y - 9) + 5x - 2y = 8$
$3(x + 3) - 4(2y + 1) - 2x = 4$

51) $10(x + 3) - 7(y + 4) = 2(4x - 3y) + 3$
$10 - 3(2x - 1) + 5y = 3y - 7x - 9$

52) $7x + 3(y - 2) = 7y + 6x - 1$
$18 + 2(x - y) = 4(x + 2) - 5y$

53) $-(y + 3) = 5(2x + 1) - 7x$
$x + 12 - 8(y + 2) = 6(2 - y)$

54) $9y - 4(2y + 3) = -2(4x + 1)$
$16 - 5(2x + 3) = 2(4 - y)$

55) Jamari wants to rent a cargo trailer to move his son into an apartment when he returns to college. A+ Rental charges $0.60 per mile while Rock Bottom Rental charges $70 plus $0.25 per mile. Let $x$ = the number of miles driven and let $y$ = the cost of the rental. The cost of renting a cargo trailer from each company can be expressed with the following equations:

$$\text{A+ Rental:} \quad y = 0.60x$$

$$\text{Rock Bottom Rental:} \quad y = 0.25x + 70$$

a) How much would it cost Jamari to rent a cargo trailer from each company if he will drive a total of 160 miles?

b) How much would it cost Jamari to rent a trailer from each company if he planned to drive 300 miles?

c) Solve the system of equations using the substitution method, and explain the meaning of the solution.

d) Graph the system of equations, and explain when it is cheaper to rent a cargo trailer from A+ and when it is cheaper to rent it from Rock Bottom Rental. When is the cost the same?

56) To rent a pressure washer, Walsh Rentals charges $16.00 per hour while Discount Company charges $24.00 plus $12.00 per hour. Let $x$ = the number of hours, and let $y$ = the cost of the rental. The cost of renting a pressure washer from each company can be expressed with the following equations:

$$\text{Walsh Rentals:} \quad y = 16.00x$$

$$\text{Discount company:} \quad y = 12.00x + 24$$

a) How much would it cost to rent a pressure washer from each company if it would be used for 4 hours?

b) How much would it cost to rent a pressure washer from each company if it would be rented for 9 hours?

c) Solve the system of equations using the substitution method, and explain the meaning of the solution.

d) Graph the system of equations, and explain when it is cheaper to rent a pressure washer from Walsh and when it is cheaper to rent it from Discount. When is the cost the same?

## Section 5.3 Solving Systems by the Elimination Method

### Objectives

1. Solve a Linear System Using the Elimination Method
2. Solve a Linear System Using the Elimination Method: Special Cases
3. Use the Elimination Method Twice to Solve a Linear System

### 1. Solve a Linear System Using the Elimination Method

The next technique we will learn for solving a system of equations is the **elimination method.** (This is also called the **addition method.**) It is based on the addition property of equality that says that we can add the *same* quantity to each side of an equation and preserve the equality.

$$\text{If } a = b, \text{ then } a + c = b + c.$$

We can extend this idea by saying that we can add *equal* quantities to each side of an equation and still preserve the equality.

$$\text{If } a = b \text{ and } c = d, \text{ then } a + c = b + d.$$

The object of the elimination method is to add the equations (or multiples of one or both of the equations) so that one variable is eliminated. Then, we can solve for the remaining variable.

### Example 1

Solve the system using the elimination method.

$$\begin{aligned} x + y &= 11 \quad (1) \\ x - y &= -5 \quad (2) \end{aligned}$$

**Solution**

The left side of each equation is equal to the right side of each equation. Therefore, if we add the left sides together and add the right sides together, we can set them equal. We will add these equations vertically. The $y$-terms are eliminated, enabling us to solve for $x$.

$$\begin{aligned} x + \phantom{}y &= 11 \quad (1) \\ + \quad x - \phantom{}y &= -5 \quad (2) \\ \hline 2x + 0y &= 6 \\ 2x &= 6 \\ x &= 3 \end{aligned}$$

Add equations (1) and (2).
Simplify.
Divide by 2.

Now we substitute $x = 3$ into either equation to find the value of $y$. Here, we will use equation (1).

$$x + y = 11 \qquad \text{Equation (1)}$$
$$3 + y = 11 \qquad \text{Substitute 3 for } x.$$
$$y = 8 \qquad \text{Subtract 3.}$$

Check $x = 3$ and $y = 8$ in *both* equations.

$$x + y = 11 \qquad\qquad\qquad x - y = -5$$
$$3 + 8 \overset{?}{=} 11 \quad \text{Substitute.} \qquad 3 - 8 \overset{?}{=} -5 \quad \text{Substitute.}$$
$$11 = 11 \quad \text{True} \qquad\qquad -5 = -5 \quad \text{True}$$

The solution is (3, 8). ■

**You Try 1**

Solve the system using the elimination method.

$$3x + y = 10$$
$$x - y = 6$$

In Example 1, simply adding the equations eliminated a variable. But what can we do if we *cannot* eliminate a variable just by adding the equations together?

**Example 2**

Solve the system using the elimination method.

$$2x + 5y = 5 \qquad (1)$$
$$x + 4y = 7 \qquad (2)$$

**Solution**

Just adding these equations will *not* eliminate a variable. The multiplication property of equality tells us that multiplying both sides of an equation by the same quantity results in an equivalent equation. If we multiply equation (2) by $-2$, the coefficient of $x$ will be $-2$.

$$-2(x + 4y) = -2(7) \qquad \text{Multiply equation (2) by } -2.$$
$$-2x - 8y = -14 \qquad \text{New, equivalent equation}$$

**Original System**          **Rewrite the System**

$$2x + 5y = 5 \qquad\qquad\qquad 2x + 5y = 5$$
$$x + 4y = 7 \quad \longrightarrow \quad -2x - 8y = -14$$

Add the equations in the rewritten system. The $x$ is eliminated.

$$\begin{array}{r} 2x + 5y = 5 \\ + \ -2x - 8y = -14 \\ \hline 0x - 3y = -9 \\ -3y = -9 \\ y = 3 \end{array}$$

       Add equations.
       Simplify.
       Solve for $y$.

Substitute $y = 3$ into (1) or (2) to find $x$. We will use equation (2).

$$x + 4y = 7 \qquad \text{Equation (2)}$$
$$x + 4(3) = 7 \qquad \text{Substitute 3 for } y.$$
$$x + 12 = 7$$
$$x = -5$$

The solution is $(-5, 3)$. Check the solution in equations (1) and (2). ■

**You Try 2**

Solve the system using the elimination method.
$$8x - y = -5$$
$$-6x + 2y = 15$$

Next we summarize the steps for solving a system using the elimination method.

> **Procedure** Solving a System of Two Linear Equations by the Elimination Method
>
> **Step 1:** Write each equation in the form $Ax + By = C$.
>
> **Step 2:** Determine which variable to eliminate. If necessary, multiply one or both of the equations by a number so that the coefficients of the variable to be eliminated are negatives of one another.
>
> **Step 3:** Add the equations, and solve for the remaining variable.
>
> **Step 4:** Substitute the value found in *Step 3* into either of the original equations to find the value of the other variable.
>
> **Step 5:** Check the solution in each of the original equations.

**Example 3**

Solve the system using the elimination method.
$$2x = 9y + 4 \qquad (1)$$
$$3x - 7 = 12y \qquad (2)$$

**Solution**

1) **Write each equation in the form $Ax + By = C$.**

$$2x = 9y + 4 \qquad (1)$$
$$2x - 9y = 4 \qquad \text{Subtract } 9y.$$

$$3x - 7 = 12y \qquad (2)$$
$$3x - 12y = 7 \qquad \text{Subtract } 12y \text{ and add } 7.$$

When we rewrite the equations in the form $Ax + By = C$, we get

$$2x - 9y = 4 \qquad (3)$$
$$3x - 12y = 7 \qquad (4)$$

2) **Determine which variable to eliminate from equations (3) and (4).** Often, it is easier to eliminate the variable with the smaller coefficients. Therefore, *we will eliminate x.*

   The least common multiple of 2 and 3 (the *x*-coefficients) is 6. Before we add the equations, one *x*-coefficient should be 6, and the other should be −6. Multiply equation (3) by 3 and equation (4) by −2.

**Rewrite the System**

$$3(2x - 9y) = 3(4) \qquad \text{3 times (3)}$$
$$-2(3x - 12y) = -2(7) \qquad -2 \text{ times (4)}$$

$$\longrightarrow$$

$$6x - 27y = 12$$
$$-6x + 24y = -14$$

3) **Add the resulting equations to eliminate *x*. Solve for *y*.**

$$\begin{array}{r} 6x - 27y = 12 \\ + \ \underline{-6x + 24y = -14} \\ -3y = -2 \\ y = \dfrac{2}{3} \end{array}$$

4)  **Substitute** $y = \dfrac{2}{3}$ **into equation (1) and solve for** $x$.

$$
\begin{aligned}
2x &= 9y + 4 && \text{Equation (1)} \\
2x &= 9\left(\frac{2}{3}\right) + 4 && \text{Substitute.} \\
2x &= 6 + 4 \\
2x &= 10 \\
x &= 5
\end{aligned}
$$

5)  **Check** to verify that $\left(5, \dfrac{2}{3}\right)$ satisfies each of the original equations. The solution is $\left(5, \dfrac{2}{3}\right)$.

**You Try 3**

Solve the system using the elimination method.

$$
\begin{aligned}
5x &= 2y - 14 \\
4x + 3y &= 21
\end{aligned}
$$

## 2. Solve a Linear System Using the Elimination Method: Special Cases

We have seen in Sections 5.1 and 5.2 that some systems have no solution, and some have an infinite number of solutions. How does the elimination method illustrate these results?

**Example 4**

Solve the system using the elimination method.

$$
\begin{aligned}
4y &= 10x + 3 && (1) \\
6y - 15x &= -8 && (2)
\end{aligned}
$$

**Solution**

1)  **Write each equation in the form** $Ax + By = C$.

$$
\begin{aligned}
4y &= 10x + 3 \\
6y - 15x &= -8
\end{aligned}
\qquad \longrightarrow \qquad
\begin{aligned}
-10x + 4y &= 3 && (3) \\
-15x + 6y &= -8 && (4)
\end{aligned}
$$

2)  **Determine which variable to eliminate from equations (3) and (4).** Eliminate $y$.
    The least common multiple of 4 and 6, the $y$-coefficients, is 12. One $y$-coefficient must be 12, and the other must be $-12$.

**Rewrite the System**

$$
\begin{aligned}
-3(-10x + 4y) &= -3(3) \\
2(-15x + 6y) &= 2(-8)
\end{aligned}
\qquad \longrightarrow \qquad
\begin{aligned}
30x - 12y &= -9 \\
-30x + 12y &= -16
\end{aligned}
$$

3)  **Add the equations.**

$$
\begin{array}{r}
30x - 12y = -9 \\
+ \ -30x + 12y = -16 \\
\hline
0 = -25 \quad \text{False}
\end{array}
$$

The variables drop out, and we get a false statement. Therefore, the system is inconsistent, and the solution set is $\varnothing$.

**You Try 4**

Solve the system using the elimination method.

$$24x + 6y = -7$$
$$4y + 3 = -16x$$

---

**Example 5**

Solve the system using the elimination method.

$$12x - 18y = 9 \qquad (1)$$
$$y = \frac{2}{3}x - \frac{1}{2} \qquad (2)$$

***Solution***

1) **Write equation (2) in the form $Ax + By = C$.**

$$y = \frac{2}{3}x - \frac{1}{2} \qquad\qquad \text{Equation (2)}$$
$$6y = 6\left(\frac{2}{3}x - \frac{1}{2}\right) \qquad \text{Multiply by 6 to eliminate fractions.}$$
$$6y = 4x - 3$$
$$-4x + 6y = -3 \qquad (3) \qquad \text{Rewrite as } Ax + By = C.$$

We can rewrite $y = \frac{2}{3}x - \frac{1}{2}$ as $-4x + 6y = -3$, equation (3).

2) **Determine which variable to eliminate from equations (1) and (3).**

$$12x - 18y = 9 \qquad (1)$$
$$-4x + 6y = -3 \qquad (3)$$

Eliminate $x$. Multiply equation (3) by 3.

$$12x - 18y = 9 \qquad (1)$$
$$-12x + 18y = -9 \qquad \text{3 times (3)}$$

3) **Add the equations.**

$$\begin{array}{r} 12x - 18y = 9 \\ + \underline{-12x + 18y = -9} \\ 0 = 0 \qquad \text{True} \end{array}$$

The variables drop out, and we get a true statement. The equations are dependent, so there are an infinite number of solutions. The solution set is $\left\{ (x, y) \,\middle|\, y = \frac{2}{3}x - \frac{1}{2} \right\}$. ■

---

**You Try 5**

Solve the system using the elimination method.

$$-6x + 8y = 4$$
$$3x - 4y = -2$$

### 3. Use the Elimination Method Twice to Solve a Linear System

Sometimes, applying the elimination method *twice* is the best strategy.

**Example 6**

Solve using the elimination method.

$$5x - 6y = 2 \quad (1)$$
$$9x + 4y = -3 \quad (2)$$

**Solution**

Each equation is written in the form $Ax + By = C$, so we begin with Step 2.

2)  We will eliminate $y$ from equations (1) and (2).

**Rewrite the System**

$$2(5x - 6y) = 2(2)$$
$$3(9x + 4y) = 3(-3)$$

$\longrightarrow$

$$10x - 12y = 4$$
$$27x + 12y = -9$$

3)  Add the resulting equations to eliminate $y$. Solve for $x$.

$$
\begin{array}{r}
10x - 12y = \phantom{-}4 \\
+ \quad 27x + 12y = -9 \\
\hline
37x = -5
\end{array}
$$

$$x = -\frac{5}{37} \qquad \text{Solve for } x.$$

Normally, we would substitute $x = -\dfrac{5}{37}$ into equation (1) or equation (2) and solve for $y$.

This time, however, working with a number like $-\dfrac{5}{37}$ would be difficult, so *we will use the elimination method a second time.*

Go back to the original equations, (1) and (2), and use the elimination method again but eliminate the other variable, $x$. Then, solve for $y$.

Eliminate $x$ from 
$$5x - 6y = 2 \quad (1)$$
$$9x + 4y = -3 \quad (2)$$

**Rewrite the System**

$$-9(5x - 6y) = -9(2)$$
$$5(9x + 4y) = 5(-3)$$

$\longrightarrow$

$$-45x + 54y = -18$$
$$45x + 20y = -15$$

Add the equations
$$
\begin{array}{r}
-45x + 54y = -18 \\
+ \quad 45x + 20y = -15 \\
\hline
74y = -33
\end{array}
$$

$$y = -\frac{33}{74} \qquad \text{Solve for } y.$$

Check to verify that the solution is $\left( -\dfrac{5}{37}, -\dfrac{33}{74} \right)$.

**You Try 6**

Solve using the elimination method.

$$-9x + 2y = -3$$
$$2x - 5y = 4$$

**Answers to You Try Exercises**

1)  $(4, -2)$    2)  $\left(\dfrac{1}{2}, 9\right)$    3)  $(0, 7)$    4)  $\varnothing$

5)  infinite number of solutions of the form $\{(x, y)\,|\,3x - 4y = -2\}$    6)  $\left(\dfrac{7}{41}, -\dfrac{30}{41}\right)$

# 5.3 Exercises

**Mixed Exercises: Objectives 1 and 2**

1)  What is the first step you would use to solve this system by elimination if you wanted to eliminate $y$?

$$5x + y = 2$$
$$3x - y = 6$$

2)  What is the first step you would use to solve this system by elimination if you wanted to eliminate $x$?

$$4x - 3y = 14$$
$$8x - 11y = 18$$

Solve each system using the elimination method.

3)  $x - y = -3$
    $2x + y = 18$

4)  $x + 3y = 1$
    $-x + 5y = -9$

5)  $-x + 2y = 2$
    $x - 7y = 8$

6)  $4x - y = -15$
    $3x + y = -6$

7)  $x + 4y = 1$
    $3x - 4y = -29$

8)  $5x - 4y = -10$
    $-5x + 7y = 25$

9)  $-8x + 5y = -16$
    $4x - 7y = 8$

10)  $7x + 6y = 3$
     $3x + 2y = -1$

11)  $4x + 15y = 13$
     $3x + 5y = 16$

12)  $12x + 7y = 7$
     $-3x + 8y = 8$

VIDEO 13)  $9x - 7y = -14$
     $4x + 3y = 6$

14)  $5x - 2y = -6$
     $4x + 5y = -18$

15)  $-9x + 2y = -4$
     $6x - 3y = 11$

16)  $12x - 2y = 3$
     $8x - 5y = -9$

17)  $9x - y = 2$
     $18x - 2y = 4$

18)  $-4x + 7y = 13$
     $12x - 21y = -5$

19)  $x = 12 - 4y$
     $2x - 7 = 9y$

20)  $5x + 3y = -11$
     $y = 6x + 4$

21)  $4y = 9 - 3x$
     $5x - 16 = -6y$

22)  $8x = 6y - 1$
     $10y - 6 = -4x$

23)  $2x - 9 = 8y$
     $20y - 5x = 6$

24)  $3x + 2y = 6$
     $4y = 12 - 6x$

25)  $6x - 11y = -1$
     $-7x + 13y = 2$

26)  $10x - 4y = 7$
     $12x - 3y = -15$

27)  $9x + 6y = -2$
     $-6x - 4y = 11$

28)  $4x - 9y = -3$
     $36y - 16x = 12$

29)  What is the first step in solving this system by the elimination method? DO NOT SOLVE.

$$\frac{x}{4} + \frac{y}{2} = -1$$
$$\frac{3}{8}x + \frac{5}{3}y = -\frac{7}{12}$$

30)  What is the first step in solving this system by the elimination method? DO NOT SOLVE.

$$0.1x + 2y = -0.8$$
$$0.03x + 0.10y = 0.26$$

Solve each system by elimination.

31)  $\dfrac{4}{5}x - \dfrac{1}{2}y = -\dfrac{3}{2}$
     $2x - \dfrac{1}{4}y = \dfrac{1}{4}$

32)  $\dfrac{1}{3}x - \dfrac{4}{5}y = \dfrac{13}{15}$
     $\dfrac{1}{6}x - \dfrac{3}{4}y = -\dfrac{1}{2}$

33)  $\dfrac{5}{4}x - \dfrac{1}{2}y = \dfrac{7}{8}$
     $\dfrac{2}{5}x - \dfrac{1}{10}y = -\dfrac{1}{2}$

34)  $\dfrac{1}{2}x - \dfrac{11}{8}y = -1$
     $-\dfrac{2}{5}x + \dfrac{3}{10}y = \dfrac{4}{5}$

VIDEO 35)  $\dfrac{x}{4} + \dfrac{y}{2} = -1$
     $\dfrac{3}{8}x + \dfrac{5}{3}y = -\dfrac{7}{12}$

36)  $\dfrac{x}{12} - \dfrac{y}{6} = \dfrac{2}{3}$
     $\dfrac{x}{4} + \dfrac{y}{3} = 2$

37)  $\dfrac{x}{12} - \dfrac{y}{8} = \dfrac{7}{8}$
     $y = \dfrac{2}{3}x - 7$

38)  $\dfrac{5}{3}x + \dfrac{1}{3}y = \dfrac{2}{3}$
     $\dfrac{3}{4}x + \dfrac{3}{20}y = -\dfrac{5}{4}$

39)  $-\dfrac{1}{2}x + \dfrac{5}{4}y = \dfrac{3}{4}$
     $\dfrac{2}{5}x - \dfrac{1}{2}y = -\dfrac{1}{10}$

40)  $y = 2 - 4x$
     $\dfrac{1}{3}x - \dfrac{3}{8}y = \dfrac{5}{8}$

41)  $0.08x + 0.07y = -0.84$
     $0.32x - 0.06y = -2$

42)  $0.06x + 0.05y = 0.58$
     $0.18x - 0.13y = 1.18$

VIDEO 43)  $0.1x + 2y = -0.8$
     $0.03x + 0.10y = 0.26$

44)  $0.6x - 0.1y = 0.5$
     $0.1x - 0.03y = -0.01$

45) $-0.4x + 0.2y = 0.1$
    $0.6x - 0.3y = 1.5$

46) $x - 0.5y = 0.2$
    $-0.3x + 0.15y = -0.06$

47) $0.04x + 0.03y = 0.16$
    $0.6x + 0.2y = 1.15$

48) $-0.5x + 0.8y = 0.3$
    $0.03x + 0.1y = -0.24$

49) $17x - 16(y + 1) = 4(x - y)$
    $19 - 10(x + 2) = -4(x + 6) - y + 2$

50) $28 - 4(y + 1) = 3(x - y) + 4$
    $-5(x + 4) - y + 3 = 28 - 5(2x + 5)$

51) $5 - 3y = 6(3x + 4) - 8(x + 2)$
    $6x - 2(5y + 2) = -7(2y - 1) - 4$

52) $5(y + 3) = 6(x + 1) + 6x$
    $7 - 3(2 - 3x) - y = 2(3y + 8) - 5$

53) $6(x - 3) + x - 4y = 1 + 2(x - 9)$
    $4(2y - 3) + 10x = 5(x + 1) - 4$

54) $8y + 2(4x + 5) - 5x = 7y - 11$
    $11y - 3(x + 2) = 16 + 2(3y - 4) - x$

### Objective 3: Use the Elimination Method Twice to Solve a Linear System

Solve each system using the elimination method twice.

VIDEO 55) $4x + 5y = -6$
    $3x + 8y = 15$

56) $8x - 4y = -21$
    $-5x + 6y = 12$

57) $4x + 9y = 7$
    $6x + 11y = -14$

58) $10x + 3y = 18$
    $9x - 4y = 5$

Find $k$ so that the given ordered pair is a solution of the given system.

59) $x + ky = 17$;  (5, 4)
    $2x - 3y = -2$

60) $kx + y = -13$;  (-1, -8)
    $9x - 2y = 7$

61) $3x + 4y = -9$;  (-7, 3)
    $kx - 5y = 41$

62) $4x + 3y = -7$;  (2, -5)
    $3x + ky = 16$

63) Given the following system of equations,

$$x - y = 5$$
$$x - y = c$$

find $c$ so that the system has

a) an infinite number of solutions.

b) no solution.

64) Given the following system of equations,

$$2x + y = 9$$
$$2x + y = c$$

find $c$ so that the system has

a) an infinite number of solutions.

b) no solution.

65) Given the following system of equations,

$$9x + 12y = -15$$
$$ax + 4y = -5$$

find $a$ so that the system has

a) an infinite number of solutions.

b) exactly one solution.

66) Given the following system of equations,

$$-2x + 7y = 3$$
$$4x + by = -6$$

find $b$ so that the system has

a) an infinite number of solutions.

b) exactly one solution.

### Extension

Let $a$, $b$, and $c$ represent nonzero constants. Solve each system for $x$ and $y$.

67) $-5x + 4by = 6$
    $5x + 3by = 8$

68) $ax - 6y = 4$
    $-ax + 9y = 2$

69) $3ax + by = 4$
    $ax - by = -5$

70) $2ax + by = c$
    $ax + 3by = 4c$

## Putting It All Together

### Objective

1. Choose the Best Method for Solving a System of Linear Equations

## 1. Choose the Best Method for Solving a System of Linear Equations

We have learned three methods for solving systems of linear equations:

1) Graphing      2) Substitution      3) Elimination

How do we know which method is best for a particular system? We will answer this question by looking at a few examples, and then we will summarize our findings.

First, solving a system by graphing is the least desirable of the methods. The point of intersection can be difficult to read, especially if one of the numbers is a fraction. But, the graphing method is important in certain situations and is one you should know.

**Example 1**

Decide which method to use to solve each system, substitution or elimination, and explain why this method was chosen. Then, solve the system.

a)  $-5x + 2y = -8$
    $x = 4y + 16$

b)  $4x + y = 10$
    $-3x - 8y = 7$

c)  $4x - 5y = -3$
    $6x + 8y = 11$

**Solution**

a)  $-5x + 2y = -8$
    $x = 4y + 16$

The second equation in this system is solved for $x$, *and* there are no fractions in this equation. *Solve this system by substitution.*

$$-5x + 2y = -8 \qquad \text{First equation}$$
$$-5(4y + 16) + 2y = -8 \qquad \text{Substitute } 4y + 16 \text{ for } x.$$
$$-20y - 80 + 2y = -8 \qquad \text{Distribute.}$$
$$-18y - 80 = -8 \qquad \text{Combine like terms.}$$
$$-18y = 72 \qquad \text{Add 80.}$$
$$y = -4 \qquad \text{Solve for } y.$$

Substitute $y = -4$ into $x = 4y + 16$:

$$x = 4(-4) + 16 \qquad \text{Substitute.}$$
$$x = 0 \qquad \text{Solve for } x.$$

The solution is $(0, -4)$. The check is left to the student.

b)  $4x + y = 10$
    $-3x - 8y = 7$

In the first equation, $y$ has a coefficient of 1, so we can easily solve for $y$ and *substitute* the expression into the second equation. (Solving for another variable would result in having fractions in the equation.) *Or*, since each equation is in the form $Ax + By = C$, elimination would work well too. *Either substitution or elimination would be a good choice to solve this system.* The student should choose whichever method he or she prefers. Here, we use substitution.

$$y = -4x + 10 \qquad \text{Solve the first equation for } y.$$
$$-3x - 8y = 7 \qquad \text{Second equation}$$
$$-3x - 8(-4x + 10) = 7 \qquad \text{Substitute } -4x + 10 \text{ for } y.$$
$$-3x + 32x - 80 = 7 \qquad \text{Distribute.}$$
$$29x = 87 \qquad \text{Simplify.}$$
$$x = 3 \qquad \text{Solve for } x.$$

Substitute $x = 3$ into $y = -4x + 10$.

$$y = -4(3) + 10 \qquad \text{Substitute.}$$
$$y = -2 \qquad \text{Solve for } y.$$

The check is left to the student. The solution is $(3, -2)$.

c)  $4x - 5y = -3$
    $6x + 8y = 11$

None of the variables has a coefficient of 1 or $-1$. Therefore, we do *not* want to use substitution because we would have to work with fractions in the equation. *To solve a system like this, where none of the coefficients are 1 or −1, use the elimination method.*

Eliminate $x$.

**Rewrite the System**

$$-3(4x - 5y) = -3(-3)$$
$$2(6x + 8y) = 2(11)$$

$\longrightarrow$

$$-12x + 15y = 9$$
$$+\ \underline{12x + 16y = 22}$$
$$31y = 31$$
$$y = 1 \qquad \text{Solve for } y.$$

Substitute $y = 1$ into $4x - 5y = -3$.

$$4x - 5(1) = -3 \qquad \text{Substitute.}$$
$$4x - 5 = -3 \qquad \text{Multiply.}$$
$$4x = 2 \qquad \text{Add 5.}$$
$$x = \frac{2}{4} = \frac{1}{2} \qquad \text{Solve for } x.$$

The check is left to the student. The solution is $\left(\dfrac{1}{2}, 1\right)$. ∎

---

**Procedure**   Choosing Between Substitution and the Elimination Method to Solve a System

1) If at least one equation is solved for a variable and contains no fractions, *use substitution*.

$$-5x + 2y = -8$$
$$x = 4y + 16 \qquad \text{Example 1(a)}$$

2) If a variable has a coefficient of 1 or −1, you can solve for that variable and *use substitution*.

$$4x + y = 10$$
$$-3x - 8y = 7 \qquad \text{Example 1(b)}$$

*Or,* leave each equation in the form $Ax + By = C$ and use *elimination*. Either approach is good and is a matter of personal preference.

3) If no variable has a coefficient of 1 or −1, *use elimination*.

$$4x - 5y = -3$$
$$6x + 8y = 11 \qquad \text{Example 1(c)}$$

Remember, if an equation contains fractions or decimals, begin by eliminating them. Then, decide which method to use following the guidelines listed here.

---

**You Try 1**

Decide which method to use to solve each system, substitution or elimination, and explain why this method was chosen. Then, solve the system.

a)   $9x - 7y = -9$
     $2x + 9y = -2$

b)   $9x - 2y = 0$
     $x = y - 7$

c)   $4x + y = 13$
     $-3x - 2y = 4$

---

**Answers to You Try Exercises**

1)  a) elimination; $(-1, 0)$   b) substitution; $(2, 9)$   c) substitution or elimination; $(6, -11)$

## Putting It All Together
## Summary Exercises

### Objective 1: Choose the Best Method for Solving a System of Linear Equations

Decide which method to use to solve each system, substitution or addition, and explain why this method was chosen. Then solve the system.

1) $8x - 5y = 10$
   $2x - 3y = -8$

2) $x = 2y - 7$
   $8x - 3y = 9$

3) $12x - 5y = 18$
   $8x + y = -1$

4) $11x + 10y = -4$
   $9x - 5y = 2$

5) $y - 4x = -11$
   $x = y + 8$

6) $4x - 5y = 4$
   $y = \dfrac{3}{4}x - \dfrac{1}{2}$

Solve each system using either the substitution or elimination method.

7) $4x + 5y = 24$
   $x - 3y = 6$

8) $6y - 5x = 22$
   $-9x - 8y = 2$

9) $6x + 15y = -1$
   $9x = 10y - 8$

10) $x + 2y = 9$
    $7x - y = 3$

11) $10x + 4y = 7$
    $15x + 6y = -2$

12) $y = -6x + 5$
    $12x + 2y = 10$

13) $10x + 9y = 4$
    $x = -\dfrac{1}{2}$

14) $6x - 4y = 11$
    $\dfrac{3}{2}x + \dfrac{1}{4}y = \dfrac{7}{8}$

15) $7y - 2x = 13$
    $3x - 2y = 6$

16) $y = 6$
    $12x + y = 8$

17) $\dfrac{2}{5}x + \dfrac{4}{5}y = -2$
    $\dfrac{1}{6}x + \dfrac{1}{6}y = \dfrac{1}{3}$

18) $5x + 4y = 14$
    $y = -\dfrac{8}{5}x + 7$

19) $-0.3x + 0.1y = 0.4$
    $0.01x + 0.05y = 0.2$

20) $0.01x - 0.06y = 0.03$
    $0.4x + 0.3y = -1.5$

21) $-6x + 2y = -10$
    $21x - 7y = 35$

22) $\dfrac{5}{3}x + \dfrac{4}{3}y = \dfrac{2}{3}$
    $10x + 8y = -5$

23) $2 = 5y - 8x$
    $y = \dfrac{3}{2}x - \dfrac{1}{2}$

24) $\dfrac{5}{6}x - \dfrac{3}{4}y = \dfrac{2}{3}$
    $\dfrac{1}{3}x + 2y = \dfrac{10}{3}$

25) $2x - 3y = -8$
    $7x + 10y = 4$

26) $6x = 9 - 13y$
    $4x + 3y = -2$

27) $6(2x - 3) = y + 4(x - 3)$
    $5(3x + 4) + 4y = 11 - 3y + 27x$

28) $3 - 5(x - 4) = 2(1 - 4y) + 2$
    $2(x + 10) + y + 1 = 3x + 5(y + 6) - 17$

29) $2y - 2(3x + 4) = -5(y - 2) - 17$
    $4(2x + 3) = 10 + 5(y + 1)$

30) $x - y + 23 = 2y + 3(2x + 7)$
    $9y - 8 + 4(x + 2) = 2(4x - 1) - 3x + 10y$

31) $y = -4x$
    $10x + 2y = -5$

32) $x = \dfrac{2}{3}y$
    $9x - 5y = -6$

Solve each system by graphing.

33) $y = \dfrac{1}{2}x + 1$
    $x + y = 4$

34) $x + y = -3$
    $y = 3x + 1$

35) $x + y = 0$
    $x - 2y = -12$

36) $2y - x = 6$
    $y = 2x$

37) $2x - 3y = 3$
    $y = \dfrac{2}{3}x + 1$

38) $y = -\dfrac{5}{2}x - 3$
    $10x + 4y = -12$

Solve each system using a graphing calculator.

39) $8x - 6y = -7$
    $4x - 16y = 3$

40) $4x + 3y = -9$
    $2x + y = 2$

# Section 5.4 Applications of Systems of Two Equations

## Objectives

1. **Solve Problems Involving General Quantities**
2. **Solve Geometry Problems**
3. **Solve Problems Involving Cost**
4. **Solve Mixture Problems**
5. **Solve Distance, Rate, and Time Problems**

In Section 3.3, we introduced the five-step method for solving applied problems. Here, we modify the method for problems with *two* unknowns and *two* equations.

---

**Procedure**   Solving an Applied Problem Using a System of Equations

**Step 1:**   **Read** the problem carefully, more than once if necessary. Draw a picture, if applicable. Identify what you are being asked to find.

**Step 2:**   **Choose variables** to represent the unknown quantities. Label any pictures with the variables.

**Step 3:**   **Write a system of equations using two variables.** It may be helpful to begin by writing the equations in words.

**Step 4:**   **Solve** the system.

**Step 5:**   **Check** the answer in the original problem, and **interpret** the solution as it relates to the problem. Be sure your answer makes sense in the context of the problem.

---

## 1. Solve Problems Involving General Quantities

### Example 1

Write a system of equations and solve.

Pink Floyd's album, *The Dark Side of the Moon,* spent more weeks on the Billboard 200 chart for top-selling albums than any other album in history. It was on the chart 251 more weeks than the second-place album, *Johnny's Greatest Hits*, by Johnny Mathis. If they were on the charts for a total of 1231 weeks, how many weeks did each album spend on the Billboard 200 chart? (www.billboard.com)

**Solution**

**Step 1:**   **Read** the problem carefully, and identify what we are being asked to find.

We must find the number of weeks each album was on the chart.

**Step 2:**   **Choose variables** to represent the unknown quantities.

$x$ = the number of weeks *The Dark Side of the Moon* was on the Billboard 200 chart

$y$ = the number of weeks *Johnny's Greatest Hits* was on the Billboard 200 chart

**Step 3:**   **Write a system of equations using two variables.** First, let's think of the equations in English. Then we will translate them into algebraic equations.

*To get one equation,* use the information that says these two albums were on the Billboard 200 chart for a total of 1231 weeks. Write an equation in words, then translate it into an algebraic equation.

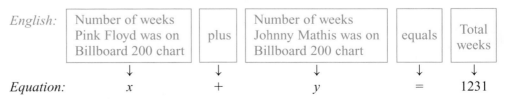

| *English:* | Number of weeks Pink Floyd was on Billboard 200 chart | plus | Number of weeks Johnny Mathis was on Billboard 200 chart | equals | Total weeks |
|---|---|---|---|---|---|
| | ↓ | ↓ | ↓ | ↓ | ↓ |
| *Equation:* | $x$ | $+$ | $y$ | $=$ | 1231 |

The first equation is $x + y = 1231$.

*To get the second equation,* use the information that says the Pink Floyd album was on the chart 251 weeks more than the Johnny Mathis album.

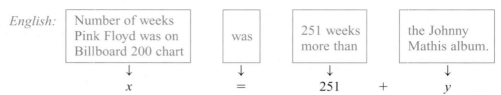

*English:*

| Number of weeks Pink Floyd was on Billboard 200 chart | was | 251 weeks more than | the Johnny Mathis album. |
|---|---|---|---|
| ↓ | ↓ | ↓ | ↓ |
| $x$ | $=$ | $251$ $+$ | $y$ |

The second equation is $x = 251 + y$.

The system of equations is $x + y = 1231$

$$x = 251 + y.$$

**Step 4:**   **Solve** the system.

$$x + y = 1231$$
$$(251 + y) + y = 1231 \qquad \text{Substitute.}$$
$$251 + 2y = 1231 \qquad \text{Combine like terms.}$$
$$2y = 980 \qquad \text{Subtract 251.}$$
$$\frac{2y}{2} = \frac{980}{2} \qquad \text{Divide each side by 2.}$$
$$y = 490 \qquad \text{Simplify}$$

Find $x$ by substituting $y = 490$ into $x = 251 + y$.

$$x = 251 + 490 = 741$$

The solution to the system is (741, 490).

**Step 5:**   **Check** the answer and **interpret** the solution as it relates to the problem.

*The Dark Side of the Moon* was on the Billboard 200 for 741 weeks, and *Johnny's Greatest Hits* was on the chart for 490 weeks.

They were on the chart for a total of $741 + 490 = 1231$ weeks, and the Pink Floyd album was on there $741 - 490 = 251$ weeks longer than the other album.  ■

**You Try 1**

Write a system of equations and solve.
In 2007, Carson City, NV, had about 33,000 fewer citizens than Elmira, NY. Find the population of each city if together they had approximately 143,000 residents. (www.census.gov)

Next we will see how we can use two variables and a system of equations to solve geometry problems.

## 2. Solve Geometry Problems

**Example 2**

Write a system of equations and solve.
    A builder installed a rectangular window in a new house and needs 182 in. of trim to go around it on the inside of the house. Find the dimensions of the window if the width is 23 in. less than the length.

### Solution

**Step 1:**   **Read** the problem carefully, and identify what we are being asked to find. Draw a picture.

We must find the length and width of the window.

**Step 2:**   **Choose variables** to represent the unknown quantities.

$$w = \text{the width of the window}$$
$$l = \text{the length of the window}$$

Label the picture with the variables.

**Step 3:**   **Write a system of equations using two variables.**

*To get one equation,* we know that the width is 23 in. less than the length. We can write the equation $w = l - 23$.

If it takes 182 in. of trim to go around the window, this is the *perimeter* of the rectangular window. Use the equation for the perimeter of a rectangle.

$$2l + 2w = 182$$

The system of equations is        $w = l - 23$
$$2l + 2w = 182.$$

**Step 4:**   **Solve** the system.

$$
\begin{aligned}
2l + 2w &= 182 \\
2l + 2(l - 23) &= 182 &&\text{Substitute.} \\
2l + 2l - 46 &= 182 &&\text{Distribute.} \\
4l - 46 &= 182 &&\text{Combine like terms.} \\
4l &= 228 &&\text{Add 46.} \\
\frac{4l}{4} &= \frac{228}{4} &&\text{Divide each side by 4.} \\
l &= 57 &&\text{Simplify.}
\end{aligned}
$$

Find $w$ by substituting $l = 57$ into $w = l - 23$.

$$w = 57 - 23 = 34$$

The solution to the system is $(57, 34)$. (The ordered pair is written as $(l, w)$, in alphabetical order.)

**Step 5:**   **Check** the answer and **interpret** the solution as it relates to the problem.

The length of the window is 57 in., and the width is 34 in. The check is left to the student.  ■

### You Try 2

Write a system of equations and solve.

The top of a desk is twice as long as it is wide. If the perimeter of the desk is 162 in., find its dimensions.

### 3. Solve Problems Involving Cost

| Example 3 |

Write a system of equations and solve.

Ari buys two mezzanine tickets to a Broadway play and four tickets to the top of the Empire State Building for $352. Lloyd spends $609 on four mezzanine tickets and three tickets to the top of the Empire State Building. Find the cost of a ticket to each attraction.

**Solution**

*Step 1:* **Read** the problem carefully, and identify what we are being asked to find.

We must find the cost of a ticket to a Broadway play and to the top of the Empire State Building.

*Step 2:* **Choose variables** to represent the unknown quantities.

$x$ = the cost of a ticket to a Broadway play

$y$ = the cost of a ticket to the Empire State Building

*Step 3:* **Write a system of equations using two variables.** First, let's think of the equations in English. Then we will translate them into algebraic equations.

First, use the information about Ari's purchase.

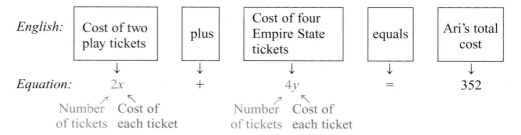

One equation is $2x + 4y = 352$.

Next, use the information about Lloyd's purchase.

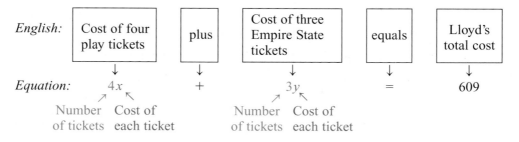

The other equation is $4x + 3y = 609$.

The system of equations is $2x + 4y = 352$

$$4x + 3y = 609.$$

*Step 4:* **Solve** the system.

Use the elimination method. Multiply the first equation by $-2$ to eliminate $x$.

$$-4x - 8y = -704$$
$$+ \quad \underline{4x + 3y = 609}$$
$$-5y = -95 \quad \text{Add the equations.}$$
$$y = 19$$

Find $x$. We will substitute $y = 19$ into $2x + 4y = 352$.

$$2x + 4(19) = 352 \qquad \text{Substitute.}$$
$$2x + 76 = 352$$
$$2x = 276$$
$$x = 138$$

The solution to the system is (138, 19).

**Step 5:**   **Check** the answer and **interpret** the solution as it relates to the problem.

A Broadway play ticket costs $138.00, and a ticket to the top of the Empire State Building costs $19.00.

The check is left to the student.

**You Try 3**

Write a system of equations and solve.
   Torie bought three scarves and a belt for $105 while Liz bought two scarves and two belts for $98. Find the cost of a scarf and a belt.

In Chapter 3, we learned how to solve mixture problems by writing an equation in one variable. Now we will learn how to solve the same type of problem using two variables and a system of equations.

## 4. Solve Mixture Problems

**Example 4**

A pharmacist needs to make 200 mL of a 10% hydrogen peroxide solution. She will make it from some 8% hydrogen peroxide solution and some 16% hydrogen peroxide solution that are in the storeroom. How much of the 8% solution and 16% solution should she use?

**Solution**

**Step 1:**   **Read** the problem carefully, and identify what we are being asked to find. Draw a picture.

We must find the amount of 8% solution and 16% solution she should use.

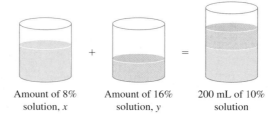

Amount of 8%         Amount of 16%         200 mL of 10%
solution, $x$          solution, $y$           solution

**Step 2:**   **Choose variables** to represent the unknown quantities.

$$x = \text{amount of 8\% solution needed}$$
$$y = \text{amount of 16\% solution needed}$$

***Step 3:*** **Write a system of equations using two variables.**

Let's begin by arranging the information in a table. Remember, to obtain the expression in the last column, multiply the percent of hydrogen peroxide in the solution by the amount of solution to get the amount of pure hydrogen peroxide in the solution.

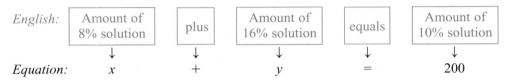

| | Percent of Hydrogen Peroxide in Solution (as a decimal) | Amount of Solution | Amount of Pure Hydrogen Peroxide in Solution |
|---|---|---|---|
| Mix these | 0.08 | $x$ | $0.08x$ |
| | 0.16 | $y$ | $0.16y$ |
| to make → | 0.10 | 200 | $0.10(200)$ |

*To get one equation*, use the information in the second column. It tells us that

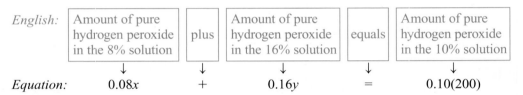

| | English: | Amount of 8% solution | plus | Amount of 16% solution | equals | Amount of 10% solution |
|---|---|---|---|---|---|---|
| | | ↓ | ↓ | ↓ | ↓ | ↓ |
| | Equation: | $x$ | $+$ | $y$ | $=$ | 200 |

The equation is $x + y = 200$.

*To get the second equation*, use the information in the third column. It tells us that

| | English: | Amount of pure hydrogen peroxide in the 8% solution | plus | Amount of pure hydrogen peroxide in the 16% solution | equals | Amount of pure hydrogen peroxide in the 10% solution |
|---|---|---|---|---|---|---|
| | | ↓ | ↓ | ↓ | ↓ | ↓ |
| | Equation: | $0.08x$ | $+$ | $0.16y$ | $=$ | $0.10(200)$ |

The equation is $0.08x + 0.16y = 0.10(200)$.

The system of equations is
$$x + y = 200$$
$$0.08x + 0.16y = 0.10(200).$$

***Step 4:*** **Solve** the system.

Multiply the second equation by 100 to eliminate the decimals. Our system becomes

$$x + y = 200$$
$$8x + 16y = 2000$$

Use the elimination method. Multiply the first equation by $-8$ to eliminate $x$.

$$\begin{array}{r} -8x - 8y = -1600 \\ +\ \underline{8x + 16y = \ \ \ 2000} \\ 8y = 400 \\ y = 50 \end{array}$$

Find $x$. Substitute $y = 50$ into $x + y = 200$.

$$x + 50 = 200$$
$$x = 150$$

The solution to the system is (150, 50).

***Step 5:*** **Check** the answer and **interpret** the solution as it relates to the problem.

The pharmacist needs 150 mL of the 8% solution and 50 mL of the 16% solution. Check the answers in the original problem to verify that they are correct. ∎

**You Try 4**

Write an equation and solve.
  How many milliliters of a 5% acid solution and how many milliliters of a 17% acid solution must be mixed to obtain 60 mL of a 13% acid solution?

## 5. Solve Distance, Rate, and Time Problems

**Example 5**

Write an equation and solve.
  Two cars leave Kearney, Nebraska, one driving east and the other heading west. The eastbound car travels 4 mph faster than the westbound car, and after 2.5 hours they are 330 miles apart. Find the speed of each car.

### Solution

**Step 1:** **Read** the problem carefully, and identify what we are being asked to find.

  We must find the speed of the eastbound and westbound cars. We will draw a picture to help us see what is happening in this problem. After 2.5 hours, the position of the cars looks like this:

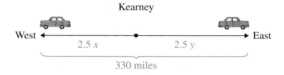

**Step 2:** **Choose variables** to represent the unknown quantities.

$$x = \text{the speed of the westbound car}$$
$$y = \text{the speed of the eastbound car}$$

**Step 3:** **Write a system of equations using two variables.**

  Let's make a table using the equation $d = rt$. Fill in the time, 2.5 hr, and the rates first, then multiply those together to fill in the values for the distance.

|  | $d$ | $r$ | $t$ |
|---|---|---|---|
| Westbound | $2.5x$ | $x$ | $2.5$ |
| Eastbound | $2.5y$ | $y$ | $2.5$ |

  Label the picture with the expressions for the distances.

  *To get one equation,* look at the picture and think about the distance between the cars after 2.5 hr.

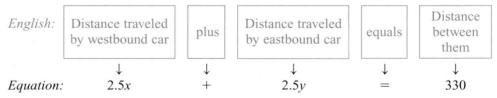

  The equation is $2.5x + 2.5y = 330$.

*To get the second equation*, use the information that says the eastbound car goes 4 mph faster than the westbound car.

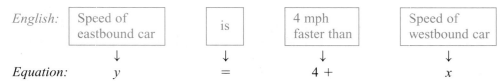

*English:*    | Speed of eastbound car | is | 4 mph faster than | Speed of westbound car |

↓       ↓       ↓       ↓

*Equation:*     $y$       $=$       $4 +$       $x$

The equation is $y = 4 + x$.

The system of equations is $2.5x + 2.5y = 330$
$$y = 4 + x.$$

**Step 4:**    **Solve** the system.

Use substitution.

$$2.5x + 2.5y = 330$$
$$2.5x + 2.5(4 + x) = 330 \qquad \text{Substitute } 4 + x \text{ for } y.$$
$$2.5x + 10 + 2.5x = 330 \qquad \text{Distribute.}$$
$$5x + 10 = 330 \qquad \text{Combine like terms.}$$
$$5x = 320$$
$$x = 64$$

Find $y$ by substituting $x = 64$ into $y = 4 + x$.
$$y = 4 + 64 = 68$$

The solution to the system is (64, 68).

**Step 5:**    **Check** the answer and **interpret** the solution as it relates to the problem.

The speed of the westbound car is 64 mph, and the speed of the eastbound car is 68 mph.

**Check.**

Distance of westbound car      Distance of eastbound car
↓                ↓

$$2.5(64) \quad + \quad 2.5(68) = 160 + 170 = 330 \text{ mi}$$

**You Try 5**

Write an equation and solve.

Two planes leave the same airport, one headed north and the other headed south. The northbound plane goes 100 mph slower than the southbound plane. Find each of their speeds if they are 1240 miles apart after 2 hours.

**Answers to You Try Exercises**

1)   Carson City: 55,000; Elmira: 88,000     2)   width: 27 in.; length: 54 in.

3)   scarf: $28; belt: $21     4)   20 mL of 5% solution; 40 mL of 17% solution

5)   northbound plane: 260 mph; southbound plane: 360 mph

## 5.4 Exercises

**Objective 1: Solve Problems Involving General Quantities**

Write a system of equations and solve.

1) The sum of two numbers is 87, and one number is eleven more than the other. Find the numbers.

2) One number is half another number. The sum of the two numbers is 141. Find the numbers.

3) Through the summer of 2009, *The Dark Knight* and *Transformers: Revenge of the Fallen* earned more money on their opening days than any other movies. *The Dark Knight* grossed $6.6 million more than *Transformers*. Together, they brought in $127.8 million. How much did each film earn on opening day? (http://hollywoodinsider.ew.com)

4) In the 1976–1977 season, Kareem Abdul-Jabbar led all players in blocked shots. He blocked 50 more shots than Bill Walton, who finished in second place. How many shots did each man block if they rejected a total of 472 shots? (www.nba.com)

5) Through 2009, Beyonce had been nominated for five more BET Awards than T.I. Determine how many nominations each performer received if they got a total of 27 nominations. (http://en.wikipedia.org)

6) From 1965 to 2000, twice as many people immigrated to the United States from The Philippines as from Vietnam. The total number of immigrants from these two countries was 2,100,000. How many people came to the United States from each country? (www.ellisisland.org)

7) According to a U.S. Census Bureau survey in 2006, about half as many people in the United States spoke Urdu at home as spoke Polish. If a total of about 975,000 people spoke these languages in their homes, how many spoke Urdu and how many spoke Polish? (www.census.gov)

8) During one week, a hardware store sold 27 fewer "regular" incandescent lightbulbs than energy-efficient compact fluorescent light (CFL) bulbs. How many of each type of bulb was sold if the store sold a total of 79 of these two types of lightbulbs?

9) On April 12, 1961, Yuri Gagarin of the Soviet Union became the first person in space when he piloted the

Vostok 1 mission. The next month, Alan B. Shepard became the first American in space in the Freedom 7 space capsule. The two of them spent a total of about 123 minutes in space, with Gagarin logging 93 more minutes than Shepard. How long did each man spend in space? (www-pao.ksc.nasa.gov, www.enchantedlearning.com)

10) Mr. Monet has 85 students in his Art History lecture. For their assignment on impressionists, one-fourth as many students chose to recreate an impressionist painting as chose to write a paper. How many students will be painting, and how many will be writing papers?

**Objective 2: Solve Geometry Problems**

11) Find the dimensions of a rectangular door that has a perimeter of 220 in. if the width is 50 in. less than the height of the door.

12) The length of a rectangle is 3.5 in. more than its width. If the perimeter is 23 in., what are the dimensions of the rectangle?

13) An iPod Touch is rectangular in shape and has a perimeter of 343.6 mm. Find its length and width given that it is 48.2 mm longer than it is wide.

14) Eliza needs 332 in. of a decorative border to sew around a rectangular quilt she just made. Its width is 26 in. less than its length. Find the dimensions of the quilt.

15) A rectangular horse corral is bordered on one side by a barn as pictured here. The length of the corral is 1.5 times the width. If 119 ft of fencing was used to make the corral, what are its dimensions?

16) The length of a rectangular mirror is twice its width. Find the dimensions of the mirror if its perimeter is 246 cm.

17) Find the measures of angles $x$ and $y$ if the measure of angle $x$ is three-fifths the measure of angle $y$ and if the angles are related according to the figure.

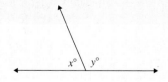

18) Find the measures of angles $x$ and $y$ if the measure of angle $y$ is two-thirds the measure of angle $x$ and if the angles are related according to the figure.

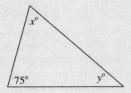

### Objective 3: Solve Problems Involving Cost

19) Kenny and Kyle are huge Colorado Avalanche fans. Kenny buys a T-shirt and two souvenir hockey pucks for $36.00, and Kyle spends $64.00 on two T-shirts and three pucks. Find the price of a T-shirt and the price of a puck.

20) Bruce Springsteen and Jimmy Buffett each played in Chicago in 2009. Four Springsteen tickets and four Buffett tickets cost $908.00, while three Springsteen tickets and two Buffett tickets cost $552.00. Find the cost of a ticket to each concert. (www.ticketmaster.com)

21) Angela and Andy watch *The Office* every week with their friends and decide to buy them some gifts. Angela buys three Dwight bobbleheads and four star mugs for $105.00, while Andy spends $74.00 on two bobbleheads and three mugs. Find the cost of each item. (www.nbcuniversalstore.com)

22) Manny and Hiroki buy tickets in advance to some Los Angeles Dodgers games. Manny buys three left-field pavilion seats and six club seats for $423.00. Hiroki spends $413.00 on eight left-field pavilion seats and five club seats. Find the cost of each type of ticket. (www.dodgers.com)

23) Carol orders five White Castle hamburgers and a small order of french fries for $4.44, and Momar orders four hamburgers and two small fries for $5.22. Find the cost of a hamburger and the cost of a small order of french fries at White Castle. (White Castle menu)

24) Phuong buys New Jersey lottery tickets every Friday. One day she spent $17.00 on four Gold Strike tickets and three Super Cashout tickets. The next Friday, she bought three Gold Strike tickets and six Super Cashout tickets for $24.00. How much did she pay for each type of lottery ticket?

25) Lakeisha is selling wrapping paper products for a school fund-raiser. Her mom buys four rolls of wrapping paper and three packages of gift bags for $52.00. Her grandmother spends $29.00 on three rolls of wrapping paper and one package of gift bags. Find the cost of a roll of wrapping paper and a package of gift bags.

26) Alberto is selling popcorn to raise money for his Cub Scout den. His dad spends $86.00 on two tins of popcorn and three tins of caramel corn. His neighbor buys two tins of popcorn and one tin of caramel corn for $48.00. How much does each type of treat cost?

### Objective 4: Solve Mixture Problems

27) How many ounces of a 9% alcohol solution and how many ounces of a 17% alcohol solution must be mixed to obtain 12 oz of a 15% alcohol solution?

28) How many milliliters of a 15% acid solution and how many milliliters of a 3% acid solution must be mixed to get 45 mL of a 7% alcohol solution?

29) How many liters of pure acid and how many liters of a 25% acid solution should be mixed to get 10 L of a 40% acid solution?

30) How many ounces of pure cranberry juice and how many ounces of a citrus fruit drink containing 10% fruit juice should be mixed to get 120 oz of a fruit drink that is 25% fruit juice?

31) How many ounces of Asian Treasure tea that sells for $7.50/oz should be mixed with Pearadise tea that sells for $5.00/oz so that a 60-oz mixture is obtained that will sell for $6.00/oz?

32) How many pounds of peanuts that sell for $1.80 per pound should be mixed with cashews that sell for $4.50 per pound so that a 10-pound mixture is obtained that will sell for $2.61 per pound?

33) During a late-night visit to Taco Bell, Giovanni orders three Crunchy Tacos and a chicken chalupa supreme. His order contains 1640 mg of sodium. Jurgis orders two Crunchy Tacos and two chicken chalupa supremes, and his order contains 1960 mg of sodium. How much sodium is in each item? (www.tacobell.com)

34) Five White Castle hamburgers and one small order of french fries contain 1010 calories. Four hamburgers and two orders of fries contain 1180 calories. Determine how many calories are in a White Castle hamburger and in a small order of french fries. (www.whitecastle.com)

35) Mahmud invested $6000 in two accounts, some of it at 2% simple interest, the rest in an account earning 4% simple interest. How much did he invest in each account if he earned $190 in interest after 1 year?

36) Marijke inherited $15,000 and puts some of it into an account earning 5% simple interest and the rest in an account earning 4% simple interest. She earns a total of $660 in interest after 1 year. How much did she deposit into each account?

37) Oscar purchased 16 stamps. He bought some $0.44 stamps and some $0.28 stamps and spent $6.40. How many of each type of stamp did he buy?

38) Kelly saves all of her dimes and nickels in a jar on her desk. When she counts her money, she finds that she has 133 coins worth a total of $10.45. How many dimes and how many nickels does she have?

### Objective 5: Solve Distance, Rate, and Time Problems

39) Michael and Jan leave the same location but head in opposite directions on their bikes. Michael rides 1 mph faster than Jan, and after 3 hr they are 51 miles apart. How fast was each of them riding?

40) A passenger train and a freight train leave cities 400 miles apart and travel toward each other. The passenger train is traveling 16 mph faster than the freight train. Find the speed of each train if they pass each other after 5 hours.

41) A small plane leaves an airport and heads south, while a jet takes off at the same time heading north. The speed of the small plane is 160 mph less than the speed of the jet. If they are 1280 miles apart after 2 hours, find the speeds of both planes.

42) Tyreese and Justine start jogging toward each other from opposite ends of a trail 6.5 miles apart. They meet after 30 minutes. Find their speeds if Tyreese jogs 3 mph faster than Justine.

43) Pam and Jim leave from opposite ends of a bike trail 9 miles apart and travel toward each other. Pam is traveling 2 mph slower than Jim. Find each of their speeds if they meet after 30 minutes.

44) Stanley and Phyllis leave the office and travel in opposite directions. Stanley drives 6 mph slower than Phyllis, and after 1 hr they are 76 miles apart. How fast was each person driving?

Other types of distance, rate, and time problems involve a boat traveling upstream and downstream, and a plane traveling with and against the wind. To solve problems like these, we will still use a table to help us organize our information, but we must understand what is happening in such problems.

Let's think about the case of a boat traveling upstream and downstream.

Downstream

Upstream

Let $x$ = the speed of the boat in still water and let $y$ = the speed of the current.

When the boat is going *downstream* (*with* the current), the boat is being pushed along by the current so that

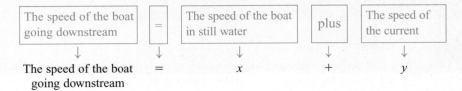

| The speed of the boat going downstream | = | The speed of the boat in still water | plus | The speed of the current |

$$\text{The speed of the boat going downstream} \quad = \quad x \quad + \quad y$$

When the boat is going *upstream* (*against* the current), the boat is being slowed down by the current so that

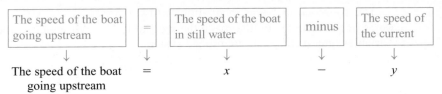

| The speed of the boat going upstream | = | The speed of the boat in still water | minus | The speed of the current |

$$\text{The speed of the boat going upstream} \quad = \quad x \quad - \quad y$$

Use this idea to solve Exercises 45–50.

45) It takes 2 hours for a boat to travel 14 miles downstream. The boat can travel 10 miles upstream in the same amount of time. Find the speed of the boat in still water and the speed of the current. (Hint: Use the information in the following table, and write a system of equations.)

|  | d | r | t |
|---|---|---|---|
| Downstream | 14 | $x + y$ | 2 |
| Upstream | 10 | $x - y$ | 2 |

46) A boat can travel 15 miles downstream in 0.75 hours. It takes the same amount of time for the boat to travel 9 miles upstream. Find the speed of the boat in still water and the speed of the current. (Hint: Use the information in the following table, and write a system of equations.)

|  | d | r | t |
|---|---|---|---|
| Downstream | 15 | $x + y$ | 0.75 |
| Upstream | 9 | $x - y$ | 0.75 |

47) It takes 5 hours for a boat to travel 80 miles downstream. The boat can travel the same distance back upstream in 8 hours. Find the speed of the boat in still water and the speed of the current.

48) A boat can travel 12 miles downstream in 1.5 hours. It takes 3 hours for the boat to travel back to the same spot going upstream. Find the speed of the boat in still water and the speed of the current.

49) A jet can travel 1000 miles against the wind in 2.5 hours. Going with the wind, the jet could travel 1250 miles in the same amount of time. Find the speed of the jet in still air and speed of the wind.

50) It takes 2 hours for a small plane to travel 390 miles with the wind. Going against the wind, the plane can travel 330 miles in the same amount of time. Find the speed of the plane in still air and the speed of the wind.

## Section 5.5 Linear Inequalities in Two Variables

**Objectives**

1. **Define a Linear Inequality in Two Variables**
2. **Graph a Linear Inequality in Two Variables**
3. **Solve a System of Linear Inequalities by Graphing**

In Chapter 3, we first learned how to solve linear inequalities in *one variable* such as $4x + 9 \geq 7$. In this section, we will first learn how to graph the solution set of linear inequalities in *two variables*. Then we will learn how to graph the solution set of *systems* of linear inequalities in two variables.

### 1. Define a Linear Inequality in Two Variables

**Definition**

A **linear equality in two variables** is an inequality that can be written in the form $Ax + By \geq C$ or $Ax + By \leq C$, where A, B, and C are real numbers and where A and B are not both zero. ($>$ and $<$ may be substituted for $\geq$ and $\leq$.)

Here are some examples of linear inequalities in two variables.

$$2x - 5y \geq 3, \qquad y < \frac{1}{5}x + 2, \qquad x \leq 6, \qquad y > -1$$

**Note**

$x \leq 6$ can be considered a linear inequality in two variables because it can be written as $x + 0y \leq 6$. Likewise, $y > -1$ can be written as $0x + y > -1$.

The solutions to linear inequalities in two variables, such as $x + y \geq 1$, are *ordered pairs* of the form $(x, y)$ that make the inequality true. We graph a linear inequality in two variables on a rectangular coordinate system.

**Example 1**

Shown here is the graph of $x + y \geq 1$. Find three points that solve $x + y \geq 1$, and find three points that are not in the solution set.

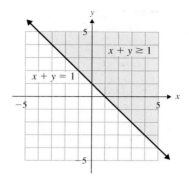

**Solution**

The solution set of $x + y \geq 1$ consists of all points either on the line or in the shaded region. *Any* point on the line or in the shaded region will make $x + y \geq 1$ true.

| Solutions of $x + y \geq 1$ | Check by Substituting into $x + y \geq 1$ |
|---|---|
| $(-1, 3)$ | $-1 + 3 \geq 1$ ✓ |
| $(4, 1)$ | $4 + 1 \geq 1$ ✓ |
| $(2, -1)$ (on the line) | $2 + (-1) \geq 1$ ✓ |

$(-1, 3)$, $(4, 1)$, and $(2, -1)$ are just some points that satisfy $x + y \geq 1$. There are infinitely many solutions.

| Not in the Solution Set of $x + y \geq 1$ | Verify by Substituting into $x + y \geq 1$ |
|---|---|
| $(0, 0)$ | $0 + 0 \geq 1$   False |
| $(-4, 1)$ | $-4 + 1 \geq 1$   False |
| $(2, -3)$ | $2 + (-3) \geq 1$   False |

$(0, 0)$, $(-4, 1)$, and $(2, -3)$ are just three points that do not satisfy $x + y \geq 1$. There are infinitely many such points.

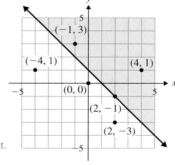

Points in the shaded region and on the line are in the solution set.

The points in the unshaded region are *not* in the solution set.

**Note**

If the inequality in Example 1 had been $x + y > 1$, then the line would have been drawn as a *dotted line* and all points on the line would *not* be part of the solution set.

**You Try 1**

Shown here is the graph of $2x - 3y \geq 6$. Find three points that solve $2x - 3y \geq 6$, and find three points that are not in the solution set.

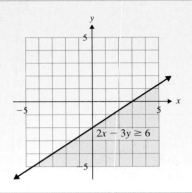

## 2. Graph a Linear Inequality in Two Variables

As you saw in the graph in Example 1, the line divides the *plane* into two regions or **half planes.** The line $x + y = 1$ is the **boundary line** between the two half planes. We can use this boundary and two different methods to graph a linear inequality in two variables. The first method we will discuss is the **test point** method.

---

**Procedure    Graphing a Linear Inequality in Two Variables Using the Test Point Method**

1) **Graph the boundary line.** If the inequality contains $\geq$ or $\leq$, make it a *solid line.* If the inequality contains $>$ or $<$, make it a *dotted line.*

2) **Choose a test point not on the line, and shade the appropriate region.** Substitute the test point into the inequality. If $(0, 0)$ is not on the line, it is an easy point to test in the inequality.

    a)   If it *makes the inequality true,* shade the side of the line *containing* the test point. All points in the shaded region are part of the solution set.

    b)   If the test point *does not satisfy the inequality,* shade the *other* side of the line. All points in the shaded region are part of the solution set.

---

**Example 2**

Graph $3x + 2y \leq -6$.

**Solution**

1) Graph the boundary line $3x + 2y = -6$ as a solid line.

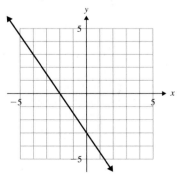

2) Choose a test point not on the line and substitute it into the inequality to determine whether or not it makes the inequality true.

| Test Point | Substituting into $3x + 2y \leq -6$ |
|---|---|
| $(0, 0)$ | $3(0) + 2(0) \leq -6$ |
| | $0 + 0 \leq -6$ |
| | $0 \leq -6$   False |

Since the test point $(0, 0)$ does *not* satisfy the inequality, we will shade the side of the line that does *not* contain the point $(0, 0)$.

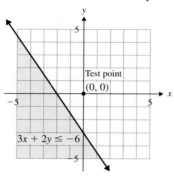

All points on the line and in the shaded region satisfy the inequality $3x + 2y \leq -6$.

### Example 3

Graph $-x + 3y > -3$.

**Solution**

1) Since the inequality symbol is $>$ and not $\geq$, graph the boundary line $-x + 3y = -3$ as a dotted line. (This means that the points *on* the line are not part of the solution set.)

2) Choose a test point not on the line and substitute it into the inequality to determine whether or not it makes the inequality true.

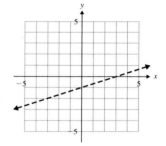

| Test Point | Substitute into $-x + 3y > -3$ |
|---|---|
| $(0, 0)$ | $-(0) + 3(0) > -3$ |
| | $0 > -3$   True |

Since the test point $(0, 0)$ satisfies the inequality, shade the side of the line containing the point $(0, 0)$.

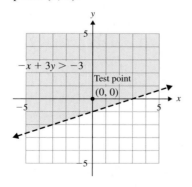

All points in the shaded region satisfy the inequality $-x + 3y > -3$.

### You Try 2

Graph each inequality.

a)   $-x + 3y \geq 1$        b)   $x + 2y < 4$

Another method we can use to graph linear inequalities in two variables involves writing the boundary line in *slope-intercept form.*

---

**Procedure    Using Slope-Intercept Form to Graph a Linear Inequality in Two Variables**

1)  **Write the inequality in the form $y \geq mx + b$ ($y > mx + b$) or $y \leq mx + b$ ($y < mx + b$).**

2)  **Graph the boundary line $y = mx + b$.**

   a)  If the inequality contains $\geq$ or $\leq$, make it a *solid line.*

   b)  If the inequality contains $>$ or $<$, make it a *dotted line.*

3)  **Shade the appropriate side of the line.**

   a)  If the inequality is in the form $y \geq mx + b$ or $y > mx + b$, shade above the line.

   b)  If the inequality is in the form $y \leq mx + b$ or $y < mx + b$, shade below the line.

---

**Example 4**

Graph each inequality using the slope-intercept method.

a)  $y < -\dfrac{1}{3}x + 2$        b)  $2x - y \leq 1$

**Solution**

a)  1) $y < -\dfrac{1}{3}x + 2$ is already in the correct form.

    2) Graph the boundary line $y = -\dfrac{1}{3}x + 2$ as a *dotted line.*

    3) Since $y < -\dfrac{1}{3}x + 2$ has a *less than* symbol, shade *below* the line.

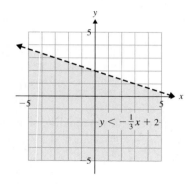

All points in the shaded region satisfy $y < -\dfrac{1}{3}x + 2$. We can choose a point, such as $(0, 0)$, in the shaded region as a check. Substitute this point into the inequality.

$$0 < -\frac{1}{3}(0) + 2$$
$$0 < 0 + 2$$
$$0 < 2 \qquad \text{True}$$

b)  1) Solve $2x - y \leq 1$ for $y$.

$$-y \leq -2x + 1 \qquad \text{Subtract } 2x.$$
$$y \geq 2x - 1 \qquad \text{Divide by } -1 \text{ and change the direction of the inequality symbol.}$$

    2) Graph $y = 2x - 1$ as a *solid line.*

    3) Since $y \geq 2x - 1$ has a *greater than or equal to* symbol, shade *above* the line.

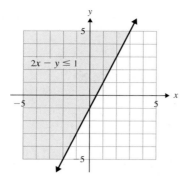

All points on the line and in the shaded region satisfy $2x - y \leq 1$.

 **You Try 3**

Graph each inequality using the slope-intercept method.

a)  $y \leq -\dfrac{3}{2}x + 2$          b)  $3x - y < -1$

### 3. Solve a System of Linear Inequalities by Graphing

A **system of linear inequalities** contains two or more linear inequalities. The **solution set of a system of linear inequalities** consists of all the ordered pairs that make all the inequalities in the system true. To solve such a system, we will graph each inequality on the same axes. The region where the solutions intersect, or overlap, is the solution to the system. We use the following steps to solve a system of linear inequalities:

> **Procedure**   Solving a System of Linear Inequalities by Graphing
>
> 1)  Graph each inequality separately on the same axes. Shade lightly.
> 2)  The solution set is the *intersection* (overlap) of the shaded regions. Heavily shade this region to indicate this is the solution set.

**Example 5**

Graph the solution set of the system.

$$x \leq 1$$
$$2x + 3y \geq -3$$

**Solution**

To graph $x \leq 1$, graph the boundary line $x = 1$ as a solid line. The $x$-values are *less than* 1 to the *left* of 1, so shade to the left of the line $x = 1$.

Graph $2x + 3y \geq -3$ as shown in the middle graph. The region shaded purple in the third graph is the *intersection* of the shaded regions and the solution set of the system.

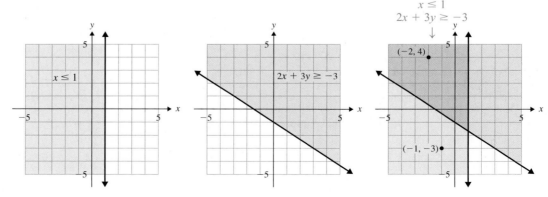

Any point inside the purple area will satisfy *both* inequalities. For example, the point $(-2, 4)$ is in this region (see the graph).

| Test Point | Verify | | |
|---|---|---|---|
| $(-2, 4)$ | $x \leq 1$: | $-2 \leq 1$ ✓ | True |
| | $2x + 3y \geq -3$: $\quad 2(-2) + 3(4) \geq -3$ | | |
| | $-4 + 12 \geq -3$ | | |
| | $8 \geq -3$ ✓ | | True |

Any point outside this region of intersection will not satisfy *both* inequalities and is *not* part of the system's solution set. One such point is $(-1, -3)$.

| Test Point | Verify | | |
|---|---|---|---|
| $(-1, -3)$ | $x \leq 1$: | $-1 \leq 1$ ✓ | True |
| | $2x + 3y \geq -3$: $\quad 2(-1) + 3(-3) \geq -3$ | | |
| | $-2 - 9 \geq -3$ | | |
| | $-11 \geq -3$ | | False |

Although we show three separate graphs in this example, it is customary to graph everything on the same axes, shading lightly at first, then to heavily shade the region that is the graph of the solution set. ∎

## You Try 4

Graph the solution set of the system.

$$y \geq 2$$
$$3x + 2y \leq -2$$

## Example 6

Graph the solution set of the system.

$$y < x + 2$$
$$-2x - y > -1$$

### Solution

Graph each inequality separately, and lightly shade the solution set of each of them. (Notice that we use dotted lines for the boundary lines because of the inequality symbols.) The solution set of the system is the intersection (overlap) of the shaded regions. This is the purple region. The points on the boundary lines are not included in the solution set.

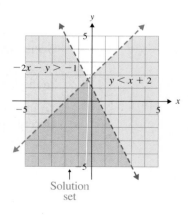

∎

## You Try 5

Graph the solution set of the system.

$$y > -\frac{1}{2}x + 2$$
$$x - y > -3$$

Next we will solve a system that contains more than two inequalities.

**Example 7**

Graph the solution set of the system.

$$x \geq 0$$
$$y \geq 0$$
$$y \leq -\frac{4}{3}x + 4$$

### Solution

The inequalities $x \geq 0$ and $y \geq 0$ tell us that the solution set will be in the first quadrant since this is where both $x$ and $y$ are positive. (The axes may be included since $x = 0$ is the $y$-axis and $y = 0$ is the $x$-axis.)

Graph the third inequality. The solution set of the system is the region shaded in purple. The points on the boundary lines are included in the solution set.

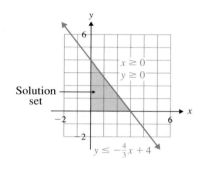

**You Try 6**

Graph the solution set of the system.

$$x \geq 0$$
$$y \geq 0$$
$$y \leq -\frac{1}{2}x + 1$$

### Using Technology

A graphing calculator can be used to graph one or more linear inequalities in two variables.

To graph a linear inequality in two variables using a graphing calculator, first solve the inequality for y. Then graph the boundary line found by replacing the inequality symbol with an = symbol. For example, to graph the inequality $2x - y \leq 5$, solve the inequality for y giving $y \geq 2x - 5$. Next, graph the boundary equation $y = 2x - 5$ using a solid line since the inequality symbol is ≤. Enter $2x - 5$ in Y1, press $\boxed{\text{ZOOM}}$, and select 6:ZStandard to graph the equation.

To shade above the line, press $\boxed{\text{Y=}}$ and move the cursor to the left of Y1 using the left arrow key. Press $\boxed{\text{ENTER}}$ twice and then move the cursor to the next line as shown next on the left. Press $\boxed{\text{GRAPH}}$ to see the inequality as shown next on the right.

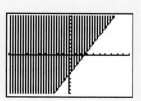

To shade below the line, press ☐Y=☐ and move the cursor to the left of Y1 using the left arrow key. Press ☐ENTER☐ three times and then move the cursor to the next line as shown next on the left. Press ☐GRAPH☐ to see the inequality $y \leq 2x - 5$, as shown next on the right.

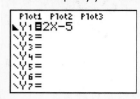
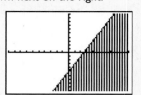

Graph the linear inequalities in two variables.

1)  $y \leq 3x - 1$       2)  $y \geq x + 2$       3)  $2x - y \leq 4$

4)  $x - y \geq 3$       5)  $y \leq -5x - 3$       6)  $4 - y \leq 3x$

---

## Answers to You Try Exercises

1)  Answers may vary; in solution set: $(-2, -4), (0, -2), (4, -1)$; not in solution set: $(0, 0), (-3, 1), (2, 3)$

2)  a)

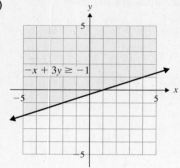

$-x + 3y \geq -1$

b)

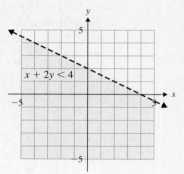

$x + 2y < 4$

3)  a)

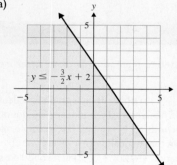

$y \leq -\frac{3}{2}x + 2$

b)

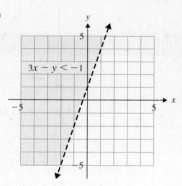

$3x - y < -1$

4)

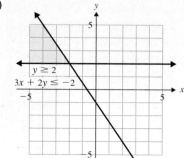

$y \geq 2$

$3x + 2y \leq -2$

5)

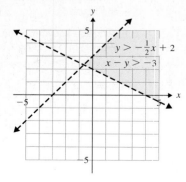

$y > -\frac{1}{2}x + 2$

$x - y > -3$

6)

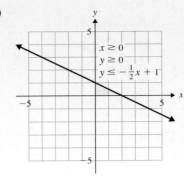

$x \geq 0$
$y \geq 0$
$y \leq -\frac{1}{2}x + 1$

**Answers to Technology Exercises**

1)

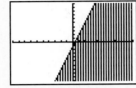

2)

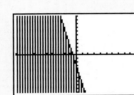

3)

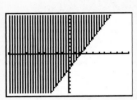

4)

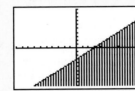

5)

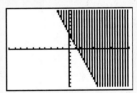

6)

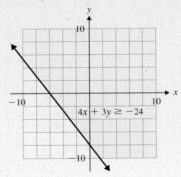

## 5.5 Exercises

### Objective 1: Define a Linear Inequality in Two Variables

The graphs of linear inequalities are given next. For each, find three points that satisfy the inequality and three that are not in the solution set.

1)  $3x - y \geq -1$

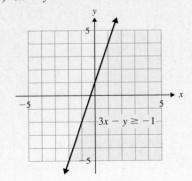

$3x - y \geq -1$

2)  $4x + 3y \geq -24$

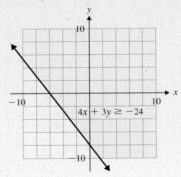

$4x + 3y \geq -24$

3) $y > -\dfrac{4}{3}x + 4$

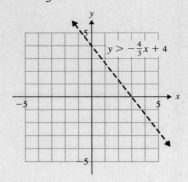

4) $y < \dfrac{1}{4}x - 2$

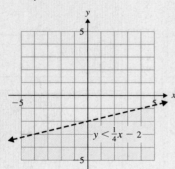

5) $y < x$

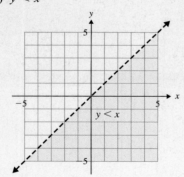

6) $x \geq 3$

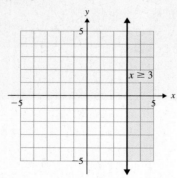

## Objective 2:  Graph a Linear Inequality in Two Variables

Graph using the test point method.

7) $2x + 3y \geq 6$        8) $x + 2y \leq 2$

9) $y < -x + 4$        10) $y > \dfrac{1}{2}x + 3$

11) $2x - 7y \leq 14$        12) $4x + 3y < 15$

(VIDEO) 13) $y < x$        14) $y \geq -\dfrac{2}{3}x$

15) $x \leq 4$        16) $y > -3$

Use the slope-intercept form to graph each inequality.

17) $y \leq x + 3$        18) $y \geq -2x + 1$

19) $y > \dfrac{2}{5}x - 4$        20) $y < \dfrac{1}{4}x + 1$

21) $4x + y < 2$        22) $-3x + y > -4$

(VIDEO) 23) $9x - 3y \leq -21$        24) $3x + 5y < -20$

25) $x \leq \dfrac{2}{3}y$        26) $x - 2y \geq 0$

27) To graph an inequality like $y \leq \dfrac{3}{4}x - 5$, which method, test point or slope-intercept, would you prefer? Why?

28) To graph an inequality like $5x + 8y > 12$, which method, test point or slope-intercept, would you prefer? Why?

Graph using either the test point or slope-intercept method.

29) $y > \dfrac{1}{3}x + 1$        30) $y \leq -\dfrac{3}{4}x - 2$

31) $5x + 2y < -8$        32) $4x + y < 7$

33) $9x - 3y \leq 21$        34) $5x - 3y \geq -9$

35) $y > \dfrac{5}{2}$        36) $x \leq -\dfrac{1}{2}$

37) $x - 2y \geq -5$        38) $3x + 2y > -8$

## Objective 3:  Solve a System of Linear Inequalities by Graphing

The graphs of systems of linear inequalities are given next. For each, find three points that are in the solution set and three that are not.

39) $y \geq \dfrac{4}{5}x + 2$

$y < 5$

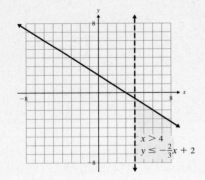

40) $x > 4$

$y \leq -\dfrac{2}{3}x + 2$

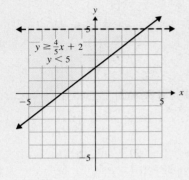

41) $x + 3y \leq 2$

$y - x \leq 0$

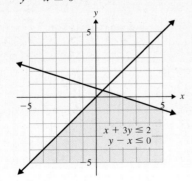

42) $3x - 2y \leq 6$

$3x + y \geq 1$

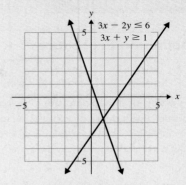

43) Is $(9, -2)$ in the solution set of the system $4x + 3y \geq 27$ and $x - 4y \geq -4$? Why or why not?

44) Is $(3, 7)$ in the solution set of the system $4x + 3y \geq 27$ and $x - 4y \geq -4$? Why or why not?

Graph the solution set of each system.

45) $x \leq 4$

$y \geq -\dfrac{3}{2}x + 3$

46) $y \leq \dfrac{1}{4}x + 2$

$y \geq -1$

47) $y < x + 3$

$y \geq -1$

48) $x < 2$

$y > \dfrac{3}{2}x - 4$

49) $y < 4$

$2x - y < -2$

50) $x - 3y \geq 0$

$x \leq -3$

51) $y \leq 4x - 1$

$y \leq -3x + 3$

52) $y > -x + 2$

$y > 2x + 1$

53) $y > -\dfrac{3}{2}x + 2$

$y < 2x - 3$

54) $y \leq \dfrac{1}{4}x + 3$

$y \geq -\dfrac{1}{3}x + 4$

55) $y \geq \dfrac{2}{3}x - 4$

$4x + y \leq 3$

56) $y - 2x \leq 1$

$y \geq -\dfrac{1}{5}x - 2$

57) $5x - 3y > 9$

$2x + 3y \leq 12$

58) $2x - 3y < -9$

$x + 6y < 12$

59) $x \leq 6$

$y \geq 1$

60) $x \geq 5$

$y \leq -3$

61) $x \geq 0$

$y \geq 0$

$y \leq -x + 2$

62) $x \geq 0$

$y \geq 0$

$y \leq -\dfrac{2}{3}x + 3$

63) $x \geq 0$

$y \geq 0$

$y \leq -3x + 4$

64) $x \geq 0$

$y \geq 0$

$y \leq -x + 1$

65) $x \geq 0$

$y \geq 0$

$2x - y \geq 3$

66) $x \geq 0$

$y \geq 0$

$2x - 3y \leq -3$

| Definition/Procedure | Example |
|---|---|

## 5.1 Solving Systems by Graphing

A **system of linear equations** consists of two or more linear equations with the same variables. A **solution of a system** of two equations in two variables is an ordered pair that is a solution of each equation in the system. **(p. 282)**

Determine whether $(4, 2)$ is a solution of the system

$$x + 2y = 8$$
$$-3x + 4y = -4$$

| | |
|---|---|
| $x + 2y = 8$ | $-3x + 4y = -4$ |
| $4 + 2(2) \stackrel{?}{=} 8$   Substitute. | $-3(4) + 4(2) \stackrel{?}{=} -4$   Substitute. |
| $4 + 4 \stackrel{?}{=} 8$ | $-12 + 8 \stackrel{?}{=} -4$ |
| $8 = 8$   True | $-4 = -4$   True |

Since $(4, 2)$ is a solution of each equation in the system, **yes,** it is a solution of the system.

---

To **solve a system by graphing,** graph each line on the same axes.
a) If the lines intersect at a single point, then this point is the solution of the system. The system is **consistent.**
b) If the lines are parallel, then the system has **no solution.** We write the solution set as $\varnothing$. The system is **inconsistent.**
c) If the graphs are the same line, then the system has an **infinite number of solutions.** The system is **dependent. (p. 283)**

Solve by graphing.
$$y = -\frac{1}{2}x + 2$$
$$5x + 3y = -1$$

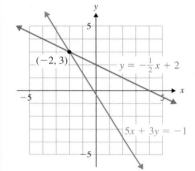

The solution of the system is $(-2, 3)$.

The system is **consistent.**

## 5.2 Solving Systems by the Substitution Method

**Steps for Solving a System by the Substitution Method**
1) Solve one of the equations for one of the variables. If possible, solve for a variable that has a coefficient of 1 or $-1$.

Solve by the substitution method.
$$7x - 3y = 8$$
$$x + 2y = -11$$

1) Solve for $x$ in the second equation since its coefficient is 1.
$$x = -2y - 11$$

2) Substitute the expression in *Step 1* into the *other* equation. The equation you obtain should contain only one variable.

2) Substitute $-2y - 11$ for the $x$ in the first equation.
$$7(-2y - 11) - 3y = 8$$

3) Solve the equation in *Step 2*.

3) Solve the equation above for $y$.

| | |
|---|---|
| $7(-2y-11) - 3y = 8$ | |
| $-14y - 77 - 3y = 8$ | Distribute. |
| $-17y - 77 = 8$ | Combine like terms. |
| $-17y = 85$ | Add 77. |
| $y = -5$ | Divide by $-17$. |

4) Substitute the value found in *Step 3* into either of the equations to obtain the value of the other variable.

4) Substitute $y = -5$ into the equation in *Step 1* to find $x$.

| | |
|---|---|
| $x = -2(-5) - 11$ | Substitute $-5$ for $y$. |
| $x = 10 - 11$ | Multiply. |
| $x = -1$ | |

5) Check the values in the original equations. **(p. 293)**

5) The solution is $(-1, -5)$. Verify this by substituting $(-1, -5)$ into each of the original equations.

| Definition/Procedure | Example |
|---|---|
| If the variables drop out and a false equation is obtained, the system has **no solution. The system is inconsistent, and the solution set is** $\varnothing$. (p. 295) | Solve by the substitution method. $\quad 2x - 8y = 9$<br>$\qquad\qquad\qquad\qquad\qquad\qquad\qquad\qquad x = 4y + 2$<br>1) The second equation is solved for $x$.<br>2) Substitute $4y + 2$ for $x$ in the first equation.<br>$\qquad\quad 2(4y + 2) - 8y = 9$<br>3) Solve the above equation for $y$.<br><br>$\qquad\qquad 2(4y + 2) - 8y = 9$<br>$\qquad\qquad\quad 8y + 4 - 8y = 9 \qquad$ Distribute.<br>$\qquad\qquad\qquad\qquad\qquad 4 = 9 \qquad$ False<br><br>4) The system has no solution. The solution set is $\varnothing$. |
| If the variables drop out and a true equation is obtained, the system has an **infinite number of solutions. The equations are dependent.** (p. 296) | Solve by the substitution method. $\qquad y = x - 3$<br>$\qquad\qquad\qquad\qquad\qquad\qquad\qquad 3x - 3y = 9$<br>1) The first equation is solved for $y$.<br>2) Substitute $x - 3$ for $y$ in the second equation.<br>$\qquad\qquad 3x - 3(x - 3) = 9$<br>3) Solve the above equation for $x$.<br><br>$\qquad\qquad\quad 3x - 3(x - 3) = 9$<br>$\qquad\qquad\quad 3x - 3x + 9 = 9 \qquad$ Distribute.<br>$\qquad\qquad\qquad\qquad\qquad 9 = 9 \qquad$ True<br><br>4) The system has an infinite number of solutions of the form $\{(x, y) \mid y = x - 3\}$. |

## 5.3 Solving Systems by the Elimination Method

**Steps for Solving a System of Two Linear Equations by the Elimination Method**

1) Write each equation in the form $Ax + By = C$.
2) Determine which variable to eliminate. If necessary, multiply one or both of the equations by a number so that the coefficients of the variable to be eliminated are negatives of one another.
3) Add the equations, and solve for the remaining variable.
4) Substitute the value found in *Step 3* into either of the original equations to find the value of the other variable.
5) Check the solution in each of the original equations. **(p. 298)**

Solve using the elimination method. $\qquad 4x + 5y = -7$
$$-5x - 6y = 8$$

Eliminate $x$. Multiply the first equation by 5, and multiply the second equation by 4 to rewrite the system with equivalent equations.

$$\qquad\qquad\qquad\qquad\qquad\qquad \text{Rewrite the system.}$$
$$5(4x + 5y) = 5(-7) \quad \rightarrow \quad 20x + 25y = -35$$
$$4(-5x - 6y) = 4(8) \quad \rightarrow \quad -20x - 24y = 32$$

Add the equations:
$$\begin{aligned} 20x + 25y &= -35 \\ + \quad -20x - 24y &= 32 \\ \hline y &= -3 \end{aligned}$$

Substitute $y = -3$ into either of the original equations and solve for $x$.

$$4x + 5y = -7$$
$$4x + 5(-3) = -7$$
$$4x - 15 = -7$$
$$4x = 8$$
$$x = 2$$

The solution is $(2, -3)$. Verify this by substituting $(2, -3)$ into each of the original equations.

| Definition/Procedure | Example |
|---|---|

## 5.4 Applications of Systems of Two Equations

Use the **Five Steps for Solving Applied Problems** outlined in the section to solve an applied problem.

1) **Read** the problem carefully. Draw a picture, if applicable.
2) **Choose variables** to represent the unknown quantities. If applicable, label the picture with the variables.
3) **Write a system of equations using two variables.** It may be helpful to begin by writing an equation in words.
4) **Solve the system.**
5) **Check** the answer in the original problem, and **interpret** the solution as it relates to the problem. **(p. 309)**

Amana spent $40.20 at a second-hand movie and music store when she purchased some DVDs and CDs. Each DVD cost $6.30, and each CD cost $2.50. How many DVDs and CDs did she buy if she purchased 10 items all together?
1) Read the problem carefully.
2) Choose variables.

$$x = \text{number of DVDs she bought}$$
$$y = \text{number of CDs she bought}$$

3) One equation involves the *cost* of the items:

$$\text{Cost DVDs} + \text{Cost CDs} = \text{Total Cost}$$
$$6.30x \quad + \quad 2.50y \quad = \quad 40.20$$

The second equation involves the number of items:

$$\underset{\text{DVDs}}{\text{Number of}} + \underset{\text{CDs}}{\text{Number of}} = \underset{\text{of items}}{\text{Total number}}$$
$$x \quad + \quad y \quad = \quad 10$$

The system is $6.30x + 2.50y = 40.20$.
$$x + y = 10$$

4) Multiply by 10 to eliminate the decimals in the first equation, and then solve the system using substitution.

$$10(6.30x + 2.50y) = 10(40.20) \qquad \text{Eliminate decimals.}$$
$$63x + 25y = 402$$

Solve the system $63x + 25y = 402$ to determine that the
$$x + y = 10$$
solution is (4, 6).
5) Amana bought 4 DVDs and 6 CDs. Verify the solution.

## 5.5 Linear Inequalities in Two Variables

A **linear inequality in two variables** is an inequality that can be written in the form $Ax + By \geq C$ or $Ax + By \leq C$, where $A$, $B$, and $C$ are real numbers and where $A$ and $B$ are not both zero. ( $>$ and $<$ may be substituted for $\geq$ and $\leq$.) **(p. 320)**

Some examples of linear inequalities in two variables are

$$x + 2y \leq 3, \qquad y > -\frac{1}{2}x + 4, \qquad y \geq -5, \qquad x < 7$$

**Graphing a Linear Inequality in Two Variables Using the Test Point Method**

1) **Graph the boundary line.**
   a) If the inequality contains $\geq$ or $\leq$, make it a *solid line*.
   b) If the inequality contains $>$ or $<$, make it a *dotted line*.

2) **Choose a test point not on the line, and shade the appropriate region.** Substitute the test point into the inequality.
   a) If it *makes the inequality true,* shade the side of the line *containing* the test point. All points in the shaded region are part of the solution set.
   b) If the test point *does not satisfy the inequality*, shade the *other* side of the line. All points in the shaded region are part of the solution set.

Graph using the test point method.

$$2x - y > -3$$

1) Graph the boundary line as a *dotted* line.

2) Choose a test point not on the line and substitute it into the inequality to determine whether or not it makes the inequality true.

| Test Point | Substitute into $2x - y > -3$ |
|---|---|
| (0, 0) | $2(0) - (0) > -3$ |
| | $0 - 0 > -3$ |
| | $0 > -3 \quad$ True |

Since the test point satisfies the inequality, shade the side of the line containing (0, 0).

| Definition/Procedure | Example |
|---|---|

**Note:** If $(0, 0)$ is not on the line, it is an easy point to test in the inequality. **(p. 322)**

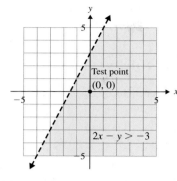

All points in the shaded region satisfy $2x - y > -3$.

**Use the Slope-Intercept Form to Graph a Linear Inequality in Two Variables**

1) **Write the inequality in the form**
$y \geq mx + b$ ($y > mx + b$) or
$y \leq mx + b$ ($y < mx + b$)

2) **Graph the boundary line $y = mx + b$.**
   a) If the inequality contains $\geq$ or $\leq$, make it a *solid line*.
   b) If the inequality contains $>$ or $<$, make it a *dotted line*.

3) **Shade the appropriate side of the line.**
   a) If the inequality is in the form $y \geq mx + b$ or $y > mx + b$, shade *above* the line.
   b) If the inequality is in the form $y \leq mx + b$ or $y < mx + b$, shade *below* the line. **(p. 324)**

Graph using the slope-intercept method.

$$-x + 4y \leq 2$$

1) Solve $-x + 4y \leq 2$ for $y$.   $4y \leq x + 2$

$$y \leq \frac{1}{4}x + \frac{1}{2}$$

2) Graph $y = \frac{1}{4}x + \frac{1}{2}$ as a *solid* line.

3) Since $y \leq \frac{1}{4}x + \frac{1}{2}$ has a *less than or equal to* symbol, shade *below* the line.

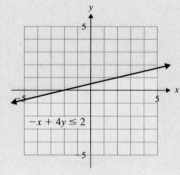

All points on the line and in the shaded region satisfy $-x + 4y \leq 2$.

**Solve a System of Linear Inequalities by Graphing**

1) Graph each inequality separately on the same axes. Shade lightly.

2) The solution set is the *intersection* (overlap) of the shaded regions. Heavily shade this region to indicate this is the solution set. **(p. 325)**

Graph the solution set of the system.   $y \geq -2x + 1$
$y \geq -2$

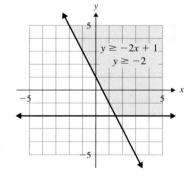

Any point in the shaded region will satisfy *both* inequalities.

**(5.1) Determine whether the ordered pair is a solution of the system of equations.**

1) $x - 5y = 13$
$2x + 7y = 20$
$(-4, -5)$

2) $8x + 3y = 16$
$10x - 6y = 7$
$\left(\dfrac{3}{2}, \dfrac{4}{3}\right)$

3) If you are solving a system of equations by graphing, how do you know whether the system has no solution?

**Solve each system by graphing.**

4) $y = \dfrac{1}{2}x + 1$
$x + y = 4$

5) $x - 3y = 9$
$-x + 3y = 6$

6) $6x - 3y = 12$
$-2x + y = -4$

7) $-x + 2y = 1$
$2x + 3y = -9$

**Without graphing, determine whether each system has no solution, one solution, or an infinite number of solutions.**

8) $8x + 9y = -2$
$x - 4y = 1$

9) $y = -\dfrac{5}{2}x + 3$
$5x + 2y = 6$

10) The graph shows the number of millions of barrels of crude oil reserves in Alabama and Michigan from 2002 to 2006. (www.census.gov)

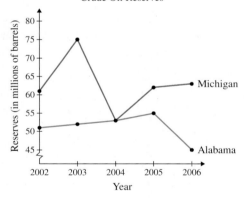

Crude Oil Reserves

a) In 2006, approximately how much more crude oil did Michigan have in reserve than Alabama?

b) Write the point of intersection as an ordered pair in the form (year, reserves) and explain its meaning.

c) Which line segment has the most negative slope? How can this be explained in the context of the problem.

**(5.2) Solve each system by the substitution method.**

11) $9x - 2y = 8$
$y = 2x + 1$

12) $y = -6x + 5$
$12x + 2y = 10$

13) $-x + 8y = 19$
$4x - 3y = 11$

14) $-12x + 7y = 9$
$8x - y = -6$

**(5.3) Solve each system using the elimination method.**

15) $x - 7y = 3$
$-x + 5y = -1$

16) $5x + 4y = -23$
$3x - 8y = -19$

17) $-10x + 4y = -8$
$5x - 2y = 4$

18) $7x - 4y = 13$
$6x - 5y = 8$

**Solve each system using the elimination method twice.**

19) $2x + 9y = -6$
$5x + y = 3$

20) $7x - 4y = 10$
$6x + 3y = 8$

**(5.2–5.3)**

21) When is the best time to use substitution to solve a system?

22) If an equation in a system contains fractions, what should you do first to make the system easier to solve?

**Solve each system by either the substitution or elimination method.**

23) $6x + y = -8$
$9x + 7y = -1$

24) $4y - 5x = -23$
$2x + 3y = -23$

25) $\dfrac{1}{3}x - \dfrac{2}{9}y = -\dfrac{2}{3}$
$\dfrac{5}{12}x + \dfrac{1}{3}y = 1$

26) $0.02x - 0.01y = 0.13$
$-0.1x + 0.4y = 1.8$

27) $6(2x - 3) = y + 4(x - 3)$
$5(3x + 4) + 4y = 11 - 3y + 27x$

28) $x - 3y = 36$
$y = \dfrac{5}{3}x$

29) $\dfrac{3}{4}x - \dfrac{5}{4}y = \dfrac{7}{8}$
$4 - 2(x + 5) - y = 3(1 - 2y) + x$

30) $y = -\dfrac{9}{7}x + \dfrac{6}{7}$
$18x + 14y = 12$

**(5.4) Write a system of equations and solve.**

31) One day in the school cafeteria, the number of children who bought white milk was twice the number who bought chocolate milk. How many cartons of each type of milk were sold if the cafeteria sold a total of 141 cartons of milk?

32) How many ounces of a 7% acid solution and how many ounces of a 23% acid solution must be mixed to obtain 20 oz of a 17% acid solution?

33) Edwin and Camille leave from opposite ends of a jogging trail 7 miles apart and travel toward each other. Edwin jogs 2 mph faster than Camille, and they meet after half an hour. How fast does each of them jog?

34) At a movie theater concession stand, three candy bars and two small sodas cost $14.00. Four candy bars and three small sodas cost $19.50. Find the cost of a candy bar and the cost of a small soda.

35) The width of a rectangle is 5 cm less than the length. Find the dimensions of the rectangle if the perimeter is 38 cm.

36) Two planes leave the same airport and travel in opposite directions. The northbound plane flies 40 mph slower than the southbound plane. After 1.5 hours they are 1320 miles apart. Find the speed of each plane.

37) Shawanna saves her quarters and dimes in a piggy bank. When she opens it, she has 63 coins worth a total of $11.55. How many of each type of coin does she have?

38) Find the measure of angles $x$ and $y$ if the measure of angle $x$ is half the measure of angle $y$.

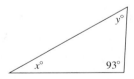

39) At a ski shop, two packs of hand warmers and one pair of socks cost $27.50. Five packs of hand warmers and three pairs of socks cost $78.00. Find the cost of a pack of hand warmers and a pair of socks.

40) A 7 P.M. spinning class has 9 more members than a 10 A.M. spinning class. The two classes have a total of 71 students. How many are in each class?

**(5.5)**

41) Find three points that satisfy the inequality and three that are not in the solution set.

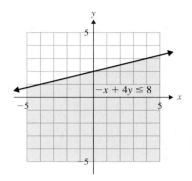

**Graph each inequality.**

42) $y \le -x + 4$

43) $y > \dfrac{1}{3}x + 1$

44) $x > 2$

45) $2x + 3y \le 9$

46) $3x - y \le 2$

47) $y > 0$

48) Find three points that are in the solution set of the system and three that are not.

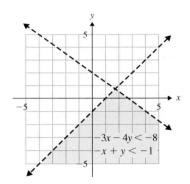

**Graph the solution set of each system.**

49) $x \le 1$
$y \ge -x + 1$

50) $y > 2x - 4$
$2x + y < -1$

51) $-3x + y < 0$
$3x + 2y > 4$

52) $x + 3y \le 3$
$x - y \ge 2$

53) $x \ge 0$
$y \ge 0$
$y \le -2x + 4$

54) $x \ge 0$
$y \ge 0$
$2x + 5y \le 5$

1) Determine whether $\left(-\dfrac{2}{3}, 4\right)$ is a solution of
the system $9x + 5y = 14$
$-6x - y = 0$.

**Solve each system by graphing.**

2) $y = -x + 2$
$3x - 4y = 20$

3) $3y - 6x = 6$
$2x - y = 1$

4) The graph shows the unemployment rate in the civilian labor force in Hawaii and New Hampshire in various years from 2001 to 2007. (www.bls.gov)

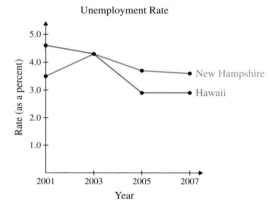

Unemployment Rate

a) When were more people unemployed in Hawaii? Approximately what percent of the state's population was unemployed at that time?

b) Write the point of intersection of the graphs as an ordered pair in the form (year, percentage) and explain its meaning.

c) Which line segment has the most negative slope? How can this be explained in the context of the problem?

**Solve each system by the substitution method.**

5) $3x - 10y = -10$
$x + 8y = -9$

6) $y = \dfrac{1}{2}x - 3$
$4x - 8y = 24$

**Solve each system by the elimination method.**

7) $2x + 5y = 11$
$7x - 5y = 16$

8) $3x + 4y = 24$
$7x - 3y = -18$

9) $-6x + 9y = 14$
$4x - 6y = 5$

**Solve each system using any method.**

10) $11x - y = -14$
$-9x + 7y = -38$

11) $\dfrac{5}{8}x + \dfrac{1}{4}y = \dfrac{1}{4}$
$\dfrac{1}{3}x + \dfrac{1}{2}y = -\dfrac{4}{3}$

12) $7 - 4(2x + 3) = x + 7 - y$
$5(x - y) + 20 = 8(2 - x) + x - 12$

13) Write a system of equations in two variables that has $(5, -1)$ as its only solution.

**Write a system of equations and solve.**

14) The area of Yellowstone National Park is about 1.1 million fewer acres than the area of Death Valley National Park. If they cover a total of 5.5 million acres, how big is each park? (www.nps.gov)

15) The Mahmood and Kuchar families take their kids to an amusement park. The Mahmoods buy one adult ticket and two children's tickets for $85.00. The Kuchars spend $150.00 on two adult and three children's tickets. How much did they pay for each type of ticket?

16) The width of a rectangle is half its length. Find the dimensions of the rectangle if the perimeter is 114 cm.

17) How many milliliters of a 12% alcohol solution and how many milliliters of a 30% alcohol solution must be mixed to obtain 72 mL of a 20% alcohol solution?

18) Rory and Lorelai leave Stars Hollow, Connecticut, and travel in opposite directions. Rory drives 4 mph faster than Lorelai, and after 1.5 hr they are 120 miles apart. How fast was each driving?

**Graph each inequality.**

19) $y > -3x - 1$

20) $x - 4y \geq -12$

**Graph the solution set of each system of inequalities.**

21) $x \geq -3$
$y \leq \dfrac{1}{2}x + 1$

22) $2x + y < 1$
$x - y < 0$

**Perform the operations and simplify.**

1) $\dfrac{7}{15} + \dfrac{9}{10}$

2) $4\dfrac{1}{5} \div \dfrac{9}{20}$

3) $3(5 - 7)^3 + 18 \div 6 - 8$

4) Find the area of the triangle.

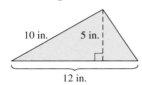

10 in.    5 in.

12 in.

5) Simplify $-3(4x^2 + 5x - 1)$.

**Simplify. The answer should not contain any negative exponents.**

6) $(2p^4)^5$

7) $9x^2 \cdot 7x^{-6}$

8) $\dfrac{36m^{-7}n^5}{24m^3n}$

9) Write 0.0007319 in scientific notation.

10) Solve $0.04(3p - 2) - 0.02p = 0.1(p + 3)$.

11) Solve $11 - 3(2k - 1) = 2(6 - k)$.

12) Solve. Write the answer in interval notation.

$$-5 < 4v - 9 < 15$$

13) Write an equation and solve.

During the first week of the "Cash for Clunkers" program, the average increase in gas mileage for the new car purchased versus the car traded in was 61%. If the average gas mileage of the new cars was 25.4 miles per gallon, what was the average gas mileage of the cars traded in? Round the answer to the nearest tenth. (www.yahoo.com)

14) The area, $A$ of a trapezoid is $A = \dfrac{1}{2}h(b_1 + b_2)$, where $h$ = height of the trapezoid, $b_1$ = length of one base of the trapezoid, and $b_2$ = length of the second base of the trapezoid.

   a) Solve the equation for $h$.

   b) Find the height of the trapezoid that has an area of $39 \text{ cm}^2$ and bases of length 8 cm and 5 cm.

15) Graph $2x - 3y = 9$.

16) Find the $x$- and $y$-intercepts of the graph of $x - 8y = 16$.

17) Write the slope-intercept form of the equation of the line containing $(3, 2)$ and $(-9, -1)$

18) Determine whether the lines are parallel, perpendicular, or neither.    $10x + 18y = 9$
$$9x - 5y = 17$$

**Solve each system of equations.**

19) $9x - 3y = 6$
$\phantom{0}3x - 2y = -8$

20) $3(2x - 1) - (y + 10) = 2(2x - 3) - 2y$
$\phantom{3(2x-1)}3x + 13 = 4x - 5(y - 3)$

21) $\phantom{-0}x + 2y = 4$
$-3x - 6y = 6$

22) $-\dfrac{1}{4}x - \dfrac{3}{4}y = \dfrac{1}{6}$
$\phantom{-}\dfrac{1}{2}x + \dfrac{3}{2}y = -\dfrac{1}{3}$

23) $\phantom{2x-}y = 4x + 1$
$2x - y = 3$

**Write a system of equations and solve.**

24) Dhaval used twice as many 6-foot boards as 4-foot boards when he made a treehouse for his children. If he used a total of 48 boards, how many of each size did he use?

25) Graph the solution set of the system:
$$y \le -2x + 1$$
$$y < 4$$

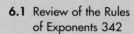

# Polynomials

## Algebra at Work: Custom Motorcycles

This is a final example of how algebra is used to build motorcycles in a custom chopper shop.

The support bracket for the fender of a custom motorcycle must be fabricated. To save money, Jim's boss told him to use a piece of scrap metal and not a new piece. So, he has to figure out how big a piece of scrap metal he needs to be able to cut out the shape needed to make the fender.

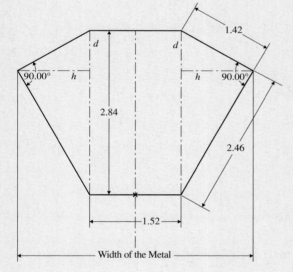

Jim drew the sketch on the left that showed the shape and dimension of the piece of metal to be cut so that it could be bent into the correct shape and size for the fender. He knows that the height of the piece must be 2.84 in.

To determine the width of the piece of metal that he needs, Jim analyzes the sketch and writes the equation

$$[(1.42)^2 - d^2] + (2.84 - d)^2 = (2.46)^2$$

In order to solve this equation, he must know how to square the binomial $(2.84 - d)^2$, something we will learn in this chapter. When he solves the equation, Jim determines that $d \approx 0.71$ in. He uses this value of $d$ to find that the width of the piece of metal that he must use to cut the correct shape for the fender is 3.98 in.

We will learn how to square binomials and perform other operations with polynomials in this chapter.

## Section 6.1 Review of the Rules of Exponents

**Objective**

1. **Review the Rules of Exponents**

In Chapter 2, we learned the rules of exponents. We will review them here to prepare us for the topics in the rest of this chapter—adding, subtracting, multiplying, and dividing polynomials.

### 1. Review of the Rules of Exponents

**Rules of Exponents**

For real numbers $a$ and $b$ and integers $m$ and $n$, the following rules apply:

---

**Summary    The Rules of Exponents**

| Rule | Example |
|---|---|
| **Product Rule:** $a^m \cdot a^n = a^{m+n}$ | $y^6 \cdot y^9 = y^{6+9} = y^{15}$ |
| **Basic Power Rule:** $(a^m)^n = a^{mn}$ | $(k^4)^7 = k^{28}$ |
| **Power Rule for a Product:** $(ab)^n = a^n b^n$ | $(9t)^2 = 9^2 t^2 = 81t^2$ |
| **Power Rule for a Quotient:** | |
| $\left(\dfrac{a}{b}\right)^n = \dfrac{a^n}{b^n}$, where $b \neq 0$. | $\left(\dfrac{2}{r}\right)^5 = \dfrac{2^5}{r^5} = \dfrac{32}{r^5}$ |
| **Zero Exponent:** If $a \neq 0$, then $a^0 = 1$. | $(7)^0 = 1$ |
| **Negative Exponent:** | |
| For $a \neq 0$, $a^{-n} = \left(\dfrac{1}{a}\right)^n = \dfrac{1}{a^n}$. | $\left(\dfrac{3}{4}\right)^{-3} = \left(\dfrac{4}{3}\right)^3 = \dfrac{4^3}{3^3} = \dfrac{64}{27}$ |
| If $a \neq 0$ and $b \neq 0$, then $\dfrac{a^{-m}}{b^{-n}} = \dfrac{b^n}{a^m}$; | $\dfrac{x^{-6}}{y^{-3}} = \dfrac{y^3}{x^6}$ |
| also, $\left(\dfrac{a}{b}\right)^{-m} = \left(\dfrac{b}{a}\right)^m$. | $\left(\dfrac{8c}{d}\right)^{-2} = \left(\dfrac{d}{8c}\right)^2 = \dfrac{d^2}{64c^2}$ |
| **Quotient Rule:** If $a \neq 0$, then $\dfrac{a^m}{a^n} = a^{m-n}$. | $\dfrac{2^9}{2^4} = 2^{9-4} = 2^5 = 32$ |

---

When we use the rules of exponents to simplify an expression, we must remember to use the order of operations.

---

**Example 1**

Simplify. Assume all variables represent nonzero real numbers. The answer should contain only positive exponents.

a) $(7k^{10})(-2k)$

b) $\dfrac{(-4)^5 \cdot (-4)^2}{(-4)^4}$

c) $\dfrac{10x^5 y^{-3}}{2x^2 y^5}$

d) $\left(\dfrac{c^2}{2d^4}\right)^{-5}$

e) $4(3p^5 q)^2$

**Solution**

a) $(7k^{10})(-2k) = -14k^{10+1}$     Multiply coefficients and add the exponents.

$\qquad\qquad\qquad = -14k^{11}$     Simplify.

b)   $\dfrac{(-4)^5 \cdot (-4)^2}{(-4)^4} = \dfrac{(-4)^{5+2}}{(-4)^4} = \dfrac{(-4)^7}{(-4)^4}$     Product rule—the bases in the numerator are the same, so add the exponents.

$= (-4)^{7-4}$     Quotient rule

$= (-4)^3$

$= -64$     Evaluate.

c)   $\dfrac{10x^5y^{-3}}{2x^2y^5} = 5x^{5-2}y^{-3-5}$     Divide coefficients and subtract the exponents.

$= 5x^3y^{-8}$     Simplify.

$= \dfrac{5x^3}{y^8}$     Write the answer with only positive exponents.

d)   $\left(\dfrac{c^2}{2d^4}\right)^{-5} = \left(\dfrac{2d^4}{c^2}\right)^5$     Take the reciprocal of the base and make the exponent positive.

$= \dfrac{2^5 d^{20}}{c^{10}}$     Power rule

$= \dfrac{32d^{20}}{c^{10}}$     Simplify.

e)   In this expression, a quantity is raised to a power and that quantity is being multiplied by 4. Remember what the order of operations says: Perform exponents before multiplication.

$4(3p^5q)^2 = 4(9p^{10}q^2)$     Apply the power rule *before* multiplying factors.

$= 36p^{10}q^2$     Multiply.

## You Try 1

Simplify. Assume all variables represent nonzero real numbers. The answer should contain only positive exponents.

a)   $(-6u^2)(-4u^3)$

b)   $\dfrac{8^3 \cdot 8^4}{8^5}$

c)   $\dfrac{8n^9}{12n^5}$

d)   $(3y^{-9})^2(2y^7)$

e)   $\left(\dfrac{3a^3b^{-4}}{2ab^6}\right)^{-4}$

---

### Answers to You Try Exercises

1)   a) $24u^5$     b) $64$     c) $\dfrac{2n^4}{3}$     d) $\dfrac{18}{y^{11}}$     e) $\dfrac{16b^{40}}{81a^8}$

## 6.1 Exercises

**Objective 1: Review the Rules of Exponents**

State which exponent rule must be used to simplify each exercise. Then simplify.

1) $\dfrac{k^{10}}{k^4}$

2) $p^5 \cdot p^2$

3) $(2h)^4$

4) $\left(\dfrac{5}{w}\right)^3$

Evaluate using the rules of exponents.

5) $2^2 \cdot 2^4$

6) $(-3)^2 \cdot (-3)$

7) $\dfrac{(-4)^8}{(-4)^5}$

8) $\dfrac{2^{10}}{2^6}$

9) $6^{-1}$

10) $(12)^{-2}$

11) $\left(\dfrac{1}{9}\right)^{-2}$

12) $\left(-\dfrac{1}{5}\right)^{-3}$

13) $\left(\dfrac{3}{2}\right)^{-4}$

14) $\left(\dfrac{7}{9}\right)^{-2}$

15) $6^0 + \left(-\dfrac{1}{2}\right)^{-5}$

16) $\left(\dfrac{1}{4}\right)^{-2} + \left(\dfrac{1}{4}\right)^0$

17) $\dfrac{8^5}{8^7}$

18) $\dfrac{2^7}{2^{12}}$

Simplify. Assume all variables represent nonzero real numbers. The answer should not contain negative exponents.

19) $t^5 \cdot t^8$

20) $n^{10} \cdot n^6$

21) $(-8c^4)(2c^5)$

22) $(3w^9)(-7w)$

23) $(z^6)^4$

24) $(y^3)^2$

25) $(5p^{10})^3$

26) $(-6m^4)^2$

27) $\left(-\dfrac{2}{3}a^7b\right)^3$

28) $\left(\dfrac{7}{10}r^2s^5\right)^2$

29) $\dfrac{f^{11}}{f^7}$

30) $\dfrac{u^9}{u^8}$

31) $\dfrac{35v^9}{5v^8}$

32) $\dfrac{36k^8}{12k^5}$

33) $\dfrac{9d^{10}}{54d^6}$

34) $\dfrac{7m^4}{56m^2}$

35) $\dfrac{x^3}{x^9}$

36) $\dfrac{v^2}{v^5}$

37) $\dfrac{m^2}{m^3}$

38) $\dfrac{t^3}{t^3}$

39) $\dfrac{45k^{-2}}{30k^2}$

40) $\dfrac{22n^{-9}}{55n^{-3}}$

41) $5(2m^4n^7)^2$

42) $2(-3a^8b)^3$

43) $(6y^2)(2y^3)^2$

44) $(-c^4)(5c^9)^3$

45) $\left(\dfrac{7a^4}{b^{-1}}\right)^{-2}$

46) $\left(\dfrac{3t^{-3}}{2u}\right)^{-4}$

47) $\dfrac{a^{-12}b^7}{a^{-9}b^2}$

48) $\dfrac{mn^{-4}}{m^9n^7}$

49) $\dfrac{(x^2y^{-3})^4}{x^5y^8}$

50) $\dfrac{10r^{-6}t}{(4r^{-5}t^4)^3}$

51) $\dfrac{12a^6bc^{-9}}{(3a^2b^{-7}c^4)^2}$

52) $\dfrac{(-7k^2m^{-3}n^{-1})^2}{14km^{-2}n^2}$

53) $(xy^{-3})^{-5}$

54) $-(s^{-6}t^2)^{-4}$

55) $\left(\dfrac{a^2b}{4c^2}\right)^{-3}$

56) $\left(\dfrac{2s^3}{rt^4}\right)^{-5}$

57) $\left(\dfrac{7h^{-1}k^9}{21h^{-5}k^5}\right)^{-2}$

58) $\left(\dfrac{24m^8n^{-3}}{16mn}\right)^{-3}$

59) $\left(\dfrac{15cd^{-4}}{5c^3d^{-10}}\right)^{-3}$

60) $\left(\dfrac{10x^{-5}y}{20x^5y^{-3}}\right)^{-2}$

61) $\dfrac{(2u^{-5}v^2w^4)^{-5}}{(u^6v^{-7}w^{-10})^2}$

62) $\dfrac{(a^{-10}b^{-5}c^2)^4}{6(a^9bc^{-4})^{-2}}$

Write expressions for the area and perimeter for each rectangle.

63)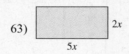

64)

65)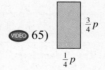

66)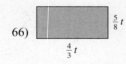

Simplify. Assume that the variables represent nonzero integers.

67) $k^{4a} \cdot k^{2a}$

68) $r^{9y} \cdot r^y$

69) $(g^{2x})^4$

70) $(t^{5c})^3$

71) $\dfrac{x^{7b}}{x^{4b}}$

72) $\dfrac{m^{10u}}{m^{3u}}$

73) $(2r^{6m})^{-3}$

74) $(5a^{-2x})^{-2}$

## Section 6.2 Addition and Subtraction of Polynomials; Graphing

**Objectives**

1. Learn the Vocabulary Associated with Polynomials
2. Evaluate Polynomials
3. Add Polynomials
4. Subtract Polynomials
5. Add and Subtract Polynomials in More Than One Variable
6. Define and Evaluate a Polynomial Function
7. Graph Basic Polynomials

### 1. Learn the Vocabulary Associated with Polynomials

In Section 1.7, we defined an *algebraic expression* as a collection of numbers, variables, and grouping symbols connected by operation symbols such as $+$, $-$, $\times$, and $\div$.

An example of an algebraic expression is

$$5x^3 + \frac{7}{4}x^2 - x + 9$$

The *terms* of this algebraic expression are $5x^3$, $\frac{7}{4}x^2$, $-x$, and 9. A *term* is a number or a variable or a product or quotient of numbers and variables.

The expression $5x^3 + \frac{7}{4}x^2 - x + 9$ is also a *polynomial*.

---

**Definition**

A **polynomial in $x$** is the sum of a finite number of terms of the form $ax^n$, where $n$ is a whole number and $a$ is a real number. (The exponents must be whole numbers.)

---

Let's look more closely at the polynomial $5x^3 + \frac{7}{4}x^2 - x + 9$.

1) The polynomial is written **in descending powers of $x$** since the powers of $x$ decrease from left to right. Generally, we write polynomials in descending powers of the variable.

2) Recall that the term without a variable is called a **constant.** The constant is 9. The **degree of a term** equals the exponent on its variable. (If a term has more than one variable, the degree equals the *sum* of the exponents on the variables.) We will list each term, its coefficient, and its degree.

| Term | Coefficient | Degree |
|------|-------------|--------|
| $5x^3$ | 5 | 3 |
| $\frac{7}{4}x^2$ | $\frac{7}{4}$ | 2 |
| $-x$ | $-1$ | 1 |
| 9 | 9 | $0\,(9 = 9x^0)$ |

3) The **degree of the polynomial** equals the highest degree of any nonzero term. The degree of $5x^3 + \frac{7}{4}x^2 - x + 9$ is 3. Or, we say that this is a **third-degree polynomial.**

**Example 1**

Decide whether each expression *is* or *is not* a polynomial. If it is a polynomial, identify each term and the degree of each term. Then, find the degree of the polynomial.

a)   $-8p^4 + 5.7p^3 - 9p^2 - 13$     b)   $4c^2 - \dfrac{2}{5}c + 3 + \dfrac{6}{c^2}$

c)   $a^3b^3 + 4a^3b^2 - ab + 1$     d)   $7n^6$

**Solution**

a)   The expression $-8p^4 + 5.7p^3 - 9p^2 - 13$ is a polynomial in $p$. Its terms have whole-number exponents and real coefficients. The term with the highest degree is $8p^4$, so the degree of the polynomial is 4.

| Term | Degree |
|------|--------|
| $-8p^4$ | 4 |
| $5.7p^3$ | 3 |
| $-9p^2$ | 2 |
| $-13$ | 0 |

b)   The expression $4c^2 - \dfrac{2}{5}c + 3 + \dfrac{6}{c^2}$ is *not* a polynomial because one of its terms has a variable in the denominator. ($\dfrac{6}{c^2} = 6c^{-2}$; the exponent $-2$ is not a whole number.)

c)   The expression $a^3b^3 + 4a^3b^2 - ab + 1$ *is* a polynomial because the variables have whole-number exponents and the coefficients are real numbers. Since this is a polynomial in two variables, we find the degree of each term by adding the exponents. The first term, $a^3b^3$, has the highest degree, 6, so the polynomial has degree 6.

| Term | Degree |
|------|--------|
| $a^3b^3$ | 6 |
| $4a^3b^2$ | 5 |
| $-ab$ | 2 |
| $1$ | 0 |

Add the exponents to get the degree.

d)   The expression $7n^6$ *is* a polynomial even though it has only one term. The degree of the term is 6, and that is the degree of the polynomial as well.

**You Try 1**

Decide whether each expression *is* or *is not* a polynomial. If it is a polynomial, identify each term and the degree of each term. Then, find the degree of the polynomial.

a)   $d^4 + 7d^3 + \dfrac{3}{d}$     b)   $k^3 - k^2 - 3.8k + 10$

c)   $5x^2y^2 + \dfrac{1}{2}xy - 6x - 1$     d)   $2r + 3r^{1/2} - 7$

The polynomial in Example 1d) is $7n^6$ and has one term. We call $7n^6$ a *monomial*. A **monomial** is a polynomial that consists of one term ("mono" means one). Some other examples of monomials are

$$y^2, \qquad -4t^5, \qquad x, \qquad m^2n^2, \qquad \text{and} \qquad -3$$

A **binomial** is a polynomial that consists of exactly two terms ("bi" means two). Some examples are

$$w + 2, \qquad 4z^2 - 11, \qquad a^4 - b^4, \qquad \text{and} \qquad -8c^3d^2 + 3cd$$

A **trinomial** is a polynomial that consists of exactly three terms ("tri" means three). Here are some examples:

$$x^2 - 3x - 40, \qquad 2q^4 - 18q^2 + 10q, \qquad \text{and} \qquad 6a^4 + 29a^2b + 28b^2$$

In Section 1.7, we saw that expressions have different values depending on the value of the variable(s). The same is true for polynomials.

## 2. Evaluate Polynomials

**Example 2**

Evaluate the trinomial $n^2 - 7n + 4$ when

a)  $n = 3$        and             b)  $n = -2$

**Solution**

a)   Substitute 3 for $n$ in $n^2 - 7n + 4$. Remember to put 3 in parentheses.

$$\begin{aligned} n^2 - 7n + 4 &= (3)^2 - 7(3) + 4 \qquad \text{Substitute.} \\ &= 9 - 21 + 4 \\ &= -8 \qquad\qquad\qquad \text{Add.} \end{aligned}$$

b)   Substitute $-2$ for $n$ in $n^2 - 7n + 4$. Put $-2$ in parentheses.

$$\begin{aligned} n^2 - 7n + 4 &= (-2)^2 - 7(-2) + 4 \qquad \text{Substitute.} \\ &= 4 + 14 + 4 \\ &= 22 \qquad\qquad\qquad\qquad \text{Add.} \end{aligned}$$

 **You Try 2**

Evaluate $t^2 - 9t - 6$ when

a)  $t = 5$        b)  $t = -4$

### Adding and Subtracting Polynomials

Recall in Section 1.7 we said that **like terms** contain the same variables with the same exponents. We add or subtract like terms by adding or subtracting the coefficients and leaving the variable(s) and exponent(s) the same. We use the same idea for adding and subtracting polynomials.

## 3. Add Polynomials

**Procedure**   Adding Polynomials

To add polynomials, add like terms.

We can set up an addition problem vertically or horizontally.

### Example 3

Add $(m^3 - 9m^2 + 5m - 4) + (2m^3 + 3m^2 - 1)$.

**Solution**

We will add these vertically. Line up like terms in columns and add.

$$
\begin{array}{r}
m^3 - 9m^2 + 5m - 4 \\
+\ \ 2m^3 + 3m^2\ \ \quad - 1 \\
\hline
3m^3 - 6m^2 + 5m - 5
\end{array}
$$

■

### You Try 3

Add $(6b^3 - 11b^2 + 3b + 3) + (-2b^3 - 6b^2 + b - 8)$.

### Example 4

Add $10k^2 + 2k - 1$ and $5k^2 + 7k + 9$.

**Solution**

Let's add these horizontally. Put the polynomials in parentheses since each contains more than one term. Use the associative and commutative properties to rewrite like terms together.

$$
\begin{aligned}
(10k^2 + 2k - 1) + (5k^2 + 7k + 9) &= (10k^2 + 5k^2) + (2k + 7k) + (-1 + 9) \\
&= 15k^2 + 9k + 8 \quad \text{Combine like terms.}
\end{aligned}
$$

■

## 4. Subtract Polynomials

To subtract two polynomials such as $(8x + 3) - (5x - 4)$ we will be using the distributive property to clear the parentheses in the second polynomial.

### Example 5

Subtract $(8x + 3) - (5x - 4)$.

**Solution**

$$
\begin{aligned}
(8x + 3) - (5x - 4) &= (8x + 3)\ -1\,(5x - 4) \\
&= (8x + 3) + (-1)(5x - 4) \quad \text{Change } -1 \text{ to } + (-1). \\
&= (8x + 3) + (-5x + 4) \quad \text{Distribute.} \\
&= 3x + 7 \quad \text{Combine like terms.}
\end{aligned}
$$

■

In Example 5, notice that we changed the sign of each term in the second polynomial and then added it to the first.

---

**Procedure**　Subtracting Polynomials

To subtract two polynomials, change the sign of each term in the second polynomial. Then, add the polynomials.

Let's see how we use this rule to subtract polynomials both horizontally and vertically.

**Example 6**

Subtract $(-6w^3 - w^2 + 10w + 1) - (2w^3 - 4w^2 + 9w - 5)$ vertically.

**Solution**

To subtract vertically, line up like terms in columns.

$$\begin{array}{r} -6w^3 - w^2 + 10w + 1 \\ -\ (2w^3 - 4w^2 + 9w - 5) \end{array}$$

Change the signs in the second polynomial and add the polynomials.

$$\begin{array}{r} -6w^3 - w^2 + 10w + 1 \\ +\ -2w^3 + 4w^2 - 9w + 5 \\ \hline -8w^3 + 3w^2 +\ \ w + 6 \end{array}$$  ■

**You Try 4**

Subtract $(-7h^2 - 8h + 1) - (-3h^2 + h - 4)$.

## 5. Add and Subtract Polynomials in More Than One Variable

To add and subtract polynomials in more than one variable, remember that like terms contain the same variables with the same exponents.

**Example 7**

Perform the indicated operation.

a)  $(a^2b^2 + 2a^2b - 13ab - 4) + (9a^2b^2 - 5a^2b - ab + 17)$

b)  $(6tu - t + 2u + 5) - (4tu + 8t - 2)$

**Solution**

a)  $(a^2b^2 + 2a^2b - 13ab - 4) + (9a^2b^2 - 5a^2b - ab + 17)$
    $= 10a^2b^2 - 3a^2b - 14ab + 13$    Combine like terms.

b)  $(6tu - t + 2u + 5) - (4tu + 8t - 2) = (6tu - t + 2u + 5) - 4tu - 8t + 2$
    $= 2tu - 9t + 2u + 7$    Combine like terms.  ■

**You Try 5**

Perform the indicated operation.

a)  $(-12x^2y^2 + xy - 6y + 1) - (-4x^2y^2 - 10xy + 3y + 6)$

b)  $(3.6m^3n^2 + 8.1mn - 10n) + (8.5m^3n^2 - 11.2mn + 4.3)$

## 6. Define and Evaluate a Polynomial Function

Look at the polynomial $3x^2 - x + 5$. If we substitute 2 for $x$, the *only* value of the expression is 15:

$$3(2)^2 - (2) + 5 = 3(4) - 2 + 5$$
$$= 12 - 2 + 5$$
$$= 15$$

For any value we substitute for $x$ in a polynomial like $3x^2 - x + 5$, there will be *only one value* of the expression. Since each value substituted for the variable produces *only one value* of the expression, we can use function notation to represent a polynomial like $3x^2 - x + 5$.

**Note**

$f(x) = 3x^2 - x + 5$ is a **polynomial function** since $3x^2 - x + 5$ is a polynomial.

Therefore, finding $f(2)$ when $f(x) = 3x^2 - x + 5$ is the same as evaluating $3x^2 - x + 5$ when $x = 2$.

### Example 8

If $f(x) = x^3 - 2x^2 + 4x + 19$, find $f(-3)$.

**Solution**

Substitute $-3$ for $x$.

$$f(x) = x^3 - 2x^2 + 4x + 19$$
$$f(-3) = (-3)^3 - 2(-3)^2 + 4(-3) + 19 \qquad \text{Substitute } -3 \text{ for } x.$$
$$f(-3) = -27 - 2(9) - 12 + 19$$
$$f(-3) = -27 - 18 - 12 + 19$$
$$f(-3) = -38$$

### You Try 6

If $h(t) = 5t^4 + 8t^3 - t^2 + 7$, find $h(-1)$.

## 7. Graph Basic Polynomials

In Chapter 4, we learned that the graph of a linear equation is a line. The graphs of polynomial equations are not all the same, but different types of polynomials result in graphs that, in general, have the same shape.

A **quadratic equation** is a special type of polynomial. A quadratic equation can be written in the form $y = ax^2 + bx + c$, where $a$, $b$, and $c$ are real numbers and $a \neq 0$. Its graph is called a **parabola.** The simplest form of a quadratic equation is $y = x^2$. Let's look at its graph.

### Example 9

Graph $y = x^2$.

**Solution**

Make a table of values. Plot the points and connect them with a smooth curve.

| $y = x^2$ | |
|---|---|
| **x** | **y** |
| 0 | 0 |
| 1 | 1 |
| 2 | 4 |
| -1 | 1 |
| -2 | 4 |

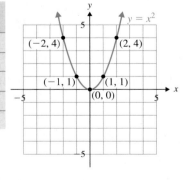

The graphs of quadratic equations where $y$ is defined in terms of $x$ open either upward or downward. Notice that this graph opens upward. The lowest point on a parabola that opens upward or the highest point on a parabola that opens downward is called the **vertex.** The vertex of the graph of $y = x^2$ is $(0, 0)$. When graphing a quadratic equation by plotting points, it is important to locate the vertex.

### You Try 7

Graph $y = x^2 - 3$.

We can also use function notation to represent a polynomial equation *if* it is a function. So let's see how we can graph a quadratic function. Its graph will be a parabola.

**Example 10**

Graph $f(x) = -(x - 2)^2$.

**Solution**

In Example 9, we said that it is important to locate the vertex of a parabola. Let's do that first. Begin by finding the *x*-coordinate of the vertex of this graph. *The x-coordinate of the vertex is the value of x that makes the expression being squared equal to zero.* The *x*-coordinate of the vertex of this parabola is 2.

Make a table of values, and use the *x*-coordinate of the vertex as the first *x*-value in the table. Then, we will choose a couple of values of *x* that are larger than 2 and a couple of values that are smaller than 2. Find the corresponding *y*-values, plot the points, and connect them with a smooth curve.

| $f(x) = -(x-2)^2$ | |
|---|---|
| **x** | **f(x)** |
| 2 | 0 |
| 3 | -1 |
| 4 | -4 |
| 1 | -1 |
| 0 | -4 |

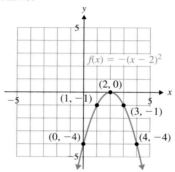

The vertex of this parabola is (2, 0). Notice that this time it is the *highest* point on the parabola because the graph opens downward.

We will study quadratic functions and parabolas in more detail in Chapter 10.

**You Try 8**

Graph $g(x) = -(x + 3)^2$.

**Answers to You Try Exercises**

1)  a) not a polynomial      b) polynomial of degree 3          c) polynomial of degree 4

| Term | Degree |
|---|---|
| $k^3$ | 3 |
| $-k^2$ | 2 |
| $-3.8k$ | 1 |
| 10 | 0 |

| Term | Degree |
|---|---|
| $5x^2y^2$ | 4 |
| $\frac{1}{2}xy$ | 2 |
| $-6x$ | 1 |
| $-1$ | 0 |

d) not a polynomial      2) a) $-26$  b) 46      3) $4b^3 - 17b^2 + 4b - 5$      4) $10h^2 - 9h + 5$
5) a) $-8x^2y^2 + 11xy - 9y - 5$      b) $12.1m^3n^2 - 3.1mn - 10n + 4.3$      6) 3
7)                          8)

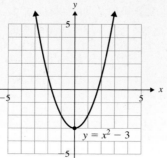

$y = x^2 - 3$

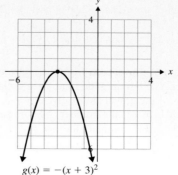

$g(x) = -(x + 3)^2$

## 6.2 Exercises

**Objective 1:  Learn the Vocabulary Associated with Polynomials**

Is the given expression a polynomial? Why or why not?

1) $-2p^2 - 5p + 6$

2) $8r^3 + 7r^2 - t + \dfrac{4}{5}$

3) $c^3 + 5c^2 + 4c^{-1} - 8$

4) $9a^5$

5) $f^{3/4} + 6f^{2/3} + 1$

6) $7y - 1 + \dfrac{3}{y}$

Determine whether each is a monomial, a binomial, or a trinomial.

7) $4x - 1$

8) $-5q^2$

9) $m^2n^2 - mn + 13$

10) $11c^2 + 3c$

11) $8$

12) $k^5 + 2k^3 + 8k$

13) How do you determine the degree of a polynomial in one variable?

14) Write a third-degree polynomial in one variable.

15) How do you determine the degree of a term in a polynomial in more than one variable?

16) Write a fourth-degree monomial in $x$ and $y$.

For each polynomial, identify each term in the polynomial, the coefficient and degree of each term, and the degree of the polynomial.

17) $3y^4 + 7y^3 - 2y + 8$

18) $6a^2 + 2a - 11$

19) $-4x^2y^3 - x^2y^2 + \dfrac{2}{3}xy + 5y$

20) $3c^2d^2 + 0.7c^2d + cd - 1$

**Objective 2:  Evaluate Polynomials**

Evaluate each polynomial when a) $r = 3$ and b) $r = -1$.

21) $2r^2 - 7r + 4$

22) $2r^3 + 5r - 6$

Evaluate each polynomial when $x = 5$ and $y = -2$.

23) $9x + 4y$

24) $-2x + 3y + 16$

25) $x^2y^2 - 5xy + 2y$

26) $-2xy^2 + 7xy + 12y - 6$

27) $\dfrac{1}{2}xy - 4x - y$

28) $x^2 - y^2$

29) Bob will make a new gravel road from the highway to his house. The cost of building the road, $y$ (in dollars), includes the cost of the gravel and is given by $y = 60x + 380$, where $x$ is the number of hours he rents the equipment needed to complete the job.

a) Evaluate the binomial when $x = 5$, and explain what it means in the context of the problem.

b) If he keeps the equipment for 9 hours, how much will it cost to build the road?

c) If it cost \$860.00 to build the road, for how long did Bob rent the equipment?

30) An object is thrown upward so that its height, $y$ (in feet), $x$ seconds after being thrown is given by
$$y = -16x^2 + 48x + 64.$$

a) Evaluate the polynomial when $x = 2$, and explain what it means in the context of the problem.

b) What is the height of the object 3 seconds after it is thrown?

c) Evaluate the polynomial when $x = 4$, and explain what it means in the context of the problem.

**Objective 3:  Add Polynomials**

Add like terms.

31) $-6z + 8z + 11z$

32) $m^2 + 7m^2 - 14m^2$

33) $5c^2 + 9c - 16c^2 + c - 3c$

34) $-4y^3 + 3y^5 + 17y^5 + 6y^3 - 5y^5$

35) $6.7t^2 - 9.1t^6 - 2.5t^2 + 4.8t^6$

36) $\dfrac{5}{4}w^3 + \dfrac{3}{8}w^4 - \dfrac{2}{3}w^4 - \dfrac{5}{6}w^3$

37) $7a^2b^2 + 4ab^2 - 16ab^2 - a^2b^2 + 5ab^2$

38) $x^5y^2 - 14xy + 6xy + 5x^5y^2 + 8xy$

Add the polynomials.

39)
$$\begin{array}{r} 5n - 8 \\ + \ 4n + 3 \\ \hline \end{array}$$

40)
$$\begin{array}{r} 9d + 14 \\ + \ 2d + 5 \\ \hline \end{array}$$

41)
$$\begin{array}{r} -7a^3 + 11a \\ + \ 2a^3 - 4a \\ \hline \end{array}$$

42)
$$\begin{array}{r} -h^4 + 6h^2 \\ + \ 5h^4 - 3h^2 \\ \hline \end{array}$$

43)
$$\begin{array}{r} 9r^2 + 16r + 2 \\ + \ 3r^2 - 10r + 9 \\ \hline \end{array}$$

44)
$$\begin{array}{r} m^2 - 3m - 8 \\ + \ 2m^2 + 7m + 1 \\ \hline \end{array}$$

45)
$$\begin{array}{r} b^2 - 8b - 14 \\ + \ 3b^2 + 8b + 11 \\ \hline \end{array}$$

46)
$$\begin{array}{r} 8g^2 + g + 5 \\ + \ 5g^2 - 6g - 5 \\ \hline \end{array}$$

47)
$$\begin{array}{r} \dfrac{5}{6}w^4 - \dfrac{2}{3}w^2 \qquad\quad + \dfrac{1}{2} \\ + \ -\dfrac{4}{9}w^4 + \dfrac{1}{6}w^2 - \dfrac{3}{8}w - 2 \\ \hline \end{array}$$

48) $\begin{array}{r} -1.7p^3 - 2p^2 + 3.8p - 6 \\ + \ \ 6.2p^3 \quad\quad\ \ - 1.2p + 14 \\ \hline \end{array}$

49) $(6m^2 - 5m + 10) + (-4m^2 + 8m + 9)$

50) $(3t^4 - 2t^2 + 11) + (t^4 + t^2 - 7)$

51) $\left(-2c^4 - \dfrac{7}{10}c^3 + \dfrac{3}{4}c - \dfrac{2}{9}\right)$

$+ \left(12c^4 + \dfrac{1}{2}c^3 - c + 3\right)$

52) $\left(\dfrac{7}{4}y^3 - \dfrac{3}{8}\right) + \left(\dfrac{5}{6}y^3 + \dfrac{7}{6}y^2 - \dfrac{9}{16}\right)$

53) $\begin{array}{l} (2.7d^3 + 5.6d^2 - 7d + 3.1) \\ + (-1.5d^3 + 2.1d^2 - 4.3d - 2.5) \end{array}$

54) $\begin{array}{l} (0.2t^4 - 3.2t + 4.1) \\ + (-2.7t^4 + 0.8t^3 - 6.4t + 3.9) \end{array}$

## Objective 4: Subtract Polynomials

Subtract the polynomials.

55) $\begin{array}{r} 15w + 7 \\ - \ \ 3w + 11 \\ \hline \end{array}$

56) $\begin{array}{r} 12a - 8 \\ - \ \ 2a + 9 \\ \hline \end{array}$

57) $\begin{array}{r} y - 6 \\ - \ \ 2y - 8 \\ \hline \end{array}$

58) $\begin{array}{r} 6p + 1 \\ - \ \ 9p - 17 \\ \hline \end{array}$

59) $\begin{array}{r} 3b^2 - 8b + 12 \\ - \ \ 5b^2 + 2b - 7 \\ \hline \end{array}$

60) $\begin{array}{r} -7d^2 + 15d + 6 \\ - \ \ 8d^2 + 3d - 9 \\ \hline \end{array}$

61) $\begin{array}{r} f^4 - 6f^3 + 5f^2 - 8f + 13 \\ - \ -3f^4 + 8f^3 - f^2 \quad\quad + 4 \\ \hline \end{array}$

62) $\begin{array}{r} 11x^4 + x^3 - 9x^2 + 2x - 4 \\ - \ -3x^4 + x^3 \quad\quad\ - x + 1 \\ \hline \end{array}$

63) $\begin{array}{r} 10.7r^2 + 1.2r + \ \ 9 \\ - \ 4.9r^2 - 5.3r - 2.8 \\ \hline \end{array}$

64) $\begin{array}{r} -\dfrac{11}{10}m^3 + \dfrac{1}{2}m - \dfrac{5}{8} \\[2mm] - \ \dfrac{2}{5}m^3 + \dfrac{1}{7}m - \dfrac{5}{6} \\ \hline \end{array}$

65) $(j^2 + 16j) - (-6j^2 + 7j + 5)$

66) $(-3p^2 + p + 4) - (4p^2 + p + 1)$

67) $(17s^5 - 12s^2) - (9s^5 + 4s^4 - 8s^2 - 1)$

68) $(-5d^4 - 8d^2 + d + 3) - (-3d^4 + 17d^3 - 6d^2 - 20)$

69) $\left(-\dfrac{3}{8}r^2 + \dfrac{2}{9}r + \dfrac{1}{3}\right) - \left(-\dfrac{7}{16}r^2 - \dfrac{5}{9}r + \dfrac{7}{6}\right)$

70) $(3.8t^5 + 7.5t - 9.6) - (-1.5t^5 + 2.9t^2 - 1.1t + 3.4)$

71) Explain, in your own words, how to subtract two polynomials.

72) Do you prefer adding and subtracting polynomials vertically or horizontally? Why?

73) Will the sum of two trinomials always be a trinomial? Why or why not? Give an example.

74) Write a third-degree polynomial in $x$ that does not contain a second-degree term.

### Mixed Exercises: Objectives 3 and 4

Perform the indicated operations.

75) $(8a^4 - 9a^2 + 17) - (15a^4 + 3a^2 + 3)$

76) $(-x + 15) + (-5x - 12)$

77) $(-11n^2 - 8n + 21) + (4n^2 + 15n - 3) + (7n^2 - 10)$

78) $(-15a^3 + 8) - (-7a^3 + 3a + 5) + (10a^3 - a + 17)$

79) $(w^3 + 5w^2 + 3) - (6w^3 - 2w^2 + w + 12) + (9w^3 + 7)$

80) $(3r + 2) - (r^2 + 5r - 1) - (-9r^3 - r + 6)$

81) $\left(y^3 - \dfrac{3}{4}y^2 - 5y + \dfrac{3}{7}\right) + \left(\dfrac{1}{3}y^3 - y^2 + 8y - \dfrac{1}{2}\right)$

82) $\left(\dfrac{3}{5}c^4 + c^3 - \dfrac{3}{2}c^2 + 1\right)$

$+ \left(c^4 - 6c^3 - \dfrac{1}{4}c^2 + 6c - 1\right)$

83) $\begin{array}{l}(3m^3 - 5m^2 + m + 12) - [(7m^3 + 4m^2 - m + 11) \\ + (-5m^3 - 2m^2 + 6m + 8)]\end{array}$

84) $(j^2 - 13j - 9) - [(-7j^2 + 10j - 2) + (4j^2 - 11j - 6)]$

Perform the indicated operations.

85) Find the sum of $p^2 - 7$ and $8p^2 + 2p - 1$.

86) Add $12n - 15$ to $5n + 4$.

87) Subtract $z^6 - 8z^2 + 13$ from $6z^6 + z^2 + 9$.

88) Subtract $-7x^2 + 8x + 2$ from $2x^2 + x$.

89) Subtract the sum of $6p^2 + 1$ and $3p^2 - 8p + 4$ from $2p^2 + p + 5$.

90) Subtract $17g^3 + 2g - 10$ from the sum of $5g^3 + g^2 + g$ and $3g^3 - 2g - 7$.

### Objective 5: Add and Subtract Polynomials in More Than One Variable

Each of the polynomials is a polynomial in two variables. Perform the indicated operations.

91) $(5w + 17z) - (w + 3z)$

92) $(-4g - 7h) + (9g + h)$

93) $(ac + 8a + 6c) + (-6ac + 4a - c)$

94) $(11rt - 6r + 2) - (10rt - 7r + 12t + 2)$

95) $(-6u^2v^2 + 11uv + 14)$
$- (-10u^2v^2 - 20uv + 18)$

96) $(-7j^2k^2 + 9j^2k - 23jk^2 + 13)$
$+ (10j^2k^2 + 5j^2k - 17)$

97) $(12x^3y^2 - 5x^2y^2 + 9x^2y - 17) + (5x^3y^2 + x^2y - 1)$
$- (6x^2y^2 + 10x^2y + 2)$

98) $(r^3s^2 + 2r^2s^2 + 10) - (7r^3s^2 + 18r^2s^2 - 9)$
$+ (11r^3s^2 - 3r^2s^2 - 4)$

Find the polynomial that represents the perimeter of each rectangle.

99) 
2x + 7

x − 4

100)
a² + 3a − 4

a² − 5a + 1

101)
5p² − 2p + 3

p − 6

102)
⅔m + 4

⅔m + 4

## Objective 6: Define and Evaluate a Polynomial Function

103) If $f(x) = 5x^2 + 7x - 8$, find
   a) $f(-3)$        b) $f(1)$

104) If $h(a) = -a^2 - 4a + 11$, find
   a) $h(4)$        b) $h(-5)$

105) If $P(t) = t^3 - 2t^2 + 5t + 8$, find
   a) $P(3)$        b) $P(0)$

106) If $G(c) = 4c^4 + c^2 - 3c - 5$, find
   a) $G(0)$        b) $G(-1)$

107) $H(z) = -3z + 11$. Find $z$ so that $H(z) = 13$.

108) $f(x) = \dfrac{1}{4}x + 7$. Find $x$ so that $f(x) = 9$.

109) $r(k) = \dfrac{3}{5}k - 4$. Find $k$ so that $r(k) = 14$.

110) $Q(a) = 4a - 3$. Find $a$ so that $Q(a) = -9$.

## Objective 7: Graph Basic Polynomials

Graph.

111) $y = x^2 + 1$            112) $y = x^2 + 2$

113) $y = -x^2$             114) $y = (x + 2)^2$

115) $f(x) = -(x + 1)^2$     116) $h(x) = -(x - 1)^2$

117) $g(x) = x^2 - 5$        118) $g(x) = -x^2 + 4$

119) $h(x) = -(x - 3)^2$     120) $f(x) = \dfrac{1}{2}x^2$

---

## Section 6.3 Multiplication of Polynomials

### Objectives

1. Multiply a Monomial and a Polynomial
2. Multiply Two Polynomials
3. Multiply Two Binomials Using FOIL
4. Find the Product of More Than Two Polynomials
5. Find the Product of Binomials of the Form $(a + b)(a - b)$
6. Square a Binomial
7. Find Higher Powers of a Binomial

We have already learned that when multiplying two monomials, we multiply the coefficients and add the exponents of the same bases:

$$4c^5 \cdot 3c^6 = 12c^{11} \qquad\qquad -3x^2y^4 \cdot 7xy^3 = -21x^3y^7$$

In this section, we will discuss how to multiply other types of polynomials.

### 1. Multiply a Monomial and a Polynomial

To multiply a monomial and a polynomial, we use the distributive property.

### Example 1

Multiply $2k^2(6k^2 + 5k - 3)$.

### Solution

$$2k^2(6k^2 + 5k - 3) = (2k^2)(6k^2) + (2k^2)(5k) + (2k^2)(-3) \qquad \text{Distribute.}$$
$$= 12k^4 + 10k^3 - 6k^2 \qquad\qquad\qquad\qquad \text{Multiply.} \quad \blacksquare$$

**You Try 1**

Multiply $5z^4(4z^3 - 7z^2 - z + 8)$.

## 2. Multiply Two Polynomials

To multiply two polynomials, we use the distributive property repeatedly. For example, to multiply $(2x - 3)(x^2 + 7x + 4)$, we multiply each term in the second polynomial by $(2x - 3)$.

$$(2x - 3)(x^2 + 7x + 4) = (2x - 3)(x^2) + (2x - 3)(7x) + (2x - 3)(4) \qquad \text{Distribute.}$$

Next, we distribute again.

$$(2x - 3)(x^2) + (2x - 3)(7x) + (2x - 3)(4)$$
$$= (2x)(x^2) - (3)(x^2) + (2x)(7x) - (3)(7x) + (2x)(4) - (3)(4)$$
$$= 2x^3 - 3x^2 + 14x^2 - 21x + 8x - 12 \qquad \text{Multiply.}$$
$$= 2x^3 + 11x^2 - 13x - 12 \qquad \text{Combine like terms.}$$

This process of repeated distribution leads us to the following rule.

**Procedure**   Multiplying Polynomials

To multiply two polynomials, multiply each term in the second polynomial by each term in the first polynomial. Then combine like terms. The answer should be written in descending powers.

Let's use this rule to multiply the polynomials in Example 2.

**Example 2**

Multiply $(n^2 + 5)(2n^3 + n - 9)$.

**Solution**

Multiply each term in the second polynomial by each term in the first.

$$(n^2 + 5)(2n^3 + n - 9)$$
$$= (n^2)(2n^3) + (n^2)(n) + (n^2)(-9) + (5)(2n^3) + (5)(n) + (5)(-9) \qquad \text{Distribute.}$$
$$= 2n^5 + n^3 - 9n^2 + 10n^3 + 5n - 45 \qquad \text{Multiply.}$$
$$= 2n^5 + 11n^3 - 9n^2 + 5n - 45 \qquad \text{Combine like terms.}$$

Polynomials can be multiplied vertically as well. The process is very similar to the way we multiply whole numbers, so let's review a multiplication problem here.

$$
\begin{array}{r}
271 \\
\times \ 53 \\
\hline
813 \\
13\ 55 \\
\hline
14{,}363
\end{array}
$$

Multiply the 271 by 3.
Multiply the 271 by 5.
Add.

In the next example, we will find a product of polynomials by multiplying vertically.

## Example 3

Multiply vertically. $(a^3 - 4a^2 + 5a - 1)(3a + 7)$

**Solution**

Set up the multiplication problem like you would for whole numbers:

$$
\begin{array}{r}
a^3 - 4a^2 + 5a - 1 \\
\times \qquad\qquad 3a + 7 \\
\hline
7a^3 - 28a^2 + 35a - 7 \\
3a^4 - 12a^3 + 15a^2 - 3a \qquad\quad \\
\hline
3a^4 - 5a^3 - 13a^2 + 32a - 7
\end{array}
$$

Multiply each term in $a^3 - 4a^2 + 5a - 1$ by 7.

Multiply each term in $a^3 - 4a^2 + 5a - 1$ by $3a$.

Line up like terms in the same column. Add.    ■

## You Try 2

Multiply.

a)   $(9x + 5)(2x^2 - x - 3)$

b)   $\left(t^2 - \dfrac{2}{3}t - 4\right)(4t^2 + 6t - 5)$

## 3. Multiply Two Binomials Using FOIL

Multiplying two binomials is one of the most common types of polynomial multiplication used in algebra. A method called **FOIL** is one that is often used to multiply two binomials, and it comes from using the distributive property.

Let's use the distributive property to multiply $(x + 6)(x + 4)$.

$$
\begin{aligned}
(x + 6)(x + 4) &= (x + 6)(x) + (x + 6)(4) && \text{Distribute.} \\
&= x(x) + 6(x) + x(4) + 6(4) && \text{Distribute.} \\
&= x^2 + 6x + 4x + 24 && \text{Multiply.} \\
&= x^2 + 10x + 24 && \text{Combine like terms.}
\end{aligned}
$$

To be sure that each term in the first binomial has been multiplied by each term in the second binomial, we can use FOIL. FOIL stands for **F**irst **O**uter **I**nner **L**ast. Let's see how we can apply FOIL to the example above:

$$
(x + 6)(x + 4) = (x + 6)(x + 4) = \overset{F}{x \cdot x} + \overset{O}{x \cdot 4} + \overset{I}{6 \cdot x} + \overset{L}{6 \cdot 4}
$$

$$
\begin{aligned}
&= x^2 + 4x + 6x + 24 && \text{Multiply.} \\
&= x^2 + 10x + 24 && \text{Combine like terms.}
\end{aligned}
$$

You can see that we get the same result.

## Example 4

Use FOIL to multiply the binomials.

a)   $(p + 5)(p - 2)$          b)   $(4r - 3)(r - 1)$

c)   $(a + 4b)(a - 3b)$        d)   $(2x + 9)(3y + 5)$

**Solution**

a)   $(p + 5)(p - 2) = (p + 5)(p - 2) = \overset{F}{p(p)} + \overset{O}{p(-2)} + \overset{I}{5(p)} + \overset{L}{5(-2)}$    Use FOIL.

$$
\begin{aligned}
&= p^2 - 2p + 5p - 10 && \text{Multiply.} \\
&= p^2 + 3p - 10 && \text{Combine like terms.}
\end{aligned}
$$

Notice that the middle terms are like terms, so we can combine them.

b) $(4r - 3)(r - 1) = (4r - 3)(r - 1) = 4r(r) + 4r(-1) - 3(r) - 3(-1)$   Use FOIL.

$\phantom{b) (4r - 3)(r - 1)} = 4r^2 - 4r - 3r + 3$   Multiply.

$\phantom{b) (4r - 3)(r - 1)} = 4r^2 - 7r + 3$   Combine like terms.

The middle terms are like terms, so we can combine them.

c) $(a + 4b)(a - 3b) = a(a) + a(-3b) + 4b(a) + 4b(-3b)$   Use FOIL.

$\phantom{c) (a + 4b)(a - 3b)} = a^2 - 3ab + 4ab - 12b^2$   Multiply.

$\phantom{c) (a + 4b)(a - 3b)} = a^2 + ab - 12b^2$   Combine like terms.

As in parts a) and b), we combined the middle terms.

d) $(2x + 9)(3y + 5) = 2x(3y) + 2x(5) + 9(3y) + 9(5)$   Use FOIL.

$\phantom{d) (2x + 9)(3y + 5)} = 6xy + 10x + 27y + 45$   Multiply.

In this case the middle terms were not like terms, so we could not combine them.   ■

### You Try 3

Use FOIL to multiply the binomials.

a)   $(n + 8)(n + 5)$       b)   $(3k + 7)(k - 4)$

c)   $(x - 2y)(x - 6y)$       d)   $(5c - 8)(2d + 1)$

With practice, you should get to the point where you can find the product of two binomials "in your head." Remember that, as in the case of parts a) – c) in Example 4, it is often possible to combine the middle terms.

## 4. Find the Product of More Than Two Polynomials

Sometimes we must find the product of more than two polynomials.

### Example 5

Multiply $3t^2(5t + 7)(t - 2)$.

#### Solution

We can approach this problem a couple of ways.

**Method 1**
Begin by multiplying the binomials, *then* multiply by the monomial.

$3t^2(5t + 7)(t - 2) = 3t^2(5t^2 - 10t + 7t - 14)$   Use FOIL to multiply the binomials.

$\phantom{3t^2(5t + 7)(t - 2)} = 3t^2(5t^2 - 3t - 14)$   Combine like terms.

$\phantom{3t^2(5t + 7)(t - 2)} = 15t^4 - 9t^3 - 42t^2$   Distribute.

**Method 2**
Begin by multiplying $3t^2$ and $(5t + 7)$, then multiply *that* product by $(t - 2)$.

$3t^2(5t + 7)(t - 2) = (15t^3 + 21t^2)(t - 2)$   Multiply $3t^2$ and $(5t + 7)$.

$\phantom{3t^2(5t + 7)(t - 2)} = 15t^4 - 30t^3 + 21t^3 - 42t^2$   Use FOIL to multiply.

$\phantom{3t^2(5t + 7)(t - 2)} = 15t^4 - 9t^3 - 42t^2$   Combine like terms.

The result is the same. These may be multiplied by whichever method you prefer.   ■

**You Try 4**

Multiply $-7m^3(m-1)(2m-3)$.

There are special types of binomial products that come up often in algebra. We will look at these next.

### 5. Find the Product of Binomials of the Form $(a + b)(a - b)$

Let's find the product $(y + 6)(y - 6)$. Using FOIL, we get

$$(y + 6)(y - 6) = y^2 - 6y + 6y - 36$$
$$= y^2 - 36$$

Notice that the middle terms, the $y$-terms, drop out. In the result, $y^2 - 36$, the first term $(y^2)$ is the square of $y$ and the last term $(36)$ is the square of 6. The resulting polynomial is a *difference of squares*. This pattern always holds when multiplying two binomials of the form $(a + b)(a - b)$.

---

**Procedure   The Product of the Sum and Difference of Two Terms**

$$(a + b)(a - b) = a^2 - b^2$$

---

**Example 6**

Multiply.

a)   $(z + 9)(z - 9)$              b)   $(2 + c)(2 - c)$

c)   $(5n - 8)(5n + 8)$          d)   $\left(\dfrac{3}{4}t + u\right)\left(\dfrac{3}{4}t - u\right)$

**Solution**

a)   The product $(z + 9)(z - 9)$ is in the form $(a + b)(a - b)$, where $a = z$ and $b = 9$. Use the rule that says $(a + b)(a - b) = a^2 - b^2$.

$$(z + 9)(z - 9) = z^2 - 9^2$$
$$= z^2 - 81$$

b)   $(2 + c)(2 - c) = 2^2 - c^2$        $a = 2$ and $b = c$
$$= 4 - c^2$$

Be very careful on a problem like this. The answer is $4 - c^2$, NOT $c^2 - 4$; subtraction is not commutative.

c)   $(5n - 8)(5n + 8) = (5n + 8)(5n - 8)$        Commutative property
$$= (5n)^2 - 8^2$$        $a = 5n$ and $b = 8$; put $5n$ in parentheses.
$$= 25n^2 - 64$$

d)   $\left(\dfrac{3}{4}t + u\right)\left(\dfrac{3}{4}t - u\right) = \left(\dfrac{3}{4}t\right)^2 - u^2$        $a = \dfrac{3}{4}t$ and $b = u$; put $\dfrac{3}{4}t$ in parentheses.
$$= \dfrac{9}{16}t^2 - u^2$$

**You Try 5**

Multiply.

a)  $(k + 7)(k - 7)$

b)  $(3c + 4)(3c - 4)$

c)  $(8 - p)(8 + p)$

d)  $\left(\dfrac{5}{2}m + n\right)\left(\dfrac{5}{2}m - n\right)$

## 6. Square a Binomial

Another type of special binomial product is a **binomial square** such as $(x + 5)^2$. $(x + 5)^2$ means $(x + 5)(x + 5)$. Therefore, we can use FOIL to multiply.

$$(x + 5)^2 = (x + 5)(x + 5) = x^2 + 5x + 5x + 25 \qquad \text{Use FOIL.}$$
$$= x^2 + 10x + 25 \qquad \text{Note that } 10x = 2(x)(5).$$

Let's square the binomial $(y - 3)$.

$$(y - 3)^2 = (y - 3)(y - 3) = y^2 - 3y - 3y + 9 \qquad \text{Use FOIL.}$$
$$= y^2 - 6y + 9 \qquad \text{Note that } -6y = 2(y)(-3).$$

In each case, notice that the outer and inner products are the same. When we add those terms, we see that the middle term of the result is *twice* the product of the terms in each binomial.

In the expansion of $(x + 5)^2$,   $10x$   is   $2(x)(5)$.

In the expansion of $(y - 3)^2$,   $-6y$   is   $2(y)(-3)$.

The *first* term in the result is the square of the *first* term in the binomial, and the *last* term in the result is the square of the *last* term in the binomial. We can express these relationships with the following formulas:

**Procedure   The Square of a Binomial**

$$(a + b)^2 = a^2 + 2ab + b^2$$
$$(a - b)^2 = a^2 - 2ab + b^2$$

We can think of the formulas in words as:

*To square a binomial, you square the first term, square the second term, then multiply 2 times the first term times the second term and add.*

Finding the products $(a + b)^2 = a^2 + 2ab + b^2$ and $(a - b)^2 = a^2 - 2ab + b^2$ is also called *expanding* the binomial squares $(a + b)^2$ and $(a - b)^2$.

**BE CAREFUL**

$$(a + b)^2 \neq a^2 + b^2 \quad \text{and} \quad (a - b)^2 \neq a^2 - b^2.$$

## Example 7

Expand.

a)  $(d + 7)^2$          b)  $(m - 9)^2$          c)  $(2x - 5y)^2$          d)  $\left(\dfrac{1}{3}t + 4\right)^2$

### Solution

a)  $(d + 7)^2 = d^2 \quad + 2(d)(7) \quad + (7)^2 \qquad a = d, b = 7$

$\qquad\qquad\qquad\uparrow\qquad\qquad\uparrow\qquad\qquad\uparrow$

$\qquad$ Square the      Two times      Square the

$\qquad$ first term      first term      second term

$\qquad\qquad\qquad$ times second

$\qquad\qquad\qquad\qquad$ term

$\qquad\qquad = d^2 + 14d + 49$

Notice, $(d + 7)^2 \neq d^2 + 49$. Do not "distribute" the power of 2 to each term in the binomial!

b)  $(m - 9)^2 = m^2 \quad -2(m)(9) \quad + (9)^2 \qquad a = m, b = 9$

$\qquad\qquad\qquad\uparrow\qquad\qquad\uparrow\qquad\qquad\uparrow$

$\qquad$ Square the      Two times      Square the

$\qquad$ first term      first term      second term

$\qquad\qquad\qquad$ times second

$\qquad\qquad\qquad\qquad$ term

$\qquad\qquad = m^2 - 18m + 81$

c)  $(2x - 5y)^2 = (2x)^2 - 2(2x)(5y) + (5y)^2 \qquad a = 2x, b = 5y$

$\qquad\qquad\qquad = 4x^2 - 20xy + 25y^2$

d)  $\left(\dfrac{1}{3}t + 4\right)^2 = \left(\dfrac{1}{3}t\right)^2 + 2\left(\dfrac{1}{3}t\right)(4) + (4)^2 \qquad a = \dfrac{1}{3}t, b = 4$

$\qquad\qquad\qquad = \dfrac{1}{9}t^2 + \dfrac{8}{3}t + 16$

## You Try 6

Expand.

a)  $(r + 10)^2$      b)  $(h - 1)^2$      c)  $(2p + 3q)^2$      d)  $\left(\dfrac{3}{4}y - 5\right)^2$

## 7. Find Higher Powers of a Binomial

To find higher powers of binomials, we use techniques we have already discussed.

## Example 8

Expand.

a)  $(n + 2)^3$          b)  $(3v - 2)^4$

### Solution

a) Just as $x^2 \cdot x = x^3$, it is true that $(n + 2)^2 \cdot (n + 2) = (n + 2)^3$. So we can think of $(n + 2)^3$ as $(n + 2)^2(n + 2)$.

$$\begin{aligned}
(n + 2)^3 &= (n + 2)^2(n + 2) \\
&= (n^2 + 4n + 4)(n + 2) & \text{Square the binomial.} \\
&= n^3 + 2n^2 + 4n^2 + 8n + 4n + 8 & \text{Multiply.} \\
&= n^3 + 6n^2 + 12n + 8 & \text{Combine like terms.}
\end{aligned}$$

b) Since we can write $x^4 = x^2 \cdot x^2$, we can write $(3v - 2)^4 = (3v - 2)^2 \cdot (3v - 2)^2$.

$$\begin{aligned}
(3v - 2)^4 &= (3v - 2)^2 \cdot (3v - 2)^2 \\
&= (9v^2 - 12v + 4)(9v^2 - 12v + 4) & \text{Square each binomial.} \\
&= 81v^4 - 108v^3 + 36v^2 - 108v^3 + 144v^2 - 48v & \text{Multiply.} \\
&\quad + 36v^2 - 48v + 16 \\
&= 81v^4 - 216v^3 + 216v^2 - 96v + 16 & \text{Combine like terms.} \quad \blacksquare
\end{aligned}$$

### You Try 7

Expand.   a)  $(k - 3)^3$      b)  $(2h + 1)^4$

---

### Answers to You Try Exercises

1) $20z^7 - 35z^6 - 5z^5 + 40z^4$    2)  a) $18x^3 + x^2 - 32x - 15$   b) $4t^4 + \dfrac{10}{3}t^3 - 25t^2 - \dfrac{62}{3}t + 20$

3)  a) $n^2 + 13n + 40$   b) $3k^2 - 5k - 28$   c) $x^2 - 8xy + 12y^2$   d) $10cd + 5c - 16d - 8$

4)  $-14m^5 + 35m^4 - 21m^3$    5)  a) $k^2 - 49$   b) $9c^2 - 16$   c) $64 - p^2$   d) $\dfrac{25}{4}m^2 - n^2$

6)  a) $r^2 + 20r + 100$   b) $h^2 - 2h + 1$   c) $4p^2 + 12pq + 9q^2$   d) $\dfrac{9}{16}y^2 - \dfrac{15}{2}y + 25$

7)  a) $k^3 - 9k^2 + 27k - 27$   b) $16h^4 + 32h^3 + 24h^2 + 8h + 1$

---

## 6.3 Exercises

**Objective 1: Multiply a Monomial and a Polynomial**

1) Explain how to multiply a monomial and a binomial.

2) Explain how to multiply two binomials.

Multiply.

3) $(3m^5)(8m^3)$      4) $(2k^6)(7k^3)$

5) $(-8c)(4c^5)$      6) $\left(-\dfrac{2}{9}z^3\right)\left(\dfrac{3}{4}z^9\right)$

Multiply.

7) $5a(2a - 7)$      8) $3y(10y - 1)$

9) $-6c(7c + 2)$      10) $-15d(11d - 2)$

 11) $6v^3(v^2 - 4v - 2)$      12) $8f^5(f^2 - 3f - 6)$

13) $-9b^2(4b^3 - 2b^2 - 6b - 9)$

14) $-4h^7(5h^6 + 4h^3 + 11h - 3)$

15) $3a^2b(ab^2 + 6ab - 13b + 7)$

16) $4x^6y^2(-5x^2y + 11xy^2 - xy + 2y - 1)$

17) $-\dfrac{3}{5}k^4(15k^2 + 20k - 3)$

18) $\dfrac{3}{4}t^5(12t^3 - 20t^2 + 9)$

**Objective 2: Multiply Two Polynomials**

Multiply.

19) $(c + 4)(6c^2 - 13c + 7)$

20) $(d + 8)(7d^2 + 3d - 9)$

21) $(f - 5)(3f^2 + 2f - 4)$

22) $(k - 2)(9k^2 - 4k - 12)$

23) $(4x^3 - x^2 + 6x + 2)(2x - 5)$

24) $(3m^3 + 3m^2 - 4m - 9)(4m - 7)$

25) $\left(\frac{1}{3}y^2 + 4\right)(12y^2 + 7y - 9)$

26) $\left(\frac{3}{5}q^2 - 1\right)(10q^2 - 7q + 20)$

27) $(s^2 - s + 2)(s^2 + 4s - 3)$

28) $(t^2 + 4t + 1)(2t^2 - t - 5)$

29) $(4h^2 - h + 2)(-6h^3 + 5h^2 - 9h)$

30) $(n^4 + 8n^2 - 5)(n^2 - 3n - 4)$

Multiply both horizontally and vertically. Which method do you prefer and why?

31) $(3y - 2)(5y^2 - 4y + 3)$

32) $(2p^2 + p - 4)(5p + 3)$

### Objective 3: Multiply Two Binomials Using FOIL

33) What do the letters in the word FOIL represent?

34) Can FOIL be used to expand $(x + 8)^2$? Explain your answer.

Use FOIL to multiply.

35) $(w + 5)(w + 7)$

36) $(u + 5)(u + 3)$

37) $(r - 3)(r + 9)$

38) $(w - 12)(w - 4)$

39) $(y - 7)(y - 1)$

40) $(g + 4)(g - 8)$

41) $(3p + 7)(p - 2)$

42) $(5u + 1)(u + 7)$

43) $(7n + 4)(3n + 1)$

44) $(4y - 3)(7y + 6)$

45) $(5 - 4w)(3 - w)$

46) $(2 - 3r)(4 - 5r)$

47) $(4a - 5b)(3a + 4b)$

48) $(3c + 2d)(c - 5d)$

49) $(6x + 7y)(8x + 3y)$

50) $(0.5p - 0.3q)(0.7p - 0.4q)$

51) $\left(v + \frac{1}{3}\right)\left(v + \frac{3}{4}\right)$

52) $\left(t + \frac{5}{2}\right)\left(t + \frac{6}{5}\right)$

53) $\left(\frac{1}{2}a + 5b\right)\left(\frac{2}{3}a - b\right)$

54) $\left(\frac{3}{4}x - y\right)\left(\frac{1}{3}x + 4y\right)$

Write an expression for a) the perimeter of each figure and b) the area of each figure.

55)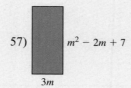
$y - 3$
$y + 5$

56)
$6w$
$5w + 4$

57)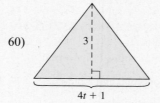
$m^2 - 2m + 7$
$3m$

58)
$3x^2 - 1$
$3x^2 - 1$

Express the area of each triangle as a polynomial.

59)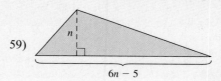
$n$
$6n - 5$

60)
$3$
$4t + 1$

### Objective 4: Find the Product of More Than Two Polynomials

61) To find the product $3(c + 5)(c - 1)$, Parth begins by multiplying $3(c + 5)$ and then he multiplies that result by $(c - 1)$. Yolanda begins by multiplying $(c + 5)(c - 1)$ and multiplies that result by 3. Who is right?

62) Find the product $(2y + 3)(y - 5)(y - 4)$

   a) by first multiplying $(2y + 3)(y - 5)$ and then multiplying that result by $(y - 4)$.

   b) by first multiplying $(y - 5)(y - 4)$ and then multiplying that result by $(2y + 3)$.

   c) What do you notice about the results?

Multiply.

63) $2(n + 3)(4n - 5)$

64) $-13(3p - 1)(p + 4)$

65) $-5z^2(z - 8)(z - 2)$

66) $11r^2(2r + 7)(-r + 1)$

VIDEO 67) $(c + 3)(c + 4)(c - 1)$

68) $(2t + 3)(t + 1)(t + 4)$

69) $(3x - 1)(x - 2)(x - 6)$

70) $(2m + 7)(m + 1)(m + 5)$

71) $8p\left(\dfrac{1}{4}p^2 + 3\right)(p^2 + 5)$

72) $10c\left(\dfrac{3}{5}c^2 - \dfrac{1}{2}\right)(c^2 + 1)$

**Mixed Exercises: Objectives 5 and 6**
Find the following special products.

73) $(y + 5)(y - 5)$

74) $(w + 2)(w - 2)$

75) $(a - 7)(a + 7)$

76) $(f - 11)(f + 11)$

77) $(3 - p)(3 + p)$

78) $(12 + d)(12 - d)$

79) $\left(u + \dfrac{1}{5}\right)\left(u - \dfrac{1}{5}\right)$

80) $\left(g - \dfrac{1}{4}\right)\left(g + \dfrac{1}{4}\right)$

VIDEO 81) $\left(\dfrac{2}{3} - k\right)\left(\dfrac{2}{3} + k\right)$

82) $\left(\dfrac{7}{4} + c\right)\left(\dfrac{7}{4} - c\right)$

83) $(2r + 7)(2r - 7)$

84) $(4h - 3)(4h + 3)$

85) $-(8j - k)(8j + k)$

86) $-(3m + 5n)(3m - 5n)$

87) $(d + 4)^2$

88) $(g + 2)^2$

89) $(n - 13)^2$

90) $(b - 3)^2$

91) $(h - 6)^2$

92) $(q - 7)^2$

93) $(3u + 1)^2$

94) $(5n + 4)^2$

VIDEO 95) $(2d - 5)^2$

96) $(4p - 3)^2$

97) $(3c + 2d)^2$

98) $(2a - 5b)^2$

99) $\left(\dfrac{3}{2}k + 8m\right)^2$

100) $\left(\dfrac{4}{5}x - 3y\right)^2$

101) $[(2a + b) + 3]^2$

102) $[(3c - d) + 5]^2$

103) $[(f - 3g) + 4][(f - 3g) - 4]$

104) $[(5j + 2k) + 1][(5j + 2k) - 1]$

105) Does $3(r + 2)^2 = (3r + 6)^2$? Why or why not?

106) Explain how to find the product $4(a - 5)^2$, then find the product.

Find each product.

107) $7(y + 2)^2$

108) $3(m + 4)^2$

109) $4c(c + 3)^2$

110) $-3a(a + 1)^2$

**Objective 7: Find Higher Powers of a Binomial**
Expand.

VIDEO 111) $(r + 5)^3$

112) $(u + 3)^3$

113) $(g - 4)^3$

114) $(w - 5)^3$

115) $(2a - 1)^3$

116) $(3x + 4)^3$

117) $(h + 3)^4$

118) $(y + 5)^4$

119) $(5t - 2)^4$

120) $(4c - 1)^4$

121) Does $(x + 2)^2 = x^2 + 4$ ? Why or why not?

122) Does $(y - 1)^3 = y^3 - 1$ ? Why or why not?

**Mixed Exercises: Objectives 1–7**

Find each product.

123) $(c - 12)(c + 7)$

124) $(11t^4)(-2t^6)$

125) $4(6 - 5a)(2a - 1)$

126) $(3y - 8z)(3y + 8z)$

127) $(2k - 9)(5k^2 + 4k - 1)$

128) $(m + 12)^2$

129) $\left(\dfrac{1}{6} - h\right)\left(\dfrac{1}{6} + h\right)$

130) $3(7p^3 + 4p^2 + 2) - (5p^3 - 18p - 11)$

131) $(3c + 1)^3$

132) $(4w - 5)(2w - 3)$

133) $\left(\dfrac{3}{8}p^7\right)\left(\dfrac{3}{4}p^4\right)$

134) $xy(2x - y)(x - 3y)(x - 2y)$

135) $(a + 7b)^2$

136) $(r - 6)^3$

137) $-5z(z - 3)^2$

138) $(2n^2 + 9n - 3)(4n^2 - n - 8)$

139) Express the volume of the cube as a polynomial.

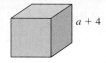
$a + 4$

140) Express the area of the square as a polynomial.

$4r - 1$

141) Express the area of the circle as a polynomial. Leave $\pi$ in your answer.

$k + 5$

Express the area of the shaded region as a polynomial. Leave $\pi$ in your answer where appropriate.

142)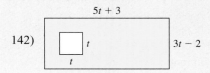
$5t + 3$
$t$
$t$
$3t - 2$

143)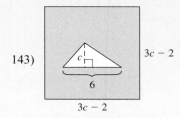
$3c - 2$
$c$
$6$
$3c - 2$

144)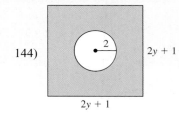
$2$
$2y + 1$
$2y + 1$

---

## Section 6.4 Division of Polynomials

**Objectives**

1. **Divide a Polynomial by a Monomial**
2. **Divide a Polynomial by a Polynomial**

The last operation with polynomials we need to discuss is division of polynomials. We will consider this in two parts:

    dividing a polynomial by a monomial and

    dividing a polynomial by a polynomial.

Let's begin with dividing a polynomial by a monomial.

### 1. Divide a Polynomial by a Monomial

The procedure for dividing a polynomial by a monomial is based on the procedure for adding or subtracting fractions.

To add $\dfrac{2}{9} + \dfrac{5}{9}$, we add the numerators and keep the denominator.

$$\frac{2}{9} + \frac{5}{9} = \frac{2 + 5}{9}$$

$$= \frac{7}{9}$$

Reversing the process above, we can write $\dfrac{7}{9} = \dfrac{2 + 5}{9} = \dfrac{2}{9} + \dfrac{5}{9}$.

We can generalize this result and say that

$$\frac{a + b}{c} = \frac{a}{c} + \frac{b}{c} \; (c \neq 0)$$

This leads us to the following rule.

---

**Procedure**   Dividing a Polynomial by a Monomial

To divide a polynomial by a monomial, divide *each term* in the polynomial by the monomial and simplify.

---

**Example 1**

Divide.

a) $\dfrac{24x^2 - 8x + 20}{4}$   b) $\dfrac{12c^3 + 54c^2 + 6c}{6c}$

**Solution**

a)  First, note that the polynomial is being divided by a *monomial*. Divide each term in the numerator by the monomial 4.

$$\frac{24x^2 - 8x + 20}{4} = \frac{24x^2}{4} - \frac{8x}{4} + \frac{20}{4}$$

$$= 6x^2 - 2x + 5$$

Let's label the components of our division problem the same way as when we divide with integers.

$$\text{Dividend} \rightarrow \frac{24x^2 - 8x + 20}{4} = 6x^2 - 2x + 5 \leftarrow \text{Quotient}$$
$$\text{Divisor} \rightarrow$$

We can check our answer by multiplying the quotient by the divisor. The answer should be the dividend.

Check: $4(6x^2 - 2x + 5) = 24x^2 - 8x + 20$ ✓   The quotient is correct.

b)  $\dfrac{12c^3 + 54c^2 + 6c}{6c} = \dfrac{12c^3}{6c} + \dfrac{54c^2}{6c} + \dfrac{6c}{6c}$   Divide each term in the numerator by $6c$.

$$= 2c^2 + 9c + 1$$   Apply the quotient rule for exponents.

**BE CAREFUL**

Students will often incorrectly "cancel out" $\dfrac{6c}{6c}$ and get nothing. But $\dfrac{6c}{6c} = 1$ since a quantity divided by itself equals one.

Check: $6c(2c^2 + 9c + 1) = 12c^3 + 54c^2 + 6c$ ✓    The quotient is correct.    ■

**Note**

In Example 1b), $c$ cannot equal zero because then the denominator of $\dfrac{12c^3 + 54c^2 + 6c}{6c}$ would equal zero. Remember, a fraction is undefined when its denominator equals zero!

**You Try 1**

Divide $\dfrac{35n^5 + 20n^4 - 5n^2}{5n^2}$.

**Example 2**

Divide $(6a - 7 - 36a^2 + 27a^3) \div (9a^2)$.

**Solution**

This is another example of a polynomial divided by a monomial. Notice, however, the terms in the numerator are not written in descending powers. Rewrite them in descending powers before dividing.

$$\frac{6a - 7 - 36a^2 + 27a^3}{9a^2} = \frac{27a^3 - 36a^2 + 6a - 7}{9a^2}$$

$$= \frac{27a^3}{9a^2} - \frac{36a^2}{9a^2} + \frac{6a}{9a^2} - \frac{7}{9a^2}$$

$$= 3a - 4 + \frac{2}{3a} - \frac{7}{9a^2} \qquad \text{Apply quotient rule and simplify.}$$

The quotient is *not* a polynomial since $a$ and $a^2$ appear in denominators. The quotient of polynomials is not necessarily a polynomial.    ■

**You Try 2**

Divide $(30z^2 + 3 - 50z^3 + 18z) \div (10z^2)$.

## 2. Divide a Polynomial by a Polynomial

When dividing a polynomial by a polynomial containing two or more terms, we use long division of polynomials. This method is similar to long division of whole numbers, so let's look at a long division problem and compare the procedure with polynomial long division.

## Example 3

Divide 854 by 3.

**Solution**

$$
\begin{array}{r}
2 \\
3{\overline{)854}} \\
-6\downarrow \\
\hline
25
\end{array}
$$

1) How many times does 3 divide into 8 evenly?  2
2) Multiply $2 \times 3 = 6$.
3) Subtract $8 - 6 = 2$.
4) Bring down the 5.

Start the process again.

$$
\begin{array}{r}
28 \\
3{\overline{)854}} \\
-6 \\
\hline
25 \\
-24\downarrow \\
\hline
14
\end{array}
$$

1) How many times does 3 divide into 25 evenly?  8
2) Multiply $8 \times 3 = 24$.
3) Subtract $25 - 24 = 1$.
4) Bring down the 4.

Do the procedure again.

$$
\begin{array}{r}
284 \\
3{\overline{)854}} \\
-6 \\
\hline
25 \\
-24 \\
\hline
14 \\
-12 \\
\hline
2
\end{array}
$$

1) How many times does 3 divide into 14 evenly?  4
2) Multiply $4 \times 3 = 12$.
3) Subtract $14 - 12 = 2$.
4) There are no more numbers to bring down, so the remainder is 2.

Write the result.

$$
854 \div 3 = 284\frac{2}{3} \quad \begin{array}{l} \leftarrow \text{Remainder} \\ \leftarrow \text{Divisor} \end{array}
$$

Check:  $(3 \times 284) + 2 = 852 + 2 = 854$ ✓

**You Try 3**

Divide 638 by 5.

Next we will divide two polynomials using a long division process similar to that of Example 3.

## Example 4

Divide $\dfrac{5x^2 + 13x + 6}{x + 2}$.

**Solution**

First, notice that we are dividing by a polynomial containing more than one term. That tells us to use long division of polynomials.

We will work with the $x$ in $x + 2$ like we worked with the 3 in Example 3.

$$\begin{array}{r} 5x \phantom{xxxxxxx} \\ x + 2 \overline{)\ 5x^2 + 13x + 6} \\ \underline{-(5x^2 + 10x)} \ \downarrow \phantom{x} \\ 3x + 6 \end{array}$$

1) By what do we multiply $x$ to get $5x^2$?   $5x$
   Line up terms in the quotient according to exponents, so write $5x$ above $13x$.

2) Multiply $5x$ by $(x + 2)$:  $5x(x + 2) = 5x^2 + 10x$

3) Subtract $(5x^2 + 13x) - (5x^2 + 10x) = 3x$.

4) Bring down the $+6$.

Start the process again. Remember, work with the $x$ in $x + 2$ like we worked with the 3 in Example 3.

$$\begin{array}{r} 5x + 3 \phantom{xxx} \\ x + 2 \overline{)\ \ 5x^2 + 13x + 6} \\ \underline{-(5x^2 + 10x)} \phantom{xx} \\ 3x + 6 \\ \underline{-(3x + 6)} \\ 0 \end{array}$$

1) By what do we multiply $x$ to get $3x$?   3
   Write $+3$ above $+6$.

2) Multiply 3 by $(x + 2)$:  $3(x + 2) = 3x + 6$

3) Subtract $(3x + 6) - (3x + 6) = 0$.

4) There are no more terms. The remainder is 0.

Write the result.

$$\frac{5x^2 + 13x + 6}{x + 2} = 5x + 3$$

Check:  $(x + 2)(5x + 3) = 5x^2 + 3x + 10x + 6 = 5x^2 + 13x + 6$  ✓  ■

## You Try 4

Divide.

a) $\dfrac{r^2 + 11r + 28}{r + 4}$     b) $\dfrac{3k^2 + 17k + 10}{k + 5}$

Next, we will look at a division problem with a remainder.

## Example 5

Divide $\dfrac{-28n + 15n^3 + 41 - 17n^2}{3n - 4}$.

### Solution

When we write our long division problem, the polynomial in the numerator must be rewritten so that the exponents are in descending order. Then, perform the long division.

$$\begin{array}{r} 5n^2 \phantom{xxxxxxxxxxx} \\ 3n - 4 \overline{)\ 15n^3 - 17n^2 - 28n + 41} \\ \underline{-(15n^3 - 20n^2)} \ \downarrow \phantom{xxxx} \\ 3n^2 - 28n \end{array}$$

1) By what do we multiply $3n$ to get $15n^3$?   $5n^2$

2) Multiply $5n^2(3n - 4) = 15n^3 - 20n^2$

3) Subtract. $(15n^3 - 17n^2) - (15n^3 - 20n^2)$
   $= 15n^3 - 17n^2 - 15n^3 + 20n^2$
   $= 3n^2$

4) Bring down the $-28n$.

Repeat the process.

$$\begin{array}{r} 5n^2 + \phantom{xx} n \phantom{xxxxxxxxxxxx} \\ 3n - 4 \overline{)\, 15n^3 - 17n^2 - 28n + 41} \\ \underline{-(15n^3 - 20n^2)} \phantom{xxxxxxxxxx} \\ 3n^2 - 28n \phantom{xxxx} \\ \underline{-(3n^2 - \phantom{x}4n)} \phantom{xxxx} \\ -24n + 41 \end{array}$$

1) By what do we multiply $3n$ to get $3n^2$?   $n$

2) Multiply $n(3n - 4) = 3n^2 - 4n$.

3) Subtract. $(3n^2 - 28n) - (3n^2 - 4n)$
   $= 3n^2 - 28n - 3n^2 + 4n$
   $= -24n$

4) Bring down the $+41$.

Continue.

$$\begin{array}{r} 5n^2 + \phantom{xx} n - \phantom{x}8 \phantom{xxxxxxx} \\ 3n - 4 \overline{)\, 15n^3 - 17n^2 - 28n + 41} \\ \underline{-(15n^3 - 20n^2)} \phantom{xxxxxxxxxx} \\ 3n^2 - 28n \phantom{xxxx} \\ \underline{-(3n^2 - \phantom{x}4n)} \phantom{xxxx} \\ -24n + 41 \phantom{x} \\ \underline{-(-24n + 32)} \phantom{x} \\ 9 \end{array}$$

1) By what do we multiply $3n$ to get $-24n$?   $-8$

2) Multiply $-8(3n - 4) = -24n + 32$.

3) Subtract. $(-24n + 41) - (-24n + 32) = 9$

We are done with the long division process. How do we know that? Since the degree of 9 (degree zero) is less than the degree of $3n - 4$ (degree one), we cannot divide anymore. *The remainder is 9.*

$$\frac{15n^3 - 17n^2 - 28n + 41}{3n - 4} = 5n^2 + n - 8 + \frac{9}{3n - 4}$$

Check: $(3n - 4)(5n^2 + n - 8) + 9 = 15n^3 + 3n^2 - 24n - 20n^2 - 4n + 32 + 9$
$= 15n^3 - 17n^2 - 28n + 41$ ✓ ■

**You Try 5**

Divide $-34a^2 + 57 + 8a^3 - 21a$ by $2a - 9$.

In Example 5, we saw that we must write our polynomials in descending order. We have to watch out for something else as well—missing terms. If a polynomial is missing one or more terms, we put them into the polynomial with coefficients of zero.

**Example 6**

Divide.

a) $x^3 + 64$ by $x + 4$

b) $t^4 + 3t^3 + 6t^2 + 11t + 5$ by $t^2 + 2$

**Solution**

a) The polynomial $x^3 + 64$ is missing the $x^2$-term and the $x$-term. We will insert these terms into the polynomial by giving them coefficients of zero.

$$x^3 + 64 = x^3 + 0x^2 + 0x + 64$$

Divide.

$$\begin{array}{r} x^2 \;-\; 4x + 16 \\ x + 4{\overline{)\,x^3 \;+\; 0x^2 \;+\; 0x \;+\; 64}} \\ \underline{-(x^3 + 4x^2)} \\ -4x^2 + \phantom{1}0x \\ \underline{-(-4x^2 - 16x)} \\ 16x + 64 \\ \underline{-(16x + 64)} \\ 0 \end{array}$$

$$(x^3 + 64) \div (x + 4) = x^2 - 4x + 16$$

Check: $(x + 4)(x^2 - 4x + 16) = x^3 - 4x^2 + 16x + 4x^2 - 16x + 64$

$$= x^3 + 64 \checkmark$$

b)   In this case, the divisor, $t^2 + 2$, is missing a $t$-term. Rewrite it as $t^2 + 0t + 2$ and divide.

$$\begin{array}{r} t^2 \;+\; 3t \;+\; \phantom{1}4 \\ t^2 + 0t + 2{\overline{)\,t^4 + 3t^3 + 6t^2 + 11t \;+\; 5}} \\ \underline{-(t^4 + 0t^3 + 2t^2)} \\ 3t^3 + 4t^2 + 11t \\ \underline{-(3t^3 + 0t^2 + \phantom{1}6t)} \\ 4t^2 + \phantom{1}5t + 5 \\ \underline{-(4t^2 + \phantom{1}0t + 8)} \\ 5t - 3 \quad \leftarrow \text{Remainder} \end{array}$$

We are done with the long division process because the degree of $5t - 3$ (degree one) is less than the degree of the divisior, $t^2 + 0t + 2$ (degree two).

Write the answer as $t^2 + 3t + 4 + \dfrac{5t - 3}{t^2 + 2}$. The check is left to the student. ∎

## You Try 6

Divide.

a)  $\dfrac{4m^3 + 17m^2 - 38}{m + 3}$

b)  $\dfrac{p^4 + 6p^3 + 3p^2 + 10p + 1}{p^2 + 1}$

## Summary    Dividing Polynomials

Remember, when asked to divide two polynomials, first identify which type of division problem it is.

1)   To divide a *polynomial* by a *monomial*, divide *each term* in the polynomial by the monomial and simplify.

Monomial →    $\dfrac{56k^3 + 24k^2 - 8k + 2}{8k} = \dfrac{56k^3}{8k} + \dfrac{24k^2}{8k} - \dfrac{8k}{8k} + \dfrac{2}{8k}$

$$= 7k^2 + 3k - 1 + \dfrac{1}{4k}$$

2) To divide a *polynomial* by a *polynomial* containing two or more terms, use *long division*.

$$\frac{15x^3 + 34x^2 - 11x - 2}{5x - 2}$$

Binomial $\rightarrow$

$$
\begin{array}{r}
3x^2 + 8x + 1 \\
5x - 2 \overline{)\,15x^3 + 34x^2 - 11x - 2} \\
\underline{-(15x^3 - 6x^2)} \\
40x^2 - 11x \\
\underline{-(40x^2 - 16x)} \\
5x - 2 \\
\underline{-(5x - 2)} \\
0
\end{array}
$$

$$\frac{15x^3 + 34x^2 - 11x - 2}{5x - 2} = 3x^2 + 8x + 1$$

**Answers to You Try Exercises**

1) $7n^3 + 4n^2 - 1$   2) $-5z + 3 + \dfrac{9}{5z} + \dfrac{3}{10z^2}$   3) $127\dfrac{3}{5}$   4) a) $r + 7$  b) $3k + 2$

5) $4a^2 + a - 6 + \dfrac{3}{2a - 9}$   6) a) $4m^2 + 5m - 15 + \dfrac{7}{m + 3}$  b) $p^2 + 6p + 2 + \dfrac{4p - 1}{p^2 + 2}$

# 6.4 Exercises

**Objective 1: Divide a Polynomial by a Monomial**

1) Label the dividend, divisor, and quotient of
$$\frac{6c^3 + 15c^2 - 9c}{3c} = 2c^2 + 5c - 3.$$

2) How would you check the answer to the division problem in Exercise 1?

3) Explain, in your own words, how to divide a polynomial by a monomial.

4) Without dividing, determine what the degree of the quotient will be when performing the division
$$\frac{48y^5 - 16y^4 + 5y^3 - 32y^2}{16y^2}.$$

Divide.

5) $\dfrac{49p^4 + 21p^3 + 28p^2}{7}$

6) $\dfrac{10m^3 + 45m^2 + 30m}{5}$

7) $\dfrac{12w^3 - 40w^2 - 36w}{4w}$

8) $\dfrac{3a^5 - 27a^4 + 12a^3}{3a}$

9) $\dfrac{22z^6 + 14z^5 - 38z^3 + 2z}{2z}$

10) $\dfrac{48u^7 - 18u^4 - 90u^2 + 6u}{6u}$

11) $\dfrac{9h^8 + 54h^6 - 108h^3}{9h^2}$

12) $\dfrac{72x^9 - 24x^7 - 56x^4}{8x^2}$

13) $\dfrac{36r^7 - 12r^4 + 6}{12r}$

14) $\dfrac{20t^6 + 130t^2 + 2}{10t}$

15) $\dfrac{8d^6 - 12d^5 + 18d^4}{2d^4}$

16) $\dfrac{21p^4 - 15p^3 + 6p^2}{3p^2}$

17) $\dfrac{28k^7 + 8k^5 - 44k^4 - 36k^2}{4k^2}$

18) $\dfrac{42n^7 + 14n^6 - 49n^4 + 63n^3}{7n^3}$

19) $(35d^5 - 7d^2) \div (-7d^2)$

20) $(-30h^7 - 8h^5 + 2h^3) \div (-2h^3)$

21) $\dfrac{10w^5 + 12w^3 - 6w^2 + 2w}{6w^2}$

22) $\dfrac{-48r^5 + 14r^3 - 4r^2 + 10}{4r}$

23) $(12k^8 - 4k^6 - 15k^5 - 3k^4 + 1) \div (2k^5)$

24) $(56m^6 + 4m^5 - 21m^2) \div (7m^3)$

Divide.

25) $\dfrac{48p^5q^3 + 60p^4q^2 - 54p^3q + 18p^2q}{6p^2q}$

26) $(-45x^5y^4 - 27x^4y^5 + 9x^3y^5 + 63x^3y^3) \div (-9x^3y^2)$

27) $\dfrac{14s^6t^6 - 28s^5t^4 - s^3t^3 + 21st}{7s^2t}$

28) $(4a^5b^4 - 32a^4b^4 - 48a^3b^4 + a^2b^3) \div (4ab^2)$

29) Chandra divides $40p^3 - 10p^2 + 5p$ by $5p$ and gets a quotient of $8p^2 - 2p$. Is this correct? Why or why not?

30) Ryan divides $\dfrac{20y^2 + 15y}{15y}$ and gets a quotient of $20y^2$. What was his mistake? What is the correct answer?

## Objective 2: Divide a Polynomial by a Polynomial

31) Label the dividend, divisor, and quotient of
$$3w + 1 \overline{)\phantom{1}} \begin{array}{c} 4w^2 - 2w - 7 \\ \hline 12w^3 - 2w^2 - 23w - 7 \end{array}.$$

32) When do you use long division to divide polynomials?

33) If a polynomial of degree 3 is divided by a binomial of degree 1, determine the degree of the quotient.

34) How would you check the answer to the division problem in Exercise 31?

Divide.

35) $6\overline{)949}$     36) $4\overline{)857}$

37) $9\overline{)3937}$     38) $8\overline{)4189}$

39) $\dfrac{g^2 + 9g + 20}{g + 5}$     40) $\dfrac{m^2 + 8m + 15}{m + 3}$

41) $\dfrac{a^2 + 13a + 42}{a + 7}$     42) $\dfrac{w^2 + 5w + 4}{w + 1}$

43) $\dfrac{k^2 - k - 30}{k + 5}$     44) $\dfrac{v^2 - 6v - 16}{v + 2}$

45) $\dfrac{x^2 + 3x - 40}{x - 5}$     46) $\dfrac{c^2 - 13c + 36}{c - 4}$

47) $\dfrac{6h^3 + 7h^2 - 17h - 4}{3h - 4}$

48) $\dfrac{20f^3 - 23f^2 + 41f - 14}{5f - 2}$

49) $(p + 23p^2 - 1 + 12p^3) \div (4p + 1)$

50) $(16y^2 + 6 + 15y^3 + 13y) \div (3y + 2)$

51) $(7m^2 - 16m - 41) \div (m - 4)$

52) $(2t^2 + 5t + 8) \div (t + 7)$

53) $\dfrac{24a + 20a^3 - 12 - 43a^2}{5a - 2}$

54) $\dfrac{17v^3 + 33v - 18 + 9v^4 - 56v^2}{9v - 1}$

55) $\dfrac{n^3 + 27}{n + 3}$     56) $\dfrac{d^3 - 8}{d - 2}$

57) $(8r^3 + 6r^2 - 25) \div (4r - 5)$

58) $(12c^3 + 23c^2 + 61) \div (3c + 8)$

59) $\dfrac{12x^3 - 17x + 4}{2x + 3}$

60) $\dfrac{16h^3 - 106h + 15}{2h - 5}$

61) $\dfrac{k^4 + k^3 + 9k^2 + 4k + 20}{k^2 + 4}$

62) $\dfrac{a^4 + 7a^3 + 6a^2 + 21a + 9}{a^2 + 3}$

63) $\dfrac{15t^4 - 40t^3 - 33t^2 + 10t + 2}{5t^2 - 1}$

64) $\dfrac{18v^4 - 15v^3 - 18v^2 + 13v - 10}{3v^2 - 4}$

65) Is the quotient of two polynomials always a polynomial? Explain.

66) Write a division problem that has a divisor of $2c - 1$ and a quotient of $8c - 5$.

## Mixed Exercises: Objectives 1 and 2
Divide.

67) $\dfrac{50a^4b^4 + 30a^4b^3 - a^2b^2 + 2ab}{10a^2b^2}$

68) $\dfrac{12n^2 - 37n + 16}{4n - 3}$

69) $\dfrac{-15f^4 - 22f^2 + 5 + 49f + 36f^3}{5f - 2}$

70) $(8p^2 + 4p - 32p^3 + 36p^4) \div (-4p)$

71) $\dfrac{8t^2 - 19t - 4}{t - 3}$

72) $\dfrac{-27x^3y^3 + 9x^2y^3 + 36xy + 72}{9x^2y}$

73) $(64p^3 - 27) \div (4p - 3)$

74) $(6g^4 - g^3 - 43g^2 + 32g - 30) \div (2g - 5)$

75) $(11x^2 + x^3 - 21 + 6x^4 + 3x) \div (x^2 + 3)$

76) $(12a^4 - 9a^3 - 10a^2 + 3a + 2) \div (3a^2 - 1)$

77) $\dfrac{-20v^3 + 35v^2 - 15v}{-5v}$

78) $\dfrac{125t^3 + 8}{5t + 2}$

VIDEO 79) $\dfrac{10h^4 - 6h^3 - 49h^2 + 27h + 19}{2h^2 - 9}$

80) $\dfrac{24n^2 - 70 + 27n + 8n^4 + 6n^3}{2n^2 + 9}$

81) $\dfrac{m^4 - 16}{m^2 + 4}$      82) $\dfrac{j^4 - 1}{j^2 - 1}$

83) $\dfrac{45t^4 - 81t^2 - 27t^3 + 8t^6}{-9t^3}$

84) $\dfrac{9q^2 + 42q^4 - 9 + 6q - 8q^3}{3q^2}$

85) $\left(x^2 + \dfrac{17}{2}x - 15\right) \div (2x - 3)$

86) $\left(3m^2 + \dfrac{44}{5}m - 11\right) \div (5m - 2)$

87) $\dfrac{21p^4 - 29p^3 - 15p^2 + 28p + 16}{7p^2 + 2p - 4}$

88) $\dfrac{8c^4 + 26c^3 + 29c^2 + 14c + 3}{2c^2 + 5c + 3}$

For each rectangle, find a polynomial that represents the missing side.

89)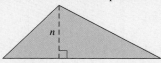
Find the width if the area is given by $6x^2 + 23x + 21$ sq units.
$2x + 3$

90)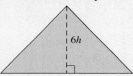
Find the width if the area is given by $20k^3 + 8k^5$ sq units.
$8k$

91) Find the base of the triangle if the area is given by $15n^3 - 18n^2 + 6n$ sq units.

92) Find the base of the triangle if the area is given by $8h^3 + 7h^2 + 2h$ sq units.

93) If Joelle travels $(3x^3 + 5x^2 - 26x + 8)$ miles in $(x + 4)$ hours, find an expression for her rate.

94) If Lorenzo spent $(4a^3 - 11a^2 + 3a - 18)$ dollars on chocolates that cost $(a - 3)$ dollars per pound, find an expression for the number of pounds of chocolates he purchased.

| Definition/Procedure | Example |
|---|---|

## 6.1  Review of the Rules of Exponents

For real numbers $a$ and $b$ and integers $m$ and $n$, the following rules apply:

**Product rule:**  $a^m \cdot a^n = a^{m+n}$   **(p. 342)**

$p^3 \cdot p^5 = p^{3+5} = p^8$

**Power rules:**

a) $(a^m)^n = a^{mn}$

b) $(ab)^n = a^n b^n$

c) $\left(\dfrac{a}{b}\right)^n = \dfrac{a^n}{b^n}$   $(b \neq 0)$    **(p. 342)**

a) $(c^4)^3 = c^{4 \cdot 3} = c^{12}$

b) $(2g)^5 = 2^5 g^5 = 32 g^5$

c) $\left(\dfrac{t}{4}\right)^3 = \dfrac{t^3}{4^3} = \dfrac{t^3}{64}$

**Zero exponent:**  $a^0 = 1$ if $a \neq 0$   **(p. 342)**

$9^0 = 1$

**Negative exponent:**

a) $a^{-n} = \left(\dfrac{1}{a}\right)^n = \dfrac{1}{a^n}$   $(a \neq 0)$

b) $\dfrac{a^{-m}}{b^{-n}} = \dfrac{b^n}{a^m}$   $(a \neq 0, b \neq 0)$    **(p. 342)**

a) $6^{-2} = \left(\dfrac{1}{6}\right)^2 = \dfrac{1}{6^2} = \dfrac{1}{36}$

b) $\dfrac{a^{-5}}{b^{-3}} = \dfrac{b^3}{a^5}$

**Quotient rule:**  a) $\dfrac{a^m}{a^n} = a^{m-n}$  $(a \neq 0)$  **(p. 342)**

$\dfrac{k^{14}}{k^4} = k^{14-4} = k^{10}$

## 6.2  Addition and Subtraction of Polynomials; Graphing

A **polynomial in x** is the sum of a finite number of terms of the form $ax^n$

where $n$ is a whole number and $a$ is a real number.

The **degree of a term** equals the exponent on its variable. If a term has more than one variable, the degree equals the *sum* of the exponents on the variables.

The **degree of the polynomial** equals the highest degree of any nonzero term.    **(p. 345)**

Identify each term in the polynomial, the coefficient and degree of each term, and the degree of the polynomial.

$3m^4 n^2 - m^3 n^2 + 2m^2 n^3 + mn - 5n$

| Term | Coeff. | Degree |
|---|---|---|
| $3m^4 n^2$ | 3 | 6 |
| $-m^3 n^2$ | $-1$ | 5 |
| $2m^2 n^3$ | 2 | 5 |
| $mn$ | 1 | 2 |
| $-5n$ | $-5$ | 1 |

The degree of the polynomial is 6.

To **add polynomials,** add like terms. Polynomials may be added horizontally or vertically.    **(p. 347)**

Add the polynomials.

$(4q^2 + 2q - 12) + (-5q^2 + 3q + 8)$
$= (4q^2 + (-5q^2)) + (2q + 3q) + (-12 + 8) = -q^2 + 5q - 4$

To **subtract two polynomials,** change the sign of each term in the second polynomial. Then add the polynomials.    **(p. 348)**

Subtract the polynomials.

$(4t^3 - 7t^2 + 4t + 4) - (12t^3 - 8t^2 + 3t + 9)$
$= (4t^3 - 7t^2 + 4t + 4) + (-12t^3 + 8t^2 - 3t - 9)$
$= -8t^3 + t^2 + t - 5$

$f(x) = 2x^2 + 9x - 5$ is an example of a **polynomial function** because $2x^2 + 9x - 5$ is a polynomial and each real number that is substituted for $x$ produces only one value for the expression. Finding $f(3)$ is the same as evaluating $2x^2 + 9x - 5$ when $x = 3$. **(p. 349)**

If  $f(x) = 2x^2 + 9x - 5$, find $f(3)$.

$f(3) = 2(3)^2 + 9(3) - 5$
$f(3) = 2(9) + 27 - 5$
$f(3) = 18 + 27 - 5$
$f(3) = 40$

| Definition/Procedure | Example |
|---|---|

**Graphing Polynomial Equations**

The graph of $y = x^2$ is a **parabola.** **(p. 350)**

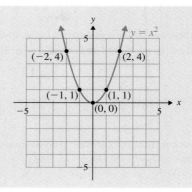

## 6.3 Multiplication of Polynomials

When multiplying a **monomial** and a **polynomial,** use the distributive property. **(p. 354)**

Multiply. $5y^3(-2y^2 + 8y - 3)$
$= (5y^3)(-2y^2) + (5y^3)(8y) + (5y^3)(-3)$
$= -10y^5 + 40y^4 - 15y^3$

To **multiply two polynomials,** multiply each term in the second polynomial by each term in the first polynomial. Then combine like terms. **(p. 355)**

Multiply.
$(5p + 2)(p^2 - 3p + 6)$
$= (5p)(p^2) + (5p)(-3p) + (5p)(6)$
$\quad + (2)(p^2) + (2)(-3p) + (2)(6)$
$= 5p^3 - 15p^2 + 30p + 2p^2 - 6p + 12$
$= 5p^3 - 13p^2 + 24p + 12$

**Multiplying Two Binomials**
We can use FOIL to multiply two binomials. **FOIL** stands for **F**irst, **O**uter, **I**nner, **L**ast. Then add like terms. **(p. 356)**

Use FOIL to multiply $(4a - 5)(a + 3)$.

First        Last
$(4a - 5)(a + 3)$
     Inner
    Outer

$(4a - 5)(a + 3) = 4a^2 + 7a - 15$

**Special Products**

a) $(a + b)(a - b) = a^2 - b^2$

a) Multiply. $(c + 9)(c - 9) = c^2 - 9^2 = c^2 - 81$

b) $(a + b)^2 = a^2 + 2ab + b^2$

b) Expand. $(x + 4)^2 = x^2 + 2(x)(4) + 4^2$
$\qquad\qquad\qquad = x^2 + 8x + 16$

c) $(a - b)^2 = a^2 - 2ab + b^2$     **(pp. 358–359)**

c) Expand. $(6v - 7)^2 = (6v)^2 - 2(6v)(7) + 7^2$
$\qquad\qquad\qquad = 36v^2 - 84v + 49$

| Definition/Procedure | Example |
|---|---|

| Definition/Procedure | Example |
|---|---|
| To **divide a polynomial by a monomial,** divide *each term* in the polynomial by the monomial and simplify.   **(p. 365)** | Divide. $\dfrac{22s^4 + 6s^3 - 7s^2 + 3s - 8}{4s^2}$ $= \dfrac{22s^4}{4s^2} + \dfrac{6s^3}{4s^2} - \dfrac{7s^2}{4s^2} + \dfrac{3s}{4s^2} - \dfrac{8}{4s^2}$ $= \dfrac{11s^2}{2} + \dfrac{3s}{2} - \dfrac{7}{4} + \dfrac{3}{4s} - \dfrac{2}{s^2}$ |
| To **divide a polynomial by another polynomial** containing two or more terms, use *long division*.   **(p. 366)** | Divide. $\dfrac{10w^3 + 2w^2 + 13w + 18}{5w + 6}$ |

$$
\begin{array}{r}
2w^2 - 2w + 5 \\
5w + 6 \overline{)\ 10w^3 + 2w^2 + 13w + 18} \\
\underline{-(10w^3 + 12w^2)} \quad\downarrow \\
-10w^2 + 13w \\
\underline{-(-10w^2 - 12w)}\ \downarrow \\
25w + 18 \\
\underline{-(25w + 30)} \\
-12 \rightarrow \text{Remainder}
\end{array}
$$

$$\frac{10w^3 + 2w^2 + 13w + 18}{5w + 6} = 2w^2 - 2w + 5 - \frac{12}{5w + 6}$$

**(6.1)**

**Evaluate using the rules of exponents.**

1) $\dfrac{2^{11}}{2^6}$

2) $6^{-2}$

3) $\left(\dfrac{2}{5}\right)^{-3}$

4) $-3^0 + 8^0$

**Simplify. Assume all variables represent nonzero real numbers. The answer should not contain negative exponents.**

5) $(p^7)^4$

6) $(5m^3)(-3m^8)$

7) $\dfrac{60t^9}{12t^3}$

8) $(-3a^5)^4$

9) $(-7c)(6c^8)$

10) $\dfrac{4p^{13}}{32p^9}$

11) $\dfrac{k^7}{k^{12}}$

12) $\dfrac{f^{-3}}{f^7}$

13) $(-2r^2s)^3(6r^{-9}s)$

14) $\dfrac{a^4b^{-2}}{a^7b^{-5}}$

15) $\left(\dfrac{2xy^{-8}}{3x^{-2}y^6}\right)^{-2}$

16) $(8p^{-7}q^3)(5p^{-2}q)^2$

17) $\dfrac{m^{-1}n^8}{mn^{14}}$

18) $\dfrac{(4b^{-3}c^4d)^2}{(2b^5cd^{-2})^{-3}}$

**Write expressions for the area and perimeter of each rectangle.**

19) [rectangle with sides $4f$ and $3f$]

20) [rectangle with sides $\frac{7}{3}c$ and $15c$]

**Simplify. Assume that the variables represent nonzero integers. Write the final answer so that the exponents have positive coefficients.**

21) $y^{4a} \cdot y^{3a}$

22) $d^{5n} \cdot d^n$

23) $\dfrac{r^{11x}}{r^{2x}}$

24) $\dfrac{g^{13h}}{g^{5h}}$

**(6.2)**

**Identify each term in the polynomial, the coefficient and degree of each term, and the degree of the polynomial.**

25) $7s^3 - 9s^2 + s + 6$

26) $a^2b^3 + 7ab^2 + 2ab + 9b$

27) Evaluate $2r^2 - 8r - 11$ for $r = -3$.

28) Evaluate $p^3q^2 + 4pq^2 - pq - 2q + 9$ for $p = -1$ and $q = 4$.

29) If $h(x) = 5x^2 - 3x - 6$, find

a) $h(-2)$

b) $h(0)$

30) $f(t) = \dfrac{2}{5}t + 4$. Find $t$ so that $f(t) = \dfrac{4}{5}$.

**Add or subtract as indicated.**

31) $(6c^2 + 2c - 8) - (8c^2 + c - 13)$

32) $(-2m^2 - m + 11) + (6m^2 - 12m + 1)$

33) $\begin{array}{r} 6.7j^3 - 1.4j^2 + \phantom{5.7}j - 5.3 \\ + \ 3.1j^3 + 5.7j^2 + 2.4j + 4.8 \\ \hline \end{array}$

34) $\begin{array}{r} -4.2p^3 + 12.5p^2 - 7.2p + 6.1 \\ - \phantom{-}1.3p^3 - \phantom{1}3.3p^2 + 2.5p + 4.3 \\ \hline \end{array}$

35) $\left(\dfrac{3}{5}k^2 + \dfrac{1}{2}k + 4\right) - \left(\dfrac{1}{10}k^2 + \dfrac{3}{2}k - 2\right)$

36) $\left(\dfrac{2}{7}u^2 - \dfrac{5}{8}u + \dfrac{4}{3}\right) + \left(\dfrac{3}{7}u^2 + \dfrac{3}{8}u - \dfrac{11}{12}\right)$

37) Subtract $4x^2y^2 - 7x^2y + xy + 5$ from $x^2y^2 + 2x^2y - 4xy + 11$.

38) Find the sum of $3c^3d^3 - 7c^2d^2 - c^2d + 8d + 1$ and $14c^3d^3 + 3c^2d^2 - 12cd - 2d - 6$.

39) Find the sum of $6m + 2n - 17$ and $-3m + 2n + 14$.

40) Subtract $-h^4 + 8j^4 - 2$ from $12h^4 - 3j^4 + 19$.

41) Subtract $2x^2 + 3x + 18$ from the sum of $7x - 16$ and $8x^2 - 15x + 6$.

42) Find the sum of $7xy + 2x - 3y - 11$ and $-3xy + 5y + 1$ and subtract it from $-5xy - 9x + y + 4$.

**Find the polynomial that represents the perimeter of each rectangle.**

43) [rectangle with top $d^2 + 6d + 2$ and side $d^2 - 3d + 1$]

44) [rectangle with top $7m - 1$ and side $3m + 5$]

**Graph.**

45) $y = x^2 + 3$

46) $y = (x + 1)^2$

47) $f(x) = -(x - 3)^2$

48) $g(x) = -x^2 + 2$

**(6.3)**
**Multiply.**

49) $3r(8r - 13)$

50) $-5m^2(7m^2 - 4m + 8)$

51) $(4w + 3)(-8w^3 - 2w + 1)$

52) $\left(2t^2 - \dfrac{1}{3}\right)(-9t^2 + 7t - 12)$

53) $(y - 3)(y - 9)$

54) $(f - 5)(f - 8)$

55) $(3n - 4)(2n - 7)$

56) $(3p + 4)(3p + 1)$

57) $-(a - 13)(a + 10)$

58) $-(5d + 2)(6d + 5)$

59) $6pq^2(7p^3q^2 + 11p^2q^2 - pq + 4)$

60) $9x^3y^4(-6x^2y + 2xy^2 + 8x - 1)$

61) $(2x - 9y)(2x + y)$

62) $(7r + 3s)(r - s)$

63) $(x^2 + 5x - 12)(10x^4 - 3x^2 + 6)$

64) $(3m^2 - 4m + 2)(m^2 + m - 5)$

65) $4f^2(2f - 7)(f - 6)$

66) $-3(5u - 11)(u + 4)$

67) $(z + 3)(z + 1)(z + 4)$

68) $(p + 2)(p + 5)(p + 4)$

69) $\left(\dfrac{2}{7}d + 3\right)\left(\dfrac{1}{2}d - 8\right)$

70) $\left(\dfrac{3}{10}t - 6\right)\left(\dfrac{2}{9}t - 5\right)$

**Expand.**

71) $(c + 4)^2$

72) $(x - 12)^2$

73) $(4p - 3)^2$

74) $(9 - 2y)^2$

75) $(x - 3)^3$

76) $(p + 4)^3$

77) $[(m - 3) + n]^2$

**Find the special products.**

78) $(z + 7)(z - 7)$

79) $(p - 13)(p + 13)$

80) $\left(\dfrac{1}{4}n - 5\right)\left(\dfrac{1}{4}n + 5\right)$

81) $\left(\dfrac{9}{2} + \dfrac{5}{6}x\right)\left(\dfrac{9}{2} - \dfrac{5}{6}x\right)$

82) $\left(\dfrac{6}{11} - r^2\right)\left(\dfrac{6}{11} + r^2\right)$

83) $\left(3a - \dfrac{1}{2}b\right)\left(3a + \dfrac{1}{2}b\right)$

84) $-4(2d - 7)^2$

85) $3u(u + 4)^2$

86) $[(2p + 5) + q][(2p + 5) - q]$

87) Write an expression for the a) area and b) perimeter of the rectangle.

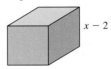

$n - 2$

$2n + 11$

88) Express the volume of the cube as a polynomial.

$x - 2$

**(6.4)**

**Divide.**

89) $\dfrac{12t^6 - 30t^5 - 15t^4}{3t^4}$

90) $\dfrac{42p^4 + 12p^3 - 18p^2 + 6p}{-6p}$

91) $\dfrac{w^2 + 9w + 20}{w + 4}$

92) $\dfrac{a^2 - 2a - 24}{a - 6}$

93) $\dfrac{8r^3 + 22r^2 - r - 15}{2r + 5}$

94) $\dfrac{-36h^3 + 99h^2 + 4h + 1}{12h - 1}$

95) $\dfrac{14t^4 + 28t^3 - 21t^2 + 20t}{14t^3}$

96) $\dfrac{48w^4 - 30w^3 + 24w^2 + 3w}{6w^2}$

97) $(14v + 8v^2 - 3) \div (4v + 9)$

98) $(-8 + 12r^2 - 19r) \div (3r - 1)$

99) $\dfrac{6v^4 - 14v^3 + 25v^2 - 21v + 24}{2v^2 + 3}$

100) $\dfrac{8t^4 + 20t^3 - 30t^2 - 65t + 13}{4t^2 - 13}$

101) $\dfrac{c^3 - 8}{c - 2}$

102) $\dfrac{g^3 + 64}{g + 4}$

103) $\dfrac{-4 + 13k + 18k^3}{3k + 2}$

104) $\dfrac{10 + 12m^3 - 34m^2}{6m + 1}$

105) $(20x^4y^4 - 48x^2y^4 - 12xy^2 + 15x) \div (-12xy^2)$

106) $(3u^4 - 31u^3 - 13u^2 + 76u - 30) \div (u^2 - 11u + 5)$

107) Find the base of the triangle if the area is given by $12a^2 + 3a$ sq units.

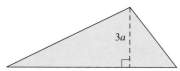

108) Find the length of the rectangle if the area is given by $28x^3 - 51x^2 + 34x - 8$ sq units.

$7x - 4$

**Mixed Exercises**

**Perform the operations and simplify. Assume all variables represent nonzero real numbers. The answer should not contain negative exponents.**

109) $\begin{array}{r} 18c^3 + \phantom{0}7c^2 - 11c + 2 \\ + \phantom{0}2c^3 - 19c^2 \phantom{- 11c} - 1 \\ \hline \end{array}$

110) $\dfrac{15a - 11 + 14a^2}{7a - 3}$

111) $(12 - 7w)(12 + 7w)$

112) $(5p - 9)(2p^2 - 4p - 7)$

113) $5(-2r^7t^9)^3$

114) $(7k^2 + k - 9) - (-4k^2 + 8k - 3)$

115) $(39a^6b^6 + 21a^4b^5 - 5a^3b^4 + a^2b) \div (3a^3b^3)$

116) $\dfrac{(6x^{-4}y^5)^{-2}}{(3xy^{-2})^{-4}}$

117) $(h - 5)^3$

118) $\left(\dfrac{1}{8}m - \dfrac{2}{3}n\right)^2$

119) $\dfrac{-23c + 41 + 2c^3}{c + 4}$

120) $-7d^3(5d^2 + 12d - 8)$

121) $\left(\dfrac{5}{y^4}\right)^{-3}$

122) $(27q^3 + 8) \div (3q + 2)$

123) $(6p^4 + 11p^3 - 20p^2 - 17p + 20) \div (3p^2 + p - 4)$

124) $\left(\dfrac{3b^{-2}c}{a^5}\right)^{-3}\left(\dfrac{4a^{-2}b}{c^4}\right)\left(\dfrac{2ab^3}{c^2}\right)$

**Evaluate.**

1) $\left(\dfrac{3}{4}\right)^{-3}$

2) $\dfrac{5^7}{5^3}$

**Simplify. Assume all variables represent nonzero real numbers. The answer should not contain negative exponents.**

3) $(8p^3)(-4p^6)$

4) $(2t^3)^5$

5) $\dfrac{g^{11}h^{-4}}{g^7h^6}$

6) $\left(\dfrac{54ab^7}{90a^4b^{-2}}\right)^{-2}$

7) Given the polynomial $6n^3 + 6n^2 - n - 7$,

   a) what is the coefficient of $n$?

   b) what is the degree of the polynomial?

8) What is the degree of the polynomial
   $6a^4b^5 + 11a^4b^3 - 2a^3b + 5ab^2 - 3$?

9) Evaluate $-2r^2 + 7s$ when $r = -4$ and $s = 5$.

**Perform the indicated operation(s).**

10) $4h^3(6h^2 - 3h + 1)$

11) $(7a^3b^2 + 9a^2b^2 - 4ab + 8)$
    $\quad + (5a^3b^2 - 12a^2b^2 + ab + 1)$

12) Subtract $6y^2 - 5y + 13$ from $15y^2 - 8y + 6$.

13) $3(-c^3 + 3c - 6) - 4(2c^3 + 3c^2 + 7c - 1)$

14) $(u - 5)(u - 9)$

15) $(4g + 3)(2g + 1)$

16) $\left(v + \dfrac{2}{5}\right)\left(v - \dfrac{2}{5}\right)$

17) $(3x - 7y)(2x + y)$

18) $(5 - 6n)(2n^2 + 3n - 8)$

19) $2y(y + 6)^2$

**Expand.**

20) $(3m - 4)^2$

21) $\left(\dfrac{4}{3}x + y\right)^2$

22) $(t - 2)^3$

23) Graph $y = x^2 - 4$.

**Divide.**

24) $\dfrac{w^2 + 9w + 18}{w + 6}$

25) $\dfrac{24m^6 - 40m^5 + 8m^4 - 6m^3}{8m^4}$

26) $(22p - 50 + 18p^3 - 45p^2) \div (3p - 7)$

27) $\dfrac{y^3 - 27}{y - 3}$

28) $(2r^4 + 3r^3 + 6r^2 + 15r - 20) \div (r^2 + 5)$

29) Write an expression for a) the area and b) the perimeter of the rectangle.

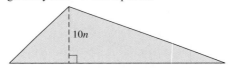

30) Write an expression for the base of the triangle if the area is given by $20n^2 + 15n$ sq units.

1) Given the set of numbers

$$\left\{\frac{3}{8}, -15, 2.1, \sqrt{17}, 41, 0.\overline{52}, 0, 9.32087326...\right\}$$

list the

a) whole numbers

b) integers

c) rational numbers

2) Evaluate $-3^4 + 2 \cdot 9 \div (-3)$.

3) Divide $3\frac{1}{8} \div 1\frac{7}{24}$.

**Simplify. The answers should not contain negative exponents.**

4) $-8(2a^7)^2$

5) $c^{10} \cdot c^7$

6) $\left(\dfrac{4p^{-12}}{p^{-5}}\right)^3$

**Solve.**

7) $-\dfrac{18}{7}m - 9 = 21$

8) $5(u + 3) + 2u = 1 + 7(u - 2)$

9) Solve $5y - 16 \geq 8y - 1$. Write the answer in interval notation.

10) *Write an equation in one variable and solve.* How many milliliters of a 15% alcohol solution and how many milliliters of an 8% alcohol solution must be mixed to obtain 70 mL of a 12% alcohol solution?

11) Find the $x$- and $y$- intercepts of $3x - 8y = 24$ and sketch a graph of the equation.

12) Graph $y = -4$.

13) Write an equation of the line containing the points $(-4, 7)$ and $(2, -11)$. Express the answer in standard form.

14) Write an equation of the line perpendicular to $4x - y = 1$ containing the point $(8, 1)$. Express the answer in slope-intercept form.

15) Solve this system using the elimination method.

$$3x - 4y = -17$$
$$x + 2y = -4$$

16) *Write a system of two equations in two variables and solve.* The length of a rectangle is 1 cm less than three times its width. The perimeter of the rectangle is 94 cm. What are the dimensions of the figure?

**Perform the indicated operations.**

17) $(6q^2 + 7q - 1) - 4(2q^2 - 5q + 8)$ $+ 3(-9q - 4)$

18) $(n - 7)(n + 8)$

19) $(3a - 11)(3a + 11)$

20) $\dfrac{12a^4b^4 - 18a^3b + 60ab + 6b}{12a^3b^2}$

21) $(5p^3 - 14p^2 - 10p + 5) \div (p - 3)$

22) $(4n^2 - 9)(3n^2 + n - 2)$

23) $5c(c - 4)^2$

24) $\dfrac{8z^3 + 1}{2z + 1}$

25) Let $g(x) = -3x^2 + 2x + 9$, find $g(-2)$.

# Factoring Polynomials

## Algebra at Work: Ophthalmology

Mark is an ophthalmologist, a doctor specializing in the treatment of diseases of the eye. He says that he could not do his job without a background in mathematics. While formulas are very important in his work, he says that the thinking skills learned in math courses are the same kinds of thinking skills he uses to treat his patients on a daily basis.

As a physician, Mark follows a very logical, analytical progression to form an accurate diagnosis and treatment plan. He examines a patient, performs tests, and then analyzes the results to form a diagnosis. Next, he must think of different ways to solve the problem and decide on the treatment plan that is best for that patient. He says that the skills he learned in his mathematics courses prepared him for the kind of problem solving he must do every day as an ophthalmologist. Factoring requires the kinds of skills that are so important to Mark in his job—the ability to think through and solve a problem in an analytical and logical manner.

In this chapter, we will learn different techniques for factoring polynomials.

## Section 7.1 The Greatest Common Factor and Factoring by Grouping

### Objectives

1. **Find the GCF of a Group of Monomials**
2. **Factoring vs. Multiplying Polynomials**
3. **Factor Out the Greatest Common Monomial Factor**
4. **Factor Out the Greatest Common Binomial Factor**
5. **Factor by Grouping**

In Section 1.1, we discussed writing a number as the product of factors:

$$18 = 3 \cdot 6$$

$$\downarrow \qquad \downarrow \qquad \downarrow$$

Product  Factor  Factor

To **factor** an integer is to write it as the product of two or more integers. Therefore, 18 can also be factored in other ways:

$$18 = 1 \cdot 18 \qquad 18 = 2 \cdot 9 \qquad 18 = -1 \cdot (-18)$$
$$18 = -2 \cdot (-9) \qquad 18 = -3 \cdot (-6) \qquad 18 = 2 \cdot 3 \cdot 3$$

The last **factorization**, $2 \cdot 3 \cdot 3$ or $2 \cdot 3^2$, is called the **prime factorization** of 18 since all of the factors are prime numbers. (See Section 1.1.) The factors of 18 are 1, 2, 3, 6, 9, 18, $-1$, $-2$, $-3$, $-6$, $-9$, and $-18$.

We can also write the factors as $\pm 1$, $\pm 2$, $\pm 3$, $\pm 6$, $\pm 9$, and $\pm 18$. (Read $\pm 1$ as "plus or minus 1.")

In this chapter, we will learn how to factor polynomials, a skill that is used in many ways throughout algebra.

### 1. Find the GCF of a Group of Monomials

> **Definition**
>
> The **greatest common factor (GCF)** of a group of two or more integers is the *largest* common factor of the numbers in the group.

For example, if we want to find the GCF of 18 and 24, we can list their positive factors.

18: 1, 2, 3, 6, 9, 18
24: 1, 2, 3, 4, 6, 8, 12, 24

The greatest common factor of 18 and 24 is 6. We can also use prime factors.

We begin our study of factoring polynomials by discussing how to find the greatest common factor of a group of monomials.

**Example 1**

Find the greatest common factor of $x^4$ and $x^6$.

**Solution**

We can write $x^4$ as $1 \cdot x^4$, and we can write $x^6$ as $x^4 \cdot x^2$. The largest power of $x$ that is a factor of both $x^4$ and $x^6$ is $x^4$. Therefore, the GCF is $x^4$.

In Example 1, notice that the power of 4 in the GCF is the smallest of the powers when comparing $x^4$ and $x^6$. This will always be true. ■

> **Note**
>
> The exponent on the variable in the GCF will be the *smallest* exponent appearing on the variable in the group of terms.

**You Try 1**

Find the greatest common factor of $y^5$ and $y^8$.

## Example 2

Find the greatest common factor for each group of terms.

a)   $24n^5, 8n^9, 16n^3$          b)   $-15x^{10}y, 25x^6y^8$

c)   $49a^4b^5, 21a^3, 35a^2b^4$

**Solution**

a)   The GCF of the coefficients, 24, 8, and 16, is 8. The smallest exponent on $n$ is 3, so $n^3$ is part of the GCF.

$$\text{The GCF is } 8n^3.$$

b)   The GCF of the coefficients, $-15$ and 25, is 5. The smallest exponent on $x$ is 6, so $x^6$ is part of the GCF. The smallest exponent on $y$ is 1, so $y$ is part of the GCF.

$$\text{The GCF is } 5x^6y.$$

c)   The GCF of the coefficients is 7. The smallest exponent on $a$ is 2, so $a^2$ is part of the GCF. There is no $b$ in the term $21a^3$, so there will be no $b$ in the GCF.

$$\text{The GCF is } 7a^2.$$   ■

## You Try 2

Find the greatest common factor for each group of terms.

a)   $18w^6, 45w^{10}, 27w^5$          b)   $-14hk^3, 18h^4k^2$          c)   $54c^5d^5, 66c^8d^3, 24c^2$

### Factoring Out the Greatest Common Factor

Earlier we said that to **factor an integer** is to write it as the product of two or more integers. To **factor a polynomial** is to write it as a product of two or more polynomials.

Throughout this chapter, we will study different factoring techniques. We will begin by discussing how to factor out the greatest common factor.

## 2. Factoring vs. Multiplying Polynomials

Factoring a polynomial is the opposite of multiplying polynomials. Let's see how factoring and multiplying are related.

## Example 3

a)   Multiply $3y(y + 4)$.          b)   Factor out the GCF from $3y^2 + 12y$.

**Solution**

a)   Use the distributive property to multiply.

$$3y(y + 4) = (3y)y + (3y)(4)$$
$$= 3y^2 + 12y$$

b)   Use the distributive property to factor out the greatest common factor from $3y^2 + 12y$.

First, identify the GCF of $3y^2$ and $12y$. The GCF is $3y$.

Then, rewrite each term as a product of two factors with one factor being $3y$.

$$3y^2 = (3y)(y) \text{ and } 12y = (3y)(4)$$
$$3y^2 + 12y = (3y)(y) + (3y)(4)$$
$$= 3y(y + 4) \qquad \text{Distributive property}$$

When we factor $3y^2 + 2y$, we get $3y(y + 4)$. We can check our result by multiplying.

$$3y(y + 4) = 3y^2 + 12y$$    ∎

---

**Procedure**    Steps for Factoring Out the Greatest Common Factor

1) Identify the GCF of all of the terms of the polynomial.
2) Rewrite each term as the product of the GCF and another factor.
3) Use the distributive property to factor out the GCF from the terms of the polynomial.
4) Check the answer by multiplying the factors. The result should be the original polynomial.

---

### 3. Factor Out the Greatest Common Monomial Factor

**Example 4**

Factor out the greatest common factor.

a)   $28p^5 + 12p^4 + 4p^3$                    b)   $w^8 - 7w^6$

c)   $6a^5b^3 + 30a^5b^2 - 54a^4b^2 - 6a^3b$

**Solution**

a)   Identify the GCF of all of the terms: GCF $= 4p^3$.

$$28p^5 + 12p^4 + 4p^3 = (4p^3)(7p^2) + (4p^3)(3p) + (4p^3)(1)$$    Rewrite each term using the GCF as one of the factors.

$$= 4p^3(7p^2 + 3p + 1)$$    Distributive property

*Check*: $4p^3(7p^2 + 3p + 1) = 28p^5 + 12p^4 + 4p^3$  ✓

b)   The GCF of all of the terms is $w^6$.

$$w^8 - 7w^6 = (w^6)(w^2) - (w^6)(7)$$    Rewrite each term using the GCF as one of the factors.

$$= w^6(w^2 - 7)$$    Distributive property

*Check*: $w^6(w^2 - 7) = w^8 - 7w^6$  ✓

c)   The GCF of all of the terms is $6a^3b$.

$$6a^5b^3 + 30a^5b^2 - 54a^4b^2 - 6a^3b$$

$$= (6a^3b)(a^2b^2) + (6a^3b)(5a^2b) - (6a^3b)(9ab) - (6a^3b)(1)$$    Rewrite using the GCF.

$$= 6a^3b(a^2b^2 + 5a^2b - 9ab - 1)$$    Distributive property

*Check*: $6a^3b(a^2b^2 + 5a^2b - 9ab - 1) = 6a^5b^3 + 30a^5b^2 - 54a^4b^2 - 6a^3b$  ✓  ∎

**You Try 3**

Factor out the greatest common factor.

a)   $3u^6 + 36u^5 + 15u^4$    b)   $z^5 - 9z^2$    c)   $45r^4t^3 + 36r^4t^2 + 18r^3t^2 - 9r^2t$

Sometimes we need to take out a negative factor.

### Example 5

Factor out $-2k$ from $-6k^4 + 10k^3 - 8k^2 + 2k$.

**Solution**

$-6k^4 + 10k^3 - 8k^2 + 2k$

$= (-2k)(3k^3) + (-2k)(-5k^2) + (-2k)(4k) + (-2k)(-1)$    Rewrite using $-2k$ as one of the factors.

$= -2k[3k^3 + (-5k^2) + 4k + (-1)]$    Distributive property

$= -2k(3k^3 - 5k^2 + 4k - 1)$    Rewrite $+(-5k^2)$ as $-5k^2$ and $+(-1)$ as $-1$.

$Check: -2k(3k^3 - 5k^2 + 4k - 1) = -6k^4 + 10k^3 - 8k^2 + 2k$   ✓   ∎

**BE CAREFUL**

When taking out a negative factor, be very careful with the signs!

**You Try 4**

Factor out $-y^2$ from $-y^4 + 10y^3 - 8y^2$.

## 4. Factor Out the Greatest Common Binomial Factor

Until now, all of the GCFs have been monomials. Sometimes, however, the greatest common factor is a *binomial*.

### Example 6

Factor out the greatest common factor.

a)   $a(b + 5) + 8(b + 5)$      b)   $c^2(c + 9) - 2(c + 9)$      c)   $x(y + 3) - (y + 3)$

**Solution**

a)   In the polynomial $\underbrace{a(b + 5)}_{\text{Term}} + \underbrace{8(b + 5)}_{\text{Term}}$, $a(b + 5)$ is a term and $8(b + 5)$ is a term. What do these terms have in common? $b + 5$

The GCF of $a(b + 5)$ and $8(b + 5)$ is $(b + 5)$. Use the distributive property to factor out $b + 5$.

$$a(b + 5) + 8(b + 5) = (b + 5)(a + 8)$$    Distributive property

$Check: (b + 5)(a + 8) = (b + 5)a + (b + 5)8$    Distribute.

$\qquad\qquad\qquad = a(b + 5) + 8(b + 5)$    Commutative property

b)   The GCF of $c^2(c + 9) - 2(c + 9)$ is $c + 9$.

$$c^2(c + 9) - 2(c + 9) = (c + 9)(c^2 - 2)$$    Distributive property

$Check: (c + 9)(c^2 - 2) = (c + 9)c^2 + (c + 9)(-2)$    Distributive property

$\qquad\qquad\qquad = c^2(c + 9) - 2(c + 9)$   ✓   Commutative property

c)   Begin by writing $x(y + 3) - (y + 3)$ as $x(y + 3) - 1(y + 3)$. The GCF is $y + 3$.

$$x(y + 3) - 1(y + 3) = (y + 3)(x - 1)$$    Distributive property

The check is left to the student.      ∎

**BE CAREFUL**

It is important to write $-1$ in front of $(y + 3)$. Otherwise, the following mistake is often made:

$$x(y + 3) - (y + 3) = (y + 3)x \quad \text{This is incorrect!}$$

The correct factor is $x - 1$, *not x.*

**You Try 5**

Factor out the GCF.

a)   $c(d - 8) + 2(d - 8)$        b)   $k(k^2 + 15) - 7(k^2 + 15)$        c)   $u(v + 2) - (v + 2)$

Taking out a binomial factor leads us to our next method of factoring—factoring by grouping.

## 5. Factor by Grouping

When we are asked to factor a polynomial containing four terms, we often try to **factor by grouping**.

**Example 7**

Factor by grouping.

a)   $rt + 7r + 2t + 14$                       b)   $3xz - 4yz + 18x - 24y$

c)   $n^3 + 8n^2 - 5n - 40$

**Solution**

a)   Begin by grouping terms together so that each group has a common factor.

$$\underbrace{rt + 7r} + \underbrace{2t + 14}$$

Factor out $r$ to        $= r(t + 7) + 2(t + 7)$        Factor out 2 to get $2(t + 7)$.
get $r(t + 7)$.            $= (t + 7)(r + 2)$              Factor out $(t + 7)$.

Check:  $(t + 7)(r + 2) = rt + 7r + 2t + 14$  ✓

b)   Group terms together so that each group has a common factor.

$$\underbrace{3xz - 4yz} + \underbrace{18x - 24y}$$

Factor out $z$          $= z(3x - 4y) + 6(3x - 4y)$        Factor out 6 to get $6(3x - 4y)$.
to get $z(3x - 4y)$.    $= (3x - 4y)(z + 6)$                Factor out $(3x - 4y)$.

Check:  $(3x - 4y)(z + 6) = 3xz - 4yz + 18x - 24y$  ✓

c)   Group terms together so that each group has a common factor.

$$\underbrace{n^3 + 8n^2} \underbrace{-5n - 40}$$

Factor out $n^2$        $= n^2(n + 8) - 5(n + 8)$        Factor out $-5$ to get $-5(n + 8)$.
to get $n^2(n + 8)$.    $= (n + 8)(n^2 - 5)$              Factor out $(n + 8)$.

We must factor out –5, *not* 5, from the second group so that the binomial factors for both groups are the same! [If we had factored out 5, then the factorization of the second group would have been $5(-n-8)$.]

*Check*: $(n + 8)(n^2 - 5) = n^3 + 8n^2 - 5n - 40$ ✓   ■

**You Try 6**

Factor by grouping.

a)   $xy + 4x + 10y + 40$        b)   $5pr - 8qr + 10p - 16q$        c)   $w^3 + 9w^2 - 6w - 54$

Sometimes we have to rearrange the terms before we can factor.

**Example 8**

Factor completely. $12c^2 - 2d + 3c - 8cd$

### Solution

Group terms together so that each group has a common factor.

$$12c^2 - 2d + 3c - 8cd$$

Factor out 2 to get $2(6c^2 - d)$.   $= 2(6c^2 - d) + c(3 - 8d)$   Factor out $c$ to get $c(3 - 8d)$.

These groups do not have common factors! Let's rearrange the terms in the original polynomial and group the terms differently.

$$12c^2 + 3c \quad - \quad 8cd - 2d$$

Factor out $3c$ to get $3c(4c + 1)$.   $= 3c(4c + 1) - 2d(4c + 1)$   Factor out $-2d$ to get $-2d(4c + 1)$.

$= (4c + 1)(3c - 2d)$   Factor out $(4c + 1)$.

*Check*: $(4c + 1)(3c - 2d) = 12c^2 - 2d + 3c - 8cd$ ✓   ■

**Note**

Often, there is more than one way that the terms can be rearranged so that the polynomial can be factored by grouping.

**You Try 7**

Factor completely. $3k^2 - 48m + 8k - 18km$

Often, we have to combine the two factoring techniques we have learned here. That is, we begin by factoring out the GCF and then we factor by grouping. Let's summarize how to factor a polynomial by grouping and then look at another example.

---

**Procedure**    Steps for Factoring by Grouping

1) Before trying to factor by grouping, look at each term in the polynomial and ask yourself, "*Can I factor out a GCF first?*" If so, factor out the GCF from all of the terms.

2) Make two groups of two terms so that each group has a common factor.

3) Take out the common factor in each group of terms.

4) Factor out the common binomial factor using the distributive property.

5) Check the answer by multiplying the factors.

---

### Example 9

Factor completely.  $7h^4 + 7h^3 - 42h^2 - 42h$

#### Solution

Notice that this polynomial has four terms. This is a clue for us to try factoring by grouping. *However*, look at the polynomial carefully and ask yourself, "*Can I factor out a GCF?*" Yes! *Therefore, the first step in factoring this polynomial is to factor out $7h$.*

$$7h^4 + 7h^3 - 42h^2 - 42h = 7h(h^3 + h^2 - 6h - 6) \qquad \text{Factor out the GCF, } 7h.$$

The polynomial in parentheses has 4 terms. Try to factor it by grouping.

$$7h(\underbrace{h^3 + h^2}\,\underbrace{- 6h - 6})$$

$$= 7h[h^2(h + 1) - 6(h + 1)] \qquad \text{Take out the common factor in each group.}$$
$$= 7h(h + 1)(h^2 - 6) \qquad \text{Factor out } (h + 1) \text{ using the distributive property.}$$

*Check*:  $7h(h + 1)(h^2 - 6) = 7h(h^3 + h^2 - 6h - 6)$
$$= 7h^4 + 7h^3 - 42h^2 - 42h \; \checkmark$$

---

 **You Try 8**

Factor completely.  $12t^3 + 12t^2 - 3t^2u - 3tu$

---

Remember, seeing a polynomial with four terms is a clue to try factoring by grouping. Not all polynomials will factor this way, however. We will learn other techniques later, and some polynomials must be factored using methods learned in later courses.

---

### Answers to You Try Exercises

1)  $y^5$    2)  a) $9w^5$  b) $2hk^2$  c) $6c^2$    3)  a) $3u^4(u^2 + 12u + 5)$  b) $z^2(z^3 - 9)$
c) $9r^2t(5r^2t^2 + 4r^2t + 2rt - 1)$    4)  $-y^2(y^2 - 10y + 8)$    5)  a) $(d - 8)(c + 2)$
b) $(k^2 + 15)(k - 7)$  c) $(v + 2)(u - 1)$    6)  a) $(y + 4)(x + 10)$  b) $(5p - 8q)(r + 2)$
c) $(w + 9)(w^2 - 6)$    7)  $(3k + 8)(k - 6m)$    8)  $3t(t + 1)(4t - u)$

## 7.1 Exercises

**Objective 1: Find the GCF of a Group of Monomials**

Find the greatest common factor of each group of terms.

1) $28, 21c$

2) $9t, 36$

3) $18p^3, 12p^2$

4) $32z^5, 56z^3$

5) $12n^6, 28n^{10}, 36n^7$

6) $63b^4, 45b^7, 27b$

7) $35a^3b^2, 15a^2b$

8) $10x^5y^4, 2x^4y^4$

9) $21r^3s^6, 63r^3s^2, -42r^4s^5$

10) $-60p^2q^2, 36pq^5, 96p^3q^3$

11) $a^2b^2, 3ab^2, 6a^2b$

12) $n^3m^4, -n^3m^4, -n^4$

13) $c(k - 9), 5(k - 9)$

14) $a^2(h + 8), b^2(h + 8)$

15) Explain how to find the GCF of a group of terms.

16) What does it mean to factor a polynomial?

**Objective 2: Factoring vs. Multiplying Polynomials**

Determine whether each expression is written in factored form.

17) $5p(p + 9)$

18) $8h^2 - 24h$

19) $18w^2 + 30w$

20) $-3z^2(2z + 7)$

21) $a^2b^2(-4ab)$

22) $c^3d - (2c + d)$

**Objective 3: Factor Out the Greatest Common Monomial Factor**

Factor out the greatest common factor. Be sure to check your answer.

23) $2w + 10$

24) $3y + 18$

25) $18z^2 - 9$

26) $14h - 12h^2$

27) $100m^3 - 30m$

28) $t^5 - t^4$

29) $r^9 + r^2$

30) $\frac{1}{2}a^2 + \frac{3}{2}a$

31) $\frac{1}{5}y^2 + \frac{4}{5}y$

32) $9a^3 + 2b^2$

33) $s^7 - 4t^3$

34) $14u^7 + 63u^6 - 42u^5$

(VIDEO) 35) $10n^5 - 5n^4 + 40n^3$

36) $3d^8 - 33d^7 - 24d^6 + 3d^5$

37) $40p^6 + 40p^5 - 8p^4 + 8p^3$

38) $44m^3n^3 - 24mn^4$

39) $63a^3b^3 - 36a^3b^2 + 9a^2b$

40) $8p^4q^3 + 8p^3q^3 - 72p^2q^2$

41) Factor out $-6$ from $-30n - 42$.

42) Factor out $-c$ from $-9c^3 + 2c^2 - c$.

43) Factor out $-4w^3$ from $-12w^5 - 16w^3$.

44) Factor out $-m$ from $-6m^3 - 3m^2 + m$.

45) Factor out $-1$ from $-k + 3$.

46) Factor out $-1$ from $-p - 10$.

**Objective 4: Factor Out the Greatest Common Binomial Factor**

Factor out the common binomial factor.

47) $u(t - 5) + 6(t - 5)$

48) $c(b + 9) + 2(b + 9)$

49) $y(6x + 1) - z(6x + 1)$

50) $s(4r - 3) - t(4r - 3)$

51) $p(q + 12) + (q + 12)$

52) $8x(y - 2) + (y - 2)$

53) $5h^2(9k + 8) - (9k + 8)$

54) $3a(4b + 1) - (4b + 1)$

**Objective 5: Factor by Grouping**

Factor by grouping.

55) $ab + 2a + 7b + 14$

56) $cd + 8c - 5d - 40$

57) $3rt + 4r - 27t - 36$

58) $5pq + 15p - 6q - 18$

59) $8b^2 + 20bc + 2bc^2 + 5c^3$

60) $4a^3 - 12ab + a^2b - 3b^2$

(VIDEO) 61) $fg - 7f + 4g - 28$

62) $xy - 8y - 7x + 56$

63) $st - 10s - 6t + 60$

64) $cd + 3c - 11d - 33$

(VIDEO) 65) $5tu + 6t - 5u - 6$

66) $qr + 3q - r - 3$

67) $36g^4 + 3gh - 96g^3h - 8h^2$

68) $40j^3 + 72jk - 55j^2k - 99k^2$

69) Explain, in your own words, how to factor by grouping.

70) What should be the first step in factoring $6ab + 24a + 18b + 54$?

Factor completely. You may need to begin by factoring out the GCF first or by rearranging terms.

**Fill It In**

Fill in the blanks with either the missing mathematical step or the reason for the given step.

71) $4xy + 12x + 20y + 60$

$4xy + 12x + 20y + 60$

$=$ _____        Factor out the GCF.

$= 4[x(y + 3) + 5(y + 3)]$   _____

_____

_____

$=$ _____        Take out the binomial factor.

72) $2m^2n - 4m^2 - 18mn + 36m$

$2m^2n - 4m^2 - 18mn + 36m$

$=$ _____        Factor out the GCF.

$= 2m[m(n - 2) - 9(n - 2)]$  _____

_____

_____

$=$ _____        Take out the binomial factor.

73) $3cd + 6c + 21d + 42$

74) $5xy + 15x - 5y - 15$

75) $2p^2q - 10p^2 - 8pq + 40p$

76) $3uv^2 - 24uv + 3v^2 - 24v$

77) $10st + 5s - 12t - 6$

78) $8pq + 12p + 10q + 15$

79) $3a^3 - 21a^2b - 2ab + 14b^2$

80) $8c^3 + 32c^2d + cd + 4d^2$

81) $8u^2v^2 + 16u^2v + 10uv^2 + 20uv$

82) $10x^2y^2 - 5x^2y - 60xy^2 + 30xy$

**Mixed Exercises: Objectives 1–5**

Factor completely.

83) $3mn + 21m + 10n + 70$

84) $4yz + 7z - 20y - 35$

85) $16b - 24$

86) $2yz^3 + 14yz^2 + 3z^3 + 21z^2$

87) $cd + 6c - 4d - 24$

88) $5x^3 - 30x^2y^2 + xy - 6y^3$

89) $6a^4b + 12a^4 - 8a^3b - 16a^3$

90) $6x^2 + 48x^3$

91) $7cd + 12 + 28c + 3d$

92) $2uv + 12u - 7v - 42$

93) $dg - d + g - 1$

94) $2ab - 10a - 12b + 60$

95) $x^4y^2 + 12x^3y^3$

96) $8u^2 - 16uv^2 + 3uv - 6v^3$

97) $4mn + 8m + 12n + 24$

98) $5c^2 - 20$

99) Factor out $-2$ from $-6p^2 - 20p + 2$.

100) Factor out $-5g$ from $-5g^3 + 50g^2 - 25g$.

## Section 7.2  Factoring Trinomials of the Form $x^2 + bx + c$

### Objectives

1. **Practice Arithmetic Skills Needed for Factoring Trinomials**
2. **Factor a Trinomial of the Form $x^2 + bx + c$**
3. **More on Factoring a Trinomial of the Form $x^2 + bx + c$**
4. **Factor a Trinomial Containing Two Variables**

One of the factoring problems encountered most often in algebra is the factoring of trinomials. In this section, we will discuss how to factor a trinomial of the form $x^2 + bx + c$; notice that the coefficient of the squared term is 1.

We will begin with arithmetic skills we need to be able to factor a trinomial of the form $x^2 + bx + c$.

## 1. Practice Arithmetic Skills Needed for Factoring Trinomials

| Example 1 |

Find two integers whose

a)   product is 15 and sum is 8.

b)   product is 24 and sum is $-10$.

c)   product is $-28$ and sum is 3.

**Solution**

a)   If the product of two numbers is *positive* 15 and the sum of the numbers is *positive* 8, *then the two numbers will be positive.* (The product of two positive numbers is positive, and their sum is positive as well.)

First, list the pairs of *positive* integers whose product is 15—the *factors* of 15. Then, find the *sum* of those factors.

| Factors of 15 | Sum of the Factors |
|:---:|:---:|
| $1 \cdot 15$ | $1 + 15 = 16$ |
| $3 \cdot 5$ | $3 + 5 = 8$ |

The product of 3 and 5 is 15, and their sum is 8.

b)   If the product of two numbers is *positive* 24 and the sum of those numbers is *negative* 10, *then the two numbers will be negative.* (The product of two negative numbers is positive, while the sum of two negative numbers is negative.)

First, list the pairs of negative numbers that are factors of 24. Then, find the sum of those factors. You can stop making your list when you find the pair that works.

| Factors of 24 | Sum of the Factors |
|:---:|:---:|
| $-1 \cdot (-24)$ | $-1 + (-24) = -25$ |
| $-2 \cdot (-12)$ | $-2 + (-12) = -14$ |
| $-3 \cdot (-8)$ | $-3 + (-8) = -11$ |
| $-4 \cdot (-6)$ | $-4 + (-6) = -10$ |

The product of $-4$ and $-6$ is 24, and their sum is $-10$.

c)   If two numbers have a product of *negative* 28 and their sum is *positive* 3, *one number must be positive and one number must be negative.* (The product of a positive number and a negative number is negative, while the sum of the numbers can be either positive *or* negative.)

First, list pairs of factors of $-28$. Then, find the sum of those factors.

| Factors of $-28$ | Sum of the Factors |
|:---:|:---:|
| $-1 \cdot 28$ | $-1 + 28 = 27$ |
| $1 \cdot (-28)$ | $1 + (-28) = -27$ |
| $-4 \cdot 7$ | $-4 + 7 = 3$ |

The product of $-4$ and 7 is $-28$ and their sum is 3.                                      ∎

**You Try 1**

Find two integers whose

a)   product is 21 and sum is 10.

b)   product is $-18$ and sum is $-3$.

c)   product is 20 and sum is $-12$.

**Note**

You should try to get to the point where you can come up with the correct numbers *in your head* without making a list.

## 2. Factor a Trinomial of the Form $x^2 + bx + c$

In Section 7.1, we said that the process of factoring is the opposite of multiplying. Let's see how this will help us understand how to factor a trinomial of the form $x^2 + bx + c$.

*Multiply* $(x + 4)(x + 5)$ using FOIL.

$$\begin{aligned}
(x + 4)(x + 5) &= x^2 + 5x + 4x + 4 \cdot 5 &&\text{Multiply using FOIL.}\\
&= x^2 + (5 + 4)x + 20 &&\text{Use the distributive property}\\
&= x^2 + 9x + 20 &&\text{and multiply } 4 \cdot 5.
\end{aligned}$$

$(x + 4)(x + 5) = x^2 + 9x + 20$ ⟵——— The *product* of 4 and 5 is 20.

↑
The *sum*
of 4 and 5
is 9.

So, if we were asked to *factor* $x^2 + 9x + 20$, we need to think of two integers whose *product* is 20 and whose *sum* is 9. Those numbers are 4 and 5. The *factored form* of $x^2 + 9x + 20$ is $(x + 4)(x + 5)$.

---

**Procedure    Factoring a Polynomial of the Form $x^2 + bx + c$**

To factor a polynomial of the form $x^2 + bx + c$, find two integers $m$ and $n$ whose product is $c$ and whose sum is $b$. Then, $x^2 + bx + c = (x + m)(x + n)$.

1)   If $b$ and $c$ are positive, then both $m$ and $n$ must be positive.

2)   If $c$ is positive and $b$ is negative, then both $m$ and $n$ must be negative.

3)   If $c$ is negative, then one integer, $m$, must be positive and the other integer, $n$, must be negative.

You can check the answer by multiplying the binomials. The result should be the original polynomial.

---

**Example 2**

Factor, if possible.

a)   $x^2 + 7x + 12$           b)   $p^2 - 9p + 14$           c)   $w^2 + w - 30$

d)   $a^2 - 3a - 54$           e)   $c^2 - 6c + 9$           f)   $y^2 + 11y + 35$

### Solution

a)   To factor $x^2 + 7x + 12$, we must find two integers whose *product* is 12 and whose *sum* is 7. Both integers will be positive.

| Factors of 12 | Sum of the Factors |
|:---:|:---:|
| $1 \cdot 12$ | $1 + 12 = 13$ |
| $2 \cdot 6$ | $2 + 6 = 8$ |
| $3 \cdot 4$ | $3 + 4 = 7$ |

The numbers are 3 and 4:   $x^2 + 7x + 12 = (x + 3)(x + 4)$

*Check*:   $(x + 3)(x + 4) = x^2 + 4x + 3x + 12 = x^2 + 7x + 12$   ✓

**Note**

The order in which the factors are written does not matter. In this example,
$(x + 3)(x + 4) = (x + 4)(x + 3)$.

b)   To factor $p^2 - 9p + 14$, find two integers whose *product* is 14 and whose *sum* is $-9$. Since 14 is positive and the coefficient of $p$ is a negative number, $-9$, both integers will be negative.

| Factors of 14 | Sum of the Factors |
|:---:|:---:|
| $-1 \cdot (-14)$ | $-1 + (-14) = -15$ |
| $-2 \cdot (-7)$ | $-2 + (-7) = -9$ |

The numbers are $-2$ and $-7$:   $p^2 - 9p + 14 = (p - 2)(p - 7)$.

*Check*:   $(p - 2)(p - 7) = p^2 - 7p - 2p + 14 = p^2 - 9p + 14$   ✓

c)   $w^2 + w - 30$

The coefficient of $w$ is 1, so we can think of this trinomial as $w^2 + 1w - 30$.

Find two integers whose *product* is $-30$ and whose *sum* is 1. Since the last term in the trinomial is negative, one of the integers must be positive and the other must be negative.

Try to find these integers mentally. Two numbers with a product of *positive* 30 are 5 and 6. We need a product of $-30$, so either the 5 is negative or the 6 is negative.

| Factors of $-30$ | Sum of the Factors |
|:---:|:---:|
| $-5 \cdot 6$ | $-5 + 6 = 1$ |

The numbers are $-5$ and 6:   $w^2 + w - 30 = (w - 5)(w + 6)$.

*Check*:   $(w - 5)(w + 6) = w^2 + 6w - 5w - 30 = w^2 + w - 30$   ✓

d) To factor $a^2 - 3a - 54$, find two integers whose *product* is $-54$ and whose *sum* is $-3$. Since the last term in the trinomial is negative, one of the integers must be positive and the other must be negative.

Find the integers mentally. First, think about two integers whose product is *positive* 54: 1 and 54, 2 and 27, 3 and 18, 6 and 9. One number must be positive and the other negative, however, to get our product of $-54$, and they must add up to $-3$.

| Factors of $-54$ | Sum of the Factors |
|---|---|
| $-6 \cdot 9$ | $-6 + 9 = 3$ |
| $6 \cdot (-9)$ | $6 + (-9) = -3$ |

The numbers are 6 and $-9$:   $a^2 - 3a - 54 = (a + 6)(a - 9)$.

The check is left to the student.

e) To factor $c^2 - 6c + 9$, notice that the *product*, 9, is positive and the *sum*, $-6$, is negative, so both integers must be negative. The numbers that multiply to 9 and add to $-6$ are the same number, $-3$ and $-3$: $(-3) \cdot (-3) = 9$ and $-3 + (-3) = -6$.

$$\text{So } c^2 - 6c + 9 = (c - 3)(c - 3) \text{ or } (c - 3)^2.$$

Either form of the factorization is correct.

f) To factor $y^2 + 11y + 35$, find two integers whose *product* is 35 and whose *sum* is 11. We are looking for two positive numbers.

| Factors of 35 | Sum of the Factors |
|---|---|
| $1 \cdot 35$ | $1 + 35 = 36$ |
| $5 \cdot 7$ | $5 + 7 = 12$ |

There are no such factors! Therefore, $y^2 + 11y + 35$ does not factor using the methods we have learned here. We say that it is **prime**. ■

**Note**

We say that trinomials like $y^2 + 11y + 35$ are **prime** if they cannot be factored using the method presented here.

In later mathematics courses you may learn how to factor such polynomials using other methods so that they are not considered prime.

**You Try 2**

Factor, if possible.

a)  $m^2 + 11m + 28$

b)  $c^2 - 16c + 48$

c)  $a^2 - 5a - 21$

d)  $r^2 - 4r - 45$

e)  $r^2 + 5r - 24$

f)  $h^2 + 12h + 36$

### 3. More on Factoring a Trinomial of the Form $x^2 + bx + c$

Sometimes it is necessary to factor out the GCF before applying this method for factoring trinomials.

**Note**

From this point on, the *first* step in factoring *any* polynomial should be to ask yourself, *"Can I factor out a greatest common factor?"*

Since some polynomials can be factored more than once, after performing one factorization, ask yourself, *"Can I factor again?"* If so, factor again. If not, you know that the polynomial has been completely factored.

**Example 3**

Factor completely. $4n^3 - 12n^2 - 40n$

**Solution**

Ask yourself, *"Can I factor out a GCF?"* Yes. The GCF is $4n$.

$$4n^3 - 12n^2 - 40n = 4n(n^2 - 3n - 10)$$

Look at the trinomial and ask yourself, *"Can I factor again?"* Yes. The integers whose product is $-10$ and whose sum is $-3$ are $-5$ and $2$. Therefore,

$$4n^3 - 12n^2 - 40n = 4n(n^2 - 3n - 10)$$
$$= 4n(n - 5)(n + 2)$$

We cannot factor again.

$$\begin{aligned} \textit{Check: } \quad 4n(n - 5)(n + 2) &= 4n(n^2 + 2n - 5n - 10) \\ &= 4n(n^2 - 3n - 10) \\ &= 4n^3 - 12n^2 - 40n \quad \checkmark \end{aligned}$$

**You Try 3**

Factor completely.

a)  $7p^4 + 42p^3 + 56p^2$            b)  $3a^2b - 33ab + 90b$

### 4. Factor a Trinomial Containing Two Variables

If a trinomial contains two variables and we cannot take out a GCF, the trinomial may still be factored according to the method outlined in this section.

**Example 4**

Factor completely. $x^2 + 12xy + 32y^2$

**Solution**

Ask yourself, *"Can I factor out a GCF?"* No. Notice that the first term is $x^2$. Let's rewrite the trinomial as

$$x^2 + 12yx + 32y^2$$

so that we can think of $12y$ as the coefficient of $x$. Find two expressions whose product is $32y^2$ and whose sum is $12y$. They are $4y$ and $8y$ since $4y \cdot 8y = 32y^2$ and $4y + 8y = 12y$.

$$x^2 + 12xy + 32y^2 = (x + 4y)(x + 8y)$$

We cannot factor $(x + 4y)(x + 8y)$ any more, so this is the complete factorization. The check is left to the student.  ■

**You Try 4**

Factor completely.

a)  $m^2 + 10mn + 16n^2$        b)  $5a^3 + 40a^2b - 45ab^2$

---

**Answers to You Try Exercises**

1) a) $3, 7$  b) $-6, 3$  c) $-2, -10$     2) a) $(m + 4)(m + 7)$  b) $(c - 12)(c - 4)$  c) prime
d) $(r - 9)(r + 5)$  e) $(r + 8)(r - 3)$  f) $(h + 6)(h + 6)$ or $(h + 6)^2$     3) a) $7p^2(p + 4)(p + 2)$
b) $3b(a - 5)(a - 6)$     4) a) $(m + 2n)(m + 8n)$  b) $5a(a - b)(a + 9b)$

---

## 7.2 Exercises

**Objective 1: Practice Arithmetic Skills Needed for Factoring Trinomials**

1) Find two integers whose

| | PRODUCT IS | and whose SUM IS | ANSWER |
|---|---|---|---|
| a) | 10 | 7 | |
| b) | $-56$ | $-1$ | |
| c) | $-5$ | 4 | |
| d) | 36 | $-13$ | |

2) Find two integers whose

| | PRODUCT IS | and whose SUM IS | ANSWER |
|---|---|---|---|
| a) | 42 | $-13$ | |
| b) | $-14$ | 13 | |
| c) | 54 | 15 | |
| d) | $-21$ | $-4$ | |

**Objective 2: Factor a Trinomial of the Form $x^2 + bx + c$**

3) If $x^2 + bx + c$ factors to $(x + m)(x + n)$ and if $c$ is positive and $b$ is negative, what do you know about the signs of $m$ and $n$?

4) If $x^2 + bx + c$ factors to $(x + m)(x + n)$ and if $b$ and $c$ are positive, what do you know about the signs of $m$ and $n$?

5) When asked to factor a polynomial, what is the first question you should ask yourself?

6) What does it mean to say that a polynomial is prime?

7) After factoring a polynomial, what should you ask yourself to be sure that the polynomial is completely factored?

8) How do you check the factorization of a polynomial?

Complete the factorization.

9) $n^2 + 7n + 10 = (n + 5)(\quad)$

10) $p^2 + 11p + 28 = (p + 4)(\quad)$

11) $c^2 - 16c + 60 = (c - 6)(\quad)$

12) $t^2 - 12t + 27 = (t - 9)(\quad)$

13) $x^2 + x - 12 = (x - 3)(\quad)$

14) $r^2 - 8r - 9 = (r + 1)(\quad)$

Factor completely, if possible. Check your answer.

VIDEO 15) $g^2 + 8g + 12$        16) $p^2 + 9p + 14$

17) $y^2 + 10y + 16$        18) $a^2 + 11a + 30$

19) $w^2 - 17w + 72$        20) $d^2 - 14d + 33$

21) $b^2 - 3b - 4$        22) $t^2 + 2t - 48$

23) $z^2 + 6z - 11$        24) $x^2 - 7x - 15$

VIDEO 25) $c^2 - 13c + 36$      26) $h^2 - 13h + 12$

27) $m^2 + 4m - 60$      28) $v^2 - 4v - 45$

29) $r^2 - 4r - 96$      30) $a^2 - 21a + 110$

31) $q^2 + 12q + 42$      32) $d^2 - 15d + 32$

33) $x^2 + 16x + 64$      34) $c^2 - 10c + 25$

35) $n^2 - 2n + 1$      36) $w^2 + 20w + 100$

37) $24 + 14d + d^2$      38) $10 + 7k + k^2$

39) $-56 + 12a + a^2$      40) $63 + 21w + w^2$

### Objective 3: More on Factoring a Trinomial of the Form $x^2 + bx + c$

Factor completely, if possible. Check your answer.

41) $2k^2 - 22k + 48$      42) $6v^2 + 54v + 120$

43) $50h + 35h^2 + 5h^3$      44) $3d^3 - 33d^2 - 36d$

45) $r^4 + r^3 - 132r^2$      46) $2n^4 - 40n^3 + 200n^2$

47) $7q^3 - 49q^2 - 42q$      48) $8b^4 + 24b^3 + 16b^2$

49) $3z^4 + 24z^3 + 48z^2$      50) $-36w + 6w^2 + 2w^3$

51) $xy^3 - 2xy^2 - 63xy$      52) $2c^3d - 14c^2d - 24cd$

Factor completely by first taking out $-1$ and then by factoring the trinomial, if possible. Check your answer.

53) $-m^2 - 12m - 35$      54) $-x^2 - 15x - 36$

55) $-c^2 - 3c + 28$      56) $-t^2 + 2t + 48$

57) $-z^2 + 13z - 30$      58) $-n^2 + 16n - 55$

59) $-p^2 + p + 56$      60) $-w^2 - 2w + 3$

### Objective 4: Factor a Trinomial Containing Two Variables

Factor completely. Check your answer.

61) $x^2 + 7xy + 12y^2$      62) $a^2 + 11ab + 18b^2$

63) $c^2 - 7cd - 8d^2$      64) $p^2 + 6pq - 72q^2$

65) $u^2 - 14uv + 45v^2$      66) $h^2 - 8hk + 7k^2$

VIDEO 67) $m^2 + 4mn - 21n^2$      68) $a^2 - 6ab - 40b^2$

69) $a^2 + 24ab + 144b^2$      70) $g^2 + 6gh + 5h^2$

Determine whether each polynomial is factored completely. If it is not, explain why and factor it completely.

71) $3x^2 + 21x + 30 = (3x + 6)(x + 5)$

72) $6a^2 + 24a - 72 = 6(a + 6)(a - 2)$

73) $n^4 - 3n^3 - 108n^2 = n^2(n - 12)(n + 9)$

74) $9y^3 - 45y^2 + 54y = (y - 2)(9y^2 - 27y)$

### Mixed Exercises: Objectives 2–4

Factor completely. Begin by asking yourself, *"Can I factor out a GCF?"*

75) $2x^2 + 16x + 30$

76) $3c^2 + 21c + 18$

77) $n^2 - 6n + 8$

78) $a^2 + a - 6$

79) $m^2 + 7mn - 44n^2$

80) $a^2 + 10ab + 24b^2$

81) $h^2 - 10h + 32$

82) $z^2 + 9z + 36$

VIDEO 83) $4q^3 - 28q^2 + 48q$

84) $3w^3 - 9w^2 - 120w$

85) $-k^2 - 18k - 81$

86) $-y^2 + 8y - 16$

87) $4h^5 + 32h^4 + 28h^3$

88) $3r^4 - 6r^3 - 45r^2$

89) $k^2 + 21k + 108$

90) $j^2 - 14j - 15$

91) $p^3q - 17p^2q^2 + 70pq^3$

92) $u^3v^2 - 2u^2v^3 - 15uv^4$

93) $a^2 + 9ab + 24b^2$

94) $m^2 - 8mn - 35n^2$

95) $x^2 - 13xy + 12y^2$

96) $p^2 - 3pq - 40q^2$

97) $5v^5 + 55v^4 - 45v^3$

98) $6t^4 + 42t^3 + 48t^2$

99) $6x^3y^2 - 48x^2y^2 - 54xy^2$

100) $2c^2d^4 - 18c^2d^3 + 28c^2d^2$

101) $36 - 13z + z^2$

102) $121 + 22w + w^2$

103) $a^2b^2 + 13ab + 42$

104) $h^2k^2 + 8hk - 20$

105) $(x + y)z^2 + 7(x + y)z - 30(x + y)$

106) $(m + n)k^2 + 17(m + n)k + 66(m + n)$

107) $(a - b)c^2 - 11(a - b)c + 28(a - b)$

108) $(r - t)u^2 - 4(r - t)u - 45(r - t)$

109) $(p + q)r^2 + 24r(p + q) + 144(p + q)$

110) $(a + b)d^2 - 8(a + b)d + 16(a + b)$

## Section 7.3 Factoring Trinomials of the Form $ax^2 + bx + c$ ($a \neq 1$)

**Objectives**

1. **Factor $ax^2 + bx + c$ ($a \neq 1$) by Grouping**
2. **Factor $ax^2 + bx + c$ ($a \neq 1$) by Trial and Error**

In the previous section, we learned that we could factor $2x^2 + 10x + 8$ by first taking out the GCF of 2 and then factoring the trinomial.

$$2x^2 + 10x + 8 = 2(x^2 + 5x + 4) = 2(x + 4)(x + 1)$$

In this section, we will learn how to factor a trinomial like $2x^2 + 11x + 15$ where we *cannot* factor out the leading coefficient of 2.

### 1. Factor $ax^2 + bx + c$ ($a \neq 1$) by Grouping

To factor $2x^2 + 11x + 15$, first find the product of 2 and 15. Then, find two integers *(Sum is 11. Product: 2 · 15 = 30)* whose *product* is 30 and whose *sum* is 11. The numbers are 6 and 5. Rewrite the middle term, $11x$, as $6x + 5x$, then factor by grouping.

$$2x^2 + 11x + 15 = 2x^2 + 6x + 5x + 15 \qquad \text{Take out the common factor from each group.}$$
$$= 2x(x + 3) + 5(x + 3) \qquad \text{Factor out } (x + 3).$$
$$= (x + 3)(2x + 5)$$

Check: $(x + 3)(2x + 5) = 2x^2 + 5x + 6x + 15 = 2x^2 + 11x + 15$ ✓

---

### Example 1

Factor completely.

a) $8k^2 + 14k + 3$    b) $6c^2 - 17c + 12$    c) $7x^2 - 34xy - 5y^2$

**Solution**

a) Since we cannot factor out a GCF (the GCF = 1), we begin with a new method.

*(Sum is 14.)*
$$8k^2 + 14k + 3$$
*(Product: 8 · 3 = 24)*

Think of two integers whose *product* is 24 and whose *sum* is 14. The integers are 2 and 12. Rewrite the middle term, $14k$, as $2k + 12k$. Factor by grouping.

$$8k^2 + 14k + 3 = 8k^2 + 2k + 12k + 3$$
$$= 2k(4k + 1) + 3(4k + 1) \qquad \text{Take out the common factor from each group.}$$
$$= (4k + 1)(2k + 3) \qquad \text{Factor out } (4k + 1).$$

Check by multiplying: $(4k + 1)(2k + 3) = 8k^2 + 14k + 3$ ✓

b)
*(Sum is −17.)*
$$6c^2 - 17c + 12$$
*(Product: 6 · 12 = 72)*

Think of two integers whose *product* is 72 and whose *sum* is −17. (Both numbers will be negative.) The integers are −9 and −8.
Rewrite the middle term, $-17c$, as $-9c - 8c$. Factor by grouping.

$$6c^2 - 17c + 12 = 6c^2 - 9c - 8c + 12$$
$$= 3c(2c - 3) - 4(2c - 3) \qquad \text{Take out the common factor from each group.}$$
$$= (2c - 3)(3c - 4) \qquad \text{Factor out } (2c - 3).$$

Check: $(2c - 3)(3c - 4) = 6c^2 - 17c + 12$ ✓

c)    *Sum is* $-34$.

$\downarrow$

$7x^2 - 34xy - 5y^2$

*Product*: $7 \cdot (-5) = -35$

The integers whose *product* is $-35$ and whose *sum* is $-34$ are $-35$ and $1$.
Rewrite the middle term, $-34xy$, as $-35xy + xy$. Factor by grouping.

$$7x^2 - 34xy - 5y^2 = \underbrace{7x^2 - 35xy}_{} + \underbrace{xy - 5y^2}_{}$$

$$= 7x(x - 5y) + y(x - 5y) \qquad \text{Take out the common factor from each group.}$$

$$= (x - 5y)(7x + y) \qquad \text{Factor out } (x - 5y).$$

*Check*: $(x - 5y)(7x + y) = 7x^2 - 34xy - 5y^2$ ✓    ∎

## You Try 1

Factor completely.

a)  $4p^2 + 16p + 15$    b)  $10y^2 - 13y + 4$    c)  $5a^2 - 29ab - 6b^2$

## Example 2

Factor completely.  $12n^2 + 64n - 48$

### Solution

It is tempting to jump right in and multiply $12 \cdot (-48) = -576$ and try to think of two integers with a product of $-576$ and a sum of $64$. However, first ask yourself, *"Can I factor out a GCF?"* Yes! We can factor out 4.

$$12n^2 + 64n - 48 = 4\underbrace{(3n^2 + 16n - 12)}_{} \qquad \text{Factor out 4.}$$

Product: $3 \cdot (-12) = -36$

Now factor $3n^2 + 16n - 12$ by finding two integers whose *product* is $-36$ and whose *sum* is $16$. The numbers are $18$ and $-2$.

$$= 4(\underbrace{3n^2 + 18n}_{} - \underbrace{2n - 12}_{})$$

$$= 4[3n(n + 6) - 2(n + 6)] \qquad \text{Take out the common factor from each group.}$$

$$= 4(n + 6)(3n - 2) \qquad \text{Factor out } (n + 6).$$

Check by multiplying: $4(n + 6)(3n - 2) = 4(3n^2 + 16n - 12)$

$$= 12n^2 + 64n - 48 \text{ ✓} \qquad ∎$$

## You Try 2

Factor completely.

a)  $24h^2 - 54h - 15$    b)  $20d^3 + 38d^2 + 12d$

## 2. Factor $ax^2 + bx + c$ ($a \neq 1$) by Trial and Error

At the beginning of this section, we factored $2x^2 + 11x + 15$ by grouping. Now we will factor it by trial and error, which is just reversing the process of FOIL.

## Example 3

Factor $2x^2 + 11x + 15$ completely.

### Solution

Can we factor out a GCF? No. So try to factor $2x^2 + 11x + 15$ as the product of two binomials. Notice that all terms are positive, so all factors will be positive.

Begin with the squared term, $2x^2$. Which two expressions with integer coefficients can we multiply to get $2x^2$? $2x$ and $x$. Put these in the binomials.

$$2x^2 + 11x + 15 = (2x \qquad )(x \qquad ) \qquad 2x \cdot x = 2x^2$$

Next, look at the last term, 15. What are the pairs of positive integers that multiply to 15? They are 15 and 1 as well as 5 and 3.

Try these numbers as the last terms of the binomials. The middle term, $11x$, comes from finding the sum of the products of the outer terms and inner terms.

First Try

$$2x^2 + 11x + 15 \overset{?}{=} (2x + 15)(x + 1) \qquad \text{Incorrect!}$$

These must both be $11x$.  $\qquad 15x$
$\qquad +\ 2x$
$\longrightarrow\ 17x$

Switch the 15 and the 1

$$2x^2 + 11x + 15 \overset{?}{=} (2x + 1)(x + 15) \qquad \text{Incorrect!}$$

These must both be $11x$.  $\qquad 1x$
$\qquad +\ 30x$
$\longrightarrow\ 31x$

Try using 5 and 3.  $\quad 2x^2 + 11x + 15 \overset{?}{=} (2x + 5)(x + 3) \qquad \text{Correct!}$

These must both be $11x$.  $\qquad 5x$
$\qquad +\ 6x$
$\longrightarrow\ 11x$

Therefore, $2x^2 + 11x + 15 = (2x + 5)(x + 3)$. Check by multiplying.  ∎

## Example 4

Factor $3t^2 - 29t + 18$ completely.

### Solution

Can we factor out a GCF? No. To get a product of $3t^2$, we will use $3t$ and $t$.

$$3t^2 - 29t + 18 = (3t \qquad )(t \qquad ) \qquad 3t \cdot t = 3t^2$$

Since the last term is positive and the middle term is negative, we want pairs of negative integers that multiply to 18. The pairs are $-1$ and $-18$, $-2$ and $-9$, and $-3$ and $-6$. Try these numbers as the last terms of the binomials. The middle term, $-29t$, comes from finding the sum of the products of the outer terms and inner terms.

$$3t^2 - 29t + 18 \overset{?}{=} (3t - 1)(t - 18) \qquad \text{Incorrect!}$$

These must both be $-29t$.  $\qquad -t$
$\qquad +\ (-54t)$
$\longrightarrow\ -55t$

Switch the $-1$ and the $-18$:   $3t^2 - 29t + 18 \stackrel{?}{=} (3t - 18)(t - 1)$

Without multiplying, we know that this choice is incorrect. How? In the factor $(3t - 18)$, a 3 can be factored out to get $3(t - 6)$. But, we concluded that we could not factor out a GCF from the original polynomial, $3t^2 - 29t + 18$. Therefore, it will not be possible to take out a common factor from one of the binomial factors.

### Note

If you cannot factor out a GCF from the original polynomial, then you cannot take out a factor from one of the binomial factors either.

Try using $-2$ and $-9$.   $3t^2 - 29t + 18 = (3t - 2)(t - 9)$   Correct !

These must both be $-29t$.

$-2t$

$+ (-27t)$

$-29t$

So, $3t^2 - 29t + 18 = (3t - 2)(t - 9)$. Check by multiplying.   ■

## You Try 3

Factor completely.

a)  $2k^2 + 17k + 8$      b)  $6z^2 - 23z + 20$

## Example 5

Factor completely.

a)  $16a^2 + 62a - 8$          b)  $-2c^2 + 3c + 20$

### Solution

a)  Ask yourself, *"Can I take out a common factor?"* Yes!

$$16a^2 + 62a - 8 = 2(8a^2 + 31a - 4)$$

Now, try to factor $8a^2 + 31a - 4$. To get a product of $8a^2$, we can try either $8a$ *and* $a$ or $4a$ *and* $2a$. Let's start by trying $8a$ and $a$.

$$8a^2 + 31a - 4 = (8a\qquad)(a\qquad)$$

List pairs of integers that multiply to $-4$: 4 and $-1$, $-4$ and 1, 2 and $-2$.
   Try 4 and $-1$. Do not put 4 in the same binomial as $8a$ since then it would be possible to factor out 2. But, 2 does not factor out of $8a^2 + 31a - 4$. Put the 4 in the same binomial as $a$.

$$8a^2 + 31a - 4 \stackrel{?}{=} (8a - 1)(a + 4)$$

$-a$

$+ 32a$

$31a$      Correct

Don't forget that the very first step was to factor out a 2. Therefore,

$$16a^2 + 62a - 8 = 2(8a^2 + 31a - 4) = 2(8a - 1)(a + 4)$$

Check by multiplying.

b)   Since the coefficient of the squared term is negative, begin by factoring out $-1$. (There is no other common factor except 1.)

$$-2c^2 + 3c + 20 = -1(2c^2 - 3c - 20)$$

Try to factor $2c^2 - 3c - 20$. To get a product of $2c^2$, we will use $2c$ and $c$ in the binomials.

$$2c^2 - 3c - 20 = (2c\quad)(c\quad)$$

We need pairs of integers whose product is $-20$. They are 1 and $-20$, $-1$ and 20, 2 and $-10$, $-2$ and 10, 4 and $-5$, $-4$ and 5.

Do *not* start with 1 and $-20$ or $-1$ and 20 because the middle term, $-3c$, is not very large. Using 1 and $-20$ or $-1$ and 20 would likely result in a larger middle term.

Think about 2 and $-10$ *and* $-2$ and 10. *These will not work because if we put any of these numbers in the factor containing 2c, then it will be possible to factor out 2.*

Try 4 and $-5$. Do not put 4 in the same binomial as $2c$ since then it would be possible to factor out 2.

$$2c^2 - 3c - 20 \stackrel{?}{=} (2c - 5)(c + 4)$$

$$-5c$$
$$\underline{+\ 8c}$$
$$3c \quad \text{This must equal } -3c. \quad \text{Incorrect}$$

Only the sign of the sum is incorrect. *Change the signs in the binomials to get the correct sum.*

$$2c^2 - 3c - 20 \stackrel{?}{=} (2c + 5)(c - 4)$$

$$5c$$
$$\underline{+\ (-8c)}$$
$$-3c \quad \text{Correct}$$

Remember that we factored out $-1$ to begin the problem.

$$-2c^2 + 3c + 20 = -1(2c^2 - 3c - 20) = -(2c + 5)(c - 4)$$

Check by multiplying.   ◼

**You Try 4**

Factor completely.

a)   $10y^2 - 58y + 40$          b)   $-4n^2 - 5n + 6$

We have seen two methods for factoring $ax^2 + bx + c$ ($a \neq 1$): factoring by grouping and factoring by trial and error. In either case, remember to begin by taking out a common factor from all terms whenever possible.

## Using Technology

We found some ways to narrow down the possibilities when factoring $ax^2 + bx + c$ $(a \neq 1)$ using the trial and error method.

We can also use a graphing calculator to help with the process. Consider the trinomial $2x^2 - 9x - 35$. Enter the trinomial into $Y_1$ and press ZOOM ; then enter 6 to display the graph in the standard viewing window.

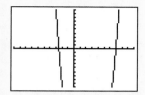

Look on the graph for the *x*-intercept (if any) that appears to be an integer. It appears that 7 is an *x*-intercept.

To check whether 7 is an *x*-intercept, press TRACE then 7 and press ENTER . As shown on the graph, when $x = 7$, $y = 0$, so 7 is an *x*-intercept.

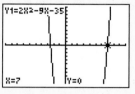

When an *x*-intercept is an integer, then *x* minus that *x*-intercept is a factor of the trinomial. In this case, $x - 7$ is a factor of $2x^2 - 9x - 35$. We can then complete the factoring as $(2x + 5)(x - 7)$, since we must multiply $-7$ by 5 to obtain $-35$.

Find an *x*-intercept using a graphing calculator and factor the trinomial.

1) $3x^2 + 11x - 4$          2) $2x^2 + x - 15$          3) $5x^2 + 6x - 8$

4) $2x^2 - 5x + 3$          5) $4x^2 - 3x - 10$          6) $14x^2 - x - 4$

### Answers to You Try Exercises

1)   a) $(2p + 5)(2p + 3)$   b) $(5y - 4)(2y - 1)$   c) $(5a + b)(a - 6b)$     2)   a) $3(2h - 5)(4h + 1)$
b) $2d(5d + 2)(2d + 3)$     3)   a) $(2k + 1)(k + 8)$   b) $(3z - 4)(2z - 5)$     4)   a) $2(5y - 4)(y - 5)$
b) $-(4n - 3)(n + 2)$

### Answers to Technology Exercises

1)   $(3x - 1)(x + 4)$     2)   $(2x - 5)(x + 3)$     3)   $(x + 2)(5x - 4)$     4)   $(x - 1)(2x - 3)$
5)   $(x - 2)(4x + 5)$     6)   $(2x + 1)(7x - 4)$

# 7.3 Exercises

**Objective 1: Factor $ax^2 + bx + c$ $(a \neq 1)$ by Grouping**

1) Find two integers whose

|  | PRODUCT IS | and whose SUM IS | ANSWER |
|---|---|---|---|
| a) | $-50$ | 5 |  |
| b) | 27 | $-28$ |  |
| c) | 12 | 8 |  |
| d) | $-72$ | $-6$ |  |

2) Find two integers whose

|  | PRODUCT IS | and whose SUM IS | ANSWER |
|---|---|---|---|
| a) | 18 | 19 |  |
| b) | $-132$ | 1 |  |
| c) | $-30$ | $-13$ |  |
| d) | 63 | $-16$ |  |

Factor by grouping.

3) $3c^2 + 12c + 8c + 32$

4) $5y^2 + 15y + 2y + 6$

5) $6k^2 - 6k - 7k + 7$

6) $4r^2 - 4r + 9r - 9$

7) $6x^2 - 27xy + 8xy - 36y^2$

8) $14p^2 - 8pq - 7pq + 4q^2$

9) When asked to factor a polynomial, what is the first question you should ask yourself?

10) After factoring a polynomial, what should you ask yourself to be sure that the polynomial is factored completely?

11) Find the polynomial that factors to $(4k + 9)(k + 2)$.

12) Find the polynomial that factors to $(6m - 5)(2m - 3)$.

Complete the factorization.

13) $5t^2 + 13t + 6 = (5t + 3)($        $)$

14) $4z^2 + 29z + 30 = (4z + 5)($        $)$

15) $6a^2 - 11a - 10 = (2a - 5)($        $)$

16) $15c^2 - 23c + 4 = (3c - 4)($        $)$

17) $12x^2 - 25xy + 7y^2 = (4x - 7y)($        $)$

18) $12r^2 - 52rt - 9t^2 = (6r + t)($        $)$

Factor by grouping. See Example 1.

19) $2h^2 + 13h + 15$

20) $3z^2 + 13z + 14$

21) $7y^2 - 11y + 4$

22) $5a^2 - 21a + 18$

23) $5b^2 + 9b - 18$

24) $11m^2 - 18m - 8$

25) $6p^2 + p - 2$

26) $8c^2 - 22c + 5$

27) $4t^2 + 16t + 15$

28) $10k^2 + 23k + 12$

29) $9x^2 - 13xy + 4y^2$

30) $6a^2 + ab - 5b^2$

### Objective 2: Factor $ax^2 + bx + c$ ($a \neq 1$) by Trial and Error

31) How do we know that $(2x - 4)$ cannot be a factor of $2x^2 + 13x - 24$?

32) How do we know that $(3p + 2)$ cannot be a factor of $6p^2 - 25p + 14$?

Factor by trial and error. See Examples 3 and 4.

33) $2r^2 + 9r + 10$

34) $3q^2 + 10q + 8$

VIDEO 35) $3u^2 - 23u + 30$

36) $7m^2 - 15m + 8$

37) $7a^2 + 31a - 20$

38) $5x^2 - 11x - 36$

39) $6y^2 + 23y + 10$

40) $8u^2 + 18u + 7$

41) $9w^2 + 20w - 21$

42) $10h^2 - 59h - 6$

43) $8c^2 - 42c + 27$

44) $15v^2 - 16v + 4$

45) $4k^2 + 40k + 99$

46) $4n^2 - 41n + 10$

47) $20b^2 - 32b - 45$

48) $14g^2 + 27g - 20$

49) $2r^2 + 13rt - 24t^2$

50) $3c^2 - 17cd - 6d^2$

51) $6a^2 - 25ab + 4b^2$

52) $6x^2 + 31xy + 18y^2$

### Mixed Exercises: Objectives 1 and 2

53) Factor $4z^2 + 5z - 6$ using each method. Do you get the same answer? Which method do you prefer? Why?

54) Factor $10a^2 + 27a + 18$ using each method. Do you get the same answer? Which method do you prefer? Why?

Factor completely.

55) $3p^2 - 16p - 12$

56) $2t^2 - 19t + 24$

57) $4k^2 + 15k + 9$

58) $12x^3 + 15x^2 - 18x$

59) $30w^3 + 76w^2 + 14w$

60) $12d^2 - 28d - 5$

61) $21r^2 - 90r + 24$

62) $45q^2 + 57q + 18$

63) $6y^2 - 10y + 3$

64) $9z^2 + 14z + 8$

65) $42b^2 + 11b - 3$

66) $13u^2 + 17u - 18$

67) $7x^2 - 17xy + 6y^2$

68) $5a^2 + 23ab + 12b^2$

VIDEO 69) $2d^2 + 2d - 40$

70) $6c^2 + 42c + 72$

71) $30r^4t^2 + 23r^3t^2 + 3r^2t^2$

72) $8m^2n^3 + 4m^2n^2 - 60m^2n$

73) $9k^2 - 42k + 49$

74) $25p^2 + 20p + 4$

75) $2m^2(n + 9) - 5m(n + 9) - 7(n + 9)$

76) $3s^2(t - 8)^2 + 19s(t - 8)^2 + 20(t - 8)^2$

77) $6v^2(u + 4)^2 + 23v(u + 4)^2 + 20(u + 4)^2$

78) $8c^2(d - 7)^3 + 69c(d - 7)^3 - 27(d - 7)^3$

79) $15b^2(2a - 1)^4 - 28b(2a - 1)^4 + 12(2a - 1)^4$

80) $12x^2(3y + 2)^3 - 28x(3y + 2)^3 + 15(3y + 2)^3$

Factor completely by first taking out a negative common factor. See Example 5.

81) $-n^2 - 8n + 48$

82) $-c^2 - 16c - 63$

83) $-7a^2 + 4a + 3$

84) $-3p^2 + 14p - 16$

VIDEO 85) $-10z^2 + 19z - 6$

86) $-16k^3 + 48k^2 - 36k$

87) $-20m^3 - 120m^2 - 135m$

88) $-3z^3 - 15z^2 + 198z$

89) $-6a^3b + 11a^2b^2 + 2ab^3$

90) $-35u^4 - 203u^3v - 140u^2v^2$

## Section 7.4 Factoring Special Trinomials and Binomials

**Objectives**

1. **Factor a Perfect Square Trinomial**
2. **Factor the Difference of Two Squares**
3. **Factor the Sum and Difference of Two Cubes**

### 1. Factor a Perfect Square Trinomial

Recall that we can square a binomial using the formulas

$$(a + b)^2 = a^2 + 2ab + b^2$$
$$(a - b)^2 = a^2 - 2ab + b^2$$

For example, $(x + 3)^2 = x^2 + 2x(3) + 3^2 = x^2 + 6x + 9$.

Since factoring a polynomial means writing the polynomial as a product of its factors, $x^2 + 6x + 9$ factors to $(x + 3)^2$.

The expression $x^2 + 6x + 9$ is a *perfect square trinomial*. A **perfect square trinomial** is a trinomial that results from squaring a binomial.

We can use the factoring method presented in Section 7.2 to factor a perfect square trinomial, or we can learn to recognize the special pattern that appears in these trinomials.

How are the terms of $x^2 + 6x + 9$ and $(x + 3)^2$ related?

$x^2$ is the square of $x$, the first term in the binomial.

9 is the square of 3, the last term in the binomial.

We get the term $6x$ by doing the following:

$$6x = \quad 2 \quad \cdot \quad x \quad \cdot \quad 3$$

Two times     First term in binomial     Last term in binomial

This follows directly from how we found $(x + 3)^2$ using the formula.

---

**Formula   Factoring a Perfect Square Trinomial**

$$a^2 + 2ab + b^2 = (a + b)^2$$
$$a^2 - 2ab + b^2 = (a - b)^2$$

---

**Note**

In order for a trinomial to be a perfect square, two of its terms must be perfect squares.

---

**Example 1**

Factor completely.  $k^2 + 18k + 81$

**Solution**

We cannot take out a common factor, so let's see whether this follows the pattern of a perfect square trinomial.

$$k^2 + 18k + 81$$

What do you square to get $k^2$? $k$    $(k)^2$      $(9)^2$    What do you square to get 81? 9

Does the middle term equal $2 \cdot k \cdot 9$? Yes.

$$2 \cdot k \cdot 9 = 18k$$

Therefore, $k^2 + 18k + 81 = (k + 9)^2$. Check by multiplying. ∎

## Example 2

Factor completely.

a) $c^2 - 16c + 64$  b) $4t^3 + 32t^2 + 64t$

c) $9w^2 + 12w + 4$  d) $4c^2 + 20c + 9$

**Solution**

a)  We cannot take out a common factor. However, since the middle term is negative and the first and last terms are positive, the sign in the binomial will be a minus $(-)$ sign. Does this fit the pattern of a perfect square trinomial?

$$c^2 - 16c + 64$$
$$\downarrow \qquad \downarrow$$

What do you square to get $c^2$? $c$ $(c)^2$  $(8)^2$ What do you square to get 64? 8

Does the middle term equal $2 \cdot c \cdot 8$? Yes. $2 \cdot c \cdot 8 = 16c$.

Notice that $c^2 - 16c + 64$ fits the pattern of $a^2 - 2ab + b^2 = (a - b)^2$ with $a = c$ and $b = 8$.

Therefore, $c^2 - 16c + 64 = (c - 8)^2$. Check by multiplying.

b)  From $4t^3 + 32t^2 + 64t$ we *can* begin by taking out the GCF of $4t$.

$$4t^3 + 32t^2 + 64t = 4t(t^2 + 8t + 16)$$
$$\downarrow \qquad \downarrow$$

What do you square to get $t^2$? $t$ $(t)^2$  $(4)^2$ What do you square to get 16? 4

Does the middle term of the trinomial in parentheses equal $2 \cdot t \cdot 4$? Yes. $2 \cdot t \cdot 4 = 8t$.

$$4t^3 + 32t^2 + 64t = 4t(t^2 + 8t + 16) = 4t(t + 4)^2$$

Check by multiplying.

c)  We cannot take out a common factor. Since the first and last terms of $9w^2 + 12w + 4$ are perfect squares, let's see whether this is a perfect square trinomial.

$$9w^2 + 12w + 4$$
$$\downarrow \qquad \downarrow$$

What do you square to get $9w^2$? $3w$ $(3w)^2$  $(2)^2$ What do you square to get 4? 2

Does the middle term equal $2 \cdot 3w \cdot 2$? Yes. $2 \cdot 3w \cdot 2 = 12w$.

Therefore, $9w^2 + 12w + 4 = (3w + 2)^2$. Check by multiplying.

d)  We cannot take out a common factor. The first and last terms of $4c^2 + 20c + 9$ are perfect squares. Is this a perfect square trinomial?

$$4c^2 + 20c + 9$$
$$\downarrow \qquad \downarrow$$

What do you square to get $4c^2$? $2c$ $(2c)^2$  $(3)^2$ What do you square to get 9? 3

Does the middle term equal $2 \cdot 2c \cdot 3$? No! $2 \cdot 2c \cdot 3 = 12c$.

This is *not* a perfect square trinomial. Applying a method from Section 7.3, we find that the trinomial does factor, however.

$$4c^2 + 20c + 9 = (2c + 9)(2c + 1).$$ Check by multiplying.    ■

**You Try 1**

Factor completely.

a)  $g^2 + 14g + 49$

b)  $6y^3 - 36y^2 + 54y$

c)  $25v^2 - 10v + 1$

d)  $9b^2 + 15b + 4$

## 2. Factor the Difference of Two Squares

Another common factoring problem is a **difference of two squares**. Some examples of these types of binomials are

$$c^2 - 36 \qquad 49x^2 - 25y^2 \qquad 64 - t^2 \qquad h^4 - 16$$

Notice that in each binomial, the terms are being *subtracted*, and each term is a perfect square.

In Section 6.3, Multiplication of Polynomials, we saw that

$$(a + b)(a - b) = a^2 - b^2$$

If we reverse the procedure, we get the factorization of the difference of two squares.

> **Formula**   **Factoring the Difference of Two Squares**
>
> $$a^2 - b^2 = (a + b)(a - b)$$

Don't forget that we can check all factorizations by multiplying.

---

**Example 3**

Factor completely.

a)  $c^2 - 36$

b)  $49x^2 - 25y^2$

c)  $t^2 - \dfrac{4}{9}$

d)  $k^2 + 81$

**Solution**

a)  First, notice that $c^2 - 36$ is the difference of two terms *and* those terms are perfect squares. We can use the formula $a^2 - b^2 = (a + b)(a - b)$.

Identify $a$ and $b$.

$$c^2 - 36$$

What do you square to get $c^2$? $c$ $(c)^2 \ (6)^2$ What do you square to get 36? 6

Then, $a = c$ and $b = 6$. Therefore, $c^2 - 36 = (c + 6)(c - 6)$.

b)  Look carefully at $49x^2 - 25y^2$. Each term *is* a perfect square, and they are being subtracted.

Identify $a$ and $b$.

$$49x^2 - 25y^2$$

What do you square to get $49x^2$? $7x$ $(7x)^2 \ (5y)^2$ What do you square to get $25y^2$? $5y$

Then, $a = 7x$ and $b = 5y$. So, $49x^2 - 25y^2 = (7x + 5y)(7x - 5y)$.

c)   Each term in $t^2 - \dfrac{4}{9}$ is a perfect square, and they are being subtracted.

$$t^2 - \frac{4}{9}$$

$$\downarrow \qquad \downarrow$$

What do you square to get $t^2$? $t$ $\quad (t)^2 \left(\dfrac{2}{3}\right)^2 \quad$ What do you square to get $\dfrac{4}{9}$? $\dfrac{2}{3}$

So, $a = t$ and $b = \dfrac{2}{3}$. Therefore, $t^2 - \dfrac{4}{9} = \left(t + \dfrac{2}{3}\right)\left(t - \dfrac{2}{3}\right)$.

d)   Each term in $k^2 + 81$ is a perfect square, but the expression is the *sum* of two squares. This polynomial does not factor.

$$k^2 + 81 \neq (k + 9)(k - 9) \text{ since } (k + 9)(k - 9) = k^2 - 81.$$
$$k^2 + 81 \neq (k + 9)(k + 9) \text{ since } (k + 9)(k + 9) = k^2 + 18k + 81.$$

So, $k^2 + 81$ is prime. ∎

**Note**

If the sum of two squares does not contain a common factor, then it cannot be factored.

 **You Try 2**

Factor completely.

a)   $m^2 - 100$      b)   $4c^2 - 81d^2$      c)   $h^2 - \dfrac{64}{25}$      d)   $p^2 + 49$

Remember that sometimes we can factor out a GCF first. And, after factoring once, ask yourself, *"Can I factor again?"*

**Example 4**

Factor completely.

a)   $300p - 3p^3$      b)   $7w^2 + 28$      c)   $x^4 - 81$

**Solution**

a)   Ask yourself, *"Can I take out a common factor?"* Yes. Factor out $3p$.

$$300p - 3p^3 = 3p(100 - p^2)$$

Now ask yourself, *"Can I factor again?"* Yes. $100 - p^2$ is the difference of two squares. Identify $a$ and $b$.

$$100 - p^2$$
$$\downarrow \qquad \downarrow$$
$$(10)^2 \ (p)^2$$

So, $a = 10$ and $b = p$. $100 - p^2 = (10 + p)(10 - p)$.

Therefore, $300p - 3p^3 = 3p(10 + p)(10 - p)$.

**BE CAREFUL**

(10 + p) (10 − p) is *not* the same as (p + 10) (p − 10) because subtraction is not commutative. While 10 + p = p + 10, 10 − p *does not equal* p − 10. You must write the terms in the correct order. Another way to see that they are not equivalent is to multiply (p + 10) (p − 10). (p + 10) (p − 10) = $p^2$ − 100. This is not the same as 100 − $p^2$.

b)   Look at $7w^2 + 28$. Ask yourself, *"Can I take out a common factor?"* Yes. Factor out 7.

$$7(w^2 + 4)$$

*"Can I factor again?"* No, because $w^2 + 4$ is the *sum* of two squares.

Therefore, $7w^2 + 28 = 7(w^2 + 4)$.

c)   The terms in $x^4 - 81$ have no common factors, but they are perfect squares. Identify *a* and *b*.

$$x^4 - 81$$

What do you square to get $x^4$? $x^2$       $(x^2)^2$  $(9)^2$    What do you square to get 81? 9

So, $a = x^2$ and $b = 9$. $x^4 - 81 = (x^2 + 9)(x^2 - 9)$.

Can we factor again?

$x^2 + 9$ is the *sum* of two squares. It will not factor.

$x^2 - 9$ is the difference of two squares, so it *will* factor.

$$x^2 - 9$$
$(x)^2$  $(3)^2$          $x^2 - 9 = (x + 3)(x - 3)$
$a = x$ and $b = 3$

Therefore,

$$x^4 - 81 = (x^2 + 9)(x^2 - 9)$$
$$= (x^2 + 9)(x + 3)(x - 3)$$

■

 **You Try 3**

Factor completely.

a)   $125d - 5d^3$           b)   $3r^2 + 48$           c)   $z^4 - 1$

## 3. Factor the Sum and Difference of Two Cubes

Before we give the formulas for factoring the sum and difference of two cubes, let's look at two products.

$(a + b)(a^2 - ab + b^2) = a(a^2 - ab + b^2) + b(a^2 - ab + b^2)$     Distributive property
$\qquad\qquad\qquad\quad = a^3 - a^2b + ab^2 + a^2b - ab^2 + b^3$     Distribute.
$\qquad\qquad\qquad\quad = a^3 + b^3$     Combine like terms.

So, $(a + b)(a^2 - ab + b^2) = a^3 + b^3$, the sum of two cubes.

Let's look at another product:

$$(a - b)(a^2 + ab + b^2) = a(a^2 + ab + b^2) - b(a^2 + ab + b^2) \qquad \text{Distributive property}$$
$$= a^3 + a^2b + ab^2 - a^2b - ab^2 - b^3 \qquad \text{Distribute.}$$
$$= a^3 - b^3 \qquad \text{Combine like terms.}$$

So, $(a - b)(a^2 + ab + b^2) = a^3 - b^3$, the difference of two cubes.

The formulas for factoring the sum and difference of two cubes, then, are as follows:

---

**Formula    Factoring the Sum and Difference of Two Cubes**

$$a^3 + b^3 = (a + b)(a^2 - ab + b^2)$$
$$a^3 - b^3 = (a - b)(a^2 + ab + b^2)$$

Notice that each factorization is the product of a binomial and a trinomial.

---

**Procedure    To factor the sum and difference of two cubes**

**Step 1:**  Identify $a$ and $b$.

**Step 2:**  Place them in the binomial factor and write the trinomial based on $a$ and $b$.

**Step 3:**  Simplify.

---

### Example 5

Factor completely.

a)  $k^3 + 27$        b)  $n^3 - 125$        c)  $64c^3 + 125d^3$

**Solution**

a)  Use steps 1–3 to factor.

**Step 1:**  Identify $a$ and $b$.

$$k^3 + 27$$
$$\downarrow \quad \downarrow$$

What do you cube to get $k^3$? $k$    $(k)^3$   $(3)^3$    What do you cube to get 27? 3

So, $a = k$ and $b = 3$.

**Step 2:**  Remember, $a^3 + b^3 = (a + b)(a^2 - ab + b^2)$.

Write the binomial factor, then write the trinomial.

                                     Square $a$.     Product     Square $b$.
                                            of $a$ and $b$

Same sign

$$k^3 + 27 = (k + 3)[(k)^2 - (k)(3) + (3)^2]$$

Opposite sign

**Step 3:**  Simplify:  $k^3 + 27 = (k + 3)(k^2 - 3k + 9)$.

b) **Step 1:** Identify $a$ and $b$.

$$n^3 - 125$$

$$\downarrow \qquad \downarrow$$

What do you cube    $(n)^3 \quad (5)^3$    What do you cube
to get $n^3$? $n$                      to get 125? 5

So, $a = n$ and $b = 5$.

**Step 2:** Write the binomial factor, then write the trinomial. Remember, $a^3 - b^3 = (a - b)(a^2 + ab + b^2)$.

Square $a$.    Square $b$.
Product
Same sign       of $a$ and $b$       ↓

$$n^3 - 125 = (n - 5)[(n)^2 + (n)(5) + (5)^2]$$
Opposite
sign

**Step 3:** Simplify:  $n^3 - 125 = (n - 5)(n^2 + 5n + 25)$.

c) $64c^3 + 125d^3$

**Step 1:** Identify $a$ and $b$.

$$64c^3 + 125d^3$$

$$\downarrow \qquad \downarrow$$

What do you cube    $(4c)^3 \quad (5d)^3$    What do you cube
to get $64c^3$? $4c$                      to get $125d^3$? $5d$

So, $a = 4c$ and $b = 5d$.

**Step 2:** Write the binomial factor, then write the trinomial. Remember, $a^3 + b^3 = (a + b)(a^2 - ab + b^2)$.

Square $a$.    Square $b$.
Product
Same sign       of $a$ and $b$       ↓

$$64c^3 + 125d^3 = (4c + 5d)[(4c)^2 - (4c)(5d) + (5d)^2]$$
Opposite
sign

**Step 3:** Simplify:  $64c^3 + 125d^3 = (4c + 5d)(16c^2 - 20cd + 25d^2)$.  ■

**You Try 4**

Factor completely.

a) $m^3 + 1000$          b) $h^3 - 1$          c) $27p^3 - 64q^3$

Just as in the other factoring problems we've studied so far, the first step in factoring *any* polynomial should be to ask ourselves, *"Can I factor out a GCF?"*

**Example 6**

Factor $5z^3 - 40$ completely.

**Solution**

*"Can I factor out a GCF?"* Yes. The GCF is 5.

$$5z^3 - 40 = 5(z^3 - 8)$$

Factor $z^3 - 8$. Use $a^3 - b^3 = (a - b)(a^2 + ab + b^2)$.

$$z^3 - 8 = (z - 2)[(z)^2 + (z)(2) + (2)^2]$$

$$(z)^3 - (2)^3 = (z - 2)(z^2 + 2z + 4)$$

Therefore, $5z^3 - 40 = 5(z^3 - 8)$
$$= 5(z - 2)(z^2 + 2z + 4)$$ ∎

### You Try 5

Factor completely.

a)  $2t^3 - 54$          b)  $2a^7 + 128ab^3$

As always, the first thing you should do when factoring is ask yourself, *"Can I factor out a GCF?"* and the last thing you should do is ask yourself, *"Can I factor again?"* Now we will summarize the factoring methods discussed in this section.

---

**Summary**   Special Factoring Rules

Perfect square trinomials:   $a^2 + 2ab + b^2 = (a + b)^2$
$a^2 - 2ab + b^2 = (a - b)^2$

Difference of two squares:   $a^2 - b^2 = (a + b)(a - b)$

Sum of two cubes:   $a^3 + b^3 = (a + b)(a^2 - ab + b^2)$

Difference of two cubes:   $a^3 - b^3 = (a - b)(a^2 + ab + b^2)$

---

**Answers to You Try Exercises**

1)  a) $(g + 7)^2$   b) $6y(y - 3)^2$   c) $(5v - 1)^2$   d) $(3b + 4)(3b + 1)$        2)  a) $(m + 10)(m - 10)$

b) $(2c + 9d)(2c - 9d)$   c) $\left(h + \dfrac{8}{5}\right)\left(h - \dfrac{8}{5}\right)$   d) prime

3)  a) $5d(5 + d)(5 - d)$   b) $3(r^2 + 16)$   c) $(z^2 + 1)(z + 1)(z - 1)$

4)  a) $(m + 10)(m^2 - 10m + 100)$   b) $(h - 1)(h^2 + h + 1)$   c) $(3p - 4q)(9p^2 + 12pq + 16q^2)$

5)  a) $2(t - 3)(t^2 + 3t + 9)$   b) $2a(a^2 + 4b)(a^4 - 4a^2b + 16b^2)$

---

## 7.4 Exercises

**Objective 1: Factor a Perfect Square Trinomial**

1) Find the following.

a) $7^2$                    b) $9^2$

c) $6^2$                    d) $10^2$

e) $5^2$                    f) $4^2$

g) $11^2$                   h) $\left(\dfrac{1}{3}\right)^2$

i) $\left(\dfrac{3}{8}\right)^2$

2) What is perfect square trinomial?

3) Fill in the blank with a term that has a positive coefficient.

a) $(\_\_)^2 = c^4$                 b) $(\_\_)^2 = 9r^2$

c) $(\_\_)^2 = 81p^2$               d) $(\_\_)^2 = 36m^4$

e) $(\_\_)^2 = \dfrac{1}{4}$        f) $(\_\_)^2 = \dfrac{144}{25}$

4) If $x^n$ is a perfect square, then $n$ is divisible by what number?

5) What perfect square trinomial factors to $(y + 6)^2$?

6) What perfect square trinomial factors to $(3k - 8)^2$?

7) Why isn't $4a^2 - 10a + 9$ a perfect square trinomial?

8) Why isn't $x^2 + 5x + 12$ a perfect square trinomial?

Factor completely.

9) $h^2 + 10h + 25$

10) $q^2 + 8q + 16$

11) $b^2 - 14b + 49$

12) $t^2 - 24t + 144$

13) $4w^2 + 4w + 1$

14) $25m^2 + 20m + 4$

VIDEO 15) $9k^2 - 24k + 16$

16) $16a^2 - 56a + 49$

17) $c^2 + c + \dfrac{1}{4}$

18) $h^2 + \dfrac{1}{3}h + \dfrac{1}{36}$

19) $k^2 - \dfrac{14}{5}k + \dfrac{49}{25}$

20) $p^2 - \dfrac{4}{3}h + \dfrac{4}{9}$

21) $a^2 + 8ab + 16b^2$

22) $4x^2 - 12xy + 9y^2$

23) $25m^2 - 30mn + 9n^2$

24) $49p^2 + 14pq + q^2$

VIDEO 25) $4f^2 + 24f + 36$

26) $8r^2 - 16r + 8$

27) $5a^4 + 30a^3 + 45a^2$

28) $3k^3 - 42k^2 + 147k$

29) $-16y^2 - 80y - 100$

30) $-81n^2 + 54n - 9$

31) $75h^3 - 6h^2 + 12h$

32) $98b^5 + 42b^4 + 18b^3$

**Objective 2: Factor the Difference of Two Squares**

33) What binomial factors to

a) $(x + 9)(x - 9)$?

b) $(9 + x)(9 - x)$?

34) What binomial factors to

a) $(y - 10)(y + 10)$?

b) $(10 - y)(10 + y)$?

Complete the factorization.

35) $w^2 - 64 = (w + 8)\,(\quad)$

36) $t^2 - 1 = (t - 1)\,(\quad)$

37) $121 - p^2 = (11 + p)(\quad)$

38) $9h^2 - 49 = (3h + 7)(\quad)$

39) $64c^2 - 25b^2 = (8c + 5b)(\quad)$

40) Does $n^2 + 9 = (n + 3)^2$ ? Explain.

Factor completely.

41) $k^2 - 4$

42) $z^2 - 100$

43) $c^2 - 25$

44) $y^2 - 81$

45) $w^2 + 49$

46) $b^2 + 64$

47) $x^2 - \dfrac{1}{9}$

48) $p^2 - \dfrac{1}{4}$

49) $a^2 - \dfrac{4}{49}$

50) $t^2 - \dfrac{121}{64}$

51) $144 - v^2$

52) $36 - r^2$

53) $1 - h^2$

54) $169 - d^2$

55) $\dfrac{36}{25} - b^2$

56) $\dfrac{9}{100} - q^2$

VIDEO 57) $100m^2 - 49$

58) $25a^2 - 121$

59) $169k^2 - 1$

60) $36p^2 - 1$

61) $4y^2 + 49$

62) $9d^2 + 25$

63) $\dfrac{1}{9}t^2 - \dfrac{25}{4}$

64) $\dfrac{16}{9}x^2 - \dfrac{1}{49}$

65) $u^4 - 100$

66) $a^4 - 4$

67) $36c^2 - d^4$

68) $25y^2 - 144z^4$

VIDEO 69) $r^4 - 1$

70) $h^4 - 16$

71) $r^4 - 81t^4$

72) $y^4 - x^4$

73) $5u^2 - 45$

74) $3k^2 - 300$

75) $2n^2 - 288$

76) $11p^2 - 11$

77) $12z^4 - 75z^2$

78) $45b^5 - 245b^3$

**Objective 3: Factor the Sum and Difference of Two Cubes**

79) Find the following.

a) $4^3$

b) $1^3$

c) $10^3$

d) $3^3$

e) $5^3$

f) $2^3$

80) If $x^n$ is a perfect cube, then $n$ is divisible by what number?

81) Fill in the blank.

a) $(\_)^3 = m^3$

b) $(\_)^3 = 27t^3$

c) $(\_)^3 = 8b^3$

d) $(\_)^3 = h^6$

82) If $x^n$ is a perfect square *and* a perfect cube, then $n$ is divisible by what number?

Complete the factorization.

83) $y^3 + 8 = (y + 2)\,(\quad)$

84) $p^3 - 1000 = (p - 10)\,(\quad)$

Factor completely.

85) $t^3 + 64$

86) $d^3 - 125$

87) $z^3 - 1$

88) $r^3 + 27$

VIDEO 89) $27m^3 - 125$

90) $64c^3 + 1$

91) $125y^3 - 8$

92) $27a^3 + 64$

93) $1000c^3 - d^3$

94) $125v^3 + w^3$

95) $8j^3 + 27k^3$

96) $125m^3 - 27n^3$

97) $64x^3 + 125y^3$

98) $27a^3 - 1000b^3$

VIDEO 99) $6c^3 + 48$

100) $9k^3 - 9$

101) $7v^3 - 7000w^3$

102) $216a^3 + 64b^3$

103) $h^6 - 64$

104) $p^6 - 1$

Extend the concepts of this section to factor completely.

105) $(d + 4)^2 - (d - 3)^2$

106) $(w - 9)^2 - (w + 2)^2$

107) $(3k + 1)^2 - (k + 5)^2$

108) $(2m - 3)^2 - (m - 1)^2$

109) $(r - 2)^3 + 27$

110) $(x + 7)^3 + 8$

111) $(c + 4)^3 - 125$

112) $(p - 3)^3 - 1$

## Putting It All Together

### Objective

1. Learn Strategies for Factoring a Given Polynomial

### 1. Learn Strategies for Factoring a Given Polynomial

In this chapter, we have discussed several different types of factoring problems:

1) Factoring out a GCF (Section 7.1)
2) Factoring by grouping (Section 7.1)
3) Factoring a trinomial of the form $x^2 + bx + c$ (Section 7.2)
4) Factoring a trinomial of the form $ax^2 + bx + c$ (Section 7.3)
5) Factoring a perfect square trinomial (Section 7.4)
6) Factoring the difference of two squares (Section 7.4)
7) Factoring the sum and difference of two cubes (Section 7.4)

We have practiced the factoring methods separately in each section, but how do we know which factoring method to use given many different types of polynomials together? We will discuss some strategies in this section. First, recall the steps for factoring *any* polynomial:

---

**Summary** To Factor a Polynomial

1) *Always* begin by asking yourself, *"Can I factor out a GCF?"* If so, factor it out.

2) Look at the expression to decide whether it will factor further. Apply the appropriate method to factor. If there are

    a) *two terms*, see whether it is a difference of two squares or the sum or difference of two cubes as in Section 7.4.

    b) *three terms*, see whether it can be factored using the methods of Section 7.2 or Section 7.3 or determine whether it is a perfect square trinomial (Section 7.4).

    c) *four terms*, see whether it can be factored by grouping as in Section 7.1.

3) After factoring, *always* look carefully at the result and ask yourself, *"Can I factor it again?"* If so, factor again.

---

Next, we will discuss how to decide which factoring method should be used to factor a particular polynomial.

### Example 1

Factor completely.

a)   $8x^2 - 50y^2$       b)   $t^2 - t - 56$       c)   $a^2b - 9b + 4a^2 - 36$

d)   $k^2 - 12k + 36$       e)   $15p^2 + 51p + 18$       f)   $27k^3 + 8$

g)   $c^2 + 4$

**Solution**

a)   *"Can I factor out a GCF?"* is the first thing you should ask yourself. Yes. Factor out 2.

$$8x^2 - 50y^2 = 2(4x^2 - 25y^2)$$

Ask yourself, *"Can I factor again?"* Examine $4x^2 - 25y^2$. It has two terms that are being subtracted and each term is a perfect square. $4x^2 - 25y^2$ is the difference of squares.

$$4x^2 - 25y^2 = (2x + 5y)(2x - 5y)$$
$$\quad\downarrow\qquad\downarrow$$
$$(2x)^2\ (5y)^2$$
$$8x^2 - 50y^2 = 2(4x^2 - 25y^2) = 2(2x + 5y)(2x - 5y)$$

*"Can I factor again?"* No. It is completely factored.

b)  Look at $t^2 - t - 56$. *"Can I factor out a GCF?"* No. Think of two numbers whose *product* is $-56$ and *sum* is $-1$. The numbers are $-8$ and $7$.

$$t^2 - t - 56 = (t - 8)(t + 7)$$

*"Can I factor again?"* No. It is completely factored.

c)  We have to factor $a^2b - 9b + 4a^2 - 36$. *"Can I factor out a GCF?"* No. Notice that this polynomial has *four terms*. When a polynomial has *four terms*, think about *factoring by grouping*.

$$\underbrace{a^2b - 9b}_{\downarrow} + \underbrace{4a^2 - 36}_{\downarrow}$$

$$= b(a^2 - 9) + 4(a^2 - 9) \quad \text{Take out the common factor from each pair of terms.}$$
$$= (a^2 - 9)(b + 4) \quad \text{Factor out } (a^2 - 9) \text{ using the distributive property.}$$

Examine $(a^2 - 9)(b + 4)$ and ask yourself, *"Can I factor again?"* Yes! $(a^2 - 9)$ is the difference of two squares. Factor again.

$$(a^2 - 9)(b + 4) = (a + 3)(a - 3)(b + 4)$$

*"Can I factor again?"* No. So, $a^2b - 9b + 4a^2 - 36 = (a + 3)(a - 3)(b + 4)$.

**Note**

Seeing four terms is a clue to try factoring by grouping.

d)  We cannot take out a GCF from $k^2 - 12k + 36$. It is a trinomial, and notice that the first and last terms are perfect squares. *Is this a perfect square trinomial?*

$$\underset{\downarrow}{k^2} - 12k + \underset{\downarrow}{36}$$
$$(k)^2 \qquad (6)^2$$

Does the middle term equal $2 \cdot k \cdot 6$? Yes. $2 \cdot k \cdot 6 = 12k$

Use $a^2 - 2ab + b^2 = (a - b)^2$ with $a = k$ and $b = 6$.

Then, $k^2 - 12k + 36 = (k - 6)^2$.

*"Can I factor again?"* No. It is completely factored.

e)  It is tempting to jump right in and try to factor $15p^2 + 51p + 18$ as the product of two binomials, but ask yourself, *"Can I take out a GCF?"* Yes! Factor out 3.

$$15p^2 + 51p + 18 = 3(5p^2 + 17p + 6)$$

*"Can I factor again?"* Yes.

$$3(5p^2 + 17p + 6) = 3(5p + 2)(p + 3)$$

*"Can I factor again?"* No. So, $15p^2 + 51p + 18 = 3(5p + 2)(p + 3)$.

f)  We cannot take out a GCF from $27k^3 + 8$. Notice that $27k^3 + 8$ has two terms, so think about squares and cubes. Neither term is a perfect square *and* the positive terms are being added, so this *cannot* be the difference of squares.

Is each term a perfect cube? Yes! $27k^3 + 8$ is the sum of two cubes. We will factor $27k^3 + 8$ using $a^3 + b^3 = (a + b)(a^2 - ab + b^2)$ with $a = 3k$ and $b = 2$.

$$27k^3 + 8 = (3k + 2)[(3k)^2 - (3k)(2) + (2)^2]$$
$$\underset{(3k)^3}{\downarrow} \quad \underset{(2)^3}{\downarrow}$$
$$= (3k + 2)(9k^2 - 6k + 4)$$

*"Can I factor again?"* No. It is completely factored.

g)  Look at $c^2 + 4$ and ask yourself, *"Can I factor out a GCF?"* No.

The binomial $c^2 + 4$ is the sum of two squares, so it does not factor. This polynomial is prime. ■

## You Try 1

Factor completely.

a)  $6a^2 + 27a - 54$    b)  $5h^3 - h^2 + 15h - 3$    c)  $d^2 - 11d + 24$

d)  $8 - 8t^4$    e)  $1000x^3 - y^3$    f)  $4m^3 + 9$    g)  $w^2 + 22w + 121$

---

**Answers to You Try Exercises**

1)  a) $3(2a - 3)(a + 6)$   b) $(h^2 + 3)(5h - 1)$   c) $(d - 3)(d - 8)$   d) $8(1 + t^2)(1 + t)(1 - t)$
    e) $(10x - y)(100x^2 + 10xy + y^2)$   f) prime   g) $(w + 11)^2$

---

## Putting It All Together
## Summary Exercises

### Objective 1: Learn Strategies for Factoring a Given Polynomial

Factor completely.

1)  $c^2 + 15c + 56$

2)  $r^2 - 100$

VIDEO 3)  $uv + 6u + 9v + 54$

4)  $5t^2 - 36t - 32$

5)  $2p^2 - 13p + 21$

6)  $h^2 - 22h + 121$

7)  $9v^5 + 90v^4 - 54v^3$

8)  $m^2 + 6mn - 40n^2$

9)  $24q^3 + 52q^2 - 32q$

10)  $5k^3 - 40$

11)  $g^3 + 125$

12)  $xy - x - 9y + 9$

13)  $144 - w^2$

14)  $z^2 - 11z + 42$

15)  $9r^2 + 12rt + 4t^2$

16)  $40b - 35$

17)  $7n^4 - 63n^3 - 70n^2$

18)  $4x^2 + 4x - 15$

19)  $9h^2 + 25$

20)  $4abc - 24ab + 12ac - 72a$

VIDEO 21)  $40x^3 - 135$

22)  $49c^2 + 56c + 16$

23)  $m^2 - \dfrac{1}{100}$

24)  $p^2 + 10p + 14$

25)  $20x^2y + 6 - 24x^2 - 5y$

26)  $100a^5b - 36ab^3$

27)  $p^2 + 17pq + 30q^2$

28)  $8k^3 + 64$

29)  $t^2 - 2t - 16$

30)  $12g^4h^3 + 54g^3h + 30g^2h$

31)  $50n^2 - 40n + 8$

32)  $8a^2 - a - 9$

33)  $36r^2 + 57rs + 21s^2$

34)  $t^2 - \dfrac{81}{169}$

35)  $81x^4 - y^4$

36)  $v^2 - 23v + 132$

VIDEO 37)  $2a^2 - 10a - 72$

38)  $p^2q - q - 6p^2 + 6$

39)  $h^2 - \dfrac{2}{5}h + \dfrac{1}{25}$

40)  $m^3 - 64$

41)  $16uv + 24u - 10v - 15$

42)  $-27r^3 + 144r^2 - 180r$

VIDEO 43)  $8b^2 - 14b - 15$

44)  $12b^2 + 36b + 27$

45)  $8y^4z^3 - 28y^3z^3 - 40y^3z^2 + 4y^2z^2$

46)  $49 - p^2$

47)  $2a^2 - 7a + 8$

48)  $6h^3k + 54h^2k^2 + 48hk^3$

49)  $16u^2 + 40uv + 25v^2$

50)  $b^4 - 16$

51)  $24k^2 + 31k - 15$

52)  $r^2 + 81$

53)  $5s^3 - 320t^3$

54)  $36w^6 - 84w^4 + 12w^3$

55)  $ab - a - b + 1$

56)  $d^2 + 16d + 64$

57)  $7h^2 - 7$

58)  $9p^2 - 18pq + 8p^2$

59)  $6m^2 - 60m + 150$

60)  $100x^4 + 49y^2$

61)  $121z^2 - 169$

62)  $64a^3 - 125b^3$

63)  $-12r^2 - 75r - 18$

64)  $9c^2 + 54c + 72$

65)  $n^3 + 1$

66)  $16t^2 + 8t + 1$

VIDEO 67)  $81u^4 - v^4$

68) $14v^3 + 12u^2 + 28uv^2 + 6uv$

69) $13h^2 + 15h + 2$

70) $2g^3 - 2g^2 - 112g$

71) $5t^7 - 8t^4$

72) $m^2 - \dfrac{144}{25}$

73) $d^2 - 7d - 30$

74) $25k^2 - 60k + 36$

75) $z^2 + 144$

76) $54w^3 + 16$

77) $r^2 + 2r + 1$

78) $b^2 - 19b + 84$

79) $49n^2 - 100$

80) $9y^4 - 81y^2$

Extend the concepts of Sections 7.1–7.4 to factor completely.

VIDEO 81) $(2z + 1)y^2 + 6(2z + 1)y - 55(2z + 1)$

82) $(2k - 7)h^2 - 4(2k - 7)h - 45(2k - 7)$

83) $(t - 3)^2 + 3(t - 3) - 4$

84) $(v + 8)^2 - 14(v + 8) + 48$

85) $(z + 7)^2 - 11(z + 7) + 28$

86) $(3n - 1)^2 - (3n - 1) - 72$

87) $(a + b)^2 - (a - b)^2$

88) $(x + y)^2 - (x + 3y)^2$

89) $(5p - 2q)^2 - (2p + q)^2$

90) $(4s + t)^2 - (3s - 2t)^2$

91) $(r + 2)^3 + 27$

92) $(d - 5)^3 + 8$

93) $(k - 7)^3 - 1$

94) $(2w + 3)^3 - 125$

95) $a^2 - 8a + 16 - b^2$

96) $x^2 + 6x + 9 - y^2$

97) $s^2 + 18s + 81 - t^2$

98) $m^2 - 2m + 1 - n^2$

## Section 7.5 Solving Quadratic Equations by Factoring

### Objectives

1. **Solve a Quadratic Equation of the Form $ab = 0$**

2. **Solve Quadratic Equations by Factoring**

3. **Solve Higher-Degree Equations by Factoring**

Earlier, we learned that a *linear equation in one variable* is an equation that can be written in the form $ax + b = 0$, where $a$ and $b$ are real numbers and $a \neq 0$.

In this section, we will learn how to solve *quadratic equations*.

### Definition

A **quadratic equation** can be written in the form $ax^2 + bx + c = 0$ where $a, b,$ and $c$ are real numbers and $a \neq 0$.

We say that a quadratic equation written in the form $ax^2 + bx + c = 0$ is in **standard form**. But quadratic equations can be written in other forms, too.

Some examples of quadratic equations are

$$x^2 + 13x + 36 = 0, \quad 5n(n - 3) = 0, \quad \text{and} \quad (z + 4)(z - 7) = -10.$$

Quadratic equations are also called *second-degree equations* because the highest power on the variable is 2.

There are many different ways to solve quadratic equations. In this section, we will learn how to solve them by factoring; other methods will be discussed later in this book.

Solving a quadratic equation by factoring is based on the *zero product rule:* If the product of two quantities is zero, then one or both of the quantities is zero.

For example, if $5y = 0$, then $y = 0$. If $p \cdot 4 = 0$, then $p = 0$. If $ab = 0$, then either $a = 0$, $b = 0$, or *both* $a$ and $b$ equal zero.

### Definition

**Zero product rule:** If $ab = 0$, then $a = 0$ or $b = 0$.

We will use this idea to solve quadratic equations by factoring.

## 1. Solve a Quadratic Equation of the Form $ab = 0$

**Example 1**

Solve.   a)   $p(p + 4) = 0$        b)   $(3x + 1)(x - 7) = 0$

**Solution**

a)   The zero product rule says that at least one of the factors on the left must equal zero in order for the *product* to equal zero.

$$p(p + 4) = 0$$

$p = 0$    or    $p + 4 = 0$         Set each factor equal to 0.
                $p = -4$         Solve.

Check the solutions in the original equation:

If $p = 0$:                                If $p = -4$:
$0(0 + 4) \overset{?}{=} 0$                                $-4(-4 + 4) \overset{?}{=} 0$
     $0(4) = 0$  ✓                                     $-4(0) = 0$  ✓

The solution set is $\{-4, 0\}$.

**Note**

It is important to remember that the factor $p$ gives us the solution 0.

b)   At least one of the factors on the left must equal zero for the *product* to equal zero.

$$(3x + 1)(x - 7) = 0$$

$3x + 1 = 0$                     $x - 7 = 0$     Set each factor equal to 0.
$3x = -1$          or
$x = -\dfrac{1}{3}$                     $x = 7$     Solve each equation.

Check in the original equation:

If $x = -\dfrac{1}{3}$                                If $x = 7$:

$\left[3\left(-\dfrac{1}{3}\right) + 1\right]\left(-\dfrac{1}{3} - 7\right) \overset{?}{=} 0$          $[3(7) + 1](7 - 7) \overset{?}{=} 0$

$(-1 + 1)\left(-\dfrac{22}{3}\right) \overset{?}{=} 0$                     $22(0) = 0$  ✓

$0\left(-\dfrac{22}{3}\right) = 0$  ✓

The solution set is $\left\{-\dfrac{1}{3}, 7\right\}$.

**You Try 1**

Solve

a)   $k(k + 2) = 0$          b)   $(2r - 3)(r + 6) = 0$

## 2. Solve Quadratic Equations by Factoring

If the equation is in standard form, $ax^2 + bx + c = 0$, begin by factoring the expression.

**Example 2**

Solve $y^2 - 6y - 16 = 0$.

**Solution**

$$y^2 - 6y - 16 = 0$$
$$(y - 8)(y + 2) = 0 \qquad \text{Factor.}$$

$$y - 8 = 0 \quad \text{or} \quad y + 2 = 0 \qquad \text{Set each factor equal to zero.}$$
$$y = 8 \quad \text{or} \qquad y = -2 \qquad \text{Solve.}$$

Check in the original equation:

If $y = 8$:
$$(8)^2 - 6(8) - 16 \stackrel{?}{=} 0$$
$$64 - 48 - 16 = 0 \ \checkmark$$

If $y = -2$:
$$(-2)^2 - 6(-2) - 16 \stackrel{?}{=} 0$$
$$4 + 12 - 16 = 0 \ \checkmark$$

The solution set is $\{-2, 8\}$.

Here are the steps to use to solve a quadratic equation by factoring:

**Procedure** Solving a Quadratic Equation by Factoring

1) Write the equation in the form $ax^2 + bx + c = 0$ (standard form) so that all terms are on one side of the equal sign and zero is on the other side.

2) Factor the expression.

3) Set each factor equal to zero, and solve for the variable. (Use the zero product rule.)

4) Check the answer(s).

**You Try 2**

Solve $r^2 + 7r + 6 = 0$.

**Example 3**

Solve each equation by factoring.

a)  $2t^2 + 7t = 15$          b)  $9v^2 = -54v$          c)  $h^2 = -5(2h + 5)$

d)  $4(a^2 + 3) - 11a = 7a(a - 3) + 20$          e)  $(w - 4)(w - 5) = 2$

**Solution**

a)  Begin by writing $2t^2 + 7t = 15$ in standard form, $at^2 + bt + c = 0$.

$$2t^2 + 7t - 15 = 0 \qquad \text{Standard form}$$
$$(2t - 3)(t + 5) = 0 \qquad \text{Factor.}$$

$$2t - 3 = 0 \quad \text{or} \quad t + 5 = 0 \qquad \text{Set each factor equal to zero.}$$
$$2t = 3$$

$$t = \frac{3}{2} \quad \text{or} \qquad t = -5 \qquad \text{Solve.}$$

Check in the original equation:

If $t = \dfrac{3}{2}$:

$$2\left(\dfrac{3}{2}\right)^2 + 7\left(\dfrac{3}{2}\right) \stackrel{?}{=} 15$$

$$2\left(\dfrac{9}{4}\right) + \dfrac{21}{2} \stackrel{?}{=} 15$$

$$\dfrac{9}{2} + \dfrac{21}{2} \stackrel{?}{=} 15$$

$$\dfrac{30}{2} = 15 \checkmark$$

If $t = -5$:

$$2(-5)^2 + 7(-5) \stackrel{?}{=} 15$$

$$2(25) - 35 \stackrel{?}{=} 15$$

$$50 - 35 = 15 \checkmark$$

The solution set is $\left\{-5, \dfrac{3}{2}\right\}$.

b)  Write $9v^2 = -54v$ in standard form.

$$9v^2 + 54v = 0 \qquad \text{Standard form}$$
$$9v(v + 6) = 0 \qquad \text{Factor.}$$

$$9v = 0 \quad \text{or} \quad v + 6 = 0 \qquad \text{Set each factor equal to zero.}$$
$$v = 0 \quad \text{or} \qquad v = -6 \qquad \text{Solve.}$$

Check. The solution set is $\{-6, 0\}$.

**Note**

Since both terms in $9v^2 = -54v$ are divisible by 9, we could have started part b) by dividing by 9:

$$\dfrac{9v^2}{9} = \dfrac{-54v}{9} \qquad \text{Divide by 9.}$$
$$v^2 = -6v$$
$$v^2 + 6v = 0 \qquad \text{Write in standard form.}$$
$$v(v + 6) = 0 \qquad \text{Factor.}$$

$$v = 0 \quad \text{or} \quad v + 6 = 0 \qquad \text{Set each factor equal to zero.}$$
$$d = -6 \qquad \text{Solve.}$$

The solution set is $\{-6, 0\}$. We get the same result.

**BE CAREFUL**

We cannot divide by $v$ even though each term contains a factor of $v$. Doing so would eliminate the solution of zero. *In general, we can divide an equation by a nonzero real number but we cannot divide an equation by a variable because we may eliminate a solution, and we may be dividing by zero.*

c)  To solve $h^2 = -5(2h + 5)$, begin by writing the equation in standard form.

$$h^2 = -10h - 25 \qquad \text{Distribute.}$$
$$h^2 + 10h + 25 = 0 \qquad \text{Write in standard form.}$$
$$(h + 5)^2 = 0 \qquad \text{Factor.}$$

Since $(h + 5)^2 = 0$ means $(h + 5)(h + 5) = 0$, setting each factor equal to zero will result in the same value for $h$.

$$h + 5 = 0 \qquad \text{Set } h + 5 = 0.$$
$$h = -5 \qquad \text{Solve.}$$

Check. The solution set is $\{-5\}$.

d)  We will have to perform several steps to write the equation in standard form.

$$4(a^2 + 3) - 11a = 7a(a - 3) + 20$$

$$4a^2 + 12 - 11a = 7a^2 - 21a + 20 \qquad \text{Distribute.}$$

Move the terms on the left side of the equation to the right side so that the coefficient of $a^2$ is positive.

$$0 = 3a^2 - 10a + 8 \qquad \text{Write in standard form.}$$

$$0 = (3a - 4)(a - 2) \qquad \text{Factor.}$$

$$3a - 4 = 0 \quad \text{or} \quad a - 2 = 0 \qquad \text{Set each factor equal to zero.}$$

$$3a = 4$$

$$a = \frac{4}{3} \quad \text{or} \quad a = 2 \qquad \text{Solve.}$$

Check. The solution set is $\left\{\dfrac{4}{3}, 2\right\}$.

e)  It is tempting to solve $(w - 4)(w - 5) = 2$ like this:

$$(w - 4)(w - 5) = 2$$

$$w - 4 = 2 \quad \text{or} \quad w - 5 = 2 \qquad \text{This is incorrect!}$$

*One side of the equation must equal zero in order to use the zero product rule.* Begin by multiplying on the left.

$$(w - 4)(w - 5) = 2$$

$$w^2 - 9w + 20 = 2 \qquad \text{Multiply using FOIL.}$$

$$w^2 - 9w + 18 = 0 \qquad \text{Standard form.}$$

$$(w - 6)(w - 3) = 0 \qquad \text{Factor.}$$

$$w - 6 = 0 \quad \text{or} \quad w - 3 = 0 \qquad \text{Set each factor equal to zero.}$$

$$w = 6 \quad \text{or} \quad w = 3 \qquad \text{Solve.}$$

The check is left to the student. The solution set is $\{3, 6\}$. ■

**You Try 3**

Solve.

a)  $5c^2 = 6c + 8$      b)  $3q^2 = -18q$      c)  $6n(n + 2) = -7(n - 1)$

d)  $(m - 5)(m - 10) = 6$      e)  $z(z + 3) = 40$

## 3. Solve Higher-Degree Equations by Factoring

Sometimes, equations that are not quadratics can be solved by factoring as well.

**Example 4**

Solve each equation.

a)  $(4x - 1)(x^2 - 8x - 20) = 0$      b)  $12n^3 - 108n = 0$

### Solution

a)  This is *not* a quadratic equation because if we multiplied the factors on the left we would get $4x^3 - 33x^2 - 72x + 20 = 0$. This is a *cubic* equation because the degree of the polynomial on the left is 3.

   The original equation is the product of two factors so we can use the zero product rule.

$$(4x - 1)(x^2 - 8x - 20) = 0$$
$$(4x - 1)(x - 10)(x + 2) = 0 \qquad \text{Factor.}$$

$$4x - 1 = 0 \quad \text{or} \quad x - 10 = 0 \quad \text{or} \quad x + 2 = 0 \qquad \text{Set each factor equal to zero.}$$
$$4x = 1$$
$$x = \frac{1}{4} \quad \text{or} \qquad x = 10 \quad \text{or} \qquad x = -2 \qquad \text{Solve.}$$

   The check is left to the student. The solution set is $\left\{ -2, \dfrac{1}{4}, 10 \right\}$.

b)  The GCF of the terms in the equation $12n^3 - 108n = 0$ is $12n$. Remember, however, that *we can divide an equation by a constant but we cannot divide an equation by a variable.* Dividing by a variable may eliminate a solution and may mean we are dividing by zero. So let's begin by dividing each term by 12.

$$\frac{12n^3}{12} - \frac{108n}{12} = \frac{0}{12} \qquad \text{Divide by 12.}$$
$$n^3 - 9n = 0 \qquad \text{Simplify.}$$
$$n(n^2 - 9) = 0 \qquad \text{Factor out } n.$$
$$n(n + 3)(n - 3) = 0 \qquad \text{Factor } n^2 - 9.$$
$$n = 0 \quad \text{or} \quad n + 3 = 0 \quad \text{or} \quad n - 3 = 0 \qquad \text{Set each factor equal to zero.}$$
$$n = -3 \qquad \qquad n = 3 \qquad \text{Solve.}$$

   Check. The solution set is $\{0, -3, 3\}$.    ■

**You Try 4**

Solve.

a)  $(5y + 3)(y^2 - 10y + 21) = 0$       b)  $8k^3 - 32k = 0$

In this section, it was possible to solve all of the equations by factoring. Below we show the relationship between solving a quadratic equation by factoring and solving it using a graphing calculator. In Chapter 10, we will learn other methods for solving quadratic equations.

### Using Technology

Solve $x^2 - x - 6 = 0$ using a graphing calculator.

Recall from Chapter 4 that to find the *x*-intercepts of the graph of an equation, we let $y = 0$ and solve the equation for *x*. If we let $y = x^2 - x - 6$, then solving $x^2 - x - 6 = 0$ is the same as finding the *x*-intercepts of the graph of $y = x^2 - x - 6$. The *x*-intercepts are also called zeros of the equation since they are the values of *x* that make $y = 0$. Enter $y = x^2 - x - 6$ into the calculator and display the graph using the standard viewing window. We obtain a graph called a *parabola*, and we can see that it has two *x*-intercepts.

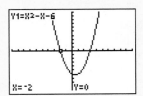

Since the scale for each tick mark on the graph is 1, it appears that the $x$-intercepts are $-2$ and $3$. To verify this, press $\boxed{\text{TRACE}}$, type $-2$, and press $\boxed{\text{ENTER}}$ as shown on the first screen. Since $x = -2$ and $y = 0$, $x = -2$ is an $x$-intercept. While still in "Trace" mode, type $3$ and press $\boxed{\text{ENTER}}$ as shown. Since $x = 3$ and $y = 0$, $x = 3$ is an $x$-intercept.

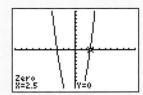

Sometimes an $x$-intercept is not an integer. Solve $2x^2 + x - 15 = 0$ using a graphing calculator.

Enter $2x^2 + x - 15$ into the calculator and press $\boxed{\text{GRAPH}}$. The $x$-intercept on the right side of the graph is between two tick marks, so it is not an integer. To find this $x$-intercept press $\boxed{\text{2nd}}$ $\boxed{\text{TRACE}}$ and select 2: zero. Now move the cursor to the left of one of the intercepts and press $\boxed{\text{ENTER}}$, then move the cursor again, so that it is to the right of the same intercept and press $\boxed{\text{ENTER}}$. Press $\boxed{\text{ENTER}}$ one more time, and the calculator will reveal the intercept and, therefore, one solution to the equation as shown on the third screen.

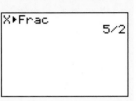

Press $\boxed{\text{2nd}}$ $\boxed{\text{MODE}}$ to return to the home screen. Press $\boxed{\text{X,T,}\Theta\text{,n}}$ $\boxed{\text{MATH}}$ $\boxed{\text{ENTER}}$ $\boxed{\text{ENTER}}$ to display the $x$-intercept in fraction form: $x = \dfrac{5}{2}$, as

shown on the final screen. Since the other $x$-intercept appears to be $-3$, press $\boxed{\text{TRACE}}$ $-3$ $\boxed{\text{ENTER}}$ to reveal that $x = -3$ and $y = 0$.

Solve using a graphing calculator.

1)  $x^2 - 5x - 6 = 0$

2)  $2x^2 - 9x - 5 = 0$

3)  $x^2 + 4x - 21 = 0$

4)  $5x^2 - 12x + 4 = 0$

5)  $x^2 + 2x - 35 = 0$

6)  $2x^2 - 11x + 12 = 0$

---

**Answers to You Try Exercises**

1)  a) $\{-2, 0\}$  b) $\left\{-6, \dfrac{3}{2}\right\}$     2)  $\{-6, -1\}$     3)  a) $\left\{-\dfrac{4}{5}, 2\right\}$  b) $\{-6, 0\}$

c) $\left\{-\dfrac{7}{2}, \dfrac{1}{3}\right\}$   d) $\{4, 11\}$   e) $\{-8, 5\}$     4)  a) $\left\{-\dfrac{3}{5}, 3, 7\right\}$  b) $\{0, -2, 2\}$

---

**Answers to Technology Exercises**

1)  $\{-1, 6\}$     2)  $\left\{-\dfrac{1}{2}, 5\right\}$     3)  $\{-7, 3\}$     4)  $\left\{\dfrac{2}{5}, 2\right\}$     5)  $\{5, -7\}$     6)  $\left\{4, \dfrac{3}{2}\right\}$

---

## 7.5 Exercises

**Objective 1: Solve a Quadratic Equation of the Form $ab = 0$**

1)  What is the standard form of an equation that is quadratic in $x$?

2)  A quadratic equation is also called a _____-degree equation.

3)  Identify the following equations as linear or quadratic.

a) $5x^2 + 3x - 7 = 0$

b) $6(p + 1) = 0$

c) $(n + 4)(n - 9) = 8$

d) $2w + 3(w - 5) = 4w + 9$

4) Which of the following equations are quadratic?

   a) $t^3 - 6t^2 - 4t + 24 = 0$

   b) $2(y^2 - 7) + 3y = 6y + 1$

   c) $3a(a - 11) = 0$

   d) $(c + 4)(2c^2 - 5c - 3) = 0$

5) Explain the zero product rule.

6) When Stephanie solves $m(m - 8) = 0$, she gets a solution set of $\{8\}$. Is this correct? Why or why not?

Solve each equation.

7) $(z + 11)(z - 4) = 0$    8) $(b + 1)(b + 7) = 0$

9) $(2r - 3)(r - 10) = 0$    10) $(5k - 4)(k + 9) = 0$

11) $d(d - 12) = 0$    12) $6w(w + 2) = 0$

13) $(3x + 5)^2 = 0$    14) $(c - 14)^2 = 0$

15) $(9h + 2)(2h + 1) = 0$    16) $(6q - 5)(2q - 3) = 0$

17) $\left(m + \dfrac{1}{4}\right)\left(m - \dfrac{2}{5}\right) = 0$    18) $\left(v + \dfrac{7}{3}\right)\left(v + \dfrac{4}{3}\right) = 0$

19) $n(n - 4.6) = 0$    20) $g(g + 0.7) = 0$

### Objective 2: Solve Quadratic Equations by Factoring

21) Can we solve $(k - 4)(k - 8) = 5$ by setting each factor equal to 5 like this: $k - 4 = 5$ or $k - 8 = 5$? Why or why not?

22) State two ways you could begin to solve $3x^2 + 18x + 24 = 0$.

Solve each equation.

23) $p^2 + 8p + 12 = 0$    24) $c^2 + 3c - 28 = 0$

25) $t^2 - t - 110 = 0$    26) $w^2 - 17w + 72 = 0$

27) $3a^2 - 10a + 8 = 0$    28) $2y^2 + 7y + 5 = 0$

29) $12z^2 + z - 6 = 0$    30) $8b^2 - 18b - 5 = 0$

31) $r^2 = 60 - 7r$    32) $h^2 + 20 = 12h$

33) $d^2 - 15d = -54$    34) $h^2 + 17h = -66$

35) $x^2 - 64 = 0$    36) $n^2 - 144 = 0$

37) $49 = 100u^2$    38) $81 = 4a^2$

39) $22k = -10k^2 - 12$    40) $4m - 48 = -24m^2$

41) $v^2 = 4v$    42) $x^2 = x$

43) $(z + 3)(z + 1) = 15$    44) $(c - 10)(c - 1) = -14$

45) $t(19 - t) = 84$    46) $48 = w(w - 2)$

47) $6k(k + 4) + 3 = 5(k^2 - 12) + 8k$

48) $7b(b + 1) + 15 = 6(b^2 + 2) + 11b$

49) $3(n^2 - 15) + 4n = 4n(n - 3) + 19$

50) $8(p^2 - 6) + 9p = 3p(3p + 7) - 13$

51) $\dfrac{1}{2}(m + 1)^2 = -\dfrac{3}{4}m(m + 5) - \dfrac{5}{2}$

52) $\dfrac{1}{8}(2y - 3)^2 + \dfrac{1}{8}y = \dfrac{1}{8}(y - 5)^2 - \dfrac{3}{4}$

### Objective 3: Solve Higher-Degree Equations by Factoring

53) To solve $5t^3 - 20t = 0$, Julio begins by dividing the equation by $5t$ to get $t^2 - 4 = 0$. Is this correct? Why or why not?

54) What are two possible first steps for solving $5t^3 - 20t = 0$?

The following equations are not quadratic but can be solved by factoring and applying the zero product rule. Solve each equation.

55) $7w(8w - 9)(w + 6) = 0$

56) $-5q(4q - 7)(q + 3) = 0$

57) $(6m + 7)(m^2 - 5m + 6) = 0$

58) $(9c - 2)(c^2 + 9c + 8) = 0$

59) $49h = h^3$    60) $r^3 = 36r$

61) $5w^2 + 36w = w^3$    62) $10p^2 - 25p = p^3$

63) $60a = 44a^2 - 8a^3$    64) $6z^3 + 16z = -50z^2$

65) $162b^3 - 8b = 0$    66) $75x = 27x^3$

### Mixed Exercises: Objectives 1–3

Solve each equation.

67) $-63 = 4y(y - 8)$    68) $-84 = g(g + 19)$

69) $\dfrac{1}{2}d(2 - d) - \dfrac{3}{2} = \dfrac{2}{5}d(d + 1) - \dfrac{7}{5}$

70) $(9p - 2)(p^2 - 10p - 11) = 0$

71) $a^2 - a = 30$    72) $45k + 27 = 18k^2$

73) $48t = 3t^3$

74) $\dfrac{1}{2}c(c + 3) - \dfrac{5}{4} = \dfrac{5}{8}(c^2 + 6) - \dfrac{1}{4}c$

75) $104r + 36 = 12r^2$

76) $3t(t - 5) + 14 = 5 - t(t + 3)$

77) $w^2 - 121 = 0$    78) $h^2 + 15h + 54 = 0$

79) $(2n - 5)(n^2 - 6n + 9) = 0$    80) $36b^2 + 60b = 0$

Extend the concepts of this section to solve each of the following equations.

81) $(2d - 5)^2 - (d + 6)^2 = 0$

82) $(3x + 8)^2 - (x + 4)^2 = 0$

83) $(11z - 4)^2 - (2z + 5)^2 = 0$

84) $(8g - 7)^2 - (g - 6)^2 = 0$

85) $2p^2(p - 4) + 9p(p - 4) + 9(p - 4) = 0$

86) $4a^2(a + 5) - 23a(a + 5) - 6(a + 5) = 0$

87) $10c^2(2c - 7) + 7(2c - 7) = 37c(2c - 7)$

88) $14y^2(5y - 8) = 4(5y - 8) - y(5y - 8)$

89) $h^3 + 8h^2 - h - 8 = 0$

90) $r^3 - 7r^2 - 4r + 28 = 0$

Find the indicated values for the following polynomial functions.

91) $f(x) = x^2 + 10x + 16$. Find $x$ so that $f(x) = 0$.

92) $g(t) = t^2 - 9t - 36$. Find $t$ so that $g(t) = 0$.

93) $g(a) = 2a^2 - 13a + 24$. Find $a$ so that $g(a) = 4$.

94) $h(c) = 2c^2 - 11c + 11$. Find $c$ so that $h(c) = -3$.

95) $P(a) = a^2 - 12$. Find $a$ so that $P(a) = 13$.

96) $Q(x) = 4x^2 + 19$. Find $x$ so that $Q(x) = 20$.

97) $h(t) = 3t^3 - 21t^2 + 18t$. Find $t$ so that $h(t) = 0$.

98) $F(x) = 2x^3 - 72x^2$. Find $x$ so that $F(x) = 0$.

## Section 7.6 Applications of Quadratic Equations

### Objectives

1. Solve Problems Involving Geometry
2. Solve Problems Involving Consecutive Integers
3. Solve Problems Using the Pythagorean Theorem
4. Solve Applied Problems Using Given Quadratic Equations

In Chapters 3 and 5, we learned how to solve applied problems involving linear equations. In this section, we will learn how to solve applications involving quadratic equations. Let's begin by reviewing the five steps for solving applied problems.

---

**Procedure**   Steps for Solving Applied Problems

**Step 1:** **Read** the problem carefully, more than once if necessary, until you understand it. Draw a picture, if applicable. Identify what you are being asked to find.

**Step 2:** **Choose a variable** to represent an unknown quantity. If there are any other unknowns, define them in terms of the variable.

**Step 3:** **Translate** the problem from English into an equation using the chosen variable.

**Step 4:** **Solve** the equation.

**Step 5:** **Check** the answer in the original problem, and **interpret** the solution as it relates to the problem. Be sure your answer makes sense in the context of the problem.

---

### 1. Solve Problems Involving Geometry

**Example 1**

A builder must cut a piece of tile into a right triangle. The tile will have an area of 40 in$^2$, and the height will be 2 in. shorter than the base. Find the base and height.

**Solution**

**Step 1:** **Read** the problem carefully. Draw a picture.

**Step 2:** **Choose a variable** to represent the unknown.

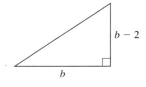

Let         $b$ = the base

$b - 2$ = the height

**Step 3:** **Translate** the information that appears in English into an algebraic equation. We are given the area of a triangular-shaped tile, so let's use the formula for the area of a triangle and substitute the expressions above for the base and the height and 40 for the area.

$$\text{Area} = \frac{1}{2}(\text{base})(\text{height})$$

$$40 = \frac{1}{2}(b)(b - 2) \qquad \text{Substitute Area} = 40, \text{base} = b, \text{height} = b - 2.$$

**Step 4:** **Solve** the equation. Eliminate the fraction first.

$$80 = (b)(b - 2) \qquad \text{Multiply by 2.}$$
$$80 = b^2 - 2b \qquad \text{Distribute.}$$
$$0 = b^2 - 2b - 80 \qquad \text{Write the equation in standard form.}$$
$$0 = (b - 10)(b + 8) \qquad \text{Factor.}$$

$$b - 10 = 0 \quad \text{or} \quad b + 8 = 0 \qquad \text{Set each factor equal to zero.}$$
$$b = 10 \quad \text{or} \qquad b = -8 \qquad \text{Solve.}$$

**Step 5:** **Check** the answer and **interpret** the solution as it relates to the problem. Since $b$ represents the length of the base of the triangle, it cannot be a negative number. So, $b = -8$ cannot be a solution. Therefore, the length of the base is 10 in., which will make the height $10 - 2 = 8$ in. The area, then, is $\frac{1}{2}(10)(8) = 40 \text{ in}^2$.

**You Try 1**

The height of a triangle is 3 cm more than its base. Find the height and base if its area is 35 cm$^2$.

## 2. Solve Problems Involving Consecutive Integers

In Chapter 3, we solved problems involving consecutive integers. Some applications involving consecutive integers lead to quadratic equations.

**Example 2**

Twice the sum of three consecutive odd integers is five less than the product of the two larger integers. Find the numbers.

**Solution**

**Step 1:** **Read** the problem carefully, and identify what we are being asked to find.

We must find three consecutive odd integers.

**Step 2:** **Choose a variable** to represent an unknown, and define the other unknowns in terms of the variable.

$$x = \text{the first odd integer}$$
$$x + 2 = \text{the second odd integer}$$
$$x + 4 = \text{the third odd integer}$$

**Step 3:** **Translate** the information that appears in English into an algebraic equation. Read the problem slowly and carefully, breaking it into small parts.

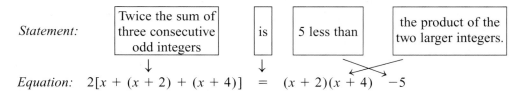

***Step 4:*** **Solve** the equation.

$$2[x + (x + 2) + (x + 4)] = (x + 2)(x + 4) - 5$$
$$2(3x + 6) = x^2 + 6x + 8 - 5 \qquad \text{Combine like terms; distribute.}$$
$$6x + 12 = x^2 + 6x + 3 \qquad \text{Combine like terms; distribute.}$$
$$0 = x^2 - 9 \qquad \text{Write in standard form.}$$
$$0 = (x + 3)(x - 3) \qquad \text{Factor.}$$

$$x + 3 = 0 \quad \text{or} \quad x - 3 = 0 \qquad \text{Set each factor equal to zero.}$$
$$x = -3 \qquad\qquad x = 3 \qquad \text{Solve.}$$

***Step 5:*** **Check** the answer and **interpret** the solution as it relates to the problem.

We get two sets of solutions. If $x = -3$, then the other odd integers are $-1$ and $1$. If $x = 3$, the other odd integers are $5$ and $7$.

Check these numbers in the original statement of the problem.

$$2[-3 + (-1) + 1] = (-1)(1) - 5 \qquad 2(3 + 5 + 7) = (5)(7) - 5$$
$$2(-3) = -1 - 5 \qquad\qquad 2(15) = 35 - 5$$
$$-6 = -6 \qquad\qquad\qquad 30 = 30 \qquad ■$$

**You Try 2**

Find three consecutive odd integers such that the product of the smaller two is 15 more than four times the sum of the three integers.

## 3. Solve Problems Using the Pythagorean Theorem

Recall that a **right triangle** contains a 90° angle. We can label it this way.

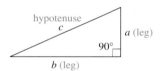

The side opposite the 90° angle is the longest side of the triangle and is called the **hypotenuse**. The other two sides are called the **legs**. The Pythagorean theorem states a relationship between the lengths of the sides of a right triangle. This is a very important relationship in mathematics and is used in many different ways.

---

**Definition** Pythagorean Theorem

Given a right triangle with legs of length $a$ and $b$ and hypotenuse of length $c$, the **Pythagorean theorem** states that $a^2 + b^2 = c^2$ [or $(\text{leg})^2 + (\text{leg})^2 = (\text{hypotenuse})^2$].

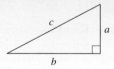

The Pythagorean theorem is true *only* for right triangles.

---

**Example 3**

Find the length of the missing side.

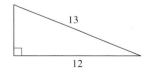

### Solution

Since this is a right triangle, we can use the Pythagorean theorem to find the length of the side. Let $a$ represent its length, and label the triangle.

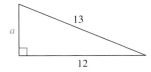

The length of the hypotenuse is 13, so $c = 13$. $a$ and 12 are legs. Let $b = 12$.

$$a^2 + b^2 = c^2 \qquad \text{Pythagorean theorem}$$
$$a^2 + (12)^2 = (13)^2 \qquad \text{Substitute values.}$$
$$a^2 + 144 = 169$$
$$a^2 - 25 = 0 \qquad \text{Write the equation in standard form.}$$
$$(a + 5)(a - 5) = 0 \qquad \text{Factor.}$$

$$a + 5 = 0 \quad \text{or} \quad a - 5 = 0 \qquad \text{Set each factor equal to 0.}$$
$$a = -5 \quad \text{or} \quad a = 5 \qquad \text{Solve.}$$

$a = -5$ does not make sense as an answer because the length of a side of a triangle cannot be negative. Therefore, $a = 5$.

*Check*:   $5^2 + (12)^2 \stackrel{?}{=} (13)^2$
$25 + 144 = 169$ ✓

### You Try 3

Find the length of the missing side.

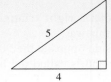

### Example 4

A community garden sits on a corner lot and is in the shape of a right triangle. One side is 10 ft longer than the shortest side, while the longest side is 20 ft longer than the shortest side. Find the lengths of the sides of the garden.

### Solution

**Step 1:**   **Read** the problem carefully, and identify what we are being asked to find. Draw a picture.

We must find the lengths of the sides of the garden.

**Step 2:**   **Choose a variable** to represent an unknown, and define the other unknowns in terms of this variable. Draw and label the picture.

$x$ = length of the shortest side (a leg)
$x + 10$ = length of the second side (a leg)
$x + 20$ = length of the third side (hypotenuse)

**Step 3:**   **Translate** the information that appears in English into an algebraic equation. We will use the Pythagorean theorem.

$$a^2 + b^2 = c^2 \qquad \text{Pythagorean theorem}$$
$$x^2 + (x + 10)^2 = (x + 20)^2 \qquad \text{Substitute.}$$

***Step 4:*** **Solve** the equation.

$$x^2 + (x + 10)^2 = (x + 20)^2$$

$$x^2 + x^2 + 20x + 100 = x^2 + 40x + 400 \qquad \text{Multiply using FOIL.}$$

$$2x^2 + 20x + 100 = x^2 + 40x + 400$$

$$x^2 - 20x - 300 = 0 \qquad\qquad \text{Write in standard form.}$$

$$(x - 30)(x + 10) = 0 \qquad\qquad \text{Factor.}$$

| $x - 30 = 0$ | or | $x + 10 = 0$ | Set each factor equal to 0. |
|---|---|---|---|
| $x = 30$ | or | $x = -10$ | Solve. |

***Step 5:*** **Check** the answer and **interpret** the solution as it relates to the problem.

The length of the shortest side, $x$, cannot be a negative number, so $x$ cannot equal $-10$. Therefore, the length of the shortest side must be 30 ft.

The length of the second side $= x + 10$, so $30 + 10 = 40$ ft.

The length of the longest side $= x + 20$, so $30 + 20 = 50$ ft.

Do these lengths satisfy the Pythagorean theorem? Yes.

$$a^2 + b^2 = c^2$$

$$(30)^2 + (40)^2 \overset{?}{=} (50)^2$$

$$900 + 1600 = 2500 \quad \checkmark$$

Therefore, the lengths of the sides are 30 ft, 40 ft, and 50 ft. ∎

**You Try 4**

A wire is attached to the top of a pole. The pole is 2 ft shorter than the wire, and the distance from the wire on the ground to the bottom of the pole is 9 ft less than the length of the wire. Find the length of the wire and the height of the pole.

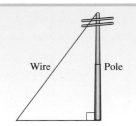

Wire        Pole

## 4. Solve Applied Problems Using Given Quadratic Equations

Let's see how to use a quadratic equation to model a real-life situation.

**Example 5**

An object is launched from a platform with an initial velocity of 32 ft/sec. The height $h$ (in feet) of the object $t$ seconds after it is released is given by the quadratic equation

$$h = -16t^2 + 32t + 20.$$

a) What is the initial height of the ball?

b) How long does it take the ball to reach a height of 32 feet?

c) How long does it take for the ball to hit the ground?

### Solution

a)  Since $t$ represents the number of seconds after the ball is thrown, $t = 0$ at the time of release.

Let $t = 0$ and solve for $h$.

$$h = -16(0)^2 + 32(0) + 20 \qquad \text{Substitute 0 for } t.$$
$$h = 0 + 0 + 20$$
$$h = 20$$

The initial height of the ball is 20 ft.

b)  We must find the *time* it takes for the ball to reach a height of 32 feet. Find $t$ when $h = 32$.

$$h = -16t^2 + 32t + 20$$
$$32 = -16t^2 + 32t + 20 \qquad \text{Substitute 32 for } h.$$
$$0 = -16t^2 + 32t - 12 \qquad \text{Write in standard form.}$$
$$0 = 4t^2 - 8t + 3 \qquad \text{Divide by } -4.$$
$$0 = (2t - 1)(2t - 3) \qquad \text{Factor.}$$

| $2t - 1 = 0$ | or | $2t - 3 = 0$ | Set each factor equal to 0. |
| $2t = 1$ | | $2t = 3$ | |
| $t = \dfrac{1}{2}$ | or | $t = \dfrac{3}{2}$ | Solve. |

How can two answers be possible? After $\dfrac{1}{2}$ sec, the ball is 32 feet above the ground *on its way up*, and after $\dfrac{3}{2}$ sec, the ball is 32 feet above the ground *on its way down*.

The ball reaches a height of 32 feet after $\dfrac{1}{2}$ sec and after $\dfrac{3}{2}$ sec.

c)  When the ball hits the ground, how high off the ground is it? *It is 0 feet high.*

Find $t$ when $h = 0$.

$$h = -16t^2 + 32t + 20$$
$$0 = -16t^2 + 32t + 20 \qquad \text{Substitute 0 for } h.$$
$$0 = 4t^2 - 8t - 5 \qquad \text{Divide by } -4.$$
$$0 = (2t + 1)(2t - 5) \qquad \text{Factor.}$$

| $2t + 1 = 0$ | or | $2t - 5 = 0$ | Set each factor equal to 0. |
| $2t = -1$ | | $2t = 5$ | |
| $t = -\dfrac{1}{2}$ | or | $t = \dfrac{5}{2}$ | Solve. |

Since $t$ represents time, $t$ cannot equal $-\dfrac{1}{2}$. Therefore, $t = \dfrac{5}{2}$.

The ball will hit the ground after $\dfrac{5}{2}$ sec (or 2.5 sec).

### Note

In Example 5, the equation can also be written using function notation $h(t) = -16t^2 + 32t + 20$ since the expression $-16t^2 + 32t + 20$ is a polynomial. Furthermore, $h(t) = -16t^2 + 32t + 20$ is a *quadratic function,* and we say that the height, $h$, is a function of the time, $t$. We will study quadratic functions in more detail in Chapter 10.

**You Try 5**

An object is thrown upward from a building. The height $h$ of the object (in feet) $t$ sec after the object is released is given by the quadratic equation

$$h = -16t^2 + 36t + 36.$$

a)   What is the initial height of the object?

b)   How long does it take the object to reach a height of 44 ft?

c)   How long does it take for the object to hit the ground?

---

**Answers to You Try Exercises**

1)   base = 7 cm; height = 10 cm       2)   13, 15, 17 or −3, −1, 1       3)   3

4)   length of wire = 17 ft; height of pole = 15 ft

5)   a) 36 ft   b) 0.25 sec and 2 sec   c) 3 sec

---

## 7.6 Exercises

**Objective 1: Solve Problems Involving Geometry**

Find the length and width of each rectangle.

1)   Area = 28 in$^2$

$x$
$x + 3$

2)   Area = 96 cm$^2$

$x - 3$
$x + 1$

Find the base and height of each triangle.

3)   Area = 44 cm$^2$

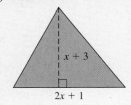

$x + 3$
$2x + 1$

4)   Area = 12 ft$^2$

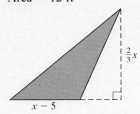

$\frac{2}{3}x$
$x - 5$

Find the base and height of each parallelogram.

5)   Area = 36 in$^2$

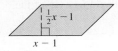

$\frac{1}{2}x - 1$
$x - 1$

6)   Area = 240 mm$^2$

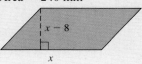

$x - 8$
$x$

7)   The volume of the box is 648 in$^3$. Find its height and width.

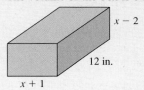

$x - 2$
12 in.
$x + 1$

8)   The volume of the box is 6 ft$^3$. Find its width and length.

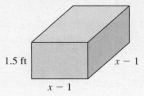

1.5 ft
$x - 1$
$x - 1$

Write an equation and solve.

9)   A rectangular sign is twice as long as it is wide. If its area is 8 ft$^2$, what are its length and width?

10)   An ad in a magazine is in the shape of a rectangle and occupies 88 in$^2$. The length is three inches longer than the width. Find the dimensions of the ad.

11) The top of a kitchen island is a piece of granite that has an area of 15 ft². It is 3.5 ft longer than it is wide. Find the dimensions of the surface.

12) To install an exhaust fan, a builder cuts a rectangular hole in the ceiling so that the width is 3 in. less than the length. The area of the hole is 180 in². Find the length and width of the hole.

13) A rectangular make-up case is 3 in. high and has a volume of 90 in³. The width is 1 in. less than the length. Find the length and width of the case.

14) An artist's sketchbox is 4 in. high and shaped like a rectangular solid. The width is three-fourths as long as the length. Find the length and width of the box if its volume is 768 in³.

15) The height of a triangle is 1 cm more than its base. Find the height and base if its area is 21 cm².

16) The area of a triangle is 24 in². Find the height and base if its height is one-third the length of the base.

## Objective 2: Solve Problems Involving Consecutive Integers

Write an equation and solve.

17) The product of two consecutive integers is 13 less than five times their sum. Find the integers.

18) The product of two consecutive integers is 10 less than four times their sum. Find the integers.

19) Find three consecutive even integers such that the product of the two smaller numbers is the same as twice the sum of all three integers.

20) Find three consecutive even integers such that five times the sum of the smallest and largest integers is the same as the square of the middle number.

21) Find three consecutive odd integers such that the product of the two larger numbers is 18 more than three times the sum of all three numbers.

22) Find three consecutive odd integers such that the square of the largest integer is 9 more than six times the sum of the two smaller numbers.

## Objective 3: Solve Problems Using the Pythagorean Theorem

23) In your own words, explain the Pythagorean theorem.

24) Can the Pythagorean theorem be used to find $a$ in this triangle? Why or why not?

Use the Pythagorean theorem to find the length of the missing side.

VIDEO 25)    26)

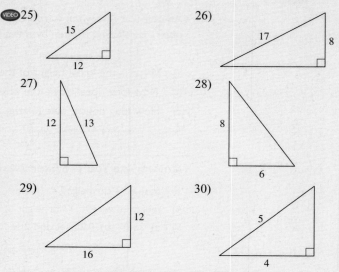

27)    28)

29)    30)

Find the lengths of the sides of each right triangle.

31)

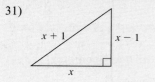

32)

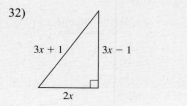

33)

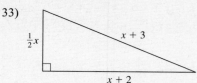

34)

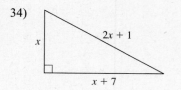

Write an equation and solve.

35) The longer leg of a right triangle is 2 cm more than the shorter leg. The length of the hypotenuse is 4 cm more than the shorter leg. Find the length of the hypotenuse.

36) The hypotenuse of a right triangle is 1 in. longer than the longer leg. The shorter leg measures 7 in. less than the longer leg. Find the measure of the longer leg of the triangle.

37) A 13-ft ladder is leaning against a wall. The distance from the top of the ladder to the bottom of the wall is 7 ft more than the distance from the bottom of the ladder to the wall. Find the distance from the bottom of the ladder to the wall.

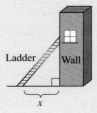

38) A wire is attached to the top of a pole. The wire is 4 ft longer than the pole, and the distance from the wire on the ground to the bottom of the pole is 4 ft less than the height of the pole. Find the length of the wire and the height of the pole.

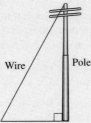

Write an equation and solve.

39) Lance and Alberto leave the same location with Lance heading due north and Alberto riding due east. When Alberto has ridden 4 miles, the distance between him and Lance is 2 miles more than Lance's distance from the starting point. Find the distance between Lance and Alberto.

40) A car heads east from an intersection while a motorcycle travels south. After 20 minutes, the car is 2 miles farther from the intersection than the motorcycle. The distance between the two vehicles is 4 miles more than the motorcycle's distance from the intersection. What is the distance between the car and the motorcycle?

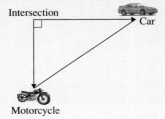

**Objective 4: Solve Applied Problems Using Given Quadratic Equations**

Solve.

41) A rock is dropped from a cliff and into the ocean. The height $h$ (in feet) of the rock after $t$ sec is given by $h = -16t^2 + 144$.

a) What is the initial height of the rock?

b) When is the rock 80 ft above the water?

c) How long does it take the rock to hit the water?

42) A Little League baseball player throws a ball upward. The height $h$ of the ball (in feet) $t$ seconds after the ball is released is given by $h = -16t^2 + 30t + 4$.

a) What is the initial height of the ball?

b) When is the ball 18 feet above the ground?

c) How long does it take for the ball to hit the ground?

Organizers of fireworks shows use quadratic and linear equations to help them design their programs. *Shells* contain the chemicals that produce the bursts we see in the sky. At a fireworks show, the shells are shot from *mortars* and when the chemicals inside the shells ignite, they explode, producing the brilliant bursts we see in the night sky.

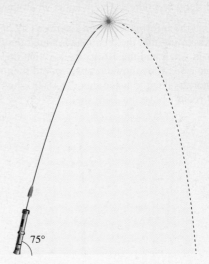

43) At a fireworks show, a 3-in. shell is shot from a mortar at an angle of 75°. The height, $y$ (in feet), of the shell $t$ sec after being shot from the mortar is given by the quadratic equation

$$y = -16t^2 + 144t$$

and the horizontal distance of the shell from the mortar, $x$ (in feet), is given by the linear equation

$$x = 39t.$$

(http://library.thinkquest.org/15384/physics/physics.html)

a) How high is the shell after 3 sec?

b) What is the shell's horizontal distance from the mortar after 3 sec?

c) The maximum height is reached when the shell explodes. How high is the shell when it bursts after 4.5 sec?

d) What is the shell's horizontal distance from its launching point when it explodes? (Round to the nearest foot.)

44) When a 10-in. shell is shot from a mortar at an angle of 75°, the height, $y$ (in feet), of the shell $t$ sec after being shot from the mortar is given by

$$y = -16t^2 + 264t$$

and the horizontal distance of the shell from the mortar, $x$ (in feet), is given by

$$x = 71t.$$

a) How high is the shell after 3 sec?

b) Find the shell's horizontal distance from the mortar after 3 sec.

c) The shell explodes after 8.25 sec. What is its height when it bursts?

d) What is the shell's horizontal distance from its launching point when it explodes? (Round to the nearest foot.)

e) Compare your answers to 43) a) and 44) a). What is the difference in their heights after 3 sec?

f) Compare your answers to 43) c) and 44) c). What is the difference in the shells' heights when they burst?

g) Use the information from Exercises 43 and 44. Assuming that the technicians timed the firings of the 3-in. shell and the 10-in. shell so that they exploded at the same time, how far apart would their respective mortars need to be so that the 10-in. shell would burst directly above the 3-in. shell?

45) An object is launched upward with an initial velocity of 96 ft/sec. The height $h$ (in feet) of the object after $t$ seconds is given by $h(t) = -16t^2 + 96t$.

a) From what height is the object launched?

b) When does the object reach a height of 128 ft?

c) How high is the object after 3 sec?

d) When does the object hit the ground?

46) An object is launched upward with an initial velocity of 128 ft/sec. The height $h$ (in feet) of the object after $t$ seconds is given by $h(t) = -16t^2 + 128t$.

a) From what height is the object launched?

b) Find the height of the object after 2 sec.

c) When does the object hit the ground?

47) The equation $R(p) = -9p^2 + 324p$ describes the relationship between the price of a ticket, $p$, in dollars, and the revenue, $R$, in dollars, from ticket sales at a music club. That is, the revenue is a function of price.

a) Determine the club's revenue from ticket sales if the price of a ticket is $15.

b) Determine the club's revenue from ticket sales if the price of a ticket is $20.

c) If the club is expecting its revenue from ticket sales to be $2916, how much should it charge for each ticket?

48) The equation $R(p) = -7p^2 + 700p$ describes the revenue, $R$, from ticket sales, in dollars, as a function of the price, $p$, in dollars, of a ticket to a fund-raising dinner. That is, the revenue is a function of price.

a) Determine the revenue if ticket price is $40.

b) Determine the revenue if the ticket price is $70.

c) If the goal of the organizers is to have ticket revenue of $17,500, how much should it charge for each ticket?

| **Definition/Procedure** | **Example** |
|---|---|

## 7.1 The Greatest Common Factor and Factoring by Grouping

| | |
|---|---|
| To **factor a polynomial** is to write it as a product of two or more polynomials.<br><br>To factor out a greatest common factor (GCF):<br><br>1) Identify the GCF of all of the terms of the polynomial.<br>2) Rewrite each term as the product of the GCF and another factor.<br>3) Use the distributive property to factor out the GCF from the terms of the polynomial.<br>4) Check the answer by multiplying the factors. **(p. 386)** | Factor out the greatest common factor.<br>$$6k^6 - 27k^5 + 15k^4$$<br>The GCF is $3k^4$.<br>$$6k^6 - 27k^5 + 15k^4 = (3k^4)(2k^2) - (3k^4)(9k) + (3k^4)(5)$$<br>$$= 3k^4(2k^2 - 9k + 5)$$<br>Check: $3k^4(2k^2 - 9k + 5) = 6k^6 - 27k^5 + 15k^4$ ✓ |
| The first step in factoring any polynomial is to ask yourself, *"Can I factor out a GCF?"*<br><br>The last step in factoring any polynomial is to ask yourself, *"Can I factor again?"*<br><br>Try to **factor by grouping** when you are asked to factor a polynomial containing four terms.<br><br>1) Make two groups of two terms so that each group has a common factor.<br>2) Take out the common factor from each group of terms.<br>3) Factor out the common factor using the distributive property.<br>4) Check the answer by multiplying the factors. **(p. 390)** | Factor completely. $10xy + 5y - 8x - 4$<br><br>Since the four terms have a GCF of 1, we will not factor out a GCF. Begin by grouping two terms together so that each group has a common factor.<br><br>$$\underbrace{10xy + 5y}\;\; \underbrace{-8x - 4}$$<br>$$= 5y(2x + 1) - 4(2x + 1) \quad \text{Take out the common factor.}$$<br>$$= (2x + 1)(5y - 4) \quad \text{Factor out } (2x + 1).$$<br>**Check:** $(2x + 1)(5y - 4) = 10xy + 5y - 8x - 4$ ✓ |

## 7.2 Factoring Trinomials of the Form $x^2 + bx + c$

| | |
|---|---|
| **Factoring $x^2 + bx + c$**<br><br>If $x^2 + bx + c = (x + m)(x + n)$, then<br><br>1) if $b$ and $c$ are positive, then both $m$ and $n$ must be positive.<br>2) if $c$ is positive and $b$ is negative, then both $m$ and $n$ must be negative.<br>3) if $c$ is negative, then one integer, $m$, must be positive and the other integer, $n$, must be negative. **(p. 394)** | Factor completely.<br>a) $t^2 + 9t + 20$<br><br>Think of two numbers whose **product** is 20 and whose **sum** is 9. **4 and 5** Then,<br>$$t^2 + 9t + 20 = (t + 4)(t + 5)$$<br>b) $3s^3 - 33s^2 + 54s$<br><br>Begin by factoring out the GCF of $3s$.<br>$$3s^3 - 33s^2 + 54s = 3s(s^2 - 11s + 18) = 3s(s - 2)(s - 9)$$ |

## 7.3 Factoring Trinomials of the Form $ax^2 + bx + c$ ($a \neq 1$)

| | |
|---|---|
| Factoring $ax^2 + bx + c$ by **grouping. (p. 400)** | Factor completely. $5t^2 + 18t - 8$<br><br>$$\text{Sum is 18.}$$<br>$$\downarrow$$<br>$$5t^2 + 18t - 8$$<br>$$\text{Product: } 5 \cdot (-8) = -40$$<br><br>Think of two integers whose **product** is $-40$ and whose **sum** is 18. **20 and $-2$** |

| Definition/Procedure | Example |
|---|---|
| | Factor by grouping.<br><br>$$5t^2 + 18t - 8 = \underbrace{5t^2 + 20t}_{\text{Group}} \underbrace{- 2t - 8}_{\text{Group}} \qquad \text{Write } 18t \text{ as } 20t - 2t.$$<br>$$= 5t(t + 4) - 2(t + 4)$$<br>$$= (t + 4)(5t - 2)$$ |
| **Factoring $ax^2 + bx + c$ by trial and error.**<br><br>When approaching a problem in this way, we must keep in mind that we are reversing the FOIL process. **(p. 401)** | Factor completely. $4x^2 - 16x + 15$<br><br>$$4x^2 - 16x + 15 = (2x - 3)(2x - 5)$$<br><br>$$4x^2 - 16x + 15 = (2x - 3)(2x - 5)$$ |

## 7.4 Factoring Special Trinomials and Binomials

| A **perfect square trinomial** is a trinomial that results from squaring a binomial.<br><br>**Factoring a Perfect Square Trinomial**<br><br>$$a^2 + 2ab + b^2 = (a + b)^2$$<br>$$a^2 - 2ab + b^2 = (a - b)^2 \quad \textbf{(p. 407)}$$ | Factor completely.<br>a) $g^2 + 22g + 121 = (g + 11)^2$<br>$\quad a = g \qquad b = 11$<br><br>b) $16d^2 - 8d + 1 = (4d - 1)^2$<br>$\quad a = 4d \qquad b = 1$ |
| **Factoring the Difference of Two Squares**<br><br>$$a^2 - b^2 = (a + b)(a - b) \quad \textbf{(p. 409)}$$ | Factor completely.<br>$$w^2 - 64 = (w + 8)(w - 8)$$<br>$\quad\downarrow \qquad \downarrow$<br>$(w)^2 \quad (8)^2 \quad a = w, b = 8$ |
| **Factoring the Sum and Difference of Two Cubes**<br><br>$$a^3 + b^3 = (a + b)(a^2 - ab + b^2)$$<br>$$a^3 - b^3 = (a - b)(a^2 + ab + b^2) \quad \textbf{(p. 412)}$$ | Factor completely.<br><br>$$r^3 + 64 = (r + 4)\left[(r)^2 - (r)(4) + (4)^2\right]$$<br>$\quad\downarrow \quad \downarrow$<br>$(r)^3 \ (4)^3 \quad a = r, b = 4$<br><br>$$r^3 + 64 = (r + 4)(r^2 - 4r + 16)$$ |

## 7.5 Solving Quadratic Equations by Factoring

| A **quadratic equation** can be written in the form $ax^2 + bx + c = 0$, where $a, b,$ and $c$ are real numbers and $a \neq 0$. **(p. 419)** | Some examples of quadratic equations are<br><br>$$5x^2 + 9 = 0, \quad y^2 = 4y + 21, \quad 4(p - 2)^2 = 8 - 7p$$ |
| To solve a quadratic equation by factoring, use the **zero product rule:** If $ab = 0$, then $a = 0$ or $b = 0$. **(p. 419)** | Solve $(y + 7)(y - 4) = 0$<br><br>$y + 7 = 0 \quad$ or $\quad y - 4 = 0 \qquad$ Set each factor equal to zero.<br>$\quad\ y = -7 \quad$ or $\qquad\ y = 4 \qquad$ Solve.<br><br>The solution set is $\{-7, 4\}$. |

| Definition/Procedure | Example |
|---|---|

**Steps for Solving a Quadratic Equation by Factoring**

1) Write the equation in the form $ax^2 + bx + c = 0$.
2) Factor the expression.
3) Set each factor equal to zero, and solve for the variable.
4) Check the answer(s). **(p. 421)**

Solve $5p^2 - 11 = 3p^2 + 3p - 9$.

$$2p^2 - 3p - 2 = 0 \qquad \text{Standard form}$$
$$(2p + 1)(p - 2) = 0 \qquad \text{Factor.}$$

$$2p + 1 = 0 \qquad \text{or} \qquad p - 2 = 0$$
$$2p = -1$$
$$p = -\frac{1}{2} \qquad \text{or} \qquad p = 2$$

The solution set is $\left\{ -\dfrac{1}{2}, 2 \right\}$. Check the answers.

### 7.6 Applications of Quadratic Equations

**Pythagorean Theorem**
Given a right triangle with legs of length $a$ and $b$ and hypotenuse of length $c$,

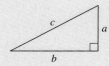

the Pythagorean theorem states that

$$a^2 + b^2 = c^2. \text{ (p. 429)}$$

Find the length of side $a$.

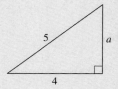

Let $b = 4$ and $c = 5$ in $\boldsymbol{a^2 + b^2 = c^2}$.

$$a^2 + (4)^2 = (5)^2$$
$$a^2 + 16 = 25$$
$$a^2 - 9 = 0$$
$$(a + 3)(a - 3) = 0$$

$$a + 3 = 0 \qquad \text{or} \qquad a - 3 = 0$$
$$a = 3 \qquad \text{or} \qquad a = -3$$

Reject $-3$ as a solution since the length of a side cannot be negative.
Therefore, $a = 3$.

**(7.1)  Find the greatest common factor of each group of terms.**

1) $40, 56$

2) $36y, 12y^2, 54y^2$

3) $15h^4, 45h^5, 20h^3$

4) $4c^4d^3, 20c^4d^2, 28c^2d$

**Factor out the greatest common factor.**

5) $63t + 45$

6) $21w^5 - 56w$

7) $2p^6 - 20p^5 + 2p^4$

8) $18a^3b^3 - 3a^2b^3 - 24ab^3$

9) $n(m + 8) - 5(m + 8)$

10) $x(9y - 4) + w(9y - 4)$

11) Factor out $-5r$ from $-15r^3 - 40r^2 + 5r$.

12) Factor out $-1$ from $-z^2 + 9z - 4$.

**Factor by grouping.**

13) $ab + 2a + 9b + 18$

14) $cd - 3c + 8d - 24$

15) $4xy - 28y - 3x + 21$

16) $hk^2 + 6h - k^2 - 6$

**(7.2)  Factor completely.**

17) $q^2 + 10q + 24$

18) $t^2 - 12t + 27$

19) $z^2 - 6z - 72$

20) $h^2 + 6h - 7$

21) $m^2 - 13mn + 30n^2$

22) $a^2 + 11ab + 30b^2$

23) $4v^2 - 24v - 64$

24) $7c^2 - 7c - 84$

25) $9w^4 + 9w^3 - 18w^2$

26) $5x^3y - 25x^2y^2 + 20xy^3$

**(7.3)  Factor completely.**

27) $3r^2 - 23r + 14$

28) $5k^2 + 11k + 6$

29) $4p^2 - 8p - 5$

30) $8d^2 + 29d - 12$

31) $12c^2 + 38c + 20$

32) $21n^2 - 54n + 24$

33) $10x^2 + 39xy - 27y^2$

34) $6g^2(h - 8)^2 + 23g(h - 8)^2 + 20(h - 8)^2$

**(7.4)  Factor completely.**

35) $w^2 - 49$

36) $121 - p^2$

37) $64t^2 - 25u^2$

38) $y^4 - 81$

39) $4b^2 + 9$

40) $12c^2 - 48d^2$

41) $64x - 4x^3$

42) $\dfrac{25}{9} - h^2$

43) $r^2 + 12r + 36$

44) $9z^2 - 24z + 16$

45) $20k^2 - 60k + 45$

46) $25a^2 + 20ab + 4b^2$

47) $v^3 - 27$

48) $8m^3 + 125$

49) $125x^3 + 64y^3$

50) $81p^4 - 3pq^3$

**(7.1–7.4)  Mixed Exercises**
**Factor completely.**

51) $10z^2 - 7z - 12$

52) $4c^2 + 24c + 36$

53) $9k^4 - 16k^2$

54) $14m^5 + 63m^4 + 21m^3$

55) $d^2 - 17d + 60$

56) $\dfrac{4}{25}t^2 - \dfrac{1}{9}u^2$

57) $3a^2b + a^2 - 12b - 4$

58) $h^2 + 100$

59) $48p^3 - 6q^3$

60) $t^4 + t$

61) $(x + 4)^2 - (y - 5)^2$

62) $8mn - 8m + 56n - 56$

63) $25c^2 - 20c + 4$

64) $12v^2 + 32v + 5$

**(7.5) Solve each equation.**

65) $y(3y + 7) = 0$

66) $(2n - 3)^2 = 0$

67) $2k^2 + 18 = 13k$

68) $3t^2 - 75 = 0$

69) $h^2 + 17h + 72 = 0$

70) $21 = 8p^2 - 2p$

71) $121 = 81r^2$

72) $12c = -c^2$

73) $3m^2 - 120 = 18m$

74) $x(16 - x) = 63$

75) $(w + 3)(w + 8) = -6$

76) $(2a + 3)^2 + 3a = a(a - 10) + 1$

77) $(5z + 4)(3z^2 - 7z + 4) = 0$

78) $18 = 9b^2 + 9b$

79) $3v + (v - 3)^2 = 5(v^2 - 4v + 1) + 8$

80) $6d^3 + 45d = 33d^2$

81) $45p^3 - 20p = 0$

82) $(r + 6)^2 - (4r - 1)^2 = 0$

**(7.6)**

83) Find the base and height if the area of the triangle is 18 cm$^2$.

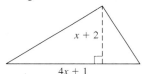

84) Find the length and width of the rectangle if its area is 60 in$^2$.

85) Find the base and height of the parallelogram if its area is 12 ft$^2$.

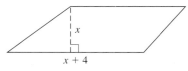

86) Find the height and length of the box if its volume is 480 in$^3$.

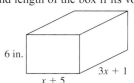

87) Use the Pythagorean theorem to find the length of the missing side.

88) Find the length of the hypotenuse.

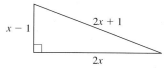

**Write an equation and solve.**

89) A rectangular mirror has an area of 10 ft², and it is 1.5 ft longer than it is wide. Find the dimensions of the mirror.

90) The base of a triangular banner is 1 ft less than its height. If the area is 3 ft², find the base and height.

91) The sum of three consecutive integers is one less than the square of the smallest number. Find the integers.

92) The product of two consecutive odd integers is 18 more than the square of the smaller number. Find the integers.

93) Desmond and Marcus leave an intersection with Desmond jogging north and Marcus jogging west. When Marcus is 1 mile farther from the intersection than Desmond, the distance between them is 2 miles more than Desmond's distance from the intersection. How far is Desmond from the intersection?

94) An object is thrown upward with an initial velocity of 68 ft/sec. The height $h$ (in feet) of the object $t$ seconds after it is thrown is given by

$$h = -16t^2 + 68t + 60.$$

a) How long does it take for the object to reach a height of 120 ft?

b) What is the initial height of the object?

c) What is the height of the object after 2 seconds?

d) How long does it take the object to hit the ground?

1) What is the first thing you should do when you are asked to factor a polynomial?

**Factor completely.**

2) $h^2 - 14h + 48$

3) $36 - v^2$

4) $7p^2 + 6p - 16$

5) $20a^3b^4 + 36a^2b^3 + 4ab^2$

6) $k^2 + 81$

7) $64t^3 - 27u^3$

8) $4z^3 + 28z^2 + 48z$

9) $36r^2 - 60r + 25$

10) $n^3 + 7n^2 - 4n - 28$

11) $x^2 - 3xy - 18y^2$

12) $81c^4 - d^4$

13) $p^2(q - 4)^2 + 17p(q - 4)^2 + 30(q - 4)^2$

14) $32w^2 + 28w - 30$

15) $k^8 + k^5$

**Solve each equation.**

16) $t^2 + 5t - 36 = 0$

17) $144r = r^3$

18) $49a^2 = 16$

19) $(y - 7)(v - 5) = 3$

20) $x - 2x(x + 4) = (2x + 3)^2 - 1$

21) $20k^2 - 52k = 24$

**Write an equation and solve.**

22) Find the length and width of a rectangular ice cream sandwich if its volume is 9 in$^3$.

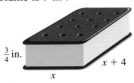

$\frac{3}{4}$ in.

$x + 4$

$x$

23) Find three consecutive odd integers such that the sum of the three numbers is 110 less than the product of the larger two integers.

24) Eric and Katrina leave home to go to work. Eric drives due east while his wife drives due south. At 8:30 A.M., Katrina is 3 miles farther from home than Eric, and the distance between them is 6 miles more than Eric's distance from home. Find Eric's distance from his house.

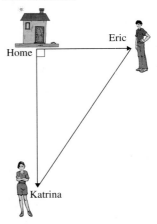

Home

Eric

Katrina

25) A rectangular cheerleading mat has an area of 252 ft$^2$. It is 7 times longer than it is wide. Find the dimensions of the mat.

26) An object is launched upward from the ground with an initial velocity of 200 ft/sec. The height $h$ (in feet) of the object after $t$ seconds is given by

$$h = -16t^2 + 200t.$$

a) Find the height of the object after 1 second.

b) Find the height of the object after 4 seconds.

c) When is the object 400 feet above the ground?

d) How long does it take for the object to hit the ground?

**Perform the indicated operation(s) and simplify.**

1) $\dfrac{2}{9} - \dfrac{5}{6} + \dfrac{1}{3}$

2) $\dfrac{35}{48} \cdot \dfrac{32}{63}$

**Simplify. The answer should not contain any negative exponents.**

3) $2(-3p^4 q)^2$

4) $\dfrac{28a^8 b^4}{40a^{-3} b^9}$

5) Write 0.0000839 in scientific notation.

6) Solve $\dfrac{3}{8}t + \dfrac{1}{2} = \dfrac{1}{4}(4t + 1) - \dfrac{3}{4}t$

7) Solve for $b_2$. $A = \dfrac{1}{2}h(b_1 + b_2)$

8) Solve $5 - \dfrac{2}{3}k \le 13$.

9) Graph $y = -\dfrac{1}{4}x + 2$.

10) Write the equation of the line parallel to $5x - 3y = 7$ containing the point $(-5, -7)$. Express the answer in slope-intercept form.

11) Use any method to solve this system of equations.

$$4 + 2(1 - 3x) + 7y = 3x - 5y$$
$$10y + 9 = 3(2x + 5) + 2y$$

12) Write a system of equations and solve.
Tickets to a high school football game cost $6.00 for an adult and $2.00 for a student. The revenue from the sale of 465 tickets was $1410. Find the number of adult tickets and number of student tickets sold.

**Multiply and simplify.**

13) $(4w - 7)(2w + 3)$

14) $(3n + 4)^2$

15) $(6z - 5)(2z^2 + 7z - 3)$

16) Add $(11v^2 - 16v + 4) + (2v^2 + 3v - 9)$.

**Divide.**

17) $\dfrac{16x^3 - 57x + 14}{4x - 7}$

18) $\dfrac{24m^2 - 8m + 4}{8m}$

**Factor completely.**

19) $6c^2 + 15c - 54$

20) $r^2 + 16$

21) $xy^2 + 4y^2 - x - 4$

22) $\dfrac{1}{4} - b^2$

23) $h^3 + 125$

**Solve.**

24) $12 = 13t - 3t^2$

25) $24n^3 = 54n$

# Rational Expressions

## Algebra at Work: Ophthalmology

At the beginning of Chapter 7, we saw how an ophthalmologist, a doctor specializing in diseases of the eye, uses mathematics every day to treat his patients. Here we will see another example of how math is used in this branch of medicine.

Some formulas in optics involve rational expressions. If Calvin determines that one of his patients needs glasses, he would use the following formula to figure out the proper prescription:

$$P = \frac{1}{f}$$

where $f$ is the focal length, in meters, and $P$ is the power of the lens, in diopters.

While computers now aid in these calculations, physicians believe that it is still important to double-check the calculations by hand.

In this chapter, we will learn how to perform operations with rational expressions and how to solve equations, like the one above, for a specific variable.

# Section 8.1 Simplifying Rational Expressions

## Objectives

1. Evaluate a Rational Expression
2. Find the Values of the Variable That Make a Rational Expression Undefined or Equal to Zero
3. Write a Rational Expression in Lowest Terms
4. Simplify a Rational Expression of the Form $\dfrac{a-b}{b-a}$
5. Write Equivalent Forms of a Rational Expression

## 1. Evaluate a Rational Expression

In Section 1.4, we defined a **rational number** as the quotient of two integers where the denominator does not equal zero. Some examples of rational numbers are

$$\frac{7}{8}, \qquad -\frac{2}{5}, \qquad 18\left(\text{since } 18 = \frac{18}{1}\right)$$

We can define a rational expression in a similar way. A rational expression is a quotient of two polynomials provided that the denominator does not equal zero. We state the definition formally next.

> **Definition**
>
> A **rational expression** is an expression of the form $\dfrac{P}{Q}$, where $P$ and $Q$ are polynomials and where $Q \neq 0$.

Some examples of rational expressions are

$$\frac{4k^3}{7}, \quad \frac{2n-1}{n+6}, \quad \frac{5}{t^2-3t-28}, \quad \frac{9x+2y}{x^2+y^2}$$

We can *evaluate* rational expressions for given values of the variables as long as the values do not make any denominators equal zero.

---

**Example 1**

Evaluate $\dfrac{x^2-9}{x+1}$ (if possible) for each value of $x$.

a) $x = 7$       b) $x = 3$       c) $x = -1$

**Solution**

a) $\dfrac{x^2-9}{x+1} = \dfrac{(7)^2-9}{(7)+1}$      Substitute 7 for $x$.

$= \dfrac{49-9}{7+1} = \dfrac{40}{8} = 5$

b) $\dfrac{x^2-9}{x+1} = \dfrac{(3)^2-9}{(3)+1}$      Substitute 3 for $x$.

$= \dfrac{9-9}{3+1} = \dfrac{0}{4} = 0$

c) $\dfrac{x^2-9}{x+1} = \dfrac{(-1)^2-9}{(-1)+1}$      Substitute $-1$ for $x$.

$= \dfrac{1-9}{0} = \dfrac{-8}{0}$

Undefined

Remember, a fraction is **undefined** when its denominator equals zero. Therefore, we say that $\dfrac{x^2-9}{x+1}$ is *undefined* when $x = -1$ since this value of $x$ makes the denominator equal zero. So, $x$ *cannot equal* $-1$ in this expression.

 **You Try 1**

Evaluate $\dfrac{k^2 - 1}{k + 11}$ (if possible) for each value of $k$.

a) $k = 9$        b) $k = -2$        c) $k = -11$        d) $k = 1$

## 2. Find the Values of the Variable That Make a Rational Expression Undefined or Equal to Zero

Parts b) and c) in Example 1 remind us about two important aspects of fractions and rational expressions.

**Note**

1) A fraction (rational expression) *equals zero* when its numerator equals zero.

2) A fraction (rational expression) is *undefined* when its denominator equals zero.

**Example 2**

For each rational expression, for what values of the variable

i)   does the expression equal zero?
ii)  is the expression undefined?

a) $\dfrac{m + 8}{m - 3}$        b) $\dfrac{2z}{z^2 - 5z - 36}$        c) $\dfrac{9c^2 - 49}{6}$        d) $\dfrac{4}{2w + 1}$

**Solution**

a)   i)   $\dfrac{m + 8}{m - 3} = 0$ when its *numerator* equals zero. Set the numerator equal to zero, and solve for $m$.

$$m + 8 = 0$$
$$m = -8$$

Therefore, $\dfrac{m + 8}{m - 3} = 0$ when $m = -8$.

ii)   $\dfrac{m + 8}{m - 3}$ is *undefined* when its *denominator* equals zero. Set the denominator equal to zero, and solve for $m$.

$$m - 3 = 0$$
$$m = 3$$

$\dfrac{m + 8}{m - 3}$ is *undefined* when $m = 3$. This means that any real number *except* 3 can be substituted for $m$ in this expression.

b)   i)   $\dfrac{2z}{z^2 - 5z - 36} = 0$ when its *numerator* equals zero. Set the numerator equal to zero, and solve for $z$.

$$2z = 0$$
$$z = \dfrac{0}{2} = 0$$

So, $\dfrac{2z}{z^2 - 5z - 36} = 0$ when $z = 0$.

ii) $\dfrac{2z}{z^2 - 5z - 36}$ is *undefined* when its *denominator* equals zero. Set the denominator equal to zero, and solve for $z$.

$$z^2 - 5z - 36 = 0$$
$$(z + 4)(z - 9) = 0 \qquad \text{Factor.}$$
$$z + 4 = 0 \quad \text{or} \quad z - 9 = 0 \qquad \text{Set each factor equal to 0.}$$
$$z = -4 \quad \text{or} \qquad z = 9 \qquad \text{Solve.}$$

$\dfrac{2z}{z^2 - 5z - 36}$ is undefined when $z = -4$ or $z = 9$. All real numbers *except* $-4$ and 9 can be substituted for $z$ in this expression.

c)   i)   To determine the values of $c$ that make $\dfrac{9c^2 - 49}{6} = 0$, set $9c^2 - 49 = 0$ and solve.

$$9c^2 - 49 = 0$$
$$(3c + 7)(3c - 7) = 0 \qquad \text{Factor.}$$
$$3c + 7 = 0 \quad \text{or} \quad 3c - 7 = 0$$
$$3c = -7 \qquad\qquad 3c = 7 \qquad \text{Set each factor equal to 0.}$$
$$c = -\frac{7}{3} \quad \text{or} \qquad c = \frac{7}{3} \qquad \text{Solve.}$$

So, $\dfrac{9c^2 - 49}{6} = 0$ when $c = -\dfrac{7}{3}$ or $c = \dfrac{7}{3}$.

ii) $\dfrac{9c^2 - 49}{6}$ is *undefined* when the denominator equals zero. However, the denominator is 6 and $6 \neq 0$. Therefore, there *is no value of $c$* that makes $\dfrac{9c^2 - 49}{6}$ undefined. *We say that* $\dfrac{9c^2 - 49}{6}$ *is defined for all real numbers.*

d)   i)   $\dfrac{4}{2w + 1} = 0$ when the numerator equals zero. The numerator is 4, and $4 \neq 0$. Therefore, $\dfrac{4}{2w + 1}$ will *never* equal zero.

ii) $\dfrac{4}{2w + 1}$ is *undefined* when $2w + 1 = 0$. Solve for $w$.

$$2w + 1 = 0$$
$$2w = -1$$
$$w = -\frac{1}{2}$$

So, $\dfrac{4}{2w + 1}$ is undefined when $w = -\dfrac{1}{2}$. Therefore, $w \neq -\dfrac{1}{2}$ in the expression.

**You Try 2**

For each rational expression, for what values of the variable

i) does the expression equal zero?

ii) is the expression undefined?

a) $\dfrac{v - 6}{v + 11}$   b) $\dfrac{9w}{w^2 - 12w + 20}$   c) $\dfrac{x^2 - 25}{8}$   d) $\dfrac{1}{5q + 4}$

All of the operations that can be performed with fractions can also be done with rational expressions. We begin our study of these operations with rational expressions by learning how to write a rational expression in lowest terms.

## 3. Write a Rational Expression in Lowest Terms

One way to think about writing a fraction such as $\dfrac{8}{12}$ in lowest terms is

$$\frac{8}{12} = \frac{2 \cdot 4}{3 \cdot 4} = \frac{2}{3} \cdot \frac{4}{4} = \frac{2}{3} \cdot 1 = \frac{2}{3}$$

Since $\dfrac{4}{4} = 1$, we can also think of reducing $\dfrac{8}{12}$ as $\dfrac{8}{12} = \dfrac{2 \cdot \cancel{4}}{3 \cdot \cancel{4}} = \dfrac{2}{3}$.

To write $\dfrac{8}{12}$ in lowest terms we can *factor* the numerator and denominator, then *divide* the numerator and denominator by the common factor, 4. This is the approach we use to write a rational expression in lowest terms.

---

**Definition**   **Fundamental Property of Rational Expressions**

If $P$, $Q$, and $C$ are polynomials such that $Q \neq 0$ and $C \neq 0$, then

$$\frac{PC}{QC} = \frac{P}{Q}$$

---

This property mirrors the example above since

$$\frac{PC}{QC} = \frac{P}{Q} \cdot \frac{C}{C} = \frac{P}{Q} \cdot 1 = \frac{P}{Q}$$

Or, we can also think of the reducing procedure as dividing the numerator and denominator by the common factor, $C$.

$$\frac{P\cancel{C}}{Q\cancel{C}} = \frac{P}{Q}$$

---

**Procedure**   **Writing a Rational Expression in Lowest Terms**

1) Completely **factor** the numerator and denominator.

2) **Divide** the numerator and denominator by the greatest common factor.

---

## Example 3

Write each rational expression in lowest terms.

a) $\dfrac{21r^{10}}{3r^4}$     b) $\dfrac{8a + 40}{3a + 15}$     c) $\dfrac{5n^2 - 20}{n^2 + 5n + 6}$

### Solution

a)   We can simplify $\dfrac{21r^{10}}{3r^4}$ using the quotient rule presented in Chapter 2.

$$\frac{21r^{10}}{3r^4} = 7r^6$$

Divide 21 by 3 and use the quotient rule:
$\dfrac{r^{10}}{r^4} = r^{10-4} = r^6.$

b) $\dfrac{8a + 40}{3a + 15} = \dfrac{8\cancel{(a+5)}}{3\cancel{(a+5)}}$        Factor.

$= \dfrac{8}{3}$        Divide out the common factor, $a + 5$.

c) $\dfrac{5n^2 - 20}{n^2 + 5n + 6} = \dfrac{5(n^2 - 4)}{(n + 2)(n + 3)}$        Factor.

$= \dfrac{5\cancel{(n+2)}(n - 2)}{\cancel{(n+2)}(n + 3)}$        Factor completely.

$= \dfrac{5(n - 2)}{n + 3}$        Divide out the common factor, $n + 2$.

■

 **BE CAREFUL**

Notice that we divided by *factors* not *terms*.

$\dfrac{\cancel{x+5}}{2\cancel{(x+5)}} = \dfrac{1}{2}$                                   $\dfrac{\cancel{x}}{x + 5} \neq \dfrac{1}{5}$

Divide by the *factor* $x + 5$.        We cannot divide by $x$ because the $x$ in the denominator is a *term* in a sum.

 **You Try 3**

Write each rational expression in lowest terms.

a) $\dfrac{12b^8}{18b^3}$        b) $\dfrac{2h + 8}{7h + 28}$        c) $\dfrac{y^3 - 8}{9y^4 - 9y^3 - 18y^2}$

## 4. Simplify a Rational Expression of the Form $\dfrac{a - b}{b - a}$

Do you think that $\dfrac{x - 4}{4 - x}$ is in lowest terms? Let's look at it more closely to understand the answer.

$\dfrac{x - 4}{4 - x} = \dfrac{x - 4}{-1(-4 + x)}$        Factor $-1$ out of the denominator.

$= \dfrac{1\cancel{(x - 4)}}{-1\cancel{(x - 4)}}$        Rewrite $-4 + x$ as $x - 4$.

$= -1$        Divide out the common factor, $x - 4$.

$\dfrac{x - 4}{4 - x} = -1$

We can generalize this result as

 **Note**

1) $b - a = -1(a - b)$    and    2) $\dfrac{a - b}{b - a} = -1$

The terms in the numerator and denominator in 2) differ only in sign. The rational expression simplifies to $-1$.

**Example 4**

Write each rational expression in lowest terms.

a) $\dfrac{7 - t}{t - 7}$       b) $\dfrac{4h^2 - 25}{5 - 2h}$       c) $\dfrac{2x^3 - 12x^2 + 18x}{9 - x^2}$

**Solution**

a) $\dfrac{7 - t}{t - 7} = -1$ since $\dfrac{7 - t}{t - 7} = \dfrac{-1(t - 7)}{(t - 7)} = -1.$

b) $\dfrac{4h^2 - 25}{5 - 2h} = \dfrac{(2h + 5)\overset{-1}{\cancel{(2h - 5)}}}{\cancel{5 - 2h}}$        Factor.

$= -1(2h + 5)$        $\dfrac{2h - 5}{5 - 2h} = -1$

$= -2h - 5$        Distribute.

c) $\dfrac{2x^3 - 12x^2 + 18x}{9 - x^2} = \dfrac{2x(x^2 - 6x + 9)}{(3 + x)(3 - x)}$        Factor $2x$ out of the numerator.

$= \dfrac{2x(x - 3)^2}{(3 + x)(3 - x)}$        Factor the trinomial.

$= \dfrac{2x\overset{-1}{\cancel{(x - 3)}}(x - 3)}{(3 + x)\cancel{(3 - x)}}$        $\dfrac{x - 3}{3 - x} = -1$

$= -\dfrac{2x(x - 3)}{3 + x}$        The negative sign can be written in front of the fraction.

**You Try 4**

Write each rational expression in lowest terms.

a) $\dfrac{10 - m}{m - 10}$       b) $\dfrac{8a - 4b}{6b - 12a}$       c) $\dfrac{100 - 4k^2}{k^2 - 8k + 15}$

## 5. Write Equivalent Forms of a Rational Expression

The answer to Example 4c) can be written in several different ways. You should be able to recognize equivalent forms of rational expressions because there isn't always just one way to write the correct answer.

**Example 5**

Write the answer to Example 4c), $-\dfrac{2x(x - 3)}{3 + x}$, in three different ways.

**Solution**

The negative sign in front of a fraction can be applied to the numerator or to the denominator. $\left(\text{For example, } -\dfrac{4}{9} = \dfrac{-4}{9} = \dfrac{4}{-9}.\right)$ Applying this concept to rational expressions can result in expressions that look quite different but that are, actually, equivalent.

i)  Apply the negative sign to the denominator.

$$-\frac{2x(x-3)}{3+x} = \frac{2x(x-3)}{-1(3+x)}$$

$$= \frac{2x(x-3)}{-3-x} \qquad \text{Distribute.}$$

ii)  Apply the negative sign to the numerator.

$$-\frac{2x(x-3)}{3+x} = \frac{-2x(x-3)}{3+x}$$

iii)  Apply the negative sign to the numerator, but distribute the $-1$.

$$-\frac{2x(x-3)}{3+x} = \frac{(2x)(-1)(x-3)}{x+3}$$

$$= \frac{2x(-x+3)}{3+x} \qquad \text{Distribute.}$$

$$= \frac{2x(3-x)}{3+x} \qquad \text{Rewrite } -x+3 \text{ as } 3-x.$$

Therefore, $\dfrac{2x(x-3)}{-3-x}$, $\dfrac{-2x(x-3)}{3+x}$, and $\dfrac{2x(3-x)}{3+x}$ are *all* equivalent forms

of $-\dfrac{2x(x-3)}{3+x}$.

Keep this idea of equivalent forms of rational expressions in mind when checking your answers against the answers in the back of the book. Sometimes students believe their answer is wrong because it "looks different" when, in fact, it is an *equivalent form* of the given answer!

**You Try 5**

Write $\dfrac{-(1-t)}{5t-8}$ in three different ways.

**Answers to You Try Exercises**

1)  a) 4   b) $\dfrac{1}{3}$   c) undefined   d) 0      2)  a) i) 6   ii) $-11$   b) i) 0   ii) 2, 10

c)  i) $-5$, 5   ii) defined for all real numbers      d) i) never equals zero   ii) $-\dfrac{4}{5}$

3)  a) $\dfrac{2b^5}{3}$   b) $\dfrac{2}{7}$   c) $\dfrac{y^2+2y+4}{9y^2(y+1)}$      4)  a) $-1$   b) $-\dfrac{2}{3}$   c) $-\dfrac{4(5+k)}{k-3}$

5)  Some possibilities are $\dfrac{t-1}{5t-8}$, $-\dfrac{1-t}{5t-8}$, $\dfrac{1-t}{8-5t}$.

## 8.1 Exercises

**Objective 1: Evaluate a Rational Expression**

1) When is a fraction or a rational expression undefined?

2) When does a fraction or a rational expression equal 0?

Evaluate (if possible) for a) $x = 3$ and b) $x = -2$.

3) $\dfrac{2x - 1}{5x + 2}$

4) $\dfrac{3(x^2 + 1)}{x^2 + 2x + 1}$

Evaluate (if possible) for a) $z = 1$ and b) $z = -3$.

5) $\dfrac{(4z)^2}{z^2 - z - 12}$

6) $\dfrac{3(z^2 - 9)}{z^2 + 8}$

7) $\dfrac{15 + 5z}{16 - z^2}$

8) $\dfrac{4z - 3}{z^2 + 6z - 7}$

**Objective 2: Find the Values of the Variable That Make a Rational Expression Undefined or Equal to Zero**

9) How do you determine the values of the variable for which a rational expression is undefined?

10) If $x^2 + 9$ is the numerator of a rational expression, can that expression equal zero? Give a reason.

Determine the value(s) of the variable for which
   a) the expression equals zero.
   b) the expression is undefined.

11) $\dfrac{m + 4}{3m}$

12) $\dfrac{-y}{y + 3}$

13) $\dfrac{2w - 7}{4w + 1}$

14) $\dfrac{3x + 13}{2x + 13}$

15) $\dfrac{11v - v^2}{5v - 9}$

16) $-\dfrac{r + 5}{r^2 - 100}$

17) $\dfrac{8}{p}$

18) $\dfrac{22}{m - 1}$

19) $-\dfrac{7k}{k^2 + 9k + 20}$

20) $\dfrac{a - 9}{a^2 + 8a - 9}$

21) $\dfrac{c + 20}{2c^2 + 3c - 9}$

22) $\dfrac{4}{3f^2 - 13f + 10}$

23) $\dfrac{g^2 + 9g + 18}{9g}$

24) $\dfrac{6m - 11}{10}$

VIDEO 25) $\dfrac{4y}{y^2 + 9}$

26) $\dfrac{q^2 + 49}{7}$

**Objective 3: Write a Rational Expression in Lowest Terms**

Write each rational expression in lowest terms.

27) $\dfrac{7x(x - 11)}{3(x - 11)}$

28) $\dfrac{24(g + 3)}{6(g + 3)(g - 5)}$

29) $\dfrac{24g^2}{56g^4}$

30) $\dfrac{99d^7}{9d^3}$

31) $\dfrac{4d - 20}{5d - 25}$

32) $\dfrac{12c - 3}{8c - 2}$

33) $\dfrac{-14h - 56}{6h + 24}$

34) $\dfrac{-15v^2 + 12}{40v^2 - 32}$

35) $\dfrac{39u^2 + 26}{30u^2 + 20}$

36) $\dfrac{3q + 15}{-7q - 35}$

37) $\dfrac{g^2 - g - 56}{g + 7}$

38) $\dfrac{b^2 + 9b + 20}{b^2 + b - 12}$

39) $\dfrac{t - 5}{t^2 - 25}$

40) $\dfrac{r + 9}{r^2 + 7r - 18}$

41) $\dfrac{3c^2 + 28c + 32}{c^2 + 10c + 16}$

42) $\dfrac{3k^2 - 36k + 96}{k - 8}$

43) $\dfrac{q^2 - 25}{2q^2 - 7q - 15}$

44) $\dfrac{6p^2 + 11p - 10}{9p^2 - 4}$

VIDEO 45) $\dfrac{w^3 + 125}{5w^2 - 25w + 125}$

46) $\dfrac{6w^3 - 48}{w^2 + 2w + 4}$

47) $\dfrac{9c^2 - 27c + 81}{c^3 + 27}$

48) $\dfrac{4x + 20}{x^3 + 125}$

49) $\dfrac{4u^2 - 20u + 4uv - 20v}{13u + 13v}$

50) $\dfrac{ab + 3a - 6b - 18}{b^2 - 9}$

51) $\dfrac{m^2 - n^2}{m^3 - n^3}$

52) $\dfrac{a^3 + b^3}{a^2 - b^2}$

**Objective 4: Simplify a Rational Expression of the Form $\dfrac{a - b}{b - a}$**

53) Any rational expression of the form $\dfrac{a - b}{b - a}$ $(a \neq b)$ reduces to what?

54) Does $\dfrac{h + 4}{h - 4} = -1$?

Write each rational expression in lowest terms.

55) $\dfrac{8 - q}{q - 8}$

56) $\dfrac{m - 15}{15 - m}$

57) $\dfrac{m^2 - 121}{11 - m}$

58) $\dfrac{k - 9}{162 - 2k^2}$

59) $\dfrac{36 - 42x}{7x^2 + 8x - 12}$

60) $\dfrac{a^2 - 6a - 27}{9 - a}$

61) $\dfrac{16 - 4b^2}{b - 2}$

62) $\dfrac{45 - 9v}{v^2 - 25}$

63) $\dfrac{y^3 - 3y^2 + 2y - 6}{21 - 7y}$

64) $\dfrac{8t^3 - 27}{9 - 4t^2}$

## Mixed Exercises: Objectives 1–4

Write each rational expression in lowest terms.

65) $\dfrac{18c + 45}{12c^2 + 18c - 30}$

66) $\dfrac{36n^3}{42n^9}$

67) $\dfrac{r^3 - t^3}{t^2 - r^2}$

68) $\dfrac{k^2 - 16k + 64}{k^3 - 8k^2 + 9k - 72}$

69) $\dfrac{b^2 + 6b - 72}{4b^2 + 52b + 48}$

70) $\dfrac{5p^2 - 13p + 6}{32 - 8p^2}$

71) $\dfrac{28h^4 - 56h^3 + 7h}{7h}$

72) $\dfrac{z^2 + 5z - 36}{64 - z^3}$

73) $\dfrac{14 - 6w}{12w^3 - 28w^2}$

74) $\dfrac{54d^6 + 6d^5 - 42d^3 - 18d^2}{6d^2}$

75) $\dfrac{-5v - 10}{v^3 - v^2 - 4v + 4}$

76) $\dfrac{38x^2 + 38}{-12x^2 - 12}$

## Objective 5: Write Equivalent Forms of a Rational Expression

Find three equivalent forms of each rational expression.

77) $-\dfrac{u + 7}{u - 2}$

78) $-\dfrac{8y - 1}{2y + 5}$

79) $-\dfrac{9 - 5t}{2t - 3}$

80) $\dfrac{w - 6}{-4w + 7}$

81) $\dfrac{-12m}{m^2 - 3}$

82) $\dfrac{-9x - 11}{18 - x}$

Reduce to lowest terms
   a) using long division.
   b) using the methods of this section.

83) $\dfrac{4y^2 - 11y + 6}{y - 2}$

84) $\dfrac{2x^2 + x - 28}{x + 4}$

85) $\dfrac{8a^3 + 125}{2a + 5}$

86) $\dfrac{27t^3 - 8}{3t - 2}$

Recall that the area of a rectangle is $A = lw$, where $w$ = width and $l$ = length. Solving for the width, we get $w = \dfrac{A}{l}$ and solving for the length gives us $l = \dfrac{A}{w}$.

Find the missing side in each rectangle.

87) Area $= 3x^2 + 8x + 4$

$x + 2$

Find the length.

88) Area $= 2y^2 - 3y - 20$

$2y + 5$

Find the width.

89) Area $= 2c^3 + 4c^2 + 8c + 16$

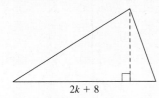
$c^2 + 4$

Find the width.

90) Area $= 3n^3 - 12n^2 - n + 4$

$n - 4$

Find the length.

Recall that the area of a triangle is $A = \dfrac{1}{2}bh$, where $b$ = length of the base and $h$ = height. Solving for the height, we get $h = \dfrac{2A}{b}$. Find the height of each triangle.

91) Area $= 3k^2 + 13k + 4$

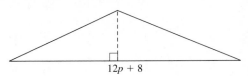
$2k + 8$

92) Area $= 6p^2 + 52p + 32$

$12p + 8$

## Section 8.2   Multiplying and Dividing Rational Expressions

**Objectives**

1. **Multiply Fractions**
2. **Multiply Rational Expressions**
3. **Divide Rational Expressions**

We multiply rational expressions the same way we multiply rational numbers. Multiply numerators, multiply denominators, and simplify.

---

**Procedure**   Multiplying Rational Expressions

If $\dfrac{P}{Q}$ and $\dfrac{R}{T}$ are rational expressions, then $\dfrac{P}{Q} \cdot \dfrac{R}{T} = \dfrac{PR}{QT}$.

To multiply two rational expressions, multiply their numerators, multiply their denominators, and simplify.

---

Let's begin by reviewing how we multiply two fractions.

### 1. Multiply Fractions

**Example 1**

Multiply $\dfrac{9}{16} \cdot \dfrac{8}{15}$.

**Solution**

We can multiply numerators, multiply denominators, then simplify by dividing out common factors, *or* we can divide out the common factors before multiplying.

$$\frac{9}{16} \cdot \frac{8}{15} = \frac{\cancel{3} \cdot 3}{2 \cdot \cancel{8}} \cdot \frac{\cancel{8}}{\cancel{3} \cdot 5}$$   Factor and divide out common factors.

$$= \frac{3}{2 \cdot 5}$$   Multiply.

$$= \frac{3}{10}$$   Simplify.

**You Try 1**

Multiply $\dfrac{8}{35} \cdot \dfrac{5}{12}$.

---

Multiplying rational expressions works the same way.

### 2. Multiply Rational Expressions

---

**Procedure**   Multiplying Rational Expressions

1) Factor.          2)   Reduce and multiply.

*All products must be written in lowest terms.*

---

**Example 2**

Multiply.

a) $\dfrac{18x^3}{y^4} \cdot \dfrac{y^7}{9x^8}$     b) $\dfrac{9c + 45}{6c^{10}} \cdot \dfrac{c^6}{c^2 - 25}$

c) $\dfrac{2n^2 - 11n - 6}{n^2 - 2n - 24} \cdot \dfrac{n^2 + 8n + 16}{2n^2 + n}$

**Solution**

a) $\dfrac{18x^3}{y^4} \cdot \dfrac{y^7}{9x^8} = \dfrac{\overset{2}{\cancel{18}}\cancel{x^3}}{\cancel{y^4}} \cdot \dfrac{\cancel{y^4} \cdot y^3}{9\cancel{x^3} \cdot x^5}$     Factor and reduce.

$= \dfrac{2y^3}{x^5}$     Multiply.

b) $\dfrac{9c + 45}{6c^{10}} \cdot \dfrac{c^6}{c^2 - 25} = \dfrac{\overset{3}{\cancel{9}}\cancel{(c + 5)}}{\underset{2}{\cancel{6}}\cancel{c^6} \cdot c^4} \cdot \dfrac{\cancel{c^6}}{\cancel{(c + 5)}(c - 5)}$     Factor and reduce.

$= \dfrac{3}{2c^4(c - 5)}$     Multiply.

c) $\dfrac{2n^2 - 11n - 6}{n^2 - 2n - 24} \cdot \dfrac{n^2 + 8n + 16}{2n^2 + n} = \dfrac{\cancel{(2n + 1)}\cancel{(n - 6)}}{\cancel{(n + 4)}\cancel{(n - 6)}} \cdot \dfrac{\overset{(n+4)}{\cancel{(n + 4)^2}}}{n\cancel{(2n + 1)}}$     Factor and reduce.

$= \dfrac{n + 4}{n}$     Multiply.

**You Try 2**

Multiply.

a) $\dfrac{n^7}{20m^9} \cdot \dfrac{10m^5}{n^2}$     b) $\dfrac{d^2}{d^2 - 4} \cdot \dfrac{4d - 8}{12d^5}$     c) $\dfrac{h^2 + 10h + 25}{3h^2 - 4h} \cdot \dfrac{3h^2 + 5h - 12}{h^2 + 8h + 15}$

### 3.  Divide Rational Expressions

When we divide rational numbers, we multiply by a reciprocal. For example,

$\dfrac{7}{4} \div \dfrac{3}{8} = \dfrac{7}{\underset{1}{\cancel{4}}} \cdot \dfrac{\overset{2}{\cancel{8}}}{3} = \dfrac{14}{3}$. We divide rational expressions the same way.

To divide rational expressions, we multiply the first rational expression by the reciprocal of the second rational expression.

**Procedure**   Dividing Rational Expressions

If $\dfrac{P}{Q}$ and $\dfrac{R}{T}$ are rational expressions with $Q$, $R$, and $T$ not equal to zero, then

$$\dfrac{P}{Q} \div \dfrac{R}{T} = \dfrac{P}{Q} \cdot \dfrac{T}{R} = \dfrac{PT}{QR}$$

Multiply the first rational expression by the reciprocal of the second rational expression.

**Example 3**

Divide.

a) $\dfrac{15a^7}{b^3} \div \dfrac{3a^4}{b^9}$  b) $\dfrac{t^2 - 16t + 63}{t^2} \div (t - 7)^2$

c) $\dfrac{x^2 - 9}{x^2 + 3x - 10} \div \dfrac{24 - 8x}{x^2 + 9x + 20}$

**Solution**

a) $\dfrac{15a^7}{b^3} \div \dfrac{3a^4}{b^9} = \dfrac{\overset{5a^3}{\cancel{15a^7}}}{b^3} \cdot \dfrac{\overset{b^6}{\cancel{b^9}}}{\cancel{3}\,\cancel{a^4}}$   Multiply by the reciprocal and reduce.

$\qquad\qquad = 5a^3 b^6$   Multiply.

Notice that we used the *quotient rule* for exponents to reduce:

$$\dfrac{a^7}{a^4} = a^3, \quad \dfrac{b^9}{b^3} = b^6$$

b) $\dfrac{t^2 - 16t + 63}{t^2} \div (t - 7)^2 = \dfrac{(t - 9)\cancel{(t-7)}}{t^2} \cdot \dfrac{1}{\underset{(t-7)}{\cancel{(t-7)^2}}}$   Since $(t - 7)^2$ can be written as $\dfrac{(t - 7)^2}{1}$, its reciprocal is $\dfrac{1}{(t - 7)^2}$.

$\qquad\qquad\qquad\qquad = \dfrac{t - 9}{t^2(t - 7)}$   Reduce and multiply.

c) $\dfrac{x^2 - 9}{x^2 + 3x - 10} \div \dfrac{24 - 8x}{x^2 + 9x + 20} = \dfrac{x^2 - 9}{x^2 + 3x - 10} \cdot \dfrac{x^2 + 9x + 20}{24 - 8x}$   Multiply by the reciprocal.

$\qquad\qquad = \dfrac{(x + 3)\overset{-1}{\cancel{(x - 3)}}}{\cancel{(x + 5)}(x - 2)} \cdot \dfrac{(x + 4)\cancel{(x + 5)}}{8\cancel{(3 - x)}}$   Factor; $\dfrac{x - 3}{3 - x} = -1$.

$\qquad\qquad = -\dfrac{(x + 3)(x + 4)}{8(x - 2)}$   Reduce and multiply. ∎

**You Try 3**

Divide.

a) $\dfrac{k^3}{12h^7} \div \dfrac{k^4}{16h^2}$  b) $\dfrac{w^2 + 4w - 45}{w} \div (w + 9)^2$

c) $\dfrac{2m^2 - m - 15}{1 - m^2} \div \dfrac{m^2 - 10m + 21}{12m - 12}$

Remember that a fraction, itself, represents division. That is, $\dfrac{30}{5} = 30 \div 5 = 6$. We can write division problems involving fractions and rational expressions in a similar way.

## Example 4

Divide.

a) $\dfrac{\dfrac{8}{35}}{\dfrac{16}{45}}$    b) $\dfrac{\dfrac{3w + 2}{5}}{\dfrac{9w^2 - 4}{10}}$

### Solution

a) $\dfrac{\dfrac{8}{35}}{\dfrac{16}{45}}$ means $\dfrac{8}{35} \div \dfrac{16}{45}$. Then,

$$\frac{8}{35} \div \frac{16}{45} = \frac{8}{35} \cdot \frac{45}{16} \qquad \text{Multiply by the reciprocal.}$$

$$= \frac{\overset{1}{\cancel{8}}}{\underset{7}{\cancel{35}}} \cdot \frac{\overset{9}{\cancel{45}}}{\underset{2}{\cancel{16}}} \qquad \text{Divide 8 and 16 by 8. Divide 45 and 35 by 5.}$$

$$= \frac{9}{14} \qquad \text{Multiply.}$$

b) $\dfrac{\dfrac{3w + 2}{5}}{\dfrac{9w^2 - 4}{10}}$ means $\dfrac{3w + 2}{5} \div \dfrac{9w^2 - 4}{10}$. Then,

$$\frac{3w + 2}{5} \div \frac{9w^2 - 4}{10} = \frac{3w + 2}{5} \cdot \frac{10}{9w^2 - 4} \qquad \text{Multiply by the reciprocal.}$$

$$= \frac{\cancel{3w + 2}}{\underset{1}{\cancel{5}}} \cdot \frac{\overset{2}{\cancel{10}}}{\cancel{(3w + 2)}(3w - 2)} \qquad \text{Factor and reduce.}$$

$$= \frac{2}{3w - 2} \qquad \text{Multiply.} \qquad \blacksquare$$

## You Try 4

Divide.

a) $\dfrac{\dfrac{4}{45}}{\dfrac{20}{27}}$    b) $\dfrac{\dfrac{25u^2 - 9}{24}}{\dfrac{5u + 3}{16}}$

### Answers to You Try Exercises

1) $\dfrac{2}{21}$    2) a) $\dfrac{n^5}{2m^4}$  b) $\dfrac{1}{3d^3(d + 2)}$  c) $\dfrac{h + 5}{h}$

3) a) $\dfrac{4}{3h^5k}$  b) $\dfrac{w - 5}{w(w + 9)}$  c) $-\dfrac{12(2m + 5)}{(m + 1)(m - 7)}$    4) a) $\dfrac{3}{25}$  b) $\dfrac{2(5u - 3)}{3}$

## 8.2 Exercises

**Objective 1: Multiply Fractions**

Multiply.

1) $\dfrac{5}{6} \cdot \dfrac{7}{9}$

2) $\dfrac{4}{11} \cdot \dfrac{2}{3}$

3) $\dfrac{6}{15} \cdot \dfrac{25}{42}$

4) $\dfrac{12}{21} \cdot \dfrac{7}{4}$

**Objective 2: Multiply Rational Expressions**

Multiply.

5) $\dfrac{16b^5}{3} \cdot \dfrac{4}{36b}$

6) $\dfrac{26}{25r^3} \cdot \dfrac{15r^6}{2}$

7) $\dfrac{21s^4}{15t^2} \cdot \dfrac{5t^4}{42s^{10}}$

8) $\dfrac{15u^4}{14v^2} \cdot \dfrac{7v^7}{20u^8}$

9) $\dfrac{9c^4}{42c} \cdot \dfrac{35}{3c^3}$

10) $-\dfrac{10}{8x^7} \cdot \dfrac{24x^9}{9x}$

VIDEO 11) $\dfrac{5t^2}{(3t-2)^2} \cdot \dfrac{3t-2}{10t^3}$

12) $\dfrac{11(z+5)^5}{6(z-4)} \cdot \dfrac{3}{22(z+5)}$

13) $\dfrac{4u-5}{9u^3} \cdot \dfrac{6u^5}{(4u-5)^3}$

14) $\dfrac{5k+6}{2k^3} \cdot \dfrac{12k^5}{(5k+6)^4}$

15) $\dfrac{6}{n+5} \cdot \dfrac{n^2+8n+15}{n+3}$

16) $\dfrac{9p^2-1}{12} \cdot \dfrac{9}{9p+3}$

17) $\dfrac{18y-12}{4y^2} \cdot \dfrac{y^2-4y-5}{3y^2+y-2}$

18) $\dfrac{12v-3}{8v+12} \cdot \dfrac{2v^2-5v-12}{3v-12}$

19) $(c-6) \cdot \dfrac{5}{c^2-6c}$

20) $(r^2+r-2) \cdot \dfrac{18r^2}{3r^2+6r}$

21) $\dfrac{7x}{11-x} \cdot (x^2-121)$

22) $\dfrac{4b}{2b^2-3b-5} \cdot (b+1)^2$

**Objective 3: Divide Rational Expressions**

Divide.

23) $\dfrac{20}{9} \div \dfrac{10}{27}$

24) $\dfrac{4}{5} \div \dfrac{12}{7}$

25) $42 \div \dfrac{9}{2}$

26) $\dfrac{18}{7} \div 6$

Divide.

27) $\dfrac{12}{5m^5} \div \dfrac{21}{8m^{12}}$

28) $\dfrac{12k^3}{35} \div \dfrac{42k^6}{25}$

29) $-\dfrac{50g}{7h^3} \div \dfrac{15g^4}{14h}$

30) $-\dfrac{c^{12}}{b} \div \dfrac{c^2}{6b}$

31) $\dfrac{2(k-2)}{21k^6} \div \dfrac{(k-2)^2}{28}$

32) $\dfrac{18}{(x+4)^3} \div \dfrac{36(x-7)}{x+4}$

33) $\dfrac{16q^5}{p+7} \div \dfrac{2q^4}{(p+7)^2}$

34) $\dfrac{(2a-5)^2}{32a^5} \div \dfrac{2a-5}{8a^3}$

35) $\dfrac{q+8}{q} \div \dfrac{q^2+q-56}{5}$

36) $\dfrac{4y^2-25}{10} \div \dfrac{18y-45}{18}$

37) $\dfrac{z^2+18z+80}{2z+1} \div (z+8)^2$

38) $\dfrac{6w^2-30w}{7} \div (w-5)^2$

39) $\dfrac{36a-12}{16} \div (9a^2-1)$

40) $\dfrac{h^2-21h+108}{4h} \div (144-h^2)$

41) $\dfrac{7n^2-14n}{8n} \div \dfrac{n^2+4n-12}{4n+24}$

42) $\dfrac{4j+24}{9} \div \dfrac{j^2-36}{9j-54}$

VIDEO 43) $\dfrac{4c-9}{2c^2-8c} \div \dfrac{12c-27}{c^2-3c-4}$

44) $\dfrac{p+13}{p+3} \div \dfrac{p^3+13p^2}{p^2-5p-24}$

45) Explain how to multiply rational expressions.

46) Explain how to divide rational expressions.

47) Find the missing polynomial in the denominator of
$$\dfrac{9h+45}{h^4} \cdot \dfrac{h^3}{} = \dfrac{9}{h(h-2)}.$$

48) Find the missing monomial in the numerator of
$$\dfrac{}{m^2-81} \cdot \dfrac{3m-27}{2m^2} = \dfrac{15m^3}{m+9}.$$

49) Find the missing binomial in the numerator of
$$\dfrac{4z^2-49}{z^2-3z-40} \div \dfrac{}{z+5} = \dfrac{2z+7}{8-z}.$$

50) Find the missing polynomial in the denominator of
$$\dfrac{12x^4}{50x^2+40x} \div \dfrac{x^3+2x^2+x+2}{} = \dfrac{6x^3}{5(x^2+1)}.$$

Divide.

51) $\dfrac{\dfrac{25}{42}}{\dfrac{8}{21}}$

52) $\dfrac{\dfrac{9}{35}}{\dfrac{4}{15}}$

53) $\dfrac{\dfrac{5}{24}}{\dfrac{15}{4}}$

54) $\dfrac{\dfrac{4}{3}}{\dfrac{2}{9}}$

55) $\dfrac{\dfrac{3d + 7}{24}}{\dfrac{3d + 7}{6}}$

56) $\dfrac{\dfrac{8s - 7}{4}}{\dfrac{8s - 7}{16}}$

57) $\dfrac{\dfrac{16r + 24}{r^3}}{\dfrac{12r + 18}{r}}$

58) $\dfrac{\dfrac{44m - 33}{3m^2}}{\dfrac{8m - 6}{m}}$

59) $\dfrac{\dfrac{a^2 - 25}{3a^{11}}}{\dfrac{4a + 20}{a^3}}$

60) $\dfrac{\dfrac{4z - 8}{z^8}}{\dfrac{z^2 - 4}{z^6}}$

61) $\dfrac{\dfrac{16x^2 - 25}{x^7}}{\dfrac{36x - 45}{6x^3}}$

62) $-\dfrac{\dfrac{16a^2}{3a^2 + 2a}}{\dfrac{12}{9a^2 - 4}}$

**Mixed Exercises: Objectives 2 and 3**

63) $\dfrac{c^2 + c - 30}{9c + 9} \cdot \dfrac{c^2 + 2c + 1}{c^2 - 25}$

64) $\dfrac{d^2 + 3d - 54}{d - 12} \cdot \dfrac{d^2 - 10d - 24}{7d + 63}$

65) $\dfrac{3x + 2}{9x^2 - 4} \div \dfrac{4x}{15x^2 - 7x - 2}$

66) $\dfrac{b^2 - 10b + 25}{8b - 40} \div \dfrac{2b^2 - 5b - 25}{2b + 5}$

67) $\dfrac{3k^2 - 12k}{12k^2 - 30k - 72} \cdot (2k + 3)^2$

68) $\dfrac{4a^3}{a^2 + a - 72} \cdot (a^2 - a - 56)$

69) $\dfrac{7t^6}{t^2 - 4} \div \dfrac{14t^2}{3t^2 - 7t + 2}$

70) $\dfrac{4n^2 - 1}{10n^3} \div \dfrac{2n^2 - 7n - 4}{6n^5}$

71) $\dfrac{4h^3}{h^2 - 64} \cdot \dfrac{8h - h^2}{12}$

72) $\dfrac{c^2 - 36}{c + 6} \div \dfrac{30 - 5c}{c - 9}$

73) $\dfrac{54x^8}{22x^3y^2} \div \dfrac{36xy^5}{11x^2y}$

74) $\dfrac{28cd^9}{2c^3d} \cdot \dfrac{5d^2}{84c^{10}d^2}$

75) $\dfrac{r^3 + 8}{r + 2} \cdot \dfrac{7}{3r^2 - 6r + 12}$

76) $\dfrac{2t^2 - 6t + 18}{5t - 5} \cdot \dfrac{t^2 - 9}{t^3 + 27}$

VIDEO  77) $\dfrac{a^2 - 4a}{6a + 54} \cdot \dfrac{a^2 + 13a + 36}{16 - a^2}$

78) $\dfrac{64 - u^2}{40 - 5u} \div \dfrac{u^2 + 10u + 16}{2u + 3}$

79) $\dfrac{2a^2}{a^2 + a - 20} \cdot \dfrac{a^3 + 5a^2 + 4a + 20}{2a^2 + 8}$

80) $\dfrac{18x^4}{x^3 + 3x^2 - 9x - 27} \cdot \dfrac{6x^2 + 19x + 3}{18x^2 + 3x}$

81) $\dfrac{30}{4y^2 - 4x^2} \div \dfrac{10x^2 + 10xy + 10y^2}{x^3 - y^3}$

82) $\dfrac{a^2 - b^2}{a^3 + b^3} \div \dfrac{8b - 8a}{9}$

83) $\dfrac{3m^2 + 8m + 4}{4} \div (12m + 8)$

84) $\dfrac{w^2 - 17w + 72}{3w} \div (w - 8)$

Perform the operations and simplify.

85) $\dfrac{4j^2 - 21j + 5}{j^3} \div \left(\dfrac{3j + 2}{j^3 - j^2} \cdot \dfrac{j^2 - 6j + 5}{j}\right)$

86) $\dfrac{2a}{a^2 + 18a + 81} \div \left(\dfrac{a^2 + 3a - 4}{a^2 + 9a + 20} \cdot \dfrac{a^2 + 5a}{a^2 + 8a - 9}\right)$

87) $\dfrac{x}{3x^2 - 15x + 75} \div \left(\dfrac{4x + 20}{x + 9} \cdot \dfrac{x^2 - 81}{x^3 + 125}\right)$

88) $\dfrac{t^3 - 8}{t - 2} \div \left(\dfrac{3t + 11}{5t + 15} \cdot \dfrac{t^2 + 2t + 4}{3t^2 + 11t}\right)$

89) If the area of a rectangle is $\dfrac{3x}{2y^2}$ and the width is $\dfrac{y}{8x^4}$, what is the length of the rectangle?

90) If the area of a triangle is $\dfrac{2n}{n^2 - 4n + 3}$ and the height is $\dfrac{n + 3}{n - 1}$, what is the length of the base of the triangle?

## Section 8.3 Finding the Least Common Denominator

**Objectives**

1. Find the Least Common Denominator for a Group of Rational Expressions
2. Rewrite Rational Expressions Using Their LCD

### 1. Find the Least Common Denominator for a Group of Rational Expressions

Recall that to add or subtract fractions, they must have a common denominator. Similarly, rational expressions must have common denominators in order to be added or subtracted. In this section, we will discuss how to find the least common denominator (LCD) of rational expressions.

We begin by looking at the fractions $\dfrac{3}{8}$ and $\dfrac{5}{12}$. By inspection, we can see that the LCD $= 24$. But, *why* is that true? Let's write each of the denominators, 8 and 12, as the product of their prime factors:

$$8 = 2 \cdot 2 \cdot 2 = 2^3$$
$$12 = 2 \cdot 2 \cdot 3 = 2^2 \cdot 3$$

The LCD will contain each factor the *greatest* number of times it appears in any single factorization.

*The LCD will contain $2^3$ because 2 appears *three* times in the factorization of 8.*

*The LCD will contain 3 because 3 appears *one* time in the factorization of 12.*

The LCD, then, is the product of the factors we have identified.

$$\text{LCD of } \frac{3}{8} \text{ and } \frac{5}{12} = 2^3 \cdot 3 = 8 \cdot 3 = 24$$

This is the same result as the one we obtained just by inspecting the two denominators.

We use the same procedure to find the least common denominator of rational expressions.

---

**Procedure** Finding the Least Common Denominator (LCD)

**Step 1:** Factor the denominators.

**Step 2:** The LCD will contain each unique factor the *greatest* number of times it appears in any single factorization.

**Step 3:** The LCD is the *product* of the factors identified in Step 2.

---

**Example 1**

Find the LCD of each pair of rational expressions.

a) $\dfrac{17}{24}, \dfrac{5}{36}$    b) $\dfrac{1}{12n}, \dfrac{10}{21n}$    c) $\dfrac{8}{49c^3}, \dfrac{13}{14c^2}$

**Solution**

a) Follow the steps for finding the least common denominator.

   **Step 1:** Factor the denominators.

$$24 = 2 \cdot 2 \cdot 2 \cdot 3 = 2^3 \cdot 3$$
$$36 = 2 \cdot 2 \cdot 3 \cdot 3 = 2^2 \cdot 3^2$$

   **Step 2:** The LCD will contain each unique factor the *greatest* number of times it appears in any factorization. *The LCD will contain $2^3$ and $3^2$.*

   **Step 3:** The LCD is the *product* of the factors in Step 2.

$$\text{LCD} = 2^3 \cdot 3^2 = 8 \cdot 9 = 72$$

b) **Step 1:** Factoring the denominators of $\dfrac{1}{12n}$ and $\dfrac{10}{21n}$ gives us

$$12n = 2 \cdot 2 \cdot 3 \cdot n = 2^2 \cdot 3 \cdot n$$
$$21n = 3 \cdot 7 \cdot n$$

**Step 2:** The LCD will contain each unique factor the *greatest* number of times it appears in any factorization. *It will contain $2^2$, 3, 7, and n.*

**Step 3:** The LCD is the *product* of the factors in Step 2.

$$\text{LCD} = 2^2 \cdot 3 \cdot 7 \cdot n = 84n$$

c) **Step 1:** Factoring the denominators of $\dfrac{8}{49c^3}$ and $\dfrac{13}{14c^2}$ gives us

$$49c^3 = 7 \cdot 7 \cdot c^3 = 7^2 \cdot c^3$$
$$14c^2 = 2 \cdot 7 \cdot c^2$$

**Step 2:** The LCD will contain each unique factor the *greatest* number of times it appears in any factorization. *It will contain 2, $7^2$, and $c^3$.*

**Step 3:** The LCD is the *product* of the factors in Step 2.

$$\text{LCD} = 2 \cdot 7^2 \cdot c^3 = 98c^3$$ ∎

**You Try I**

Find the LCD of each pair of rational expressions.

a) $\dfrac{14}{15}, \dfrac{11}{18}$    b) $\dfrac{3}{14}, \dfrac{7}{10}$    c) $\dfrac{20}{27h^2}, \dfrac{1}{6h^4}$

**Example 2**

Find the LCD of each group of rational expressions.

a) $\dfrac{4}{k}, \dfrac{6}{k + 3}$    b) $\dfrac{7}{a - 6}, \dfrac{2a}{a^2 + 3a - 54}$    c) $\dfrac{3p}{p^2 + 4p + 4}, \dfrac{1}{5p^2 + 10p}$

**Solution**

a) The denominators of $\dfrac{4}{k}$, and $\dfrac{6}{k + 3}$ are already in simplest form. It is important to recognize that $k$ and $k + 3$ are *different factors*.

The LCD will be the product of $k$ and $k + 3$:  $\text{LCD} = k(k + 3)$.

Usually, we leave the LCD in this form; we do not distribute.

b) **Step 1:** Factor the denominators of $\dfrac{7}{a - 6}$ and $\dfrac{2a}{a^2 + 3a - 54}$.

$a - 6$ cannot be factored.  $a^2 + 3a - 54 = (a - 6)(a + 9)$

**Step 2:** The LCD will contain each unique factor the *greatest* number of times it appears in any factorization. *It will contain $a - 6$ and $a + 9$.*

**Step 3:** The LCD is the *product* of the factors identified in Step 2.

$$\text{LCD} = (a - 6)(a + 9)$$

c) **Step 1:** Factor the denominators of $\dfrac{3p}{p^2 + 4p + 4}$ and $\dfrac{1}{5p^2 + 10p}$.

$$p^2 + 4p + 4 = (p + 2)^2 \qquad 5p^2 + 10p = 5p(p + 2)$$

**Step 2:** The unique factors are 5, $p$, and $p + 2$, with $p + 2$ *appearing at most twice. The factors we will use in the LCD are $5, p,$ and $(p + 2)^2$.*

**Step 3:** LCD $= 5p(p + 2)^2$ ∎

## You Try 2

Find the LCD of each group of rational expressions.

a) $\dfrac{6}{w}, \dfrac{9w}{w + 1}$
   b) $\dfrac{1}{r + 8}, \dfrac{5r}{r^2 + r - 56}$
   c) $\dfrac{4b}{b^2 - 18b + 81}, \dfrac{3}{8b^2 - 72b}$

At first glance it may appear that the least common denominator of $\dfrac{9}{x - 7}$ and $\dfrac{5}{7 - x}$ is $(x - 7)(7 - x)$. This is *not* the case. Recall from Section 8.1 that $a - b = -1(b - a)$. We will use this idea to find the LCD of $\dfrac{9}{x - 7}$ and $\dfrac{5}{7 - x}$.

## Example 3

Find the LCD of $\dfrac{9}{x - 7}$ and $\dfrac{5}{7 - x}$.

**Solution**

Since $7 - x = -(x - 7)$, we can rewrite $\dfrac{5}{7 - x}$ as $\dfrac{5}{-(x - 7)} = -\dfrac{5}{x - 7}$.

Therefore, we can now think of our task as finding the LCD of $\dfrac{9}{x - 7}$ and $-\dfrac{5}{x - 7}$.

The least common denominator is $x - 7$. ∎

## You Try 3

Find the LCD of $\dfrac{2}{k - 5}$ and $\dfrac{13}{5 - k}$.

## 2. Rewrite Rational Expressions Using Their LCD

As we know from our previous work with fractions, after *determining* the least common denominator, we must *rewrite* those fractions as equivalent fractions with the LCD so that they can be added or subtracted.

**Example 4**

Identify the LCD of $\dfrac{1}{6}$ and $\dfrac{8}{9}$, and rewrite each as an equivalent fraction with the LCD as its denominator.

**Solution**

The LCD of $\dfrac{1}{6}$ and $\dfrac{8}{9}$ is 18. We must rewrite each fraction with a denominator of 18.

$\dfrac{1}{6}$:   By what number should we multiply 6 to get 18?   3

$$\frac{1}{6} \cdot \frac{3}{3} = \frac{3}{18}$$   Multiply the numerator *and* denominator by 3 to obtain an equivalent fraction.

$\dfrac{8}{9}$:   By what number should we multiply 9 to get 18?   2

$$\frac{8}{9} \cdot \frac{2}{2} = \frac{16}{18}$$   Multiply the numerator *and* denominator by 2 to obtain an equivalent fraction.

Therefore, $\dfrac{1}{6} = \dfrac{3}{18}$ and $\dfrac{8}{9} = \dfrac{16}{18}$.

**You Try 4**

Identify the LCD of $\dfrac{7}{12}$ and $\dfrac{2}{9}$, and rewrite each as an equivalent fraction with the LCD as its denominator.

The procedure for rewriting rational expressions as equivalent expressions with the least common denominator is very similar to the process used in Example 4.

---

**Procedure**   Writing Rational Expressions as Equivalent Expressions with the Least Common Denominator

**Step 1:**   Identify and write down the LCD.

**Step 2:**   Look at each rational expression (with its denominator in factored form) and compare its denominator with the LCD. Ask yourself, "What factors are missing?"

**Step 3:**   Multiply the numerator and denominator by the "missing" factors to obtain an equivalent rational expression with the desired LCD.

**Note**

Use the distributive property to multiply the terms in the numerator, but leave the denominator as the product of factors. (We will see why this is done in Section 8.4.)

## Example 5

Identify the LCD of each pair of rational expressions, and rewrite each as an equivalent expression with the LCD as its denominator.

a) $\dfrac{5}{12z}, \dfrac{4}{9z^3}$     b) $\dfrac{m}{m-4}, \dfrac{3}{m+8}$     c) $\dfrac{2}{3x^2-6x}, \dfrac{9x}{x^2-7x+10}$

d) $\dfrac{t}{t-3}, \dfrac{8}{3-t}$

**Solution**

a) **Step 1:**  Identify and write down the LCD of $\dfrac{5}{12z}$ and $\dfrac{4}{9z^3}$: LCD $= 36z^3$.

**Step 2:**  Compare the denominators of $\dfrac{5}{12z}$ and $\dfrac{4}{9z^3}$ to the LCD and ask yourself, "What's missing?"

$\dfrac{5}{12z}$: $12z$ is "missing" the       |   $\dfrac{4}{9z^3}$: $9z^3$ is "missing"

*factors* $3$ *and* $z^2$.                       |   *the factor* $4$.

**Step 3:**  Multiply the numerator and denominator by $3z^2$.          |   Multiply the numerator and denominator by $4$.

$\dfrac{5}{12z} \cdot \dfrac{3z^2}{3z^2} = \dfrac{15z^2}{36z^3}$       |   $\dfrac{4}{9z^3} \cdot \dfrac{4}{4} = \dfrac{16}{36z^3}$

$$\dfrac{5}{12z} = \dfrac{15z^2}{36z^3} \quad \text{and} \quad \dfrac{4}{9z^3} = \dfrac{16}{36z^3}$$

b) **Step 1:**  Identify and write down the LCD of $\dfrac{m}{m-4}$ and $\dfrac{3}{m+8}$:
LCD $= (m-4)(m+8)$.

**Step 2:**  Compare the denominators of $\dfrac{m}{m-4}$ and $\dfrac{3}{m+8}$ to the LCD and ask yourself, "What's missing?"

$\dfrac{m}{m-4}$: $m-4$ is "missing"       |   $\dfrac{3}{m+8}$: $m+8$ is "missing"

*the factor* $m+8$.                         |   *the factor* $m-4$.

**Step 3:**  Multiply the numerator and denominator by $m+8$.          |   Multiply the numerator and denominator by $m-4$.

$\dfrac{m}{m-4} \cdot \dfrac{m+8}{m+8} = \dfrac{m(m+8)}{(m-4)(m+8)}$  |  $\dfrac{3}{m+8} \cdot \dfrac{m-4}{m-4} = \dfrac{3(m-4)}{(m+8)(m-4)}$

$= \dfrac{m^2+8m}{(m-4)(m+8)}$       |       $= \dfrac{3m-12}{(m-4)(m+8)}$

Notice that we multiplied the factors in the numerator but left the denominator in factored form.

$$\dfrac{m}{m-4} = \dfrac{m^2+8m}{(m-4)(m+8)} \quad \text{and} \quad \dfrac{3}{m+8} = \dfrac{3m-12}{(m-4)(m+8)}$$

c) **Step 1:** Identify and write down the LCD of $\dfrac{2}{3x^2 - 6x}$ and $\dfrac{9x}{x^2 - 7x + 10}$.

First, we must factor the denominators.

$$\frac{2}{3x^2 - 6x} = \frac{2}{3x(x-2)}, \frac{9x}{x^2 - 7x + 10} = \frac{9x}{(x-2)(x-5)}$$

We will work with the factored forms of the expressions.

$$\text{LCD} = 3x(x-2)(x-5)$$

**Step 2:** Compare the denominators of $\dfrac{2}{3x(x-2)}$ and $\dfrac{9x}{(x-2)(x-5)}$ to the LCD and ask yourself, "What's missing?"

$\dfrac{2}{3x(x-2)}$: $3x(x-2)$ is "missing" the factor $x - 5$. | $\dfrac{9x}{(x-2)(x-5)}$: $(x-2)(x-5)$ is "missing" $3x$.

**Step 3:** Multiply the numerator and denominator by $x - 5$. | Multiply the numerator and denominator by $3x$.

$$\frac{2}{3x(x-2)} \cdot \frac{x-5}{x-5} = \frac{2(x-5)}{3x(x-2)(x-5)} \qquad \frac{9x}{(x-2)(x-5)} \cdot \frac{3x}{3x} = \frac{27x^2}{3x(x-2)(x-5)}$$

$$= \frac{2x - 10}{3x(x-2)(x-5)}$$

$$\frac{2}{3x^2 - 6x} = \frac{2x - 10}{3x(x-2)(x-5)} \quad \text{and} \quad \frac{9x}{x^2 - 7x + 10} = \frac{27x^2}{3x(x-2)(x-5)}$$

d) To find the LCD of $\dfrac{t}{t-3}$ and $\dfrac{8}{3-t}$, recall that $3 - t$ can be rewritten as $-(t-3)$. So,

$$\frac{8}{3 - t} = \frac{8}{-(t-3)} = -\frac{8}{t-3}$$

Therefore, the LCD of $\dfrac{t}{t-3}$ and $-\dfrac{8}{t-3}$ is $t - 3$.

$\dfrac{t}{t-3}$ already has the LCD, and $\dfrac{8}{3-t} = -\dfrac{8}{t-3}$. ∎

**You Try 5**

Identify the least common denominator of each pair of rational expressions, and rewrite each as an equivalent expression with the LCD as its denominator.

a) $\dfrac{3}{10a^6}, \dfrac{7}{8a^5}$    b) $\dfrac{6}{n+10}, \dfrac{n}{2n-3}$    c) $\dfrac{v-9}{v^2+15v+44}, \dfrac{8}{5v^2+55v}$

d) $\dfrac{c}{10-c}, \dfrac{5}{c-10}$

### Answers to You Try Exercises

1) a) 90  b) 70  c) $54h^4$    2) a) $w(w + 1)$  b) $(r + 8)(r - 7)$  c) $8b(b - 9)^2$

3) $k - 5$    4) LCD = 36; $\dfrac{7}{12} = \dfrac{21}{36}, \dfrac{2}{9} = \dfrac{8}{36}$    5) a) LCD = $40a^6$; $\dfrac{3}{10a^6} = \dfrac{12}{40a^6}, \dfrac{7}{8a^5} = \dfrac{35a}{40a^6}$

b) LCD = $(2n - 3)(n + 10)$; $\dfrac{6}{n + 10} = \dfrac{12n - 18}{(2n - 3)(n + 10)}, \dfrac{n}{2n - 3} = \dfrac{n^2 + 10n}{(2n - 3)(n + 10)}$

c) LCD = $5v(v + 4)(v + 11)$; $\dfrac{v - 9}{v^2 + 15v + 44} = \dfrac{5v^2 - 45v}{5v(v + 4)(v + 11)}, \dfrac{8}{5v^2 + 55v} = \dfrac{8v + 32}{5v(v + 4)(v + 11)}$

d) LCD = $c - 10$; $\dfrac{c}{10 - c} = -\dfrac{c}{c - 10}, \dfrac{5}{c - 10} = \dfrac{5}{c - 10}$

# 8.3 Exercises

## Objective 1: Find the LCD for a Group of Rational Expressions

Find the LCD of each group of fractions.

1) $\dfrac{7}{12}, \dfrac{2}{15}$

2) $\dfrac{3}{8}, \dfrac{9}{7}$

3) $\dfrac{27}{40}, \dfrac{11}{10}, \dfrac{5}{12}$

4) $\dfrac{19}{8}, \dfrac{1}{12}, \dfrac{3}{32}$

5) $\dfrac{3}{n^7}, \dfrac{5}{n^{11}}$

6) $\dfrac{4}{c^2}, \dfrac{8}{c^3}$

7) $\dfrac{13}{14r^4}, \dfrac{3}{4r^7}$

8) $\dfrac{11}{6p^4}, \dfrac{3}{10p^9}$

9) $-\dfrac{5}{6z^5}, \dfrac{7}{36z^2}$

10) $\dfrac{5}{24w^5}, -\dfrac{1}{4w^{10}}$

11) $\dfrac{7}{10m}, \dfrac{9}{22m^4}$

12) $-\dfrac{3}{2k^2}, \dfrac{5}{14k^5}$

13) $\dfrac{4}{24x^3y^2}, \dfrac{11}{6x^3y}$

14) $\dfrac{3}{10a^4b^2}, \dfrac{8}{15ab^4}$

15) $\dfrac{4}{11}, \dfrac{8}{z - 3}$

16) $\dfrac{3}{n + 8}, \dfrac{1}{5}$

VIDEO 17) $\dfrac{10}{w}, \dfrac{6}{2w + 1}$

18) $\dfrac{1}{y}, -\dfrac{6}{6y + 1}$

19) What is the first step for finding the LCD of $\dfrac{9}{8t - 10}$ and $\dfrac{3t}{20t - 25}$?

20) Is $(h - 9)(9 - h)$ the LCD of $\dfrac{2h}{h - 9}$ and $\dfrac{4}{9 - h}$? Explain your answer.

Find the LCD of each group of fractions.

21) $\dfrac{8}{5c - 5}, \dfrac{9}{2c - 2}$

22) $\dfrac{5}{7k + 14}, \dfrac{11}{4k + 8}$

23) $\dfrac{2}{9p^4 - 6p^3}, \dfrac{3}{3p^6 - 2p^5}$

24) $\dfrac{21}{6a^2 - 8a}, \dfrac{13}{18a^3 - 24a^2}$

25) $\dfrac{4m}{m - 7}, \dfrac{2}{m - 3}$

26) $\dfrac{5}{r + 9}, \dfrac{7}{r - 1}$

27) $\dfrac{11}{z^2 + 11z + 24}, \dfrac{7z}{z^2 + 5z - 24}$

28) $\dfrac{7x}{x^2 - 12x + 35}, \dfrac{x}{x^2 - x - 20}$

29) $\dfrac{14t}{t^2 - 3t - 18}, -\dfrac{6}{t^2 - 36}, \dfrac{t}{t^2 + 9t + 18}$

30) $\dfrac{6w}{w^2 - 10w + 16}, \dfrac{3}{w^2 - 7w - 8}, \dfrac{4w}{w^2 - w - 2}$

31) $\dfrac{6}{a - 8}, \dfrac{7}{8 - a}$

32) $\dfrac{6}{b - 3}, \dfrac{5}{3 - b}$

33) $\dfrac{12}{y - x}, \dfrac{5y}{x - y}$

34) $\dfrac{u}{v - u}, \dfrac{8}{u - v}$

## Objective 2: Rewrite Rational Expressions Using Their LCD

35) Explain, in your own words, how to rewrite $\dfrac{4}{x + 9}$ as an equivalent rational expression with a denominator of $(x + 9)(x - 3)$.

36) Explain, in your own words, how to rewrite $\dfrac{7}{5 - m}$ as an equivalent rational expression with a denominator of $m - 5$.

Rewrite each rational expression with the indicated denominator.

37) $\dfrac{7}{12} = \dfrac{}{48}$

38) $\dfrac{5}{7} = \dfrac{}{42}$

39) $\dfrac{8}{z} = \dfrac{}{9z}$

40) $\dfrac{-6}{b} = \dfrac{}{4b}$

41) $\dfrac{3}{8k} = \dfrac{}{56k^4}$

42) $\dfrac{5}{3p^4} = \dfrac{}{9p^6}$

43) $\dfrac{6}{5t^5u^2} = \dfrac{}{10t^7u^5}$

44) $\dfrac{13}{6cd^2} = \dfrac{}{24c^3d^3}$

45) $\dfrac{7}{3r + 4} = \dfrac{}{r(3r + 4)}$

46) $\dfrac{8}{m - 8} = \dfrac{}{m(m - 8)}$

47) $\dfrac{v}{4(v - 3)} = \dfrac{}{16v^5(v - 3)}$

48) $\dfrac{a}{5(2a + 7)} = \dfrac{}{15a(2a + 7)}$

49) $\dfrac{9x}{x + 6} = \dfrac{}{(x + 6)(x - 5)}$

50) $\dfrac{5b}{b + 3} = \dfrac{}{(b + 3)(b + 7)}$

51) $\dfrac{z - 3}{2z - 5} = \dfrac{}{(2z - 5)(z + 8)}$

52) $\dfrac{w + 2}{4w - 1} = \dfrac{}{(4w - 1)(w - 4)}$

53) $\dfrac{5}{3 - p} = \dfrac{}{p - 3}$

54) $\dfrac{10}{10 - n} = \dfrac{}{n - 10}$

55) $-\dfrac{8c}{6c - 7} = \dfrac{}{7 - 6c}$

56) $-\dfrac{g}{3g - 2} = \dfrac{}{2 - 3g}$

Identify the least common denominator of each pair of rational expressions, and rewrite each as an equivalent rational expression with the LCD as its denominator.

57) $\dfrac{8}{15}, \dfrac{1}{6}$

58) $\dfrac{3}{8}, \dfrac{5}{12}$

59) $\dfrac{4}{u}, \dfrac{8}{u^3}$

60) $\dfrac{9}{d^5}, \dfrac{7}{d^2}$

61) $\dfrac{9}{8n^6}, \dfrac{2}{3n^2}$

62) $\dfrac{5}{8a}, \dfrac{7}{10a^5}$

63) $\dfrac{6}{4a^3b^5}, \dfrac{6}{a^4b}$

64) $\dfrac{3}{x^3y}, \dfrac{6}{5xy^5}$

65) $\dfrac{r}{5}, \dfrac{2}{r - 4}$

66) $\dfrac{t}{5t - 1}, \dfrac{8}{7}$

67) $\dfrac{3}{d}, \dfrac{7}{d - 9}$

68) $\dfrac{5}{c}, \dfrac{4}{c + 2}$

69) $\dfrac{m}{m + 7}, \dfrac{3}{m}$

70) $\dfrac{z}{z - 4}, \dfrac{5}{z}$

71) $\dfrac{a}{30a - 15}, \dfrac{1}{12a - 6}$

72) $\dfrac{7}{24x - 16}, \dfrac{x}{18x - 12}$

73) $\dfrac{9}{k - 9}, \dfrac{5k}{k + 3}$

74) $\dfrac{6}{h + 1}, \dfrac{11h}{h + 7}$

75) $\dfrac{3}{a + 2}, \dfrac{2a}{3a + 4}$

76) $\dfrac{b}{6b - 5}, \dfrac{8}{b - 9}$

77) $\dfrac{9y}{y^2 - y - 42}, \dfrac{3}{2y^2 + 12y}$

78) $\dfrac{12q}{q^2 - 6q - 16}, \dfrac{4}{2q^2 - 16q}$

79) $\dfrac{c}{c^2 + 9c + 18}, \dfrac{11}{c^2 + 12c + 36}$

80) $\dfrac{z}{z^2 - 8z + 16}, \dfrac{9z}{z^2 + 4z - 32}$

81) $\dfrac{11}{g - 3}, \dfrac{4}{9 - g^2}$

82) $\dfrac{6}{g - 9}, \dfrac{1}{81 - g^2}$

83) $\dfrac{4}{3x - 4}, \dfrac{7x}{16 - 9x^2}$

84) $\dfrac{12}{5k - 2}, \dfrac{4k}{4 - 25k^2}$

85) $\dfrac{2}{z^2 + 3z}, \dfrac{6}{3z^2 + 9z}, \dfrac{8}{z^2 + 6z + 9}$

86) $\dfrac{4}{w^2 - 4w}, \dfrac{6}{7w^2 - 28w}, \dfrac{11}{w^2 - 8w + 16}$

87) $\dfrac{t}{t^2 - 13t + 30}, \dfrac{6}{t - 10}, \dfrac{7}{t^2 - 9}$

88) $-\dfrac{2}{a + 2}, \dfrac{a}{a^2 - 4}, \dfrac{15}{a^2 - 3a + 2}$

89) $-\dfrac{9}{h^3 + 8}, \dfrac{2h}{5h^2 - 10h + 20}$

90) $\dfrac{5x}{x^3 - y^3}, \dfrac{7}{8x - 8y}$

# Section 8.4 Adding and Subtracting Rational Expressions

## Objectives

1. **Add and Subtract Rational Expressions with a Common Denominator**
2. **Add and Subtract Rational Expressions with Different Denominators**
3. **Add and Subtract Rational Expressions with Special Denominators**

We know that in order to add or subtract fractions, they must have a common denominator. The same is true for rational expressions.

## 1. Add and Subtract Rational Expressions with a Common Denominator

Let's first look at fractions and rational expressions with common denominators.

 **Example 1**

Add or subtract.

a) $\dfrac{8}{11} - \dfrac{5}{11}$    b) $\dfrac{2x}{4x - 9} + \dfrac{5x + 3}{4x - 9}$

### Solution

a) Since the fractions have the same denominator, subtract the terms in the numerator and keep the common denominator.

$$\frac{8}{11} - \frac{5}{11} = \frac{8 - 5}{11} = \frac{3}{11} \qquad \text{Subtract terms in the numerator.}$$

b) Since the rational expressions have the same denominator, add the terms in the numerator and keep the common denominator.

$$\frac{2x}{4x - 9} + \frac{5x + 3}{4x - 9} = \frac{2x + (5x + 3)}{4x - 9} \qquad \text{Add terms in the numerator.}$$

$$= \frac{7x + 3}{4x - 9} \qquad \text{Combine like terms.}$$

We can generalize the procedure for adding and subtracting rational expressions that have a common denominator as follows.

---

**Procedure**   Adding and Subtracting Rational Expressions

If $\dfrac{P}{Q}$ and $\dfrac{R}{Q}$ are rational expressions with $Q \neq 0$, then

1) $\dfrac{P}{Q} + \dfrac{R}{Q} = \dfrac{P + R}{Q}$    and    2) $\dfrac{P}{Q} - \dfrac{R}{Q} = \dfrac{P - R}{Q}$

---

 **You Try 1**

Add or subtract.

a) $\dfrac{11}{12} - \dfrac{7}{12}$    b) $\dfrac{6h}{5h - 2} + \dfrac{3h + 8}{5h - 2}$

All answers to a sum or difference of rational expressions should be in lowest terms. Sometimes it is necessary to simplify our result to lowest terms by factoring the numerator and dividing the numerator and denominator by the greatest common factor.

## Example 2

Add or subtract.

a) $\dfrac{4}{15k} + \dfrac{8}{15k}$        b) $\dfrac{c^2 - 3}{c(c + 4)} - \dfrac{5 - 2c}{c(c + 4)}$

### Solution

a) $\dfrac{4}{15k} + \dfrac{8}{15k} = \dfrac{4 + 8}{15k} = \dfrac{12}{15k}$     Add terms in the numerator.

$\phantom{a) \dfrac{4}{15k} + \dfrac{8}{15k}} = \dfrac{4}{5k}$     Reduce to lowest terms.

b) $\dfrac{c^2 - 3}{c(c + 4)} - \dfrac{5 - 2c}{c(c + 4)} = \dfrac{(c^2 - 3) - (5 - 2c)}{c(c + 4)}$     Subtract terms in the numerator.

$\phantom{xxxxxxxxxxxxxxxx} = \dfrac{c^2 - 3 - 5 + 2c}{c(c + 4)}$     Distribute.

$\phantom{xxxxxxxxxxxxxxxx} = \dfrac{c^2 + 2c - 8}{c(c + 4)}$     Combine like terms.

$\phantom{xxxxxxxxxxxxxxxx} = \dfrac{(c + 4)(c - 2)}{c(c + 4)}$     Factor the numerator.

$\phantom{xxxxxxxxxxxxxxxx} = \dfrac{c - 2}{c}$     Reduce to lowest terms.

## You Try 2

Add or subtract.

a) $\dfrac{19}{32w} + \dfrac{9}{32w}$        b) $\dfrac{m^2 - 5}{m(m + 6)} - \dfrac{3m + 49}{m(m + 6)}$

> **BE CAREFUL**
>
> After combining like terms in the numerator, ask yourself, *"Can I factor the numerator?"* If so, factor it. Sometimes, the expression can be reduced by dividing the numerator and denominator by the greatest common factor.

## 2. Add and Subtract Rational Expressions with Different Denominators

If we are asked to add or subtract rational expressions with different denominators, we must begin by rewriting each expression with the least common denominator. Then, add or subtract. Simplify the result.

Using the procedure studied in Section 8.3, here are the steps to follow to add or subtract rational expressions with different denominators.

> **Procedure**   Steps for Adding and Subtracting Rational Expressions with Different Denominators
>
> 1) Factor the denominators.
> 2) Write down the LCD.
> 3) Rewrite each rational expression as an equivalent rational expression with the LCD.
> 4) Add or subtract the numerators and keep the common denominator in factored form.
> 5) After combining like terms in the numerator ask yourself, *"Can I factor it?"* If so, factor.
> 6) Reduce the rational expression, if possible.

**Example 3**

Add or subtract.

a) $\dfrac{t+6}{4} + \dfrac{t-8}{12}$    b) $\dfrac{3}{10a} - \dfrac{7}{8a^2}$    c) $\dfrac{7n-30}{n^2-36} + \dfrac{n}{n+6}$

**Solution**

a)   The LCD is 12. $\dfrac{t-8}{12}$ already has the LCD.

Rewrite $\dfrac{t+6}{4}$ with the LCD: $\dfrac{t+6}{4} \cdot \dfrac{3}{3} = \dfrac{3(t+6)}{12}$

$$\dfrac{t+6}{4} + \dfrac{t-8}{12} = \dfrac{3(t+6)}{12} + \dfrac{t-8}{12} \qquad \text{Write each expression with the LCD.}$$

$$= \dfrac{3(t+6) + (t-8)}{12} \qquad \text{Add the numerators.}$$

$$= \dfrac{3t+18+t-8}{12} \qquad \text{Distribute.}$$

$$= \dfrac{4t+10}{12} \qquad \text{Combine like terms.}$$

Ask yourself, *"Can I factor the numerator?"* Yes.

$$= \dfrac{\cancel{2}(2t+5)}{\underset{6}{\cancel{12}}} \qquad \text{Factor.}$$

$$= \dfrac{2t+5}{6} \qquad \text{Reduce.}$$

b)   The LCD of $\dfrac{3}{10a}$ and $\dfrac{7}{8a^2}$ is $40a^2$. Rewrite each expression with the LCD.

$$\dfrac{3}{10a} \cdot \dfrac{4a}{4a} = \dfrac{12a}{40a^2} \qquad \text{and} \qquad \dfrac{7}{8a^2} \cdot \dfrac{5}{5} = \dfrac{35}{40a^2}$$

$$\dfrac{3}{10a} - \dfrac{7}{8a^2} = \dfrac{12a}{40a^2} - \dfrac{35}{40a^2} \qquad \text{Write each expression with the LCD.}$$

$$= \dfrac{12a-35}{40a^2} \qquad \text{Subtract the numerators.}$$

*"Can I factor the numerator?"* No. The expression is in simplest form since the numerator and denominator have no common factors.

c)  First, factor the denominator of $\dfrac{7n - 30}{n^2 - 36}$ to get $\dfrac{7n - 30}{(n + 6)(n - 6)}$.

The LCD of $\dfrac{7n - 30}{(n + 6)(n - 6)}$ and $\dfrac{n}{n + 6}$ is $(n + 6)(n - 6)$.

Rewrite $\dfrac{n}{n + 6}$ with the LCD: $\dfrac{n}{n + 6} \cdot \dfrac{n - 6}{n - 6} = \dfrac{n(n - 6)}{(n + 6)(n - 6)}$

$$\dfrac{7n - 30}{n^2 - 36} + \dfrac{n}{n + 6} = \dfrac{7n - 30}{(n + 6)(n - 6)} + \dfrac{n}{n + 6}$$  Factor the denominator.

$$= \dfrac{7n - 30}{(n + 6)(n - 6)} + \dfrac{n(n - 6)}{(n + 6)(n - 6)}$$  Write each expression with the LCD.

$$= \dfrac{7n - 30 + n(n - 6)}{(n + 6)(n - 6)}$$  Add the numerators.

$$= \dfrac{7n - 30 + n^2 - 6n}{(n + 6)(n - 6)}$$  Distribute.

$$= \dfrac{n^2 + n - 30}{(n + 6)(n - 6)}$$  Combine like terms.

Ask yourself, *"Can I factor the numerator?"* Yes.

$$= \dfrac{\cancel{(n + 6)}(n - 5)}{\cancel{(n + 6)}(n - 6)}$$  Factor.

$$= \dfrac{n - 5}{n - 6}$$  Reduce.

## You Try 3

Add or subtract.

a)  $\dfrac{t + 4}{5} + \dfrac{2t - 7}{15}$  b)  $\dfrac{7}{12v} - \dfrac{9}{16v^2}$  c)  $\dfrac{15h - 8}{h^2 - 64} + \dfrac{h}{h + 8}$

## Example 4

Subtract $\dfrac{6r}{r^2 + 10r + 16} - \dfrac{3r + 4}{r^2 + 3r - 40}$.

### Solution

Factor the denominators, then write down the LCD.

$$\dfrac{6r}{r^2 + 10r + 16} = \dfrac{6r}{(r + 8)(r + 2)}, \qquad \dfrac{3r + 4}{r^2 + 3r - 40} = \dfrac{3r + 4}{(r + 8)(r - 5)}$$

Rewrite each expression with the LCD, $(r + 8)(r + 2)(r - 5)$.

$$\frac{6r}{(r + 8)(r + 2)} \cdot \frac{r - 5}{r - 5} = \frac{6r(r - 5)}{(r + 8)(r + 2)(r - 5)}$$

$$\frac{3r + 4}{(r + 8)(r - 5)} \cdot \frac{r + 2}{r + 2} = \frac{(3r + 4)(r + 2)}{(r + 8)(r + 2)(r - 5)}$$

$$\frac{6r}{r^2 + 10r + 16} - \frac{3r + 4}{r^2 + 3r - 40}$$

$$= \frac{6r}{(r + 8)(r + 2)} - \frac{3r + 4}{(r + 8)(r - 5)}$$  Factor the denominators.

$$= \frac{6r(r - 5)}{(r + 8)(r + 2)(r - 5)} - \frac{(3r + 4)(r + 2)}{(r + 8)(r + 2)(r - 5)}$$  Write each expression with the LCD.

$$= \frac{6r(r - 5) - (3r + 4)(r + 2)}{(r + 8)(r + 2)(r - 5)}$$  Subtract the numerators.

$$= \frac{6r^2 - 30r - (3r^2 + 10r + 8)}{(r + 8)(r + 2)(r - 5)}$$  Distribute. You must use parentheses.

$$= \frac{6r^2 - 30r - 3r^2 - 10r - 8}{(r + 8)(r + 2)(r - 5)}$$  Distribute.

$$= \frac{3r^2 - 40r - 8}{(r + 8)(r + 2)(r - 5)}$$  Combine like terms.

Ask yourself, *"Can I factor the numerator?"* No. The expression is in simplest form since the numerator and denominator have no common factors. ■

---

**BE CAREFUL**

In Example 4, when you move from

$$\frac{6r(r - 5) - (3r + 4)(r + 2)}{(r + 8)(r + 2)(r - 5)} \quad \text{to} \quad \frac{6r^2 - 30r - (3r^2 + 10r + 8)}{(r + 8)(r + 2)(r - 5)}$$

you *must* use parentheses since the entire quantity $3r^2 + 10r + 8$ is being subtracted from $6r^2 - 30r$.

---

**You Try 4**

Subtract $\dfrac{4z}{z^2 + 10z + 21} - \dfrac{3z + 5}{z^2 - z - 12}$.

---

## 3. Add and Subtract Rational Expressions with Special Denominators

**Example 5**

Add or subtract.

a) $\dfrac{z}{z - 9} - \dfrac{8}{9 - z}$    b) $\dfrac{4}{7 - w} + \dfrac{10}{w^2 - 49}$

**Solution**

a)  Recall that $a - b = -(b - a)$. The least common denominator of $\dfrac{z}{z - 9}$ and $\dfrac{8}{9 - z}$

is $z - 9$ or $9 - z$. We will use LCD $= z - 9$.

Rewrite $\dfrac{8}{9-z}$ with the LCD:

$$\dfrac{8}{9-z} = \dfrac{8}{-(z-9)} = -\dfrac{8}{z-9}$$

$$\dfrac{z}{z-9} - \dfrac{8}{9-z} = \dfrac{z}{z-9} - \left(-\dfrac{8}{z-9}\right) \quad \text{Write each expression with the LCD.}$$

$$= \dfrac{z}{z-9} + \dfrac{8}{z-9} \quad \text{Distribute.}$$

$$= \dfrac{z+8}{z-9} \quad \text{Add the numerators.}$$

b)   Factor the denominator of $\dfrac{10}{w^2-49}$:   $\dfrac{10}{w^2-49} = \dfrac{10}{(w+7)(w-7)}$

Rewrite $\dfrac{4}{7-w}$ with a denominator of $w-7$:

$$\dfrac{4}{7-w} = \dfrac{4}{-(w-7)} = -\dfrac{4}{w-7}$$

Now we must find the LCD of $\dfrac{10}{(w+7)(w-7)}$ and $-\dfrac{4}{w-7}$.

$$\text{LCD} = (w+7)(w-7)$$

Rewrite $-\dfrac{4}{w-7}$ with the LCD.

$$-\dfrac{4}{w-7} \cdot \dfrac{w+7}{w+7} = -\dfrac{4(w+7)}{(w+7)(w-7)} = \dfrac{-4(w+7)}{(w+7)(w-7)}$$

$$\dfrac{4}{7-w} + \dfrac{10}{w^2-49} = -\dfrac{4}{w-7} + \dfrac{10}{(w+7)(w-7)}$$

$$= \dfrac{-4(w+7)}{(w+7)(w-7)} + \dfrac{10}{(w+7)(w-7)} \quad \begin{array}{l}\text{Write each expression}\\ \text{with the LCD.}\end{array}$$

$$= \dfrac{-4(w+7)+10}{(w+7)(w-7)} \quad \text{Add the numerators.}$$

$$= \dfrac{-4w-28+10}{(w+7)(w-7)} \quad \text{Distribute.}$$

$$= \dfrac{-4w-18}{(w+7)(w-7)} \quad \text{Combine like terms.}$$

Ask yourself, *"Can I factor the numerator?"* Yes.

$$= \dfrac{-2(2w+9)}{(w+7)(w-7)} \quad \text{Factor.}$$

Although the numerator factors, the numerator and denominator do not contain any common factors. The result, $\dfrac{-2(2w+9)}{(w+7)(w-7)}$, is in simplest form.   ■

**You Try 5**

Add or subtract.

a)   $\dfrac{n}{n-12} - \dfrac{7}{12-n}$      b)   $\dfrac{15}{4-t} + \dfrac{20}{t^2-16}$

## Answers to You Try Exercises

1) a) $\dfrac{1}{3}$  b) $\dfrac{9h+8}{5h-2}$   2) a) $\dfrac{7}{8w}$  b) $\dfrac{m-9}{m}$   3) a) $\dfrac{t+1}{3}$  b) $\dfrac{28v-27}{48v^2}$

c) $\dfrac{h-1}{h-8}$   4) $\dfrac{z^2-42z-35}{(z+7)(z+3)(z-4)}$   5) a) $\dfrac{n+7}{n-12}$  b) $\dfrac{-5(3t+8)}{(t+4)(t-4)}$

# 8.4 Exercises

## Objective 1:  Add and Subtract Rational Expressions with a Common Denominator

Add or subtract.

1) $\dfrac{5}{16}+\dfrac{9}{16}$

2) $\dfrac{5}{7}-\dfrac{3}{7}$

3) $\dfrac{11}{14}-\dfrac{3}{14}$

4) $\dfrac{1}{10}+\dfrac{9}{10}$

5) $\dfrac{5}{p}-\dfrac{23}{p}$

6) $\dfrac{6}{a}+\dfrac{3}{a}$

7) $\dfrac{7}{3c}+\dfrac{8}{3c}$

8) $\dfrac{10}{3k^2}-\dfrac{2}{3k^2}$

9) $\dfrac{6}{z-1}+\dfrac{z}{z-1}$

10) $\dfrac{4n}{n+9}-\dfrac{6}{n+9}$

11) $\dfrac{8}{x+4}+\dfrac{2x}{x+4}$

12) $\dfrac{5m}{m+7}+\dfrac{35}{m+7}$

13) $\dfrac{25t+17}{t(4t+3)}-\dfrac{5t+2}{t(4t+3)}$

14) $\dfrac{9w-20}{w(2w-5)}-\dfrac{20-7w}{w(2w-5)}$

15) $\dfrac{d^2+15}{(d+5)(d+2)}+\dfrac{8d-3}{(d+5)(d+2)}$

16) $\dfrac{2r+15}{(r-5)(r+4)}+\dfrac{r^2-10r}{(r-5)(r+4)}$

## Objective 2:  Add and Subtract Rational Expressions with Different Denominators

17) For $\dfrac{4}{9b^2}$ and $\dfrac{5}{6b^4}$,

a) find the LCD.

b) explain, in your own words, how to rewrite each expression with the LCD.

c) rewrite each expression with the LCD.

18) For $\dfrac{8}{x-3}$ and $\dfrac{2}{x}$,

a) find the LCD.

b) explain, in your own words, how to rewrite each expression with the LCD.

c) rewrite each expression with the LCD.

Add or subtract.

19) $\dfrac{3}{8}+\dfrac{2}{5}$

20) $\dfrac{5}{12}-\dfrac{1}{8}$

21) $\dfrac{4t}{3}+\dfrac{3}{2}$

22) $\dfrac{14x}{15}-\dfrac{5x}{6}$

23) $\dfrac{10}{3h^3}+\dfrac{2}{5h}$

24) $\dfrac{5}{8u}-\dfrac{2}{3u^2}$

25) $\dfrac{3}{2f^2}-\dfrac{7}{f}$

26) $\dfrac{8}{5a}+\dfrac{2}{5a^2}$

27) $\dfrac{13}{y+3}+\dfrac{3}{y}$

28) $\dfrac{3}{k}+\dfrac{11}{k+9}$

29) $\dfrac{15}{d-8}-\dfrac{4}{d}$

30) $\dfrac{14}{r-5}-\dfrac{3}{r}$

31) $\dfrac{9}{c-4}+\dfrac{6}{c+8}$

32) $\dfrac{2}{z+5}+\dfrac{1}{z+2}$

33) $\dfrac{m}{3m+5}-\dfrac{2}{m-10}$

34) $\dfrac{x}{x+4}-\dfrac{3}{2x+1}$

35) $\dfrac{8u+2}{u^2-1}+\dfrac{3u}{u+1}$

36) $\dfrac{t}{t+7}+\dfrac{9t-35}{t^2-49}$

37) $\dfrac{7g}{g^2 + 10g + 16} + \dfrac{3}{g^2 - 64}$

38) $\dfrac{b}{b^2 - 25} + \dfrac{8}{b^2 - 3b - 40}$

39) $\dfrac{5a}{a^2 - 6a - 27} - \dfrac{2a + 1}{a^2 + 2a - 3}$

40) $\dfrac{3c}{c^2 + 4c - 12} - \dfrac{2c - 5}{c^2 + 2c - 24}$

41) $\dfrac{2x}{x^2 + x - 20} - \dfrac{4}{x^2 + 2x - 15}$

42) $\dfrac{3m}{m^2 + 10m + 24} - \dfrac{2}{m^2 + 3m - 4}$

(VIDEO) 43) $\dfrac{4b + 1}{3b - 12} + \dfrac{5b}{b^2 - b - 12}$

44) $\dfrac{k + 12}{2k - 18} + \dfrac{3k}{k^2 - 12k + 27}$

**Objective 3: Add and Subtract Rational Expressions with Special Denominators**

45) Is $(x - 6)(6 - x)$ the LCD for $\dfrac{9}{x - 6} + \dfrac{4}{6 - x}$? Why or why not?

46) When Lamar adds $\dfrac{u}{7 - 2u} + \dfrac{5}{2u - 7}$, he gets $\dfrac{u - 5}{7 - 2u}$, but when he checks his answer in the back of the book, it says that the answer is $\dfrac{5 - u}{2u - 7}$. Which is the correct answer?

Add or subtract.

47) $\dfrac{16}{q - 4} + \dfrac{10}{4 - q}$

48) $\dfrac{8}{z - 9} + \dfrac{4}{9 - z}$

49) $\dfrac{11}{f - 7} - \dfrac{15}{7 - f}$

50) $\dfrac{5}{a - b} - \dfrac{4}{b - a}$

51) $\dfrac{7}{x - 4} + \dfrac{x - 1}{4 - x}$

52) $\dfrac{10}{m - 5} + \dfrac{m + 21}{5 - m}$

53) $\dfrac{8}{3 - a} + \dfrac{a + 5}{a - 3}$

54) $\dfrac{9}{6 - n} + \dfrac{n + 3}{n - 6}$

55) $\dfrac{3}{2u - 3v} - \dfrac{6u}{3v - 2u}$

56) $\dfrac{3c}{11b - 5c} - \dfrac{9}{5c - 11b}$

57) $\dfrac{8}{x^2 - 9} + \dfrac{2}{3 - x}$

58) $\dfrac{4}{8 - y} + \dfrac{12}{y^2 - 64}$

(VIDEO) 59) $\dfrac{a}{4a^2 - 9} - \dfrac{4}{3 - 2a}$

60) $\dfrac{3b}{9b^2 - 25} - \dfrac{3}{5 - 3b}$

**Mixed Exercises: Objectives 2 and 3**

Perform the indicated operations.

61) $\dfrac{5}{a^2 - 2a} + \dfrac{8}{a} - \dfrac{10a}{a - 2}$

62) $\dfrac{3}{j^2 + 6j} + \dfrac{2j}{j + 6} - \dfrac{2}{3j}$

63) $\dfrac{3b - 1}{b^2 + 8b} + \dfrac{b}{3b^2 + 25b + 8} + \dfrac{2}{3b^2 + b}$

64) $\dfrac{2k + 7}{k^2 - 4k} + \dfrac{9k}{2k^2 - 15k + 28} + \dfrac{15}{2k^2 - 7k}$

(VIDEO) 65) $\dfrac{c}{c^2 - 8c + 16} - \dfrac{5}{c^2 - c - 12}$

66) $\dfrac{n}{n^2 + 11n + 30} - \dfrac{6}{n^2 + 10n + 25}$

67) $\dfrac{9}{4a + 4b} + \dfrac{8}{a - b} - \dfrac{6a}{a^2 - b^2}$

68) $\dfrac{1}{x + y} + \dfrac{x}{x^2 - y^2} - \dfrac{10}{5x - 5y}$

69) $\dfrac{2v + 1}{6v^2 - 29v - 5} - \dfrac{v - 2}{3v^2 - 13v - 10}$

70) $\dfrac{n + 2}{4n^2 + 11n - 3} - \dfrac{n - 3}{2n^2 + 7n + 3}$

71) $\dfrac{g - 5}{5g^2 - 30g} + \dfrac{g}{2g^2 - 17g + 30} - \dfrac{6}{2g^2 - 5g}$

72) $\dfrac{y + 6}{y^2 - 4y} + \dfrac{y}{2y^2 - 13y + 20} - \dfrac{1}{2y^2 - 5y}$

For each rectangle, find a rational expression in simplest form to represent its a) area and b) perimeter.

73) $\dfrac{k - 4}{4}$ ; $\dfrac{8}{k + 1}$

74) $\dfrac{4}{r - 3}$ ; $\dfrac{r + 1}{6}$

75) $\dfrac{6}{h^2 + 9h + 20}$ ; $\dfrac{h}{h + 5}$

76) $\dfrac{1}{d^2 - 9}$ ; $\dfrac{d}{d + 3}$

77) Find a rational expression in simplest form to represent the perimeter of the triangle.

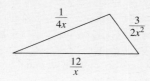

78) The total resistance of a set of resistors in parallel in an electrical circuit can be found using the formula
$$\frac{1}{R_T} = \frac{1}{R_1} + \frac{1}{R_2},$$ where $R_1 =$ the resistance in resistor 1,

$R_2 =$ the resistance in resistor 2, and $R_T =$ the total resistance in the circuit. (Resistance is measured in ohms.) For the given circuit,

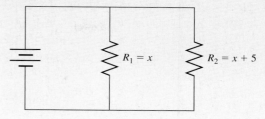

a) find the sum $\dfrac{1}{x} + \dfrac{1}{x + 5}$.

b) find an expression for the total resistance, $R_T$, in the circuit.

c) if $R_1 = 10$ ohms, what is the total resistance in the circuit?

---

## Putting It All Together

**Objective**

1. Review the Concepts Presented in Sections 8.1–8.4

### 1. Review the Concepts Presented in Sections 8.1–8.4

In Section 8.1, we defined a rational expression, and we evaluated expressions. We also discussed how to write a rational expression in lowest terms.

**Example 1**

Write in lowest terms: $\dfrac{5n^2 - 45n}{n^2 - 11n + 18}$

**Solution**

$$\frac{5n^2 - 45n}{n^2 - 11n + 18} = \frac{5n(n - 9)}{(n - 2)(n - 9)} \qquad \text{Factor.}$$

$$= \frac{5n}{n - 2} \qquad \text{Divide by } n - 9.$$

Recall that a rational expression *equals zero* when its *numerator equals zero*. A rational expression is *undefined* when its *denominator equals zero*.

**Example 2**

Determine the values of $c$ for which $\dfrac{c + 8}{c^2 - 25}$

a) equals zero.          b) is undefined.

**Solution**

a) $\dfrac{c + 8}{c^2 - 25}$ equals zero when its numerator equals zero.

Let $c + 8 = 0$, and solve for $c$.

$$c + 8 = 0$$
$$c = -8$$

$\dfrac{c + 8}{c^2 - 25}$ equals zero when $c = -8$.

b)  $\dfrac{c + 8}{c^2 - 25}$ is undefined when its denominator equals zero.

Let $c^2 - 25 = 0$, and solve for $c$.

$$c^2 - 25 = 0$$
$$(c + 5)(c - 5) = 0 \qquad \text{Factor.}$$
$$c + 5 = 0 \quad \text{or} \quad c - 5 = 0 \qquad \text{Set each factor equal to zero.}$$
$$c = -5 \quad \text{or} \qquad c = 5 \qquad \text{Solve.}$$

$\dfrac{c + 8}{c^2 - 25}$ is undefined when $c = 5$ or $c = -5$. So, $c \neq 5$ and $c \neq -5$ in the expression.

In Sections 8.2–8.4, we learned how to multiply, divide, add, and subtract rational expressions. Now we will practice these operations together so that we will learn to recognize the techniques needed to perform these operations.

## Example 3

Divide $\dfrac{t^2 - 3t - 28}{16t^2 - 81} \div \dfrac{t^2 - 7t}{54 - 24t}$.

### Solution

Do we need a common denominator to divide? *No.* A common denominator is needed to add or subtract but not to multiply or divide.

To divide, multiply the first rational expression by the reciprocal of the second expression, then factor, reduce, and multiply.

$$\frac{t^2 - 3t - 28}{16t^2 - 81} \div \frac{t^2 - 7t}{54 - 24t} = \frac{t^2 - 3t - 28}{16t^2 - 81} \cdot \frac{54 - 24t}{t^2 - 7t} \qquad \text{Multiply by the reciprocal.}$$

$$= \frac{(t + 4)\cancel{(t - 7)}}{(4t + 9)\cancel{(4t - 9)}} \cdot \frac{\overset{-1}{6\cancel{(9 - 4t)}}}{t\cancel{(t - 7)}} \qquad \text{Factor and reduce.}$$

$$= -\frac{6(t + 4)}{t(4t + 9)} \qquad \text{Multiply.}$$

Recall that $\dfrac{9 - 4t}{4t - 9} = -1$.

## Example 4

Add $\dfrac{x}{x + 2} + \dfrac{4}{3x - 1}$.

### Solution

To add or subtract rational expressions, we need a common denominator. We do not need to factor these denominators, so we are ready to identify the LCD.

$$\text{LCD} = (x + 2)(3x - 1)$$

Rewrite each expression with the LCD.

$$\frac{x}{x+2}\cdot\frac{3x-1}{3x-1}=\frac{x(3x-1)}{(x+2)(3x-1)}, \quad \frac{4}{3x-1}\cdot\frac{x+2}{x+2}=\frac{4(x+2)}{(x+2)(3x-1)}$$

$$\frac{x}{x+2}+\frac{4}{3x-1}=\frac{x(3x-1)}{(x+2)(3x-1)}+\frac{4(x+2)}{(x+2)(3x-1)} \qquad \text{Write each expression with the LCD.}$$

$$=\frac{x(3x-1)+4(x+2)}{(x+2)(3x-1)} \qquad \text{Add the numerators.}$$

$$=\frac{3x^2-x+4x+8}{(x+2)(3x-1)} \qquad \text{Distribute.}$$

$$=\frac{3x^2+3x+8}{(x+2)(3x-1)} \qquad \text{Combine like terms.}$$

Although this numerator will not factor, remember that sometimes it *is* possible to factor the numerator and simplify the result. ∎

## You Try 1

a) Write in lowest terms: $\dfrac{3k^2-14k+16}{4-k^2}$

b) Subtract $\dfrac{b}{b+10}-\dfrac{3}{b}$.

c) Multiply $\dfrac{r^3+9r^2-r-9}{12r+12}\cdot\dfrac{5}{r^2+3r-54}$.

d) Determine the values of $w$ for which $\dfrac{6w-1}{2w^2+16w}$   i) equals zero.   ii) is undefined.

### Answers to You Try Exercises

1)   a) $\dfrac{8-3k}{k+2}$   b) $\dfrac{b^2-3b-30}{b(b+10)}$   c) $\dfrac{5(r-1)}{12(r-6)}$   d) i) $\dfrac{1}{6}$   ii) $-8, 0$

## Putting It All Together
## Summary Exercises

**Objective 1: Review the Concepts Presented in Sections 8.1–8.4**

Evaluate, if possible, for a) $x=-3$ and b) $x=2$.

1) $\dfrac{x+3}{3x+4}$

2) $\dfrac{x}{x-2}$

3) $\dfrac{5x-3}{x^2+10x+21}$

4) $-\dfrac{x^2}{x^2-12}$

Determine the values of the variable for which

   a) the expression is undefined.

   b) the expression equals zero.

5) $-\dfrac{5w}{w^2-36}$

6) $\dfrac{m-4}{2m^2+11m+15}$

VIDEO 7) $\dfrac{3-5b}{b^2+2b-8}$

8) $\dfrac{5k-8}{64-k^2}$

9) $\dfrac{12}{5r}$

10) $\dfrac{t-15}{t^2+4}$

Write each rational expression in lowest terms.

11) $\dfrac{12w^{16}}{3w^5}$

12) $\dfrac{42n^3}{18n^8}$

13) $\dfrac{m^2+6m-27}{2m^2+2m-24}$

14) $\dfrac{2j+20}{2j^2+10j-100}$

VIDEO 15) $\dfrac{12-15n}{5n^2+6n-8}$

16) $\dfrac{-x-y}{xy+y^2+5x+5y}$

Perform the operations and simplify.

17) $\dfrac{4c^2 + 4c - 24}{c + 3} \div \dfrac{3c - 6}{8}$    18) $\dfrac{6}{f + 11} - \dfrac{2}{f}$

19) $\dfrac{4j}{j^2 - 81} + \dfrac{3}{j^2 - 3j - 54}$

20) $\dfrac{27a^4}{8b} \cdot \dfrac{40b^2}{81a^2}$

21) $\dfrac{12y^7}{4z^6} \cdot \dfrac{8z^4}{72y^6}$

22) $\dfrac{3}{q^2 - q - 20} + \dfrac{8q}{q^2 + 11q + 28}$

23) $\dfrac{x}{2x^2 - 7x - 4} - \dfrac{x + 3}{4x^2 + 4x + 1}$

24) $\dfrac{n - 4}{4n - 44} \cdot \dfrac{121 - n^2}{n + 11}$

25) $\dfrac{16 - m^2}{m + 4} \div \dfrac{8m - 32}{m + 7}$

26) $\dfrac{16}{r - 7} + \dfrac{4}{7 - r}$

27) $\dfrac{xy - 5x + 2y - 10}{y^2 - 25} \div \dfrac{x^3 + 8}{19x}$

28) $\dfrac{\dfrac{10d}{d + 11}}{\dfrac{5d^7}{3d + 33}}$

29) $\dfrac{9}{d + 3} + \dfrac{8}{d^2}$

30) $\dfrac{3a^2 - 6a + 12}{5a - 10} \cdot \dfrac{a^2 - 4}{a^3 + 8}$

31) $\dfrac{\dfrac{9k^2 - 1}{14k}}{\dfrac{3k - 1}{21k^4}}$    32) $\dfrac{13}{4z} - \dfrac{1}{3z}$

33) $\dfrac{2w}{25 - w^2} + \dfrac{w - 3}{w^2 - 12w + 35}$

34) $\dfrac{12a^4}{10a - 30} \div \dfrac{4a}{a^3 - 3a^2 + 5a - 15}$

35) $\dfrac{10}{x - 8} + \dfrac{4}{x + 3}$

36) $\dfrac{1}{4y} + \dfrac{8}{6y^4}$

37) $\dfrac{2h^2 + 11h + 5}{8} \div (2h + 1)^2$

38) $\dfrac{b^2 - 15b + 36}{b^2 - 8b - 48} \cdot (b + 4)^2$

39) $\dfrac{3m}{7m - 4n} - \dfrac{20n}{4n - 7m}$

40) $\dfrac{10d}{8c - 10d} + \dfrac{8c}{10d - 8c}$

41) $\dfrac{2p + 3}{p^2 + 7p} - \dfrac{4p}{p^2 - p - 56} + \dfrac{5}{p^2 - 8p}$

42) $\dfrac{6u + 1}{3u^2 - 2u} - \dfrac{u}{3u^2 + u - 2} + \dfrac{10}{u^2 + u}$

43) $\dfrac{6t + 6}{3t^2 - 24t} \cdot (t^2 - 7t - 8)$

44) $\dfrac{3r^2 + r - 14}{5r^3 - 10r^2} \div (9r^2 - 49)$

45) $\dfrac{\dfrac{3c^3}{8c + 40}}{\dfrac{9c}{c + 5}}$

46) $\dfrac{\dfrac{6v - 30}{4}}{\dfrac{v - 5}{3}}$

47) $\dfrac{f - 8}{f - 4} - \dfrac{4}{4 - f}$

48) $\dfrac{12p}{4p^2 + 11p + 6} - \dfrac{5}{p^2 - 4p - 12}$

49) $\left( \dfrac{3m}{3m - 1} - \dfrac{4}{m + 4} \right) \cdot \dfrac{9m^2 - 1}{21m^2 + 28}$

50) $\left( \dfrac{2c}{c + 8} + \dfrac{4}{c - 2} \right) \div \dfrac{6}{4c + 32}$

51) $\dfrac{3}{k^2 + 3k} - \dfrac{4}{3k} + \dfrac{1}{k + 3}$

52) $\dfrac{3}{w^2 - w} + \dfrac{4}{5w} - \dfrac{3}{w - 1}$

53) Find a rational expression in simplest form to represent the a) area and b) perimeter of the rectangle.

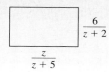

54) Find a rational expression in simplest form to represent the perimeter of the triangle.

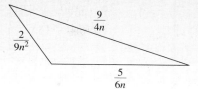

## Section 8.5  Simplifying Complex Fractions

### Objectives

1. Simplify a Complex Fraction with One Term in the Numerator and One Term in the Denominator
2. Simplify a Complex Fraction with More Than One Term in the Numerator and/or Denominator by Rewriting It as a Division Problem
3. Simplify a Complex Fraction with More Than One Term in the Numerator and/or Denominator by Multiplying by the LCD

In algebra, we sometimes encounter fractions that contain fractions in their numerators, denominators, or both. These are called **complex fractions.** Some examples of complex fractions are

$$\frac{\dfrac{3}{7}}{\dfrac{9}{2}}, \qquad \frac{\dfrac{1}{8}+\dfrac{5}{6}}{2-\dfrac{2}{3}}, \qquad \frac{\dfrac{4}{xy^2}}{\dfrac{1}{x}-\dfrac{1}{y}}, \qquad \frac{\dfrac{5a-15}{4}}{\dfrac{a-3}{a}}$$

### Definition

A **complex fraction** is a rational expression that contains one or more fractions in its numerator, its denominator, or both.

A complex fraction is not considered to be an expression in simplest form. In this section, we will learn how to simplify two different types of complex fractions:

1) Complex fractions with *one term* in the numerator and *one term* in the denominator
2) Complex fractions that have *more than one term* in their numerators, their denominators or both

### 1. Simplify a Complex Fraction with One Term in the Numerator and One Term in the Denominator

We studied these expressions in Section 8.2 when we learned how to divide fractions. We will look at another example.

---

**Example 1**

Simplify. $\dfrac{\dfrac{5a-15}{4}}{\dfrac{a-3}{a}}$

$\left.\dfrac{5a-15}{4}\right\}$ ⟵ This is the numerator.

⟵ This is the main fraction bar.

$\left.\dfrac{a-3}{a}\right\}$ ⟵ This is the denominator.

### Solution

This complex fraction contains one term in the numerator and one term in the denominator. To simplify, rewrite as a division problem, multiply by the reciprocal, and simplify.

$$\frac{\dfrac{5a-15}{4}}{\dfrac{a-3}{a}} = \frac{5a-15}{4} \div \frac{a-3}{a} \qquad \text{Rewrite as a division problem.}$$

$$= \frac{5a-15}{4} \cdot \frac{a}{a-3} = \frac{5(a-3)}{4} \cdot \frac{a}{a-3} = \frac{5\cancel{(a-3)}}{4} \cdot \frac{a}{\cancel{a-3}} = \frac{5a}{4} \quad \blacksquare$$

## You Try 1

Simplify.  $\dfrac{\dfrac{9}{z^2-64}}{\dfrac{3z+3}{z+8}}$

Let's summarize how to simplify this first type of complex fraction.

---

**Procedure**  Simplify a Complex Fraction Containing One Term in the Numerator and One Term in the Denominator

To simplify a complex fraction containing one term in the numerator and one term in the denominator:

1)  Rewrite the complex fraction as a division problem.

2)  Perform the division by multiplying the first fraction by the reciprocal of the second (that is, multiply the numerator of the complex fraction by the reciprocal of the denominator).

---

## 2. Simplify a Complex Fraction with More Than One Term in the Numerator and/or Denominator by Rewriting It as a Division Problem

When a complex fraction has more than one term in the numerator and/or the denominator, we can use one of two methods to simplify.

---

**Procedure**  Simplify a Complex Fraction Using Method 1

1)  Combine the terms in the numerator and combine the terms in the denominator so that each contains only one fraction.

2)  Rewrite as a division problem.

3)  Perform the division by multiplying the first fraction by the reciprocal of the second.

---

## Example 2

Simplify.

a)  $\dfrac{\dfrac{1}{4}+\dfrac{2}{3}}{2-\dfrac{1}{2}}$      b)  $\dfrac{\dfrac{5}{a^2b}}{\dfrac{a}{b}+\dfrac{1}{a}}$

**Solution**

a)  The numerator, $\frac{1}{4} + \frac{2}{3}$, contains two terms; the denominator, $2 - \frac{1}{2}$, contains two terms. We will add the terms in the numerator and subtract the terms in the denominator so that the numerator and denominator will each contain one fraction.

$$\frac{\dfrac{1}{4} + \dfrac{2}{3}}{2 - \dfrac{1}{2}} = \frac{\dfrac{3}{12} + \dfrac{8}{12}}{\dfrac{4}{2} - \dfrac{1}{2}} = \frac{\dfrac{11}{12}}{\dfrac{3}{2}}$$

Add the fractions in the numerator.
Subtract the fractions in the denominator.

Rewrite as a division problem, multiply by the reciprocal, and simplify.

$$\frac{11}{12} \div \frac{3}{2} = \frac{11}{\overset{6}{\cancel{12}}} \cdot \frac{\overset{1}{\cancel{2}}}{3} = \frac{11}{18}$$

b)  The numerator, $\frac{5}{a^2b}$, contains one term; the denominator, $\frac{a}{b} + \frac{1}{a}$, contains two terms.

We will add the terms in the denominator so that it, like the numerator, will contain only one term. The LCD of the expressions in the denominator is $ab$.

$$\frac{\dfrac{5}{a^2b}}{\dfrac{a}{b} + \dfrac{1}{a}} = \frac{\dfrac{5}{a^2b}}{\dfrac{a}{b} \cdot \dfrac{a}{a} + \dfrac{1}{a} \cdot \dfrac{b}{b}} = \frac{\dfrac{5}{a^2b}}{\dfrac{a^2}{ab} + \dfrac{b}{ab}} = \frac{\dfrac{5}{a^2b}}{\dfrac{a^2 + b}{ab}}$$

Rewrite as a division problem, multiply by the reciprocal, and simplify.

$$\frac{5}{a^2b} \div \frac{a^2 + b}{ab} = \frac{5}{\underset{a}{\cancel{a^2b}}} \cdot \frac{\cancel{ab}}{a^2 + b} = \frac{5}{a(a^2 + b)}$$  ∎

**You Try 2**

Simplify.

a) $\dfrac{\dfrac{5}{8} - \dfrac{1}{2}}{1 + \dfrac{1}{4}}$     b) $\dfrac{\dfrac{6}{r^2t^2}}{\dfrac{2}{r} - \dfrac{r}{t}}$

### 3. Simplify a Complex Fraction with More than One Term in the Numerator and/or Denominator by Multiplying by the LCD

Another method we can use to simplify complex fractions involves multiplying the numerator and denominator of the complex fraction by the LCD of *all* of the fractions in the expression.

**Procedure**   Simplify a Complex Fraction Using Method 2

1)  Identify and write down the LCD of *all* of the fractions in the complex fraction.

2)  Multiply the numerator and denominator of the complex fraction by the LCD.

3)  Simplify.

We will simplify the complex fractions we simplified in Example 2 using Method 2.

**Example 3**

Simplify using Method 2.

a) $\dfrac{\dfrac{1}{4} + \dfrac{2}{3}}{2 - \dfrac{1}{2}}$       b) $\dfrac{\dfrac{5}{a^2b}}{\dfrac{a}{b} + \dfrac{1}{a}}$

### Solution

a)   List all of the fractions in the complex fraction: $\dfrac{1}{4}, \dfrac{2}{3}, \dfrac{1}{2}$. Write down their

LCD:  LCD = 12.

Multiply the numerator and denominator of the complex fraction by the LCD, 12, then simplify.

$$\dfrac{12\left(\dfrac{1}{4} + \dfrac{2}{3}\right)}{12\left(2 - \dfrac{1}{2}\right)}$$      We are multiplying the expression by $\dfrac{12}{12}$, which equals 1.

$$= \dfrac{12 \cdot \dfrac{1}{4} + 12 \cdot \dfrac{2}{3}}{12 \cdot 2 - 12 \cdot \dfrac{1}{2}}$$      Distribute.

$$= \dfrac{3 + 8}{24 - 6} = \dfrac{11}{18}$$      Simplify.

This is the same result we obtained in Example 2 using Method 1.

**Note**

In the denominator, we multiplied the 2 by 12 even though 2 is not a fraction. Remember, *all* terms, not just the fractions, must be multiplied by the LCD.

b)   List all of the fractions in the complex fraction: $\dfrac{5}{a^2b}, \dfrac{a}{b}, \dfrac{1}{a}$. Write down their

LCD:  LCD = $a^2b$.

Multiply the numerator and denominator of the complex fraction by the LCD, $a^2b$, then simplify.

$$\dfrac{a^2b\left(\dfrac{5}{a^2b}\right)}{a^2b\left(\dfrac{a}{b} + \dfrac{1}{a}\right)}$$      We are multiplying the expression by $\dfrac{a^2b}{a^2b}$, which equals 1.

$$= \dfrac{a^2b \cdot \dfrac{5}{a^2b}}{a^2b \cdot \dfrac{a}{b} + a^2b \cdot \dfrac{1}{a}}$$      Distribute.

$$= \dfrac{5}{a^3 + ab} = \dfrac{5}{a(a^2 + b)}$$      Simplify.

If the numerator and denominator factor, factor them. Sometimes, you can divide by a common factor to simplify.

Notice that the result is the same as what was obtained in Example 2 using Method 1. ∎

**You Try 3**

Simplify using Method 2.

a) $\dfrac{\dfrac{5}{8} - \dfrac{1}{2}}{1 + \dfrac{1}{4}}$

b) $\dfrac{\dfrac{6}{r^2 t^2}}{\dfrac{2}{r} - \dfrac{r}{t}}$

You should be familiar with both methods for simplifying complex fractions containing two terms in the numerator or denominator. After a lot of practice, you will be able to decide which method is better for a particular problem.

**Example 4**

Determine which method to use to simplify each complex fraction, then simplify.

a) $\dfrac{\dfrac{4}{x} + \dfrac{1}{x + 3}}{\dfrac{2}{x + 3} - \dfrac{1}{x}}$

b) $\dfrac{\dfrac{n^2 - 1}{7n + 28}}{\dfrac{6n - 6}{n^2 - 16}}$

**Solution**

a) This complex fraction contains two terms in the numerator and two terms in the denominator. Let's use Method 2: multiply the numerator and denominator by the LCD of all of the fractions.

List all of the fractions in the complex fraction: $\dfrac{4}{x}, \dfrac{1}{x + 3}, \dfrac{2}{x + 3}, \dfrac{1}{x}$. Write down their LCD: LCD $= x(x + 3)$.

Multiply the numerator and denominator of the complex fraction by the LCD, $x(x + 3)$, then simplify.

$$\dfrac{x(x + 3)\left(\dfrac{4}{x} + \dfrac{1}{x + 3}\right)}{x(x + 3)\left(\dfrac{2}{x + 3} - \dfrac{1}{x}\right)} = \dfrac{x(x + 3) \cdot \dfrac{4}{x} + x(x + 3) \cdot \dfrac{1}{x + 3}}{x(x + 3) \cdot \dfrac{2}{x + 3} - x(x + 3) \cdot \dfrac{1}{x}}$$

Multiply the numerator and denominator by $x(x + 3)$ and distribute.

$$= \dfrac{4(x + 3) + x}{2x - (x + 3)}$$

Multiply.

$$= \dfrac{4x + 12 + x}{2x - x - 3} = \dfrac{5x + 12}{x - 3}$$

Distribute and combine like terms.

b)   The complex fraction $\dfrac{\dfrac{n^2 - 1}{7n + 28}}{\dfrac{6n - 6}{n^2 - 16}}$ contains one term in the numerator, $\dfrac{n^2 - 1}{7n + 28}$, and one

term in the denominator, $\dfrac{6n - 6}{n^2 - 16}$. To simplify, use Method 1 rewrite as a division

problem, multiply by the reciprocal, and simplify.

$$\dfrac{\dfrac{n^2 - 1}{7n + 28}}{\dfrac{6n - 6}{n^2 - 16}} = \dfrac{n^2 - 1}{7n + 28} \div \dfrac{6n - 6}{n^2 - 16} \qquad \text{Rewrite as a division problem.}$$

$$= \dfrac{n^2 - 1}{7n + 28} \cdot \dfrac{n^2 - 16}{6n - 6} \qquad \text{Multiply by the reciprocal.}$$

$$= \dfrac{(n + 1)(n - 1)}{7(n + 4)} \cdot \dfrac{(n + 4)(n - 4)}{6(n - 1)} \qquad \text{Factor and reduce.}$$

$$= \dfrac{(n + 1)(n - 4)}{42} \qquad \text{Multiply.}$$    ■

## You Try 4

Determine which method to use to simplify each complex fraction, then simplify.

a)  $\dfrac{\dfrac{8}{k} - \dfrac{1}{k + 5}}{\dfrac{3}{k + 5} + \dfrac{5}{k}}$    b)  $\dfrac{\dfrac{c^2 - 9}{8c - 56}}{\dfrac{2c + 6}{c^2 - 49}}$

---

### Answers to You Try Exercises

1)  $\dfrac{3}{(z - 8)(z + 1)}$    2)  a) $\dfrac{1}{10}$  b) $\dfrac{6}{rt(2t - r^2)}$    3)  a) $\dfrac{1}{10}$  b) $\dfrac{6}{rt(2t - r^2)}$

4)  a) $\dfrac{7k + 40}{8k + 25}$  b) $\dfrac{(c - 3)(c + 7)}{16}$

---

## 8.5 Exercises

**Objective 1: Simplify a Complex Fraction with One Term in the Numerator and One Term in the Denominator**

1)  Explain, in your own words, two ways to simplify $\dfrac{\dfrac{2}{9}}{\dfrac{5}{18}}$.

Then simplify it both ways. Which method do you prefer and why?

2)  Explain, in your own words, two ways to simplify $\dfrac{\dfrac{3}{2} - \dfrac{1}{5}}{\dfrac{1}{10} + \dfrac{3}{5}}$. Then simplify it both ways. Which method do you prefer and why?

Simplify completely.

3) $\dfrac{\dfrac{5}{9}}{\dfrac{7}{4}}$

4) $\dfrac{\dfrac{3}{10}}{\dfrac{5}{6}}$

5) $\dfrac{\dfrac{u^4}{v^2}}{\dfrac{u^3}{v}}$

6) $\dfrac{\dfrac{a^3}{b}}{\dfrac{a}{b^3}}$

7) $\dfrac{\dfrac{x^4}{y}}{\dfrac{x^2}{y^2}}$

8) $\dfrac{\dfrac{s^3}{t^3}}{\dfrac{s^4}{t}}$

 9) $\dfrac{\dfrac{14m^5n^4}{9}}{\dfrac{35mn^6}{3}}$

10) $\dfrac{\dfrac{11b^4c^2}{4}}{\dfrac{55bc}{12}}$

11) $\dfrac{\dfrac{m-7}{m}}{\dfrac{m-7}{18}}$

12) $\dfrac{\dfrac{t-4}{9}}{\dfrac{t-4}{t^2}}$

13) $\dfrac{\dfrac{g^2-36}{20}}{\dfrac{g+6}{60}}$

14) $\dfrac{\dfrac{6}{y^2-49}}{\dfrac{8}{y+7}}$

15) $\dfrac{\dfrac{d^3}{16d-24}}{\dfrac{d}{40d-60}}$

16) $\dfrac{\dfrac{45w-63}{w^5}}{\dfrac{30w-42}{w^2}}$

17) $\dfrac{\dfrac{c^2-7c-8}{11c}}{\dfrac{c+1}{c}}$

18) $\dfrac{\dfrac{5x}{x-3}}{\dfrac{5}{x^2+4x-21}}$

## Objective 2: Simplify a Complex Fraction with More Than One Term in the Numerator and/or Denominator by Rewriting It as a Division Problem

Simplify using Method 1.

19) $\dfrac{\dfrac{7}{9}-\dfrac{2}{3}}{3+\dfrac{1}{9}}$

20) $\dfrac{\dfrac{1}{2}+\dfrac{3}{4}}{\dfrac{2}{3}+\dfrac{3}{2}}$

21) $\dfrac{\dfrac{r}{s}-4}{\dfrac{3}{s}+\dfrac{1}{r}}$

22) $\dfrac{\dfrac{4}{c}-c^2}{1+\dfrac{8}{c}}$

23) $\dfrac{\dfrac{8}{r^2t}}{\dfrac{3}{r}-\dfrac{r}{t}}$

24) $\dfrac{\dfrac{9}{h^2k^3}}{\dfrac{6}{hk}-\dfrac{24}{k^2}}$

25) $\dfrac{\dfrac{5}{w-1}+\dfrac{3}{w+4}}{\dfrac{6}{w+4}+\dfrac{4}{w-1}}$

26) $\dfrac{\dfrac{5}{z-2}-\dfrac{2}{z+3}}{\dfrac{4}{z-2}+\dfrac{1}{z+3}}$

## Objective 3: Simplify a Complex Fraction with More Than One Term in the Numerator and/or Denominator by Multiplying by the LCD

Simplify the complex fractions in Exercises 19–26 using Method 2. Think about which method you prefer. (You will discuss your preference in Exercises 35 and 36.)

27) Rework Exercise 19.

28) Rework Exercise 20.

29) Rework Exercise 21.

30) Rework Exercise 22.

31) Rework Exercise 23.

32) Rework Exercise 24.

33) Rework Exercise 25.

34) Rework Exercise 26.

35) In Exercises 19–34, which types of complex fractions did you prefer to simplify using Method 1? Why?

36) In Exercises 19–34, which types of complex fractions did you prefer to simplify using Method 2? Why?

## Mixed Exercises: Objectives 1–3

Simplify completely using any method.

37) $\dfrac{\dfrac{a-4}{12}}{\dfrac{a-4}{a}}$

38) $\dfrac{\dfrac{z^2+1}{5}}{z+\dfrac{1}{z}}$

39) $\dfrac{\dfrac{3}{n}-\dfrac{4}{n-2}}{\dfrac{1}{n-2}+\dfrac{5}{n}}$

40) $\dfrac{\dfrac{1}{6}-\dfrac{5}{4}}{\dfrac{3}{5}+\dfrac{1}{3}}$

41) $\dfrac{\dfrac{6}{w}-w}{1+\dfrac{6}{w}}$

42) $\dfrac{\dfrac{6t+48}{t}}{\dfrac{9t+72}{7}}$

43) $\dfrac{\dfrac{6}{5}}{\dfrac{9}{15}}$

44) $\dfrac{\dfrac{8}{k+7}+\dfrac{1}{k}}{\dfrac{9}{k}+\dfrac{2}{k+7}}$

59) $\dfrac{\dfrac{x^2-x-42}{2x-14}}{\dfrac{x^2-36}{8x+16}}$

60) $\dfrac{3b+\dfrac{1}{b}}{b-\dfrac{13}{b}}$

45) $\dfrac{1-\dfrac{4}{t+5}}{\dfrac{4}{t^2-25}+\dfrac{t}{t-5}}$

46) $\dfrac{\dfrac{c^2}{d}+\dfrac{2}{c^2d}}{\dfrac{d}{c}-\dfrac{c}{d}}$

61) $\dfrac{\dfrac{y^4}{z^3}}{\dfrac{y^6}{z^4}}$

62) $\dfrac{\dfrac{k+6}{k}}{\dfrac{k+6}{5}}$

47) $\dfrac{\dfrac{9}{x}-\dfrac{9}{y}}{\dfrac{2}{x^2}-\dfrac{2}{y^2}}$

48) $\dfrac{\dfrac{m^2}{n^2}}{\dfrac{m^5}{n}}$

63) $\dfrac{7-\dfrac{8}{m}}{\dfrac{7m-8}{11}}$

64) $\dfrac{\dfrac{7}{a}-\dfrac{7}{b}}{\dfrac{1}{a^2}-\dfrac{1}{b^2}}$

49) $\dfrac{\dfrac{24c-60}{5}}{\dfrac{8c-20}{c^2}}$

50) $\dfrac{\dfrac{3}{x^2y^2}}{\dfrac{x}{y}+\dfrac{1}{x}}$

65) $\dfrac{\dfrac{1}{h^2-4}+\dfrac{2}{h+2}}{h-\dfrac{3}{2}}$

51) $\dfrac{\dfrac{4}{9}+\dfrac{2}{5}}{\dfrac{1}{5}-\dfrac{2}{3}}$

52) $\dfrac{1+\dfrac{4}{t-3}}{\dfrac{t}{t-3}+\dfrac{2}{t^2-9}}$

66) $\dfrac{\dfrac{w^2+10w+25}{25-w^2}}{\dfrac{w^3+125}{4w-20}}$

53) $\dfrac{\dfrac{1}{10}}{\dfrac{7}{8}}$

54) $\dfrac{\dfrac{r^2-6}{40}}{r-\dfrac{6}{r}}$

67) $\dfrac{\dfrac{6}{v+3}-\dfrac{4}{v-1}}{\dfrac{2}{v-1}+\dfrac{1}{v+2}}$

55) $\dfrac{\dfrac{2}{uv^2}}{\dfrac{6}{v}-\dfrac{4v}{u}}$

56) $\dfrac{\dfrac{y^2-9}{3y+15}}{\dfrac{7y-21}{y^2-25}}$

68) $\dfrac{\dfrac{5}{r+2}+\dfrac{7}{2r-3}}{\dfrac{1}{r-3}+\dfrac{3}{2r-3}}$

57) $\dfrac{1+\dfrac{b}{a-b}}{\dfrac{b}{a^2-b^2}+\dfrac{1}{a+b}}$

58) $\dfrac{\dfrac{c}{c+2}+\dfrac{1}{c^2-4}}{1-\dfrac{3}{c+2}}$

## Section 8.6 Solving Rational Equations

### Objectives

1. **Differentiate Between Rational Expressions and Rational Equations**
2. **Solve Rational Equations**
3. **Solve a Proportion**
4. **Solve an Equation for a Specific Variable**

A **rational equation** is an equation that contains a rational expression. Some examples of rational equations are

$$\frac{1}{2}a+\frac{7}{10}=\frac{3}{5}a-4, \qquad \frac{8}{p+7}-\frac{p}{p-10}=2, \qquad \frac{3n}{n^2+10n+16}+\frac{5}{n+8}=\frac{1}{n+2}$$

### 1. Differentiate Between Rational Expressions and Rational Equations

In Chapter 3, we solved rational equations like the first one above, and we learned how to add and subtract rational expressions in Section 8.4. Let's summarize the difference between the two because this is often a point of confusion for students.

---

**Summary**  Expressions Versus Equations

1) *The sum or difference of rational expressions does* not *contain an = sign.* To add or subtract, rewrite each expression with the LCD, and *keep the denominator* while performing the operations.

2) *An equation contains an = sign.* To solve an equation containing rational expressions, *multiply the equation by the LCD of all fractions to eliminate the denominators, then solve.*

---

### Example 1

Determine whether each is an equation or is a sum or difference of expressions. Then, solve the equation or find the sum or difference.

a) $\dfrac{c-5}{6} + \dfrac{c}{8} = \dfrac{3}{2}$     b) $\dfrac{c-5}{6} + \dfrac{c}{8}$

**Solution**

a) This is an *equation* because it contains an = sign. We will *solve* for $c$ using the method we learned in Chapter 3: eliminate the denominators by multiplying by the LCD of all of the expressions.  LCD = 24

$$24\left(\dfrac{c-5}{6} + \dfrac{c}{8}\right) = 24 \cdot \dfrac{3}{2} \qquad \text{Multiply by LCD of 24 to eliminate the denominators.}$$
$$4(c-5) + 3c = 36 \qquad \text{Distribute and eliminate denominators.}$$
$$4c - 20 + 3c = 36 \qquad \text{Distribute.}$$
$$7c - 20 = 36 \qquad \text{Combine like terms.}$$
$$7c = 56$$
$$c = 8$$

Check to verify that the solution set is $\{8\}$.

b) $\dfrac{c-5}{6} + \dfrac{c}{8}$ is *not* an equation to be solved because it does *not* contain an = sign.

*It is a sum of rational expressions.* Rewrite each expression with the LCD, then subtract, *keeping the denominators* while performing the operations.

LCD = 24

$$\dfrac{(c-5)}{6} \cdot \dfrac{4}{4} = \dfrac{4(c-5)}{24} \qquad\qquad \dfrac{c}{8} \cdot \dfrac{3}{3} = \dfrac{3c}{24}$$

$$\dfrac{c-5}{6} + \dfrac{c}{8} = \dfrac{4(c-5)}{24} + \dfrac{3c}{24} \qquad \text{Rewrite each expression with a denominator of 24.}$$
$$= \dfrac{4(c-5) + 3c}{24} \qquad \text{Add the numerators.}$$
$$= \dfrac{4c - 20 + 3c}{24} \qquad \text{Distribute.}$$
$$= \dfrac{7c - 20}{24} \qquad \text{Combine like terms.}$$

---

### You Try 1

Determine whether each is an equation or is a sum or difference of expressions. Then solve the equation or find the sum or difference.

a) $\dfrac{m+1}{6} - \dfrac{m}{2}$     b) $\dfrac{m+1}{6} - \dfrac{m}{2} = \dfrac{5}{6}$

## 2. Solve Rational Equations

Let's list the steps we use to solve a rational equation. Then we will look at more examples.

---

**Procedure    How to Solve a Rational Equation**

1) If possible, factor all denominators.

2) Write down the LCD of all of the expressions.

3) Multiply both sides of the equation by the LCD to *eliminate* the denominators.

4) Solve the equation.

5) Check the solution(s) in the original equation. If a proposed solution makes a denominator equal 0, then it is rejected as a solution.

---

**Example 2**

Solve $\dfrac{t}{16} + \dfrac{2}{t} = \dfrac{3}{4}$.

**Solution**

Since this is an equation, we will eliminate the denominators by multiplying the equation by the LCD of all of the expressions. LCD $= 16t$

$$16t\left(\frac{t}{16} + \frac{2}{t}\right) = 16t\left(\frac{3}{4}\right)$$    Multiply both sides of the equation by the LCD, $16t$.

$$\cancel{16t}\left(\frac{t}{\cancel{16}}\right) + 16\cancel{t}\left(\frac{2}{\cancel{t}}\right) = \overset{4}{\cancel{16t}}\left(\frac{3}{\cancel{4}}\right)$$    Distribute and divide out common factors.

$$t^2 + 32 = 12t$$

$$t^2 - 12t + 32 = 0$$    Subtract $12t$.

$$(t - 8)(t - 4) = 0$$    Factor.

$$t - 8 = 0 \quad \text{or} \quad t - 4 = 0$$
$$t = 8 \quad \text{or} \qquad t = 4$$

*Check:*    $t = 8$

$$\frac{t}{16} + \frac{2}{t} \overset{?}{=} \frac{3}{4}$$

$$\frac{8}{16} + \frac{2}{8} \overset{?}{=} \frac{3}{4}$$

$$\frac{2}{4} + \frac{1}{4} = \frac{3}{4} \quad ✓$$

$t = 4$

$$\frac{t}{16} + \frac{2}{t} \overset{?}{=} \frac{3}{4}$$

$$\frac{4}{16} + \frac{2}{4} \overset{?}{=} \frac{3}{4}$$

$$\frac{1}{4} + \frac{2}{4} = \frac{3}{4} \quad ✓$$

The solution set is $\{4, 8\}$.

**You Try 2**

Solve $\dfrac{d}{3} + \dfrac{4}{d} = \dfrac{13}{3}$.

It is *very* important to check the proposed solution. Sometimes, what appears to be a solution actually is not.

## Example 3

Solve $2 - \dfrac{9}{k + 9} = \dfrac{k}{k + 9}$.

### Solution

Since this is an equation, we will eliminate the denominators by multiplying the equation by the LCD of all of the expressions. LCD = $k + 9$

$$(k + 9)\left(2 - \frac{9}{k + 9}\right) = (k + 9)\left(\frac{k}{k + 9}\right)$$  Multiply both sides of the equation by the LCD, $k + 9$.

$$(k + 9)2 - \cancel{(k + 9)} \cdot \frac{9}{\cancel{k + 9}} = \cancel{(k + 9)}\left(\frac{k}{\cancel{k + 9}}\right)$$  Distribute and divide out common factors.

$$2k + 18 - 9 = k$$  Multiply.
$$2k + 9 = k$$
$$9 = -k$$  Subtract $2k$.
$$-9 = k$$  Divide by $-1$.

Check: $2 - \dfrac{9}{(-9) + 9} \overset{?}{=} \dfrac{-9}{(-9) + 9}$  Substitute $-9$ for $k$ in the original equation.

$$2 - \frac{9}{0} = \frac{-9}{0}$$

Since $k = -9$ makes the denominator equal zero, $-9$ cannot be a solution to the equation. Therefore, this equation has no solution. The solution set is $\varnothing$. ∎

**BE CAREFUL**

*Always* check what *appears* to be the solution or solutions to an equation containing rational expressions. If one of these values makes a denominator zero, then it *cannot* be a solution to the equation.

## You Try 3

Solve $\dfrac{3m}{m - 4} - 1 = \dfrac{12}{m - 4}$.

## Example 4

Solve $\dfrac{1}{4} - \dfrac{1}{a + 2} = \dfrac{a + 18}{4a^2 - 16}$.

### Solution

This is an equation. Eliminate the denominators by multiplying by the LCD. Begin by factoring the denominator of $\dfrac{a + 18}{4a^2 - 16}$.

$$\frac{1}{4} - \frac{1}{a + 2} = \frac{a + 18}{4(a + 2)(a - 2)}$$  Factor the denominator.

$$\text{LCD} = 4(a + 2)(a - 2)$$  Write down the LCD of all of the expressions.

$$4(a + 2)(a - 2)\left(\frac{1}{4} - \frac{1}{a + 2}\right) = 4(a + 2)(a - 2)\left(\frac{a + 18}{4(a + 2)(a - 2)}\right)$$  Multiply both sides of the equation by the LCD.

$$\cancel{4}(a + 2)(a - 2)\left(\frac{1}{\cancel{4}}\right) - 4\cancel{(a + 2)}(a - 2)\left(\frac{1}{\cancel{a + 2}}\right) = \cancel{4}\cancel{(a + 2)}\cancel{(a - 2)}\left(\frac{a + 18}{\cancel{4(a + 2)(a - 2)}}\right)$$

Distribute and divide out common factors.

$$(a + 2)(a - 2) - 4(a - 2) = a + 18 \qquad \text{Multiply.}$$
$$a^2 - 4 - 4a + 8 = a + 18 \qquad \text{Distribute.}$$
$$a^2 - 4a + 4 = a + 18 \qquad \text{Combine like terms.}$$
$$a^2 - 5a - 14 = 0 \qquad \text{Subtract } a \text{ and subtract 18.}$$
$$(a - 7)(a + 2) = 0 \qquad \text{Factor.}$$
$$a - 7 = 0 \quad \text{or} \quad a + 2 = 0 \qquad \text{Set each factor equal to zero.}$$
$$a = 7 \quad \text{or} \quad a = -2 \qquad \text{Solve.}$$

Look at the factored form of the equation. If $a = 7$, no denominator will equal zero. *If $a = -2$, however, two of the denominators will equal zero. Therefore, we must reject $a = -2$ as a solution.* Check only $a = 7$.

$$Check: \frac{1}{4} - \frac{1}{7 + 2} \stackrel{?}{=} \frac{7 + 18}{4(7)^2 - 16} \qquad \text{Substitute } a = 7 \text{ into the original equation.}$$

$$\frac{1}{4} - \frac{1}{9} \stackrel{?}{=} \frac{25}{180} \qquad \text{Simplify.}$$

$$\frac{9}{36} - \frac{4}{36} \stackrel{?}{=} \frac{5}{36} \qquad \text{Get a common denominator and reduce } \frac{25}{180}.$$

$$\frac{5}{36} = \frac{5}{36} \quad \checkmark \qquad \text{Subtract.}$$

The solution set is $\{7\}$.

The previous problem is a good example of why it is necessary to check all "solutions" to equations containing rational expressions.

## You Try 4

Solve $\dfrac{1}{3} - \dfrac{1}{z + 2} = \dfrac{z + 14}{3z^2 - 12}$.

## Example 5

Solve $\dfrac{11}{6h^2 + 48h + 90} = \dfrac{h}{3h + 15} + \dfrac{1}{2h + 6}$.

### Solution

Since this is an equation, we will eliminate the denominators by multiplying by the LCD. Begin by factoring all denominators, then identify the LCD.

$$\frac{11}{6(h + 5)(h + 3)} = \frac{h}{3(h + 5)} + \frac{1}{2(h + 3)} \qquad \text{LCD} = 6(h + 5)(h + 3)$$

$$6(h + 5)(h + 3)\left(\frac{11}{6(h + 5)(h + 3)}\right) = 6(h + 5)(h + 3)\left(\frac{h}{3(h + 5)} + \frac{1}{2(h + 3)}\right) \qquad \text{Multiply by the LCD.}$$

$$\cancel{6(h + 5)(h + 3)}\left(\frac{11}{\cancel{6(h + 5)(h + 3)}}\right) = \overset{2}{\cancel{6}}\cancel{(h + 5)}(h + 3)\left(\frac{h}{\cancel{3(h + 5)}}\right) + \overset{3}{\cancel{6}}(h + 5)\cancel{(h + 3)}\left(\frac{1}{\cancel{2(h + 3)}}\right) \qquad \text{Distribute.}$$

$$11 = 2h(h + 3) + 3(h + 5) \qquad \text{Multiply.}$$
$$11 = 2h^2 + 6h + 3h + 15 \qquad \text{Distribute.}$$
$$11 = 2h^2 + 9h + 15 \qquad \text{Combine like terms.}$$
$$0 = 2h^2 + 9h + 4 \qquad \text{Subtract 11.}$$
$$0 = (2h + 1)(h + 4) \qquad \text{Factor.}$$

$$2h + 1 = 0 \quad \text{or} \quad h + 4 = 0$$

$$h = -\frac{1}{2} \quad \text{or} \quad h = -4 \qquad \text{Solve.}$$

You can see from the factored form of the equation that neither $h = -\dfrac{1}{2}$ nor $h = -4$ will make a denominator zero. Check the values in the original equation to verify that the solution set is $\left\{ -4, -\dfrac{1}{2} \right\}$.

 **You Try 5**

Solve $\dfrac{5}{6n^2 + 18n + 12} = \dfrac{n}{2n + 2} + \dfrac{1}{3n + 6}$.

## 3. Solve a Proportion

**Example 6**

Solve $\dfrac{18}{r + 7} = \dfrac{6}{r - 1}$.

### Solution

This rational equation is also a *proportion*. A **proportion** is a statement that two ratios are equal. We can solve this proportion as we have solved the other equations in this section, by multiplying both sides of the equation by the LCD. Or, recall from Section 3.7 that *we can solve a proportion by setting the cross products equal to each other.*

$$\dfrac{18}{r + 7} \diagdown \diagup \dfrac{6}{r - 1}$$

Multiply.          Multiply.

$$18(r - 1) = 6(r + 7) \qquad \text{Set the cross products equal to each other.}$$
$$18r - 18 = 6r + 42 \qquad \text{Distribute.}$$
$$12r = 60$$
$$r = 5 \qquad \text{Solve.}$$

The proposed solution, $r = 5$, does *not* make a denominator equal zero. Check to verify that the solution set is $\{5\}$.

 **You Try 6**

Solve $\dfrac{7}{d + 3} = \dfrac{14}{3d + 5}$.

## 4. Solve an Equation for a Specific Variable

In Section 3.6, we learned how to solve an equation for a specific variable. For example, to solve $2l + 2w = P$ for $w$, we do the following:

$$2l + 2\boxed{w} = P \qquad \text{Put a box around } w, \text{ the variable for which we are solving.}$$
$$2\boxed{w} = P - 2l \qquad \text{Subtract } 2l.$$
$$w = \dfrac{P - 2l}{2} \qquad \text{Divide by 2.}$$

Next we discuss how to solve for a specific variable in a rational expression.

## Example 7

Solve $z = \dfrac{n}{d - D}$ for $d$.

### Solution

Note that the equation contains a lowercase $d$ and an uppercase $D$. These represent different quantities, so be sure to write them correctly. Put $d$ in a box.

Since $d$ is in the denominator of the rational expression, multiply both sides of the equation by $d - D$ to eliminate the denominator.

$$z = \frac{n}{\boxed{d} - D} \qquad \text{Put } d \text{ in a box.}$$

$$(\boxed{d} - D)z = (\boxed{d} - D)\left(\frac{n}{\boxed{d} - D}\right) \qquad \begin{array}{l}\text{Multiply both sides by } d - D \\ \text{to eliminate the denominator.}\end{array}$$

$$\boxed{d}z - Dz = n \qquad \text{Distribute.}$$

$$\boxed{d}z = n + Dz \qquad \text{Add } Dz.$$

$$d = \frac{n + Dz}{z} \qquad \text{Divide by } z.$$

### You Try 7

Solve $v = \dfrac{k}{m + M}$ for $m$.

## Example 8

Solve $\dfrac{1}{x} + \dfrac{1}{y} = \dfrac{1}{z}$ for $y$.

### Solution

Put the $y$ in a box. The LCD of all of the fractions is $xyz$. Multiply both sides of the equation by $xyz$.

$$\frac{1}{x} + \frac{1}{\boxed{y}} = \frac{1}{z} \qquad \text{Put } y \text{ in a box.}$$

$$x\boxed{y}z\left(\frac{1}{x} + \frac{1}{\boxed{y}}\right) = x\boxed{y}z\left(\frac{1}{z}\right) \qquad \begin{array}{l}\text{Multiply both sides by } xyz \text{ to} \\ \text{eliminate the denominator.}\end{array}$$

$$\cancel{x}\boxed{y}z \cdot \frac{1}{\cancel{x}} + x\cancel{\boxed{y}}z \cdot \frac{1}{\cancel{\boxed{y}}} = x\boxed{y}\cancel{z} \cdot \left(\frac{1}{\cancel{z}}\right) \qquad \text{Distribute.}$$

$$\boxed{y}z + xz = x\boxed{y} \qquad \text{Divide out common factors.}$$

Since we are solving for $y$ and there are terms containing $y$ on each side of the equation, we must get $yz$ and $xy$ on one side of the equation and $xz$ on the other side.

$$xz = x\boxed{y} - \boxed{y}z \qquad \text{Subtract } yz \text{ from each side.}$$

To isolate $y$, we will *factor* $y$ out of each term on the right-hand side of the equation.

$$xz = \boxed{y}(x - z) \qquad \text{Factor out } y.$$

$$\frac{xz}{x - z} = y \qquad \text{Divide by } x - z.$$

### You Try 8

Solve $\dfrac{1}{x} + \dfrac{1}{y} = \dfrac{1}{z}$ for $z$.

## Using Technology

We can use a graphing calculator to solve a rational equation in one variable. First, enter the left side of the equation in $Y_1$ and the right side of the equation in $Y_2$. Then enter $Y_1 - Y_2$ in $Y_3$. Then graph the equation in $Y_3$. The zeros or $x$-intercepts of the graph are the solutions to the equation.

We will solve $\dfrac{2}{x+5} - \dfrac{3}{x-2} = \dfrac{4x}{x^2+3x-10}$ using a graphing calculator.

1) Enter $\dfrac{2}{x+5} - \dfrac{3}{x-2}$ by entering $2/(x+5) - 3/(x-2)$ in $Y_1$.

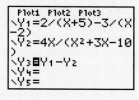

2) Enter $\dfrac{4x}{x^2+3x-10}$ by entering $4x/(x^2+3x-10)$ in $Y_2$.

3) Enter $Y_1 - Y_2$ in $Y_3$ as follows: press [VARS], select Y-VARS using the right arrow key, and press [ENTER] [ENTER] to select $Y_1$. Then press [ − ]. Press [VARS], select Y-VARS using the right arrow key, press [ENTER] [ 2 ] to select $Y_2$. Then press [ENTER].

4) Move the cursor onto the = sign just right of \$Y_1$ and press [ENTER] to deselect $Y_1$. Repeat to deselect $Y_2$. Press [GRAPH] to graph $Y_1 - Y_2$.

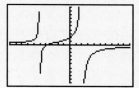

5) Press [2nd] [TRACE] 2:zero, move the cursor to the left of the zero and press [ENTER], move the cursor to the right of the zero and press [ENTER], and move the cursor close to the zero and press [ENTER] to display the zero.

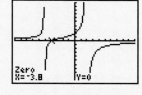

6) Press [X,T,θ,n] [MATH] [ENTER] [ENTER] to display the zero $x = -\dfrac{19}{5}$.

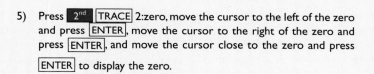

If there is more than one zero, repeat steps 5 and 6 above for each zero.

Solve each equation using a graphing calculator.

1) $\dfrac{2x}{x-3} + \dfrac{1}{x+5} = \dfrac{4-2x}{x^2+2x-15}$

2) $\dfrac{4}{x-3} + \dfrac{5}{x+3} = \dfrac{15}{x^2-9}$

3) $\dfrac{2}{x+2} + \dfrac{4}{x-5} = \dfrac{16}{x^2-3x-10}$

4) $\dfrac{1}{x-7} + \dfrac{3}{x+4} = \dfrac{7}{x^2-3x-28}$

5) $\dfrac{6}{x+1} = \dfrac{5x+3}{x^2-x-2} - \dfrac{x}{x-2}$

6) $\dfrac{4}{x+3} - \dfrac{x}{x+2} = \dfrac{3x}{x^2+5x+6}$

**Answers to You Try Exercises**

1) a) difference; $\dfrac{1-2m}{6}$   b) equation; $\{-2\}$   2) $\{1, 12\}$   3) $\varnothing$   4) $\{6\}$   5) $\left\{-3, \dfrac{1}{3}\right\}$

6) $\{1\}$   7) $m = \dfrac{k - Mv}{v}$   8) $z = \dfrac{xy}{x+y}$

**Answers to Technology Exercises**

1) $\left\{-7, \dfrac{1}{2}\right\}$   2) $\{2\}$   3) $\{3\}$   4) $\{6\}$   5) $\{-5, 3\}$   6) $\{-4, 2\}$

## 8.6 Exercises

**Objective 1: Differentiate Between Rational Expressions and Rational Equations**

1) When solving an equation containing rational expressions, do you keep the LCD throughout the problem or do you eliminate the denominators?

2) When adding or subtracting two rational expressions, do you keep the LCD throughout the problem or do you eliminate the denominators?

Determine whether each is an equation or a sum or difference of expressions. Then solve the equation or find the sum or difference.

3) $\dfrac{3r + 5}{2} - \dfrac{r}{6}$

4) $\dfrac{m}{12} + \dfrac{m - 8}{3}$

5) $\dfrac{3h}{2} + \dfrac{4}{3} = \dfrac{2h + 3}{3}$

6) $\dfrac{7f - 24}{12} = f + \dfrac{1}{2}$

7) $\dfrac{3}{a^2} + \dfrac{1}{a + 11}$

8) $\dfrac{z}{z - 5} - \dfrac{4}{z}$

9) $\dfrac{8}{b - 11} - 5 = \dfrac{3}{b - 11}$

10) $1 + \dfrac{2}{c + 5} = \dfrac{11}{c + 5}$

**Mixed Exercises: Objectives 2 and 3**

Values that make the denominators equal zero cannot be solutions of an equation. Find *all* of the values that make the

denominators zero and which, therefore, cannot be solutions of each equation. Do NOT solve the equation.

11) $\dfrac{k + 3}{k - 2} + 1 = \dfrac{7}{k}$

12) $\dfrac{t}{t + 12} - \dfrac{5}{t} = 3$

13) $\dfrac{8}{p + 3} - \dfrac{6}{p} = \dfrac{p}{p^2 - 9}$

14) $\dfrac{7}{d^2 - 64} + \dfrac{6}{d} = \dfrac{8}{d + 8}$

15) $\dfrac{9h}{h^2 - 5h - 36} + \dfrac{1}{h + 4} = \dfrac{h + 7}{3h - 27}$

16) $\dfrac{v + 8}{v^2 - 8v + 12} - \dfrac{5}{3v - 4} = \dfrac{2v}{v - 6}$

Solve each equation.

17) $\dfrac{a}{3} + \dfrac{7}{12} = \dfrac{1}{4}$

18) $\dfrac{y}{2} - \dfrac{4}{3} = \dfrac{1}{6}$

19) $\dfrac{1}{4}j - j = -4$

20) $\dfrac{1}{3}h + h = -4$

21) $\dfrac{8m - 5}{24} = \dfrac{m}{6} - \dfrac{7}{8}$

22) $\dfrac{13u - 1}{20} = \dfrac{3u}{5} - 1$

23) $\dfrac{8}{3x + 1} = \dfrac{2}{x + 3}$

24) $\dfrac{4}{5t + 2} = \dfrac{2}{2t - 1}$

25) $\dfrac{r + 1}{2} = \dfrac{4r + 1}{5}$

26) $\dfrac{w}{3} = \dfrac{6w - 4}{9}$

27) $\dfrac{23}{z} + 8 = -\dfrac{25}{z}$

28) $\dfrac{18}{a} - 2 = \dfrac{10}{a}$

29) $\dfrac{5q}{q + 1} - 2 = \dfrac{5}{q + 1}$

30) $\dfrac{n}{n + 3} + 5 = \dfrac{12}{n + 3}$

VIDEO 31) $\dfrac{2}{s+6} + 4 = \dfrac{2}{s+6}$

32) $\dfrac{u}{u-5} + 3 = \dfrac{5}{u-5}$

33) $\dfrac{3b}{b+7} - 6 = \dfrac{3}{b+7}$

34) $\dfrac{c}{c-5} - 5 = \dfrac{20}{c-5}$

35) $\dfrac{8}{r} - 1 = \dfrac{6}{r}$

36) $\dfrac{11}{g} + 3 = -\dfrac{10}{g}$

37) $z + \dfrac{12}{z} = -8$

38) $y - \dfrac{28}{y} = 3$

39) $\dfrac{15}{b} = 8 - b$

40) $n = 13 - \dfrac{12}{n}$

41) $\dfrac{8}{c+2} - \dfrac{12}{c-4} = \dfrac{2}{c+2}$

42) $\dfrac{2}{m-1} + \dfrac{1}{m+4} = \dfrac{4}{m+4}$

43) $\dfrac{9}{c-8} - \dfrac{15}{c} = 1$

44) $\dfrac{6}{r+5} - \dfrac{2}{r} = -1$

45) $\dfrac{3}{p-4} + \dfrac{8}{p+4} = \dfrac{13}{p^2 - 16}$

46) $\dfrac{5}{w-7} - \dfrac{8}{w+7} = \dfrac{52}{w^2 - 49}$

47) $\dfrac{9}{k+5} - \dfrac{4}{k+1} = \dfrac{10}{k^2 + 6k + 5}$

48) $\dfrac{3}{a+2} + \dfrac{10}{a^2 - 6a - 16} = \dfrac{5}{a-8}$

49) $\dfrac{12}{g^2 - 9} + \dfrac{2}{g+3} = \dfrac{7}{g-3}$

50) $\dfrac{9}{t+4} + \dfrac{8}{t^2 - 16} = \dfrac{1}{t-4}$

51) $\dfrac{5}{p-3} - \dfrac{7}{p^2 - 7p + 12} = \dfrac{8}{p-4}$

52) $\dfrac{8}{x^2 + 2x - 15} = \dfrac{6}{x-3} + \dfrac{4}{x+5}$

53) $\dfrac{x^2}{2} = \dfrac{x^2 - 6x}{3}$

54) $\dfrac{k^2}{3} = \dfrac{k^2 + 3k}{4}$

55) $\dfrac{3}{t^2} = \dfrac{6}{t^2 + 8t}$

56) $\dfrac{5}{m^2 - 36} = \dfrac{4}{m^2 + 6m}$

57) $\dfrac{b+3}{3b - 18} - \dfrac{b+2}{b-6} = \dfrac{b}{3}$

58) $\dfrac{3y - 2}{y+2} = \dfrac{y}{4} + \dfrac{1}{4y+8}$

59) $\dfrac{4}{n+1} = \dfrac{10}{n^2 - 1} - \dfrac{5}{n-1}$

60) $\dfrac{2}{c-6} - \dfrac{24}{c^2 - 36} = -\dfrac{3}{c+6}$

61) $-\dfrac{a}{5} = \dfrac{3}{a+8}$

62) $\dfrac{u}{7} = \dfrac{2}{9-u}$

VIDEO 63) $\dfrac{8}{p+2} + \dfrac{p}{p+1} = \dfrac{5p+2}{p^2 + 3p + 2}$

64) $\dfrac{6}{x-1} + \dfrac{x}{x+3} = \dfrac{2x+28}{x^2 + 2x - 3}$

65) $\dfrac{-14}{3a^2 + 15a - 18} = \dfrac{a}{a-1} + \dfrac{2}{3a+18}$

66) $\dfrac{3}{2n^2 + 10n + 8} = \dfrac{n}{2n+2} + \dfrac{1}{n+1}$

67) $\dfrac{3}{f+4} = \dfrac{f}{f+6} - \dfrac{2}{f^2 + 10f + 24}$

68) $\dfrac{11}{c+9} = \dfrac{c}{c-4} - \dfrac{36 - 8c}{c^2 + 5c - 36}$

69) $\dfrac{b}{b^2 + b - 6} + \dfrac{3}{b^2 + 9b + 18} = \dfrac{8}{b^2 + 4b - 12}$

70) $\dfrac{h}{h^2 + 2h - 8} + \dfrac{4}{h^2 + 8h - 20} = \dfrac{4}{h^2 + 14h + 40}$

71) $\dfrac{r}{r^2 + 8r + 15} - \dfrac{2}{r^2 + r - 6} = \dfrac{2}{r^2 + 3r - 10}$

72) $\dfrac{5}{t^2 + 5t - 6} - \dfrac{t}{t^2 + 10t + 24} = \dfrac{1}{t^2 + 3t - 4}$

73) $\dfrac{k}{k^2 - 6k - 16} - \dfrac{12}{5k^2 - 65k + 200} = \dfrac{28}{5k^2 - 15k - 50}$

74) $\dfrac{q}{q^2 + 4q - 32} + \dfrac{2}{q^2 - 14q + 40} = \dfrac{6}{q^2 - 2q - 80}$

## Objective 4: Solve an Equation for a Specific Variable

Solve for the indicated variable.

75) $W = \dfrac{CA}{m}$ for $m$

76) $V = \dfrac{nRT}{P}$ for $P$

77) $a = \dfrac{rt}{2b}$ for $b$

78) $y = \dfrac{kx}{z}$ for $z$

VIDEO 79) $B = \dfrac{t+u}{3x}$ for $x$

80) $Q = \dfrac{n-k}{5r}$ for $r$

81) $d = \dfrac{t}{z-n}$ for $n$

82) $z = \dfrac{a}{b+c}$ for $b$

83) $h = \dfrac{3A}{r+s}$ for $s$

84) $A = \dfrac{4r}{q-t}$ for $t$

85) $r = \dfrac{kx}{y-az}$ for $y$

86) $w = \dfrac{na}{kc+b}$ for $c$

VIDEO 87) $\dfrac{1}{t} = \dfrac{1}{r} - \dfrac{1}{s}$ for $r$

88) $\dfrac{1}{R_1} + \dfrac{1}{R_2} = \dfrac{1}{R_3}$ for $R_2$

89) $\dfrac{5}{x} = \dfrac{1}{y} - \dfrac{4}{z}$ for $z$

90) $\dfrac{2}{A} + \dfrac{1}{C} = \dfrac{3}{B}$ for $C$

## Section 8.7 Applications of Rational Equations and Variation

### Objectives

1. **Solve Problems Involving Proportions**
2. **Solve Problems Involving Distance, Rate, and Time**
3. **Solve Problems Involving Work**
4. **Solve Direct Variation Problems**
5. **Solve Inverse Variation Problems**

We have studied applications of linear and quadratic equations. Now we turn our attention to applications involving equations with rational expressions. We will continue to use the five-step problem-solving method outlined in Section 3.3.

### 1. Solve Problems Involving Proportions

We first solved application problems involving proportions in Section 3.7. We begin this section with a problem involving a proportion.

---

**Example 1**

Write an equation and solve.

At a small business, the ratio of employees who ride their bikes to work to those who drive a car is 4 to 3. The number of people who bike to work is three more than the number who drive. How many people bike to work, and how many drive their cars?

**Solution**

**Step 1:** **Read** the problem carefully, and identify what we are being asked to find.

We must find the number of people who bike to work and the number who drive.

**Step 2:** **Choose a variable** to represent the unknown, and define the other unknown in terms of this variable.

$$x = \text{the number of people who drive}$$
$$x + 3 = \text{the number of people who bike}$$

**Step 3:** **Translate** the information that appears in English into an algebraic equation.

Write a proportion. We will write our ratios in the form of $\dfrac{\text{number who bike}}{\text{number who drive}}$ so that the numerators contain the same quantities and the denominators contain the same quantities.

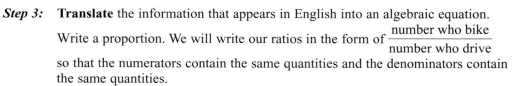

Number who bike → $\dfrac{4}{3}$ = $\dfrac{x+3}{x}$ ← Number who bike
Number who drive →              ← Number who drive

The equation is $\dfrac{4}{3} = \dfrac{x+3}{x}$.

**Step 4:** **Solve** the equation.

$$\frac{4}{3} \quad \frac{x+3}{x}$$

Multiply.
Multiply.

$4x = 3(x + 3)$    Set the cross products equal.
$4x = 3x + 9$     Distribute.
$x = 9$        Subtract $3x$.

**Step 5:** **Check** the answer and **interpret** the solution as it relates to the problem.

Therefore, 9 people drive to work and $9 + 3 = 12$ people ride their bikes. The check is left to the student.

**You Try 1**

Write an equation and solve.

In a classroom of college students, the ratio of students who have Internet access on their cell phones to those who do not is 3 to 5. The number of students who have Internet access is eight less than the number who do not. How many students have Internet access on their phones?

## 2. Solve Problems Involving Distance, Rate, and Time

In Section 3.7, we solved problems involving distance ($d$), rate ($r$), and time ($t$).

The basic formula is $d = rt$. We can solve this formula for $r$ and then for $t$ to obtain

$$r = \frac{d}{t} \quad \text{and} \quad t = \frac{d}{r}$$

The problems in this section involve boats going with and against a current, and planes going with and against the wind. Both situations use the same idea.

Suppose a boat's speed is 18 mph in still water. If that same boat had a 4-mph current pushing *against* it, how fast would it be traveling? (The current will cause the boat to slow down.)

Speed *against* the current = 18 mph − 4 mph = 14 mph

$$\frac{\text{Speed } against}{\text{the current}} = \frac{\text{Speed in}}{\text{still water}} - \frac{\text{Speed of}}{\text{the current}}$$

If the speed of the boat in still water is 18 mph and a 4-mph current is *pushing* the boat, how fast would the boat be traveling *with* the current? (The current will cause the boat to travel faster.)

Speed *with* the current = 18 mph + 4 mph = 22 mph

$$\frac{\text{Speed } with}{\text{the current}} = \frac{\text{Speed in}}{\text{still water}} + \frac{\text{Speed of}}{\text{the current}}$$

A boat traveling *against* the current is said to be traveling *upstream*. A boat traveling *with* the current is said to be traveling *downstream*.

We will use these ideas in Example 2.

**Example 2**

Write an equation and solve.

A boat can travel 15 mi downstream in the same amount of time it can travel 9 mi upstream. If the speed of the current is 4 mph, what is the speed of the boat in still water?

### Solution

*Step 1:* **Read** the problem carefully, and identify what we are being asked to find.

First, we must understand that "15 mi downstream" means *15 mi with the current*, and "9 mi upstream" means *9 miles against the current*.

We must find the speed of the boat in still water.

*Step 2:* **Choose a variable** to represent the unknown, and define the other unknowns in terms of this variable.

$x$ = the speed of the boat in still water
$x + 4$ = the speed of the boat *with* the current (downstream)
$x - 4$ = the speed of the boat *against* the current (upstream)

**Step 3:**  **Translate** from English into an algebraic equation. Use a table to organize the information.

First, fill in the distances and the rates (or speeds).

|  | d | r | t |
|---|---|---|---|
| Downstream | 15 | $x + 4$ |  |
| Upstream | 9 | $x - 4$ |  |

Next we must write expressions for the times it takes the boat to go downstream and upstream. We know that $d = rt$, so if we solve for $t$ we get $t = \dfrac{d}{r}$. Substitute the information from the table to get the expressions for the time.

$$\text{Downstream: } t = \frac{d}{r} = \frac{15}{x + 4} \qquad\qquad \text{Upstream: } t = \frac{d}{r} = \frac{9}{x - 4}$$

Put these values into the table.

|  | d | r | t |
|---|---|---|---|
| Downstream | 15 | $x + 4$ | $\dfrac{15}{x + 4}$ |
| Upstream | 9 | $x - 4$ | $\dfrac{9}{x - 4}$ |

The problem states that it takes the boat the *same amount of time* to travel 15 mi downstream as it does to go 9 mi upstream. We can write an equation in English:

$$\begin{matrix} \text{Time for boat to go} \\ \text{15 miles downstream} \end{matrix} = \begin{matrix} \text{Time for boat to go} \\ \text{9 miles upstream} \end{matrix}$$

Looking at the table, we can write the algebraic equation using the expressions for time. The equation is $\dfrac{15}{x + 4} = \dfrac{9}{x - 4}$.

**Step 4:**  **Solve** the equation.

$$\frac{15}{x + 4} \diagdown\!\!\!\!\diagup \frac{9}{x - 4} \quad \text{Multiply.}$$

$$
\begin{aligned}
15(x - 4) &= 9(x + 4) &&\text{Set the cross products equal.} \\
15x - 60 &= 9x + 36 &&\text{Distribute.} \\
6x &= 96 \\
x &= 16 &&\text{Solve.}
\end{aligned}
$$

**Step 5:**  **Check** the answer and **interpret** the solution as it relates to the problem.

The speed of the boat in still water is 16 mph.

*Check:* The speed of the boat going downstream is $16 + 4 = 20$ mph, so the time to travel downstream is

$$t = \frac{d}{r} = \frac{15}{20} = \frac{3}{4} \text{ hr}$$

The speed of the boat going upstream is $16 - 4 = 12$ mph, so the time to travel upstream is

$$t = \frac{d}{r} = \frac{9}{12} = \frac{3}{4} \text{ hr}$$

So, time upstream = time downstream. ✓                                    ■

**You Try 2**

> Write an equation and solve.
>
> It takes a boat the same amount of time to travel 12 mi downstream as it does to travel 6 mi upstream. Find the speed of the boat in still water if the speed of the current is 3 mph.

### 3. Solve Problems Involving Work

Suppose it takes Tara 3 hr to paint a fence. What is the *rate* at which she does the job?

$$\text{rate} = \frac{1 \text{ fence}}{3 \text{ hr}} = \frac{1}{3} \text{ fence/hr}$$

Tara works at a rate of $\frac{1}{3}$ of a fence per hour.

In general, we can say that if it takes $t$ units of time to do a job, then the *rate* at which the job is done is $\frac{1}{t}$ job per unit of time.

This idea of *rate* is what we use to determine how long it can take for 2 or more people or things to do a job.

Let's assume, again, that Tara can paint the fence in 3 hr. At this rate, how much of the job can she do in 2 hr?

$$\begin{array}{c}\text{Fractional part} \\ \text{of the job done}\end{array} = \begin{array}{c}\text{Rate of} \\ \text{work}\end{array} \cdot \begin{array}{c}\text{Amount of} \\ \text{time worked}\end{array}$$

$$= \frac{1}{3} \cdot 2$$

$$= \frac{2}{3}$$

She can paint $\frac{2}{3}$ of the fence in 2 hr.

---

**Procedure**  Solving Work Problems

The basic equation used to solve work problems is:

$$\begin{array}{c}\text{Fractional part of a job} \\ \text{done by one person or thing}\end{array} + \begin{array}{c}\text{Fractional part of a job} \\ \text{done by another person or thing}\end{array} = 1 \text{ (whole job)}$$

### Example 3

Write an equation and solve.

If Tara can paint the backyard fence in 3 hr but her sister, Grace, could paint the fence in 2 hr, how long would it take them to paint the fence together?

**Solution**

**Step 1:**  **Read** the problem carefully, and identify what we are being asked to find.

We must determine how long it would take Tara and Grace to paint the fence together.

**Step 2:**  **Choose a variable** to represent the unknown.

$$t = \text{the number of hours to paint the fence together}$$

**Step 3:**  **Translate** the information that appears in English into an algebraic equation.

Let's write down their rates:

$$\text{Tara's rate} \ = \frac{1}{3} \text{ fence/hr (since the job takes her 3 hours)}$$

$$\text{Grace's rate} \ = \frac{1}{2} \text{ fence/hr (since the job takes her 2 hours)}$$

It takes them $t$ hours to paint the room together. Recall that

$$\begin{array}{ccc} \text{Fractional part} & = & \text{Rate of} & \cdot & \text{Amount of} \\ \text{of job done} & & \text{work} & & \text{time worked} \end{array}$$

$$\text{Tara's fractional part} \ = \ \frac{1}{3} \ \cdot \ t \ = \frac{1}{3}t$$

$$\text{Grace's fractional part} \ = \ \frac{1}{2} \ \cdot \ t \ = \frac{1}{2}t$$

The equation we can write comes from

$$\begin{array}{ccc} \text{Fractional part of the} & + & \text{Fractional part of the} & = & \text{1 whole job} \\ \text{job done by Tara} & & \text{job done by Grace} & & \end{array}$$

$$\frac{1}{3}t \quad + \quad \frac{1}{2}t \quad = \quad 1$$

The equation is $\dfrac{1}{3}t + \dfrac{1}{2}t = 1$.

**Step 4:**  **Solve** the equation.

$$6\left(\frac{1}{3}t + \frac{1}{2}t\right) = 6(1) \qquad \text{Multiply by the LCD of 6 to eliminate the fractions.}$$

$$6\left(\frac{1}{3}t\right) + 6\left(\frac{1}{2}t\right) = 6(1) \qquad \text{Distribute.}$$

$$\begin{aligned} 2t + 3t &= 6 & \text{Multiply.} \\ 5t &= 6 & \text{Combine like terms.} \\ t &= \frac{6}{5} & \text{Divide by 5.} \end{aligned}$$

***Step 5:***   **Check** the answer and **interpret** the solution as it relates to the problem.

Tara and Grace could paint the fence together in $\frac{6}{5}$ hr or $1\frac{1}{5}$ hr.

Check:

| Fractional part of the job done by Tara | + | Fractional part of the job done by Grace | = | 1 whole job |
|:---:|:---:|:---:|:---:|:---:|
| $\frac{1}{3} \cdot \left(\frac{6}{5}\right)$ | + | $\frac{1}{2} \cdot \left(\frac{6}{5}\right)$ | $\stackrel{?}{=}$ | 1 |
| $\frac{2}{5}$ | + | $\frac{3}{5}$ | = | 1   ■ |

**You Try 3**

Write an equation and solve.

Javier can put up drywall in a house in 6 hr while it would take his coworker, Frank, 8 hr to drywall the same space. How long would it take them to install the drywall if they worked together?

## 4. Solve Direct Variation Problems

Suppose you are driving on a highway at a constant speed of 60 miles per hour. The distance you travel depends on the amount of time you drive.

Let $y$ = the distance traveled, in miles, and let $x$ = the number of hours you drive. An equation relating $x$ and $y$ is $y = 60x$.

A table relating $x$ and $y$ is shown here.

| $x$ | $y$ |
|:---:|:---:|
| 1 | 60 |
| 1.5 | 90 |
| 2 | 120 |
| 3 | 180 |

Notice that as the value of $x$ increases, the value of $y$ also increases. (The more hours you drive, the farther you will go.) Likewise, as the value of $x$ decreases, the value of $y$ also decreases.

We can say that the distance traveled, $y$, is *directly proportional to* the time spent traveling, $x$. Or *y varies directly as x.*

> **Definition**
>
> **Direct Variation:** For $k > 0$, **y varies directly as x** (or **y is directly proportional to x**) means $y = kx$. $k$ is called the **constant of variation.**

If two quantities vary directly, then as one quantity increases, the other increases as well. And, as one quantity decreases, the other decreases.
    In our example of driving distance, $y = 60x$, 60 *is the constant of variation.*

Given information about how variables are related, we can write an equation and solve a variation problem.

## Example 4

Suppose $y$ varies directly as $x$. If $y = 36$ when $x = 9$,

a) find the constant of variation, $k$.
b) write a variation equation relating $x$ and $y$ using the value of $k$ found in Part a).
c) find $y$ when $x = 7$.

### Solution

a) To find the constant of variation, write a *general* variation equation relating $x$ and $y$.
   *y varies directly as x means* $y = kx$.

   We are told that $y = 36$ when $x = 9$. Substitute these values into the equation and solve for $k$.

$$y = kx$$
$$36 = k(9) \qquad \text{Substitute 9 for } x \text{ and 36 for } y.$$
$$4 = k \qquad \text{Divide by 9.}$$

b) The *specific* variation equation is the equation obtained when we substitute 4 for $k$ in $y = kx$:   $y = 4x$.

c) To find $y$ when $x = 7$, substitute 7 for $x$ in $y = 4x$ and evaluate.

$$y = 4x$$
$$= 4(7) \qquad \text{Substitute 7 for } x.$$
$$= 28 \qquad \text{Multiply.}$$

---

### Procedure   Steps for Solving a Variation Problem

**Step 1:**  Write the *general* variation equation.

**Step 2:**  Find $k$ by substituting the known values into the equation and solving for $k$.

**Step 3:**  Write the *specific* variation equation by substituting the value of $k$ into the *general* variation equation.

**Step 4:**  Use the specific variation equation to solve the problem.

---

### You Try 4

Suppose $y$ varies directly as $x$. If $y = 24$ when $x = 8$,

a) find the constant of variation, $k$.

b) write the specific variation equation relating $x$ and $y$.

c) find $y$ when $x = 5$.

---

## Example 5

Suppose $h$ varies directly as the square of $m$. If $h = 80$ when $m = 4$, find $h$ when $m = 3$.

### Solution

**Step 1:**  Write the *general* variation equation.

   $h$ varies directly as the *square* of $m$ means $h = km.^2$

**Step 2:**   Find $k$ using the known values: $h = 80$ when $m = 4$.

$$h = km^2 \qquad \text{General variation equation}$$
$$80 = k(4)^2 \qquad \text{Substitute 4 for } m \text{ and 80 for } h.$$
$$80 = k(16)$$
$$5 = k \qquad \text{Divide by 16.}$$

**Step 3:**   Substitute $k = 5$ into $h = km^2$ to get the *specific* variation equation, $h = 5m.^2$

**Step 4:**   We are asked to find $h$ when $m = 3$. Substitute $m = 3$ into $h = 5m^2$ to get $h$.

$$h = 5m^2 \qquad \text{Specific variation equation}$$
$$= 5(3)^2 \qquad \text{Substitute 3 for } m.$$
$$= 5(9)$$
$$= 45$$

### You Try 5

Suppose $v$ varies directly as the cube of $z$. If $v = 54$ when $z = 3$, find $v$ when $z = 5$.

## Example 6

Elzbieta's weekly earnings as a personal trainer vary directly as the number of hours she spends training clients each week. If she trains clients 25 hr in a week, she earns \$875. How much would she earn if she trained clients for 30 hr?

**Solution**

Let $h$ = the number of hours Elzbieta trains clients
$E$ = earnings

We will follow the four steps for solving a variation problem.

**Step 1:**   Write the *general* variation equation, $E = kh$.

**Step 2:**   Find $k$ using the known values $E = 875$ when $h = 25$.

$$E = kh \qquad \text{General variation equation}$$
$$875 = k(25) \qquad \text{Substitute 25 for } h \text{ and 875 for } E.$$
$$35 = k \qquad \text{Divide by 25.}$$

**Step 3:**   Substitute $k = 35$ into $E = kh$ to get the *specific* variation equation, $E = 35h$.

**Step 4:**   We must find Elzbieta's earnings when she trains clients 30 hr in a week. Substitute $h = 30$ into $E = 35h$ to find $E$.

$$E = 35h \qquad \text{Specific variation equation}$$
$$E = 35(30)$$
$$E = 1050$$

Elzbieta would earn \$1050 if she trained clients for 30 hr.

### You Try 6

The amount of money raised by a booster club for its high school sports teams is directly proportional to the number of participants in its golf tournament. If they raised \$2240 with 56 golfers, how much would they have raised if 80 people had participated?

## 5. Solve Inverse Variation Problems

If two quantities vary *inversely* (are *inversely* proportional) then as one value increases, the other decreases. Likewise, as one value decreases, the other increases.

---

### Definition

**Inverse Variation:** For $k > 0$, **y varies inversely as x** (or **y is inversely proportional to x**) means $y = \dfrac{k}{x}$.

$k$ is the **constant of variation.**

---

A good example of inverse variation is the relationship between the time, $t$, it takes to travel a given distance, $d$, when driving at a certain rate (or speed), $r$. We can define this relationship as $t = \dfrac{d}{r}$. As the rate, $r$, increases, the time, $t$, that it takes to travel $d$ mi decreases. Likewise, as $r$ decreases, the time, $t$, that it takes to travel $d$ mi increases. Therefore, $t$ varies *inversely* as $r$.

---

### Example 7

Suppose $y$ varies inversely as $x$. If $y = 8$ when $x = 7$, find $y$ when $x = 4$.

**Solution**

**Step 1:**  Write the *general* variation equation, $y = \dfrac{k}{x}$.

**Step 2:**  Find $k$ using the known values: $y = 8$ when $x = 7$.

$$y = \frac{k}{x}$$

$$8 = \frac{k}{7} \qquad \text{Substitute 7 for } x \text{ and 8 for } y.$$

$$56 = k \qquad \text{Multiply by 7.}$$

**Step 3:**  Substitute $k = 56$ into $y = \dfrac{k}{x}$ to get the *specific* variation equation, $y = \dfrac{56}{x}$.

**Step 4:**  Substitute 4 for $x$ in $y = \dfrac{56}{x}$ to find $y$.

$$y = \frac{56}{4}$$

$$y = 14$$

---

### You Try 7

Suppose $p$ varies inversely as the square of $n$. If $p = 7.5$ when $n = 4$, find $p$ when $n = 2$.

## Example 8

If the voltage in an electrical circuit is held constant (stays the same), then the current in the circuit varies inversely as the resistance. If the current is 40 amps when the resistance is 3 ohms, find the current when the resistance is 8 ohms.

### Solution

Let $C$ = the current (in amps)

$\quad\ R$ = the resistance (in ohms)

We will follow the four steps for solving a variation problem.

**Step 1:** Write the *general* variation equation, $C = \dfrac{k}{R}$.

**Step 2:** Find $k$ using the known values $C = 40$ when $R = 3$.

$$C = \frac{k}{R} \qquad \text{General variation equation}$$

$$40 = \frac{k}{3} \qquad \text{Substitute 40 for } C \text{ and 3 for } R.$$

$$120 = k \qquad \text{Multiply by 3.}$$

**Step 3:** Substitute $k = 120$ into $C = \dfrac{k}{R}$ to get the *specific* variation equation, $C = \dfrac{120}{R}$.

**Step 4:** We must find the current when the resistance is 8 ohms. Substitute $R = 8$ into $C = \dfrac{120}{R}$ to find $C$.

$$C = \frac{120}{R} \qquad \text{Specific variation equation}$$

$$C = \frac{120}{8}$$

$$C = 15$$

The current is 15 amps.

## You Try 8

The time it takes to drive between two cities taking the same route varies inversely as the speed of the car. If it takes Elisha 4 hr to drive between the cities when he goes 60 mph, how long would it take him to drive between the cities if he drives 50 mph?

---

### Answers to You Try Exercises

1)  12     2)  9 mph     3)  $3\frac{3}{7}$ hr     4)  a) 3   b) $y = 3x$   c) 15     5)  250

6)  $3200     7)  30     8)  4.8 hr

## 8.7 Exercises

**Objective 1: Solve Problems Involving Proportions**

Solve the following proportions.

1) $\dfrac{8}{15} = \dfrac{32}{x}$

2) $\dfrac{9}{12} = \dfrac{6}{a}$

3) $\dfrac{4}{7} = \dfrac{n}{n + 9}$

4) $\dfrac{5}{3} = \dfrac{c}{c - 10}$

Write an equation for each and solve. See Example 1.

5) The scale on a blueprint is 2.5 in. to 10 ft in actual room length. Find the length of a room that is 3 in. long on the blueprint.

6) A survey conducted by the U.S. Centers for Disease Control revealed that, in Michigan, approximately 2 out of 5 adults between the ages of 18 and 24 smoke cigarettes. In a group of 400 Michigan citizens in this age group, how many would be expected to be smokers? (www.cdc.gov)

7) The ratio of employees at a small company who have their paychecks directly deposited into their bank accounts to those who do not is 9 to 2. If the number of people who have direct deposit is 14 more than the number who do not, how many employees do not have direct deposit?

8) In a gluten-free flour mixture, the ratio of potato-starch flour to tapioca flour is 2 to 1. If a mixture contains 3 more cups of potato-starch flour than tapioca flour, how much of each type of flour is in the mixture?

9) At a state university, the ratio of the number of freshmen who graduated in four years to those who took longer was about 2 to 5. If the number of students who graduated in four years was 1200 less than the number who graduated in more than four years, how many students graduated in four years?

10) Francesca makes her own ricotta cheese for her restaurant. The ratio of buttermilk to whole milk in her recipe is 1 to 4. How much of each type of milk will she need if she will use 18 more cups of whole milk than buttermilk?

11) The ancient Greeks believed that the rectangle most pleasing to the eye, the golden rectangle, had sides in which the ratio of its length to its width was approximately 8 to 5. They erected many buildings, including the Parthenon, using this golden ratio. The marble floor of a museum foyer is to be designed as a golden rectangle. If its width is to be 18 feet less than its length, find the length and width of the foyer.

12) The ratio of seniors at Central High School who drive to school to those who take the school bus is 7 to 2. If the number of students who drive is 320 more than the number who take the bus, how many students drive and how many take the bus?

13) A math professor surveyed her class and found that the ratio of students who used the school's tutoring service to those who did not was 3 to 8. The number of students who did not use the tutoring lab was 15 more than the number who did. How many students used the tutoring service and how many did not?

14) An industrial cleaning solution calls for 5 parts water to 2 parts concentrated cleaner. If a worker uses 15 more quarts of water than concentrated cleaner to make a solution,

   a) how much concentrated cleaner did she use?

   b) how much water did she use?

   c) how much solution did she make?

**Objective 2: Solve Problems Involving Distance, Rate, and Time**

Answer the following questions about rates.

15) If the speed of a boat in still water is 8 mph,

   a) what is its speed going *against* a 2-mph current?

   b) what is its speed *with* a 2-mph current?

16) If an airplane travels at a constant rate of 350 mph,

   a) what is its speed going *into* a 50-mph wind?

   b) what is its speed going *with* a 25-mph wind?

17) If an airplane travels at a constant rate of *x* mph,

   a) what is its speed going *with* a 40-mph wind?

   b) what is its speed going *against* a 30-mph wind?

18) If the speed of a boat in still water is 11 mph,

   a) what is its speed going *against* a current with a rate of *x* mph?

   b) what is its speed going *with* a current with a rate of *x* mph?

Write an equation for each and solve. See Example 2.

19) A boat can travel 4 mi upstream in the same amount of time it can travel 6 mi downstream. If the speed of the current is 2 mph, what is the speed of the boat in still water?

20) Flying at a constant speed, a plane can travel 800 miles with the wind in the same amount of time it can fly 650 miles against the wind. If the wind blows at 30 mph, what is the speed of the plane?

21) When the wind is blowing at 25 mph, a plane flying at a constant speed can travel 500 miles with the wind in the same amount of time it can fly 400 miles against the wind. Find the speed of the plane.

22) With a current flowing at 3 mph, a boat can travel 9 mi downstream in the same amount of time it can travel 6 mi upstream. What is the speed of the boat in still water?

23) The speed of a boat in still water is 28 mph. The boat can travel 32 mi with the current in the same amount of time it can travel 24 mi against the current. Find the speed of the current.

24) A boat can travel 20 mi downstream in the same amount of time it can travel 12 mi upstream. The speed of the boat in still water is 20 mph. Find the speed of the current.

25) The speed of a plane in still air is 280 mph. Flying against the wind, it can fly 600 mi in the same amount of time it takes to go 800 mi with the wind. What is the speed of the wind?

26) The speed of a boat in still water is 10 mph. If the boat can travel 9 mi downstream in the same amount of time it can travel 6 mi upstream, find the speed of the current.

27) Bill drives 120 miles from his house in San Diego to Los Angeles for a business meeting. Afterward, he drives from LA to Las Vegas, a distance of 240 miles. If he averages the same speed on both legs of the trip and it takes him 2 hours less to go from San Diego to Los Angeles, what is his average driving speed?

28) Rashard drives 80 miles from Detroit to Lansing, and later drives 60 miles more from Lansing to Grand Rapids. The trip from Lansing to Grand Rapids takes him a half hour less than the drive from Detroit to Lansing. Find his average driving speed if it is the same on both parts of the trip.

## Objective 3: Solve Problems Involving Work

Answer the following questions about work rate.

29) It takes Midori 3 hr to do her homework. What is her rate?

30) It takes Signe 20 hr to complete her self-portrait for art class. How much of the job does she do in 12 hr?

31) Tomasz can set up his new computer in $t$ hours. What is the rate at which he does this job?

32) It takes Jesse twice as long to edit a chapter in a book as it takes Curtis. If it takes Curtis $t$ hours to edit the chapter, at what rate does Jesse do the job?

Write an equation for each and solve. See Example 3.

33) It takes Rupinderjeet 4 hr to paint a room while the same job takes Sana 5 hr. How long would it take for them to paint the room together?

34) A hot-water faucet can fill a sink in 9 min while it takes the cold-water faucet only 7 min. How long would it take to fill the sink if both faucets were on?

35) Wayne can clean the carpets in his house in 4 hr but it would take his son, Garth, 6 hr to clean them on his own. How long would it take them to clean the carpets together?

36) Janice and Blanca have to type a report on a project they did together. Janice could type it in 40 min, and Blanca could type it in 1 hr. How long would it take them if they worked together?

37) A faucet can fill a tub in 12 minutes. The leaky drain can empty the tub in 30 minutes. If the faucet is on and the drain is leaking, how long would it take to fill the tub?

38) A pipe can fill a pool in 8 hr, and another pipe can empty a pool in 12 hr. If both pipes are accidentally left open, how long would it take to fill the pool?

39) A new machine in a factory can do a job in 5 hr. When it is working together with an older machine, the job can be done in 3 hr. How long would it take the old machine to do the job by itself?

40) It takes Lily 75 minutes to mow the lawn. When she works with her brother, Preston, it takes only 30 minutes. How long would it take Preston to mow the lawn himself?

41) It would take Mei twice as long as Ting to make decorations for a party. If they worked together, they could make the decorations in 40 min. How long would it take Mei to make the decorations by herself?

42) It takes Lemar three times as long as his boss, Emilio, to build a custom shelving unit. Together they can build the unit in 4.5 hr. How long would it take Lemar to build the shelves by himself?

## Objectives 4 and 5: Solve Direct and Inverse Variation Problems

43) If $u$ varies directly as $v$, then as $v$ increases, the value of $u$ _____.

44) If $m$ varies inversely as $n$, then as $n$ increases, the value of $m$ _____.

Decide whether each equation represents direct or inverse variation.

45) $y = 10x$

46) $r = \dfrac{7}{s}$

47) $z = \dfrac{5}{a^3}$

48) $p = 8q^2$

Write a general variation equation using $k$ as the constant of variation.

49) $A$ varies directly as $w$.

50) $h$ varies directly as $c$.

51) $x$ varies inversely as $g$.

52) $Q$ varies inversely as $T$.

53) $C$ varies directly as the cube of $d$.

54) $m$ varies directly as the square of $p$.

55) $b$ varies inversely as the square of $z$.

56) $R$ varies inversely as the cube of $x$.

Solve each step-by-step variation problem.

57) Suppose $z$ varies directly as $x$. If $z = 63$ when $x = 7$,

a) find the constant of variation.

b) write the specific variation equation relating $z$ and $x$.

c) find $z$ when $x = 6$.

58) Suppose $w$ varies directly as $q$. If $w = 30$ when $q = 5$,

a) find the constant of variation.

b) write the specific variation equation relating $w$ and $q$.

c) find $w$ when $q = 7$.

59) Suppose $T$ varies inversely as $n$. If $T = 10$ when $n = 6$,

a) find the constant of variation.

b) write the specific variation equation relating $T$ and $n$.

c) find $T$ when $n = 5$.

60) Suppose $y$ varies inversely as $x$. If $y = 15$ when $x = 5$,

a) find the constant of variation.

b) write the specific variation equation relating $x$ and $y$.

c) find $y$ when $x = 3$.

61) Suppose $u$ varies inversely as the square of $v$. If $u = 36$ when $v = 4$,

a) find the constant of variation.

b) write the specific variation equation relating $u$ and $v$.

c) find $u$ and $v = 3$.

62) Suppose $H$ varies directly as the cube of $J$. If $H = 500$ when $J = 5$,

a) find the constant of variation.

b) write the specific variation equation relating $H$ and $J$.

c) find $H$ and $J = 2$.

Solve.

63) If $N$ varies directly as $d$, and $N = 28$ when $d = 4$, find $N$ when $d = 11$.

64) If $p$ varies directly as $r$, and $p = 30$ when $r = 6$, find $p$ when $r = 2$.

65) If $b$ varies inversely as $a$, and $b = 18$ when $a = 5$, find $b$ when $a = 10$.

66) If $W$ varies inversely as $z$, and $W = 6$ when $z = 9$, find $W$ when $z = 27$.

67) If $Q$ varies inversely as the cube of $T$, and $Q = 216$ when $T = 2$, find $Q$ when $T = 3$.

68) If $y$ varies directly as the cube of $x$, and $y = 192$ when $x = 4$, find $y$ when $x = 5$.

69) If $h$ varies directly as the square of $v$, and $h = 175$ when $v = 5$, find $h$ when $v = 2$.

70) If $C$ varies inversely as the square of $D$, and $C = 16$ when $D = 3$, find $C$ when $D = 2$.

Write a variation equation and solve.

71) Hassan is paid hourly, and his weekly earnings vary directly as the number of hours he works. If Hassan earned $576.00 when he worked 32 hr, how much would he earn if he worked 40 hr?

72) If distance is held constant, the time it takes to travel that distance varies inversely as the speed at which one travels. If it takes 6 hr to travel the given distance at 70 mph, how long would it take to travel the same distance at 60 mph?

73) If the area is held constant, the width of a rectangle varies inversely as its length. If the width of a rectangle is 12 cm when the length is 24 cm, find the width of a rectangle with the same area when the length is 36 cm.

74) The circumference of a circle is directly proportional to its radius. A circle of radius 5 in. has a circumference of approximately 31.4 in. What is the circumference of a circle of radius 7 in.?

75) The weight of a ball varies directly as the cube of its radius. If a ball with radius 3 in. weighs 3.24 lb, how much would a ball made out of the same material weigh if it had a radius of 2 in.?

76) If the force (weight) is held constant, then the work done by a bodybuilder in lifting a barbell varies directly as the distance the barbell is lifted. If Jay does 300 ft-lb of work lifting a barbell 2 ft off of the ground, how much work would it take for him to lift the same barbell 3 ft off the ground?

77) If the area is held constant, the height of a triangle varies inversely as the length of its base. If the height of a triangle is 10 in. when the base is 18 in., find the height of a triangle with the same area when the base is 12 in.

78) The loudness of sound is inversely proportional to the square of the distance between the source of the sound and the listener. If the sound level measures 18 dB (decibels) 10 ft from a speaker, how loud is the sound 4 ft from the speaker?

79) Hooke's law states that the force required to stretch a spring is proportional to the distance that the spring is stretched from its original length. A force of 120 lb is required to stretch a spring 4 in. from its natural length. How much force is needed to stretch the spring 6 in. beyond its natural length?

80) The amount of garbage produced is proportional to the population. If a city of 60,000 people produces 9600 tons of garbage in a week, how much garbage would a city of 100,000 people produce in a week?

81) The intensity of light, in lumens, varies inversely as the square of the distance from the source. If the intensity of the light is 40 lumens 5 ft from the source, what is the intensity of the light 4 ft from the source?

82) The weight of an object on Earth varies inversely as the square of its distance from the center of the Earth. If an object weighs 210 lb on the surface of the Earth (4000 mi from the center), what is the weight of the object if it is 200 mi above the Earth? (Round to the nearest pound.)

| **Definition/Procedure** | **Example** |
|---|---|

### 8.1 Simplifying Rational Expressions

A **rational expression** is an expression of the form $\frac{P}{Q}$, where $P$ and $Q$ are polynomials and where $Q \neq 0$.

We can *evaluate* rational expressions. **(p. 446)**

Evaluate $\frac{5a - 8}{a + 3}$ for $a = 2$.

$$\frac{5(2) - 8}{2 + 3} = \frac{10 - 8}{5} = \frac{2}{5}$$

**How to Determine When a Rational Expression Equals Zero and When It Is Undefined**

1) To determine what values of the variable make the expression equal zero, set the numerator equal to zero and solve for the variable.

2) To determine what values of the variable make the expression undefined, set the denominator equal to zero and solve for the variable. **(p. 447)**

For what value(s) of $x$ is $\frac{x - 7}{x + 9}$

a) equal to zero?    b) undefined?

a) $\frac{x - 7}{x + 9} = 0$ when $x - 7 = 0$.

$$x - 7 = 0$$
$$x = 7$$

When $x = 7$, the expression equals zero.

b) $\frac{x - 7}{x + 9}$ is undefined when its denominator equals zero.

Solve $x + 9 = 0$.

$$x + 9 = 0$$
$$x = -9$$

When $x = -9$, the expression is undefined.

**To Write an Expression in Lowest Terms**

1) Completely *factor* the numerator and denominator.

2) *Divide* the numerator and denominator by the greatest common factor. **(p. 449)**

Simplify $\frac{3r^2 - 10r + 8}{2r^2 - 8}$.

$$\frac{3r^2 - 10r + 8}{2r^2 - 8} = \frac{(3r - 4)(r - 2)}{2(r + 2)(r - 2)} = \frac{3r - 4}{2(r + 2)}$$

**Simplifying** $\frac{a - b}{b - a}$.

A rational expression of the form $\frac{a - b}{b - a}$ will simplify to $-1$.

**(p. 450)**

Simplify $\frac{5 - w}{w^2 - 25}$.

$$\frac{5 - w}{w^2 - 25} = \frac{\overset{-1}{5 - w}}{(w + 5)(w - 5)} = -\frac{1}{w + 5}$$

| Definition/Procedure | Example |
|---|---|

## 8.2 Multiplying and Dividing Rational Expressions

**Multiplying Rational Expressions**

1) Factor numerators and denominators.

2) Reduce and multiply. **(p. 455)**

Multiply $\dfrac{16v^4}{v^2 + 10v + 21} \cdot \dfrac{3v + 21}{4v}$.

$$\dfrac{16v^4}{v^2 + 10v + 21} \cdot \dfrac{3v + 21}{4v} = \dfrac{\overset{4}{\cancel{16}v^3} \cdot \cancel{v}}{(v + 3)\cancel{(v + 7)}} \cdot \dfrac{3\cancel{(v + 7)}}{\cancel{4v}} = \dfrac{12v^3}{v + 3}$$

**Dividing Rational Expressions**

To **divide** rational expressions, multiply the first expression by the reciprocal of the second. **(p. 456)**

Divide $\dfrac{2x^2 + 5x}{x + 4} \div \dfrac{4x^2 - 25}{12x - 30}$.

$$\dfrac{2x^2 + 5x}{x + 4} \div \dfrac{4x^2 - 25}{12x - 30} = \dfrac{2x^2 + 5x}{x + 4} \cdot \dfrac{12x - 30}{4x^2 - 25}$$

$$= \dfrac{x\cancel{(2x + 5)}}{x + 4} \cdot \dfrac{6\cancel{(2x - 5)}}{\cancel{(2x + 5)}\cancel{(2x - 5)}} = \dfrac{6x}{x + 4}$$

## 8.3 Finding the Least Common Denominator

**To Find the Least Common Denominator (LCD)**

1) Factor the denominators.
2) The LCD will contain each unique factor the greatest number of times it appears in any single factorization.
3) The LCD is the *product* of the factors identified in step 2. **(p. 461)**

Find the LCD of $\dfrac{9b}{b^2 + 8b}$ and $\dfrac{6}{b^2 + 16a + 64}$.

1) $\qquad b^2 + 8b = b(b + 8)$
$\qquad b^2 + 16a + 64 = (b + 8)^2$
2) The factors we will use in the LCD are $b$ and $(b + 8)^2$.
3) LCD $= b(b + 8)^2$

## 8.4 Adding and Subtracting Rational Expressions

**Adding and Subtracting Rational Expressions**

1) Factor the denominators.
2) Write down the LCD.
3) Rewrite each rational expression as an equivalent expression with the LCD.
4) Add or subtract the numerators and keep the common denominator in factored form.
5) After combining like terms in the numerator, ask yourself, "*Can I factor it?*" If so, factor.
6) Reduce the rational expression, if possible. **(p. 471)**

Add $\dfrac{y}{y + 7} + \dfrac{10y - 28}{y^2 - 49}$.

1) Factor the denominator of $\dfrac{10y - 28}{y^2 - 49}$.

$$\dfrac{10y - 28}{y^2 - 49} = \dfrac{10y - 28}{(y + 7)(y - 7)}$$

2) The LCD is $(y + 7)(y - 7)$.

3) Rewrite $\dfrac{y}{y + 7}$ with the LCD.

$$\dfrac{y}{y + 7} \cdot \dfrac{y - 7}{y - 7} = \dfrac{y(y - 7)}{(y + 7)(y - 7)}$$

4) $\dfrac{y}{y + 7} + \dfrac{10y - 28}{y^2 - 49} = \dfrac{y(y - 7)}{(y + 7)(y - 7)} + \dfrac{10y - 28}{(y + 7)(y - 7)}$

$$= \dfrac{y(y - 7) + 10y - 28}{(y + 7)(y - 7)}$$

$$= \dfrac{y^2 - 7y + 10y - 28}{(y + 7)(y - 7)}$$

$$= \dfrac{y^2 + 3y - 28}{(y + 7)(y - 7)}$$

5) $\qquad = \dfrac{\cancel{(y + 7)}(y - 4)}{\cancel{(y + 7)}(y - 7)}$     Factor.

6) $\qquad = \dfrac{y - 4}{y - 7}$     Reduce.

| Definition/Procedure | Example |
|---|---|

## 8.5 Simplifying Complex Fractions

A **complex fraction** is a rational expression that contains one or more fractions in its numerator, its denominator, or both. **(p. 481)**

Some examples of complex fractions are

$$\frac{\frac{9}{16}}{\frac{3}{4}}, \qquad \frac{\frac{b+3}{2}}{\frac{6b+18}{7}}, \qquad \frac{\frac{1}{x}-\frac{1}{y}}{1-\frac{x}{y}}$$

---

**To simplify a complex fraction containing one term in the numerator and one term in the denominator,**

1) Rewrite the complex fraction as a division problem.
2) Perform the division by multiplying the first fraction by the reciprocal of the second. **(p. 482)**

Simplify $\dfrac{\dfrac{b+3}{2}}{\dfrac{6b+18}{7}}$.

$$\frac{\frac{b+3}{2}}{\frac{6b+18}{7}} = \frac{b+3}{2} \div \frac{6b+18}{7}$$

$$= \frac{b+3}{2} \cdot \frac{7}{6(b+3)} = \frac{\cancel{b+3}}{2} \cdot \frac{7}{6\cancel{(b+3)}} = \frac{7}{12}$$

---

**To simplify complex fractions containing more than one term in the numerator and/or the denominator,**

**Method 1**

1) Combine the terms in the numerator and combine the terms in the denominator so that each contains only one fraction.
2) Rewrite as a division problem.
3) Perform the division. **(p. 482)**

**Method 1**

Simplify $\dfrac{\dfrac{1}{x}-\dfrac{1}{y}}{1-\dfrac{x}{y}}$.

$$\frac{\frac{1}{x}-\frac{1}{y}}{1-\frac{x}{y}} = \frac{\frac{y}{xy}-\frac{x}{xy}}{\frac{y}{y}-\frac{x}{y}} = \frac{\frac{y-x}{xy}}{\frac{y-x}{y}}$$

$$= \frac{y-x}{xy} \div \frac{y-x}{y} = \frac{\cancel{y-x}}{x\cancel{y}} \cdot \frac{\cancel{y}}{\cancel{y-x}} = \frac{1}{x}$$

---

**Method 2**

1) Write down the LCD of *all* of the fractions in the complex fraction.
2) Multiply the numerator and denominator of the complex fraction by the LCD.
3) Simplify. **(p. 483)**

**Method 2**

Simplify $\dfrac{\dfrac{1}{x}-\dfrac{1}{y}}{1-\dfrac{x}{y}}$.

**Step 1:** LCD $= xy$

**Step 2:** $\dfrac{xy\left(\dfrac{1}{x}-\dfrac{1}{y}\right)}{xy\left(1-\dfrac{x}{y}\right)}$

$$\frac{xy\left(\frac{1}{x}-\frac{1}{y}\right)}{xy\left(1-\frac{x}{y}\right)} = \frac{xy \cdot \frac{1}{x} - xy \cdot \frac{1}{y}}{xy \cdot 1 - xy \cdot \frac{x}{y}} \qquad \text{Distribute.}$$

**Step 3:** $\qquad = \dfrac{y-x}{xy-x^2} \qquad \text{Simplify.}$

$$= \frac{\cancel{y-x}}{x\cancel{(y-x)}} = \frac{1}{x}$$

| Definition/Procedure | Example |
|---|---|

## 8.6 Solving Rational Equations

An **equation** contains an $=$ sign.

To solve a rational equation, *multiply* the equation by the LCD to *eliminate* the denominators, then solve.

Always check the answer to be sure it does not make a denominator equal zero. **(p. 489)**

Solve $\dfrac{n}{n+6} + 1 = \dfrac{18}{n+6}$.

This is an equation because it contains an $=$ sign.
We must eliminate the denominators. Identify the LCD of all of the expressions in the equation.

$\text{LCD} = (n+6)$

Multiply both sides of the equation by $(n+6)$.

$$(n+6)\left(\frac{n}{n+6} + 1\right) = (n+6)\left(\frac{18}{n+6}\right)$$

$$\cancel{(n+6)} \cdot \left(\frac{n}{\cancel{n+6}}\right) + (n+6) \cdot 1 = \cancel{(n+6)} \cdot \frac{18}{\cancel{n+6}}$$

$$n + n + 6 = 18$$
$$2n + 6 = 18$$
$$2n = 12$$
$$n = 6$$

The solution set is $\{6\}$.

The check is left to the student.

**Solve an Equation for a Specific Variable (p. 493)**

Solve $x = \dfrac{3b}{n+m}$ for $n$.

Since we are solving for $n$, put it in a box.

$$x = \frac{3b}{\boxed{n} + m}$$

$$(\boxed{n} + m)x = (\cancel{\boxed{n} + m}) \cdot \frac{3b}{\cancel{\boxed{n} + m}}$$

$$(\boxed{n} + m)x = 3b$$

$$\boxed{n}x + mx = 3b$$

$$\boxed{n}x = 3b - mx$$

$$n = \frac{3b - mx}{x}$$

## 8.7 Applications of Rational Equations and Variation

Use the five steps for solving word problems outlined in Section 3.3. **(p. 498)**

*Write an equation and solve.*

Jeff can wash and wax his car in 3 hours, but it takes his dad only 2 hours to wash and wax the car. How long would it take the two of them to wash and wax together?

**Step 1:** Read the problem carefully.
**Step 2:** $t =$ number of hours to wash and wax the car together.

| Definition/Procedure | Example |
|---|---|

**Step 3:** Translate from English into an algebraic equation.

$$\text{Jeff's rate} = \frac{1}{3} \text{ wash/hr} \quad \text{Dad's rate} = \frac{1}{2} \text{ wash/hr}$$

$$\text{Fractional part} = \text{Rate} \cdot \text{Time}$$

$$\text{Jeff's part} = \frac{1}{3} \cdot t = \frac{1}{3}t$$

$$\text{Dad's part} = \frac{1}{2} \cdot t = \frac{1}{2}t$$

$$\begin{array}{ccc} \text{Fractional} & \text{Fractional} & \text{1 whole} \\ \text{job by Jeff} + & \text{job by his dad} = & \text{job} \\ \frac{1}{3}t & + \quad \frac{1}{2}t & = \quad 1 \end{array}$$

*Equation:* $\frac{1}{3}t + \frac{1}{2}t = 1$

**Step 4:** Solve the equation.

$$6\left(\frac{1}{3}t + \frac{1}{2}t\right) = 6(1) \qquad \text{Multiply by 6, the LCD.}$$

$$6 \cdot \frac{1}{3}t + 6 \cdot \frac{1}{2}t = 6(1) \qquad \text{Distribute.}$$

$$2t + 3t = 6 \qquad \text{Multiply.}$$

$$5t = 6$$

$$t = \frac{6}{5}$$

**Step 5:** Interpret the solution as it relates to the problem.

Jeff and his dad could wash and wax the car together in $\frac{6}{5}$ hours or $1\frac{1}{5}$ hours.

The check is left to the student.

---

**Direct Variation**
For $k > 0$, **$y$ varies directly as $x$ (or $y$ is directly proportional to $x$)** means

$$y = kx.$$

$k$ is called the **constant of variation. (p. 503)**

The circumference, $C$, of a circle is given by $C = 2\pi r$. $C$ varies directly as $r$ where $k = 2\pi$.

---

**Inverse Variation**
For $k > 0$, **$y$ varies inversely as $x$ (or $y$ is inversely proportional to $x$)** means

$$y = \frac{k}{x}$$

$k$ is called the **constant of variation. (p. 506)**

The time, $t$ (in hours), it takes to drive 600 miles is inversely proportional to the rate, $r$, at which you drive.

$$t = \frac{600}{r}$$

where $k = 600$.

**(8.1) Evaluate, if possible, for a) $n = 5$ and b) $n = -2$.**

1) $\dfrac{n^2 - 3n - 10}{3n + 2}$

2) $\dfrac{3n - 2}{n^2 - 4}$

**Determine the value(s) of the variable for which**
**a) the expression equals zero.**
**b) the expression is undefined.**

3) $\dfrac{2s}{4s + 11}$

4) $\dfrac{k + 3}{k - 4}$

5) $\dfrac{15}{4t^2 - 9}$

6) $\dfrac{2c^2 - 3c - 9}{c^2 - 7c}$

7) $\dfrac{3m^2 - m - 10}{m^2 + 49}$

8) $\dfrac{15 - 5d}{d^2 + 25}$

**Write each rational expression in lowest terms.**

9) $\dfrac{77k^9}{7k^3}$

10) $\dfrac{54a^3}{9a^{11}}$

11) $\dfrac{r^2 - 14r + 48}{4r^2 - 24r}$

12) $\dfrac{18c - 66}{39c - 143}$

13) $\dfrac{3z - 5}{6z^2 - 7z - 5}$

14) $\dfrac{y^2 + 8y - yz - 8z}{yz - 3y - z^2 + 3z}$

15) $\dfrac{11 - x}{x^2 - 121}$

16) $\dfrac{4t^4 - 32t}{(t - 2)(t^2 + 2t + 4)}$

**Find three equivalent forms of each rational expression.**

17) $-\dfrac{4n + 1}{5 - 3n}$

18) $-\dfrac{u - 8}{u + 2}$

**Find the missing side in each rectangle.**

19) Area $= 2b^2 + 13b + 21$

[rectangle, base labeled $2b + 7$]

Find the width.

20) Area $= 3x^2 - 8x - 3$

[rectangle, side labeled $x - 3$]

Find the length.

**(8.2) Perform the operations and simplify.**

21) $\dfrac{64}{45} \cdot \dfrac{27}{56}$

22) $\dfrac{6}{25} \div \dfrac{9}{10}$

23) $\dfrac{t + 6}{4} \cdot \dfrac{2(t + 2)}{(t + 6)^2}$

24) $\dfrac{4m^3}{30n} \div \dfrac{20m^6}{3n^5}$

25) $\dfrac{3x^2 + 11x + 8}{15x + 40} \div \dfrac{9x + 9}{x - 3}$

26) $\dfrac{6w - 1}{6w^2 + 5w - 1} \cdot \dfrac{3w + 3}{12w}$

27) $\dfrac{r^2 - 16r + 63}{2r^3 - 18r^2} \div (r - 7)^2$

28) $(h^2 + 10h + 24) \cdot \dfrac{h}{h^2 + h - 12}$

29) $\dfrac{3p^5}{20q^2} \cdot \dfrac{4q^3}{21p^7}$

30) $\dfrac{25 - a^2}{4a^2 + 12a} \div \dfrac{a^3 - 125}{a^2 + 3a}$

**Divide.**

31) $\dfrac{\dfrac{3s + 8}{12}}{\dfrac{3s + 8}{4}}$

32) $\dfrac{\dfrac{16m - 8}{m^2}}{\dfrac{12m - 6}{m^4}}$

33) $\dfrac{\dfrac{9}{8}}{\dfrac{15}{4}}$

34) $\dfrac{\dfrac{2r + 10}{r^2}}{\dfrac{r^2 - 25}{4r}}$

**(8.3) Find the LCD of each group of fractions.**

35) $\dfrac{9}{10}, \dfrac{7}{15}, \dfrac{6}{5}$

36) $\dfrac{3}{9x^2y}, \dfrac{13}{4xy^4}$

37) $\dfrac{3}{k^5}, \dfrac{11}{k^2}$

38) $\dfrac{3}{2m}, \dfrac{4}{m + 4}$

39) $\dfrac{1}{4x + 9}, \dfrac{3x}{x - 7}$

40) $\dfrac{8}{3d^2 - d}, \dfrac{11}{9d - 3}$

41) $\dfrac{w}{w - 5}, \dfrac{11}{5 - w}$

42) $\dfrac{6m}{m^2 - n^2}, \dfrac{n}{n - m}$

43) $\dfrac{3c - 11}{c^2 + 9c + 20}, \dfrac{8c}{c^2 - 2c - 35}$

44) $\dfrac{6}{x^2 + 7x}, \dfrac{1}{2x^2 + 14x}, \dfrac{13}{x^2 + 14x + 49}$

**Rewrite each rational expression with the indicated denominator.**

45) $\dfrac{3}{5y} = \dfrac{}{20y^3}$

46) $\dfrac{4k}{k - 9} = \dfrac{}{(k - 6)(k - 9)}$

47) $\dfrac{6}{2z + 5} = \dfrac{}{z(2z + 5)}$

48) $\dfrac{n}{9 - n} = \dfrac{}{n - 9}$

49) $\dfrac{t - 3}{3t + 1} = \dfrac{}{(3t + 1)(t + 4)}$

**Identify the LCD of each group of fractions, and rewrite each as an equivalent fraction with the LCD as its denominator.**

50) $\dfrac{4}{5a^3b}, \dfrac{3}{8ab^5}$

51) $\dfrac{8c}{c^2 + 5c - 24}, \dfrac{5}{c^2 - 6c + 9}$

52) $\dfrac{6}{p + 9}, \dfrac{3}{p}$

53) $\dfrac{7}{2q^2 - 12q}, \dfrac{3q}{36 - q^2}, \dfrac{q - 5}{2q^2 + 12q}$

54) $\dfrac{1}{g - 12}, \dfrac{6}{12 - g}$

**(8.4) Add or subtract.**

55) $\dfrac{5}{9c} + \dfrac{7}{9c}$

56) $\dfrac{5}{6z^2} + \dfrac{9}{12z}$

57) $\dfrac{9}{10u^2v^2} - \dfrac{1}{8u^3v}$

58) $\dfrac{3m}{m-4} - \dfrac{1}{m-4}$

59) $\dfrac{n}{3n-5} - \dfrac{4}{n}$

60) $\dfrac{8}{t+2} + \dfrac{8}{t}$

61) $\dfrac{9}{y+2} - \dfrac{5}{y-3}$

62) $\dfrac{7d-3}{d^2+3d-28} + \dfrac{3d}{5d+35}$

63) $\dfrac{k-3}{k^2+14k+49} - \dfrac{2}{k^2+7k}$

64) $\dfrac{8p+3}{2p+2} - \dfrac{6}{p^2-3p-4}$

65) $\dfrac{t+9}{t-18} - \dfrac{11}{18-t}$

66) $\dfrac{1}{12-r} + \dfrac{24}{r^2-144}$

67) $\dfrac{4w}{w^2+11w+24} - \dfrac{3w-1}{2w^2-w-21}$

68) $\dfrac{2a+7}{a^2-6a+9} + \dfrac{6}{a^2+2a-15}$

69) $\dfrac{b}{9b^2-4} + \dfrac{b+1}{6b^2-4b} - \dfrac{1}{6b+4}$

70) $\dfrac{d+4}{d^2+3d} + \dfrac{d}{5d^2+12d-9} - \dfrac{8}{5d^2-3d}$

71) Find a rational expression in simplest form to represent the a) area and b) perimeter of the rectangle.

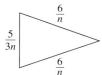

72) Find a rational expression in simplest form to represent the perimeter of the triangle.

**(8.5) Simplify completely.**

73) $\dfrac{\dfrac{x}{y}}{\dfrac{x^3}{y^2}}$

74) $\dfrac{\dfrac{a}{b} - \dfrac{2a}{b^2}}{\dfrac{4}{ab} - \dfrac{a}{b}}$

75) $\dfrac{p + \dfrac{4}{p}}{\dfrac{9}{p} + p}$

76) $\dfrac{\dfrac{6n+48}{n^2}}{\dfrac{8n+64}{}}$

77) $\dfrac{\dfrac{4}{5} - \dfrac{2}{3}}{\dfrac{1}{2} + \dfrac{1}{6}}$

78) $\dfrac{\dfrac{4q}{7q+70}}{\dfrac{q^3}{8q+80}}$

79) $\dfrac{1 - \dfrac{1}{y-8}}{\dfrac{2}{y+4} + 1}$

80) $\dfrac{\dfrac{10}{21}}{\dfrac{16}{9}}$

81) $\dfrac{1 + \dfrac{1}{r-t}}{\dfrac{1}{r^2-t^2} + \dfrac{1}{r+t}}$

82) $\dfrac{\dfrac{z}{z+2} + \dfrac{1}{z^2-4}}{1 - \dfrac{3}{z+2}}$

**(8.6) Solve each equation.**

83) $\dfrac{5a+4}{15} = \dfrac{a}{5} + \dfrac{4}{5}$

84) $\dfrac{16}{9c-27} + \dfrac{2c-4}{c-3} = \dfrac{c}{9}$

85) $\dfrac{m}{7} = \dfrac{5}{m+2}$

86) $\dfrac{2}{y-7} = \dfrac{8}{y+5}$

87) $\dfrac{r}{r+5} + 4 = \dfrac{5}{r+5}$

88) $\dfrac{3}{j+9} + \dfrac{j}{j-3} = \dfrac{2j^2+2}{j^2+6j-27}$

89) $\dfrac{5}{t^2+10t+24} + \dfrac{5}{t^2+3t-18} = \dfrac{t}{t^2+t-12}$

90) $p - \dfrac{20}{p} = 8$

91) $\dfrac{3}{x+1} = \dfrac{6x}{x^2-1} - \dfrac{4}{x-1}$

92) $\dfrac{9}{4k^2+28k+48} = \dfrac{k}{4k+16} + \dfrac{9}{8k+24}$

**Solve for the indicated variable.**

93) $R = \dfrac{s+T}{D}$ for $D$

94) $A = \dfrac{2p}{c}$ for $c$

95) $w = \dfrac{N}{c-ak}$ for $k$

96) $n = \dfrac{t}{a+b}$ for $a$

97) $\dfrac{1}{R_1} + \dfrac{1}{R_2} = \dfrac{1}{R_3}$ for $R_1$

98) $\dfrac{1}{r} = \dfrac{1}{s} + \dfrac{1}{t}$ for $s$

**(8.7) Write an equation and solve.**

99) A boat can travel 8 miles downstream in the same amount of time it can travel 6 miles upstream. If the speed of the boat in still water is 14 mph, what is the speed of the current?

100) The ratio of saturated fat to total fat in a Starbucks tall Caramel Frappuccino is 2 to 3. If there are 4 more grams of total fat in the drink than there are grams of saturated fat, how much total fat is in 2 Caramel Frappuccinos? (Starbucks brochure)

101) Crayton and Flow must put together notebooks for each person attending a conference. Working alone, it would take Crayton 5 hours while it would take Flow 8 hours. How long would it take for them to assemble the notebooks together?

102) An airplane flying at constant speed can fly 350 miles with the wind in the same amount of time it can fly 300 miles against the wind. What is the speed of the plane if the wind blows at 20 mph?

103) Suppose $c$ varies directly as $m$. If $c = 96$ when $m = 12$, find $c$ when $m = 3$.

104) Suppose $p$ varies inversely as the square of $d$. If $p = 40$ when $d = 2$, find $p$ when $d = 4$.

Solve each problem by writing a variation equation.

105) The surface area of a cube varies directly as the square of the length of one of its sides. A cube has a surface area of 54 cm$^2$ when the length of each side is 3 cm. What is the surface area of a cube with a side of length 6 centimeters?

106) The frequency of a vibrating string varies inversely as its length. If a 4-ft-long piano string vibrates at 125 cycles/second, what is the frequency of a piano string that is 2 feet long?

**Perform the operation and simplify.**

107) $\dfrac{5n}{2n - 1} - \dfrac{2n + 3}{n + 2}$

108) $\dfrac{27w^3}{3w^2 + w - 4} \cdot \dfrac{2 - 2w}{15w}$

109) $\dfrac{2a^2 + 9a + 10}{4a - 7} \div (2a + 5)^2$

110) $\dfrac{5}{8b} + \dfrac{2}{9b^4}$

111) $\dfrac{c^2}{c^2 - d^2} + \dfrac{c}{d - c}$

112) $\dfrac{\dfrac{7}{x} + \dfrac{8}{y}}{1 - \dfrac{6}{y}}$

**Solve.**

113) $\dfrac{h}{5} = \dfrac{h - 3}{h + 1} + \dfrac{12}{5h + 5}$

114) $\dfrac{5w}{6} - \dfrac{2}{3} = -\dfrac{1}{6}$

115) $\dfrac{8}{3g^2 - 7g - 6} - \dfrac{8}{3g + 2} = -\dfrac{4}{g - 3}$

116) $\dfrac{4k}{k + 16} = \dfrac{4}{k + 1}$

1) Evaluate, if possible, for $k = -4$.

$$\frac{5k + 8}{k^2 + 16}$$

**Determine the values of the variable for which**
**a) the expression is undefined.**
**b) the expression equals zero.**

2) $\dfrac{2c - 9}{c + 10}$

3) $\dfrac{n^2 + 1}{n^2 - 5n - 36}$

**Write each rational expression in lowest terms.**

4) $\dfrac{21t^8 u^2}{63t^{12} u^5}$

5) $\dfrac{3h^2 - 25h + 8}{9h^2 - 1}$

6) Write three equivalent forms of $\dfrac{7 - m}{4m - 5}$.

7) Identify the LCD of $\dfrac{2z}{z + 6}$ and $\dfrac{9}{z}$.

**Perform the operations and simplify.**

8) $\dfrac{8}{15r} + \dfrac{2}{15r}$

9) $\dfrac{28a^9}{b^2} \div \dfrac{20a^{15}}{b^3}$

10) $\dfrac{5h}{12} - \dfrac{7h}{9}$

11) $\dfrac{6}{c + 2} + \dfrac{c}{3c + 5}$

12) $\dfrac{k^3 - 9k^2 + 2k - 18}{4k - 24} \cdot \dfrac{k^2 + 3k - 54}{81 - k^2}$

13) $\dfrac{8d^2 + 24d}{20} \div (d + 3)^2$

14) $\dfrac{2t - 5}{t - 7} + \dfrac{t + 9}{7 - t}$

15) $\dfrac{3}{2v^2 - 7v + 6} - \dfrac{v + 4}{v^2 + 7v - 18}$

**Simplify completely.**

16) $\dfrac{1 - \dfrac{1}{m + 2}}{\dfrac{m}{m + 2} - \dfrac{1}{m}}$

17) $\dfrac{\dfrac{5x + 5y}{x^2 y^2}}{\dfrac{20}{xy}}$

**Solve each equation.**

18) $\dfrac{3r + 1}{2} + \dfrac{1}{10} = \dfrac{6r}{5}$

19) $\dfrac{28}{w^2 - 4} = \dfrac{7}{w - 2} - \dfrac{5}{w + 2}$

20) $\dfrac{3}{x + 8} + \dfrac{x}{x - 4} = \dfrac{7x + 9}{x^2 + 4x - 32}$

21) Solve for $b$.

$$\frac{1}{a} + \frac{1}{b} = \frac{1}{c}$$

**Write an equation for each and solve.**

22) Every Sunday night, the equipment at a restaurant must be taken apart and cleaned. Ricardo can do this job twice as fast as Michael. When they work together, they can do the cleaning in 2 hr. How long would it take each man to do the job on his own?

23) A current flows at 4 mph. If a boat can travel 12 mi downstream in the same amount of time it can go 6 mi upstream, find the speed of the boat in still water.

24) Suppose $m$ varies directly as the square of $n$. If $m = 48$ when $n = 4$, find $m$ when $n = 5$.

25) If the temperature remains the same, the volume of a gas is inversely proportional to the pressure. If the volume of a gas is 6.25 L (liters) at a pressure of 2 atm (atmospheres), what is the volume of the gas at 1.25 atm?

1) Find the area of the triangle.

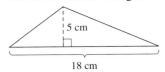

2) Evaluate $72 - 30 \div 6 + 4(3^2 - 10)$.

**Simplify. The answer should not contain negative exponents.**

3) $(2p^3)^5$

4) $(5y^2)^{-3}$

5) Write an equation and solve.

The length of a rectangular garden is 4 ft longer than the width. Find the dimensions of the garden if its perimeter is 28 ft.

**Solve each inequality. Write the answer in interval notation.**

6) $19 - 8w > 5$

7) $4 \le \dfrac{3}{5}t + 4 \le 13$

8) Find the $x$- and $y$-intercepts of $4x - 3y = 6$, and graph the equation.

9) Find the slope of the line containing the points $(4, 1)$ and $(-2, 9)$.

10) Solve the system.

$5x + 4y = 5$
$7x - 6y = 36$

**Multiply and simplify.**

11) $(2n - 3)^2$

12) $(8a + b)(8a - b)$

**Divide.**

13) $\dfrac{45h^4 - 25h^3 + 15h^2 - 10}{15h^2}$

14) $\dfrac{5k^3 + 18k^2 - 11k - 8}{k + 4}$

**Factor completely.**

15) $4d^2 + 4d - 15$

16) $3z^4 - 48$

17) $rt + 8t - r - 8$

18) Solve $x(x + 16) = x - 36$.

19) For what values of $a$ is $\dfrac{7a + 2}{a^2 - 6a}$

a) undefined?

b) equal to zero?

20) Write $\dfrac{3c^2 + 21c - 54}{c^2 + 3c - 54}$ in lowest terms.

**Perform the operations and simplify.**

21) $\dfrac{10n^2}{n^2 - 8n + 16} \cdot \dfrac{3n^2 - 14n + 8}{10n - 15n^2}$

22) $\dfrac{6}{y + 5} - \dfrac{3}{y}$

23) Simplify $\dfrac{\dfrac{2}{r - 8} + 1}{1 - \dfrac{3}{r - 8}}$.

24) Solve $\dfrac{1}{v - 1} + \dfrac{2}{5v - 3} = \dfrac{37}{5v^2 - 8v + 3}$.

25) Suppose $h$ varies inversely as the square of $p$. If $h = 12$ when $p = 2$, find $h$ when $p = 4$.

# CHAPTER 9

# Roots and Radicals

## Algebra at Work: Forensics

Forensic scientists use mathematics in many ways to help them analyze evidence and solve crimes. To help him reconstruct an accident scene, Keith can use this formula containing a radical

to estimate the minimum speed of a vehicle when the accident occurred:

$$S = \sqrt{30fd}$$

where $f$ = the drag factor, based on the type of road surface

$d$ = the length of the skid, in feet

$S$ = the speed of the vehicle in miles per hour

Keith is investigating an accident in a residential neighborhood where the speed limit is 25 mph. The car involved in the accident left skid marks 60 ft long. Tests showed that the drag factor of the asphalt road was 0.80. Was the driver speeding at the time of the accident?

Substitute the values into the equation and evaluate it to determine the minimum speed of the vehicle at the time of the accident:

$$S = \sqrt{30fd}$$
$$S = \sqrt{30(0.80)(60)}$$
$$S = \sqrt{1440} \approx 38 \text{ mph}$$

The driver was going at least 38 mph when the accident occurred. This is well over the speed limit of 25 mph.

We will learn how to simplify radicals in this chapter as well as how to work with equations like the one given here.

## Section 9.1 Finding Roots

### Objectives

1. **Find the Square Root of a Rational Number**
2. **Approximate the Square Root of a Whole Number**
3. **Use the Pythagorean Theorem**
4. **Use the Distance Formula**
5. **Find the Higher Roots of Rational Numbers**

In Section 1.2, we introduced the idea of exponents as representing repeated multiplication. For example,

$$3^2 \text{ means } 3 \cdot 3, \text{ so } 3^2 = 9$$
$$(-3)^2 \text{ means } -3 \cdot (-3), \text{ so } (-3)^2 = 9$$
$$2^4 \text{ means } 2 \cdot 2 \cdot 2 \cdot 2, \text{ so } 2^4 = 16$$

In this chapter we will study the opposite procedure, finding *roots* of numbers.

### 1. Find the Square Root of a Rational Number

The **square root** of a number, like 9, is a number that, when squared, results in the given number. So, 3 and $-3$ are square roots of 9 since $3^2 = 9$ and $(-3)^2 = 9$.

 **Example 1**

Find all square roots of 36.

**Solution**

To find a *square* root of 36, ask yourself, "What number do I *square* to get 36?" Or, "What number multiplied by itself equals 36?" One number is 6 since $6^2 = 36$. Another number is $-6$ since $(-6)^2 = 36$. So, the square roots of 36 are 6 and $-6$. ∎

 **You Try 1**

Find all square roots of 144.

The **positive** or **principal square root** of a number is represented with the $\sqrt{\phantom{x}}$ symbol. Therefore, $\sqrt{36} = 6$.

**BE CAREFUL**

$\sqrt{36} = 6$ but $\sqrt{36} \neq -6$. The $\sqrt{\phantom{x}}$ symbol represents *only* the positive square root.

To find the negative square root of a number, we must put a $-$ in front of the $\sqrt{\phantom{x}}$. For example, $-\sqrt{36} = -6$.

We call $\sqrt{\phantom{x}}$ the **square root symbol** or the **radical sign.** The number under the radical sign is the **radicand.**

$$\text{Radical sign} \rightarrow \underset{\underset{\text{Radical}}{\uparrow}}{\sqrt{25}} \leftarrow \text{Radicand}$$

The entire expression, $\sqrt{36}$, is called a **radical.**

**Example 2**

Find each square root.

a)  $\sqrt{81}$    b)  $-\sqrt{49}$    c)  $\sqrt{\dfrac{9}{16}}$    d)  $-\sqrt{\dfrac{121}{4}}$    e)  $\sqrt{0.64}$

**Solution**

a)  $\sqrt{81} = 9$ since $(9)^2 = 81$.

b)  $-\sqrt{49}$ means $-1 \cdot \sqrt{49}$. Therefore,

$$-\sqrt{49} = -1 \cdot \sqrt{49} = -1 \cdot (7) = -7$$

c)  Since $\sqrt{9} = 3$ and $\sqrt{16} = 4$, $\sqrt{\dfrac{9}{16}} = \dfrac{3}{4}$.

d)  $-\sqrt{\dfrac{121}{4}} = -1 \cdot \sqrt{\dfrac{121}{4}} = -1 \cdot \left(\dfrac{11}{2}\right) = -\dfrac{11}{2}$

e)  $\sqrt{0.64} = 0.8$    ■

**You Try 2**

Find each square root.

a)  $-\sqrt{64}$    b)  $\sqrt{\dfrac{81}{25}}$    c)  $-\sqrt{\dfrac{1}{144}}$    d)  $\sqrt{0.36}$

**Example 3**

Find $\sqrt{-4}$.

**Solution**

Recall that to find $\sqrt{-4}$, you can ask yourself, "What number do I *square* to get $-4$?" or "What number multiplied by itself equals $-4$?" *There is no such real number* since $2^2 = 4$ and $(-2)^2 = 4$. Therefore, $\sqrt{-4}$ *is not a real number*.    ■

**You Try 3**

Find $\sqrt{-49}$.

Let's review what we know about a square root and the radicand and add a third fact.

---

**Summary    Types of Square Roots**

1)  If the radicand is a perfect square, the *square* root is a *rational* number.

Example:    $\sqrt{25} = 5$        25 is a perfect square.

$\sqrt{\dfrac{4}{81}} = \dfrac{2}{9}$    $\dfrac{4}{81}$ is a perfect square.

2)  If the radicand is a negative number, the square root is *not* a real number.

Example:    $\sqrt{-9}$ is *not* a real number.

3)  If the radicand is positive and *not* a perfect square, then the square root is an *irrational* number.

Example:    $\sqrt{13}$ is irrational.        13 is not a perfect square.

The square root of such a number is a real number that is a nonrepeating, nonterminating decimal. It is important to be able to approximate such square roots because sometimes it is necessary to estimate their places on a number line or on a Cartesian coordinate system when graphing.

For the purposes of graphing, approximating a radical to the nearest tenth is sufficient. A calculator with a $\sqrt{\phantom{x}}$ key will give a better approximation of the radical.

## 2. Approximate the Square Root of a Whole Number

**Example 4**

Approximate $\sqrt{13}$ to the nearest tenth and plot it on a number line.

### Solution

What is the largest perfect square that is *less than* 13?   **9**

What is the smallest perfect square that is *greater than* 13?   **16**

Since 13 is between 9 and 16 ($9 < 13 < 16$), it is true that $\sqrt{13}$ is between $\sqrt{9}$ and $\sqrt{16}$.

$$(\sqrt{9} < \sqrt{13} < \sqrt{16})$$
$$\sqrt{9} = 3$$
$$\sqrt{13} = ?$$
$$\sqrt{16} = 4$$

$\sqrt{13}$ must be between 3 and 4. Numerically, 13 is closer to 16 than it is to 9. So, $\sqrt{13}$ will be closer to $\sqrt{16}$ than to $\sqrt{9}$. Check to see if 3.6 is a good approximation of $\sqrt{13}$. ($\approx$ means approximately equal to.)

$$\text{If } \sqrt{13} \approx 3.6, \text{ then } (3.6)^2 \approx 13$$
$$(3.6)^2 = (3.6) \cdot (3.6) = 12.96$$

Is 3.7 a better approximation of $\sqrt{13}$?

$$\text{If } \sqrt{13} \approx 3.7, \text{ then } (3.7)^2 \approx 13$$
$$(3.7)^2 = (3.7) \cdot (3.7) = 13.69$$

3.6 is a better approximation of $\sqrt{13}$.

$$\sqrt{13} \approx 3.6$$

$\sqrt{13}$

```
←─┼──┼──┼──┼─●┼──┼──┼─→
  0  1  2  3  4  5  6
```

A calculator evaluates $\sqrt{13}$ as 3.6055513. Remember that this is only an approximation. We will discuss how to approximate radicals using a calculator later in this chapter. ■

**You Try 4**

Approximate $\sqrt{29}$ to the nearest tenth and plot it on a number line.

## 3. Use the Pythagorean Theorem

Recall from Section 7.6 that we can apply the Pythagorean theorem to right triangles. If $a$ and $b$ are the lengths of the legs of a right triangle, and if $c$ is the length of the hypotenuse, then

$$a^2 + b^2 = c^2$$

**Example 5**

Let $a$ and $b$ represent the lengths of the legs of a right triangle, and let $c$ represent the length of the hypotenuse. Use the Pythagorean theorem to find the length of the missing side.

a) $a = 4, b = 3$, find $c$.  b) $a = 7, c = 10$, find $b$.

**Solution**

a) Substitute the values into the Pythagorean theorem and solve.

$$a^2 + b^2 = c^2 \qquad \text{Pythagorean theorem}$$
$$4^2 + 3^2 = c^2 \qquad \text{Let } a = 4 \text{ and } b = 3.$$
$$16 + 9 = c^2 \qquad \text{Square.}$$
$$25 = c^2 \qquad \text{Add.}$$

To solve $25 = c^2$, or $c^2 = 25$, we must find $c$ so that the square of $c$ is 25. Additionally, $c$ must be a positive number since it represents the length of the hypotenuse. In other words, one solution of $c^2 = 25$ is the *positive* square root of 25. Therefore, $c = \sqrt{25} = 5$.

**Note:**
If $k > 0$, then the *positive* solution of $x^2 = k$ is $x = \sqrt{k}$.

b) Substitute the values into the Pythagorean theorem and solve.

$$a^2 + b^2 = c^2 \qquad \text{Pythagorean theorem}$$
$$7^2 + b^2 = 10^2 \qquad \text{Let } a = 7 \text{ and } c = 10.$$
$$49 + b^2 = 100 \qquad \text{Square.}$$
$$b^2 = 51 \qquad \text{Subtract 49 from each side.}$$
$$b = \sqrt{51} \qquad \text{Solve for } b.$$

**You Try 5**

Let $a$ and $b$ represent the lengths of the legs of a right triangle, and let $c$ represent the length of the hypotenuse. Use the Pythagorean theorem to find the length of the missing side.

a) $a = 6, b = 8$, find $c$.  b) $b = 4, c = 9$, find $a$.

**Example 6**

A 17-ft wire is attached to the top of a 15-ft pole so that the wire, the pole, and the ground form a right triangle as shown in the figure. Find the distance from the base of the pole to the point where the wire is attached to the ground.

**Solution**

**Step 1:** **Read** the problem carefully, and identify what we are being asked to find.

We must find the distance from the base of the pole to the wire.

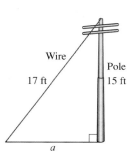

Wire
17 ft
Pole
15 ft
$a$

**Step 2:** **Choose a variable** to represent the unknown. Label the picture.

$a =$ distance from the base of the pole to the wire

Also notice that the wire is the hypotenuse of the triangle, so $c = 17$. Therefore, $b = 15$.

***Step 3:*** **Translate** the information into an algebraic equation. Use the Pythagorean theorem.

$$a^2 + b^2 = c^2 \qquad \text{Pythagorean theorem}$$
$$a^2 + 15^2 = 17^2 \qquad \text{Let } b = 15 \text{ and } c = 17.$$

***Step 4:*** **Solve** the equation.

$$a^2 + 225 = 289 \qquad \text{Square.}$$
$$a^2 = 64 \qquad \text{Subtract 225 from each side.}$$
$$a = \sqrt{64} \qquad \text{Solve.}$$
$$a = 8 \qquad \text{Simplify.}$$

When we solve $a^2 = 64$, only the *positive* square root of 64 makes sense because $a$ represents a length in this problem.

***Step 5:*** **Check** the answer and **interpret** the solution as it relates to the problem.

The distance from the base of the pole to the wire is 8 ft.

Do these lengths satisfy the Pythagorean theorem? Yes.

$$a^2 + b^2 = c^2$$
$$8^2 + 15^2 \stackrel{?}{=} 17^2$$
$$64 + 225 = 289 \checkmark$$

**You Try 6**

A 10-ft ladder is leaning against a house. The base of the ladder is 6 ft from the house. Find the distance from the ground to the top of the ladder.

We can use the Pythagorean theorem to find a formula for the distance between two points.

## 4. Use the Distance Formula

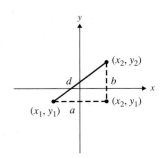

Let's say that we want to find the distance, $d$, between points $(x_1, y_1)$ and $(x_2, y_2)$ as shown in the figure. If we also include the point $(x_2, y_1)$ in our drawing, we get a right triangle.

The length of side $a$ is $a = x_2 - x_1$, the length of side $b$ is $b = y_2 - y_1$, and the length of the hypotenuse is $d$. Next, we will use the Pythagorean theorem to relate sides $a$, $b$, and $d$ and get the distance formula.

$$d^2 = a^2 + b^2 \qquad \text{Pythagorean theorem}$$
$$d^2 = (x_2 - x_1)^2 + (y_2 - y_1)^2 \qquad \text{Let } a = x_2 - x_1 \text{ and } b = y_2 - y_1.$$
$$d = \sqrt{(x_2 - x_1)^2 + (y_2 - y_1)^2} \qquad \text{Solve for } d.$$

---

**Definition    The Distance Formula**

The distance between two points with coordinates $(x_1, y_1)$ and $(x_2, y_2)$ is given by

$$d = \sqrt{(x_2 - x_1)^2 + (y_2 - y_1)^2}$$

**Example 7**

Find the distance between the points $(-5, 2)$ and $(1, 3)$.

**Solution**

Begin by labeling the points: $(\overset{x_1, y_1}{-5, 2})$ $(\overset{x_2, y_2}{1, 3})$.

Substitute the values into the distance formula.

$$d = \sqrt{(x_2 - x_1)^2 + (y_2 - y_1)^2}$$
$$= \sqrt{[1 - (-5)]^2 + (3 - 2)^2} \qquad \text{Substitute values.}$$
$$= \sqrt{(1 + 5)^2 + (1)^2}$$
$$= \sqrt{(6)^2 + (1)^2} = \sqrt{36 + 1} = \sqrt{37}$$

 **BE CAREFUL**
$\sqrt{36 + 1} \neq \sqrt{36} + \sqrt{1}$. To simplify $\sqrt{36 + 1}$, we must first simplify the radicand. Therefore, $\sqrt{36 + 1} = \sqrt{37}$.

 **You Try 7**

Find the distance between the points $(8, -1)$ and $(3, 1)$.

## 5. Find the Higher Roots of Rational Numbers

We saw in Example 2(a) that $\sqrt{81} = 9$ since $9^2 = 81$. Finding a square root is the *opposite* of squaring a number. Similarly, we can find higher roots of numbers like $\sqrt[3]{a}$ (read as "the **cube root** of $a$"), $\sqrt[4]{a}$ (read as "the **fourth root** of $a$"), $\sqrt[5]{a}$ (the **fifth root** of $a$), etc.

**Example 8**

Find each root.

a) $\sqrt[3]{125}$        b) $\sqrt[4]{81}$        c) $\sqrt[5]{32}$

**Solution**

a)  To find $\sqrt[3]{125}$ (read as "the cube root of 125") ask yourself, "What number do I *cube* to get 125?" That number is 5 since $5^3 = 125$. Therefore, $\sqrt[3]{125} = 5$.

   Finding the cube root of a number is the *opposite* of cubing a number.

b)  To find $\sqrt[4]{81}$ (read as "the fourth root of 81") ask yourself, "What number do I raise to the *fourth power* to get 81?" That number is 3 since $3^4 = 81$. Therefore, $\sqrt[4]{81} = 3$.

   Finding the fourth root of a number is the *opposite* of raising a number to the fourth power.

c)  To find $\sqrt[5]{32}$ (read as "the fifth root of 32") ask yourself, "What number do I raise to the *fifth power* to get 32?" That number is 2 since $2^5 = 32$. Therefore, $\sqrt[5]{32} = 2$.

   Finding the fifth root of a number is the *opposite* of raising a number to the fifth power.

**You Try 8**

Find each root.

a)  $\sqrt[4]{16}$          b)  $\sqrt[3]{27}$

We can use a general notation for writing roots of numbers.

**Definition**

The expression $\sqrt[n]{a}$ is read as "the *n*th root of *a*." If $\sqrt[n]{a} = b$, then $a = b^n$.

*n* is the **index** or **order** of the radical. (The plural of *index* is **indices**.)

**Note**

When finding square roots, we do not write $\sqrt[2]{a}$. The square root of *a* is written as $\sqrt{a}$, and the index is understood to be 2.

In Section 1.2, we first presented the powers of numbers that students are expected to know. ($2^2 = 4$, $2^3 = 8$, etc.) Use of these powers was first necessary in the study of the rules of exponents in Chapter 2. Knowing these powers is necessary for finding roots as well, so the student can refer to p. 18 to review this list of powers.

While it is true that the square root of a negative number is not a real number, sometimes it *is* possible to find a *higher* root of a negative number.

**Example 9**

Find each root, if possible.

a)  $\sqrt[3]{-64}$          b)  $\sqrt[5]{-32}$          c)  $-\sqrt[4]{16}$          d)  $\sqrt[4]{-16}$

**Solution**

a)  To find $\sqrt[3]{-64}$ ask yourself, "What number do I *cube* to get $-64$?" That number is $-4$.

$$\sqrt[3]{-64} = -4 \text{ since } (-4)^3 = -64$$

b)  To find $\sqrt[5]{-32}$ ask yourself, "What number do I raise to the *fifth power* to get $-32$?" That number is $-2$.

$$\sqrt[5]{-32} = -2 \text{ since } (-2)^5 = -32$$

c)  $-\sqrt[4]{16}$ means $-1 \cdot \sqrt[4]{16}$. Therefore, $-\sqrt[4]{16} = -1 \cdot \sqrt[4]{16} = -1 \cdot 2 = -2$.

d)  To find $\sqrt[4]{-16}$ ask yourself, "What number do I raise to the *fourth power* to get $-16$?" *There is no such real number* since $2^4 = 16$ and $(-2)^4 = 16$.

$$\sqrt[4]{-16} \text{ is not a real number}$$

We can summarize what we have seen in Example 9 as follows:

**Property    Roots of Negative Numbers**

1)  The *odd root* of a negative number is a negative number.

2)  The *even root* of a negative number is not a real number.

## You Try 9

Find each root, if possible.

a)   $\sqrt[6]{-64}$        b)   $-\sqrt[3]{125}$        c)   $\sqrt[4]{81}$

## Using Technology

We can evaluate square roots, cube roots, or even higher roots using a graphing calculator. A radical sometimes evaluates to an integer and sometimes must be approximated using a decimal.

**To evaluate a square root:**

For example, to evaluate $\sqrt{9}$ press [2ⁿᵈ] [x²], enter the radicand [9], and then press [)] [ENTER]. The result is 3 as shown on the screen on the left below. When the radicand is a perfect square such as 9, 16, or 25, then the square root evaluates to a whole number. For example, $\sqrt{16}$ evaluates to 4 and $\sqrt{25}$ evaluates to 5 as shown.

If the radicand of a square root is not a perfect square, then the result is a decimal approximation. For example, to evaluate $\sqrt{19}$ press [2ⁿᵈ] [x²], enter the radicand [1][9], and then press [)] [ENTER]. The result is approximately 4.3589, rounded to four decimal places.

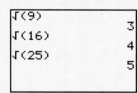   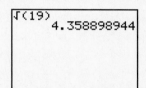

**To evaluate a cube root:**

For example, to evaluate $\sqrt[3]{27}$ press [MATH] [4], enter the radicand [2][7], and then press [)] [ENTER]. The result is 3 as shown.

If the radicand is a perfect cube such as 27, then the cube root evaluates to an integer. Since 28 is not a perfect cube, the cube root evaluates to approximately 3.0366.

**To evaluate radicals with an index greater than 3:**

For example, to evaluate $\sqrt[4]{16}$ enter the index [4], press [MATH] [5], enter the radicand [1][6], and press [ENTER]. The result is 2.

Since the fifth root of 18 evaluates to a decimal, the result is an approximation of 1.7826 rounded to four decimal places as shown.

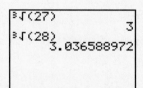

Evaluate each root using a graphing calculator. If necessary, approximate to the nearest tenth.

1)   $\sqrt{25}$        2)   $\sqrt[3]{216}$        3)   $\sqrt{29}$        4)   $\sqrt{324}$        5)   $\sqrt[5]{1024}$        6)   $\sqrt[3]{343}$

---

## Answers to You Try Exercises

1)   $-12$ and $12$      2)   a) $-8$   b) $\dfrac{9}{5}$   c) $-\dfrac{1}{12}$   d) $0.6$      3)   not a real number

4)   $5.4$                                              5)   a) $10$   b) $\sqrt{65}$      6)   8 ft      7)   $\sqrt{29}$

$\sqrt{29}$

←—+——+——+——+——+——+——+——•——+—→
     0   1   2   3   4   5   6

8)   a) $2$   b) $3$      9)   a) not a real number   b) $-5$   c) $3$

**Answers to Technology Exercises**

1) 5    2) 6    3) 5.4    4) 18    5) 4    6) 7

# 9.1    Exercises

**Objective 1: Find the Square Root of a Rational Number**

Decide whether each statement is true or false. If it is false, explain why.

1) $\sqrt{100} = 10$ and $-10$    2) $\sqrt{49} = 7$

3) The square root of a negative number is a negative number.

4) The square root of a nonnegative number is always positive.

Find all square roots of each number.

5) 81    6) 121

7) 4    8) 64

9) 900    10) 400

11) $\dfrac{1}{36}$    12) $\dfrac{25}{9}$

13) 0.25    14) 0.01

Find each square root, if possible.

15) $\sqrt{144}$    16) $\sqrt{16}$

17) $\sqrt{9}$    18) $\sqrt{25}$

19) $\sqrt{-64}$    20) $\sqrt{-1}$

VIDEO 21) $-\sqrt{36}$    22) $-\sqrt{169}$

23) $\sqrt{\dfrac{64}{81}}$    24) $\sqrt{\dfrac{1}{100}}$

25) $\sqrt{\dfrac{4}{9}}$    26) $\sqrt{\dfrac{49}{25}}$

27) $-\sqrt{\dfrac{1}{16}}$    28) $-\sqrt{\dfrac{9}{121}}$

29) $\sqrt{0.49}$    30) $\sqrt{0.04}$

31) $-\sqrt{0.0144}$    32) $-\sqrt{0.81}$

**Objective 2: Approximate the Square Root of a Whole Number**

Approximate each square root to the nearest tenth and plot it on a number line.

VIDEO 33) $\sqrt{11}$    34) $\sqrt{3}$

35) $\sqrt{2}$    36) $\sqrt{5}$

37) $\sqrt{33}$    38) $\sqrt{39}$

39) $\sqrt{55}$    40) $\sqrt{72}$

**Objective 3: Use the Pythagorean Theorem**

Use the Pythagorean theorem to find the length of the missing side.

41)
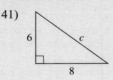

42)
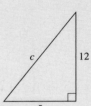

43)
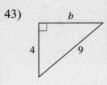

44)
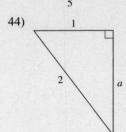

Let $a$ and $b$ represent the lengths of the legs of a right triangle, and let $c$ represent the length of the hypotenuse. Use the Pythagorean theorem to find the length of the missing side.

45) If $b = 5$ and $c = 13$, find $a$.

46) If $a = 3$ and $c = 5$, find $b$.

47) If $a = 1$ and $b = 1$, find $c$.

48) If $a = 7$ and $b = 3$, find $c$.

Write an equation and solve. (Hint: Draw a picture.)

49) The width of a rectangle is 3 in., and its diagonal is 8 in. long. What is the length of the rectangle?

50) The length of a rectangle is 7 cm, and its width is 5 cm. Find the length of the diagonal.

Write an equation and solve.

VIDEO 51) A 13-ft ladder is leaning against a wall so that the base of the ladder is 5 ft away from the wall. How high on the wall does the ladder reach?

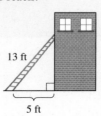

52) Salma is flying a kite. It is 30 ft from her horizontally, and it is 40 ft above her hand. How long is the kite string?

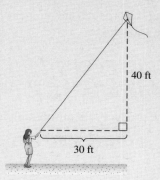

53) Martha digs a garden in the corner of her yard so that it is in the shape of a right triangle. One side of the garden is next to the garage, and the other side is against the back fence. How much fencing will she need to enclose the remaining side of the garden?

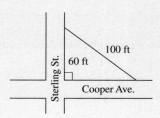

54) The Ramirez family's property is on a corner in the shape of a right triangle. How long is the property on Cooper Ave.?

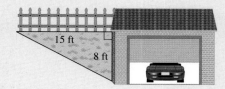

55) A laptop screen has a 13-in. diagonal and a length of 11.2 in. Find the width of the screen to the nearest tenth.

56) The screen of an LCD TV has a 40-in. diagonal and a length of 34.8 in. Find the height of the screen to the nearest tenth.

**Objective 4: Use the Distance Formula**

57) Mrs. Chang asks her students to evaluate $\sqrt{9 + 16}$. Rachel gives this answer:
$$\sqrt{9 + 16} = \sqrt{9} + \sqrt{16} = 3 + 4 = 7$$
Ivan gives this answer: $\sqrt{9 + 16} = \sqrt{25} = 5$
Who is right and why?

Evaluate.

58) $\sqrt{64 + 36}$

59) $\sqrt{25 + 144}$

60) $\sqrt{49 + 81}$

61) $\sqrt{(1)^2 + (8)^2}$

62) $\sqrt{(5)^2 + (2)^2}$

63) $\sqrt{(3)^2 + (-10)^2}$

64) $\sqrt{(-4)^2 + (-1)^2}$

65) $\sqrt{(-40)^2 + (-30)^2}$

66) $\sqrt{(15)^2 + (-8)^2}$

Find the distance between the given points.

67) (4, 2) and (8, 5)

68) (2, 13) and (7, 1)

69) (−5, −6) and (−2, −8)

70) (−3, 5) and (−1, 0)

71) (7, 3) and (2, −3)

72) (−4, −4) and (−1, −2)

73) (0, 11) and (−6, 0)

74) (2, 5) and (−3, 1)

75) $\left(\dfrac{5}{8}, \dfrac{1}{3}\right)$ and $\left(\dfrac{1}{4}, \dfrac{5}{6}\right)$

76) $\left(-\dfrac{3}{4}, -\dfrac{10}{9}\right)$ and $\left(\dfrac{1}{4}, \dfrac{8}{9}\right)$

**Objective 5: Find the Higher Roots of Rational Numbers**

Decide whether each statement is true or false. If it is false, explain why.

77) The cube root of a negative number is a negative number.

78) The even root of a negative number is a negative number.

79) The odd root of a negative number is not a real number.

80) Every nonnegative real number has two real, even roots.

81) Explain how to find $\sqrt[3]{64}$.

82) Explain how to find $\sqrt[4]{16}$.

83) Does $\sqrt[4]{-81} = -3$? Why or why not?

84) Does $\sqrt[3]{-8} = -2$? Why or why not?

Find each root, if possible.

85) $\sqrt[3]{8}$

86) $\sqrt[3]{1}$

87) $\sqrt[3]{125}$

88) $\sqrt[3]{27}$

89) $\sqrt[3]{-1}$

90) $\sqrt[3]{-8}$

91) $\sqrt[4]{81}$

92) $\sqrt[4]{16}$

93) $\sqrt[4]{-1}$

94) $\sqrt[4]{-81}$

95) $-\sqrt[4]{16}$

96) $-\sqrt[4]{1}$

97) $\sqrt[5]{-32}$

98) $-\sqrt[6]{64}$

99) $-\sqrt[3]{-27}$

100) $-\sqrt[3]{-1000}$

101) $\sqrt[6]{-64}$

102) $\sqrt[4]{-16}$

103) $\sqrt[3]{\dfrac{8}{125}}$

104) $\sqrt[4]{\dfrac{81}{16}}$

105) $\sqrt{60 - 11}$

106) $\sqrt{100 + 21}$

107) $\sqrt[3]{100 + 25}$

108) $\sqrt[3]{9 - 36}$

109) $\sqrt{1 - 9}$

110) $\sqrt{25 - 36}$

# Section 9.2 Simplifying Expressions Containing Radicals

**Objectives**

1. **Multiply Square Roots**
2. **Simplify the Square Root of a Whole Number**
3. **Use the Quotient Rule for Square Roots**
4. **Simplify Square Root Expressions Containing Variables**
5. **Simplify Higher Roots**

In this section, we will introduce rules for finding the product and quotient of square roots as well as for simplifying expressions containing square roots.

## 1. Multiply Square Roots

Let's begin with the product $\sqrt{4} \cdot \sqrt{25}$. $\sqrt{4} \cdot \sqrt{25} = 2 \cdot 5 = 10$. Also notice that $\sqrt{4} \cdot \sqrt{25} = \sqrt{4 \cdot 25} = \sqrt{100} = 10$.

We obtain the same result. This leads us to the product rule for multiplying expressions containing square roots.

---

**Definition   Product Rule for Square Roots**

Let $a$ and $b$ be nonnegative real numbers. Then,

$$\sqrt{a} \cdot \sqrt{b} = \sqrt{a \cdot b}$$

In other words, the product of two square roots equals the square root of the product.

---

**Example 1**

Multiply. Assume the variable represents a nonnegative real number.

a)  $\sqrt{7} \cdot \sqrt{2}$          b)  $\sqrt{5} \cdot \sqrt{x}$

**Solution**

a)  $\sqrt{7} \cdot \sqrt{2} = \sqrt{7 \cdot 2} = \sqrt{14}$          b)  $\sqrt{5} \cdot \sqrt{x} = \sqrt{5 \cdot x} = \sqrt{5x}$

**BE CAREFUL**

We can multiply radicals this way *only if* the indices are the same. In future math courses we will learn how to multiply radicals with different indices such as $\sqrt{5} \cdot \sqrt[3]{t}$.

**You Try 1**

Multiply. Assume the variable represents a nonnegative real number.

a)  $\sqrt{2} \cdot \sqrt{3}$          b)  $\sqrt{6} \cdot \sqrt{z}$

## 2. Simplify the Square Root of a Whole Number

Knowing how to simplify radicals is very important in the study of algebra. We begin by discussing how to simplify expressions containing square roots.

How do we know when a square root is simplified?

---

**Property   When Is a Square Root Simplified?**

An expression containing a square root is simplified when all of the following conditions are met:

1) The radicand does not contain any factors (other than 1) that are perfect squares.

2) The radicand does not contain any fractions.

3) There are no radicals in the denominator of a fraction.

Note: Condition 1) implies that the radical cannot contain variables with exponents greater than or equal to 2, the index of the square root.

We will discuss higher roots later in this section.

To simplify expressions containing square roots, we reverse the process of multiplying. That is, we use the product rule that says

$$\sqrt{a \cdot b} = \sqrt{a} \cdot \sqrt{b}$$

where $a$ or $b$ is a perfect square.

---

### Example 2

Simplify completely.

a)  $\sqrt{12}$     b)  $\sqrt{300}$     c)  $\sqrt{21}$     d)  $\sqrt{72}$

**Solution**

a)  The radical $\sqrt{12}$ is not in simplest form since 12 contains a factor (other than 1) that is a perfect square. Think of two numbers that multiply to 12 so that at least one of the numbers is a perfect square: $12 = 4 \cdot 3$.

(While it is true that $12 = 6 \cdot 2$, neither 6 nor 2 is a perfect square.)

Rewrite $\sqrt{12}$:

$$\begin{aligned}
\sqrt{12} &= \sqrt{4 \cdot 3} && \text{4 is a perfect square.}\\
&= \sqrt{4} \cdot \sqrt{3} && \text{Product rule}\\
&= 2\sqrt{3} && \sqrt{4} = 2
\end{aligned}$$

$2\sqrt{3}$ is completely simplified because 3 does not have any factors that are perfect squares.

b)  Does 300 have a factor that is a perfect square? Yes! $300 = 100 \cdot 3$. To simplify $\sqrt{300}$, rewrite it as

$$\begin{aligned}
\sqrt{300} &= \sqrt{100 \cdot 3} && \text{100 is a perfect square.}\\
&= \sqrt{100} \cdot \sqrt{3} && \text{Product rule}\\
&= 10\sqrt{3} && \sqrt{100} = 10
\end{aligned}$$

$10\sqrt{3}$ is completely simplified because 3 does not have any factors that are perfect squares.

c)  $21 = 3 \cdot 7$     Neither 3 nor 7 is a perfect square.
    $21 = 1 \cdot 21$    While 1 is a perfect square, it will not help us simplify $\sqrt{21}$.

$\sqrt{21}$ is in simplest form.

d)  There are different ways to simplify $\sqrt{72}$. We will look at two of them.

   i)  Two numbers that multiply to 72 are 36 and 2 with 36 being a perfect square. We can write

$$\begin{aligned}
\sqrt{72} &= \sqrt{36 \cdot 2} && \text{36 is a perfect square.}\\
&= \sqrt{36} \cdot \sqrt{2} && \text{Product rule}\\
&= 6\sqrt{2} && \sqrt{36} = 6
\end{aligned}$$

   ii)  We can also think of 72 as $9 \cdot 8$ since 9 is a perfect square. We can write

$$\begin{aligned}
\sqrt{72} &= \sqrt{9 \cdot 8} && \text{9 is a perfect square.}\\
&= \sqrt{9} \cdot \sqrt{8} && \text{Product rule}\\
&= 3\sqrt{8} && \sqrt{9} = 3
\end{aligned}$$

Therefore, $\sqrt{72} = 3\sqrt{8}$. Is $\sqrt{8}$ in simplest form? *No, because $8 = 4 \cdot 2$ and 4 is a perfect square.* We must continue to simplify.

$$
\begin{aligned}
\sqrt{72} &= 3\sqrt{8} \\
&= 3\sqrt{4 \cdot 2} && \text{4 is a perfect square.} \\
&= 3\sqrt{4} \cdot \sqrt{2} && \text{Product rule} \\
&= 3 \cdot 2 \cdot \sqrt{2} && \sqrt{4} = 2 \\
&= 6\sqrt{2} && \text{Multiply } 3 \cdot 2.
\end{aligned}
$$

$6\sqrt{2}$ is completely simplified because 2 does not have any factors that are perfect squares.

Example 2(d) shows that using either $\sqrt{72} = \sqrt{36 \cdot 2}$ or $\sqrt{72} = \sqrt{9 \cdot 8}$ leads us to the same result. Furthermore, this example illustrates that a radical is not always *completely* simplified after just one iteration of the simplification process. It is necessary to always examine the radical to determine whether or not it can be simplified more. ■

**Note**

After simplifying a radical, look at the result and ask yourself, "Is the radical in simplest form?" If it is not, simplify again. Asking yourself this question will help you to be sure that the radical *is* completely simplified.

**You Try 2**

Simplify completely.

a)  $\sqrt{20}$      b)  $\sqrt{150}$      c)  $\sqrt{33}$      d)  $\sqrt{24}$

**Example 3**

Multiply and simplify.

a)  $\sqrt{5} \cdot \sqrt{10}$      b)  $\sqrt{18} \cdot \sqrt{3}$

**Solution**

a)  $\begin{aligned}[t] \sqrt{5} \cdot \sqrt{10} &= \sqrt{50} && \text{Product rule} \\ &= \sqrt{25 \cdot 2} && \text{25 is a perfect square.} \\ &= \sqrt{25} \cdot \sqrt{2} = 5\sqrt{2} \end{aligned}$

b)  We can find the product $\sqrt{18} \cdot \sqrt{3}$ in two different ways.

   i)  Begin by multiplying the radicands to obtain one radical.

$$
\begin{aligned}
\sqrt{18} \cdot \sqrt{3} &= \sqrt{18 \cdot 3} && \text{Product rule} \\
&= \sqrt{54} && \text{Multiply.} \\
&= \sqrt{9 \cdot 6} && \text{9 is a perfect square.} \\
&= \sqrt{9} \cdot \sqrt{6} = 3\sqrt{6}
\end{aligned}
$$

   ii)  Simplify $\sqrt{18}$ before multiplying the radicals.

$$
\sqrt{18} = \sqrt{9 \cdot 2} = \sqrt{9} \cdot \sqrt{2} = 3\sqrt{2}
$$

   Then, substitute $3\sqrt{2}$ for $\sqrt{18}$.

$$
\begin{aligned}
\sqrt{18} \cdot \sqrt{3} &= 3\sqrt{2} \cdot \sqrt{3} && \text{Substitute } 3\sqrt{2} \text{ for } \sqrt{18}. \\
&= 3\sqrt{2 \cdot 3} && \text{Product rule} \\
&= 3\sqrt{6} && \text{Multiply.}
\end{aligned}
$$

Either way, we get the same result. ■

**You Try 3**

Multiply and simplify.

a)   $\sqrt{3} \cdot \sqrt{15}$          b)   $\sqrt{20} \cdot \sqrt{2}$

### 3.  Use the Quotient Rule for Square Roots

Let's simplify $\dfrac{\sqrt{36}}{\sqrt{9}}$. We can say $\dfrac{\sqrt{36}}{\sqrt{9}} = \dfrac{6}{3} = 2$. It is also true that $\dfrac{\sqrt{36}}{\sqrt{9}} = \sqrt{\dfrac{36}{9}} = \sqrt{4} = 2$.

This leads us to the quotient rule for dividing expressions containing square roots.

---

**Definition   Quotient Rule for Square Roots**

Let $a$ and $b$ be nonnegative real numbers such that $b \neq 0$. Then

$$\sqrt{\dfrac{a}{b}} = \dfrac{\sqrt{a}}{\sqrt{b}}$$

The square root of a quotient equals the quotient of the square roots.

---

**Example 4**

Simplify completely.

a)   $\sqrt{\dfrac{9}{25}}$          b)   $\sqrt{\dfrac{700}{7}}$          c)   $\dfrac{\sqrt{60}}{\sqrt{5}}$          d)   $\sqrt{\dfrac{2}{49}}$          e)   $\dfrac{15\sqrt{12}}{3\sqrt{2}}$

**Solution**

a)   Since 9 and 25 are each perfect squares, simplify the expression by finding the square root of each separately.

$$\sqrt{\dfrac{9}{25}} = \dfrac{\sqrt{9}}{\sqrt{25}} \qquad \text{Quotient rule}$$

$$= \dfrac{3}{5} \qquad \sqrt{9} = 3 \text{ and } \sqrt{25} = 5$$

b)   Neither 700 nor 7 is a perfect square, but if we simplify $\dfrac{700}{7}$ we get 100, which *is* a perfect square.

$$\sqrt{\dfrac{700}{7}} = \sqrt{100} \qquad \text{Simplify } \dfrac{700}{7}.$$

$$= 10$$

c)   We can simplify $\dfrac{\sqrt{60}}{\sqrt{5}}$ using two different methods.

  i)   Begin by applying the quotient rule to obtain a fraction under *one* radical and simplify the fraction.

$$\dfrac{\sqrt{60}}{\sqrt{5}} = \sqrt{\dfrac{60}{5}} \qquad \text{Quotient rule}$$

$$= \sqrt{12} \qquad \text{Simplify } \dfrac{60}{5}.$$

$$= \sqrt{4 \cdot 3} \qquad \text{4 is a perfect square.}$$

$$= \sqrt{4} \cdot \sqrt{3} \qquad \text{Product rule}$$

$$= 2\sqrt{3} \qquad \sqrt{4} = 2$$

ii)  We can apply the product rule to rewrite $\sqrt{60}$ then simplify the fraction.

$$\frac{\sqrt{60}}{\sqrt{5}} = \frac{\sqrt{5} \cdot \sqrt{12}}{\sqrt{5}} \qquad \text{Product rule}$$

$$= \frac{\overset{1}{\cancel{\sqrt{5}}} \cdot \sqrt{12}}{\underset{1}{\cancel{\sqrt{5}}}} \qquad \text{Divide out the common factor.}$$

$$= \sqrt{12} \qquad \text{Simplify.}$$
$$= \sqrt{4 \cdot 3} = \sqrt{4} \cdot \sqrt{3} = 2\sqrt{3}$$

Either method will produce the same result.

d)  We cannot simplify the fraction $\dfrac{2}{49}$, but 49 *is* a perfect square.

$$\sqrt{\frac{2}{49}} = \frac{\sqrt{2}}{\sqrt{49}} \qquad \text{Use the quotient rule.}$$

$$= \frac{\sqrt{2}}{7} \qquad \sqrt{49} = 7$$

e)  Think of $\dfrac{15\sqrt{12}}{3\sqrt{2}}$ as $\dfrac{15}{3} \cdot \dfrac{\sqrt{12}}{\sqrt{2}}$. Simplify $\dfrac{15}{3}$ and use the quotient rule.

$$\frac{15}{3} \cdot \frac{\sqrt{12}}{\sqrt{2}} = 5 \cdot \sqrt{\frac{12}{2}} = 5\sqrt{6}.$$

### You Try 4

**Simplify completely.**

a)  $\sqrt{\dfrac{49}{144}}$    b)  $\sqrt{\dfrac{360}{10}}$    c)  $\dfrac{\sqrt{120}}{\sqrt{3}}$    d)  $\sqrt{\dfrac{6}{25}}$    e)  $\dfrac{28\sqrt{30}}{4\sqrt{6}}$

Let's look at another example that requires us to use both the product and quotient rules.

### Example 5

Simplify $\sqrt{\dfrac{7}{12}} \cdot \sqrt{\dfrac{1}{3}}$.

**Solution**

Begin by using the product rule to multiply the radicands. Then simplify.

$$\sqrt{\frac{7}{12}} \cdot \sqrt{\frac{1}{3}} = \sqrt{\frac{7}{12} \cdot \frac{1}{3}} \qquad \text{Product rule}$$

$$= \sqrt{\frac{7}{36}} \qquad \text{Multiply.}$$

$$= \frac{\sqrt{7}}{\sqrt{36}} \qquad \text{Quotient rule}$$

$$= \frac{\sqrt{7}}{6} \qquad \text{Simplify.}$$

**You Try 5**

Simplify $\sqrt{\dfrac{1}{8}} \cdot \sqrt{\dfrac{3}{2}}$.

## 4. Simplify Square Root Expressions Containing Variables

Recall that a square root expression in simplified form cannot contain any factors (other than 1) that are perfect squares. Let's see what this means when the radicand contains variables by first examining some expressions containing only numbers.

It is true that $\sqrt{4^2} = \sqrt{16} = 4$ and $\sqrt{(-4)^2} = \sqrt{16} = 4$. This is just one example that shows that the square root of a squared number is nonnegative. Using a variable, we can say

$$\text{For any real number, } a, \sqrt{a^2} = |a|.$$

We can apply this to the example above: $\sqrt{4^2} = |4| = 4$ and $\sqrt{(-4)^2} = |-4| = 4$.

The first of our examples shows that if $a \geq 0$, we do not need to use absolute values to simplify square roots. **In the rest of this book, we will assume that all variables represent positive real numbers.** Then, $\sqrt{a^2} = a$ and $\sqrt{a} \cdot \sqrt{a} = a$.

Let's simplify some square root expressions containing variables.

**Example 6**

Simplify completely.

a)  $\sqrt{t^4}$    b)  $\sqrt{81p^6}$    c)  $\sqrt{45c^{18}}$    d)  $\sqrt{\dfrac{8}{w^{10}}}$    e)  $\sqrt{x^7}$

**Solution**

a)  Ask yourself, *"What do I square to get $t^4$?"* $t^2$. Therefore, $\sqrt{t^4} = t^2$.

   Or, we can think of simplifying in this way:

   $$\sqrt{t^4} = \sqrt{(t^2)^2} \qquad \text{Write } t^4 \text{ as a perfect square.}$$
   $$= t^2 \qquad \text{Simplify.}$$

b)  Ask yourself, *"What do I square to get $81p^6$?"* $9p^3$. Therefore,
   $\sqrt{81p^6} = 9p^3$.

   Or, we can begin by using the product rule:

   $$\sqrt{81p^6} = \sqrt{81} \cdot \sqrt{p^6} \qquad \text{Product rule}$$
   $$= 9p^3 \qquad \text{Simplify: } \sqrt{p^6} = \sqrt{(p^3)^2} = p^3.$$

c)  $$\sqrt{45c^{18}} = \sqrt{45} \cdot \sqrt{c^{18}} \qquad \text{Product rule}$$
   $$= \sqrt{9} \cdot \sqrt{5} \cdot \sqrt{(c^9)^2} \qquad \text{Product rule: write } \sqrt{c^{18}} \text{ as a perfect square.}$$
   $$= 3\sqrt{5} \cdot c^9 \qquad \text{Simplify.}$$
   $$= 3c^9\sqrt{5}$$

d)  $$\sqrt{\dfrac{8}{w^{10}}} = \dfrac{\sqrt{8}}{\sqrt{w^{10}}} \qquad \text{Quotient rule}$$
   $$= \dfrac{\sqrt{4} \cdot \sqrt{2}}{\sqrt{(w^5)^2}} \qquad 4 \text{ is a perfect square: } w^{10} = (w^5)^2.$$
   $$= \dfrac{2\sqrt{2}}{w^5} \qquad \text{Simplify.}$$

e)   The exponent in the radicand of $\sqrt{x^7}$ is odd, so we must take a different approach to simplifying this radical. Since this is a *square* root expression, we want to write $x^7$ as a product so that one of the factors is a perfect *square*. We will write $x^7$ as $x^6 \cdot x$ since $x^6$ is the largest perfect square that is a factor of $x^7$.

$$\sqrt{x^7} = \sqrt{x^6 \cdot x}$$
$$= \sqrt{x^6} \cdot \sqrt{x} \qquad \text{Product rule}$$
$$= x^3\sqrt{x} \qquad \sqrt{x^6} = x^3$$

### You Try 6

Simplify completely.

a)  $\sqrt{n^8}$       b)  $\sqrt{64z^{14}}$       c)  $\sqrt{63b^{10}}$       d)  $\sqrt{\dfrac{12}{r^{12}}}$       e)  $\sqrt{h^5}$

Notice in Example 6(a)–(d) that to find the square root of a variable with an *even* exponent, we can just divide the exponent by 2. For example,

$$\sqrt{t^4} = t^{4/2} = t^2 \qquad\qquad \sqrt{p^6} = p^{6/2} = p^3$$

## 5. Simplify Higher Roots

In Section 9.1, we found higher roots like $\sqrt[4]{16} = 2$ and $\sqrt[3]{-27} = -3$. Now we will extend what we learned about multiplying, dividing, and simplifying *square* roots to doing the same with higher roots.

### Definition    Product and Quotient Rules for Higher Roots

If $a$ and $b$ are real numbers such that the roots exist, then

$$\sqrt[n]{a} \cdot \sqrt[n]{b} = \sqrt[n]{a \cdot b} \qquad \text{and} \qquad \sqrt[n]{\frac{a}{b}} = \frac{\sqrt[n]{a}}{\sqrt[n]{b}} \quad (b \neq 0)$$

These rules enable us to multiply, divide, and simplify radicals with any index in a way that is similar to multiplying, dividing, and simplifying square roots.

### Example 7

Multiply.
a)  $\sqrt[3]{2} \cdot \sqrt[3]{5}$        b)  $\sqrt[4]{7} \cdot \sqrt[4]{y}$

**Solution**

a)  $\sqrt[3]{2} \cdot \sqrt[3]{5} = \sqrt[3]{2 \cdot 5} = \sqrt[3]{10}$        b)  $\sqrt[4]{7} \cdot \sqrt[4]{y} = \sqrt[4]{7 \cdot y} = \sqrt[4]{7y}$

### You Try 7

Multiply.
a)  $\sqrt[3]{4} \cdot \sqrt[3]{3}$        b)  $\sqrt[4]{9} \cdot \sqrt[4]{d}$

**BE CAREFUL**   Remember that we can apply the product rule *only* if the indices of the radicals are the same.

Let's see how we can use the product rule to simplify higher roots.

**Example 8**

Simplify completely.

a) $\sqrt[3]{40}$          b) $\sqrt[4]{48}$          c) $\dfrac{\sqrt[3]{72}}{\sqrt[3]{3}}$

**Solution**

a) Because we are simplifying a *cube* root, $\sqrt[3]{40}$, think of two numbers whose product is 40 with one of those factors being a perfect *cube*. Those numbers are 8 and 5.

$$\sqrt[3]{40} = \sqrt[3]{8 \cdot 5} \qquad \text{Rewrite 40 as } 8 \cdot 5.$$
$$= \sqrt[3]{8} \cdot \sqrt[3]{5} \qquad \text{Product rule}$$
$$= 2\sqrt[3]{5} \qquad \text{Simplify.}$$

b) To simplify a *fourth* root, $\sqrt[4]{48}$, think of two numbers whose product is 48 with one of those factors being a perfect *fourth power*. Those numbers are 16 and 3.

$$\sqrt[4]{48} = \sqrt[4]{16 \cdot 3} \qquad \text{Rewrite 48 as } 16 \cdot 3.$$
$$= \sqrt[4]{16} \cdot \sqrt[4]{3} \qquad \text{Product rule}$$
$$= 2\sqrt[4]{3} \qquad \text{Simplify.}$$

c) Begin by applying the quotient rule.

$$\frac{\sqrt[3]{72}}{\sqrt[3]{3}} = \sqrt[3]{\frac{72}{3}} \qquad \text{Quotient rule}$$
$$= \sqrt[3]{24} \qquad \text{Divide.}$$

Is $\sqrt[3]{24}$ completely simplified? If it is in simplest form, then 24 will not contain any factors that are perfect cubes. But $24 = 8 \cdot 3$, and 8 is a perfect cube. Therefore, $\sqrt[3]{24}$ is not in simplest form. We must continue to simplify.

$$= \sqrt[3]{8} \cdot \sqrt[3]{3} \qquad \text{Product rule}$$
$$= 2\sqrt[3]{3} \qquad \text{Simplify.} \qquad \blacksquare$$

 **You Try 8**

Simplify completely.

a) $\sqrt[3]{80}$          b) $\sqrt[4]{32}$          c) $\dfrac{\sqrt[3]{54}}{\sqrt[3]{2}}$

 **Note**

The expression $\sqrt[n]{a}$ is simplified when $a$ does not contain any factors (other than 1) that are perfect $n$th powers and when the radicand does not contain any fractions. This implies that the radicand cannot contain variables with exponents greater than or equal to $n$.

Next let's simplify higher roots containing variables. Recall that we are assuming that $a > 0$ so that $\sqrt[n]{a^n} = a$.

## Example 9

Simplify completely.

a) $\sqrt[3]{k^6}$     b) $\sqrt[3]{125h^{12}}$     c) $\sqrt[3]{\dfrac{z^3}{64}}$

d) $\sqrt[3]{m^5}$     e) $\sqrt[3]{54x^{13}}$     f) $\sqrt[4]{d^8}$

### Solution

a)  Since we are trying to simplify a *cube* root, ask yourself, *"What do I cube to get $k^6$?"* $k^2$. Therefore, $\sqrt[3]{k^6} = k^2$.

Or, we can think of simplifying in this way:

$$\sqrt[3]{k^6} = \sqrt[3]{(k^2)^3} \qquad \text{Write } k^6 \text{ as a perfect cube.}$$
$$= k^2 \qquad\qquad \text{Simplify.}$$

b)  $\sqrt[3]{125h^{12}} = \sqrt[3]{125} \cdot \sqrt[3]{h^{12}} \qquad$ Product rule
$$= 5 \cdot \sqrt[3]{(h^4)^3} \qquad \text{Simplify, write } h^{12} \text{ as a perfect cube.}$$
$$= 5h^4 \qquad\qquad \text{Simplify.}$$

c)  $\sqrt[3]{\dfrac{z^3}{64}} = \dfrac{\sqrt[3]{z^3}}{\sqrt[3]{64}} \qquad$ Quotient rule
$$= \dfrac{z}{4} \qquad\qquad \text{Simplify.}$$

d)  We are trying to simplify a *cube root* expression, $\sqrt[3]{m^5}$, but $m^5$ is not a perfect *cube*. Therefore, write $m^5$ as a product so that one of the factors is a perfect *cube*. We can write $m^5$ as $m^3 \cdot m^2$.

$$\sqrt[3]{m^5} = \sqrt[3]{m^3 \cdot m^2}$$
$$= \sqrt[3]{m^3} \cdot \sqrt[3]{m^2} \qquad \text{Product rule}$$
$$= m\sqrt[3]{m^2} \qquad\qquad \text{Simplify.}$$

The radicand in the expression $m\sqrt[3]{m^2}$ is completely simplified because the exponent is less than the index of the radical.

e)  In the radicand of $\sqrt[3]{54x^{13}}$, $x^{13}$ is not a perfect *cube*. We will need to write $x^{13}$ as a product so that one of the factors is a perfect *cube*. $x^{13} = x^{12} \cdot x$

$$\sqrt[3]{54x^{13}} = \sqrt[3]{54} \cdot \sqrt[3]{x^{13}} \qquad\qquad \text{Product rule}$$
$$= (\sqrt[3]{27} \cdot \sqrt[3]{2}) \cdot \sqrt[3]{x^{12} \cdot x} \qquad \text{Product rule}$$
$$= 3\sqrt[3]{2} \cdot \sqrt[3]{x^{12}} \cdot \sqrt[3]{x} \qquad\quad \text{Simplify; product rule}$$
$$= 3\sqrt[3]{2} \cdot \sqrt[3]{(x^4)^3} \cdot \sqrt[3]{x} \qquad \text{Write } x^{12} \text{ as a perfect cube.}$$
$$= 3\sqrt[3]{2} \cdot x^4 \cdot \sqrt[3]{x} \qquad\quad \text{Simplify.}$$
$$= 3x^4\sqrt[3]{2x} \qquad\qquad\qquad \text{Product rule}$$

f)  $\sqrt[4]{d^8} = \sqrt[4]{(d^2)^4} \qquad$ Write $d^8$ as a perfect fourth power.
$$= d^2 \qquad\qquad \text{Simplify.}$$

### You Try 9

Simplify completely.

a) $\sqrt[3]{y^9}$     b) $\sqrt[3]{1000t^{18}}$     c) $\sqrt[3]{\dfrac{a^6}{8}}$     d) $\sqrt[3]{w^7}$

e) $\sqrt[3]{72b^{17}}$     f) $\sqrt[4]{q^{12}}$

### Answers to You Try Exercises

1) a) $\sqrt{6}$  b) $\sqrt{6z}$    2) a) $2\sqrt{5}$  b) $5\sqrt{6}$  c) $\sqrt{33}$  d) $2\sqrt{6}$    3) a) $3\sqrt{5}$  b) $2\sqrt{10}$

4) a) $\dfrac{7}{12}$  b) 6  c) $2\sqrt{10}$  d) $\dfrac{\sqrt{6}}{5}$  e) $7\sqrt{5}$    5) $\dfrac{\sqrt{3}}{4}$

6) a) $n^4$  b) $8z^7$  c) $3b^5\sqrt{7}$  d) $\dfrac{2\sqrt{3}}{r^6}$  e) $h^2\sqrt{h}$    7) a) $\sqrt[3]{12}$  b) $\sqrt[4]{9d}$

8) a) $2\sqrt[3]{10}$  b) $2\sqrt[4]{2}$  c) 3    9) a) $y^3$  b) $10t^6$  c) $\dfrac{a^2}{2}$  d) $w^2\sqrt[3]{w}$  e) $2b^5\sqrt[3]{9b^2}$  f) $q^3$

## 9.2 Exercises

**Objective 1: Multiply Square Roots**

Multiply and simplify.

1) $\sqrt{2} \cdot \sqrt{3}$
2) $\sqrt{5} \cdot \sqrt{7}$
3) $\sqrt{13} \cdot \sqrt{10}$
4) $\sqrt{2} \cdot \sqrt{k}$
5) $\sqrt{15} \cdot \sqrt{n}$
6) $\sqrt{7} \cdot \sqrt{r}$

**Objective 2: Simplify the Square Root of a Whole Number**

Label each statement as true or false. Give a reason for your answer.

7) $\sqrt{24}$ is in simplest form.

8) $\sqrt{30}$ is in simplest form.

Simplify completely. If the radical is already simplified, then say so.

**Fill It In**

Fill in the blanks with either the missing mathematical step or reason for the given step.

9) $\sqrt{60} = \sqrt{4 \cdot 15}$       _____

   $= $ _____       Product rule

   $= 2\sqrt{15}$       _____

10) $\sqrt{200} = $ _____       Factor.

    $= \sqrt{100} \cdot \sqrt{2}$       _____

    $= $ _____       Simplify.

11) $\sqrt{20}$
12) $\sqrt{8}$
13) $\sqrt{90}$
14) $\sqrt{24}$
15) $\sqrt{21}$
16) $\sqrt{42}$
17) $-\sqrt{75}$
18) $-\sqrt{54}$
19) $3\sqrt{80}$
20) $-5\sqrt{96}$
21) $-\sqrt{1600}$
22) $\sqrt{400}$

Multiply and simplify.

23) $\sqrt{6} \cdot \sqrt{2}$
24) $\sqrt{5} \cdot \sqrt{15}$
25) $\sqrt{12} \cdot \sqrt{3}$
26) $\sqrt{8} \cdot \sqrt{2}$
27) $\sqrt{20} \cdot \sqrt{3}$
28) $\sqrt{21} \cdot \sqrt{3}$
29) $7\sqrt{6} \cdot \sqrt{12}$
30) $5\sqrt{6} \cdot \sqrt{20}$
31) $\sqrt{20} \cdot \sqrt{18}$
32) $\sqrt{50} \cdot \sqrt{12}$

**Objective 3: Use the Quotient Rule for Square Roots**

Simplify completely.

33) $\sqrt{\dfrac{9}{16}}$
34) $\sqrt{\dfrac{2}{72}}$
35) $\dfrac{\sqrt{32}}{\sqrt{2}}$
36) $\dfrac{\sqrt{54}}{\sqrt{6}}$
37) $\sqrt{\dfrac{60}{5}}$
38) $\dfrac{\sqrt{140}}{\sqrt{7}}$
39) $-\sqrt{\dfrac{3}{64}}$
40) $-\sqrt{\dfrac{6}{49}}$
41) $\sqrt{\dfrac{52}{49}}$
42) $\sqrt{\dfrac{45}{121}}$
43) $\dfrac{16\sqrt{63}}{2\sqrt{3}}$
44) $\dfrac{48\sqrt{50}}{12\sqrt{5}}$
45) $\dfrac{10\sqrt{80}}{15\sqrt{2}}$
46) $\dfrac{28\sqrt{54}}{35\sqrt{3}}$
47) $\sqrt{\dfrac{1}{2}} \cdot \sqrt{\dfrac{5}{8}}$
48) $\sqrt{\dfrac{2}{3}} \cdot \sqrt{\dfrac{1}{27}}$
49) $\sqrt{\dfrac{2}{7}} \cdot \sqrt{\dfrac{50}{7}}$
50) $\sqrt{\dfrac{12}{11}} \cdot \sqrt{\dfrac{3}{11}}$

## Objective 4: Simplify Square Root Expressions Containing Variables

Simplify completely.

51) $\sqrt{c^2}$

52) $\sqrt{a^4}$

53) $\sqrt{t^6}$

54) $\sqrt{w^8}$

55) $\sqrt{121b^{10}}$

56) $\sqrt{9r^{12}}$

57) $\sqrt{36q^{14}}$

58) $\sqrt{144n^{16}}$

VIDEO 59) $\sqrt{28r^4}$

60) $\sqrt{75n^4}$

61) $\sqrt{18z^{12}}$

62) $\sqrt{500q^{22}}$

63) $\sqrt{\dfrac{y^8}{169}}$

64) $\sqrt{\dfrac{k^6}{49}}$

65) $\sqrt{\dfrac{99}{w^2}}$

66) $\dfrac{\sqrt{45}}{\sqrt{m^{18}}}$

67) $\sqrt{r^4 s^{12}}$

68) $\sqrt{c^8 d^2}$

69) $\sqrt{36x^{10}y^2}$

70) $\sqrt{81a^6 b^{16}}$

71) $\sqrt{\dfrac{m^{18}n^8}{49}}$

72) $\sqrt{\dfrac{u^{14}v^4}{900}}$

Simplify completely.

### Fill It In

Fill in the blanks with either the missing mathematical step or reason for the given step.

73) $\sqrt{p^9} = \sqrt{p^8 \cdot p^1}$ ____

   $=$ _____    Product rule

   $= p^4\sqrt{p}$    ____

74) $\sqrt{x^{19}} = \sqrt{x^{18} \cdot x^1}$ ____

   $= \sqrt{x^{18}} \cdot \sqrt{x^1}$ _____

   $=$ ____    Simplify.

75) $\sqrt{h^3}$

76) $\sqrt{a^5}$

77) $\sqrt{g^{13}}$

78) $\sqrt{k^{11}}$

79) $\sqrt{100w^5}$

80) $\sqrt{144z^3}$

VIDEO 81) $\sqrt{75t^{11}}$

82) $\sqrt{20c^9}$

83) $\sqrt{13q^7}$

84) $\sqrt{45p^{17}}$

85) $\sqrt{x^2 y^9}$

86) $\sqrt{a^4 b^3}$

87) $\sqrt{4t^9 u^5}$

88) $\sqrt{36m^7 n^{11}}$

## Objective 5: Simplify Higher Roots

89) How do you know that a radical expression containing a cube root is completely simplified?

90) How do you know that a radical expression containing a fourth root is completely simplified?

Multiply.

91) $\sqrt[3]{2} \cdot \sqrt[3]{6}$

92) $\sqrt[3]{3} \cdot \sqrt[3]{7}$

93) $\sqrt[4]{9} \cdot \sqrt[4]{n}$

94) $\sqrt[4]{13} \cdot \sqrt[4]{p^3}$

Simplify completely.

VIDEO 95) $\sqrt[3]{24}$

96) $\sqrt[3]{48}$

97) $\sqrt[3]{72}$

98) $\sqrt[3]{81}$

99) $\sqrt[3]{108}$

100) $\sqrt[3]{4000}$

101) $\sqrt[4]{48}$

102) $\sqrt[4]{64}$

103) $\sqrt[4]{162}$

104) $\sqrt[4]{243}$

105) $\sqrt[3]{\dfrac{1}{64}}$

106) $\sqrt[3]{-\dfrac{54}{2}}$

107) $\sqrt[4]{\dfrac{1}{16}}$

108) $\sqrt[4]{\dfrac{64}{4}}$

109) $\dfrac{\sqrt[3]{500}}{\sqrt[3]{2}}$

110) $\dfrac{\sqrt[3]{120}}{\sqrt[3]{5}}$

111) $\dfrac{\sqrt[3]{8000}}{\sqrt[3]{4}}$

112) $\dfrac{\sqrt[3]{240}}{\sqrt[3]{3}}$

Simplify completely.

113) $\sqrt[3]{r^3}$

114) $\sqrt[3]{h^9}$

115) $\sqrt[3]{w^{12}}$

116) $\sqrt[3]{x^6}$

117) $\sqrt[3]{27d^{15}}$

118) $\sqrt[3]{1000p^9}$

119) $\sqrt[3]{125a^{18}b^6}$

120) $\sqrt[3]{8m^3 n^{12}}$

121) $\sqrt[3]{\dfrac{t^9}{8}}$

122) $\sqrt[3]{\dfrac{k^6}{64}}$

123) $\sqrt[3]{t^4}$

124) $\sqrt[3]{a^8}$

125) $\sqrt[3]{x^{11}}$

126) $\sqrt[3]{p^{13}}$

127) $\sqrt[3]{b^{17}}$

128) $\sqrt[3]{w^{10}}$

129) $\sqrt[3]{125g^6}$

130) $\sqrt[3]{64u^{18}}$

131) $\sqrt[3]{27x^3 y^{12}}$

132) $\sqrt[3]{1000p^{15}q^9}$

133) $\sqrt[3]{24k^7}$

134) $\sqrt[3]{81r^{11}}$

135) $\sqrt[4]{n^4}$

136) $\sqrt[4]{z^{12}}$

137) $\sqrt[4]{m^5}$

138) $\sqrt[4]{y^9}$

139) If $A$ is the area of an equilateral triangle, then the length of a side, $s$, is given by $s = \sqrt{\dfrac{4\sqrt{3}A}{3}}$. If an equilateral triangle has an area of $2\sqrt{3}$ in$^2$, how long is each side of the triangle?

140) If $V$ is the volume of a sphere, then the radius of the sphere, $r$, is given by $r = \sqrt[3]{\dfrac{3V}{4\pi}}$. If a ball has a volume of $36\pi$ in$^3$, find its radius.

# Section 9.3 Adding and Subtracting Radicals

## Objectives

1. **Add and Subtract Radical Expressions**
2. **Add and Subtract Radical Expressions Containing Variables**

Just as we can add and subtract like terms such as $2x + 5x = 7x$, we can add and subtract *like radicals* such as $2\sqrt{3} + 5\sqrt{3}$.

### Note

**Like radicals** have the same index and the same radicand.

Some examples of like radicals are

$$2\sqrt{3} \text{ and } 5\sqrt{3}, \quad -\sqrt[3]{2} \text{ and } 9\sqrt[3]{2}, \quad \sqrt{x} \text{ and } 4\sqrt{x}, \quad \text{and } 6\sqrt[3]{a^2} \text{ and } 8\sqrt[3]{a^2}$$

## 1. Add and Subtract Radical Expressions

### Note

In order to add or subtract radicals, they must be *like* radicals.

We add and subtract like radicals in the same way we add and subtract like terms—add or subtract the "coefficients" of the radicals and multiply that result by the radical. Recall that we are using the distributive property when we are combining like terms in this way.

### Example 1

Add or subtract.

a) $2x + 5x$
b) $2\sqrt{3} + 5\sqrt{3}$
c) $\sqrt{6} - 4\sqrt{6}$
d) $\sqrt{5} - \sqrt{2}$
e) $\sqrt[3]{7} + \sqrt[3]{7}$

### Solution

a) First notice that $2x$ and $5x$ are like terms. Therefore, they can be added.

$$2x + 5x = (2 + 5)x \qquad \text{Distributive property}$$
$$= 7x \qquad \text{Simplify.}$$

Or, we can say that by just adding the coefficients, $2x + 5x = 7x$.

b) We can add $2\sqrt{3}$ and $5\sqrt{3}$ because they are like radicals.

$$2\sqrt{3} + 5\sqrt{3} = (2 + 5)\sqrt{3} \qquad \text{Distributive property}$$
$$= 7\sqrt{3} \qquad \text{Simplify.}$$

c) $\sqrt{6} - 4\sqrt{6}$

$$= 1\sqrt{6} - 4\sqrt{6}$$
$$= (1 - 4)\sqrt{6}$$
$$= -3\sqrt{6}$$

d) We cannot find the difference $\sqrt{5} - \sqrt{2}$ using the distributive property because $\sqrt{5}$ and $\sqrt{2}$ are not like radicals.

e) $\sqrt[3]{7} + \sqrt[3]{7}$

$$= 1\sqrt[3]{7} + 1\sqrt[3]{7} = (1 + 1)\sqrt[3]{7} = 2\sqrt[3]{7}$$

### You Try 1

Add or subtract.

a) $4t + 6t$
b) $4\sqrt{11} + 6\sqrt{11}$
c) $\sqrt{5} - 9\sqrt{5}$
d) $\sqrt{7} - \sqrt{3}$
e) $\sqrt[3]{4} + \sqrt[3]{4}$

**Example 2**

Perform the operations and simplify. $3 + 9\sqrt{2} - 11 + \sqrt{2}$

**Solution**

Begin by writing like terms together.

$$\begin{aligned}
3 + 9\sqrt{2} - 11 + \sqrt{2} &= 3 - 11 + 9\sqrt{2} + \sqrt{2} && \text{Commutative property} \\
&= -8 + (9 + 1)\sqrt{2} && \text{Subtract; distributive property} \\
&= -8 + 10\sqrt{2} && \text{Add.}
\end{aligned}$$

Is $-8 + 10\sqrt{2}$ in simplest form? *Yes.* The terms are not like so they cannot be combined further, *and* $\sqrt{2}$ is in simplest form.  ■

**You Try 2**

Perform the operations and simplify.  $-5 + 10\sqrt{7} - 1 - \sqrt{7}$

Sometimes it looks like two radicals cannot be added or subtracted. But if the radicals can be *simplified* and they turn out to be *like* radicals, then we can add or subtract them.

---

**Procedure  Steps for Adding and Subtracting Radicals**

**Step 1:**  Write each radical expression in simplest form.

**Step 2:**  Combine like radicals.

---

**Example 3**

Perform the operations and simplify.

a)  $\sqrt{12} + 5\sqrt{3}$    b)  $6\sqrt{18} + 3\sqrt{50} - \sqrt{45}$    c)  $-7\sqrt[3]{40} + \sqrt[3]{5}$

**Solution**

a)  $\sqrt{12}$ and $5\sqrt{3}$ are not like radicals. Can either radical be simplified? *Yes.* We can simplify $\sqrt{12}$.

$$\begin{aligned}
\sqrt{12} + 5\sqrt{3} &= \sqrt{4 \cdot 3} + 5\sqrt{3} && \text{4 is a perfect square.} \\
&= \sqrt{4} \cdot \sqrt{3} + 5\sqrt{3} && \text{Product rule} \\
&= 2\sqrt{3} + 5\sqrt{3} && \sqrt{4} = 2 \\
&= 7\sqrt{3} && \text{Add like radicals.}
\end{aligned}$$

b)  $6\sqrt{18}, 3\sqrt{50}$, and $\sqrt{45}$ are not like radicals. In this case, *each* radical should be simplified to determine whether they can be combined.

$$\begin{aligned}
6\sqrt{18} + 3\sqrt{50} - \sqrt{45} &= 6\sqrt{9 \cdot 2} + 3\sqrt{25 \cdot 2} - \sqrt{9 \cdot 5} && \text{Factor.} \\
&= 6\sqrt{9} \cdot \sqrt{2} + 3\sqrt{25} \cdot \sqrt{2} - \sqrt{9} \cdot \sqrt{5} && \text{Product rule} \\
&= 6 \cdot 3 \cdot \sqrt{2} + 3 \cdot 5 \cdot \sqrt{2} - 3\sqrt{5} && \text{Simplify radicals.} \\
&= 18\sqrt{2} + 15\sqrt{2} - 3\sqrt{5} && \text{Multiply.} \\
&= 33\sqrt{2} - 3\sqrt{5} && \text{Add like radicals.}
\end{aligned}$$

$33\sqrt{2} - 3\sqrt{5}$ is in simplest form since they aren't like expressions.

c)  $$\begin{aligned}
-7\sqrt[3]{40} + \sqrt[3]{5} &= -7\sqrt[3]{8 \cdot 5} + \sqrt[3]{5} && \text{8 is a perfect cube.} \\
&= -7\sqrt[3]{8} \cdot \sqrt[3]{5} + \sqrt[3]{5} && \text{Product rule} \\
&= -7 \cdot 2 \cdot \sqrt[3]{5} + \sqrt[3]{5} && \sqrt[3]{8} = 2 \\
&= -14\sqrt[3]{5} + \sqrt[3]{5} && \text{Multiply.} \\
&= -13\sqrt[3]{5} && \text{Add like radicals.}
\end{aligned}$$  ■

**You Try 3**

Perform the operations and simplify.

a)  $7\sqrt{3} - \sqrt{12}$   b)  $2\sqrt{63} - 11\sqrt{28} + 2\sqrt{21}$   c)  $\sqrt[3]{54} + 5\sqrt[3]{16}$

## 2. Add and Subtract Radical Expressions Containing Variables

Next we will add and subtract radicals containing variables.

**Example 4**

Perform the operations and simplify.

a)  $6\sqrt{x} - \sqrt{x}$   b)  $\sqrt{8k} + 3\sqrt{2k}$   c)  $\sqrt{3c^2} - 5c\sqrt{12}$

d)  $9r\sqrt[3]{24r^3} + 2\sqrt[3]{3r^6}$

**Solution**

a)  These are like radicals, so use the distributive property.

$$6\sqrt{x} - \sqrt{x} = (6 - 1)\sqrt{x} = 5\sqrt{x}$$

b)  Simplify the first radical to see whether the two radicals can then be added.

$$\begin{aligned}
\sqrt{8k} + 3\sqrt{2k} &= \sqrt{4} \cdot \sqrt{2k} + 3\sqrt{2k} && \text{Product rule: 4 is a perfect square.} \\
&= 2\sqrt{2k} + 3\sqrt{2k} && \sqrt{4} = 2 \\
&= 5\sqrt{2k} && \text{Add like radicals.}
\end{aligned}$$

c)  Simplify each radical to determine whether they can be combined.

$$\begin{aligned}
\sqrt{3c^2} - 5c\sqrt{12} &= \sqrt{3} \cdot \sqrt{c^2} - 5c \cdot \sqrt{4} \cdot \sqrt{3} && \text{Product rule} \\
&= \sqrt{3} \cdot c - 5c \cdot 2 \cdot \sqrt{3} && \text{Simplify.} \\
&= c\sqrt{3} - 10c\sqrt{3} && \text{Multiply.} \\
&= -9c\sqrt{3} && \text{Subtract like radicals.}
\end{aligned}$$

d)  Begin by simplifying each radical.

$$\begin{aligned}
9r\sqrt[3]{24r^3} &= 9r \cdot \sqrt[3]{24} \cdot \sqrt[3]{r^3} && \text{Product rule} \\
&= 9r \cdot \sqrt[3]{8} \cdot \sqrt[3]{3} \cdot r && \text{Product rule; simplify.} \\
&= 9r \cdot 2\sqrt[3]{3} \cdot r && \sqrt[3]{8} = 2 \\
&= 18r^2\sqrt[3]{3} && \text{Multiply.}
\end{aligned}$$

$$\begin{aligned}
2\sqrt[3]{3r^6} &= 2\sqrt[3]{3} \cdot \sqrt[3]{r^6} && \text{Product rule} \\
&= 2\sqrt[3]{3} \cdot r^2 && \text{Simplify.} \\
&= 2r^2\sqrt[3]{3} && \text{Multiply.}
\end{aligned}$$

Substitute the simplified radicals into the original expression.

$$\begin{aligned}
9r\sqrt[3]{24r^3} + 2\sqrt[3]{3r^6} &= 18r^2\sqrt[3]{3} + 2r^2\sqrt[3]{3} && \text{Substitute.} \\
&= 20r^2\sqrt[3]{3} && \text{Add like radicals.}
\end{aligned}$$

**You Try 4**

Perform the operations and simplify.

a)  $3\sqrt{p} - 5\sqrt{p}$   b)  $\sqrt{45a} + 7\sqrt{5a}$

c)  $y\sqrt{24} - \sqrt{6y^2}$   d)  $10\sqrt[3]{2n^3} + 4n\sqrt[3]{16}$

**Answers to You Try Exercises**

1) a) $10t$  b) $10\sqrt{11}$  c) $-8\sqrt{5}$  d) $\sqrt{7} - \sqrt{3}$  e) $2\sqrt[3]{4}$    2) $-6 + 9\sqrt{7}$

3) a) $5\sqrt{3}$  b) $-16\sqrt{7} + 2\sqrt{21}$  c) $13\sqrt[3]{2}$    4) a) $-2\sqrt{p}$  b) $10\sqrt{5a}$  c) $y\sqrt{6}$  d) $18n\sqrt[3]{2}$

# 9.3 Exercises

**Objective 1: Add and Subtract Radical Expressions**

1) How do you know whether two radicals are *like* radicals?

2) What are the steps for adding or subtracting radicals?

Perform the operations and simplify.

3) $4\sqrt{11} + 5\sqrt{11}$

4) $14\sqrt{3} + 7\sqrt{3}$

5) $4\sqrt{2} - 9\sqrt{2}$

6) $14\sqrt{7} - 23\sqrt{7}$

7) $9\sqrt[3]{5} - 2\sqrt[3]{5}$

8) $4\sqrt[3]{4} + 6\sqrt[3]{4}$

9) $11\sqrt[3]{2} + 7\sqrt[3]{2}$

10) $\sqrt[3]{10} - 9\sqrt[3]{10}$

11) $4 - \sqrt{13} + 8 - 6\sqrt{13}$

12) $-3 + 5\sqrt{6} - 4\sqrt{6} + 9$

**Fill It In**

Fill in the blanks with either the missing mathematical step or reason for the given step.

13) $\sqrt{24} + \sqrt{6} = \sqrt{4 \cdot 6} + \sqrt{6}$    _____

   $= $ _____    Product rule

   $= 2\sqrt{6} + \sqrt{6}$    _____

   $= $ _____    Add like radicals.

14) $\sqrt{50} - 9\sqrt{2} = \sqrt{25 \cdot 2} - 9\sqrt{2}$    _____

   $= \sqrt{25} \cdot \sqrt{2} - 9\sqrt{2}$    _____

   $= $ _____    Simplify.

   $= $ _____    Subtract like radicals.

15) $6\sqrt{3} - \sqrt{12}$

16) $\sqrt{45} + 6\sqrt{5}$

17) $\sqrt{75} + \sqrt{3}$

18) $\sqrt{44} - 8\sqrt{11}$

19) $\sqrt{28} - 3\sqrt{63}$

20) $3\sqrt{45} + \sqrt{20}$

21) $3\sqrt{72} - 4\sqrt{8}$

22) $3\sqrt{98} + 4\sqrt{50}$

23) $\sqrt{32} - 3\sqrt{18}$

24) $\sqrt{96} + 4\sqrt{24}$

25) $\dfrac{5}{2}\sqrt{40} + \dfrac{1}{3}\sqrt{90}$

26) $\dfrac{2}{3}\sqrt{180} - \dfrac{6}{7}\sqrt{245}$

27) $\dfrac{4}{3}\sqrt{18} - \dfrac{5}{8}\sqrt{128}$

28) $\dfrac{4}{5}\sqrt{150} - \dfrac{4}{3}\sqrt{54}$

29) $\sqrt{50} - \sqrt{2} + \sqrt{98}$

30) $\sqrt{3} - \sqrt{12} + \sqrt{75}$

31) $\sqrt{96} + 3\sqrt{24} - \sqrt{54}$  32) $\sqrt{20} - \sqrt{45} - 2\sqrt{80}$

33) $2\sqrt[3]{11} + 5\sqrt[3]{88}$

34) $6\sqrt[3]{9} + \sqrt[3]{72}$

35) $8\sqrt[3]{3} - 3\sqrt[3]{81}$

36) $2\sqrt[3]{81} - 12\sqrt[3]{3}$

37) $11\sqrt[3]{16} + 10\sqrt[3]{2}$

38) $\sqrt[3]{6} - \sqrt[3]{48}$

**Objective 2: Add and Subtract Radical Expressions Containing Variables**

Perform the operations and simplify.

39) $12\sqrt{c} + 3\sqrt{c}$

40) $-7\sqrt{y} - 2\sqrt{y}$

41) $5\sqrt{3a} - 9\sqrt{3a}$

42) $6\sqrt{5t} + \sqrt{5t}$

43) $\sqrt{5b} + \sqrt{45b}$

44) $9\sqrt{2w} + \sqrt{32w}$

45) $\sqrt{50n} - \sqrt{18n}$

46) $\sqrt{12k} + \sqrt{48k}$

47) $11\sqrt{3z} + 2\sqrt{12z}$

48) $8\sqrt{5r} - 5\sqrt{20r}$

49) $5\sqrt{63v} + 6\sqrt{7v}$

50) $2\sqrt{8p} - 6\sqrt{2p}$

51) $\sqrt{4h} - 8\sqrt{8p} + \sqrt{4h} + 6\sqrt{8p}$

52) $6\sqrt{3m} + 10\sqrt{2m} + 8\sqrt{3m} - \sqrt{2m}$

53) $9z\sqrt{12} - \sqrt{3z^2}$

54) $2m\sqrt{45} - \sqrt{5m^2}$

55) $8\sqrt{7r^2} + 3r\sqrt{28}$

56) $\sqrt{2x^2} - 7x\sqrt{50}$

57) $6q\sqrt{q} + 7\sqrt{q^3}$

58) $9\sqrt{r^3} + 3r\sqrt{r}$

59) $16m^3\sqrt{m} - 13\sqrt{m^7}$

60) $9d^2\sqrt{d} - 25\sqrt{d^5}$

61) $8w\sqrt{w^5} + 4\sqrt{w^7}$

62) $18\sqrt{q^5} - 4q\sqrt{q^3}$

63) $\sqrt{xy^3} + 6y\sqrt{xy}$

64) $7a\sqrt{ab} + 3\sqrt{a^3b}$

65) $9v\sqrt{6u^3} - 2u\sqrt{54uv^2}$

66) $6c^2\sqrt{8d^3} - 9d\sqrt{2c^4d}$

67) $3\sqrt{75m^3n} + m\sqrt{12mn}$

68) $y\sqrt{54xy} - 6\sqrt{24xy^3}$  69) $15\sqrt[3]{t^2} - 25\sqrt[3]{t^2}$

70) $10\sqrt[3]{m} - 6\sqrt[3]{m}$

71) $5\sqrt[3]{27x^2} + \sqrt[3]{8x^2}$

72) $3\sqrt[3]{64k^2} - 9\sqrt[3]{125k^2}$  73) $7r^5\sqrt[3]{r} - 14\sqrt[3]{r^{16}}$

74) $6t^3\sqrt[3]{t} - 5\sqrt[3]{t^{10}}$

75) $7\sqrt[3]{81a^5} + 4a\sqrt[3]{3a^2}$

76) $3\sqrt[3]{40x} - 12\sqrt[3]{5x}$    77) $2c^2\sqrt[3]{108c} - 12\sqrt[3]{32c^7}$

78) $9\sqrt[3]{128h^2} + 4\sqrt[3]{16h^2}$

79) $3p^2\sqrt[3]{88pq^5} + 8p^2q\sqrt[3]{11pq^2}$

80) $18f^5\sqrt[3]{7f^2g} + 2f^3\sqrt[3]{7f^8g}$

81) $3\sqrt[4]{x^4} + \sqrt[4]{16x^4}$

82) $\sqrt[4]{81q^4} - 2\sqrt[4]{q^4}$

83) $-3\sqrt[4]{c^{15}} + 8c^3\sqrt[4]{c^3}$

84) $5a\sqrt[4]{a^7} + \sqrt[4]{a^{11}}$

85) $\sqrt[3]{y^2} + 11\sqrt[5]{y^2} - 9\sqrt[3]{y^2} + \sqrt[5]{y^2}$

86) $5\sqrt[4]{g} - 2\sqrt[3]{g} + 4\sqrt[3]{g} + 3\sqrt[4]{g}$

87) $7yz^2\sqrt[4]{11y^4z} + 3z\sqrt[4]{11y^8z^5}$

88) $15cd\sqrt[4]{9cd} - \sqrt[4]{9c^5d^5}$

Find the perimeter of each figure.

89)

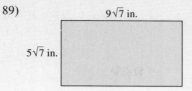

9√7 in.

5√7 in.

90)

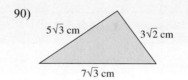

5√3 cm

3√2 cm

7√3 cm

## Section 9.4 Combining Operations on Radicals

### Objectives
1. **Multiply Radical Expressions**
2. **Multiply Radical Expressions Using FOIL**
3. **Square a Binomial Containing Radical Expressions**
4. **Multiply Two Binomials of the Form $(a + b)(a - b)$ Containing Radicals**

In Section 9.2, we learned to multiply radicals like $\sqrt{6} \cdot \sqrt{2}$. In this section, we will learn how to simplify expressions that combine multiplication, addition, and subtraction of radicals. Remember, we will assume that all variables represent positive real numbers.

### 1. Multiply Radical Expressions

> **Example 1**

Multiply and simplify.

a)  $6(\sqrt{3} - \sqrt{12})$        b)  $\sqrt{2}(\sqrt{14} + \sqrt{5})$        c)  $\sqrt{a}(\sqrt{a} + \sqrt{50b})$

**Solution**

a)  Since $\sqrt{12}$ can be simplified, we will do that first.

$$\sqrt{12} = \sqrt{4 \cdot 3} = \sqrt{4} \cdot \sqrt{3} = 2\sqrt{3}$$

Substitute $2\sqrt{3}$ for $\sqrt{12}$ in the original expression.

$$\begin{aligned} 6(\sqrt{3} - \sqrt{12}) &= 6(\sqrt{3} - 2\sqrt{3}) && \text{Substitute } 2\sqrt{3} \text{ for } \sqrt{12}. \\ &= 6(-\sqrt{3}) && \text{Subtract.} \\ &= -6\sqrt{3} && \text{Multiply.} \end{aligned}$$

b)  Neither $\sqrt{14}$ nor $\sqrt{5}$ can be simplified. Apply the distributive property.

$$\begin{aligned} \sqrt{2}(\sqrt{14} + \sqrt{5}) &= \sqrt{2} \cdot \sqrt{14} + \sqrt{2} \cdot \sqrt{5} && \text{Distribute.} \\ &= \sqrt{28} + \sqrt{10} && \text{Product rule} \end{aligned}$$

Is $\sqrt{28} + \sqrt{10}$ in simplest form? *No.* $\sqrt{28}$ can be simplified.

$$\begin{aligned} &= \sqrt{4 \cdot 7} + \sqrt{10} && \text{4 is a perfect square.} \\ &= \sqrt{4} \cdot \sqrt{7} + \sqrt{10} && \text{Product rule} \\ &= 2\sqrt{7} + \sqrt{10} && \sqrt{4} = 2 \end{aligned}$$

The sum is in simplest form. The radicals are not like, so they cannot be combined.

c)  Since $\sqrt{50b}$ can be simplified, we will do that first.

$$\begin{aligned} \sqrt{50b} &= \sqrt{50} \cdot \sqrt{b} && \text{Product rule} \\ &= \sqrt{25} \cdot \sqrt{2} \cdot \sqrt{b} && \text{Product rule; 25 is a perfect square.} \\ &= 5\sqrt{2b} && \text{Simplify; multiply } \sqrt{2} \cdot \sqrt{b}. \end{aligned}$$

Substitute $5\sqrt{2b}$ for $\sqrt{50b}$ in the original expression.

$$\sqrt{a}(\sqrt{a} + \sqrt{50b}) = \sqrt{a}(\sqrt{a} + 5\sqrt{2b}) \qquad \text{Substitute } 5\sqrt{2b} \text{ for } \sqrt{50b}.$$
$$= \sqrt{a} \cdot \sqrt{a} + \sqrt{a} \cdot 5\sqrt{2b} \qquad \text{Distribute.}$$
$$= a + 5\sqrt{2ab} \qquad \text{Multiply.} \qquad ■$$

### You Try 1

Multiply and simplify.

a)  $7(\sqrt{3} - \sqrt{75})$     b)  $\sqrt{5}(\sqrt{10} + \sqrt{6})$     c)  $\sqrt{r}(\sqrt{r} + \sqrt{72t})$

## 2. Multiply Radical Expressions Using FOIL

In Chapter 6, we first multiplied binomials using **FOIL** (**F**irst **O**uter **I**nner **L**ast).

$$(3x + 4)(x + 2) = 3x \cdot x + 3x \cdot 2 + 4 \cdot x + 4 \cdot 2$$
$$\qquad\qquad\qquad\quad \textbf{F} \qquad\quad \textbf{O} \qquad \textbf{I} \qquad \textbf{L}$$
$$= \quad 3x^2 \; + \; 6x \; + \; 4x \; + \; 8$$
$$= 3x^2 + 10x + 8$$

We can multiply binomials containing radicals the same way.

### Example 2

Multiply and simplify.

a)  $(4 + \sqrt{7})(6 + \sqrt{7})$     b)  $(2\sqrt{5} + \sqrt{3})(\sqrt{5} - 8\sqrt{3})$
c)  $(\sqrt{x} + \sqrt{2y})(\sqrt{x} + 6\sqrt{2y})$

**Solution**

a)  Since we must multiply two binomials, we will use FOIL.

$$\qquad\qquad\qquad \textbf{F} \qquad \textbf{O} \qquad \textbf{I} \qquad \textbf{L}$$
$$(4 + \sqrt{7})(6 + \sqrt{7}) = 4 \cdot 6 + 4 \cdot \sqrt{7} + 6 \cdot \sqrt{7} + \sqrt{7} \cdot \sqrt{7} \qquad \text{Use FOIL.}$$
$$= \; 24 \; + \; 4\sqrt{7} \; + \; 6\sqrt{7} \; + \; 7 \qquad \text{Multiply.}$$
$$= 31 + 10\sqrt{7} \qquad \text{Combine like terms.}$$

b)  $(2\sqrt{5} + \sqrt{3})(\sqrt{5} - 8\sqrt{3})$

$$\qquad\qquad \textbf{F} \qquad\qquad \textbf{O} \qquad\qquad \textbf{I} \qquad\qquad \textbf{L}$$
$$= 2\sqrt{5} \cdot (\sqrt{5}) + 2\sqrt{5} \cdot (-8\sqrt{3}) + \sqrt{3} \cdot (\sqrt{5}) + \sqrt{3} \cdot (-8\sqrt{3}) \qquad \text{Use FOIL.}$$
$$= \quad 2 \cdot 5 \quad + \quad (-16\sqrt{15}) \quad + \quad \sqrt{15} \quad + \quad (-8 \cdot 3) \qquad \text{Multiply.}$$
$$= \quad 10 \quad - \quad 16\sqrt{15} \quad + \quad \sqrt{15} \quad - \quad 24 \qquad \text{Multiply.}$$
$$= -14 - 15\sqrt{15} \qquad \text{Combine like terms.}$$

c)  $(\sqrt{x} + \sqrt{2y})(\sqrt{x} + 6\sqrt{2y})$

$$\qquad\qquad \textbf{F} \qquad\qquad \textbf{O} \qquad\qquad \textbf{I} \qquad\qquad \textbf{L}$$
$$= \sqrt{x} \cdot (\sqrt{x}) + \sqrt{x} \cdot (6\sqrt{2y}) + \sqrt{2y} \cdot (\sqrt{x}) + \sqrt{2y} \cdot (6\sqrt{2y}) \qquad \text{Use FOIL.}$$
$$= \quad x \quad + \quad 6\sqrt{2xy} \quad + \quad \sqrt{2xy} \quad + \quad 6 \cdot 2y \qquad \text{Multiply.}$$
$$= \quad x \quad + \quad 6\sqrt{2xy} \quad + \quad 1\sqrt{2xy} \quad + \quad 12y \qquad \text{Multiply.}$$
$$= x + 7\sqrt{2xy} + 12y \qquad \text{Combine like terms.} \qquad ■$$

### You Try 2

Multiply and simplify.

a)  $(10 + \sqrt{6})(8 + \sqrt{6})$     b)  $(6\sqrt{7} + \sqrt{5})(\sqrt{7} - 9\sqrt{5})$
c)  $(\sqrt{a} + \sqrt{10b})(\sqrt{a} + 4\sqrt{10b})$

### 3. Square a Binomial Containing Radical Expressions

Recall again, from Chapter 6, that we can use FOIL to square a binomial or we can use these special formulas:

$$(a + b)^2 = a^2 + 2ab + b^2$$
$$(a - b)^2 = a^2 - 2ab + b^2$$

For example,

$$(n + 6)^2 = (n)^2 + 2(n)(6) + (6)^2$$
$$= n^2 + 12n + 36$$

and

$$(2y - 3)^2 = (2y)^2 - 2(2y)(3) + (3)^2$$
$$= 4y^2 - 12y + 9$$

To square a binomial containing radicals, we can either use FOIL or we can use the formulas above. The formulas will help us solve radical equations in Section 9.6.

---

**Example 3**

Multiply and simplify.

a)  $(\sqrt{6} + 4)^2$     b)  $(3\sqrt{2} - 7)^2$     c)  $(\sqrt{x} + \sqrt{5})^2$

**Solution**

a)  Use $(a + b)^2 = a^2 + 2ab + b^2$.

$(\sqrt{6} + 4)^2 = (\sqrt{6})^2 + 2(\sqrt{6})(4) + (4)^2$      Substitute $\sqrt{6}$ for $a$ and 4 for $b$.
$\qquad\qquad = 6 + 8\sqrt{6} + 16$      Multiply.
$\qquad\qquad = 22 + 8\sqrt{6}$      Combine like terms.

b)  Use $(a - b)^2 = a^2 - 2ab + b^2$.

$(3\sqrt{2} - 7)^2 = (3\sqrt{2})^2 - 2(3\sqrt{2})(7) + (7)^2$      Substitute $3\sqrt{2}$ for $a$ and 7 for $b$.
$\qquad\qquad = (9 \cdot 2) - (6\sqrt{2})(7) + 49$      Multiply.
$\qquad\qquad = 18 - 42\sqrt{2} + 49$      Multiply.
$\qquad\qquad = 67 - 42\sqrt{2}$      Combine like terms.

c)  $(\sqrt{x} + \sqrt{5})^2 = (\sqrt{x})^2 + 2(\sqrt{x})(\sqrt{5}) + (\sqrt{5})^2$      Use $(a + b)^2 = a^2 + 2ab + b^2$.
$\qquad\qquad = x + 2\sqrt{5x} + 5$      Square; product rule   ∎

---

**You Try 3**

Multiply and simplify.

a) $(\sqrt{5} + 8)^2$     b)  $(2\sqrt{3} - 9)^2$     c)  $(\sqrt{h} + \sqrt{10})^2$

---

### 4. Multiply Two Binomials of the Form $(a + b)(a - b)$ Containing Radicals

We will review one last rule from Chapter 6 on multiplying binomials. We will use this in Section 9.5 when we divide radicals.

$$(a + b)(a - b) = a^2 - b^2$$

For example, $(c + 9)(c - 9) = (c)^2 - (9)^2 = c^2 - 81$. The same rule applies when we multiply binomials containing radicals.

**Example 4**

Multiply and simplify.

a)  $(4 + \sqrt{3})(4 - \sqrt{3})$

b)  $(\sqrt{x} + \sqrt{y})(\sqrt{x} - \sqrt{y})$

**Solution**

a)  Use $(a + b)(a - b) = a^2 - b^2$.

$(4 + \sqrt{3})(4 - \sqrt{3}) = (4)^2 - (\sqrt{3})^2$    Substitute 4 for $a$ and $\sqrt{3}$ for $b$.

$= 16 - 3$    Square each term.

$= 13$    Subtract.

b)  $(\sqrt{x} + \sqrt{y})(\sqrt{x} - \sqrt{y}) = (\sqrt{x})^2 - (\sqrt{y})^2$    Use $(a + b)(a - b) = a^2 - b^2$.

$= x - y$    Square each term.    ■

**Note**

Notice in Example 4 that when we multiply expressions containing square roots of the form $(a + b)(a - b)$, the radicals are eliminated. *This will always be true.*

**You Try 4**

Multiply and simplify.

a)  $(5 - \sqrt{7})(5 + \sqrt{7})$

b)  $(\sqrt{p} + \sqrt{q})(\sqrt{p} - \sqrt{q})$

**Answers to You Try Exercises**

1) a) $-28\sqrt{3}$  b) $5\sqrt{2} + \sqrt{30}$  c) $r + 6\sqrt{2rt}$

2) a) $86 + 18\sqrt{6}$  b) $-3 - 53\sqrt{35}$  c) $a + 5\sqrt{10ab} + 40b$

3) a) $69 + 16\sqrt{5}$  b) $93 - 36\sqrt{3}$  c) $h + 2\sqrt{10h} + 10$    4) a) 18  b) $p - q$

# 9.4 Exercises

**Objective 1: Multiply Radical Expressions**

Multiply and simplify.

1)  $5(m + 3)$

2)  $7(p + 4)$

3)  $7(\sqrt{2} + 6)$

4)  $4(5 - \sqrt{6})$

5)  $\sqrt{2}(\sqrt{5} - 8)$

6)  $\sqrt{10}(4 + \sqrt{3})$

7)  $-6(\sqrt{32} + \sqrt{2})$

8)  $4(\sqrt{27} - \sqrt{3})$

9)  $6(\sqrt{45} - \sqrt{20})$

10)  $-5(\sqrt{18} + \sqrt{50})$

11)  $\sqrt{5}(\sqrt{54} - \sqrt{96})$

12)  $\sqrt{3}(\sqrt{45} + \sqrt{20})$

13)  $\sqrt{2}(7 + \sqrt{6})$

14)  $\sqrt{18}(\sqrt{8} - 8)$

15)  $\sqrt{a}(\sqrt{a} - \sqrt{64b})$

16)  $\sqrt{d}(\sqrt{12c} + \sqrt{7d})$

17)  $\sqrt{xy}(\sqrt{6x} + \sqrt{27y})$

18)  $\sqrt{2pq}(\sqrt{2q} - q\sqrt{p})$

**Objective 2:  Multiply Radical Expressions Using FOIL**

19)  How are the problems *Multiply* $(x + 2)(x + 7)$ and *Multiply* $(\sqrt{10} + 2)(\sqrt{10} + 7)$ similar? What method can be used to multiply each of them?

20)  a)  Multiply $(x + 2)(x + 7)$.

b)  Multiply $(\sqrt{10} + 2)(\sqrt{10} + 7)$.

Multiply and simplify.

| Fill It In |
| --- |
| Fill in the blanks with either the missing mathematical step or reason for the given step. |

21)  $(4 + \sqrt{5})(3 + \sqrt{5})$

= _____    Use FOIL.

$= 12 + 4\sqrt{5} + 3\sqrt{5} + 5$

= _____    Combine like terms.

22) $(1 + \sqrt{6})(8 + \sqrt{6})$
$= 1 \cdot 8 + 1\sqrt{6} + 8\sqrt{6} + \sqrt{6} \cdot \sqrt{6}$
$=$ _____   Multiply.
$=$ _____   Combine like terms.

**Multiply and simplify.**

23) $(k + 3)(k + 6)$

24) $(b - 8)(b + 2)$

25) $(6 + \sqrt{7})(2 + \sqrt{7})$

26) $(5 + \sqrt{3})(1 + \sqrt{3})$

27) $(\sqrt{2} + 7)(\sqrt{2} - 4)$

28) $(\sqrt{5} - 7)(\sqrt{5} + 2)$

29) $(\sqrt{3} - 4\sqrt{5})(2\sqrt{3} - \sqrt{5})$

30) $(5\sqrt{2} - \sqrt{3})(2\sqrt{3} - \sqrt{2})$

31) $(3\sqrt{6} - 2\sqrt{2})(\sqrt{2} + 5\sqrt{6})$

32) $(2\sqrt{10} + 3\sqrt{2})(\sqrt{10} - 2\sqrt{2})$

33) $(3 + 2\sqrt{5})(\sqrt{7} + \sqrt{2})$

34) $(\sqrt{2} + 4)(\sqrt{3} - 6\sqrt{5})$

35) $(\sqrt{m} + \sqrt{7n})(\sqrt{m} + 5\sqrt{7n})$

36) $(\sqrt{x} + \sqrt{3y})(\sqrt{x} + 4\sqrt{3y})$

37) $(\sqrt{6p} - 2\sqrt{q})(8\sqrt{q} + 5\sqrt{6p})$

38) $(4\sqrt{3r} + \sqrt{s})(3\sqrt{s} - 2\sqrt{3r})$

**Objective 3: Square a Binomial Containing Radical Expressions**

39) How are the problems *Multiply* $(p - 4)^2$ and *Multiply* $(\sqrt{7} - 4)^2$ similar? What method can be used to multiply each of them?

40) a) Multiply $(p - 4)^2$.

b) Multiply $(\sqrt{7} - 4)^2$.

**Multiply and simplify.**

41) $(2b - 11)^2$

42) $(3d + 5)^2$

43) $(\sqrt{3} + 1)^2$

44) $(\sqrt{2} + 9)^2$

45) $(\sqrt{13} - \sqrt{5})^2$

46) $(\sqrt{3} + \sqrt{11})^2$

47) $(2\sqrt{7} + \sqrt{10})^2$

48) $(2\sqrt{3} - \sqrt{2})^2$

49) $(\sqrt{6} - 4\sqrt{2})^2$

50) $(\sqrt{3} + 2\sqrt{15})^2$

51) $(\sqrt{k} + \sqrt{11})^2$

52) $(\sqrt{m} + \sqrt{7})^2$

53) $(\sqrt{x} - \sqrt{y})^2$

54) $(\sqrt{b} - \sqrt{a})^2$

**Objective 4: Multiply Two Binomials of the Form $(a + b)(a - b)$ Containing Radicals**

55) What formula can be used to multiply $(5 + \sqrt{6})(5 - \sqrt{6})$?

56) What happens to the radical terms whenever we multiply $(a + b)(a - b)$ where the binomials contain square roots?

57) $(a + 9)(a - 9)$

58) $(g - 8)(g + 8)$

59) $(\sqrt{5} + 3)(\sqrt{5} - 3)$

60) $(\sqrt{3} + 2)(\sqrt{3} - 2)$

61) $(6 - \sqrt{2})(6 + \sqrt{2})$

62) $(4 - \sqrt{11})(4 + \sqrt{11})$

63) $(4\sqrt{3} + \sqrt{2})(4\sqrt{3} - \sqrt{2})$

64) $(2\sqrt{2} - 2\sqrt{7})(2\sqrt{2} + 2\sqrt{7})$

65) $(\sqrt{11} + 5\sqrt{2})(\sqrt{11} - 5\sqrt{2})$

66) $(\sqrt{15} + 5\sqrt{3})(\sqrt{15} - 5\sqrt{3})$

67) $(\sqrt{c} + \sqrt{d})(\sqrt{c} - \sqrt{d})$

68) $(\sqrt{2y} + \sqrt{z})(\sqrt{2y} - \sqrt{z})$

69) $(6 - \sqrt{t})(6 + \sqrt{t})$

70) $(4 - \sqrt{q})(4 + \sqrt{q})$

71) $(8\sqrt{f} - \sqrt{g})(8\sqrt{f} + \sqrt{g})$

72) $(\sqrt{a} + 6\sqrt{b})(\sqrt{a} - 6\sqrt{b})$

**Mixed Exercises: Objectives 1–4**

**Multiply and simplify.**

73) $(7 + \sqrt{2})(11 + \sqrt{2})$

74) $\sqrt{7}(\sqrt{14} + \sqrt{2})$

75) $\sqrt{n}(\sqrt{98m} + \sqrt{5n})$

76) $(\sqrt{x} + \sqrt{15})^2$

77) $(3\sqrt{5} + 4)(3\sqrt{5} - 4)$

78) $(1 + 4\sqrt{7})(2\sqrt{3} + \sqrt{5})$

79) $(2 - \sqrt{7})^2$

80) $-3(\sqrt{24} - \sqrt{5})$

81) $(\sqrt{5a} - \sqrt{b})(3\sqrt{5a} - \sqrt{b})$

82) $(8 - \sqrt{11})(8 + \sqrt{11})$

83) $(\sqrt{5} + \sqrt{10})(\sqrt{2} - 3\sqrt{3})$

84) $(\sqrt{3} + 2)(\sqrt{3} - 10)$

**Extension**

**Multiply and simplify.**

85) $(\sqrt[3]{2} - 4)(\sqrt[3]{2} + 4)$

86) $(1 + \sqrt[3]{7})(1 - \sqrt[3]{7})$

87) $(1 + 2\sqrt[3]{5})(1 - 2\sqrt[3]{5} + 4\sqrt[3]{25})$

88) $(3 + \sqrt[3]{2})(9 - 3\sqrt[3]{2} + \sqrt[3]{4})$

89) $[(\sqrt{5} + \sqrt{2}) + \sqrt{7}][(\sqrt{5} + \sqrt{2}) - \sqrt{7}]$

90) $[(\sqrt{2} - \sqrt{10}) + \sqrt{3}][(\sqrt{2} - \sqrt{10}) - \sqrt{3}]$

## Section 9.5 Dividing Radicals

### Objectives

1. Rationalize a Denominator Containing One Square Root Term
2. Rationalize a Denominator Containing a Higher Root
3. Rationalize a Denominator Containing Two Terms
4. Divide Out Common Factors in Radical Quotients

It is generally agreed that a radical expression is *not* in simplest form if its denominator contains a radical. For example, $\dfrac{1}{\sqrt{2}}$ is not simplified, but an equivalent form, $\dfrac{\sqrt{2}}{2}$, is simplified.

Later we will show that $\dfrac{1}{\sqrt{2}} = \dfrac{\sqrt{2}}{2}$. The process of eliminating radicals from the denominator of an expression is called **rationalizing the denominator.** We will look at two types of rationalizing problems.

    1)   Rationalizing a denominator containing one term

    2)   Rationalizing a denominator containing two terms

To rationalize a denominator, we will use the fact that multiplying the numerator and denominator of a fraction by the same quantity results in an equivalent fraction:

$$\frac{2}{3} \cdot \frac{4}{4} = \frac{8}{12} \qquad \frac{2}{3} \text{ and } \frac{8}{12} \text{ are equivalent because } \frac{4}{4} = 1.$$

We use the same idea to rationalize the denominator of a radical expression.

### 1. Rationalize a Denominator Containing One Square Root Term

The goal of rationalizing is to eliminate the radical from the denominator. With regard to square roots, recall that $\sqrt{a} \cdot \sqrt{a} = \sqrt{a^2} = a$ for $a \geq 0$. For example,

$$\sqrt{2} \cdot \sqrt{2} = \sqrt{2^2} = 2, \quad \sqrt{14} \cdot \sqrt{14} = \sqrt{(14)^2} = 14, \quad \sqrt{x} \cdot \sqrt{x} = \sqrt{x^2} = x \ (x \geq 0)$$

We will use this property to rationalize the denominators of the following expressions.

---

**Example 1**

Rationalize the denominator of each expression.

a) $\dfrac{1}{\sqrt{2}}$      b) $\dfrac{30}{\sqrt{12}}$      c) $\dfrac{7\sqrt{3}}{\sqrt{5}}$

**Solution**

a)  To eliminate the square root from the denominator of $\dfrac{1}{\sqrt{2}}$, ask yourself, "By what do I multiply $\sqrt{2}$ to get a *perfect square* under the square root?" The answer is $\sqrt{2}$ since $\sqrt{2} \cdot \sqrt{2} = \sqrt{2^2} = \sqrt{4} = 2$. Multiply by $\sqrt{2}$ in the numerator *and* denominator. (We are actually multiplying $\dfrac{1}{\sqrt{2}}$ by 1.)

$$\begin{aligned} \frac{1}{\sqrt{2}} &= \frac{1}{\sqrt{2}} \cdot \frac{\sqrt{2}}{\sqrt{2}} && \text{Rationalize the denominator.} \\[2mm] &= \frac{\sqrt{2}}{\sqrt{2^2}} && \text{Multiply.} \\[2mm] &= \frac{\sqrt{2}}{2} && \sqrt{2^2} = \sqrt{4} = 2 \end{aligned}$$

**BE CAREFUL**

$\dfrac{\sqrt{2}}{2}$ is in simplest form. We cannot reduce terms inside and outside of the radical.

**Wrong:** $\dfrac{\sqrt{2}}{2} = \dfrac{\sqrt{2}^{1}}{2_{1}} = \sqrt{1} = 1$

b)   First, simplify the denominator of $\dfrac{30}{\sqrt{12}}$: $\sqrt{12} = \sqrt{4} \cdot \sqrt{3} = 2\sqrt{3}$.

$$\dfrac{30}{\sqrt{12}} = \dfrac{30}{2\sqrt{3}} \qquad \text{Substitute } 2\sqrt{3} \text{ for } \sqrt{12}.$$

$$= \dfrac{15}{\sqrt{3}} \qquad \text{Simplify } \dfrac{30}{2}.$$

Rationalize the denominator. Ask yourself, "By what do I multiply $\sqrt{3}$ to get a *perfect square* under the square root?" The answer is $\sqrt{3}$.

$$= \dfrac{15}{\sqrt{3}} \cdot \dfrac{\sqrt{3}}{\sqrt{3}} \qquad \text{Rationalize the denominator.}$$

$$= \dfrac{15\sqrt{3}}{\sqrt{3^2}}$$

$$= \dfrac{15\sqrt{3}}{3} \qquad \sqrt{3^2} = 3$$

$$= 5\sqrt{3} \qquad \text{Simplify } \dfrac{15}{3}.$$

c)   To rationalize $\dfrac{7\sqrt{3}}{\sqrt{5}}$, multiply the numerator and denominator by $\sqrt{5}$.

$$\dfrac{7\sqrt{3}}{\sqrt{5}} = \dfrac{7\sqrt{3}}{\sqrt{5}} \cdot \dfrac{\sqrt{5}}{\sqrt{5}}$$

$$= \dfrac{7\sqrt{15}}{5}$$

**You Try 1**

Rationalize the denominator of each expression.

a)   $\dfrac{1}{\sqrt{3}}$       b)   $\dfrac{60}{\sqrt{20}}$       c)   $\dfrac{11\sqrt{6}}{\sqrt{5}}$

Sometimes we will apply the quotient or product rule before rationalizing.

**Example 2**

Simplify completely.

a)   $\sqrt{\dfrac{2}{16}}$       b)   $\sqrt{\dfrac{7}{8}} \cdot \sqrt{\dfrac{4}{3}}$

**Solution**

a)  First, simplify $\dfrac{2}{16}$ under the radical.

$$\sqrt{\dfrac{2}{16}} = \sqrt{\dfrac{1}{8}} \qquad \text{Simplify.}$$

$$= \dfrac{\sqrt{1}}{\sqrt{8}} \qquad \text{Quotient rule}$$

$$= \dfrac{1}{\sqrt{4} \cdot \sqrt{2}} \qquad \text{Product rule}$$

$$= \dfrac{1}{2\sqrt{2}} \qquad \sqrt{4} = 2$$

$$= \dfrac{1}{2\sqrt{2}} \cdot \dfrac{\sqrt{2}}{\sqrt{2}} \qquad \text{Rationalize the denominator.}$$

$$= \dfrac{\sqrt{2}}{2 \cdot 2} \qquad \sqrt{2} \cdot \sqrt{2} = 2$$

$$= \dfrac{\sqrt{2}}{4} \qquad \text{Multiply.}$$

b)  First, use the product rule to multiply the radicands.

$$\sqrt{\dfrac{7}{8}} \cdot \sqrt{\dfrac{4}{3}} = \sqrt{\dfrac{7}{8} \cdot \dfrac{4}{3}} \qquad \text{Product rule}$$

$$= \sqrt{\dfrac{7}{\underset{2}{8}} \cdot \dfrac{\overset{1}{4}}{3}} \qquad \begin{array}{l}\text{Multiply the} \\ \text{fractions} \\ \text{under the} \\ \text{radical.}\end{array}$$

$$= \sqrt{\dfrac{7}{6}} \qquad \text{Multiply.}$$

$$= \dfrac{\sqrt{7}}{\sqrt{6}} \qquad \text{Quotient rule}$$

$$= \dfrac{\sqrt{7}}{\sqrt{6}} \cdot \dfrac{\sqrt{6}}{\sqrt{6}} \qquad \begin{array}{l}\text{Rationalize the} \\ \text{denominator.}\end{array}$$

$$= \dfrac{\sqrt{42}}{6} \qquad \text{Multiply.} \quad \blacksquare$$

**You Try 2**

Simplify completely.

a)  $\sqrt{\dfrac{2}{54}}$   b)  $\sqrt{\dfrac{11}{15}} \cdot \sqrt{\dfrac{3}{2}}$

We work with radical expressions containing variables the same way. **Remember, we are assuming that all variables represent positive real numbers.**

**Example 3**

Simplify completely.

a)  $\dfrac{11}{\sqrt{x}}$   b)  $\sqrt{\dfrac{75a^3}{7b}}$   c)  $\sqrt{\dfrac{13rt^2}{rt^3}}$

**Solution**

a)  Ask yourself, "By what do I multiply $\sqrt{x}$ to get a *perfect square* under the square root?" The perfect square we want to get is $\sqrt{x^2}$.

$$\sqrt{x} \cdot \sqrt{?} = \sqrt{x^2} = x$$
$$\sqrt{x} \cdot \sqrt{x} = \sqrt{x^2} = x$$

$$\dfrac{11}{\sqrt{x}} = \dfrac{11}{\sqrt{x}} \cdot \dfrac{\sqrt{x}}{\sqrt{x}} \qquad \text{Rationalize the denominator.}$$

$$= \dfrac{11\sqrt{x}}{\sqrt{x^2}} \qquad \text{Multiply.}$$

$$= \dfrac{11\sqrt{x}}{x} \qquad \sqrt{x^2} = x$$

b)  Before rationalizing, apply the quotient rule and simplify the numerator.

$$\sqrt{\dfrac{75a^3}{7b}} = \dfrac{\sqrt{75a^3}}{\sqrt{7b}} = \dfrac{\sqrt{25} \cdot \sqrt{3} \cdot \sqrt{a^2} \cdot \sqrt{a}}{\sqrt{7b}} = \dfrac{5a\sqrt{3a}}{\sqrt{7b}}$$

Next, rationalize the denominator. "By what do I multiply $\sqrt{7b}$ to get a *perfect square* under the square root?" The perfect square we want to get is $\sqrt{7^2b^2}$ or $\sqrt{49b^2}$.

$$\sqrt{7b} \cdot \sqrt{?} = \sqrt{7^2b^2} = 7b$$
$$\sqrt{7b} \cdot \sqrt{7b} = \sqrt{7^2b^2} = 7b$$

$$\frac{5a\sqrt{3a}}{\sqrt{7b}} = \frac{5a\sqrt{3a}}{\sqrt{7b}} \cdot \frac{\sqrt{7b}}{\sqrt{7b}} \qquad \text{Rationalize the denominator.}$$

$$= \frac{5a\sqrt{21ab}}{7b} \qquad \text{Multiply.}$$

c)   First, simplify the radicand, $\dfrac{13rt^2}{rt^3}$.

$$\sqrt{\frac{13rt^2}{rt^3}} = \sqrt{\frac{13}{t}} \qquad \text{Quotient rule for exponents}$$

$$= \frac{\sqrt{13}}{\sqrt{t}} \qquad \text{Quotient rule for radicals}$$

$$= \frac{\sqrt{13}}{\sqrt{t}} \cdot \frac{\sqrt{t}}{\sqrt{t}} \qquad \text{Rationalize the denominator.}$$

$$= \frac{\sqrt{13t}}{t} \qquad \text{Multiply.}$$

 **You Try 3**

Simplify completely.

a) $\dfrac{8}{\sqrt{w}}$     b) $\sqrt{\dfrac{40h^3}{3k}}$     c) $\sqrt{\dfrac{7m^2n}{m^3n}}$

## 2. Rationalize a Denominator Containing a Higher Root

Many students assume that to rationalize denominators like we have up until this point, we always multiply the numerator and denominator of the expression by the denominator as in

$$\frac{5}{\sqrt{6}} = \frac{5}{\sqrt{6}} \cdot \frac{\sqrt{6}}{\sqrt{6}} = \frac{5\sqrt{6}}{6}$$

*We will see, however, why this reasoning is incorrect.*

To rationalize an expression like $\dfrac{5}{\sqrt{6}}$ we asked ourselves, "By what do I multiply $\sqrt{6}$ to get a *perfect square* under the *square root*?"

To rationalize an expression like $\dfrac{3}{\sqrt[3]{2}}$ we must ask ourselves, "By what do I multiply $\sqrt[3]{2}$ to get a *perfect cube* under the *cube root*?" The perfect cube we want is $2^3$ (since we began with 2) so that $\sqrt[3]{2} \cdot \sqrt[3]{2^2} = \sqrt[3]{2^3} = 2$.

We will practice some fill-in-the-blank problems to eliminate radicals before we move on to rationalizing.

**Example 4**

Fill in the blank.

a) $\sqrt[3]{5} \cdot \sqrt[3]{?} = \sqrt[3]{5^3} = 5$      b) $\sqrt[3]{3} \cdot \sqrt[3]{?} = \sqrt[3]{3^3} = 3$

c) $\sqrt[3]{x^2} \cdot \sqrt[3]{?} = \sqrt[3]{x^3} = x$

**Solution**

a) Ask yourself, "By what do I multiply $\sqrt[3]{5}$ to get $\sqrt[3]{5^3}$?" The answer is $\sqrt[3]{5^2}$.

$$\sqrt[3]{5} \cdot \sqrt[3]{?} = \sqrt[3]{5^3} = 5$$
$$\sqrt[3]{5} \cdot \sqrt[3]{5^2} = \sqrt[3]{5^3} = 5$$

b) "By what do I multiply $\sqrt[3]{3}$ to get $\sqrt[3]{3^3}$?" $\sqrt[3]{3^2}$

$$\sqrt[3]{3} \cdot \sqrt[3]{?} = \sqrt[3]{3^3} = 3$$
$$\sqrt[3]{3} \cdot \sqrt[3]{3^2} = \sqrt[3]{3^3} = 3$$

c) "By what do I multiply $\sqrt[3]{x^2}$ to get $\sqrt[3]{x^3}$?" $\sqrt[3]{x}$

$$\sqrt[3]{x^2} \cdot \sqrt[3]{?} = \sqrt[3]{x^3} = x$$
$$\sqrt[3]{x^2} \cdot \sqrt[3]{x} = \sqrt[3]{x^3} = x$$

**You Try 4**

Fill in the blank.

a) $\sqrt[3]{2} \cdot \sqrt[3]{?} = \sqrt[3]{2^3} = 2$      b) $\sqrt[3]{25} \cdot \sqrt[3]{?} = \sqrt[3]{5^2} = 5$

c) $\sqrt[3]{n} \cdot \sqrt[3]{?} = \sqrt[3]{n^3} = n$

We will use the technique presented in Example 4 to rationalize denominators with indices higher than 2.

**Example 5**

Rationalize the denominator.

a) $\dfrac{3}{\sqrt[3]{2}}$      b) $\sqrt[3]{\dfrac{5}{9}}$      c) $\dfrac{6}{\sqrt[3]{4k}}$

**Solution**

a) To rationalize the denominator of $\dfrac{3}{\sqrt[3]{2}}$, *first* identify what we want to get as the denominator *after* multiplying. **We want to obtain $\sqrt[3]{2^3}$ since $\sqrt[3]{2^3} = 2$.**

$$\frac{3}{\sqrt[3]{2}} \cdot \frac{}{\phantom{x}} = \frac{}{\sqrt[3]{2^3}} \longleftarrow \text{This is what we want to get.}$$
$$\underset{\substack{\uparrow \\ \text{What is needed here?}}}{}$$

Ask yourself, "By what do I multiply $\sqrt[3]{2}$ to get $\sqrt[3]{2^3}$?" $\sqrt[3]{2^2}$

$$\frac{3}{\sqrt[3]{2}} \cdot \frac{\sqrt[3]{2^2}}{\sqrt[3]{2^2}} = \frac{3\sqrt[3]{2^2}}{\sqrt[3]{2^3}} \qquad \text{Multiply.}$$
$$= \frac{3\sqrt[3]{4}}{2} \qquad \text{Simplify.}$$

b) $\sqrt[3]{\dfrac{5}{9}} = \dfrac{\sqrt[3]{5}}{\sqrt[3]{9}}$      Quotient rule

$\phantom{\sqrt[3]{\dfrac{5}{9}}} = \dfrac{\sqrt[3]{5}}{\sqrt[3]{3^2}}$      Rewrite 9 as $3^2$.

To rationalize the denominator, we want to obtain $\sqrt[3]{3^3}$ since $\sqrt[3]{3^3} = 3$.

"By what do I multiply $\sqrt[3]{3^2}$ to get $\sqrt[3]{3^3}$?"    $\sqrt[3]{3}$

$\dfrac{\sqrt[3]{5}}{\sqrt[3]{3^2}} \cdot \dfrac{\sqrt[3]{3}}{\sqrt[3]{3}} = \dfrac{\sqrt[3]{5} \cdot \sqrt[3]{3}}{\sqrt[3]{3^3}}$      Multiply.

$\phantom{\dfrac{\sqrt[3]{5}}{\sqrt[3]{3^2}} \cdot \dfrac{\sqrt[3]{3}}{\sqrt[3]{3}}} = \dfrac{\sqrt[3]{15}}{3}$      Multiply and simplify.

c) To rationalize the denominator of $\dfrac{6}{\sqrt[3]{4k}}$, first identify what we want to get as the denominator *after* multiplying. Remember, we need a radicand that is a perfect cube. We want to obtain $\sqrt[3]{8k^3}$ since $\sqrt[3]{8k^3} = \sqrt[3]{(2k)^3} = 2k$.

Ask yourself, "By what do I multiply $\sqrt[3]{4k}$ to get $\sqrt[3]{8k^3}$?" $\sqrt[3]{2k^2}$ Now, rationalize the denominator.

$\dfrac{6}{\sqrt[3]{4k}} \cdot \dfrac{\sqrt[3]{2k^2}}{\sqrt[3]{2k^2}} = \dfrac{6\sqrt[3]{2k^2}}{\sqrt[3]{8k^3}}$      Multiply.

$\phantom{\dfrac{6}{\sqrt[3]{4k}} \cdot \dfrac{\sqrt[3]{2k^2}}{\sqrt[3]{2k^2}}} = \dfrac{6\sqrt[3]{2k^2}}{2k}$      Simplify $\sqrt[3]{8k^3}$.

$\phantom{\dfrac{6}{\sqrt[3]{4k}} \cdot \dfrac{\sqrt[3]{2k^2}}{\sqrt[3]{2k^2}}} = \dfrac{3\sqrt[3]{2k^2}}{k}$      Simplify $\dfrac{6}{2}$.     ■

**You Try 5**

Rationalize the denominator.

a) $\dfrac{6}{\sqrt[3]{5}}$      b) $\sqrt[3]{\dfrac{11}{4}}$      c) $\dfrac{30}{\sqrt[3]{3z^2}}$

## 3. Rationalize a Denominator Containing Two Terms

To rationalize the denominator of an expression like $\dfrac{9}{4 + \sqrt{6}}$, we multiply the numerator and the denominator of the expression by the *conjugate* of $4 + \sqrt{6}$.

**Definition**

The **conjugate** of a binomial is the binomial obtained by changing the sign between the two terms.

| Expression | Conjugate |
|---|---|
| $4 + \sqrt{6}$ | $4 - \sqrt{6}$ |
| $\sqrt{10} - 2\sqrt{3}$ | $\sqrt{10} + 2\sqrt{3}$ |
| $\sqrt{a} + \sqrt{b}$ | $\sqrt{a} - \sqrt{b}$ |

In Section 9.4, we applied the formula $(a + b)(a - b) = a^2 - b^2$ to multiply binomials containing square roots. Recall that the terms containing the square roots were eliminated.

## Example 6

Multiply $7 - \sqrt{2}$ by its conjugate.

### Solution

The conjugate of $7 - \sqrt{2}$ is $7 + \sqrt{2}$. We will first multiply using FOIL to show *why* the radical drops out, then we will multiply using the formula $(a + b)(a - b) = a^2 - b^2$.

i)   Use FOIL to multiply.

$$(7 - \sqrt{2})(7 + \sqrt{2}) = \overset{F}{7 \cdot 7} + \overset{O}{7 \cdot \sqrt{2}} - \overset{I}{7 \cdot \sqrt{2}} - \overset{L}{\sqrt{2} \cdot \sqrt{2}}$$
$$= 49 - 2$$
$$= 47$$

ii)  Use $(a + b)(a - b) = a^2 - b^2$.

$$(7 - \sqrt{2})(7 + \sqrt{2}) = (7)^2 - (\sqrt{2})^2 \qquad \text{Substitute 7 for } a \text{ and } \sqrt{2} \text{ for } b.$$
$$= 49 - 2$$
$$= 47$$

Each method gives the same result.  ■

## You Try 6

Multiply $5 + \sqrt{6}$ by its conjugate.

---

### Procedure    Rationalizing a Denominator Containing Two Terms

To rationalize the denominator of an expression in which the denominator contains two terms, multiply the numerator and denominator of the expression by the conjugate of the denominator.

---

## Example 7

Rationalize the denominator and simplify completely.

a)  $\dfrac{9}{4 + \sqrt{6}}$        b)  $\dfrac{8 + \sqrt{3}}{\sqrt{3} - 5}$        c)  $\dfrac{5}{\sqrt{x} + 2}$

### Solution

a)  The denominator of $\dfrac{9}{4 + \sqrt{6}}$ has two terms, so we must multiply the numerator and denominator by the conjugate, $4 - \sqrt{6}$.

$$\frac{9}{4 + \sqrt{6}} \cdot \frac{4 - \sqrt{6}}{4 - \sqrt{6}} \qquad \text{Multiply by the conjugate.}$$
$$= \frac{9(4 - \sqrt{6})}{(4)^2 - (\sqrt{6})^2} \qquad (a + b)(a - b) = a^2 - b^2$$
$$= \frac{36 - 9\sqrt{6}}{16 - 6} \qquad \text{Simplify.}$$
$$= \frac{36 - 9\sqrt{6}}{10} \qquad \text{Subtract.}$$

b) $\dfrac{8 + \sqrt{3}}{\sqrt{3} - 5} = \dfrac{8 + \sqrt{3}}{\sqrt{3} - 5} \cdot \dfrac{\sqrt{3} + 5}{\sqrt{3} + 5}$    Multiply by the conjugate.

$= \dfrac{8\sqrt{3} + 40 + 3 + 5\sqrt{3}}{(\sqrt{3})^2 - (5)^2}$    Use FOIL to multiply the numerators. $(a + b)(a - b) = a^2 - b^2$

$= \dfrac{13\sqrt{3} + 43}{3 - 25}$    Combine like terms; simplify.

$= \dfrac{13\sqrt{3} + 43}{-22} = -\dfrac{13\sqrt{3} + 43}{22}$    Simplify.

c) $\dfrac{5}{\sqrt{x} + 2} = \dfrac{5}{\sqrt{x} + 2} \cdot \dfrac{\sqrt{x} - 2}{\sqrt{x} - 2}$    Multiply by the conjugate.

$= \dfrac{5(\sqrt{x} - 2)}{(\sqrt{x})^2 - (2)^2}$    Multiply. $(a + b)(a - b) = a^2 - b^2$

$= \dfrac{5\sqrt{x} - 10}{x - 4}$    Multiply; simplify.    ■

**You Try 7**

Rationalize the denominator and simplify completely.

a) $\dfrac{4}{7 + \sqrt{2}}$    b) $\dfrac{6 + \sqrt{5}}{\sqrt{5} - 9}$    c) $\dfrac{11}{\sqrt{n} + 8}$

## 4. Divide Out Common Factors in Radical Quotients

Sometimes it is necessary to simplify a radical expression by dividing out common factors from the numerator and denominator. This is a skill we will need in Chapter 10 when we are solving quadratic equations, so we will look at an example here.

**Example 8**

Simplify completely.  $\dfrac{5\sqrt{2} + 45}{5}$

**Solution**

It is tempting to do one of the following:

$$\dfrac{\cancel{5}\sqrt{2} + 45}{\cancel{5}} = \sqrt{2} + 45 \quad \textbf{Incorrect!}$$

or

$$\dfrac{5\sqrt{2} + \overset{9}{\cancel{45}}}{\cancel{5}} = 5\sqrt{2} + 9 \quad \textbf{Incorrect!}$$

**Each is incorrect because $5\sqrt{2}$ is a *term* in a sum and 45 is a *term* in a sum.**

The correct way to simplify $\dfrac{5\sqrt{2} + 45}{5}$ is to begin by factoring out a 5 in the numerator and *then* divide the numerator and denominator by any common factors.

$$\frac{5\sqrt{2} + 45}{5} = \frac{5(\sqrt{2} + 9)}{5}$$ Factor out 5 from the numerator.

$$= \frac{\overset{1}{\cancel{5}}(\sqrt{2} + 9)}{\underset{1}{\cancel{5}}}$$ Divide by 5, a *factor* of the numerator.

$$= \sqrt{2} + 9$$ Simplify.  ∎

### Note

In Example 8, we can divide numerator and denominator by 5 in $\dfrac{5(\sqrt{2} + 9)}{5}$ because the 5 in the numerator is part of a *product* not a sum or difference.

## You Try 8

Simplify completely. $\dfrac{4\sqrt{6} - 20}{4}$

### Answers to You Try Exercises

1) a) $\dfrac{\sqrt{3}}{3}$  b) $6\sqrt{5}$  c) $\dfrac{11\sqrt{30}}{5}$  2) a) $\dfrac{\sqrt{3}}{9}$  b) $\dfrac{\sqrt{110}}{10}$

3) a) $\dfrac{8\sqrt{w}}{w}$  b) $\dfrac{2h\sqrt{30hk}}{3k}$  c) $\dfrac{\sqrt{7m}}{m}$  4) a) $2^2$ or 4  b) 5  c) $n^2$

5) a) $\dfrac{6\sqrt[3]{25}}{5}$  b) $\dfrac{\sqrt[3]{22}}{2}$  c) $\dfrac{10\sqrt[3]{9z}}{z}$  6) 19

7) a) $\dfrac{28 - 4\sqrt{2}}{47}$  b) $-\dfrac{15\sqrt{5} + 59}{76}$  c) $\dfrac{11\sqrt{n} - 88}{n - 64}$  8) $\sqrt{6} - 5$

# 9.5 Exercises

### Objective 1: Rationalize a Denominator Containing One Square Root Term

1) What does it mean to rationalize the denominator of a radical expression?

2) In your own words, explain how to rationalize the denominator of an expression containing one term in the denominator.

Rationalize the denominator of each expression.

3) $\dfrac{1}{\sqrt{6}}$

4) $\dfrac{1}{\sqrt{10}}$

5) $\dfrac{7}{\sqrt{3}}$

6) $\dfrac{9}{\sqrt{2}}$

7) $\dfrac{8}{\sqrt{2}}$

8) $\dfrac{15}{\sqrt{5}}$

9) $\dfrac{45}{\sqrt{10}}$

10) $\dfrac{32}{\sqrt{6}}$

11) $\dfrac{4}{\sqrt{72}}$

12) $\dfrac{15}{\sqrt{75}}$

13) $-\dfrac{20}{\sqrt{8}}$

14) $-\dfrac{6}{\sqrt{24}}$

15) $\dfrac{\sqrt{11}}{\sqrt{7}}$

16) $\dfrac{\sqrt{5}}{\sqrt{3}}$

17) $\dfrac{2\sqrt{10}}{\sqrt{3}}$

18) $\dfrac{9\sqrt{7}}{\sqrt{2}}$

19) $\dfrac{\sqrt{5}}{\sqrt{18}}$

20) $-\dfrac{21\sqrt{5}}{\sqrt{98}}$

21) $\dfrac{6\sqrt{8}}{\sqrt{27}}$

22) $\dfrac{10\sqrt{12}}{\sqrt{32}}$

23) $\dfrac{35\sqrt{6}}{14\sqrt{5}}$

24) $\dfrac{27\sqrt{7}}{45\sqrt{3}}$

25) $\sqrt{\dfrac{7}{14}}$

26) $\sqrt{\dfrac{10}{30}}$

27) $\sqrt{\dfrac{12}{80}}$

28) $\sqrt{\dfrac{14}{600}}$

29) $\sqrt{\dfrac{50}{2000}}$

30) $\sqrt{\dfrac{88}{180}}$

Multiply and simplify.

31) $\sqrt{\dfrac{10}{7}} \cdot \sqrt{\dfrac{7}{3}}$

32) $\sqrt{\dfrac{11}{6}} \cdot \sqrt{\dfrac{2}{11}}$

33) $\sqrt{\dfrac{8}{15}} \cdot \sqrt{\dfrac{3}{2}}$

34) $\sqrt{\dfrac{20}{21}} \cdot \sqrt{\dfrac{14}{15}}$

35) $\sqrt{\dfrac{1}{13}} \cdot \sqrt{\dfrac{8}{3}}$

36) $\sqrt{\dfrac{7}{5}} \cdot \sqrt{\dfrac{1}{10}}$

37) $\sqrt{\dfrac{3}{8}} \cdot \sqrt{\dfrac{27}{2}}$

38) $\sqrt{\dfrac{5}{12}} \cdot \sqrt{\dfrac{20}{3}}$

Simplify completely.

39) $\dfrac{5}{\sqrt{k}}$

40) $\dfrac{7}{\sqrt{r}}$

41) $\dfrac{\sqrt{6}}{\sqrt{d}}$

42) $\dfrac{\sqrt{3}}{\sqrt{u}}$

43) $\sqrt{\dfrac{16}{a}}$

44) $\sqrt{\dfrac{81}{z}}$

45) $\sqrt{\dfrac{x}{y}}$

46) $\sqrt{\dfrac{m}{n}}$

47) $\sqrt{\dfrac{40p^3}{q}}$

48) $\sqrt{\dfrac{63r^3}{t}}$

49) $\sqrt{\dfrac{64v^7}{5w}}$

50) $\sqrt{\dfrac{49t^5}{2u}}$

51) $\dfrac{\sqrt{18h^5}}{\sqrt{2h}}$

52) $\dfrac{\sqrt{84k^7}}{\sqrt{21k}}$

53) $\dfrac{\sqrt{15x^6}}{\sqrt{9y}}$

54) $\dfrac{\sqrt{11a^4}}{\sqrt{100b}}$

55) $\sqrt{\dfrac{2mn^2}{mn^3}}$

56) $\sqrt{\dfrac{6c^4d^2}{c^5d^2}}$

57) $\sqrt{\dfrac{8u^5v^3}{24u^6v}}$

58) $\sqrt{\dfrac{14x^5y^2}{28xy^3}}$

59) $\dfrac{\sqrt{500}}{\sqrt{p^3}}$

60) $\dfrac{\sqrt{128}}{\sqrt{w^5}}$

**Objective 2: Rationalize a Denominator Containing a Higher Root**

Fill in the blank.

61) $\sqrt[3]{2} \cdot \sqrt[3]{?} = \sqrt[3]{2^3} = 2$

62) $\sqrt[3]{5} \cdot \sqrt[3]{?} = \sqrt[3]{5^3} = 5$

63) $\sqrt[3]{9} \cdot \sqrt[3]{?} = \sqrt[3]{3^3} = 3$

64) $\sqrt[3]{4} \cdot \sqrt[3]{?} = \sqrt[3]{2^3} = 2$

65) $\sqrt[3]{c} \cdot \sqrt[3]{?} = \sqrt[3]{c^3} = c$

66) $\sqrt[3]{m^2} \cdot \sqrt[3]{?} = \sqrt[3]{m^3} = m$

67) To rationalize the denominator of $\dfrac{9}{\sqrt[3]{2}}$, Jason did the

following: $\dfrac{9}{\sqrt[3]{2}} \cdot \dfrac{\sqrt[3]{2}}{\sqrt[3]{2}} = \dfrac{9\sqrt[3]{2}}{2}$.

What did he do wrong? What should he have done?

68) To rationalize the denominator of $\dfrac{9}{\sqrt[3]{25}}$, the numerator

and denominator should be multiplied by $\sqrt[3]{5}$:

$$\dfrac{9}{\sqrt[3]{25}} \cdot \dfrac{\sqrt[3]{5}}{\sqrt[3]{5}} = \dfrac{9\sqrt[3]{5}}{\sqrt[3]{125}} = \dfrac{9\sqrt[3]{5}}{5}.$$

When you multiply the numerator and denominator by $\sqrt[3]{5}$, you are actually multiplying the fraction by what number?

Rationalize the denominator of each expression.

69) $\dfrac{4}{\sqrt[3]{3}}$

70) $\dfrac{11}{\sqrt[3]{5}}$

71) $\dfrac{1}{\sqrt[3]{4}}$

72) $\dfrac{1}{\sqrt[3]{9}}$

73) $\dfrac{35}{\sqrt[3]{5}}$

74) $\dfrac{26}{\sqrt[3]{2}}$

75) $\sqrt[3]{\dfrac{8}{9}}$

76) $\sqrt[3]{\dfrac{27}{4}}$

77) $\sqrt[3]{\dfrac{11}{25}}$

78) $\sqrt[3]{\dfrac{1}{81}}$

79) $\dfrac{6}{\sqrt[3]{x}}$

80) $\dfrac{10}{\sqrt[3]{p}}$

81) $\dfrac{7}{\sqrt[3]{m^2}}$

82) $\dfrac{3}{\sqrt[3]{h^2}}$

83) $\sqrt[3]{\dfrac{1000}{c}}$

84) $\sqrt[3]{\dfrac{64}{z}}$

85) $\sqrt[3]{\dfrac{4}{w^2}}$

86) $\sqrt[3]{\dfrac{9}{r^2}}$

87) $\dfrac{16}{\sqrt[3]{2t}}$

88) $\dfrac{12}{\sqrt[3]{2v^2}}$

89) $\dfrac{\sqrt[3]{5a}}{\sqrt[3]{36b}}$

90) $\dfrac{\sqrt[3]{3v}}{\sqrt[3]{5u^2}}$

91) $\dfrac{7}{\sqrt[4]{2}}$

92) $\dfrac{5}{\sqrt[4]{3}}$

93) $\dfrac{2}{\sqrt[4]{x^3}}$

94) $\dfrac{13}{\sqrt[4]{t}}$

## Objective 3: Rationalize a Denominator Containing Two Terms

95) How do you find the conjugate of a binomial?

96) When you multiply a binomial containing a square root by its conjugate, what happens to the radical?

Find the conjugate of each binomial. Then multiply the binomial by its conjugate.

97) $(4 + \sqrt{3})$

98) $(\sqrt{7} - 9)$

99) $(\sqrt{5} - \sqrt{11})$

100) $(\sqrt{13} + \sqrt{2})$

101) $(\sqrt{p} - 6)$

102) $(\sqrt{r} + 7)$

Rationalize the denominator and simplify completely.

### Fill It In

Fill in the blanks with either the missing mathematical step or reason for the given step.

103) $\dfrac{5}{4 - \sqrt{3}} = \dfrac{5}{4 - \sqrt{3}} \cdot \dfrac{4 + \sqrt{3}}{4 + \sqrt{3}}$ _____

$= \dfrac{5(4 + \sqrt{3})}{(4)^2 - (\sqrt{3})^2}$ _____

$=$ _____ Multiply terms in numerator, square terms in denominator.

$=$ _____ Simplify.

104) $\dfrac{\sqrt{2}}{\sqrt{11} + \sqrt{6}}$

$= \dfrac{\sqrt{2}}{\sqrt{11} + \sqrt{6}} \cdot \dfrac{\sqrt{11} - \sqrt{6}}{\sqrt{11} - \sqrt{6}}$ _____

$= \dfrac{\sqrt{2}(\sqrt{11} - \sqrt{6})}{(\sqrt{11})^2 - (\sqrt{6})^2}$ _____

$=$ _____ Multiply terms in numerator, square terms in denominator.

$=$ _____ Simplify.

105) $\dfrac{6}{3 + \sqrt{2}}$

106) $\dfrac{7}{4 - \sqrt{5}}$

107) $\dfrac{5}{7 - \sqrt{3}}$

108) $\dfrac{1}{6 + \sqrt{7}}$

109) $\dfrac{10}{\sqrt{5} - 6}$

110) $\dfrac{\sqrt{6}}{\sqrt{3} + 1}$

111) $\dfrac{\sqrt{2}}{\sqrt{10} - 3}$

112) $\dfrac{9 + \sqrt{7}}{5 - \sqrt{7}}$

113) $\dfrac{5 + \sqrt{3}}{4 + \sqrt{3}}$

114) $\dfrac{6 + \sqrt{2}}{\sqrt{2} + 1}$

115) $\dfrac{1 - 2\sqrt{3}}{\sqrt{2} - \sqrt{3}}$

116) $\dfrac{3\sqrt{6} - \sqrt{2}}{\sqrt{7} - \sqrt{5}}$

117) $\dfrac{12}{\sqrt{w} + 3}$

118) $\dfrac{7}{\sqrt{h} - 1}$

119) $\dfrac{9}{5 - \sqrt{n}}$

120) $\dfrac{4}{2 + \sqrt{x}}$

121) $\dfrac{\sqrt{m}}{\sqrt{m} + \sqrt{n}}$

122) $\dfrac{\sqrt{x} + \sqrt{y}}{\sqrt{x} - \sqrt{y}}$

## Objective 4: Divide Out Common Factors in Radical Quotients

Simplify completely.

123) $\dfrac{9 + 36\sqrt{5}}{9}$

124) $\dfrac{21 - 14\sqrt{11}}{7}$

125) $\dfrac{20 - 44\sqrt{3}}{12}$

126) $\dfrac{12 + 21\sqrt{2}}{15}$

127) $\dfrac{\sqrt{45} + 6}{9}$

128) $\dfrac{\sqrt{60} - 18}{4}$

129) $\dfrac{-24 - \sqrt{800}}{4}$

130) $\dfrac{-18 + \sqrt{96}}{8}$

## Mixed Exercises: Objectives 1–4

Rationalize the denominator and simplify completely.

131) $\dfrac{\sqrt{11}}{\sqrt{50}}$

132) $\dfrac{5\sqrt{2}}{\sqrt{t}}$

133) $\sqrt[3]{\dfrac{4}{3}}$

134) $\dfrac{9}{\sqrt{2} + 4}$

135) $\sqrt{\dfrac{125a^3}{6b}}$

136) $\sqrt{\dfrac{54}{8}}$

137) $\dfrac{3 + \sqrt{10}}{\sqrt{10} - 9}$

138) $-\dfrac{12\sqrt{3}}{\sqrt{20}}$

139) $\sqrt{\dfrac{108}{63}}$

140) $\sqrt{\dfrac{20c^3}{11d}}$

141) $\dfrac{\sqrt{256}}{\sqrt{2y}}$

142) $\sqrt[3]{\dfrac{1}{81}}$

143) The equation $r = \sqrt{\dfrac{A}{\pi}}$ describes the radius of a circle, $r$, in terms of its area, $A$. If the area of a circle is measured in square inches, find $r$ when $A = 12\pi$.

144) The equation $r = \sqrt{\dfrac{V}{\pi h}}$ describes the radius of a right circular cylinder, $r$, in terms of its volume, $V$, and height, $h$. If the volume of a cylinder is measured in cubic inches, find $r$ when the volume is $20\pi$ in³ and the cylinder is 5 in. tall.

# Section 9.6 Solving Radical Equations

## Objectives

1. **Understand the Steps for Solving a Radical Equation**
2. **Solve a Radical Equation Containing One Square Root**
3. **Solve a Radical Equation Containing Two Square Roots**
4. **Solve a Radical Equation by Squaring a Binomial**
5. **Solve a Radical Equation Containing a Cube Root**

In this section, we will learn how to solve *radical equations*.

An equation containing a variable in the radicand is a **radical equation.** Some examples of radical equations are

$$\sqrt{x} = 5, \qquad \sqrt{4w - 3} + 2 = w, \qquad \text{and} \qquad \sqrt{3y + 1} - \sqrt{y - 1} = 2.$$

## 1. Understand the Steps for Solving a Radical Equation

Let's review what happens when we square a square root expression: If $x \geq 0$, then $(\sqrt{x})^2 = x$. That is, to eliminate the radical from $\sqrt{x}$, we *square* the expression.

Therefore to solve equations like those above containing *square roots*, we *square* both sides of the equation to obtain new equations. The solutions of the new equations contain all of the solutions of the original equation and may also contain *extraneous solutions*.

An **extraneous solution** is a value that satisfies one of the new equations but does not satisfy the original equation. Extraneous solutions occur frequently when solving radical equations, so we *must* check all possible solutions in the original equation and discard any that are extraneous.

---

**Procedure**   Steps for Solving Radical Equations Containing Square Roots

**Step 1:** Get a radical on a side by itself.

**Step 2:** Square both sides of the equation to eliminate a radical.

**Step 3:** Combine like terms on each side of the equation.

**Step 4:** If the equation still contains a radical, repeat Steps 1–3.

**Step 5:** Solve the equation.

**Step 6:** Check the proposed solutions *in the original equation* and discard extraneous solutions.

---

## 2. Solve a Radical Equation Containing One Square Root

### Example 1

Solve.

a)  $\sqrt{x + 5} = 4$          b)  $\sqrt{p - 9} + 6 = 8$

**Solution**

a) **Step 1:** The radical *is* on a side by itself: $\sqrt{x + 5} = 4$

**Step 2:** *Square* both sides to eliminate the *square root*.

$$(\sqrt{x + 5})^2 = 4^2 \qquad \text{Square both sides.}$$
$$x + 5 = 16$$

**Steps 3 and 4** do not apply because there are no like terms to combine and no radicals remain.

**Step 5:** Solve the equation.

$$x = 11 \qquad \text{Subtract 5 from each side.}$$

**Step 6:** Check $x = 11$ in the *original* equation.

$$\sqrt{x + 5} = 4$$
$$\sqrt{11 + 5} \stackrel{?}{=} 4$$
$$\sqrt{16} = 4 \quad \checkmark$$

The solution set is $\{11\}$.

b) The first step is to get the radical on a side by itself.

$$\sqrt{p - 9} + 6 = 8$$
$$\sqrt{p - 9} = 2 \qquad \text{Subtract 6 from each side.}$$
$$(\sqrt{p - 9})^2 = 2^2 \qquad \text{Square both sides to eliminate the radical.}$$
$$p - 9 = 4 \qquad \text{The square root has been eliminated.}$$
$$p = 13 \qquad \text{Add 9 to each side.}$$

Check $p = 13$ in the *original* equation.

$$\sqrt{p - 9} + 6 = 8$$
$$\sqrt{13 - 9} + 6 \overset{?}{=} 8$$
$$2 + 6 = 8 \quad \checkmark$$

The solution set is $\{13\}$. ∎

**You Try 1**

Solve.

a)  $\sqrt{t + 10} = 3$      b)  $\sqrt{w - 3} + 7 = 11$

**Example 2**

Solve $t = \sqrt{t^2 + 2t + 3}$.

**Solution**

The radical is already on a side by itself, so we will square both sides.

$$t = \sqrt{t^2 + 2t + 3}$$
$$t^2 = (\sqrt{t^2 + 2t + 3})^2 \qquad \text{Square both sides to eliminate the radical.}$$
$$t^2 = t^2 + 2t + 3 \qquad \text{The square root has been eliminated.}$$

Solve the equation.

$$0 = 2t + 3 \qquad \text{Subtract } t^2 \text{ from each side.}$$
$$-2t = 3 \qquad \text{Subtract } 2t \text{ from each side.}$$
$$t = -\frac{3}{2} \qquad \text{Divide by } -2.$$

Check $t = -\dfrac{3}{2}$ in the *original* equation.

$$t = \sqrt{t^2 + 2t + 3}$$
$$-\frac{3}{2} \overset{?}{=} \sqrt{\left(-\frac{3}{2}\right)^2 + 2\left(-\frac{3}{2}\right) + 3}$$
$$-\frac{3}{2} \overset{?}{=} \sqrt{\frac{9}{4} - 3 + 3}$$
$$-\frac{3}{2} \overset{?}{=} \sqrt{\frac{9}{4}}$$
$$-\frac{3}{2} = \frac{3}{2} \qquad \text{FALSE}$$

Because $t = -\dfrac{3}{2}$ does not satisfy the *original* equation, it is an extraneous solution. There is no real solution to this equation. The solution is $\varnothing$. ∎

**You Try 2**

Solve $c = \sqrt{c^2 - 8c - 4}$.

### 3. Solve a Radical Equation Containing Two Square Roots

Next, we will take our first look at solving an equation containing two square roots.

**Example 3**

Solve $\sqrt{7n + 1} - 4\sqrt{n - 5} = 0$.

**Solution**

Begin by getting a radical on a side by itself.

$$\sqrt{7n + 1} = 4\sqrt{n - 5} \qquad \text{Add } 4\sqrt{n-5} \text{ to each side.}$$
$$(\sqrt{7n + 1})^2 = (4\sqrt{n - 5})^2 \qquad \text{Square both sides to eliminate the radicals.}$$
$$7n + 1 = 16(n - 5) \qquad 4^2 = 16$$
$$7n + 1 = 16n - 80 \qquad \text{Distribute.}$$
$$-9n = -81$$
$$n = 9 \qquad \text{Solve.}$$

Check $n = 9$ in the original equation.

$$\sqrt{7n + 1} - 4\sqrt{n - 5} = 0$$
$$\sqrt{7(9) + 1} - 4\sqrt{9 - 5} = 0$$
$$\sqrt{63 + 1} - 4\sqrt{4} \stackrel{?}{=} 0$$
$$\sqrt{64} - 4(2) \stackrel{?}{=} 0$$
$$8 - 8 = 0 \quad \checkmark \qquad \text{The solution set is } \{9\}. \qquad ■$$

**You Try 3**

Solve $\sqrt{3z + 5} - 2\sqrt{z - 2} = 0$.

### 4. Solve a Radical Equation by Squaring a Binomial

Sometimes, we have to square a binomial in order to solve a radical equation.

**Example 4**

Solve $\sqrt{4w - 3} + 2 = w$.

**Solution**

As usual, start by getting the radical on a side by itself.

$$\sqrt{4w - 3} = w - 2 \qquad \text{Subtract 2 from each side.}$$
$$(\sqrt{4w - 3})^2 = (w - 2)^2 \qquad \text{Square both sides to eliminate the radicals.}$$
$$4w - 3 = w^2 - 4w + 4 \qquad \text{Simplify; square the binomial.}$$
$$0 = w^2 - 8w + 7 \qquad \text{Subtract } 4w; \text{ add 3.}$$
$$0 = (w - 1)(w - 7) \qquad \text{Factor.}$$

$$w - 1 = 0 \quad \text{or} \quad w - 7 = 0 \qquad \text{Set each factor equal to zero.}$$
$$w = 1 \quad \text{or} \qquad w = 7 \qquad \text{Solve.}$$

Check $w = 1$ and $w = 7$ in the *original* equation.

$w = 1$:    $\sqrt{4w - 3} + 2 = w$            $w = 7$:    $\sqrt{4w - 3} + 2 = w$

$\sqrt{4(1) - 3} + 2 \overset{?}{=} 1$                        $\sqrt{4(7) - 3} + 2 \overset{?}{=} 7$

$\sqrt{1} + 2 \overset{?}{=} 1$                            $\sqrt{25} + 2 \overset{?}{=} 7$

$3 \overset{?}{=} 1$    FALSE                        $5 + 2 = 7$    TRUE

$w = 7$ *is* a solution but $w = 1$ is *not* because $w = 1$ does not satisfy the original equation. The solution set is $\{7\}$.    ■

**You Try 4**

Solve $\sqrt{3r - 5} + 3 = r$.

Recall from Section 9.4 that we can square binomials containing radical expressions just like we squared the binomial $(w - 2)^2$ in Example 4. We can use the formulas

$$(a + b)^2 = a^2 + 2ab + b^2 \qquad \text{and} \qquad (a - b)^2 = a^2 - 2ab + b^2$$

**Example 5**

Square the binomial and simplify $(3 - \sqrt{h + 6})^2$.

**Solution**

Use the formula $(a - b)^2 = a^2 - 2ab + b^2$.

$(3 - \sqrt{h + 6})^2 = (3)^2 - 2(3)(\sqrt{h + 6}) + (\sqrt{h + 6})^2$    Substitute 3 for $a$ and $\sqrt{h + 6}$ for $b$.

$= 9 - 6\sqrt{h + 6} + (h + 6)$

$= h + 15 - 6\sqrt{h + 6}$    Combine like terms.    ■

**You Try 5**

Square each binomial and simplify.

a)   $(\sqrt{t} - 5)^2$            b)   $(2 + \sqrt{n - 8})^2$

To solve the next two equations, we will have to square both sides of the equation twice to eliminate the radicals. Be very careful when you are squaring the binomials that contain a radical.

**Example 6**

Solve each equation.

a)   $\sqrt{a + 7} + \sqrt{a} = 7$            b)   $\sqrt{3y + 1} - \sqrt{y - 1} = 2$

**Solution**

a)   This equation contains two radicals *and* a constant. Get one of the radicals on a side by itself, then square both sides.

$\sqrt{a + 7} = 7 - \sqrt{a}$                        Subtract $\sqrt{a}$ from each side.

$(\sqrt{a + 7})^2 = (7 - \sqrt{a})^2$                    Square both sides.

$a + 7 = (7)^2 - 2(7)(\sqrt{a}) + (\sqrt{a})^2$        Use the formula $(a - b)^2 = a^2 - 2ab + b^2$.

$a + 7 = 49 - 14\sqrt{a} + a$                    Simplify.

The equation still contains a radical. Therefore, repeat steps 1–3. Begin by getting the radical on a side by itself.

$$7 = 49 - 14\sqrt{a}$$   Subtract $a$ from each side.
$$-42 = -14\sqrt{a}$$   Subtract 49 from each side.
$$3 = \sqrt{a}$$   Divide by $-14$.
$$3^2 = (\sqrt{a})^2$$   Square both sides.
$$9 = a$$   Solve.

The check is left to the student. The solution set is $\{9\}$.

b) **Step 1:**   Get a radical on a side by itself.

$$\sqrt{3y + 1} - \sqrt{y - 1} = 2$$
$$\sqrt{3y + 1} = 2 + \sqrt{y - 1}$$   Add $\sqrt{y - 1}$ to each side.

**Step 2:**   Square both sides of the equation to eliminate a radical.

$$(\sqrt{3y + 1})^2 = (2 + \sqrt{y - 1})^2$$   Square both sides.
$$3y + 1 = (2)^2 + 2(2)(\sqrt{y - 1}) + (\sqrt{y - 1})^2$$   Use the formula $(a + b)^2 = a^2 + 2ab + b^2$.
$$3y + 1 = 4 + 4\sqrt{y - 1} + y - 1$$

**Step 3:**   Combine like terms on the right side.

$$3y + 1 = y + 3 + 4\sqrt{y - 1}$$   Combine like terms.

**Step 4:**   The equation still contains a radical, so repeat Steps 1–3.

**Step 1:**   Get the radical on a side by itself.

$$3y + 1 = y + 3 + 4\sqrt{y - 1}$$
$$2y - 2 = 4\sqrt{y - 1}$$   Subtract $y$ and subtract 3.
$$y - 1 = 2\sqrt{y - 1}$$   Divide by 2.

We do not need to eliminate the 2 from in front of the radical before squaring both sides. The radical must not be a part of a *sum* or *difference* when we square.

**Step 2:**   Square both sides of the equation to eliminate the radical.

$$(y - 1)^2 = (2\sqrt{y - 1})^2$$   Square both sides.
$$y^2 - 2y + 1 = 4(y - 1)$$   Square the binomial; $2^2 = 4$.

Steps 3 and 4 no longer apply.

**Step 5:**   Solve the equation.

$$y^2 - 2y + 1 = 4y - 4$$   Distribute.
$$y^2 - 6y + 5 = 0$$   Subtract $4y$; add 4.
$$(y - 1)(y - 5) = 0$$   Factor.

$$y - 1 = 0 \quad \text{or} \quad y - 5 = 0$$   Set each factor equal to zero.
$$y = 1 \quad \text{or} \quad y = 5$$   Solve.

**Step 6:**   The check is left to the student. Verify that $y = 1$ and $y = 5$ each satisfy the original equation. The solution set is $\{1, 5\}$. ∎

**You Try 6**

Solve each equation.

a)   $\sqrt{r + 9} + \sqrt{r} = 9$        b)   $\sqrt{5z + 6} - \sqrt{4z + 1} = 1$

Watch out for two common mistakes that students make when solving an equation like the one in Example 6b.

1) Do not square both sides before getting a radical on a side by itself.

**This is incorrect:**

$$(\sqrt{3y + 1} - \sqrt{y - 1})^2 = 2^2$$
$$3y + 1 - (y - 1) = 4$$

On the left we must multiply using FOIL or the formula $(a - b)^2 = a^2 - 2ab + b^2$.

2) The *second* time we perform Step 2, watch out for this common error:

**This is incorrect:**

$$(y - 1)^2 = (2\sqrt{y - 1})^2$$
$$y^2 - 1 = 2(y - 1)$$

On the left we must multiply using FOIL or the formula $(a - b)^2 = a^2 - 2ab + b^2$ and on the right we must remember to square the 2.

### 5. Solve a Radical Equation Containing a Cube Root

We can solve many equations containing cube roots the same way we solve equations containing square roots except, to eliminate a *cube root*, we *cube* both sides of the equation.

**Example 7**

Solve $\sqrt[3]{4x} - \sqrt[3]{x + 2} = 0$.

**Solution**

Begin by getting a radical on a side by itself.

$$\sqrt[3]{4x} = \sqrt[3]{x + 2} \qquad \text{Add } \sqrt[3]{x + 2} \text{ to each side.}$$
$$(\sqrt[3]{4x})^3 = (\sqrt[3]{x + 2})^3 \qquad \text{Cube both sides to eliminate the radicals.}$$
$$4x = x + 2 \qquad \text{Simplify.}$$
$$3x = 2 \qquad \text{Subtract } x.$$
$$x = \frac{2}{3} \qquad \text{Solve.}$$

Check $x = \frac{2}{3}$ in the original equation.

$$\sqrt[3]{4x} - \sqrt[3]{x + 2} = 0$$
$$\sqrt[3]{4\left(\frac{2}{3}\right)} - \sqrt[3]{\left(\frac{2}{3}\right) + 2} \stackrel{?}{=} 0$$
$$\sqrt[3]{\frac{8}{3}} - \sqrt[3]{\frac{2}{3} + \frac{6}{3}} \stackrel{?}{=} 0$$
$$\sqrt[3]{\frac{8}{3}} - \sqrt[3]{\frac{8}{3}} = 0 \checkmark \qquad \text{The solution set is } \left\{\frac{2}{3}\right\}.$$

**You Try 7**

Solve $\sqrt[3]{7m - 12} - \sqrt[3]{3m} = 0$.

## Using Technology

We can use a graphing calculator to solve a radical equation in one variable. First subtract every term on the right side of the equation from both sides of the equation and enter the result in $Y_1$. Graph the equation in $Y_1$. The zeros or $x$-intercepts of the graph are the solutions to the equation.

We will solve $\sqrt{x+3} = 2$ using a graphing calculator.

1. Enter $\sqrt{x+3} - 2$ in $Y_1$.

2. Press $\boxed{\text{ZOOM}}$ $\boxed{6}$ to graph the function in $Y_1$ as shown.

3. Press $\boxed{\text{2nd}}$ $\boxed{\text{TRACE}}$ 2:zero, move the cursor to the left of the zero and press $\boxed{\text{ENTER}}$, move the cursor to the right of the zero and press $\boxed{\text{ENTER}}$, and move the cursor close to the zero and press $\boxed{\text{ENTER}}$ to display the zero. The solution to the equation is $x = 1$.

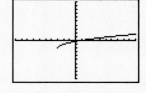
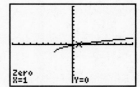

Solve each equation using a graphing calculator.

1) $\sqrt{x-2} = 1$  

2) $\sqrt{3x-2} = 5$

3) $\sqrt{3x-2} = \sqrt{x+2}$  

4) $\sqrt{4x-5} = \sqrt{x+4}$

5) $\sqrt{2x-7} = \sqrt{x}-1$  

6) $\sqrt{\sqrt{x}-1} = 1$

---

**Answers to You Try Exercises**

1) a) $\{-1\}$  b) $\{19\}$    2) $\varnothing$    3) $\{13\}$    4) $\{7\}$

5) a) $t - 10\sqrt{t} + 25$  b) $n + 4\sqrt{n-8} - 4$    6) a) $\{16\}$  b) $\{2, 6\}$

---

**Answers to Technology Exercises**

1) $\{3\}$    2) $\{9\}$    3) $\{9\}$    4) $\{3\}$    5) $\{4\}$    6) $\{4\}$

---

# 9.6 Exercises

**Objective 1: Understand the Steps for Solving a Radical Equation**

1) Why is it necessary to check the proposed solutions to a radical equation in the original equation?

2) How do you know, without actually solving and checking the solution, that $\sqrt{x} = -2$ has no solution?

**Objective 2: Solve a Radical Equation Containing One Square Root**

Solve.

3) $\sqrt{d} = 12$

4) $\sqrt{w} = 5$

5) $\sqrt{x} = \dfrac{1}{3}$

6) $\sqrt{t} = \dfrac{5}{8}$

7) $\sqrt{b-11} - 3 = 0$

8) $\sqrt{z+9} - 1 = 0$

9) $\sqrt{c-6} + 3 = 4$

10) $\sqrt{y+7} - 2 = 3$

11) $-9 = \sqrt{3k+1} - 11$

12) $8 = \sqrt{2n+1} - 1$

13) $\sqrt{5p+8} + 7 = 5$

14) $\sqrt{2b-1} + 3 = 0$

15) $\sqrt{5-2r} - 13 = -7$

16) $8 = 11 - \sqrt{4-3v}$

17) $g = \sqrt{g^2 + 8g - 24}$

18) $x = \sqrt{x^2 + 9x - 36}$

19) $\sqrt{v^2 - 7v - 42} = v$

20) $c = \sqrt{c^2 + c + 9}$

21) $2p = \sqrt{4p^2 - 3p + 6}$

22) $\sqrt{9r^2 - 2r + 10} = 3r$

**Objective 3: Solve a Radical Equation Containing Two Square Roots**

Solve.

23) $\sqrt{5y+4} = \sqrt{y+8}$

24) $\sqrt{3a+1} = \sqrt{5a-9}$

25) $\sqrt{z} = 3\sqrt{7}$

26) $4\sqrt{3} = \sqrt{a}$

27) $\sqrt{3-11h} - 3\sqrt{h+7} = 0$

28) $3\sqrt{t + 32} - 5\sqrt{t + 16} = 0$

29) $\sqrt{2p - 1} + 2\sqrt{p + 4} = 0$

30) $2\sqrt{3c + 4} + \sqrt{c - 6} = 0$

## Objective 4: Solve a Radical Equation by Squaring a Binomial

Solve.

31) $\sqrt{x} = x - 6$

32) $\sqrt{h} = h - 2$

33) $m + 4 = 5\sqrt{m}$

34) $b + 5 = 6\sqrt{b}$

35) $c = \sqrt{5c - 9} + 3$

36) $\sqrt{2k + 11} + 2 = k$

37) $6 + \sqrt{w^2 + 3w - 9} = w$

38) $\sqrt{z^2 + 5z - 8} = z + 4$

39) $\sqrt{7u - 5} - u = 1$

40) $3 = \sqrt{10z + 6} - z$

41) $a - 2\sqrt{a + 1} = -1$

42) $r - 3\sqrt{r + 2} = 2$

Square each binomial and simplify.

43) $(y + 7)^2$

44) $(t - 5)^2$

45) $(\sqrt{n} + 7)^2$

46) $(\sqrt{w} - 5)^2$

47) $(4 - \sqrt{c + 3})^2$

48) $(3 + \sqrt{x - 2})^2$

49) $(2\sqrt{3p - 4} + 1)^2$

50) $(3\sqrt{2k + 1} - 4)^2$

Solve.

51) $\sqrt{w + 5} = 5 - \sqrt{w}$

52) $\sqrt{p + 8} = 4 - \sqrt{p}$

53) $\sqrt{y} - \sqrt{y - 13} = 1$

54) $\sqrt{d} - \sqrt{d - 16} = 2$

55) $\sqrt{3t + 10} - \sqrt{2t} = 2$

56) $\sqrt{4z - 3} - \sqrt{5z + 1} = -1$

57) $\sqrt{4c + 5} + \sqrt{2c - 1} = 4$

58) $4 = \sqrt{3a} + \sqrt{a - 2}$

59) $3 - \sqrt{3h + 1} = \sqrt{3h - 14}$

60) $\sqrt{7x - 6} - 1 = \sqrt{7x - 3}$

61) $\sqrt{5a + 19} - \sqrt{a + 12} = 1$

62) $\sqrt{v + 7} - \sqrt{2v + 7} = 1$

63) $3 = \sqrt{5 - 2y} + \sqrt{y + 2}$

64) $4 + \sqrt{r - 4} = \sqrt{3r + 4}$

65) $\sqrt{2n - 5} = \sqrt{2n + 3} - 2$

66) $-3 = \sqrt{5k - 9} - \sqrt{5k + 6}$

67) $\sqrt{3b + 4} - \sqrt{2b + 9} = -1$

68) $\sqrt{z - 4} + \sqrt{5z} = 6$

## Objective 5: Solve a Radical Equation Containing a Cube Root

69) How do you eliminate the radical from an equation like $\sqrt[3]{x} = 2$?

70) Juanita says that there is no solution to the equation $\sqrt[3]{c} = -3$, while Alejandra says that the solution is $\{-27\}$. Who is correct and why?

Solve.

71) $\sqrt[3]{n} = -5$

72) $\sqrt[3]{v} = 4$

73) $\sqrt[3]{6j - 2} = \sqrt[3]{j - 7}$

74) $\sqrt[3]{3m} = \sqrt[3]{2m + 9}$

75) $\sqrt[3]{5a + 4} - \sqrt[3]{2a - 3} = 0$

76) $\sqrt[3]{3p + 10} - \sqrt[3]{5p + 11} = 0$

77) $\sqrt[3]{x^2} - \sqrt[3]{5x + 14} = 0$

78) $\sqrt[3]{33 - 8g} - \sqrt[3]{g^2} = 0$

79) $\sqrt[3]{4h - 1} = \sqrt[3]{2h^2 + 15h + 11}$

80) $\sqrt[3]{4w^2 - 19w + 8} = \sqrt[3]{w^2 + 2}$

Extend the concepts of this section to solve the following equations.

81) $\sqrt[4]{y + 10} = 3$

82) $1 = \sqrt[4]{z - 9}$

83) $2 = \sqrt[4]{b^2 - 6b}$

84) $\sqrt[4]{k^2 + 16k + 1} = 3$

85) $\sqrt[5]{4x} = -2$

86) $\sqrt[5]{3p} = -1$

Write an equation and solve.

87) The sum of 7 and the square root of a number is 12. Find the number.

88) The sum of 6 and the square root of a number equals the number. Find the number.

89) Three times the square root of a number is the sum of 4 and the square root of the number. Find the number.

90) Twice the square root of a number is 8 less than the number. Find the number.

Solve.

91) If the area of a square is $A$ and each side has length $l$, then the length of a side is given by $l = \sqrt{A}$. A square rug has an area of 36 ft$^2$. Find the dimensions of the rug.

92) If the area of a circle is $A$ and the radius is $r$, then the radius of the circle is given by $r = \sqrt{\dfrac{A}{\pi}}$. The center circle on a basketball court has an area of 113.1 ft$^2$. What is the radius of the center circle to the nearest tenth?

At the beginning of the chapter, we learned that forensic scientists can use the formula $S = \sqrt{30\,fd}$ to determine the speed, $S$ in mph, of a vehicle in an accident based on the drag factor, $f$, of the road and the length of the skid marks, $d$, in feet. Use this formula for Exercises 93 and 94.

93) Accident investigators respond to a car accident on a highway where the speed limit is 65 mph. They determine that the concrete highway has a drag factor of 0.90. The skid marks left by the car are 142 ft long. Find the speed of the car, to the nearest unit, at the time of the accident to determine whether the car was exceeding the speed limit.

94) A car involved in an accident on a busy street leaves skid marks 80 ft long. Accident investigators determine that the asphalt street has a drag factor of 0.78. If the speed limit is 35 mph, was the driver speeding at the time of the accident? Support your answer by finding the speed of the car to the nearest unit.

 Use the following information for Exercises 95 and 96.

The distance a person can see to the horizon is approximated by $d = 1.2\sqrt{h}$, where $d$ is the number of miles a person can see to the horizon from a height of $h$ feet.

95) Andrew is in a hot air balloon and can see 48 miles to the horizon. Find his height above the ground.

96) On a clear day, visitors to Chicago's Willis Tower (formerly the Sears Tower) can see about 44.1 miles to the horizon. Find the height of the observation deck at Willis Tower to the nearest foot. (www.the-skydeck.com)

## Section 9.7 Rational Exponents

### Objectives

1. Evaluate Expressions of the Form $a^{1/n}$
2. Evaluate Expressions of the Form $a^{m/n}$
3. Evaluate Expressions of the Form $a^{-m/n}$
4. Combine the Rules of Exponents to Simplify Expressions
5. Convert a Radical Expression to Exponential Form and Simplify

### 1. Evaluate Expressions of the Form $a^{1/n}$

In this section, we will explain the relationship between radicals and rational exponents (fractional exponents). Sometimes, converting between these two forms makes it easier to simplify expressions.

**Formula**   $a^{1/n} = \sqrt[n]{a}$

If $a$ is a nonnegative number and $n$ is a positive integer greater than 1, then

$$a^{1/n} = \sqrt[n]{a}$$

(The denominator of the fractional exponent is the index of the radical.)

---

**Example 1**

Write in radical form and evaluate.

a) $8^{1/3}$     b) $25^{1/2}$     c) $81^{1/4}$

**Solution**

a) The denominator of the fractional exponent is the index of the radical: $8^{1/3} = \sqrt[3]{8} = 2$.

b) The denominator in the exponent of $25^{1/2}$ is 2, so the index of the radical is 2, meaning *square* root.

$$25^{1/2} = \sqrt{25} = 5$$

c) $81^{1/4} = \sqrt[4]{81} = 3$

---

 **You Try 1**

Write in radical form and evaluate.

a) $125^{1/3}$     b) $36^{1/2}$     c) $64^{1/6}$

## 2. Evaluate Expressions of the Form $a^{m/n}$

Let's look at another relationship between rational exponents and radicals.

> **Formula**   $a^{m/n} = (a^{1/n})^m = (\sqrt[n]{a})^m$
>
> If $a$ is a nonnegative number and $m$ and $n$ are integers such that $n > 1$,
>
> $$a^{m/n} = (a^{1/n})^m = (\sqrt[n]{a})^m$$
>
> (The denominator of the fractional exponent is the index of the radical, and the numerator is the power to which we raise the radical expression.)

We can also think of $a^{m/n}$ in this way: $a^{m/n} = (a^m)^{1/n} = \sqrt[n]{a^m}$.

### Example 2

Write in radical form and evaluate.

a)   $16^{3/2}$          b)   $-125^{2/3}$

**Solution**

a)   The denominator of the fractional exponent is the index of the radical, and the numerator is the power to which we raise the radical expression.

$$16^{3/2} = (16^{1/2})^3 \qquad \text{Use the definition to rewrite the exponent.}$$
$$= (\sqrt{16})^3 \qquad \text{Rewrite as a radical.}$$
$$= 4^3 \qquad\quad \sqrt{16} = 4$$
$$= 64$$

b)   To evaluate $-125^{2/3}$, *first* evaluate $125^{2/3}$, *then* take the negative of that result.

$$-125^{2/3} = -(125^{2/3}) = -(125^{1/3})^2 \qquad \text{Use the definition to rewrite the exponent.}$$
$$= -(\sqrt[3]{125})^2 \qquad\quad \text{Rewrite as a radical.}$$
$$= -(5)^2 \qquad\qquad \sqrt[3]{125} = 5$$
$$= -25$$

### You Try 2

Write in radical form and evaluate.

a)   $36^{3/2}$          b)   $8^{5/3}$

## 3. Evaluate Expressions of the Form $a^{-m/n}$

Recall from Section 2.2 that if $n$ is any integer and $a \neq 0$, then $a^{-n} = \dfrac{1}{a^n}$.

For example, $6^{-2} = \dfrac{1}{6^2} = \dfrac{1}{36}$.

We can extend this to rational exponents.

> **Formula**   $a^{-m/n} = \dfrac{1}{a^{m/n}}$
>
> If $a$ is a positive number and $m$ and $n$ are integers such that $n > 1$,
>
> $$a^{-m/n} = \dfrac{1}{a^{m/n}}$$

## Example 3

Rewrite with a positive exponent and evaluate.

a)  $49^{-1/2}$          b)  $64^{-5/6}$

**Solution**

a)  $49^{-1/2} = \dfrac{1}{49^{1/2}}$        $a^{-m/n} = \dfrac{1}{a^{m/n}}$

$= \dfrac{1}{\sqrt{49}}$        The denominator of the fractional exponent
is the index of the radical.

$= \dfrac{1}{7}$

b)  $64^{-5/6} = \dfrac{1}{64^{5/6}}$

$= \dfrac{1}{(\sqrt[6]{64})^5}$        The denominator of the fractional exponent
is the index of the radical.

$= \dfrac{1}{2^5}$        $\sqrt[6]{64} = 2$

$= \dfrac{1}{32}$

**BE CAREFUL**

The negative exponent does not make the expression negative! For example,

$$49^{-1/2} \neq -49^{1/2}$$

## You Try 3

Rewrite with a positive exponent and evaluate.

a)  $100^{-1/2}$          b)  $32^{-2/5}$

We can combine the rules presented in this section with the rules of exponents we learned in Chapter 2 to simplify expressions containing numbers or variables.

## 4. Combine the Rules of Exponents to Simplify Expressions

## Example 4

Simplify completely. The answer should contain only positive exponents.

a)  $(7^{1/9})^2$          b)  $36^{5/8} \cdot 36^{-1/8}$          c)  $\dfrac{4^{7/6}}{4^{13/6}}$

**Solution**

a)  $(7^{1/9})^2 = 7^{2/9}$        Multiply exponents.

b)  $36^{5/8} \cdot 36^{-1/8} = 36^{\frac{5}{8}+(-\frac{1}{8})}$        Add exponents.

$= 36^{4/8} = 36^{1/2} = 6$

c)  $\dfrac{4^{7/6}}{4^{13/6}} = 4^{7/6 - 13/6}$        Subtract exponents.

$= 4^{-6/6} = 4^{-1} = \dfrac{1}{4^1} = \dfrac{1}{4}$

**You Try 4**

Simplify completely. The answer should contain only positive exponents.

a)  $(4^{1/6})^3$     b)  $125^{5/9} \cdot 125^{-2/9}$     c)  $\dfrac{8^{8/3}}{8^{10/3}}$

---

**Example 5**

Simplify completely. The answer should contain only positive exponents.

a)  $k^{1/12} \cdot k^{5/12}$     b)  $\left(\dfrac{a^{3/4}}{b^{1/6}}\right)^8$     c)  $\dfrac{x^{-5/4} \cdot x^{1/2}}{x^{-1/4}}$

**Solution**

a)  $k^{1/12} \cdot k^{5/12} = k^{\frac{1}{12}+\frac{5}{12}}$          Add exponents.
$= k^{6/12}$
$= k^{1/2}$

b)  $\left(\dfrac{a^{3/4}}{b^{1/6}}\right)^8 = \dfrac{a^{\frac{3}{4}\cdot 8}}{b^{\frac{1}{6}\cdot 8}}$          Multiply exponents.

$= \dfrac{a^6}{b^{4/3}}$          Multiply and reduce.

c)  $\dfrac{x^{-5/4} \cdot x^{1/2}}{x^{-1/4}} = \dfrac{x^{-\frac{5}{4}+\frac{1}{2}}}{x^{-1/4}}$          Add exponents.

$= \dfrac{x^{-\frac{5}{4}+\frac{2}{4}}}{x^{-1/4}}$          Get a common denominator.

$= \dfrac{x^{-3/4}}{x^{-1/4}}$          Add exponents.

$= x^{-\frac{3}{4}-\left(-\frac{1}{4}\right)}$          Subtract exponents.

$= x^{-\frac{3}{4}+\frac{1}{4}}$

$= x^{-2/4}$

$= x^{-1/2}$          Reduce.

$= \dfrac{1}{x^{1/2}}$          Write with a positive exponent.

---

**You Try 5**

Simplify completely. The answer should contain only positive exponents.

a)  $t^{4/9} \cdot t^{2/9}$     b)  $\left(\dfrac{n^{5/6}}{m^{4/3}}\right)^9$     c)  $\dfrac{r^{3/4} \cdot r^{-5/2}}{r^{-9/4}}$

## 5. Convert a Radical Expression to Exponential Form and Simplify

Some radicals can be simplified by first putting them into rational exponent form and then converting them back to radicals.

**Example 6**

Rewrite each radical in exponential form, then simplify. Write the answer in simplest (or radical) form.

a)   $\sqrt[10]{4^5}$          b)   $\sqrt[8]{c^6}$

**Solution**

a)   Since the index of the radical is the denominator of the rational exponent and the power is the numerator, we can write

$$\sqrt[10]{4^5} = 4^{5/10} \qquad \text{Write with a rational exponent.}$$
$$= 4^{1/2} \qquad \text{Reduce } \frac{5}{10}.$$
$$= 2 \qquad \text{Evaluate.}$$

b)
$$\sqrt[8]{c^6} = c^{6/8} \qquad \text{Write with a rational exponent.}$$
$$= c^{3/4} \qquad \text{Reduce } \frac{6}{8}.$$
$$= \sqrt[4]{c^3} \qquad \text{Write in radical form.}$$

$\sqrt[8]{c^6}$ is not in simplest form because the 6 and the 8 contain a common factor of 2. $\sqrt[4]{c^3}$ *is* in simplest form because 3 and 4 do not have any common factors besides 1.  ∎

**You Try 6**

Rewrite each radical in exponential form, then simplify. Write the answer in simplest (or radical) form.

a)   $\sqrt[6]{49^3}$          b)   $\sqrt[10]{h^8}$

## Using Technology

We can evaluate square roots, cube roots, and even higher roots by first rewriting the radical in exponential form and then using a graphing calculator.

For example, to evaluate $\sqrt{49}$, first rewrite the radical as $49^{1/2}$, then enter 4 9, press ∧ (, enter 1 ÷ 2, and press ) ENTER. The result is 7, as shown on the screen on the left below.

To approximate $\sqrt[3]{12^2}$ rounded to the nearest tenth, first rewrite the radical as $12^{2/3}$, then enter 1 2, press ∧ (, enter 2 ÷ 3, and press ) ENTER. The result is 5.241482788 as shown on the screen on the right. The result rounded to the nearest tenth is then 5.2.

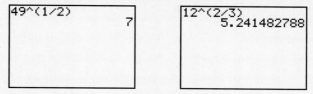

To evaluate radicals with an index greater than 3, follow the same procedure explained above.

Evaluate by rewriting in exponential form if necessary and then using a graphing calculator. If necessary, approximate to the nearest tenth.

1)  $16^{1/2}$          2)  $\sqrt[3]{512}$          3)  $\sqrt{37}$

4)  $361^{1/2}$         5)  $4096^{2/3}$         6)  $2401^{1/4}$

**Answers to You Try Exercises**

1)  a) 5  b) 6  c) 2    2)  a) 216  b) 32    3)  a) $\dfrac{1}{10}$  b) $\dfrac{1}{4}$

4)  a) 2  b) 5  c) $\dfrac{1}{4}$    5)  a) $t^{2/3}$  b) $\dfrac{n^{15/2}}{m^{12}}$  c) $r^{1/2}$    6)  a) 7  b) $\sqrt[5]{h^4}$

**Answers to Technology Exercises**

1)  4    2)  8    3)  6.1    4)  19    5)  256    6)  7

## 9.7 Exercises

**Objective 1: Evaluate Expressions of the Form $a^{1/n}$**

1)  Explain how to write $36^{1/2}$ in radical form.

2)  Explain how to write $27^{1/3}$ in radical form.

Write in radical form and evaluate.

3)  $16^{1/2}$

4)  $81^{1/2}$

5)  $1000^{1/3}$

6)  $8^{1/3}$

7)  $81^{1/4}$

8)  $32^{1/5}$

9)  $-64^{1/3}$

10)  $-64^{1/6}$

11)  $\left(\dfrac{9}{4}\right)^{1/2}$

12)  $\left(\dfrac{16}{121}\right)^{1/2}$

13)  $\left(\dfrac{125}{64}\right)^{1/3}$

14)  $\left(\dfrac{16}{81}\right)^{1/4}$

15)  $-\left(\dfrac{100}{49}\right)^{1/2}$

16)  $-\left(\dfrac{100}{9}\right)^{1/2}$

**Objective 2: Evaluate Expressions of the Form $a^{m/n}$**

17)  Explain how to write $16^{3/4}$ in radical form.

18)  Explain how to write $100^{3/2}$ in radical form.

Write in radical form and evaluate.

19)  $8^{4/3}$

20)  $125^{2/3}$

21)  $81^{3/4}$

22)  $32^{4/5}$

23)  $64^{5/6}$

24)  $1000^{2/3}$

25)  $-36^{3/2}$

26)  $-27^{4/3}$

27)  $\left(\dfrac{16}{81}\right)^{5/4}$

28)  $\left(\dfrac{64}{125}\right)^{2/3}$

29)  $-\left(\dfrac{8}{27}\right)^{4/3}$

30)  $-\left(\dfrac{1000}{27}\right)^{2/3}$

**Objective 3: Evaluate Expressions of the Form $a^{-m/n}$**

Decide whether each statement is true or false. Explain your answer.

31)  $121^{-1/2} = -11$

32)  $(100)^{-3/2} = \dfrac{1}{(100)^{3/2}}$

Rewrite with a positive exponent and evaluate.

33)  $16^{-1/2}$

34)  $121^{-1/2}$

35)  $100^{-1/2}$

36)  $64^{-1/3}$

37)  $125^{-1/3}$

38)  $1000^{-1/3}$

39)  $-81^{-1/4}$

40)  $-64^{-1/6}$

41)  $64^{-5/6}$

42)  $81^{-3/4}$

43)  $64^{-2/3}$

44)  $125^{-2/3}$

45)  $-27^{-2/3}$

46)  $-32^{-6/5}$

47)  $-16^{-5/4}$

48)  $100^{-3/2}$

**Objective 4: Combine the Rules of Exponents to Simplify Expressions**

Simplify completely. The answer should contain only positive exponents.

49)  $5^{2/3} \cdot 5^{7/3}$

50)  $3^{3/4} \cdot 3^{5/4}$

51)  $(16^{1/4})^2$

52)  $(2^{2/3})^3$

53)  $7^{7/5} \cdot 7^{-3/5}$

54)  $6^{-4/3} \cdot 6^{5/3}$

55)  $\dfrac{9^{10/3}}{9^{4/3}}$

56)  $\dfrac{2^{31/4}}{2^{11/4}}$

57)  $\dfrac{5^{7/2}}{5^{13/2}}$

58)  $\dfrac{32^{3/5}}{32^{7/5}}$

59)  $\dfrac{7^{-1}}{7^{1/2} \cdot 7^{-5/2}}$

60)  $\dfrac{10^{-5/2} \cdot 10^{3/2}}{10^{-4}}$

61)  $\dfrac{6^{4/9} \cdot 6^{1/9}}{6^{2/9}}$

62)  $\dfrac{8^{2/5}}{8^{6/5} \cdot 8^{3/5}}$

Simplify completely. The answer should contain only positive exponents.

63) $k^{9/4} \cdot k^{5/4}$

64) $y^{1/7} \cdot y^{6/7}$

65) $m^{-4/5} \cdot m^{1/10}$

66) $h^{1/6} \cdot h^{-3/4}$

67) $(-9v^{5/8})(8v^{3/4})$

68) $(-5x^{-1/3})(8x^{4/9})$

69) $\dfrac{n^{5/9}}{n^{2/9}}$

70) $\dfrac{p^{1/6}}{p^{5/6}}$

71) $\dfrac{26c^{-2/3}}{72c^{5/6}}$

72) $\dfrac{14w^{3/10}}{10w^{2/5}}$

73) $(q^{3/5})^{10}$

74) $(t^{3/4})^{16}$

75) $(x^{-5/9})^{3}$

76) $(n^{-2/11})^{3}$

77) $(z^{1/5})^{3/4}$

78) $(p^{4/3})^{5/2}$

### Objective 5: Convert a Radical Expression to Exponential Form and Simplify

Rewrite each radical in exponential form, then simplify. Write the answer in simplest (or radical) form.

79) $\sqrt[6]{49^3}$

80) $\sqrt[6]{1000^2}$

81) $\sqrt[9]{8^3}$

82) $\sqrt[4]{81^2}$

83) $\sqrt[6]{t^4}$

84) $\sqrt[9]{w^6}$

85) $\sqrt[4]{h^2}$

86) $\sqrt[8]{s^4}$

87) $\sqrt{d^4}$

88) $\sqrt{c^6}$

The period $T$, in seconds, of a simple pendulum is given by

$$T = 2\pi\left(\frac{L}{g}\right)^{1/2}$$

where $L$ is the length of the pendulum, in meters, and $g \approx 9.8$ m/sec$^2$. Use 3.14 for $\pi$. Use this information for Exercises 89 and 90.

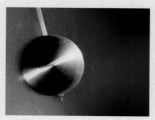

89) Find the period of a pendulum that is 0.25 m long. Round your answer to the nearest second.

90) Find the period of a pendulum that is 1 m long. Round your answer to the nearest second.

Let $N_i$ = the initial population of a group, $d$ = the time it takes for the population to double, $t$ = time elapsed, and $N_t$ = the population after time $t$. The formula $N_t = N_i \cdot 2^{t/d}$ can be used to determine the population of a group after time $t$. Use this information for Exercises 91 and 92.

91) A laboratory culture initially contains 5000 bacteria, and the population doubles every 48 minutes. How many bacteria are in the culture after

   a) 30 minutes?
   b) 2 hours?

92) The population of a particular type of fruit fly doubles every 20 days. A scientist begins his study with 600 fruit flies. How many fruit flies are in her sample after

   a) 15 days?
   b) 4 weeks?

| Definition/Procedure | Example |
|---|---|

## 9.1 Finding Roots

| | |
|---|---|
| If the radicand is a perfect square, then the square root is a *rational* number. **(p. 523)** | $\sqrt{25} = 5$ since $5^2 = 25$. |
| If the radicand is a negative number, then the square root is *not* a real number. **(p. 523)** | $\sqrt{-49}$ is not a real number. |
| If the radicand is positive and not a perfect square, then the square root is an *irrational* number. **(p. 523)** | $\sqrt{3}$ is irrational because 3 is not a perfect square. |
| The $\sqrt[n]{a}$ is read as "the *n*th root of *a*." If $\sqrt[n]{a} = b$, then $b^n = a$. *n* is the **index** of the radical. **(p. 523)** | $\sqrt[4]{81} = 3$ since $3^4 = 81$. |

**The Pythagorean Theorem**

If the lengths of the legs of a right triangle are $a$ and $b$ and the hypotenuse has length $c$, then $a^2 + b^2 = c^2$. **(p. 524)**

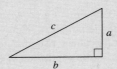

Find the length of the missing side.

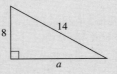

Use the Pythagorean theorem, $a^2 + b^2 = c^2$.

$$b = 8, c = 14 \qquad \text{Find } a.$$

$$a^2 + (8)^2 = (14)^2 \qquad \text{Substitute values.}$$
$$a^2 + 64 = 196$$
$$a^2 = 132 \qquad \text{Subtract 64 from each side.}$$
$$a = \pm\sqrt{132} \qquad \text{Square root property}$$
$$a = \pm 2\sqrt{33} \qquad \sqrt{132} = \sqrt{4}\sqrt{33} = 2\sqrt{33}$$

Discard $-2\sqrt{33}$ as a solution since the length of a side of a triangle cannot be negative.

$$a = 2\sqrt{33}$$

**The Distance Formula**

The **distance between two points** with coordinates $(x_1, y_1)$ and $(x_2, y_2)$ is given by $d = \sqrt{(x_2 - x_1)^2 + (y_2 - y_1)^2}$. **(p. 526)**

Find the distance between the points $(6, -2)$ and $(0, 2)$.

Label the points: $\overset{x_1 \quad y_1}{(6, -2)} \quad \overset{x_2 \quad y_2}{(0, 2)}$

Substitute the values into the distance formula.

$$d = \sqrt{(0 - 6)^2 + (2 - (-2))^2}$$
$$d = \sqrt{(-6)^2 + (4)^2}$$
$$d = \sqrt{36 + 16}$$
$$d = \sqrt{52} = 2\sqrt{13}$$

| | |
|---|---|
| The *odd root* of a negative number is a negative number. **(p. 528)** | $\sqrt[3]{-8} = -2$ since $(-2)^3 = -8$. |
| The *even root* of a negative number is not a real number. **(p. 528)** | $\sqrt[4]{-16}$ is not a real number. |

## 9.2 Simplifying Expressions Containing Radicals

| | |
|---|---|
| **Product Rule for Square Roots**<br>Let $a$ and $b$ be nonnegative real numbers. Then $\sqrt{a} \cdot \sqrt{b} = \sqrt{ab}$ **(p. 532)** | $\sqrt{5} \cdot \sqrt{6} = \sqrt{5 \cdot 6} = \sqrt{30}$ |

| Definition/Procedure | Example |
|---|---|

**An expression containing a square root is simplified when all of the following conditions are met:**

1) The radicand does not contain any factors (other than 1) which are perfect squares.
2) The radicand does not contain any fractions.
3) There are no radicals in the denominator of a fraction.

To *simplify square roots*, reverse the process of multiplying radicals, where $a$ or $b$ is a perfect square.

$$\sqrt{ab} = \sqrt{a} \cdot \sqrt{b}$$

After simplifying a radical, look at the result and ask yourself, *"Is the radical in simplest form?"* If it is not, simplify again. **(p. 532)**

---

Simplify $\sqrt{24}$.

$$\begin{aligned}\sqrt{24} &= \sqrt{4 \cdot 6} &&\text{4 is a perfect square.}\\ &= \sqrt{4} \cdot \sqrt{6} &&\text{Product rule}\\ &= 2\sqrt{6} &&\sqrt{4} = 2\end{aligned}$$

Simplify $\sqrt{12} \cdot \sqrt{2}$.

$$\begin{aligned}\sqrt{12} \cdot \sqrt{2} &= \sqrt{24}\\ &= \sqrt{4 \cdot 6}\\ &= \sqrt{4} \cdot \sqrt{6}\\ &= 2\sqrt{6}\end{aligned} \quad \text{or} \quad \begin{aligned}\sqrt{12} \cdot \sqrt{2} &= \sqrt{4 \cdot 3} \cdot \sqrt{2}\\ &= 2\sqrt{3} \cdot \sqrt{2}\\ &= 2\sqrt{3 \cdot 2}\\ &= 2\sqrt{6}\end{aligned}$$

---

**Quotient Rule for Square Roots**

Let $a$ and $b$ be nonnegative real numbers such that $b \neq 0$. Then

$$\sqrt{\frac{a}{b}} = \frac{\sqrt{a}}{\sqrt{b}}. \text{ (p. 535)}$$

---

Simplify $\sqrt{\dfrac{72}{25}}$.

$$\begin{aligned}\sqrt{\frac{72}{25}} &= \frac{\sqrt{72}}{\sqrt{25}} &&\text{Quotient rule}\\ &= \frac{\sqrt{36} \cdot \sqrt{2}}{5} &&\text{Product rule; } \sqrt{25} = 5\\ &= \frac{6\sqrt{2}}{5} &&\sqrt{36} = 6\end{aligned}$$

---

**Simplifying Square Root Expressions Containing Variables**

If $a$ is a nonnegative real number, then $\sqrt{a^2} = a$. In the rest of the book, we will assume that all variables represent positive real numbers.

To simplify a square root expression containing variables, use the product rule and factor out all perfect squares. **(p. 537)**

---

Simplify.

a) $\sqrt{w^2} = w$

b) $\sqrt{49t^8} = \sqrt{(7t^4)^2} = 7t^4$

c) $\begin{aligned}\sqrt{n^5} &= \sqrt{n^4 \cdot n}\\ &= \sqrt{n^4} \cdot \sqrt{n} &&\text{Product rule}\\ &= n^2\sqrt{n} &&\text{Simplify.}\end{aligned}$

---

**Product Rule for Higher Roots**

If $a$ and $b$ are real numbers such that the roots exist, then

$$\sqrt[n]{a} \cdot \sqrt[n]{b} = \sqrt[n]{a \cdot b}. \text{ (p. 538)}$$

---

Multiply $\sqrt[3]{3} \cdot \sqrt[3]{7}$.

$$\sqrt[3]{3} \cdot \sqrt[3]{7} = \sqrt[3]{3 \cdot 7} = \sqrt[3]{21} \quad \text{Product rule}$$

Simplify $\sqrt[3]{40}$.

$$\begin{aligned}\sqrt[3]{40} &= \sqrt[3]{8 \cdot 5}\\ &= \sqrt[3]{8} \cdot \sqrt[3]{5} &&\text{Product rule}\\ &= 2\sqrt[3]{5} &&\sqrt[3]{8} = 2\end{aligned}$$

---

**Quotient Rule for Higher Roots**

If $a$ and $b$ are real numbers such that the roots exist and $b \neq 0$, then

$$\sqrt[n]{\frac{a}{b}} = \frac{\sqrt[n]{a}}{\sqrt[n]{b}} \text{ (p. 538)}$$

---

Simplify $\sqrt[4]{\dfrac{64}{81}}$.

$$\sqrt[4]{\frac{64}{81}} = \frac{\sqrt[4]{64}}{\sqrt[4]{81}} = \frac{\sqrt[4]{16} \cdot \sqrt[4]{4}}{3} = \frac{2\sqrt[4]{4}}{3}$$

---

**Simplifying Higher Roots with Variables in the Radicand**

To simplify $\sqrt[n]{a^m}$, use the product rule and factor out all perfect $n$th powers. **(p. 539)**

---

Simplify each expression.

a) $\sqrt[3]{k^{12}} = k^4$

b) $\begin{aligned}\sqrt[3]{d^{16}} &= \sqrt[3]{d^{15} \cdot d}\\ &= \sqrt[3]{d^{15}} \cdot \sqrt[3]{d} &&\text{Product rule}\\ &= d^5\sqrt[3]{d} &&\text{Simplify.}\end{aligned}$

| Definition/Procedure | Example |
|---|---|

## 9.3 Adding and Subtracting Radicals

**Like radicals** have the same index and the same radicand. In order to add or subtract radicals, they must be like radicals.

**Steps for Adding and Subtracting Radicals**
1) Write each radical expression in simplest form.
2) Combine like radicals. **(p. 543)**

Perform the operations and simplify.

a) $6\sqrt{5} + 2\sqrt{7} - 3\sqrt{5} + 9\sqrt{7} = 3\sqrt{5} + 11\sqrt{7}$

b) $\sqrt{72} - \sqrt{45} + \sqrt{18}$
$= \sqrt{36}\cdot\sqrt{2} - \sqrt{9}\cdot\sqrt{5} + \sqrt{9}\cdot\sqrt{2}$
$= \phantom{=} 6\sqrt{2} \phantom{=} - \phantom{=} 3\sqrt{5} \phantom{=} + \phantom{=} 3\sqrt{2}$
$= 9\sqrt{2} - 3\sqrt{5}$

## 9.4 Combining Operations on Radicals

Multiply expressions containing radicals using the same techniques that are used for multiplying polynomials. **(p. 547)**

Multiply and simplify.

a) $\sqrt{2}(\sqrt{5} + 9) = \sqrt{2}\cdot\sqrt{5} + \sqrt{2}\cdot 9 = \sqrt{10} + 9\sqrt{2}$

b) $(\sqrt{g} + \sqrt{6})(\sqrt{g} - \sqrt{3})$

Use FOIL to multiply two binomials.

$(\sqrt{g} + \sqrt{6})(\sqrt{g} - \sqrt{3})$
$= \sqrt{g}\cdot\sqrt{g} - \sqrt{3}\sqrt{g} + \sqrt{6}\cdot\sqrt{g} - \sqrt{6}\cdot\sqrt{3}$

$\phantom{=}\quad\ \ \text{F}\qquad\quad\ \text{O}\qquad\quad\ \text{I}\qquad\quad\ \text{L}$

$= \phantom{xx} g^2 \phantom{xx} - \phantom{x} \sqrt{3g} + \phantom{x} \sqrt{6g} - \phantom{x} \sqrt{18}$
$= g^2 - \sqrt{3g} + \sqrt{6g} - 3\sqrt{2} \qquad \sqrt{18} = 3\sqrt{2}$

---

To *square a binomial*, we can use either FOIL or one of the special formulas from Chapter 6:

$$(a + b)^2 = a^2 + 2ab + b^2$$
$$(a - b)^2 = a^2 - 2ab + b^2 \quad \textbf{(p. 549)}$$

$(\sqrt{3} + 4)^2 = (\sqrt{3})^2 + 2(\sqrt{3})(4) + (4)^2$
$= 3 + 8\sqrt{3} + 16$
$= 19 + 8\sqrt{3}$

---

To multiply binomials of the form $(a + b)(a - b)$, use the formula $(a + b)(a - b) = a^2 - b^2$. **(p. 549)**

$(2 + \sqrt{11})(2 - \sqrt{11}) = (2)^2 - (\sqrt{11})^2$
$= 4 - 11$
$= -7$

## 9.5 Dividing Radicals

The process of eliminating radicals from the denominator of an expression is called **rationalizing the denominator.**

First, we give examples of rationalizing denominators containing one term. **(p. 552)**

Rationalize the denominator of each expression.

a) $\dfrac{7}{\sqrt{5}} = \dfrac{7}{\sqrt{5}}\cdot\dfrac{\sqrt{5}}{\sqrt{5}} = \dfrac{7\sqrt{5}}{5}$

b) $\dfrac{9}{\sqrt[3]{4}} = \dfrac{9}{\sqrt[3]{2^2}}\cdot\dfrac{\sqrt[3]{2}}{\sqrt[3]{2}} = \dfrac{9\sqrt[3]{2}}{\sqrt[3]{2^3}} = \dfrac{9\sqrt[3]{2}}{2}$

---

The **conjugate** of a binomial is the binomial obtained by changing the sign between the two terms. **(p. 557)**

The conjugate of $\sqrt{21} - 1$ is $\sqrt{21} + 1$.
The conjugate of $-14 + \sqrt{5}$ is $-14 - \sqrt{5}$.

---

**Rationalizing a Denominator with Two Terms**

To rationalize the denominator of an expression containing two terms, multiply the numerator and denominator of the expression by the conjugate of the denominator. **(p. 558)**

Rationalize the denominator of $\dfrac{7}{\sqrt{5}-3}$.

$\dfrac{7}{\sqrt{5} - 3} = \dfrac{7}{\sqrt{5} - 3}\cdot\dfrac{\sqrt{5} + 3}{\sqrt{5} + 3}$   Multiply by the conjugate.

$= \dfrac{7(\sqrt{5} + 3)}{(\sqrt{5})^2 - (3)^2}$   $(a + b)(a - b) = a^2 - b^2$

$= \dfrac{7(\sqrt{5} + 3)}{5 - 9}$   Square the terms.

$= \dfrac{7(\sqrt{5} + 3)}{-4}$   Subtract.

$= -\dfrac{7\sqrt{5} + 21}{4}$   Distribute.

| Definition/Procedure | Example |
|---|---|

## 9.6 Solving Radical Equations

**Steps for Solving Radical Equations Containing Square Roots**

**Step 1:** Get a radical on a side by itself.

**Step 2:** Square both sides of the equation to eliminate a radical.

**Step 3:** Combine like terms on each side of the equation.

**Step 4:** If the equation still contains a radical, repeat Steps 1–3.

**Step 5:** Solve the equation.

**Step 6:** Check the proposed solutions *in the original equation* and discard extraneous solutions. **(p. 563)**

Solve $n = 1 + \sqrt{2n + 1}$.

$$n - 1 = \sqrt{2n + 1} \quad \text{Get the radical by itself.}$$
$$(n - 1)^2 = (\sqrt{2n + 1})^2 \quad \text{Square both sides.}$$
$$n^2 - 2n + 1 = 2n + 1$$
$$n^2 - 4n = 0 \quad \text{Get all terms on the same side.}$$
$$n(n - 4) = 0 \quad \text{Factor.}$$
$$n = 0 \quad \text{or} \quad n - 4 = 0$$
$$\text{or} \quad n = 4$$

Check $n = 0$ and $n = 4$ in the *original* equation.

**$n = 0$:**             **$n = 4$:**

$$n = 1 + \sqrt{2n + 1} \qquad\qquad n = 1 + \sqrt{2n + 1}$$
$$0 \overset{?}{=} 1 + \sqrt{2(0) + 1} \qquad\quad 4 \overset{?}{=} 1 + \sqrt{2(4) + 1}$$
$$0 \overset{?}{=} 1 + \sqrt{1} \qquad\qquad\quad 4 \overset{?}{=} 1 + \sqrt{9}$$
$$0 = 2 \qquad\qquad\qquad\quad 4 = 1 + 3$$
$$\text{FALSE} \qquad\qquad\qquad \text{TRUE}$$

$n = 4$ *is* a solution, but $n = 0$ is *not* because $n = 0$ does not satisfy the original equation.

The solution set is $\{4\}$.

## 9.7 Rational Exponents

| | |
|---|---|
| If $a$ is a nonnegative number and $n$ is a positive integer greater than 1, then $a^{1/n} = \sqrt[n]{a}$. **(p. 571)** | $27^{1/3} = \sqrt[3]{27} = 3$ |
| If $a$ is a nonnegative number and $m$ and $n$ are integers such that $n > 1$, then $a^{m/n} = (a^{1/n})^m = (\sqrt[n]{a})^m$ **(p. 572)** | $16^{5/4} = (\sqrt[4]{16})^5 = 2^5 = 32$ |
| If $a$ is a positive number and $m$ and $n$ are integers such that $n > 1$, then $a^{-m/n} = \dfrac{1}{a^{m/n}}$. **(p. 572)** | $1000^{-2/3} = \dfrac{1}{1000^{2/3}} = \dfrac{1}{(\sqrt[3]{1000})^2} = \dfrac{1}{10^2} = \dfrac{1}{100}$ <br><br>  BE CAREFUL    The negative exponent does not make the expression negative. |

**(9.1)**
**Find each root, if possible.**

1) $\sqrt{49}$

2) $\sqrt{-25}$

3) $-\sqrt{16}$

4) $\sqrt{\dfrac{169}{9}}$

5) $\sqrt[3]{-125}$

6) $\sqrt[5]{32}$

7) $\sqrt[4]{81}$

8) Approximately $\sqrt{34}$ to the nearest tenth and plot it on a number line.

9) Use the Pythagorean theorem to find the length of the missing side.

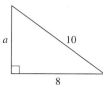

10) The length of a rectangle is 9 cm, and its width is 4 cm. Find the length of its diagonal.

**Find the distance between the given points.**

11) $(4, 1)$ and $(10, -7)$

12) $(-8, -3)$ and $(-1, 2)$

**(9.2)**
**Simplify completely.**

13) $\sqrt{45}$

14) $\sqrt{108}$

15) $\sqrt[3]{81}$

16) $\sqrt[3]{250}$

17) $\sqrt[4]{80}$

18) $-\sqrt{\dfrac{125}{5}}$

19) $\sqrt{\dfrac{48}{6}}$

20) $\dfrac{\sqrt{48}}{\sqrt{121}}$

21) $\sqrt[3]{-\dfrac{16}{2}}$

22) $\sqrt[4]{\dfrac{162}{2}}$

23) $\sqrt{h^8}$

24) $-\sqrt{\dfrac{128}{p^{12}}}$

25) $\sqrt[3]{b^{24}}$

26) $\sqrt[3]{64u^{15}}$

27) $\sqrt{w^{15}}$

28) $\sqrt[3]{r^{20}}$

29) $\sqrt{63d^4}$

30) $\sqrt{72h^3}$

31) $\sqrt{44x^{12}y^5}$

32) $\sqrt[3]{-27p^6q^{21}}$

33) $\sqrt[3]{48m^{17}}$

34) $\dfrac{\sqrt{54c^{17}}}{\sqrt{6c^9}}$

35) $\dfrac{40\sqrt{150}}{56\sqrt{2}}$

36) $\sqrt{\dfrac{10}{27}} \cdot \sqrt{\dfrac{2}{3}}$

**Perform the indicated operation and simplify.**

37) $\sqrt{6} \cdot \sqrt{5}$

38) $\sqrt{8} \cdot \sqrt{10}$

39) $\sqrt{5} \cdot \sqrt{10}$

40) $\sqrt{d^3} \cdot \sqrt{d^{11}}$

41) $\sqrt[3]{3} \cdot \sqrt[3]{10}$

**(9.3)**
**Perform the operations and simplify.**

42) $8\sqrt{6} + 4\sqrt{6}$

43) $\sqrt{80} + \sqrt{150}$

44) $\sqrt{80} - \sqrt{20} + \sqrt{48}$

45) $4\sqrt[3]{9} - 9\sqrt[3]{72}$

46) $8\sqrt{p^3} - 3p\sqrt{p}$

47) $10n^2\sqrt{8n} - 45n\sqrt{2n^3}$

48) $4y^3\sqrt[3]{y} + 6\sqrt[3]{y^{10}}$

**(9.4)**
**Multiply and simplify.**

49) $\sqrt{7}(\sqrt{7} - \sqrt{5})$

50) $4\sqrt{t}(\sqrt{20t} + \sqrt{2})$

51) $(5 - \sqrt{2})(3 + \sqrt{2})$

52) $(\sqrt{2r} + 5\sqrt{s})(3\sqrt{s} + 4\sqrt{2r})$

53) $(2\sqrt{6} - 5)^2$

54) $(1 + \sqrt{k})^2$

55) $(\sqrt{7} - \sqrt{6})(\sqrt{7} + \sqrt{6})$

56) $(3 + \sqrt{n})(3 - \sqrt{n})$

**(9.5)**
**Rationalize the denominator of each expression.**

57) $\dfrac{18}{\sqrt{3}}$

58) $\dfrac{20\sqrt{3}}{\sqrt{6}}$

59) $\sqrt{\dfrac{98}{h}}$

60) $\dfrac{\sqrt{63}}{\sqrt{2k^5}}$

61) $\dfrac{15}{\sqrt[3]{3}}$

62) $-\dfrac{9}{\sqrt[3]{4}}$

63) $\dfrac{7}{\sqrt[3]{2c}}$

64) $\dfrac{\sqrt[3]{x^2}}{\sqrt[3]{y}}$

65) $\dfrac{2}{3 + \sqrt{2}}$

66) $\dfrac{-1 + \sqrt{5}}{8 - \sqrt{5}}$

67) $\dfrac{6}{9 - \sqrt{x}}$

68) $\dfrac{z - 4}{\sqrt{z} + 2}$

**Simplify completely.**

69) $\dfrac{16 - 24\sqrt{3}}{8}$

70) $\dfrac{-\sqrt{48} - 6}{12}$

**(9.6)**
**Solve.**

71) $\sqrt{r} - 9 = 0$

72) $\sqrt{3w - 4} = 6$

73) $3\sqrt{2v + 7} - 15 = 0$

74) $m = \sqrt{m^2 + 9m + 18}$

75) $\sqrt{a + 11} - 2\sqrt{a + 8} = 0$

76) $k + \sqrt{k + 7} = 5$

77) $1 = \sqrt{5x - 1} - x$

78) $\sqrt{t + 8} + \sqrt{5 - t} = 5$

79) $\sqrt{3b + 1} + \sqrt{b} = 3$

80) $\sqrt[3]{4z + 7} - \sqrt[3]{6z - 3} = 0$

**(9.7)**

81) Explain how to write $8^{2/3}$ in radical form.

82) Explain how to eliminate the negative from the exponent in an expression like $9^{-1/2}$.

**Evaluate.**

83) $64^{1/2}$

84) $32^{1/5}$

85) $\left(\dfrac{64}{125}\right)^{1/3}$

86) $\left(\dfrac{25}{49}\right)^{1/2}$

87) $-81^{1/4}$

88) $-1000^{1/3}$

89) $125^{2/3}$

90) $32^{3/5}$

91) $\left(\dfrac{1000}{27}\right)^{2/3}$

92) $\left(\dfrac{8}{125}\right)^{2/3}$

93) $121^{-1/2}$

94) $64^{-1/3}$

95) $-81^{-3/4}$

96) $8^{-2/3}$

97) $4^{6/7} \cdot 4^{8/7}$

98) $(144^5)^{1/10}$

99) $\dfrac{64^2}{64^{8/3}}$

100) $\dfrac{9^2}{9^{5/3} \cdot 9^{1/3}}$

**Simplify completely. The answer should contain only positive exponents.**

101) $(-8p^{5/6})(3p^{-1/2})$

102) $\dfrac{c^{3/8}}{c^{3/4}}$

103) $(m^{4/3})^{-3}$

**Rewrite each radical in exponential form, then simplify. Write the answer in simplest (or radical) form.**

104) $\sqrt[4]{25^2}$

105) $\sqrt[10]{n^5}$

106) $\sqrt[9]{k^3}$

**Mixed Exercises**

Simplify each expression.

107) $9d\sqrt{d} - 4\sqrt{d^3}$

108) $\sqrt{\dfrac{3}{24}}$

109) $3(5\sqrt{8} + \sqrt{2})$

110) $\sqrt[3]{-1000y^{12}}$

111) $t^{9/5} \cdot t^{6/5}$

112) $\dfrac{32\sqrt{13} + 8}{8}$

113) $\dfrac{\sqrt{5}}{6 + \sqrt{10}}$

114) $\dfrac{4}{\sqrt[3]{25m}}$

115) $32^{-4/5}$

116) $\sqrt{\dfrac{2}{15}} \cdot \sqrt{\dfrac{6}{5}}$

117) $(4 - \sqrt{3})^2$

118) $\dfrac{\sqrt{20}}{\sqrt{6x}}$

119) $\sqrt{128p^{15}}$

120) If the area of a circle is $A$ and the radius is $r$, then the radius of the circle is given by $r = \sqrt{\dfrac{A}{\pi}}$. A circle has an area of $28\pi$ cm$^2$. What is the radius of the circle?

# Chapter 9:

**Find each root, if possible.**

1) $\sqrt{121}$

2) $\sqrt{-81}$

3) $\sqrt[3]{-1000}$

4) $\sqrt[4]{81}$

5) Find the length of the missing side.

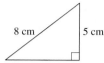

6) Find the distance between $(4, -1)$ and $(3, 1)$.

**Simplify completely.**

7) $\sqrt{125}$

8) $\sqrt[3]{48}$

9) $\sqrt{\dfrac{192}{6}}$

10) $\sqrt{m^6}$

11) $\sqrt{h^7}$

12) $\sqrt[3]{k^{12}}$

13) $\sqrt[3]{b^{14}}$

14) $\sqrt{63m^5n^8}$

15) $\sqrt[3]{\dfrac{x^{15}y^7}{8}}$

16) $\dfrac{\sqrt{120w^{15}}}{\sqrt{2w^4}}$

**Perform the operations and simplify.**

17) $\sqrt{3y} \cdot \sqrt{12y}$

18) $\sqrt[3]{z^4} \cdot \sqrt[3]{z^6}$

19) $2\sqrt{3} - \sqrt{18} + \sqrt{108}$

20) $\sqrt{18a^5} - 4a\sqrt{8a^3}$

21) $\sqrt{\dfrac{8}{3}} \cdot \sqrt{\dfrac{10}{27}}$

**Multiply and simplify.**

22) $\sqrt{5}(\sqrt{10} - 9)$

23) $(4 - 3\sqrt{7})(\sqrt{3} + 9)$

24) $(\sqrt{5} + 4)(\sqrt{5} - 4)$

25) $(7 - 2\sqrt{t})^2$

**Rationalize the denominator of each expression.**

26) $\dfrac{2}{\sqrt{6}}$

27) $\dfrac{11}{\sqrt{2} + 3}$

28) $\sqrt{\dfrac{7}{5k}}$

29) $\dfrac{12}{\sqrt[3]{4}}$

30) Simplify completely. $\dfrac{8 - \sqrt{48}}{4}$

31) How do you know, without actually solving and checking the solution, that $\sqrt{x} = -4$ has no real number solution?

**Solve.**

32) $\sqrt{4n - 1} = 3$

33) $y = 2 + \sqrt{y - 2}$

34) $\sqrt{3p} + \sqrt{p - 2} = 4$

35) In the formula $r = \sqrt{\dfrac{V}{\pi h}}$, $V$ represents the volume of a right circular cylinder, $h$ represents the height of the cylinder, and $r$ represents the radius. A cylindrical container has a volume of $64\pi$ cubic inches. It is 4 inches high. What is the radius of the container?

**Evaluate.**

36) $-16^{1/4}$

37) $169^{-1/2}$

38) Explain, in words, how to evaluate $81^{3/4}$. Then, evaluate the expression.

**Simplify completely. The answer should contain only positive exponents.**

39) $(-4g^{5/8}) \cdot (5g^{1/4})$

40) $\dfrac{42w^{1/9}}{12w^{4/9}}$

41) $(x^{3/10}y^{-2/5})^{-5}$

1) Divide $\dfrac{5}{8} \div \dfrac{7}{12}$.

2) Write an expression for "twice the sum of $-9$ and 4" and simplify.

3) Simplify $2^{-3} - 4^{-2}$.

4) Combine like terms. $5a - 7b + 3 - 3a + \dfrac{5}{2}b - 2$

5) Write in scientific notation. $0.000941$

6) Solve $4(5k - 6) + 9 = 3k + 2(k + 5)$.

7) Graph $-x + 4y = 8$.

8) Write the slope-intercept form of the line containing the points $(-4, 2)$ and $(1, -4)$.

9) Solve the system. $\begin{aligned} 3x + 2y &= 7 \\ 6x - y &= -6 \end{aligned}$

**Perform the operations and simplify.**

10) $(4t - 3)(2t^2 - 9t - 5)$

11) $(7n^2 + 10n - 1) - (8n^2 + n - 5)$

12) $\dfrac{12a^3 - 17a + 7}{2a - 1}$

**Factor completely.**

13) $p^2 - 10p + 25$

14) $6y^2 - 21y + 18$

15) $h^3 - 9h^2 + 4h - 36$

16) Solve $5(r^2 - 5) - 9r = 4r^2 - 6r + 3$.

17) *Write an equation and solve.* The width of a rectangle is 3 in. less than its length. The area is 40 in². Find the dimensions of the rectangle.

**Perform the operations and simplify.**

18) $\dfrac{m - 1}{m + 3} - \dfrac{5}{4m}$

19) $\dfrac{k^2 - 2k - 63}{64 - k^2} \cdot \dfrac{3k^2 - 24k}{k + 7}$

20) Solve $\dfrac{4}{x - 5} = \dfrac{x}{x^2 - 25} - \dfrac{2}{3}$.

21) Simplify $\dfrac{10\sqrt{80}}{15\sqrt{12}}$.

22) Simplify $\sqrt[3]{72x^6}$.

23) Rationalize the denominator of $\dfrac{\sqrt{2}}{\sqrt{6} - 4}$.

24) Evaluate $8^{4/3}$.

25) *Write an equation and solve.* Twice the square root of a number is 3 less than the number. Find the number.

# Quadratic Equations

## Algebra at Work: Ophthalmology

We have already seen two applications of mathematics to ophthalmology, and here we have a third. An ophthalmologist can use a quadratic equation to convert between a prescription for glasses and a prescription for contact lenses.

After having reexamined her patient for contact lens use, Sarah can use the following quadratic equation to double-check the prescription for the contact lenses based on the prescription her patient currently has for her glasses.

$$D_c = s(D_g)^2 + D_g$$

where $D_g$ = power of the glasses, in diopters

$s$ = distance of the glasses to the eye, in meters

$D_c$ = power of the contact lenses, in diopters

For example, if the power of a patient's eyeglasses is +9.00 diopters and the glasses rest 1 cm or 0.01 m from the eye, the power the patient would need in her contact lenses would be

$$D_c = 0.01(9)^2 + 9$$
$$D_c = 0.01(81) + 9$$
$$D_c = 0.81 + 9$$
$$D_c = 9.81 \text{ diopters}$$

An eyeglass power of +9.00 diopters would convert to a contact lens power of +9.81 diopters.

In this chapter, we will learn different ways to solve quadratic equations.

# Section 10.1 Solving Quadratic Equations Using the Square Root Property

## Objectives

1. **Solve an Equation of the Form $x^2 = k$**
2. **Solve an Equation of the Form $(ax + b)^2 = k$**
3. **Solve a Formula for a Specific Variable**

We defined a quadratic equation in Chapter 7. Let's restate the definition:

> **Definition**
>
> A **quadratic equation** can be written in the form $ax^2 + bx + c = 0$, where $a$, $b$, and $c$ are real numbers and $a \neq 0$.

In Section 7.5, we learned how to solve quadratic equations by factoring. For example, we can use the zero product rule to solve $x^2 - 3x - 28 = 0$.

$$x^2 - 3x - 28 = 0$$
$$(x - 7)(x + 4) = 0 \qquad \text{Factor.}$$
$$x - 7 = 0 \quad \text{or} \quad x + 4 = 0 \qquad \text{Set each factor equal to zero.}$$
$$x = 7 \quad \text{or} \qquad x = -4 \qquad \text{Solve.}$$

The solution set is $\{-4, 7\}$.

It is not easy to solve all quadratic equations by factoring, however. Therefore, we need to learn other methods for solving quadratic equations. In this chapter, we will discuss three more methods for solving quadratic equations. Let's begin with the square root property.

## 1. Solve an Equation of the Form $x^2 = k$

Look at the equation $x^2 = 4$, for example. We can solve by factoring like this

$$x^2 = 4$$
$$x^2 - 4 = 0 \qquad \text{Get all terms on the same side.}$$
$$(x + 2)(x - 2) = 0 \qquad \text{Factor.}$$
$$x + 2 = 0 \quad \text{or} \quad x - 2 = 0 \qquad \text{Set each factor equal to zero.}$$
$$x = -2 \quad \text{or} \qquad x = 2 \qquad \text{Solve.}$$

giving us a solution set of $\{-2, 2\}$.

Or, we can solve an equation like $x^2 = 4$ using the **square root property,** as we will see in Example 1a.

> **Definition    The Square Root Property**
>
> Let $k$ be a constant. If $x^2 = k$, then $x = \sqrt{k}$ or $x = -\sqrt{k}$.
>     (The solution is often written as $x = \pm\sqrt{k}$, read as "x equals plus or minus the square root of k.")

**Note**

We can use the square root property to solve an equation containing a squared quantity and a constant. To do so, we will get the squared quantity containing the variable on one side of the equal sign and the constant on the other side.

## Example 1

Solve using the square root property.

a)  $x^2 = 4$    b)  $c^2 - 45 = 0$    c)  $2n^2 + 7 = 19$    d)  $y^2 + 15 = 6$

**Solution**

a)
$$x^2 = 4$$
$$x = \sqrt{4} \quad \text{or} \quad x = -\sqrt{4} \qquad \text{Square root property}$$
$$x = 2 \quad \text{or} \quad x = -2$$

*Check:*

$$x = 2: \quad x^2 = 4 \qquad\qquad x = -2: \quad x^2 = 4$$
$$(2)^2 \overset{?}{=} 4 \qquad\qquad\qquad\qquad (-2)^2 \overset{?}{=} 4$$
$$4 = 4 \ \checkmark \qquad\qquad\qquad\qquad 4 = 4 \ \checkmark$$

The solution set is $\{-2, 2\}$. We can also write it as $\{\pm 2\}$.

An equivalent way to solve $x^2 = 4$ is to write it as

$$x^2 = 4$$
$$x = \pm\sqrt{4} \qquad \text{Square root property}$$
$$x = \pm 2$$

We will use this approach when solving equations using the square root property.

b)  To solve $c^2 - 45 = 0$, begin by getting $c^2$ on a side by itself.

$$c^2 - 45 = 0$$
$$c^2 = 45 \qquad\qquad \text{Add 45 to each side.}$$
$$c = \pm\sqrt{45} \qquad\quad \text{Square root property}$$
$$c = \pm\sqrt{9} \cdot \sqrt{5} \quad\ \text{Product rule for radicals}$$
$$c = \pm 3\sqrt{5} \qquad\quad \sqrt{9} = 3$$

The check is left to the student. The solution set is $\{-3\sqrt{5}, 3\sqrt{5}\}$ or $\{\pm 3\sqrt{5}\}$.

c)  To solve $2n^2 + 7 = 19$, begin by getting $2n^2$ on a side by itself.

$$2n^2 + 7 = 19$$
$$2n^2 = 12 \qquad\quad \text{Subtract 7 from each side.}$$
$$n^2 = 6 \qquad\qquad \text{Divide by 2.}$$
$$n = \pm\sqrt{6} \qquad\quad \text{Square root property}$$

The check is left to the student. The solution set is $\{-\sqrt{6}, \sqrt{6}\}$ or $\{\pm\sqrt{6}\}$.

d)
$$y^2 + 15 = 6$$
$$y^2 = -9 \qquad\qquad \text{Subtract 15 from each side.}$$
$$y = \pm\sqrt{-9} \qquad\quad \text{Square root property}$$

Since $\sqrt{-9}$ is not a real number, there is no real number solution to $y^2 + 15 = 6$. The solution set is $\varnothing$.

## You Try 1

Solve using the square root property.

a)  $m^2 = 25$    b)  $h^2 - 28 = 0$    c)  $4a^2 + 9 = 49$    d)  $p^2 + 30 = 14$

Can we solve $(a - 5)^2 = 9$ using the square root property? Yes. The equation has a *squared quantity* and a *constant*.

## 2. Solve an Equation of the Form $(ax + b)^2 = k$

### Example 2

Solve $x^2 = 9$ and $(a - 5)^2 = 9$ using the square root property.

**Solution**

While the equation $(a - 5)^2 = 9$ has a *binomial* that is being squared, the two equations are actually in the same form.

$$x^2 = 9 \qquad\qquad (a - 5)^2 = 9$$
$$\uparrow \quad \uparrow \qquad\qquad\qquad \uparrow \qquad \uparrow$$
$x$ squared = constant $\qquad$ $(a - 5)$ squared = constant

Solve $x^2 = 9$:

$$x^2 = 9$$
$$x = \pm\sqrt{9} \qquad \text{Square root property}$$
$$x = \pm 3$$

The solution set is $\{-3, 3\}$ or $\{\pm 3\}$.

We solve $(a - 5)^2 = 9$ in the same way with some additional steps.

$$(a - 5)^2 = 9$$
$$a - 5 = \pm\sqrt{9} \qquad \text{Square root property}$$
$$a - 5 = \pm 3$$

This means $a - 5 = 3$ or $a - 5 = -3$. Solve both equations.

$$a - 5 = 3 \quad \text{or} \quad a - 5 = -3$$
$$a = 8 \quad \text{or} \qquad a = 2 \qquad \text{Add 5 to each side.}$$

*Check:*

$$a = 8: \quad (a - 5)^2 = 9 \qquad\qquad a = 2: \quad (a - 5)^2 = 9$$
$$(8 - 5)^2 \stackrel{?}{=} 9 \qquad\qquad\qquad (2 - 5)^2 \stackrel{?}{=} 9$$
$$3^2 \stackrel{?}{=} 9 \qquad\qquad\qquad\qquad (-3)^2 \stackrel{?}{=} 9$$
$$9 = 9 \ \checkmark \qquad\qquad\qquad\qquad 9 = 9 \ \checkmark$$

The solution set is $\{2, 8\}$. ∎

### You Try 2

Solve $(t + 3)^2 = 4$ using the square root property.

### Example 3

Solve.

a) $(w - 4)^2 = 3$ $\qquad$ b) $(3r + 2)^2 = 25$ $\qquad$ c) $(2x - 7)^2 = 18$

**Solution**

a) $(w - 4)^2 = 3$
$$w - 4 = \pm\sqrt{3} \qquad \text{Square root property}$$
$$w = 4 \pm\sqrt{3} \qquad \text{Add 4 to each side.}$$

Check: $w = 4 + \sqrt{3}$: $(w - 4)^2 = 3$ | $w = 4 - \sqrt{3}$: $(w - 4)^2 = 3$
$(4 + \sqrt{3} - 4)^2 \overset{?}{=} 3$ | $(4 - \sqrt{3} - 4)^2 \overset{?}{=} 3$
$(\sqrt{3})^2 \overset{?}{=} 3$ | $(-\sqrt{3})^2 \overset{?}{=} 3$
$3 = 3$ ✓ | $3 = 3$ ✓

The solution set is $\{4 - \sqrt{3}, 4 + \sqrt{3}\}$ or $\{4 \pm \sqrt{3}\}$.

b) $(3r + 2)^2 = 25$

$\quad\quad 3r + 2 = \pm\sqrt{25}$    Square root property

$\quad\quad 3r + 2 = \pm 5$

This means $3r + 2 = 5$ or $3r + 2 = -5$. Solve both equations.

$\quad\quad 3r + 2 = 5$   or   $3r + 2 = -5$

$\quad\quad\quad\quad 3r = 3$ $\quad\quad\quad\quad 3r = -7$    Subtract 2 from each side.

$\quad\quad\quad\quad\quad r = 1$   or   $r = -\dfrac{7}{3}$    Divide by 3.

The check is left to the student. The solution set is $\left\{ -\dfrac{7}{3}, 1 \right\}$.

c) $(2x - 7)^2 = 18$

$\quad\quad 2x - 7 = \pm\sqrt{18}$    Square root property

$\quad\quad 2x - 7 = \pm 3\sqrt{2}$    Simplify $\sqrt{18}$.

$\quad\quad\quad\quad 2x = 7 \pm 3\sqrt{2}$    Add 7 to each side.

$\quad\quad\quad\quad\quad x = \dfrac{7 \pm 3\sqrt{2}}{2}$    Divide by 2.

One solution is $\dfrac{7 + 3\sqrt{2}}{2}$, and the other is $\dfrac{7 - 3\sqrt{2}}{2}$.

The solution set is $\left\{ \dfrac{7 - 3\sqrt{2}}{2}, \dfrac{7 + 3\sqrt{2}}{2} \right\}$. This can also be written as $\left\{ \dfrac{7 \pm 3\sqrt{2}}{2} \right\}$. The check is left to the student.  ∎

**You Try 3**

Solve.

a) $(z - 10)^2 = 7$ $\quad\quad\quad$ b) $(4d + 1)^2 = 16$ $\quad\quad\quad$ c) $(5p - 2)^2 = 27$

## 3. Solve a Formula for a Specific Variable

Sometimes, we need to use the square root property to solve a formula for a specific variable.

**Example 4**

The formula for the volume, $V$, of a right circular cone is $V = \dfrac{1}{3}\pi r^2 h$, where $r$ is the radius of the base and $h$ is the height. Find the radius of the base of a right circular cone if it is 9 in. high and its volume is $12\pi$ in.$^3$.

### Solution

We will substitute the given values into the formula and solve for $r$.

$$V = \frac{1}{3}\pi r^2 h$$

$$12\pi = \frac{1}{3}\pi r^2 (9) \qquad \text{Substitute } 12\pi \text{ for } V \text{ and } 9 \text{ for } h.$$

$$12\pi = 3\pi r^2 \qquad \text{Multiply.}$$

$$\frac{12\pi}{3\pi} = \frac{3\pi r^2}{3\pi} \qquad \text{Divide both sides by } 3\pi.$$

$$4 = r^2 \qquad \text{Simplify.}$$

$$\pm\sqrt{4} = r \qquad \text{Square root property}$$

$$\pm 2 = r \qquad \text{Simplify.}$$

The radius of the cone cannot be negative, so we discard $r = -2$. The radius is 2 in. ∎

### You Try 4

The formula for the volume, $V$, of a right circular cylinder is $V = \pi r^2 h$, where $r$ is the radius and $h$ is the height. Find the radius of a cylinder if it is 15 cm high and its volume is $240\pi$ cm$^3$.

### Answers to You Try Exercises

1) a) $\{-5, 5\}$   b) $\{-2\sqrt{7}, 2\sqrt{7}\}$   c) $\{-\sqrt{10}, \sqrt{10}\}$   d) $\varnothing$     2) $\{-5, -1\}$

3) a) $\{10 - \sqrt{7}, 10 + \sqrt{7}\}$   b) $\left\{-\dfrac{5}{4}, \dfrac{3}{4}\right\}$   c) $\left\{\dfrac{2 - 3\sqrt{3}}{5}, \dfrac{2 + 3\sqrt{3}}{5}\right\}$     4) 4 cm

# 10.1 Exercises

**Objective 1: Solve an Equation of the Form $x^2 = k$**

1) What are two methods that can be used to solve $x^2 - 81 = 0$? Solve the equation using both methods.

2) If $k$ is a negative number and $x^2 = k$, what can you conclude about the solution to the equation?

Solve using the square root property.

3) $b^2 = 16$

4) $h^2 = 100$

5) $w^2 = 11$

6) $x^2 = 23$

7) $p^2 = -49$

8) $s^2 = -81$

9) $x^2 = \dfrac{25}{9}$

10) $m^2 = \dfrac{16}{121}$

11) $y^2 = 0.04$

12) $d^2 = 0.25$

13) $r^2 - 144 = 0$

14) $a^2 - 1 = 0$

15) $c^2 - 19 = 0$

16) $a^2 - 6 = 0$

17) $v^2 - 54 = 0$

18) $g^2 - 75 = 0$

19) $t^2 - \dfrac{5}{64} = 0$

20) $c^2 - \dfrac{14}{81} = 0$

 21) $z^2 + 5 = 19$

22) $x^2 + 9 = 17$

23) $n^2 + 10 = 6$

24) $y^2 + 11 = 9$

25) $3d^2 + 14 = 41$

26) $2m^2 - 5 = 67$

27) $4p^2 - 9 = 39$

28) $3j^2 + 7 = 31$

29) $3 = 35 - 8h^2$

30) $145 = 2w^2 - 55$

31) $10 = 14 + 2x^2$

32) $6 = 24 + 3k^2$

33) $4y^2 + 15 = 24$

34) $9n^2 + 17 = 18$

35) $9w^2 - 5 = 5$

36) $16a^2 + 2 = 13$

37) $-7 = 4 - 5b^2$

38) $-1 = 13 - 6t^2$

**Objective 2: Solve an Equation of the Form $(ax + b)^2 = k$**

Solve using the square root property.

39) $(r + 6)^2 = 25$

40) $(x - 1)^2 = 16$

41) $(q - 8)^2 = 1$

42) $(c + 11)^2 = 49$

43) $(a + 2)^2 = 13$

44) $(t - 5)^2 = 7$

45) $(k - 10)^2 = 45$

46) $(b + 4)^2 = 20$

47) $(m + 7)^2 = -18$

48) $(v - 3)^2 = -100$

49) $0 = (p + 3)^2 - 68$

50) $0 = (d + 5)^2 - 72$

51) $(2z - 1)^2 = 9$

52) $(5h + 9)^2 = 36$

53) $121 = (4q + 5)^2$

54) $64 = (3c - 4)^2$

55) $(3g - 10)^2 = 24$

56) $(2w - 7)^2 = 63$

57) $125 = (5u + 8)^2$

58) $44 = (4a + 5)^2$

59) $(2x + 3)^2 - 54 = 0$

60) $(6t - 1)^2 - 90 = 0$

61) $(7h - 8)^2 + 32 = 0$

62) $(2b + 9)^2 + 18 = 0$

63) $(5y - 2)^2 + 6 = 22$

64) $29 = 4 + (3m + 1)^2$

65) $1 = (6r + 7)^2 - 8$

66) $(3 - 4k)^2 - 18 = -2$

67) $(2z - 11)^2 + 3 = 17$

68) $(5x + 8)^2 - 2 = 6$

69) $\left(1 - \dfrac{1}{2}c\right)^2 - 6 = -5$

70) $\left(\dfrac{2}{3}p + 5\right)^2 + 7 = 56$

**Objective 3: Solve a Formula for a Specific Variable**

Solve each problem.

71) The area of a circle is $81\pi$ cm$^2$. Find the radius of the circle.

72) The volume of a right circular cylinder is $28\pi$ in$^3$. Find the radius of the cylinder if it is 7 in. tall.

73) The surface area, $S$, of a sphere is given by $S = 4\pi r^2$, where $r$ is the radius. Find the radius of the sphere with a surface area of $\pi$ m$^2$.

74) The surface area, $S$, of a cube is given by $S = 6L^2$, where $L$ is the length of one of its sides. Find the length of a side of a cube with a surface area of 150 in$^2$.

75) The illuminance $E$ (the measure of light emitted, in lux) of a light source is given by $E = \dfrac{1}{d^2}$, where $I$ is the luminous intensity (measured in candela) and $d$ is the distance, in

meters, from the light source. Find the distance, $d$, from the light source when $E = 300$ lux and $I = 2700$ candela.

76) The power, $P$ (in watts), in an electrical system is given by $P = \dfrac{V^2}{R}$, where $V$ is the voltage, in volts, and $R$ is the resistance, in ohms. Find the voltage if the power is 192 watts and the resistance is 3 ohms.

The kinetic energy $K$, in joules, of an object is given by $K = \dfrac{1}{2}mv^2$, where $m$ is the mass of the object, in kg, and $v$ is the velocity of the object in m/sec. Use this formula for Exercises 77 and 78.

77) Find the velocity of a roller coaster car with a mass of 1200 kg and kinetic energy of 153,600 joules.

78) Find the velocity of a boat with a mass of 25,000 kg and kinetic energy of 28,125 joules.

---

## Section 10.2 Solving Quadratic Equations by Completing the Square

**Objectives**

1. **Complete the Square for an Expression of the Form $x^2 + bx$**
2. **Solve an Equation of the Form $ax^2 + bx + c = 0$ by Completing the Square**

The next method we will learn for solving a quadratic equation is **completing the square.** But first we need to review an idea presented in Section 7.4.

A **perfect square trinomial** is a trinomial whose factored form is the square of a binomial. Some examples of perfect square trinomials are

| Perfect Square Trinomials | Factored Form |
|---------------------------|---------------|
| $x^2 + 6x + 9$ | $(x + 3)^2$ |
| $d^2 - 14d + 49$ | $(d - 7)^2$ |

In the trinomial $x^2 + 6x + 9$, $x^2$ is called the *quadratic term*, $6x$ is called the *linear term*, and 9 is called the *constant*.

## 1. Complete the Square for an Expression of the Form $x^2 + bx$

In a perfect square trinomial where the coefficient of the quadratic term is 1, the constant term is related to the coefficient of the linear term in the following way: *If you find half of the linear coefficient and square the result, you will get the constant term.*

$x^2 + 6x + 9$: The constant, 9, is obtained by

1) finding half of the coefficient of $x$

$$\frac{1}{2}(6) = 3$$

2) then squaring the result.

$$3^2 = 9 \text{ (the constant)}$$

$d^2 - 14d + 49$: The constant, 49, is obtained by

1) finding half of the coefficient of $d$

$$\frac{1}{2}(-14) = -7$$

2) then squaring the result.

$$(-7)^2 = 49 \text{ (the constant)}$$

We can generalize this procedure so that we can find the constant needed to obtain the perfect square trinomial for any quadratic expression of the form $x^2 + bx$. Finding this perfect square trinomial is called *completing the square* because the trinomial will factor to the square of a binomial.

---

**Procedure**   Completing the Square for $x^2 + bx$

To find the constant needed to complete the square for $x^2 + bx$:

**Step 1:** Find half of the coefficient of $x$: $\dfrac{1}{2}b$

**Step 2:** Square the result: $\left(\dfrac{1}{2}b\right)^2$

**Step 3:** Then add it to $x^2 + bx$ to get $x^2 + bx + \left(\dfrac{1}{2}b\right)^2$

---

**BE CAREFUL**

The coefficient of the squared term *must* be 1 before you complete the square!

---

**Example 1**

Complete the square for each expression to obtain a perfect square trinomial. Then, factor.

a)  $k^2 + 10k$          b)  $p^2 - 8p$

**Solution**

a)  Find the constant needed to complete the square for $k^2 + 10k$.

**Step 1:** Find half of the coefficient of $k$:

$$\frac{1}{2}(10) = 5$$

b)  Find the constant needed to complete the square for $p^2 - 8p$.

**Step 1:** Find half of the coefficient of $p$:

$$\frac{1}{2}(-8) = -4$$

**Step 2:** Square the result:

$$5^2 = 25$$

**Step 3:** Add 25 to $k^2 + 10k$:

$$k^2 + 10k + 25$$

The perfect square trinomial is $k^2 + 10k + 25$. The factored form is $(k + 5)^2$.

**Step 2:** Square the result:

$$(-4)^2 = 16$$

**Step 3:** Add 16 to $p^2 - 8p$:

$$p^2 - 8p + 16$$

The perfect square trinomial is $p^2 - 8p + 16$. The factored form is $(p - 4)^2$. ∎

### You Try 1

Complete the square for each expression to obtain a perfect square trinomial. Then, factor.

a) $w^2 + 2w$ \qquad b) $r^2 - 18r$

At the beginning of the section and in Example 1, we saw the following perfect square trinomials and their factored forms. We will look at the relationship between the constant in the factored form and the coefficient of the linear term.

| Perfect Square Trinomial | | Factored Form |
|---|---|---|
| $x^2 + 6x + 9$ | 3 is $\frac{1}{2}(6)$. | $(x + 3)^2$ |
| $d^2 - 14d + 49$ | $-7$ is $\frac{1}{2}(-14)$. | $(d - 7)^2$ |
| $k^2 + 10k + 25$ | 5 is $\frac{1}{2}(10)$. | $(k + 5)^2$ |
| $p^2 - 8p + 16$ | $-4$ is $\frac{1}{2}(-16)$. | $(p - 4)^2$ |

This pattern will always hold true and can be helpful in factoring some perfect square trinomials.

### Example 2

Complete the square for $t^2 + 3t$ to obtain a perfect square trinomial. Then, factor.

**Solution**

**Step 1:** Find half of the coefficient of $t$: $\frac{1}{2}(3) = \frac{3}{2}$

**Step 2:** Square the result: $\left(\frac{3}{2}\right)^2 = \frac{9}{4}$

**Step 3:** Add $\frac{9}{4}$ to $t^2 + 3t$. The perfect square trinomial is $t^2 + 3t + \frac{9}{4}$.

The factored form is $\left(t + \frac{3}{2}\right)^2$

$\frac{3}{2}$ is $\frac{1}{2}(3)$, the coefficient of $t$.

Check: $\left(t + \frac{3}{2}\right)^2 = t^2 + 2t\left(\frac{3}{2}\right) + \left(\frac{3}{2}\right)^2 = t^2 + 3t + \frac{9}{4}$ ∎

**You Try 2**

Complete the square for $n^2 - 7n$ to obtain a perfect square trinomial. Then, factor.

## 2. Solve an Equation of the Form $ax^2 + bx + c = 0$ by Completing the Square

Any quadratic equation of the form $ax^2 + bx + c = 0$ $(a \neq 0)$ can be written in the form $(x - h)^2 = k$ by completing the square. Once an equation is in this form, we can use the square root property to solve for the variable.

---

**Procedure   Steps for Solving a Quadratic Equation ($ax^2 + bx + c = 0$) by Completing the Square**

**Step 1:**   **The coefficient of the squared term must be 1.** If it is not 1, divide both sides of the equation by $a$ to obtain a leading coefficient of 1.

**Step 2:**   **Get the variables on one side of the equal sign and the constant on the other side.**

**Step 3:**   **Complete the square.** Find half of the linear coefficient, then square the result. Add that quantity to *both* sides of the equation.

**Step 4:**   **Factor.**

**Step 5:**   **Solve using the square root property.**

---

**Example 3**

Solve by completing the square.

a)   $x^2 + 12x + 27 = 0$         b)   $k^2 - 2k + 5 = 0$

**Solution**

a)   $x^2 + 12x + 27 = 0$

  **Step 1:**   The coefficient of $x^2$ is already 1.

  **Step 2:**   Get the variables on one side of the equal sign and the constant on the other side: $x^2 + 12x = -27$

  **Step 3:**   Complete the square: $\dfrac{1}{2}(12) = 6$

$$6^2 = 36$$

  Add 36 to both sides of the equation: $x^2 + 12x + 36 = -27 + 36$
$$x^2 + 12x + 36 = 9$$

  **Step 4:**   Factor: $(x + 6)^2 = 9$

  **Step 5:**   Solve using the square root property.

$$(x + 6)^2 = 9$$
$$x + 6 = \pm\sqrt{9}$$
$$x + 6 = \pm 3$$

$$x + 6 = 3 \quad \text{or} \quad x + 6 = -3$$
$$x = -3 \quad \text{or} \quad x = -9$$

The check is left to the student. The solution set is $\{-9, -3\}$.

Notice that we would have obtained the same result if we had solved the equation by factoring.

$$x^2 + 12x + 27 = 0$$
$$(x + 9)(x + 3) = 0$$

$$x + 9 = 0 \quad \text{or} \quad x + 3 = 0$$
$$x = -9 \quad \text{or} \quad x = -3$$

b)  $k^2 - 2k + 5 = 0$

**Step 1:**  The coefficient of $k^2$ is already 1.

**Step 2:**  Get the variables on one side of the equal sign and the constant on the other side: $k^2 - 2k = -5$

**Step 3:**  Complete the square: $\dfrac{1}{2}(-2) = -1$

$$(-1)^2 = 1$$

Add 1 to both sides of the equation: $k^2 - 2k + 1 = -5 + 1$
$$k^2 - 2k + 1 = -4$$

**Step 4:**  Factor: $(k - 1)^2 = -4$

**Step 5:**  Solve using the square root property.

$$(k - 1)^2 = -4$$
$$k - 1 = \pm\sqrt{-4}$$

Since $\sqrt{-4}$ is not a real number, there is no real number solution to $k^2 - 2k + 5 = 0$. The solution set is $\varnothing$.

**You Try 3**

Solve by completing the square.

a)  $w^2 + 4w - 21 = 0$          b)  $a^2 - 6a + 34 = 0$

Remember that in order to complete the square, the coefficient of the quadratic term must be 1.

**Example 4**

Solve by completing the square.

a)  $4n^2 - 16n + 15 = 0$          b)  $10y + 2y^2 = 3$

**Solution**

a)  $4n^2 - 16n + 15 = 0$

**Step 1:**  Since the coefficient of $n^2$ is *not* 1, divide the whole equation by 4.

$$\frac{4n^2}{4} - \frac{16n}{4} + \frac{15}{4} = \frac{0}{4} \qquad \text{Divide by 4.}$$

$$n^2 - 4n + \frac{15}{4} = 0 \qquad \text{Simplify.}$$

**Step 2:**  Get the constant on a side by itself: $n^2 - 4n = -\dfrac{15}{4}$

**Step 3:** Complete the square: $\dfrac{1}{2}(-4) = -2$

$$(-2)^2 = 4$$

Add 4 to both sides of the equation.

$$n^2 - 4n + 4 = -\frac{15}{4} + 4$$

$$n^2 - 4n + 4 = -\frac{15}{4} + \frac{16}{4} \qquad \text{Get a common denominator.}$$

$$n^2 - 4n + 4 = \frac{1}{4} \qquad \text{Add.}$$

**Step 4:** Factor: $(n - 2)^2 = \dfrac{1}{4}$

**Step 5:** Solve using the square root property.

$$(n - 2)^2 = \frac{1}{4}$$

$$n - 2 = \pm\sqrt{\frac{1}{4}} \qquad \text{Square root property}$$

$$n - 2 = \pm\frac{1}{2} \qquad \text{Simplify.}$$

This means $n - 2 = \dfrac{1}{2}$ or $n - 2 = -\dfrac{1}{2}$. Solve both equations.

$$n - 2 = \frac{1}{2} \qquad\qquad\qquad n - 2 = -\frac{1}{2}$$

$$n = \frac{1}{2} + 2 \qquad \text{or} \qquad n = -\frac{1}{2} + 2 \qquad \text{Add 2 to each side.}$$

$$n = \frac{5}{2} \qquad\qquad\qquad\qquad n = \frac{3}{2}$$

The check is left to the student. The solution set is $\left\{\dfrac{3}{2}, \dfrac{5}{2}\right\}$.

b) $10y + 2y^2 = 3$

**Step 1:** Since the coefficient of $y^2$ is *not* 1, divide the whole equation by 2.

$$\frac{10y}{2} + \frac{2y^2}{2} = \frac{3}{2} \qquad \text{Divide by 2.}$$

$$5y + y^2 = \frac{3}{2} \qquad \text{Simplify.}$$

**Step 2:** The constant is on a side by itself. Rewrite the left side of the equation.

$$y^2 + 5y = \frac{3}{2}$$

**Step 3:** Complete the square: $\dfrac{1}{2}(5) = \dfrac{5}{2}$

$$\left(\frac{5}{2}\right)^2 = \frac{25}{4}$$

Add $\dfrac{25}{4}$ to both sides of the equation.

$$y^2 + 5y + \frac{25}{4} = \frac{3}{2} + \frac{25}{4}$$

$$y^2 + 5y + \frac{25}{4} = \frac{6}{4} + \frac{25}{4} \qquad \text{Get a common denominator.}$$

$$y^2 + 5y + \frac{25}{4} = \frac{31}{4} \qquad \text{Add.}$$

**Step 4:**   Factor: $\left(y + \dfrac{5}{2}\right)^2 = \dfrac{31}{4}$

↑

$\dfrac{5}{2}$ is $\dfrac{1}{2}(5)$, the coefficient of $y$.

**Step 5:**   Solve using the square root property.

$$\left(y + \frac{5}{2}\right)^2 = \frac{31}{4}$$

$$y + \frac{5}{2} = \pm\sqrt{\frac{31}{4}} \qquad \text{Square root property}$$

$$y + \frac{5}{2} = \pm\frac{\sqrt{31}}{2} \qquad \text{Simplify.}$$

$$y = -\frac{5}{2} \pm \frac{\sqrt{31}}{2} \qquad \text{Add } -\frac{5}{2} \text{ to each side.}$$

Check the answers. The solution set is $\left\{-\dfrac{5}{2} - \dfrac{\sqrt{31}}{2},\ -\dfrac{5}{2} + \dfrac{\sqrt{31}}{2}\right\}$.   ■

**You Try 4**

Solve by completing the square.

a)   $4c^2 + 8c - 45 = 0$        b)   $2b^2 - 6b = 5$

---

**Answers to You Try Exercises**

1)   a) $w^2 + 2w + 1;\ (w + 1)^2$   b) $r^2 - 18r + 81;\ (r - 9)^2$     2) $n^2 - 7n + \dfrac{49}{4};\ \left(n - \dfrac{7}{2}\right)^2$

3)   a) $\{-7, 3\}$   b) $\varnothing$     4)   a) $\left\{-\dfrac{9}{2}, \dfrac{5}{2}\right\}$   b) $\left\{\dfrac{3}{2} - \dfrac{\sqrt{19}}{2}, \dfrac{3}{2} + \dfrac{\sqrt{19}}{2}\right\}$

---

## 10.2 Exercises

**Objectives 1: Complete the Square for an Expression of the Form $x^2 + bx$**

1)   What is a perfect square trinomial? Give an example.

2)   In $x^2 - 9x + 14$, what is the

a)   quadratic term?

b)   linear term?

c)   constant?

Complete the square for each expression to obtain a perfect square trinomial. Then, factor.

**Fill It In**

Fill in the blanks with either the missing mathematical step or reason for the given step.

3) $y^2 + 18y$

_____     Find half of the coefficient of $y$.

_____     Square the result.

_____     Add the constant to the expression.

The perfect square trinomial is _____.

The factored form of the trinomial is _____.

4) $c^2 - 5c$

$\dfrac{1}{2}(-5) = -\dfrac{5}{2}$     _____

$\left(-\dfrac{5}{2}\right)^2 = \dfrac{25}{4}$     _____

$c^2 - 5c + \dfrac{25}{4}$     _____

The perfect square trinomial is _____.

The factored form of the trinomial is _____.

5) $a^2 + 12a$          6) $b^2 + 8b$

7) $k^2 - 10k$          8) $p^2 - 4p$

9) $g^2 - 24g$          10) $z^2 + 26z$

11) $h^2 + 9h$          12) $m^2 - 3m$

13) $x^2 - x$          14) $y^2 + 7y$

**Objective 2: Solve an Equation of the Form $ax^2 + bx + c = 0$ by Completing the Square**

15) What are the steps used to solve a quadratic equation by completing the square?

16) Can $x^3 + 12x + 20 = 0$ be solved by completing the square? Give a reason for your answer.

Solve by completing the square.

17) $x^2 + 6x + 8 = 0$          18) $a^2 + 10a - 24 = 0$

19) $z^2 - 14z + 45 = 0$          20) $t^2 - 12t - 45 = 0$

21) $p^2 + 8p + 20 = 0$          22) $c^2 - 2c + 37 = 0$

23) $y^2 - 11 = -4y$          24) $k^2 - 1 = 8k$

25) $x^2 - 10x = 3$          26) $-4b = b^2 - 14$

27) $2a = 22 + a^2$          28) $w^2 + 39 = 6w$

29) $m^2 + 3m - 40 = 0$          30) $h^2 - 7h + 6 = 0$

31) $c^2 - 56 = c$          32) $p^2 + 5p = -4$

33) $h^2 + 9h = -12$          34) $q^2 - 3 = q$

35) $b^2 - 5b + 27 = 6$          36) $g^2 + 3g + 11 = 4$

37) Can you complete the square on $2x^2 + 16x$ as it is given? Why or why not?

38) What is the first thing you should do if you want to solve $3n^2 - 9n = 12$ by completing the square?

Solve by completing the square.

39) $4r^2 + 32r + 55 = 0$          40) $4t^2 - 16t + 7 = 0$

41) $3x^2 + 39 = 30x$          42) $5p^2 + 30p = 10$

43) $7k^2 + 84 = 49k$          44) $10m = 2m^2 + 12$

45) $54y - 6y^2 = 72$          46) $8w^2 + 8w = 32$

47) $16z^2 + 3 = 16z$          48) $5 - 16c = 16c^2$

49) $3g^2 + 15g + 37 = 0$          50) $7t^2 - 21t + 40 = 0$

51) $-v^2 - 2v + 35 = 0$          52) $-k^2 + 12k - 32 = 0$

53) $(a - 4)(a + 10) = -17$     54) $(y + 5)(y - 3) = 5$

55) $n + 2 = 3n^2$          56) $15 = m + 2m^2$

57) $(5p + 2)(p + 4) = 1$          58) $(3c + 4)(c + 2) = 3$

Solve each problem by writing an equation and completing the square.

59) The length of a rectangular portfolio is 7 inches more than its width. Find the dimensions of the portfolio if it has an area of 170 in$^2$.

60) The area of rectangular sign is 220 ft$^2$. Its width is 9 ft less than its length. What are the dimensions of the sign?

61) The area of a triangle is 60 cm$^2$. Its base is 1 cm less than twice its height. Find the lengths of the base and the height.

62) The length of the base of a triangle is 5 in. more than twice its height. Find the lengths of the base and the height if the area is 26 in$^2$.

Find the lengths of the sides of each right triangle. Use the Pythagorean theorem, and solve by completing the square.

63)

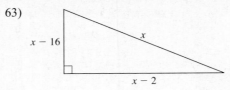

64)

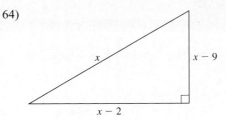

# Section 10.3 Solving Equations Using the Quadratic Formula

## Objectives

1. Derive the Quadratic Formula
2. Solve a Quadratic Equation Using the Quadratic Formula

## 1. Derive the Quadratic Formula

In Section 10.2, we saw that any quadratic equation of the form $ax^2 + bx + c = 0$ $(a \neq 0)$ can be solved by completing the square.

We can develop another method, called the *quadratic formula,* for solving quadratic equations if we complete the square on $ax^2 + bx + c = 0$ $(a \neq 0)$.

The steps we use to complete the square on $ax^2 + bx + c = 0$ are *exactly* the same steps we use to solve an equation like $2x^2 + 3x - 1 = 0$. We will do these steps side by side so that you can more easily understand the process.

### Solve Each Equation for x by Completing the Square.

$$2x^2 + 3x - 1 = 0 \qquad \qquad ax^2 + bx + c = 0$$

***Step 1:*** **The coefficient of the squared term must be 1.**

$$2x^2 + 3x - 1 = 0 \qquad\qquad ax^2 + bx + c = 0$$

$$\frac{2x^2}{2} + \frac{3x}{2} - \frac{1}{2} = \frac{0}{2} \quad \text{Divide by 2.} \qquad \frac{ax^2}{a} + \frac{bx}{a} + \frac{c}{a} = \frac{0}{a} \quad \text{Divide by } a.$$

$$x^2 + \frac{3}{2}x - \frac{1}{2} = 0 \quad \text{Simplify.} \qquad x^2 + \frac{b}{a}x + \frac{c}{a} = 0 \quad \text{Simplify.}$$

***Step 2:*** **Get the constant on the other side of the equal sign.**

$$x^2 + \frac{3}{2}x = \frac{1}{2} \quad \text{Add } \frac{1}{2}. \qquad x^2 + \frac{b}{a}x = -\frac{c}{a} \quad \text{Subtract } \frac{c}{a}.$$

***Step 3:*** **Complete the square.**

$$\frac{1}{2}\left(\frac{3}{2}\right) = \frac{3}{4} \quad \tfrac{1}{2} \text{ of } x\text{-coefficient} \qquad \frac{1}{2}\left(\frac{b}{a}\right) = \frac{b}{2a} \quad \tfrac{1}{2} \text{ of } x\text{-coefficient}$$

$$\left(\frac{3}{4}\right)^2 = \frac{9}{16} \quad \text{Square the result.} \qquad \left(\frac{b}{2a}\right)^2 = \frac{b^2}{4a^2} \quad \text{Square the result.}$$

$$\text{Add } \frac{9}{16} \text{ to both sides of the equation.} \qquad \text{Add } \frac{b^2}{4a^2} \text{ to both sides of the equation.}$$

$$x^2 + \frac{3}{2}x + \frac{9}{16} = \frac{1}{2} + \frac{9}{16} \qquad\qquad x^2 + \frac{b}{a}x + \frac{b^2}{4a^2} = -\frac{c}{a} + \frac{b^2}{4a^2}$$

$$x^2 + \frac{3}{2}x + \frac{9}{16} = \frac{8}{16} + \frac{9}{16} \quad \begin{array}{l}\text{Get a common}\\ \text{denominator.}\end{array} \qquad x^2 + \frac{b}{a}x + \frac{b^2}{4a^2} = -\frac{4ac}{4a^2} + \frac{b^2}{4a^2} \quad \begin{array}{l}\text{Get a common}\\ \text{denominator.}\end{array}$$

$$x^2 + \frac{3}{2}x + \frac{9}{16} = \frac{17}{16} \quad \text{Add.} \qquad x^2 + \frac{b}{a}x + \frac{b^2}{4a^2} = \frac{b^2 - 4ac}{4a^2} \quad \text{Add.}$$

***Step 4:*** **Factor.**

$$\left(x + \frac{3}{4}\right)^2 = \frac{17}{16} \qquad\qquad \left(x + \frac{b}{2a}\right)^2 = \frac{b^2 - 4ac}{4a^2}$$

$$\uparrow \qquad\qquad\qquad \uparrow$$

$$\frac{3}{4} \text{ is } \frac{1}{2}\left(\frac{3}{2}\right), \text{ the coefficient of } x. \qquad \frac{b}{2a} \text{ is } \frac{1}{2}\left(\frac{b}{a}\right), \text{ the coefficient of } x.$$

**Step 5:    Solve using the square root property.**

$$\left(x + \frac{3}{4}\right)^2 = \frac{17}{16}$$

$$x + \frac{3}{4} = \pm\sqrt{\frac{17}{16}}$$

$$x + \frac{3}{4} = \frac{\pm\sqrt{17}}{4} \qquad \sqrt{16} = 4$$

$$x = -\frac{3}{4} \pm \frac{\sqrt{17}}{4} \qquad \text{Subtract } \frac{3}{4}.$$

$$x = \frac{-3 \pm \sqrt{17}}{4} \qquad \begin{array}{l}\text{Same denomi-}\\\text{nators, combine}\\\text{numerators.}\end{array}$$

$$\left(x + \frac{b}{2a}\right)^2 = \frac{b^2 - 4ac}{4a^2}$$

$$x + \frac{b}{2a} = \pm\sqrt{\frac{b^2 - 4ac}{4a^2}}$$

$$x + \frac{b}{2a} = \frac{\pm\sqrt{b^2 - 4ac}}{2a} \qquad \sqrt{4a^2} = 2a$$

$$x = -\frac{b}{2a} \pm \frac{\sqrt{b^2 - 4ac}}{2a} \qquad \text{Subtract } \frac{b}{2a}.$$

$$x = \frac{-b \pm \sqrt{b^2 - 4ac}}{2a} \qquad \begin{array}{l}\text{Same denomi-}\\\text{nators, combine}\\\text{numerators.}\end{array}$$

The result on the right is called the *quadratic formula.*

---

**Definition    The Quadratic Formula**

The solutions of any quadratic equation of the form $ax^2 + bx + c = 0$ $(a \neq 0)$ are

$$x = \frac{-b \pm \sqrt{b^2 - 4ac}}{2a}$$

---

**Note**

1)   The equation to be solved *must* be written in the form $ax^2 + bx + c = 0$ so that $a, b,$ and $c$ can be identified correctly.

2)   $x = \dfrac{-b \pm \sqrt{b^2 - 4ac}}{2a}$ represents the two solutions $x = \dfrac{-b + \sqrt{b^2 - 4ac}}{2a}$ and $x = \dfrac{-b - \sqrt{b^2 - 4ac}}{2a}.$

3)   Notice that the fraction bar continues under $-b$ and does not end at the radical.

$$x = \frac{-b \pm \sqrt{b^2 - 4ac}}{2a} \qquad\qquad x = -b \pm \frac{\sqrt{b^2 - 4ac}}{2a}$$

Correct                                    Incorrect

4)   When deriving the quadratic formula, using the $\pm$ allows us to say that $\sqrt{4a^2} = 2a.$

5)   The quadratic formula is a *very* important result and is one that is used often. *It should be memorized!*

## 2. Solve a Quadratic Equation Using the Quadratic Formula

**Example 1**

Solve using the quadratic formula.

a)   $2x^2 + 3x - 1 = 0$          b)   $3n^2 - 10n + 8 = 0$

**Solution**

a)   Is $2x^2 + 3x - 1 = 0$ in the form $ax^2 + bx + c = 0$? *Yes*. Identify the values of $a$, $b$, and $c$, and substitute them into the quadratic formula.

$$a = 2 \qquad b = 3 \qquad c = -1$$

$$x = \frac{-b \pm \sqrt{b^2 - 4ac}}{2a} \qquad \text{Quadratic formula}$$

$$= \frac{-(3) \pm \sqrt{(3)^2 - 4(2)(-1)}}{2(2)} \qquad \text{Substitute } a = 2, b = 3, \text{ and } c = -1.$$

$$= \frac{-3 \pm \sqrt{9 - (-8)}}{4} \qquad \text{Perform the operations.}$$

$$= \frac{-3 \pm \sqrt{17}}{4} \qquad 9 - (-8) = 9 + 8 = 17$$

The solution set is $\left\{ \dfrac{-3 - \sqrt{17}}{4}, \dfrac{-3 + \sqrt{17}}{4} \right\}$. This is the same result we obtained when we solved this equation by completing the square at the beginning of the section.

b)   Is $3n^2 - 10n + 8 = 0$ in the form $ax^2 + bx + c = 0$? *Yes*. Identify $a$, $b$, and $c$, and substitute them into the quadratic formula.

$$a = 3 \qquad b = -10 \qquad c = 8$$

$$n = \frac{-b \pm \sqrt{b^2 - 4ac}}{2a} \qquad \text{Quadratic formula}$$

$$n = \frac{-(-10) \pm \sqrt{(-10)^2 - 4(3)(8)}}{2(3)} \qquad \text{Subsitute } a = 3, b = -10, \text{ and } c = 8.$$

$$n = \frac{10 \pm \sqrt{100 - 96}}{6} \qquad \text{Perform the operations.}$$

$$n = \frac{10 \pm \sqrt{4}}{6} \qquad \text{Simplify the radicand.}$$

$$n = \frac{10 \pm 2}{6} \qquad \text{Simplify } \sqrt{4}.$$

Find the two values of $n$, one using the plus sign and the other using the minus sign:

$$n = \frac{10 + 2}{6} = \frac{12}{6} = 2 \qquad \text{or} \qquad n = \frac{10 - 2}{6} = \frac{8}{6} = \frac{4}{3}$$

Check the values in the original equation. The solution set is $\left\{ \dfrac{4}{3}, 2 \right\}$.  ∎

  **You Try !**

Solve using the quadratic formula.

a)   $5p^2 - p - 3 = 0$          b)   $3r^2 + r - 10 = 0$

If a quadratic equation is not in standard form, we must write it that way.

## Example 2

Solve using the quadratic formula.

a) $t^2 + 1 = 4t$

b) $2w(w + 3) = -5$

**Solution**

a) Is $t^2 + 1 = 4t$ in the form $ax^2 + bx + c = 0$? *No.* Before we can apply the quadratic formula, we must write it in that form.

$$t^2 - 4t + 1 = 0 \qquad \text{Subtract } 4t.$$

Identify $a$, $b$, and $c$:   $a = 1$      $b = -4$      $c = 1$

$$t = \frac{-b \pm \sqrt{b^2 - 4ac}}{2a} \qquad \text{Quadratic formula}$$

$$t = \frac{-(-4) \pm \sqrt{(-4)^2 - 4(1)(1)}}{2(1)} \qquad \text{Substitute } a = 1, b = -4, \text{ and } c = 1.$$

$$t = \frac{4 \pm \sqrt{16 - 4}}{2} \qquad \text{Perform the operations.}$$

$$t = \frac{4 \pm \sqrt{12}}{2} \qquad \text{Simplify the radicand.}$$

$$t = \frac{4 \pm 2\sqrt{3}}{2} \qquad \text{Simplify the radical.}$$

$$t = \frac{2(2 \pm \sqrt{3})}{2} \qquad \text{Factor out 2.}$$

$$t = 2 \pm \sqrt{3} \qquad \text{Simplify.}$$

The solution set is $\{2 - \sqrt{3}, 2 + \sqrt{3}\}$.

b) Is $2w(w + 3) = -5$ in the form $ax^2 + bx + c = 0$? *No.* Distribute on the left side of the equation, then get all terms on the same side of the equal sign.

$$2w^2 + 6w = -5 \qquad \text{Distribute.}$$
$$2w^2 + 6w + 5 = 0 \qquad \text{Add 5.}$$

Identify $a$, $b$, and $c$:   $a = 2$      $b = 6$      $c = 5$

$$w = \frac{-b \pm \sqrt{b^2 - 4ac}}{2a} \qquad \text{Quadratic formula}$$

$$w = \frac{-(6) \pm \sqrt{(6)^2 - 4(2)(5)}}{2(2)} \qquad \text{Substitute } a = 2, b = 6, \text{ and } c = 5.$$

$$w = \frac{-6 \pm \sqrt{36 - 40}}{4} \qquad \text{Perform the operations.}$$

$$w = \frac{-6 \pm \sqrt{-4}}{4} \qquad \text{Simplify the radicand.}$$

Since $\sqrt{-4}$ is not a real number, there is no real number solution to $2w(w + 3) = -5$. The solution set is $\varnothing$. ∎

## You Try 2

Solve using the quadratic formula.

a) $y^2 + 4 = 6y$

b) $m(5m - 2) = -3$

The expression under the radical, $b^2 - 4ac$, in the quadratic formula is called the **discriminant**. Examples 1a and 2a show that if the discriminant is positive but not a perfect square, then the given equation has *two irrational solutions*. We see in Example 1b that if the discriminant is positive and the square of an integer, then the equation has *two rational solutions* and can be solved by factoring. If the discriminant is negative, as in Example 2b, then the equation has *no real number solution*.

What if the discriminant equals 0? What does that tell us about the solution set?

## Example 3

Solve $\dfrac{2}{9}k^2 - \dfrac{2}{3}k + \dfrac{1}{2} = 0$ using the quadratic formula.

### Solution

Is $\dfrac{2}{9}k^2 - \dfrac{2}{3}k + \dfrac{1}{2} = 0$ in the form $ax^2 + bx + c = 0$? *Yes.* However, working with fractions in the quadratic formula would be difficult. *Eliminate the fractions by multiplying the equation by 18, the least common denominator of the fractions.*

$$18\left(\frac{2}{9}k^2 - \frac{2}{3}k + \frac{1}{2}\right) = 18 \cdot 0 \qquad \text{Multiply by 18 to eliminate the fractions.}$$
$$4k^2 - 12k + 9 = 0$$

Identify $a$, $b$, and $c$:  $\quad a = 4 \qquad b = -12 \qquad c = 9$

$$k = \frac{-b \pm \sqrt{b^2 - 4ac}}{2a} \qquad \text{Quadratic formula}$$

$$k = \frac{-(-12) \pm \sqrt{(-12)^2 - 4(4)(9)}}{2(4)} \qquad \text{Substitute } a = 4,\, b = -12,\text{ and } c = 9.$$

$$k = \frac{12 \pm \sqrt{144 - 144}}{8} \qquad \text{Perform the operations.}$$

$$k = \frac{12 \pm \sqrt{0}}{8} \qquad \text{Simplify the radicand. The discriminant} = 0.$$

$$k = \frac{12 \pm 0}{8} = \frac{12}{8} = \frac{3}{2}$$

The solution set is $\left\{\dfrac{3}{2}\right\}$.

Example 3 illustrates that when the discriminant equals 0, the equation has *one rational solution*.

## You Try 3

Solve $\dfrac{3}{4}h^2 + \dfrac{1}{2}h + \dfrac{1}{12} = 0$ using the quadratic formula.

### Answers to You Try Exercises

1)  a) $\left\{\dfrac{1 - \sqrt{61}}{10}, \dfrac{1 + \sqrt{61}}{10}\right\}$  b) $\left\{-2, \dfrac{5}{3}\right\}$

2)  a) $\{3 - \sqrt{5}, 3 + \sqrt{5}\}$   b) $\varnothing$     3) $\left\{-\dfrac{1}{3}\right\}$

## 10.3 Exercises

**Objective 2: Solve a Quadratic Equation Using the Quadratic Formula**

1) To solve a quadratic equation, $ax^2 + bx + c = 0$ ($a \neq 0$), for $x$, we can use the quadratic formula. Write the quadratic formula.

Find the error in each, and correct the mistake.

2) The solution to $ax^2 + bx + c = 0$ ($a \neq 0$) can be found using the quadratic formula

$$x = -b \pm \frac{\sqrt{b^2 - 4ac}}{2a}.$$

3) In order to solve $3x^2 - 5x = 4$ using the quadratic formula, a student substitutes $a$, $b$, and $c$ into the formula in this way: $a = 3$, $b = -5$, $c = 4$.

$$x = \frac{-(-5) \pm \sqrt{(-5)^2 - 4(3)(4)}}{2(3)}$$

4) $\dfrac{-3 \pm 12\sqrt{5}}{3} = -1 \pm 12\sqrt{5}$

Solve using the quadratic formula.

5) $x^2 + 2x - 8 = 0$

6) $p^2 + 8p + 12 = 0$

7) $6z^2 - 7z + 2 = 0$

8) $2h^2 + h - 15 = 0$

9) $k^2 + 2 = 5k$

10) $d^2 = 5 - 3d$

11) $3w^2 = 2w + 4$

12) $8r = 2 - 5r^2$

13) $y = 2y^2 + 6$

14) $3v + 4v^2 = -3$

15) $m^2 + 11m = 0$

16) $w^2 - 8w = 0$

17) $2p(p - 3) = -3$

18) $3q(q + 3) = 7q + 4$

19) $(2s + 3)(s - 1) = s^2 - s + 6$

20) $2n^2 + 2n - 4 = (3n + 4)(n - 2)$

21) $k(k + 2) = -5$

22) $3r = 2(r^2 + 4)$

23) $(x - 8)(x - 3) = 3(3 - x)$

24) $2(k + 10) = (k + 10)(k - 2)$

25) $8t = 1 + 16t^2$

26) $9u^2 + 12u + 4 = 0$

27) $\dfrac{1}{8}z^2 + \dfrac{3}{4}z + \dfrac{1}{2} = 0$

28) $\dfrac{1}{6}p^2 + \dfrac{4}{3}p = \dfrac{5}{2}$

29) $\dfrac{1}{6}k + \dfrac{1}{2} = \dfrac{3}{4}k^2$

30) $\dfrac{1}{5} + \dfrac{1}{2}w = \dfrac{1}{5}w^2$

31) $0.8v^2 + 0.1 = 0.6v$

32) $0.2m^2 + 0.1m = 1.5$

33) $16g^2 - 3 = 0$

34) $49n^2 - 20 = 0$

35) $9d^2 - 4 = 0$

36) $4k^2 - 25 = 0$

37) $3(3 - 4r) = -4r^2$

38) $5y(5y + 2) = -1$

39) $6 = 7h - 3h^2$

40) $1 + t(2 + 5t) = 0$

41) $4p^2 + 6 = 20p$

42) $4x^2 = 6x + 16$

Write an equation and solve. Give an exact answer and use a calculator to round the answer to the nearest hundredth.

43) The hypotenuse of a right triangle is 1 inch less than twice the shorter leg. The longer leg is $\sqrt{23}$ inches long. Find the length of the shorter leg.

44) The hypotenuse of the right triangle is 1 inch more than twice the longer leg. The length of the shorter leg is $\sqrt{14}$ in. Find the length of the longer leg.

Solve. Give an exact answer and use a calculator to round the answer to the nearest hundredth.

45) An object is thrown upward from a height of 24 feet. The height $h$ of the object (in feet) $t$ seconds after the object is released is given by

$$h = -16t^2 + 24t + 24$$

a) How long does it take the object to reach a height of 8 feet?

b) How long does it take the object to hit the ground? (Hint: When the object hits the ground, $h = 0$.)

46) A ball is thrown upward from a height of 20 feet. The height $h$ of the ball (in feet) $t$ seconds after the ball is released is given by

$$h = -16t^2 + 16t + 20$$

a) How long does it take the ball to reach a height of 8 feet?

b) How long does it take the ball to hit the ground? (Hint: when the ball hits the ground, $h = 0$.)

## Putting It All Together

**Objective**

1. **Decide Which Method to Use to Solve a Quadratic Equation**

We have learned four methods for solving quadratic equations.

**Methods for Solving Quadratic Equations**

1) Factoring

2) Square root property

3) Completing the square

4) Quadratic formula

While it is true that the quadratic formula can be used to solve *every* quadratic equation of the form $ax^2 + bx + c = 0$ ($a \neq 0$), it is not always the most *efficient* method. In this section we will discuss how to decide which method to use to solve a quadratic equation.

### 1. Decide Which Method to Use to Solve a Quadratic Equation

**Example 1**

Solve.

a) $c^2 - 3c = 28$   b) $x^2 - 10x + 18 = 0$

c) $2k^2 + 5k + 9 = 0$   d) $(5n + 3)^2 - 12 = 0$

**Solution**

a) Write $c^2 - 3c = 28$ in standard form:   $c^2 - 3c - 28 = 0$

Does $c^2 - 3c - 28$ factor? *Yes. Solve by factoring.*

$$(c - 7)(c + 4) = 0$$

$c - 7 = 0$   or   $c + 4 = 0$      Set each factor equal to 0.

$c = 7$   or      $c = -4$      Solve.

The solution set is $\{-4, 7\}$.

b) To solve $x^2 - 10x + 18 = 0$, ask yourself, "Can I factor $x^2 - 10x + 18$?" *No.* We could solve this using the quadratic formula, but *completing the square* is also a good method for solving this equation. Why?

*Completing the square is a desirable method for solving a quadratic equation when the coefficient of the squared term is 1 or −1 and when the coefficient of the linear term is even.*
  We will solve $x^2 - 10x + 18 = 0$ by completing the square.

**Step 1:** The coefficient of $x^2$ is 1.

**Step 2:** Get the variables on one side of the equal sign and the constant on the other side:   $x^2 - 10x = -18$

**Step 3:** Complete the square:   $\dfrac{1}{2}(-10) = -5$

$(-5)^2 = 25$

Add 25 to both sides of the equation:   $x^2 - 10x + 25 = -18 + 25$

$x^2 - 10x + 25 = 7$

***Step 4:*** Factor: $(x - 5)^2 = 7$

***Step 5:*** Solve using the square root property:

$$(x - 5)^2 = 7$$
$$x - 5 = \pm\sqrt{7}$$
$$x = 5 \pm \sqrt{7}$$

The solution set is $\{5 - \sqrt{7}, 5 + \sqrt{7}\}$.

### Note

Completing the square works well when the coefficient of the squared term is 1 or −1 and when the coefficient of the linear term is *even* because when we complete the square in Step 3, we will not obtain a fraction. (Half of an even number is an integer.)

c)  Ask yourself, "Can I solve $2k^2 + 5k + 9 = 0$ by factoring?" *No.* Completing the square would not be a very efficient way to solve the equation because the coefficient of $k^2$ is 2, and dividing the equation by 2 would give us $k^2 + \dfrac{5}{2}k + \dfrac{9}{2} = 0$.

*We will solve $2k^2 + 5k + 9 = 0$ using the quadratic formula.*

Identify $a$, $b$, and $c$:    $a = 2$    $b = 5$    $c = 9$

$$k = \frac{-b \pm \sqrt{b^2 - 4ac}}{2a} \qquad \text{Quadratic formula}$$

$$k = \frac{-(5) \pm \sqrt{(5)^2 - 4(2)(9)}}{2(2)} \qquad \text{Substitute } a = 2, b = 5, \text{ and } c = 9.$$

$$k = \frac{-5 \pm \sqrt{25 - 72}}{4} = \frac{-5 \pm \sqrt{-47}}{4}$$

Since $\sqrt{-47}$ is not a real number, there is no real number solution to $2k^2 + 5k + 9 = 0$. The solution set is $\varnothing$.

d)  Which method should we use to solve $(5n + 3)^2 - 12 = 0$?

We *could* square the binomial, combine like terms, then solve, possibly, by factoring or using the quadratic formula. However, this would be very inefficient. The equation contains a squared quantity and a constant.
*We will solve $(5n + 3)^2 - 12 = 0$ using the square root property.*

$$(5n + 3)^2 - 12 = 0$$
$$(5n + 3)^2 = 12 \qquad \text{Add 12 to each side.}$$
$$5n + 3 = \pm\sqrt{12} \qquad \text{Square root property}$$
$$5n + 3 = \pm 2\sqrt{3} \qquad \text{Simplify } \sqrt{12}.$$
$$5n = -3 \pm 2\sqrt{3} \qquad \text{Add } -3 \text{ to each side.}$$
$$n = \frac{-3 \pm 2\sqrt{3}}{5} \qquad \text{Divide by 5.}$$

The solution set is $\left\{\dfrac{-3 - 2\sqrt{3}}{5}, \dfrac{-3 + 2\sqrt{3}}{5}\right\}$.

### You Try 1

Solve.

a)  $3t^2 - 2 = -8t$          b)  $m^2 + 36 = 13m$

c)  $(4r - 1)^2 + 10 = 0$      d)  $w^2 + 8w - 2 = 0$

### Answers to You Try Exercises

1) a) $\left\{\dfrac{-4 - \sqrt{22}}{3}, \dfrac{-4 + \sqrt{22}}{3}\right\}$   b) $\{4, 9\}$   c) $\varnothing$   d) $\{-4 - 3\sqrt{2}, -4 + 3\sqrt{2}\}$

---

## Putting It All Together
### Summary Exercises

**Objective 1: Decide Which Method to Use to Solve a Quadratic Equation**

Keep in mind the four methods we have learned for solving quadratic equations: *factoring, the square root property, completing the square,* and *the quadratic formula.* Solve the equations using one of these methods.

1) $f^2 - 75 = 0$

2) $t^2 - 8t = 4$

3) $a(a + 1) = 20$

4) $3m^2 + 7 = 4m$

5) $v^2 + 6v + 7 = 0$

6) $4u^2 - 5u - 6 = 0$

7) $3x(x + 3) = -5(x - 1)$

8) $5 = (p + 4)^2 + 7$

9) $n^2 + 12n + 42 = 0$

10) $\dfrac{1}{2}r^2 = \dfrac{3}{4} - \dfrac{3}{2}r$

11) $12 + (2k - 1)^2 = 3$

12) $h^2 + 7h + 15 = 0$

13) $1 = \dfrac{x^2}{40} - \dfrac{3x}{20}$

14) $72 = 2p^2$

15) $b^2 - 6b = 5$

16) $3m^3 + 42m = -27m^2$

17) $q(q + 12) = 3(q^2 + 5) + q$

18) $w^2 = 3w$

19) $\dfrac{9}{c} = 1 + \dfrac{18}{c^2}$

20) $2t(2t + 4) = 5t + 6$

21) $(3v + 4)(v - 2) = -9$

22) $y^2 + 4y = 2$

23) $2r^2 + 3r - 2 = 0$

24) $(6r + 1)(r - 4) = -2(12r + 1)$

25) $5m = m^2$

26) $6z^2 + 12z + 18 = 0$

27) $4m^3 = 9m$

28) $\dfrac{8}{x^2} = \dfrac{1}{4} + \dfrac{1}{x}$

29) $2k^2 + 3 = 9k$

30) $6v^2 + 3 = 15v$

---

## Section 10.4 Complex Numbers

### Objectives

1. Define a Complex Number
2. Find the Square Root of a Negative Number
3. Add and Subtract Complex Numbers
4. Multiply Complex Numbers
5. Multiply a Complex Number by Its Conjugate
6. Divide Complex Numbers
7. Solve Quadratic Equations with Complex Number Solutions

### 1. Define a Complex Number

We have seen that not all quadratic equations have real number solutions. For example, there is no real number solution to $x^2 = -1$ since the square of a real number will not be negative. However, the equation $x^2 = -1$ *does* have a solution under another set of numbers called *complex numbers.*

Before we define a complex number, we must define the number $i$. $i$ is called an *imaginary number.*

### Definition

The **imaginary number** $i$ is defined as $i = \sqrt{-1}$.

Therefore, squaring both sides gives us $i^2 = -1$.

**Note**

$i = \sqrt{-1}$ and $i^2 = -1$ are two *very* important facts to remember. We will be using them often!

## Definition

A **complex number** is a number of the form $a + bi$, where $a$ and $b$ are real numbers. $a$ is called the **real part** and $b$ is called the **imaginary part**.

The following table lists some examples of complex numbers and their real and imaginary parts.

| Complex Number | Real Part | Imaginary Part |
|:---:|:---:|:---:|
| $8 + 3i$ | 8 | 3 |
| $\dfrac{1}{2} - 5i$ | $\dfrac{1}{2}$ | $-5$ |
| $9i$ | 0 | 9 |
| 2 | 2 | 0 |

**Note**

$9i$ can be written in the form $a + bi$ as $0 + 9i$. Likewise, besides being a real number, 2 is a complex number since it can be written as $2 + 0i$.

Since any real number, $a$, can be written in the form $a + 0i$, all real numbers are also complex numbers.

**Note**

The set of real numbers is a subset of the set of complex numbers.

Since we defined $i$ as $i = \sqrt{-1}$, we can now evaluate square roots of negative numbers.

## 2. Find the Square Root of a Negative Number

### Example 1

Simplify.

a)  $\sqrt{-25}$          b)  $\sqrt{-2}$          c)  $\sqrt{-20}$

**Solution**

a)  $\sqrt{-25} = \sqrt{-1 \cdot 25} = \sqrt{-1} \cdot \sqrt{25} = i \cdot 5 = 5i$

b)  $\sqrt{-2} = \sqrt{-1 \cdot 2} = \sqrt{-1} \cdot \sqrt{2} = i\sqrt{2}$

c)  $\sqrt{-20} = \sqrt{-1 \cdot 20} = \sqrt{-1} \cdot \sqrt{20} = i\sqrt{4}\sqrt{5} = i \cdot 2\sqrt{5} = 2i\sqrt{5}$    ■

**Note**

In Example 1b) we wrote $i\sqrt{2}$ instead of $\sqrt{2}i$, and in Example 1c) we wrote $2i\sqrt{5}$ instead of $2\sqrt{5}i$. That is because we want to avoid the confusion of thinking that the $i$ is under the radical. It is good practice to write the $i$ *before* the radical.

### You Try 1

Simplify.

a)  $\sqrt{-4}$          b)  $\sqrt{-7}$          c)  $\sqrt{-54}$

### 3. Add and Subtract Complex Numbers

Just as we can add, subtract, multiply, and divide real numbers, we can perform all of these operations with complex numbers.

> **Procedure**    Adding and Subtracting Complex Numbers
> 1) To add complex numbers, add the real parts and add the imaginary parts.
> 2) To subtract complex numbers, apply the distributive property and combine the real parts and combine the imaginary parts.

**Example 2**

Add or subtract.

a)  $(5 + 2i) + (3 + 7i)$          b)   $(4 + i) - (10 - 3i)$

**Solution**

a)   $(5 + 2i) + (3 + 7i) = (5 + 3) + (2 + 7)i$     Add real parts; add imaginary parts.
$$= 8 + 9i$$

b)   $(4 + i) - (10 - 3i) = 4 + i - 10 + 3i$     Distributive property
$$= (4 - 10) + (1 + 3)i$$     Subtract real parts; add imaginary parts.
$$= -6 + 4i$$ ∎

 **You Try 2**

Add or subtract.
a)  $(-8 + 5i) + (2 + 9i)$          b)   $(1 - 12i) - (7 - 9i)$

### 4. Multiply Complex Numbers

We multiply complex numbers just like we would multiply polynomials. There may be an additional step, however. Remember to replace $i^2$ with $-1$.

**Example 3**

Multiply and simplify.

a)   $4(-3 + 8i)$          b)   $(6 + 5i)(-1 + 2i)$          c)   $(5 + 4i)(5 - 4i)$

**Solution**

a)   $4(-3 + 8i) = -12 + 32i$     Distributive property

b)   Look carefully at $(6 + 5i)(-1 + 2i)$. Each complex number is a *binomial* similar to, say, $(x + 5)(x + 2)$. How can we multiply two binomials? We can use FOIL.

$$(6 + 5i)(-1 + 2i) = (6)(-1) + (6)(2i) + (5i)(-1) + (5i)(2i)$$
$$\qquad\qquad\qquad\quad \text{F}\qquad\quad \text{O}\qquad\quad \text{I}\qquad\quad \text{L}$$
$$= -6\quad +\quad 12i\quad -\quad 5i\quad +\quad 10i^2$$
$$= -6 + 7i + 10(-1)$$     Replace $i^2$ with $-1$.
$$= -6 + 7i - 10 = -16 + 7i$$

c)   $(5 + 4i)$ and $(5 - 4i)$ are each binomials, so we will multiply them using FOIL.
$$(5 + 4i)(5 - 4i) = (5)(5) + (5)(-4i) + (4i)(5) + (4i)(-4i)$$
$$\qquad\qquad\qquad\quad \text{F}\qquad\quad \text{O}\qquad\quad \text{I}\qquad\quad \text{L}$$
$$= 25\quad -\quad 20i\quad +\quad 20i\quad -\quad 16i^2$$
$$= 25 - 16(-1)$$     Replace $i^2$ with $-1$.
$$= 25 + 16 = 41$$ ∎

**You Try 3**

Multiply and simplify.

a)  $-5(4 - 9i)$    b)  $(7 + 2i)(5 - 9i)$    c)  $(4 + 3i)(4 - 3i)$

### 5. Multiply a Complex Number by Its Conjugate

In Chapter 9, we learned about conjugates of radical expressions. For example, the conjugate of $3 + \sqrt{5}$ is $3 - \sqrt{5}$.

The complex numbers in Example 3c), $5 + 4i$ and $5 - 4i$, are **complex conjugates.**

---

**Definition**

The **conjugate** of $a + bi$ is $a - bi$.

---

We found that $(5 + 4i)(5 - 4i) = 41$. The result is a real number. The product of a complex number and its conjugate is *always* a real number, as illustrated next.

$$(a + bi)(a - bi) = (a)(a) + (a)(-bi) + (bi)(a) + (bi)(-bi)$$

$$\begin{array}{cccc} & F & O & I & L \\ = & a^2 & - \; abi & + \; abi & - \; b^2i^2 \end{array}$$

$$= a^2 - b^2(-1) \qquad \text{Replace } i^2 \text{ with } -1.$$
$$= a^2 + b^2$$

We can summarize these facts about complex numbers and their conjugates as follows:

---

**Property    Complex Conjugates**

1)  The conjugate of $a + bi$ is $a - bi$.

2)  The product of $a + bi$ and $a - bi$ is a real number.

3)  We can find the product $(a + bi)(a - bi)$ by using FOIL or by using $(a + bi)(a - bi) = a^2 + b^2$.

---

**Example 4**

Multiply $-7 + 2i$ by its conjugate using the formula $(a + bi)(a - bi) = a^2 + b^2$.

**Solution**

The conjugate of $-7 + 2i$ is $-7 - 2i$.

$$(-7 + 2i)(-7 - 2i) = (-7)^2 + (2)^2 \qquad a = -7, b = 2$$
$$= 49 + 4$$
$$= 53$$

**You Try 4**

Multiply $4 - 10i$ by its conjugate using the formula $(a + bi)(a - bi) = a^2 + b^2$.

### 6. Divide Complex Numbers

To rationalize the denominator of a radical expression like $\dfrac{8}{2 + \sqrt{7}}$, we multiply the numerator and denominator by $2 - \sqrt{7}$, the conjugate of the denominator. We divide complex numbers in the same way.

---

**Procedure**   Dividing Complex Numbers

To divide complex numbers, multiply the numerator and denominator by the *conjugate of the denominator*. Write the result in the form $a + bi$.

---

## Example 5

Divide. Write the result in the form $a + bi$.

a) $\dfrac{9}{3 - 5i}$        b) $\dfrac{6 - 2i}{-7 + i}$

**Solution**

a) $\dfrac{9}{3 - 5i} = \dfrac{9}{(3 - 5i)} \cdot \dfrac{(3 + 5i)}{(3 + 5i)}$   Multiply the numerator and denominator by the conjugate of the denominator.

$= \dfrac{27 + 45i}{3^2 + (5)^2}$   Multiply numerators.
$(a + bi)(a - bi) = a^2 + b^2$

$= \dfrac{27 + 45i}{9 + 25}$

$= \dfrac{27 + 45i}{34}$

$= \dfrac{27}{34} + \dfrac{45}{34}i$   Write the result in the form $a + bi$.

b) $\dfrac{6 - 2i}{-7 + i} = \dfrac{(6 - 2i)}{(-7 + i)} \cdot \dfrac{(-7 - i)}{(-7 - i)}$   Multiply the numerator and denominator by the conjugate of the denominator.

$= \dfrac{-42 - 6i + 14i + 2i^2}{(-7)^2 + (1)^2}$   Multiply using FOIL.
$(a + bi)(a - bi) = a^2 + b^2$

$= \dfrac{-42 + 8i + 2(-1)}{49 + 1}$   Replace $i^2$ with $-1$.

$= \dfrac{-42 + 8i - 2}{50} = \dfrac{-44 + 8i}{50} = -\dfrac{44}{50} + \dfrac{8}{50}i = -\dfrac{22}{25} + \dfrac{4}{25}i$   ∎

## You Try 5

Divide. Write the result in the form $a + bi$.

a) $\dfrac{12}{4 + i}$        b) $\dfrac{2 - 8i}{6 - 2i}$

### 7. Solve Quadratic Equations with Complex Number Solutions

Now that we have defined $i$ as $\sqrt{-1}$, we can find complex number solutions to quadratic equations.

**Example 6**

Solve.

a)   $(k - 6)^2 + 15 = 11$        b)   $6x^2 + 1 = -2x$

**Solution**

a)   $(k - 6)^2 + 15 = 11$

$\quad\quad (k - 6)^2 = -4$        Subtract 15 from each side.

$\quad\quad k - 6 = \pm\sqrt{-4}$        Square root property

$\quad\quad k - 6 = \pm 2i$

$\quad\quad k = 6 \pm 2i$        Add 6 to each side.

The solution set is $\{6 - 2i, 6 + 2i\}$.

b)   Write $6x^2 + 1 = -2x$ in standard form, then use the quadratic formula.

$$6x^2 + 2x + 1 = 0 \quad\quad \text{Add } 2x.$$

Identify $a$, $b$, and $c$, substitute into the quadratic formula, and simplify:

$$a = 6 \quad\quad b = 2 \quad\quad c = 1$$

$$x = \frac{-b \pm \sqrt{b^2 - 4ac}}{2a} \quad\quad \text{Quadratic formula}$$

$$x = \frac{-(2) \pm \sqrt{(2)^2 - 4(6)(1)}}{2(6)} \quad\quad \text{Substitute } a = 6, b = 2, \text{ and } c = 1.$$

$$x = \frac{-2 \pm \sqrt{4 - 24}}{12} \quad\quad \text{Perform the operations.}$$

$$x = \frac{-2 \pm \sqrt{-20}}{12} \quad\quad \text{Simplify the radicand.}$$

$$x = \frac{-2 \pm 2i\sqrt{5}}{12} \quad\quad \text{Simplify the radical.}$$

$$x = \frac{2(-1 \pm i\sqrt{5})}{12} \quad\quad \text{Factor out 2.}$$

$$x = \frac{-1 \pm i\sqrt{5}}{6} \quad\quad \text{Simplify.}$$

$$x = -\frac{1}{6} \pm \frac{\sqrt{5}}{6}i \quad\quad \text{Write in the form } a + bi.$$

The solution set is $\left\{ -\dfrac{1}{6} - \dfrac{\sqrt{5}}{6}i, -\dfrac{1}{6} + \dfrac{\sqrt{5}}{6}i \right\}$.

Did you notice that in each example a complex number *and* its conjugate were the solutions to the equation? This will always be true provided that the variables in the equation have real number coefficients.

**Note**

If $a + bi$ is a solution of a quadratic equation having only real coefficients, then $a - bi$ is also a solution.

**You Try 6**

Solve.

a)   $(t - 4)^2 + 29 = -7$        b)   $5y^2 = 6y - 3$

## Using Technology

We can use a graphing calculator to perform operations on complex numbers or to evaluate square roots of negative numbers.

If the calculator is in the default REAL mode, the result is an error message "ERR: NONREAL ANS," which indicates that $\sqrt{-4}$ is a complex number rather than a real number. Before evaluating $\sqrt{-4}$ on the home screen of your calculator, check the mode by pressing MODE and looking at row 7, change the mode to complex numbers by selecting $a + bi$, as shown at the left below. Now evaluating $\sqrt{-4}$ on the home screen results in the correct answer $2i$, as shown at the right below.

    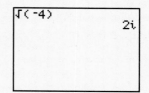

Operations can be performed on complex numbers with the calculator in either REAL or $a + bi$ mode. Simply use the arithmetic operators on the right column on your calculator. To enter the imaginary number $i$, press 2nd . . To add $2 - 5i$ and $4 + 3i$, enter $(2 - 5i) + (4 + 3i)$ on the home screen and press ENTER , as shown at the right.

To subtract $8 + 6i$ from $7 - 2i$, enter $(7 - 2i) - (8 + 6i)$ on the home screen and press ENTER as shown.

```
(2-5i)+(4+3i)
               6-2i
(7-2i)-(8+6i)
              -1-8i
```

To multiply $3 - 5i$ and $7 + 4i$, enter $(3 - 5i) * (7 + 4i)$ on the home screen and press ENTER , as shown at the right.

To divide $2 + 9i$ by $2 - i$, enter $(2 + 9i)/(2 - i)$ on the home screen and press ENTER as shown.

To raise $3 - 4i$ to the fifth power, enter $(3 - 4i)^5$ on the home screen and press ENTER as shown.

```
(3-5i)*(7+4i)
             41-23i
(2+9i)/(2-i)
               -1+4i
(3-4i)^5
         -237+3116i
```

Consider the quotient $(5 + 3i)/(4 - 7i)$. The exact answer is $-\dfrac{1}{65} + \dfrac{47}{65}i$. The calculator automatically displays the decimal result.

Press MATH 1 ENTER to convert the decimal result to the exact fractional result, as shown at the right.

```
(5+3i)/(4-7i)
-.0153846154+.7…
Ans▶Frac
        -1/65+47/65i
```

Perform the indicated operation using a graphing calculator.

1) Simplify $\sqrt{-36}$.

2) $(3 + 7i) + (5 - 8i)$

3) $(10 - 3i) - (4 + 8i)$

4) $(3 + 2i)(6 - 3i)$

5) $(4 + 3i) \div (1 - i)$

6) $(5 - 3i)^3$

---

### Answers to You Try Exercises

1) a) $2i$   b) $i\sqrt{7}$   c) $3i\sqrt{6}$     2) a) $-6 + 14i$   b) $-6 - 3i$

3) a) $-20 + 45i$   b) $53 - 53i$   c) $25$     4) $116$

5) a) $\dfrac{48}{17} - \dfrac{12}{17}i$   b) $\dfrac{7}{10} - \dfrac{11}{10}i$     6) a) $\{4 - 6i, 4 + 6i\}$   b) $\left\{\dfrac{3}{5} - \dfrac{\sqrt{6}}{5}i, \dfrac{3}{5} + \dfrac{\sqrt{6}}{5}i\right\}$

---

### Answers to Technology Exercises

1) $6i$     2) $8 - i$     3) $6 - 11i$     4) $24 + 3i$     5) $\dfrac{1}{2} + \dfrac{7}{2}i$     6) $-10 - 198i$

## 10.4 Exercises

**Mixed Exercises: Objectives 1 and 2**

Determine whether each statement is true or false.

1) Every complex number is a real number.

2) Every real number is a complex number.

3) Since $i = \sqrt{-1}$, it follows that $i^2 = -1$.

4) In the complex number $-9 + 4i$, $-9$ is the real part and $4i$ is the imaginary part.

Simplify.

5) $\sqrt{-36}$

6) $\sqrt{-49}$

7) $\sqrt{-16}$

8) $\sqrt{-100}$

9) $\sqrt{-5}$

10) $\sqrt{-17}$

VIDEO 11) $\sqrt{-27}$

12) $\sqrt{-28}$

13) $\sqrt{-45}$

14) $\sqrt{-75}$

15) $\sqrt{-32}$

16) $\sqrt{-500}$

**Objective 3: Add and Subtract Complex Numbers**

17) Explain how to add complex numbers.

18) Can the sum or difference of two complex numbers be a real number? Give an example.

Perform the indicated operations.

19) $(5 + 2i) + (-8 + 10i)$

20) $(6 + i) + (2 - 7i)$

VIDEO 21) $(13 - 8i) - (9 + i)$

22) $(-15 + 7i) - (-11 - 3i)$

23) $\left(-\dfrac{5}{8} - \dfrac{1}{3}i\right) - \left(-\dfrac{1}{2} + \dfrac{3}{4}i\right)$

24) $\left(\dfrac{3}{4} + \dfrac{1}{2}i\right) - \left(\dfrac{5}{6} + \dfrac{1}{3}i\right)$

25) $10i - (5 + 9i) + (5 + i)$

26) $(-8 + 7i) + (-1 - i) - (-9 + 6i)$

**Objective 4: Multiply Complex Numbers**

27) How is multiplying $(1 + 8i)(3 - 5i)$ similar to multiplying $(x + 8)(3x - 5)$?

28) When $i^2$ appears in an expression, it should be replaced with what?

Multiply and simplify.

29) $2(7 - 6i)$

30) $-7(4 + 8i)$

31) $-\dfrac{5}{2}(8 + 3i)$

32) $\dfrac{2}{3}(-5 + i)$

33) $3i(-2 + 7i)$

34) $8i(5 + i)$

35) $5i(4 - 3i)$

36) $-4i(-7 - 6i)$

37) $(3 + i)(2 + 8i)$

38) $(5 + 8i)(4 + 2i)$

VIDEO 39) $(-1 + 3i)(4 - 6i)$

40) $(-7 - i)(10 - 9i)$

41) $(6 - 2i)(8 - 7i)$

42) $(4 - 5i)(5 + 3i)$

43) $\left(\dfrac{2}{3} + \dfrac{2}{3}i\right)\left(\dfrac{1}{5} + \dfrac{2}{5}i\right)$

44) $\left(\dfrac{5}{2} - \dfrac{3}{2}i\right)\left(\dfrac{3}{4} + \dfrac{2}{3}i\right)$

**Objective 5: Multiply a Complex Number by Its Conjugate**

Identify the conjugate of each complex number, then multiply the number and its conjugate.

45) $3 + 5i$

46) $-6 - i$

47) $-10 - 2i$

48) $4 + 8i$

49) $-7 + 9i$

50) $7 - 3i$

**Objective 6: Divide Complex Numbers**

51) Explain how to divide complex numbers.

52) On Amy's test, she had to divide $\dfrac{12 + 30i}{4 + 5i}$. Her answer was $3 + 6i$. What did she do wrong? What is the correct way to divide these complex numbers?

Divide. Write the result in the form $a + bi$.

53) $\dfrac{2}{1 - 6i}$

54) $\dfrac{-3}{7 - 2i}$

55) $\dfrac{8}{-3 + 5i}$

56) $\dfrac{15}{8 + i}$

57) $\dfrac{-10i}{-6 + 3i}$

58) $\dfrac{i}{4 + 3i}$

VIDEO 59) $\dfrac{3 - 8i}{-6 + 7i}$

60) $\dfrac{-2 + 9i}{6 - i}$

61) $\dfrac{1 + 8i}{2 - 5i}$

62) $\dfrac{4 + 10i}{4 + 2i}$

63) $\dfrac{7}{i}$

64) $\dfrac{13 + 2i}{-i}$

**Objective 7: Solve Quadratic Equations with Complex Number Solutions**

Solve.

65) $t^2 = -9$

66) $k^2 = -144$

67) $n^2 + 11 = 0$

68) $y^2 + 23 = 8$

69) $z^2 - 5 = -33$

70) $a^2 - 12 = -30$

71) $-23 = 3p^2 + 52$

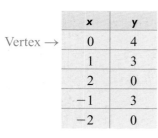 73) $(c + 3)^2 - 4 = -29$

75) $3 = 23 + (d - 8)^2$

77) $5 + (4b + 1)^2 = 2$

79) $3x^2 + x + 2 = 0$

81) $h^2 + 4 = -2h$

83) $4k = k^2 + 29$

85) $2(n^2 + 3n) = 7$

72) $11 = 83 + 2m^2$

74) $(h - 10)^2 + 17 = 13$

76) $-16 = (r + 1)^2 - 14$

78) $(3w - 4)^2 + 48 = -24$

80) $5v^2 + 3v + 1 = 0$

82) $t^2 + 6t = -10$

84) $4w - 5 = 2w^2$

86) $z^2 - 8 = 3z$

87) $4q^2 = 2q - 3$

89) $3y(y + 2) = -5$

91) $5g^2 + 9 = 12g$

88) $4m(2m + 1) = -1$

90) $5p^2 + 10p + 6 = 0$

92) $4a^2 + 28a = -97$

Solve by completing the square.

93) $n^2 - 4n + 5 = 0$

95) $z^2 + 8z + 29 = 0$

97) $2p^2 - 12p + 54 = 0$

99) $3k^2 + 9k + 12 = 0$

94) $d^2 + 6d + 13 = 0$

96) $x^2 - 2x + 11 = 0$

98) $3t^2 + 12t + 72 = 0$

100) $4y^2 - 4y + 16 = 0$

# Section 10.5  Graphs of Quadratic Equations

## Objectives

1. **Graph a Quadratic Equation of the Form** $y = ax^2 + bx + c$
2. **Use x-intercepts to Find the Solutions to Quadratic Equations**
3. **Find the Domain and Range of Quadratic Functions**

In Section 6.2, we learned that the graph of a quadratic equation, $y = ax^2 + bx + c$ where $a \neq 0$, is called a **parabola**. We graphed some basic parabolas by plotting points. For example, if we plot some points for $y = x^2$ we get its graph, shown in Figure 1.

$$y = x^2$$

| x | y |
|---|---|
| Vertex → 0 | 0 |
| 1 | 1 |
| 2 | 4 |
| −1 | 1 |
| −2 | 4 |

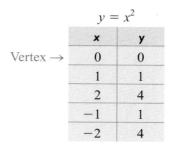

**Figure 1**

The graphs of quadratic equations where $y$ is defined in terms of $x$ open either upward or downward. The lowest point on a parabola that opens upward or the highest point on a parabola that opens downward is called the **vertex.** The vertex of the graph of $y = x^2$ is (0, 0).

The graphs of $y = -x^2 + 4$ and $y = x^2 - 1$ are shown in Figures 2 and 3.

| x | y |
|---|---|
| Vertex → 0 | 4 |
| 1 | 3 |
| 2 | 0 |
| −1 | 3 |
| −2 | 0 |

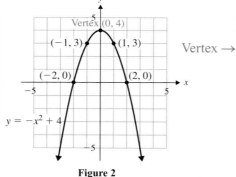

**Figure 2**

| x | y |
|---|---|
| Vertex → 0 | −1 |
| 1 | 0 |
| 2 | 3 |
| −1 | 0 |
| −2 | 3 |

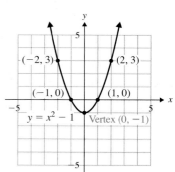

**Figure 3**

These are basic forms of quadratic equations, and it is easy to find the vertex of each parabola. But how do we find the vertex of the graph of a quadratic equation like $y = -x^2 + 4x - 3$? We need another approach because to accurately graph a quadratic equation, we must be able to locate the vertex.

Notice in Figures 2 and 3 that the $x$-coordinate of the vertex is halfway between the $x$-coordinates of the $x$-intercepts of the parabola. This will always be true and will help us derive a formula to find the vertex of a parabola with an equation like $y = -x^2 + 4x - 3$.

## 1. Graph a Quadratic Equation of the Form $y = ax^2 + bx + c$

Recall from previous chapters that to find the $x$-intercepts of the graph of an equation, we let $y = 0$ and solve for $x$. Therefore, to find the $x$-intercepts of the graph of $y = ax^2 + bx + c$, let $y = 0$ and solve for $x$.

$$0 = ax^2 + bx + c$$

We can use the quadratic formula to solve for $x$, giving us the following $x$-coordinates of the $x$-intercepts:

$$x = \frac{-b - \sqrt{b^2 - 4ac}}{2a} \quad \text{and} \quad x = \frac{-b + \sqrt{b^2 - 4ac}}{2a}$$

Since the $x$-coordinate of the vertex of a parabola is halfway between the $x$-intercepts, we find the *average* of the $x$-coordinates of the $x$-intercepts to find the $x$-coordinate of the vertex.

$$
\begin{aligned}
x\text{-coordinate of the vertex} &= \frac{1}{2}\left( \frac{-b - \sqrt{b^2 - 4ac}}{2a} + \frac{-b + \sqrt{b^2 - 4ac}}{2a} \right) \\
&= \frac{1}{2}\left( \frac{-b - \sqrt{b^2 - 4ac} - b + \sqrt{b^2 - 4ac}}{2a} \right) \\
&= \frac{1}{2}\left( \frac{-2b}{2a} \right) \\
&= -\frac{b}{2a}
\end{aligned}
$$

---

**Procedure**    Finding the Vertex of a Parabola of the Form $y = ax^2 + bx + c \ (a \neq 0)$

The **$x$-coordinate of the vertex of a parabola** written in the form $y = ax^2 + bx + c \ (a \neq 0)$ is $x = -\dfrac{b}{2a}$. To find the $y$-coordinate of the vertex, substitute the $x$-value into the equation and solve for $y$.

---

When graphing a parabola, the first point we should locate is the vertex. After finding the vertex, we will modify the procedure we learned in Chapter 6 to find other points on the parabola. We summarize the steps here.

---

**Procedure**    Graphing a Parabola of the Form $y = ax^2 + bx + c$

**Step 1:** **Find the vertex.** The $x$-coordinate of the vertex is $-\dfrac{b}{2a}$. Find the $y$-coordinate of the vertex by substituting the $x$-value into the equation.

**Step 2:** **Find the $y$-intercept** by substituting 0 for $x$ and solving for $y$.

**Step 3:** **Find the $x$-intercepts,** if they exist, by substituting 0 for $y$ and solving for $x$.

**Step 4:** **Find additional points** on the parabola using a table of values.

**Step 5:** **Plot the points and sketch the graph.**

## Example 1

Graph $y = -x^2 + 4x - 3$.

**Solution**

**Step 1:   Find the vertex.** First, find the $x$-coordinate of the vertex:

$$x = -\frac{b}{2a} = -\frac{4}{2(-1)} = \frac{4}{2} = 2 \qquad a = -1, \quad b = 4$$

Next, substitute $x = 2$ into the equation to find the $y$-coordinate of the vertex.

$$y = -x^2 + 4x - 3$$
$$y = -(2)^2 + 4(2) - 3 \qquad \text{Substitute 2 for } x.$$
$$y = -4 + 8 - 3 = 1$$

The vertex of the parabola is (2, 1).

**Step 2:   Find the $y$-intercept** by substituting 0 for $x$ and solving for $y$:

$$y = -x^2 + 4x - 3$$
$$y = -(0)^2 + 4(0) - 3 = -3$$

The $y$-intercept is $(0, -3)$.

**Step 3:   Find the $x$-intercepts** by substituting 0 for $y$ and solving for $x$:

$$0 = -x^2 + 4x - 3 \qquad \text{Substitute 0 for } y.$$
$$0 = x^2 - 4x + 3 \qquad \text{Divide both sides by } -1.$$
$$0 = (x - 1)(x - 3) \qquad \text{Factor.}$$

$$x = 1 \qquad \text{or} \qquad x = 3 \qquad \text{Solve.}$$

The $x$-intercepts are (1, 0) and (3, 0).

**Steps 4 and 5:   Find additional points, plot the points, and sketch the graph.**

|  | x | y |
|---|---|---|
| Vertex → | 2 | 1 |
| y-int. → | 0 | -3 |
| x-int. → | 1 | 0 |
| x-int. → | 3 | 0 |
|  | 4 | -3 |

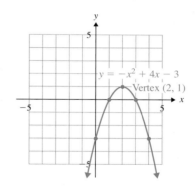

## You Try 1

Graph $y = -x^2 - 6x - 8$.

Sometimes we need to use the quadratic formula to find the $x$-intercepts of the graph.

## Example 2

Graph $y = x^2 + 2x - 2$.

**Solution**

**Step 1:** **Find the vertex.** $x = -\dfrac{b}{2a} = -\dfrac{2}{2(1)} = -\dfrac{2}{2} = -1$     $a = 1, \quad b = 2$

Substitute $x = -1$ into the equation to find the $y$-coordinate of the vertex.

$$y = (-1)^2 + 2(-1) - 2 = 1 - 2 - 2 = -3$$

The vertex of the parabola is $(-1, -3)$.

**Step 2:** **Find the $y$-intercept** by substituting 0 for $x$ and solving for $y$:

$$y = (0)^2 + 2(0) - 2 = -2$$

The $y$-intercept is $(0, -2)$.

**Step 3:** **Find the $x$-intercepts** by substituting 0 for $y$ and solving for $x$:

$$0 = x^2 + 2x - 2$$

We will solve this equation using the quadratic formula.

$$x = \frac{-(2) \pm \sqrt{(2)^2 - 4(1)(-2)}}{2(1)}$$     Let $a = 1, b = 2, c = -2$.

$$x = \frac{-2 \pm \sqrt{12}}{2} = \frac{-2 \pm 2\sqrt{3}}{2} = \frac{2(-1 \pm \sqrt{3})}{2} = -1 \pm \sqrt{3}$$

The $x$-intercepts are $(-1 - \sqrt{3}, 0)$ and $(-1 + \sqrt{3}, 0)$.

**Steps 4 and 5:** **Find additional points, plot the points, and sketch the graph.**

To plot the $x$-intercepts, we can either approximate them in our heads or use a calculator.

|  | $x$ | $y$ |
|---|---|---|
| Vertex → | $-1$ | $-3$ |
| $y$-int. → | $0$ | $-2$ |
| $x$-int. → | $-1 - \sqrt{3} \approx -2.7$ | $0$ |
| $x$-int. → | $-1 + \sqrt{3} \approx 0.7$ | $0$ |
|  | $-2$ | $-2$ |

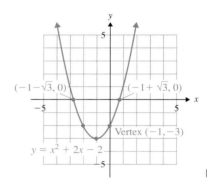

**You Try 2**

Graph $y = x^2 - 8x + 14$.

Notice in Example 1, $y = -x^2 + 4x - 3$, the value of $a$ is negative ($a = -1$) and the parabola opens downward. In Example 2, the graph of $y = x^2 + 2x - 2$ opens upward and $a$ is positive ($a = 1$). This is another characteristic of the graph of $y = ax^2 + bx + c$. If $a$ is positive, the graph opens upward. If $a$ is negative, the graph opens downward.

## 2. Use x-intercepts to Find the Solutions to Quadratic Equations

We have seen that to find the $x$-intercepts of $y = ax^2 + bx + c$, we let $y = 0$ and solve for $x$. So solving the equation $x^2 - 6x + 5 = 0$ is equivalent to finding the $x$-intercepts of the graph of $y = x^2 - 6x + 5$.

**Example 3**

Use the graph to determine the solutions to each quadratic equation.

a) $x^2 - 6x + 5 = 0$      b) $x^2 + 4x + 4 = 0$      c) $x^2 - 4x + 7 = 0$

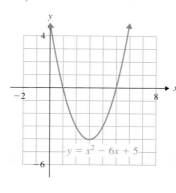 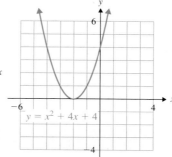 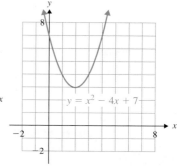

**Solution**

a) The graph of $y = x^2 - 6x + 5$ has two $x$-intercepts: $(1, 0)$ and $(5, 0)$. Therefore, the solution set of $x^2 - 6x + 5 = 0$ is $\{1, 5\}$. You can verify this result by solving $x^2 - 6x + 5 = 0$ by factoring.

b) The graph of $y = x^2 + 4x + 4$ has one $x$-intercept: $(-2, 0)$. So, the solution set of $x^2 + 4x + 4 = 0$ is $\{-2\}$.

c) The graph of $y = x^2 - 4x + 7$ has no $x$-intercepts; therefore, the equation $x^2 - 4x + 7 = 0$ has no real number solutions. (The equation has two complex solutions, $2 - i\sqrt{3}$ and $2 + i\sqrt{3}$.)

---

**Note**

Recall that in Section 7.5 we used a graphing calculator to solve equations like $x^2 - 6x + 5 = 0$ by graphing $y = x^2 - 6x + 5$ and finding the $x$-intercepts. (See p. 424). This may be particularly helpful when the solutions are irrational numbers because the calculator can give us a good approximation to the solutions.

## 3. Find the Domain and Range of Quadratic Functions

Recall from Chapter 4 that the vertical line test tells us that if every vertical line that can be drawn through a graph touches the graph only once, then the graph represents a function. Since the graph of $y = ax^2 + bx + c$ is a parabola that opens either upward or downward, it passes the vertical line test and is, therefore, a function.

**Definition**

A **quadratic function** is a function that can be written in the form $f(x) = ax^2 + bx + c$, where $a, b,$ and $c$ are real numbers and $a \neq 0$.

Since any real number may be substituted for $x$, the *domain* of a quadratic function is $(-\infty, \infty)$. The *range* of the function is the set of $y$-values that result from the corresponding $x$-values. To determine the range of a quadratic function, we look at its graph.

## Example 4

Determine the range of $f(x) = -x^2 + 3$.

### Solution

Graph the function. The set of $y$-values on the graph includes all real numbers from $-\infty$ to 3. The range is $(-\infty, 3]$.

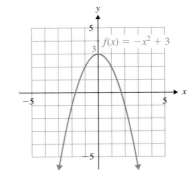

## You Try 3

Determine the range of $g(x) = x^2 - 5$.

## Using Technology

In Section 7.5, we said that the solutions of the equation $x^2 - x - 6 = 0$ are the $x$-intercepts of the graph of $y = x^2 - x - 6$. The $x$-intercepts are also called the zeros of the equation since they are the values of $x$ that make $y = 0$. We can find the $x$-intercepts shown on the graphs by pressing 2nd TRACE and then selecting 2: zero. Move the cursor to the left of an $x$-intercept using the right arrow key and press ENTER. Move the cursor to the right of the $x$-intercept using the right arrow key and press ENTER. Move the cursor close to the $x$-intercept using the left arrow key and press ENTER. The result finds the $x$-intercepts $(-2, 0)$ and $(3, 0)$ as shown on the graphs below.

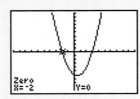

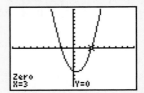

The $y$-intercept is found by graphing the function and pressing 2nd TRACE 0. As shown on the graph at the left below, the $y$-intercept for $y = x^2 - x - 6$ is $(0, -6)$.

The $x$-value of the vertex can be found using the vertex formula. In this case, $a = 1$ and $b = -1$, so $-\dfrac{b}{2a} = \dfrac{1}{2}$. To find the vertex on the graph, press TRACE, type 1/2, and press ENTER. The vertex is shown as $(0.5, -6.25)$ on the graph in the center below.

To convert the coordinates of the vertex to fractions, go to the home screen and select Frac from the Math menu. The vertex is then $\left(\dfrac{1}{2}, -\dfrac{25}{4}\right)$ as shown on the screen at the right below.

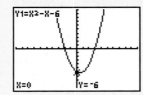

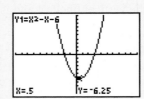

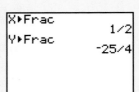

Find the $x$-intercepts, $y$-intercept, and vertex using a graphing calculator.

1)  $y = x^2 + 2x - 3$

2)  $y = x^2 - 4x + 5$

3)  $y = (x - 1)^2 - 9$

4)  $y = x^2 - 9$

5)  $y = x^2 - 4x + 4$

6)  $y = -x^2 + 6x - 10$

**Answers to You Try Exercises**

1)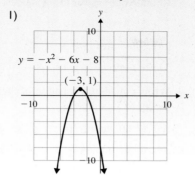

$y = -x^2 - 6x - 8$

$(-3, 1)$

2)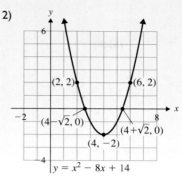

$(2, 2)$   $(6, 2)$

$(4-\sqrt{2}, 0)$   $(4+\sqrt{2}, 0)$

$(4, -2)$

$y = x^2 - 8x + 14$

3) $[-5, \infty)$

**Answers to Technology Exercises**

1) $x$-intercepts $(1, 0)$, $(-3, 0)$; $y$-intercept $(0, -3)$; vertex $(-1, -4)$
2) no $x$-intercepts; $y$-intercept $(0, 5)$; vertex $(2, 1)$
3) $x$-intercepts $(-2, 0)$, $(4, 0)$; $y$-intercept $(0, -8)$; vertex $(1, -9)$
4) $x$-intercepts $(-3, 0)$, $(3, 0)$; $y$-intercept $(0, -9)$; vertex $(0, -9)$
5) $x$-intercept $(2, 0)$; $y$-intercept $(0, 4)$; vertex $(2, 0)$
6) no $x$-intercepts; $y$-intercept $(0, -10)$; vertex $(3, -1)$

## 10.5 Exercises

**Objective 1: Graph a Quadratic Equation of the Form $y = ax^2 + bx + c$**

1) Explain how to find the vertex of a parabola with equation $y = ax^2 + bx + c$.

2) How do you know whether the graph of $y = ax^2 + bx + c$ opens upward or downward?

Sketch the graph of each equation. Identify the vertex.

3) $y = x^2 - 4$   4) $y = -x^2 + 1$

5) $y = -x^2 - 1$   6) $y = x^2 - 5$

7) $y = (x - 3)^2$   8) $y = (x + 2)^2$

9) $y = x^2 - 4x + 3$   10) $y = -x^2 + 6x - 5$

11) $y = -x^2 + 2x + 3$   12) $y = x^2 + 2x - 8$

13) $y = -x^2 - 2x + 4$   14) $y = x^2 + 4x + 1$

15) $y = x^2 + 6x + 11$   16) $y = -x^2 + 8x - 17$

17) $y = -x^2 + 2x - 1$   18) $y = x^2 + 6x + 9$

19) $y = 2x^2 - 4x$   20) $y = -2x^2 - 8x - 6$

**Objective 2: Use $x$-Intercepts to Find the Solutions to Quadratic Equations**

Use the graph to determine the real number solutions to each quadratic equation. Verify the result by solving the equation algebraically.

21) $x^2 - 2x - 3 = 0$

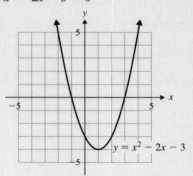

$y = x^2 - 2x - 3$

22) $x^2 + 6x + 8 = 0$

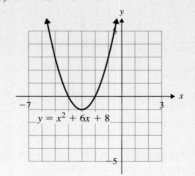

$y = x^2 + 6x + 8$

23) $x^2 + 2x + 1 = 0$

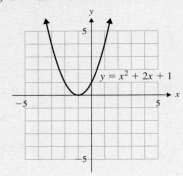

24) $-x^2 + 2x - 4 = 0$

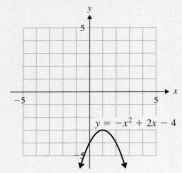

25) $x^2 - 4x + 5 = 0$

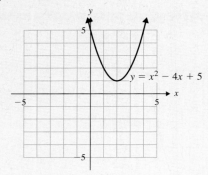

26) $x^2 - 6x + 9 = 0$

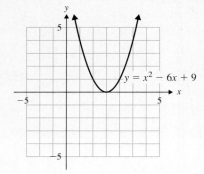

We have seen that the solutions of $ax^2 + bx + c = 0$ are the x-intercepts of the corresponding quadratic equation $y = ax^2 + bx + c$. Use a graphing calculator to solve each equation in Exercises 27–32 by finding the x-intercepts of the graph of the corresponding equation, $y = ax^2 + bx + c$. Round your answer to the nearest hundredth where appropriate.

27) $5x^2 - 17x + 6 = 0$     28) $4x^2 - 5x - 6 = 0$

29) $8x^2 + 38x + 9 = 0$     30) $20x^2 - 64x + 35 = 0$

31) $x^2 + 2x - 6 = 0$     32) $x^2 + 10x + 22 = 0$

Write a quadratic equation of the form $y = ax^2 + bx + c$ with the given x-intercepts.

33) $(2, 0)$ and $(4, 0)$     34) $(-5, 0)$ and $(-3, 0)$

35) $(-3, 0)$ and $(1, 0)$     36) $(-1, 0)$ and $(5, 0)$

## Objective 3: Find the Domain and Range of Quadratic Functions

37) What is the domain of a quadratic function?

38) How can we find the range of a quadratic function?

Determine the range of each function.

39) $f(x) = x^2 - 2$     40) $g(x) = -x^2 + 6$

41) $h(x) = -(x + 1)^2$     42) $f(x) = (x - 2)^2$

43) $k(x) = -x^2 + 6x - 5$     44) $h(x) = x^2 - 4x + 3$

45) $g(x) = x^2 + 6x + 8$     46) $k(x) = -x^2 - 8x - 7$

Recall that the graph of $f(x) = ax^2 + bx + c$ opens downward if $a < 0$. Therefore, the vertex is the highest point on the parabola so that the **maximum value of the function** is the y-coordinate of vertex. The x-coordinate of the vertex is the value at which the maximum occurs. Use this information to answer the questions in Exercises 47–50.

47) Let $f(x) = -2x^2 + 16x - 25$.

   a) Find the value of x at which the maximum value of the function occurs.

   b) What is the maximum value of the function?

48) Let $g(x) = -3x^2 + 12x - 2$.

   a) Find the value of x at which the maximum value of the function occurs.

   b) What is the maximum value of the function?

49) An object is fired upward from the ground so that its height $h$ (in feet) $t$ seconds after being fired is given by $h(t) = -16t^2 + 288t$.

   a) How long does it take the object to reach its maximum height?

   b) What is the maximum height attained by the object?

50) An object is thrown upward from a height of 12 feet so that its height $h$ (in feet) $t$ seconds after being thrown is given by $h(t) = -16t^2 + 64t + 12$.

   a) How long does it take the object to reach its maximum height?

   b) What is the maximum height attained by the object?

| Definition/Procedure | Example |
|---|---|

## 10.1 Solving Quadratic Equations Using the Square Root Property

**The Square Root Property**

Let $k$ be a constant. If $x^2 = k$, then $x = \sqrt{k}$ or $x = -\sqrt{k}$. **(p. 588)**

Solve $(t - 8)^2 = 5$.

$$t - 8 = \pm\sqrt{5} \qquad \text{Square root property}$$
$$t = 8 \pm \sqrt{5} \qquad \text{Add 8 to both sides.}$$

The solution set is $\{8 - \sqrt{5}, 8 + \sqrt{5}\}$.

## 10.2 Solving Quadratic Equations by Completing the Square

A **perfect square trinomial** is a trinomial whose factored form is the square of a binomial. **(p. 593)**

| Perfect Square Trinomial | Factored Form |
|---|---|
| $g^2 + 6g + 9$ | $(g + 3)^2$ |
| $4t^2 - 20t + 25$ | $(2t - 5)^2$ |

To find the constant needed to complete the square for $x^2 + bx$:

**Step 1:** Find half of the coefficient of $x$: $\dfrac{1}{2}b$

**Step 2:** Square the result: $\left(\dfrac{1}{2}b\right)^2$

**Step 3:** Add it to $x^2 + bx$: $x^2 + bx + \left(\dfrac{1}{2}b\right)^2$ **(p. 594)**

Complete the square for $x^2 + 10x$ to obtain a perfect square trinomial. Then, factor.

**Step 1:** Find half of the coefficient of $x$: $\dfrac{1}{2}(10) = 5$

**Step 2:** Square the result: $5^2 = 25$

**Step 3:** Add 25 to $x^2 + 10x$: $x^2 + 10x + 25$

The perfect square trinomial is $x^2 + 10x + 25$.
The factored form is $(x + 5)^2$.

**Solving a Quadratic Equation ($ax^2 + bx + c = 0$) by Completing the Square**

**Step 1:** **The coefficient of the squared term must be 1.** If it is not 1, divide both sides of the equation by $a$ to obtain a leading coefficient of 1.

**Step 2:** **Get the variables on one side of the equal sign and the constant on the other side.**

**Step 3:** **Complete the square.** Find half of the linear coefficient, then square the result. Add that quantity to *both* sides of the equation.

**Step 4:** **Factor.**

**Step 5:** **Solve using the square root property. (p. 596)**

Solve $x^2 + 8x + 9 = 0$ by completing the square.

$$x^2 + 8x + 9 = 0 \qquad \text{The coefficient of } x^2 \text{ is 1.}$$
$$x^2 + 8x = -9 \qquad \text{Get the constant on the other side of the equal sign.}$$

Complete the square.

$$\frac{1}{2}(8) = 4$$
$$(4)^2 = 16$$

Add 16 to both sides of the equation.

$$x^2 + 8x + 16 = -9 + 16$$
$$(x + 4)^2 = 7 \qquad \text{Factor.}$$
$$x + 4 = \pm\sqrt{7} \qquad \text{Square root property}$$
$$x = -4 \pm \sqrt{7}$$

The solution set is $\{-4 - \sqrt{7}, -4 + \sqrt{7}\}$.

| Definition/Procedure | Example |
|---|---|

## 10.3 Solving Equations Using the Quadratic Formula

**The Quadratic Formula**

The solutions of any quadratic equation of the form
$ax^2 + bx + c = 0$ ($a \neq 0$) are

$$x = \frac{-b \pm \sqrt{b^2 - 4ac}}{2a} \quad \text{(p. 602)}$$

Solve $2x^2 - 9x + 8 = 0$ using the quadratic formula.

$$a = 2 \qquad b = -9 \qquad c = 8$$

Substitute the values into the quadratic formula.

$$x = \frac{-(-9) \pm \sqrt{(-9)^2 - 4(2)(8)}}{2(2)}$$

$$x = \frac{9 \pm \sqrt{81 - 64}}{4} = \frac{9 \pm \sqrt{17}}{4}$$

The solution set is $\left\{ \dfrac{9 - \sqrt{17}}{4}, \dfrac{9 + \sqrt{17}}{4} \right\}$.

## 10.4 Complex Numbers

**Definition of $i$:**

$$i = \sqrt{-1}$$

Therefore,

$$i^2 = -1$$

A **complex number** is a number of the form $a + bi$, where $a$ and $b$ are real numbers. $a$ is called the **real part** and $b$ is called the **imaginary part.** The set of real numbers is a subset of the set of complex numbers. **(pp. 609–610)**

*Examples of complex numbers:*

$$-3 + 11i$$
$$4 \quad \text{(since it can be written } 4 + 0i)$$
$$7i \quad \text{(since it can be written } 0 + 7i)$$

---

**Simplifying Complex Numbers (p. 610)**

Simplify $\sqrt{-81}$. $\quad \sqrt{-81} = \sqrt{-1} \cdot \sqrt{81}$
$$= i \cdot 9$$
$$= 9i$$

---

**Adding and Subtracting Complex Numbers**

To add and subtract complex numbers, combine the real parts and combine the imaginary parts. **(p. 611)**

Subtract $(12 + 8i) - (-2 + 7i)$.

$$(12 + 8i) - (-2 + 7i) = 12 + 8i + 2 - 7i$$
$$= 14 + i$$

---

Multiply complex numbers like we multiply polynomials. Remember to replace $i^2$ with $-1$. **(p. 611)**

Multiply and simplify.

a) $3(7 + 8i) = 21 + 24i$

b)
$$\overset{\text{F} \quad \text{O} \quad \text{I} \quad \text{L}}{(-4 + 2i)(6 - 7i)} = -24 + 28i + 12i - 14i^2$$
$$= -24 + 40i - 14(-1)$$
$$= -24 + 40i + 14$$
$$= -10 + 40i$$

---

**Complex Conjugates**

1) The **conjugate** of $a + bi$ is $a - bi$.
2) The product of $a + bi$ and $a - bi$ is a real number.
3) Find the product $(a + bi)(a - bi)$ using FOIL or recall that $(a + bi)(a - bi) = a^2 + b^2$. **(p. 612)**

Multiply $-3 - 4i$ by its conjugate.
The conjugate of $-3 - 4i$ is $-3 + 4i$.

i) Multiply using FOIL:

$$\overset{\text{F} \quad \text{O} \quad \text{I} \quad \text{L}}{(-3 - 4i)(-3 + 4i)} = 9 - 12i + 12i - 16i^2$$
$$= 9 - 16(-1)$$
$$= 9 + 16$$
$$= 25$$

ii) Use $(a + bi)(a - bi) = a^2 + b^2$.

$$(-3 - 4i)(-3 + 4i) = (-3)^2 + (4)^2$$
$$= 9 + 16$$
$$= 25$$

| Definition/Procedure | Example |
|---|---|

**Dividing Complex Numbers**

To **divide complex numbers,** multiply the numerator and denominator by the *conjugate of the denominator.* Write the result in the form $a + bi$. **(p. 613)**

Divide $\dfrac{5i}{3 - 2i}$. Write the result in the form $a + bi$.

$$\frac{5i}{3 - 2i} = \frac{5i}{(3 - 2i)} \cdot \frac{(3 + 2i)}{(3 + 2i)} = \frac{15i + 10i^2}{3^2 + 2^2}$$

$$= \frac{15i + 10(-1)}{13} = \frac{15i - 10}{13} = -\frac{10}{13} + \frac{15}{13}i$$

---

We can find complex solutions to quadratic equations.

If $a + bi$ is a solution of a quadratic equation having only real coefficients, then $a - bi$ is also a solution. **(p. 614)**

Solve $(q - 3)^2 + 10 = 1$.

$$\begin{array}{ll} (q - 3)^2 = -9 & \text{Subtract 10 from each side.} \\ q - 3 = \pm\sqrt{-9} & \text{Square root property} \\ q - 3 = \pm 3i & \sqrt{-9} = 3i \\ q = 3 \pm 3i & \text{Add 3 to each side.} \end{array}$$

The solution set is $\{3 - 3i, 3 + 3i\}$.

## 10.5 Graphs of Quadratic Equations

**Graphing a Parabola of the Form $y = ax^2 + bx + c\,(a \neq 0)$**

**Step 1:** **Find the vertex.** The $x$-coordinate of the vertex is $x = -\dfrac{b}{2a}$. Find the $y$-coordinate of the vertex by substituting the $x$-value in the equation.

**Step 2:** **Find the $y$-intercept** by substituting 0 for $x$ and solving for $y$.

**Step 3:** **Find the $x$-intercepts,** if they exist, by substituting 0 for $y$ and solving for $x$.

**Step 4:** **Find additional points** on the parabola using a table of values.

**Step 5:** **Plot the points and sketch the graph. (p. 618)**

Graph $y = x^2 + 4x + 3$.

**Step 1:** **Find the vertex.**

$$x = -\frac{b}{2a} = -\frac{4}{2(1)} = -2$$
$$y = (-2)^2 + 4(-2) + 3 = 4 - 8 + 3 = -1$$

The vertex of the parabola is $(-2, -1)$.

**Step 2:** **Find the $y$-intercept:** $y = (0)^2 + 4(0) + 3 = 3$
The $y$-intercept is $(0, 3)$.

**Step 3:** **Find the $x$-intercepts:** $0 = x^2 + 4x + 3$
$$0 = (x + 3)(x + 1)$$
$$x = -3 \text{ or } x = -1$$
The $x$-intercepts are $(-3, 0)$ and $(-1, 0)$.

**Step 4:** **Find an additional point:** Let $x = -4$.
$$y = (-4)^2 + 4(-4) + 3 = 16 - 16 + 3 = 3$$

Another point on the graph is $(-4, 3)$.

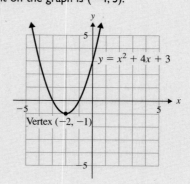

| Definition/Procedure | Example |
|---|---|
| **Use *x*-intercepts to Find the Solutions to Quadratic Equations**<br>The *x*-intercepts of $y = ax^2 + bx + c$ are the solutions of $ax^2 + bx + c = 0$. **(p. 621)** | Here is the graph of $y = x^2 - 6x + 5$.<br>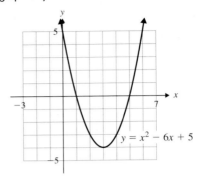<br>The *x*-intercepts are $(1, 0)$ and $(5, 0)$.<br>The solution set of $x^2 - 6x + 5 = 0$ is $\{1, 5\}$. |
| A **quadratic function** is a function that can be written in the form $f(x) = ax^2 + bx + c$, where $a, b$, and $c$ are real numbers and $a \neq 0$.<br>The **domain** of a quadratic function is $(-\infty, \infty)$.<br>The **range** of a quadratic function is the set of *y*-values that result from the corresponding *x*-values. This can be determined from its graph. **(p. 621)** | Determine the range of $f(x) = -x^2 + 2$.<br>Graph the function:<br>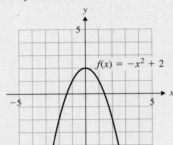<br>The set of *y*-values on the graph includes all real numbers from $-\infty$ to 2.<br><br>The range is $(-\infty, 2]$. |

**(10.1)**

**Solve using the square root property. Find only the real number solutions.**

1) $w^2 = 81$

2) $p^2 - 50 = 0$

3) $3a^2 + 7 = 40$

4) $(z + 6)^2 = 20$

5) $5 = (k - 4)^2 + 14$

6) $35 = (2y - 1)^2 + 7$

7) $\left(\frac{3}{4}c + 5\right)^2 - 2 = 14$

8) The power, $P$ (in watts), in an electrical system is given by $P = \dfrac{V^2}{R}$, where $V$ is the voltage, in volts, and $R$ is the resistance, in ohms. Find the voltage if the power is 45 watts and the resistance is 5 ohms.

**(10.2)**

**Complete the square for each expression to obtain a perfect square trinomial. Then, factor.**

9) $x^2 + 18x$

10) $n^2 - 20n$

11) $z^2 - 3z$

12) $y^2 + y$

**Solve by completing the square. Find only the real number solutions.**

13) $d^2 + 10d + 9 = 0$

14) $v^2 - 8v + 11 = 0$

15) $r^2 + 15 = 4r$

16) $x(x + 14) = -24$

17) $a^2 + 5a + 2 = 0$

18) $t^2 - 7 = 3t$

19) $4n^2 - 8n = 21$

20) $2h - 7 = -2h^2$

**(10.3)**

21) We can use the quadratic formula to solve $ax^2 + bx + c = 0$ $(a \neq 0)$ for $x$. Write the quadratic formula.

22) Correct the error in this simplification:
$$\frac{6 \pm 3\sqrt{10}}{3} = 6 \pm \sqrt{10}$$

**Solve using the quadratic formula. Find only the real number solutions.**

23) $x^2 + 7x + 12 = 0$

24) $3z^2 + z - 10 = 0$

25) $2r^2 + 3r = -6$

26) $8n^2 + 3 = 14n$

27) $t(t - 10) = -7$

28) $-1 = y(5 - 4y)$

29) $\dfrac{3}{8}p^2 + \dfrac{1}{2}p - \dfrac{1}{4} = 0$

30) $(2w + 3)(w + 4) = 6(w + 1)$

31) The hypotenuse of a right triangle is 1 inch less than three times the shorter leg. The longer leg is $\sqrt{55}$ inches long. Find the length of the shorter leg.

**Solve. Give an exact answer and use a calculator to round the answer to the nearest hundredth.**

32) An object is thrown upward from a height of 26 feet. The height $h$ of the object (in feet) $t$ seconds after the object is released is given by

$$h = -16t^2 + 22t + 26$$

a) How long does it take the object to reach a height of 6 feet?

b) How long does it take the object to hit the ground? (Hint: When the object hits the ground, $h = 0$.)

**Mixed Exercises**

**Keep in mind the four methods we have learned for solving quadratic equations:** *factoring, the square root property, completing the square,* **and** *the quadratic formula.* **Solve the equations using one of these methods. Find only the real number solutions.**

33) $d^2 - 3d - 28 = 0$

34) $2y(y - 3) = -1$

35) $(3t - 1)(t + 1) = 3$

36) $\dfrac{1}{18}p^2 + \dfrac{7}{9}p + \dfrac{8}{3} = 0$

37) $(n - 9)^2 + 7 = 12$

38) $(3a - 4)^2 + 15 = 11$

39) $6c^3 + 12c = 38c^2$

40) $4m^2 - 25 = 0$

41) $k - 2 = 5k^2 + 3k + 4$

42) $w^2 + 6w + 4 = 0$

43) $h^2 = 4h$

44) $0.01x^2 + 0.09x + 0.12 = 0$

**(10.4)**

**Simplify.**

45) $\sqrt{-144}$

46) $\sqrt{-6}$

47) $\sqrt{-40}$

48) $\sqrt{-125}$

**Perform the operations and simplify.**

49) $(-10 + 3i) + (7 + 2i)$

50) $(4 + i) - (1 - 7i)$

51) $(-2 - 11i) - (9 + 4i) + (3 - 6i)$

52) $\left(\dfrac{2}{3} - \dfrac{1}{8}i\right) + \left(\dfrac{5}{6} + \dfrac{2}{5}i\right)$

53) When $i^2$ appears in an expression, it should be replaced with what?

**Multiply and simplify.**

54) $-4(5 - 9i)$

55) $\dfrac{3}{4}(-6 + 12i)$

56) $2i(7 + i)$

57) $(2 + 5i)(1 + 9i)$

58) $(3 + 2i)(7 - 6i)$

59) Identify the conjugate of $-6 - 4i$, then multiply the number and its conjugate.

60) Explain how to divide complex numbers.

**Divide. Write the result in the form $a + bi$.**

61) $\dfrac{7}{3 + i}$

62) $\dfrac{2i}{5 - 6i}$

63) $\dfrac{8 - 2i}{-4 - 3i}$

64) $\dfrac{6 - 6i}{1 + 7i}$

**Mixed Exercises**

65) The instructions in an exercise set say, "Perform the operations and simplify." What is the difference between the two problems $(5 + 2i) + (3 - 8i)$ and $(5 + 2i)(3 - 8i)$?

66) Perform the operations and simplify.

  a) $(5 + 2i) + (3 - 8i)$     b) $(5 + 2i)(3 - 8i)$

**Perform the operations and simplify.**

67) $(-4 + 5i) + (8 + 2i)$

68) $\dfrac{5}{2}(-6 + 7i)$

69) $-6i(3 - 4i)$

70) $(9 - i)(2 + 3i)$

71) $\dfrac{6i}{-1 + 4i}$

72) $(10 + 3i) - (2 - 11i)$

73) $(-7 + 2i)(1 - 3i)$

74) $\dfrac{6 - 8i}{3 - 2i}$

**Solve the following equations with complex solutions.**

75) $r^2 = -36$

76) $4y^2 + 19 = 6$

77) $m^2 - 8m + 17 = 0$

78) $3x^2 + 2x + 5 = 0$

79) $4 = 31 + (t + 2)^2$

80) $k(k + 10) = -34$

81) $(2z + 1)(z - 3) = z - 12$

82) $(2a - 1)^2 + 11 = 9$

**(10.5)**

83) What do you call the graph of $y = ax^2 + bx + c \; (a \neq 0)$?

84) Explain how to find the vertex of the graph of $y = ax^2 + bx + c (a \neq 0)$.

**Sketch the graph of each equation. Identify the vertex.**

85) $y = -x^2 + 4$

86) $y = (x + 2)^2$

87) $y = x^2 + 4x + 3$

88) $y = -x^2 + 2x + 2$

89) $y = x^2 - 6x + 11$

90) Use the graph to determine the real number solutions to $x^2 + 6x + 8 = 0$.

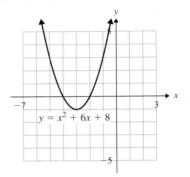

Use a graphing calculator to solve the equations in Exercises 91 and 92. Round your answer to the nearest hundredth, where appropriate.

91) $8x^2 - 2x - 3 = 0$

92) $x^2 - 4x + 2 = 0$

**Determine the domain and range of each quadratic function.**

93) $f(x) = (x - 3)^2$

94) $h(x) = -x^2 + 5$

95) $g(x) = -x^2 - 6x - 10$

96) $f(x) = x^2 + 4x + 1$

# Chapter 10: Test

1) Solve $3a^2 + 4 = 22$ using the square root property.

2) The kinetic energy $K$, in joules, of an object is given by $K = \frac{1}{2}mv^2$, where $m$ is the mass of the object, in kg, and $v$ is the velocity of the object in m/sec. Find the velocity of a motorcycle with a mass of 180 kg and kinetic energy of 1440 joules.

**Solve by completing the square.**

3) $m^2 - 12m + 26 = 0$

4) $3k^2 + 15k + 9 = 0$

5) Elaine simplified the expression as follows: $\frac{12 \pm 8\sqrt{5}}{4} = 3 \pm 8\sqrt{5}$. Explain her error, and simplify the expression correctly.

6) Solve $2p^2 + 9p + 5 = 0$ using the quadratic formula.

**Solve using any method. Find only the real number solutions.**

7) $(p - 7)^2 + 1 = 13$

8) $\frac{1}{6}c^2 + \frac{2}{3}c - 2 = 0$

9) $(2k - 3)(2k + 5) = 6(k - 2)$  10) $n^2 = 3n - 10$

**Solve. Give an exact answer and use a calculator to round the answer to the nearest hundredth.**

11) An object is thrown upward from a height of 32 feet. The height $h$ of the object (in feet) $t$ seconds after the object is released is given by $h = -16t^2 + 40t + 32$.

a) How long does it take the object to reach a height of 8 feet?

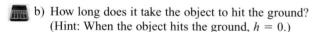 b) How long does it take the object to hit the ground? (Hint: When the object hits the ground, $h = 0$.)

**Simplify.**

12) $\sqrt{-121}$

13) $\sqrt{-48}$

**Perform the operations and simplify. Write the result in the form $a + bi$.**

14) $4i(3 + 5i)$

15) $(2 - 9i) - (6 - 10i)$

16) $(5 + 2i)(7 - 4i)$

17) $\frac{10}{9 - 2i}$

18) $\frac{4 - 5i}{-2 + 5i}$

**Solve the equations with complex solutions. Use any method.**

19) $(r + 8)^2 + 15 = 12$

20) $5y^2 - 2y + 1 = 0$

**Sketch the graph of each equation. Identify the vertex.**

21) $y = -(x - 1)^2$

22) $y = x^2 + 2x - 3$

23) $y = -x^2 + 4x - 1$

24) Use the graph to determine the real number solutions to $x^2 - 6x + 8 = 0$.

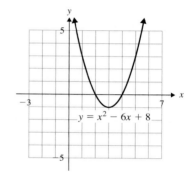

25) Determine the domain and range of $f(x) = x^2 - 4$.

**Perform the operations and simplify.**

1) $\dfrac{5}{8} - \dfrac{2}{7}$

2) $\dfrac{12 - 56 \div 8}{(1 + 5)^2 - 2^4}$

3) Find the area and perimeter.

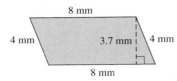

4) Given this set of numbers,

$$\left\{ \sqrt{23}, -6, 14.38, \dfrac{3}{11}, 2, 5.7, 0, 9.21743819 \ldots \right\}$$

list the

a) natural numbers

b) integers

c) rational numbers

d) irrational numbers

5) The lowest temperature on record in the state of Wyoming is $-66°$F. Georgia's record low is $49°$ higher than Wyoming's. What is the lowest temperature ever recorded in Georgia? (www.weather.com)

6) Use the rules of exponents to simplify.

a) $2^3 \cdot 2^2$

b) $\left(\dfrac{1}{3}\right)^2 \cdot \left(\dfrac{1}{3}\right)$

c) $\dfrac{x^7 \cdot (x^2)^5}{(2y^3)^4}$

d) $(6 - 8)^2$

7) Evaluate.

a) $(-12)^0$

b) $5^0 + 4^0$

c) $-6^{-2}$

d) $2^{-4}$

8) Simplify by applying one or more of the rules of exponents. The final answer should not contain negative exponents.

a) $(-3s^4t^5)^4$

b) $\left(\dfrac{d^{-4}}{d^{-9}}\right)^5$

c) $\left(\dfrac{3k^{-1}t}{5k^{-7}t^4}\right)^{-3}$

d) $\left(\dfrac{40}{21}x^{10}\right)(3x^{-12})\left(\dfrac{49}{20}x^2\right)$

9) Write 0.0000575 in scientific notation.

10) Simplify $\dfrac{8 \times 10^6}{2 \times 10^{13}}$ and write the final answer without an exponent.

**Solve each equation.**

11) $8b - 7 = 57$

12) $6 - 5(4d - 3) = 7(3 - 4d) + 8d$

**Solve using the five steps for solving applied problems.**

13) A number increased by nine is one less than twice the number. Find the number.

14) Jerome invested some money in an account earning 7% simple interest and $3000 more than that at 8% simple interest. After 1 year, he earned $915 in interest. How much money did he invest in each account?

15) Solve $-15 < 4p - 7 \le 5$ and write the answer in interval notation.

16) Complete the table of values for $4x - 6y = 8$.

| x | y |
|---|---|
|  | 0 |
| 0 |  |
| 3 |  |
|  | −4 |

17) Graph $x - 2y = 6$ by finding the intercepts and at least one other point.

18) Write the *standard form* of the equation of the line given the following information.

$$m = -\dfrac{5}{3} \text{ and } y\text{-intercept } (0, 2)$$

19) Write an equation of the line *perpendicular* to the given line and containing the given point. Write the answer in slope-intercept form.

$$y = -\dfrac{1}{2}x + 9; \quad (6, 5)$$

20) Let $f(x) = 5x - 12$, $g(x) = x^2 + 6x + 5$. Find each of the following and simplify.

a) $f(4)$

b) $f(-3)$

c) $g(0)$

d) $f(c - 2)$

21) Solve by the system of equations graphing. $2x + 3y = 5$

$$y = \dfrac{1}{2}x + 4$$

**Solve each system.**

22) $6x - y = -3$
$15x + 2y = 15$

23) $\dfrac{3}{4}x - y = \dfrac{1}{2}$
$-\dfrac{x}{3} + \dfrac{y}{2} = -\dfrac{1}{6}$

24) A car and a tour bus leave the same location and travel in opposite directions. The car's speed is 12 mph more than the speed of the bus. If they are 270 miles apart after $2\dfrac{1}{2}$ hours, how fast is each vehicle traveling?

25) Graph the solution set of $x \le 4$ and $y \ge -\dfrac{3}{2}x + 3$.

26) Simplify $\left(\dfrac{3pq^{-10}}{2p^{-2}q^5}\right)^{-2}$.

**Perform the indicated operation and simplify.**

27) $\quad 5.8p^3 - 1.2p^2 + \quad p - 7.5$
$\quad + 2.1p^3 + 6.3p^2 + 3.8p + 3.9$

28) $\left(\dfrac{7}{4}k^2 + \dfrac{1}{6}k + 5\right) - \left(\dfrac{1}{2}k^2 + \dfrac{5}{6}k - 2\right)$

29) $-5(7w - 12)(w + 3)$

30) $(6q^2 + 2q - 35) \div (3q + 7)$

**Factor completely.**

31) $t^2 - 2tu - 63u^2$

32) $3g^2 + g - 44$

33) $a^2 + 16a + 64$

34) Solve $(z + 2)^2 = -z(3z + 4) + 9$

35) Find the length and width of the rectangle if its area is 28 cm².

$x + 1$
$x + 4$

**Perform the operation and simplify.**

36) $\dfrac{3x^2 + 14x + 16}{15x + 40} \div \dfrac{11x + 22}{x - 5}$

37) $\dfrac{10p + 3}{4p + 4} - \dfrac{8}{p^2 - 6p - 7}$

38) $\dfrac{2}{5z^2} + \dfrac{9}{10z}$

39) $\dfrac{\dfrac{c}{c + 2} + \dfrac{1}{c^2 - 4}}{1 - \dfrac{3}{c + 2}}$

40) Solve $\dfrac{3k}{k + 9} = \dfrac{3}{k + 1}$.

**Simplify.**

41) $\dfrac{\sqrt{200k^{21}}}{\sqrt{2k^5}}$

42) $\sqrt{80} - \sqrt{48} + \sqrt{20}$

43) $\dfrac{z - 4}{\sqrt{z} + 2}$

44) $81^{-3/4}$

**Solve.**

45) $\sqrt{6x + 9} - \sqrt{2x + 1} = 4$

46) $2c^2 - 11 = 25$

47) $t^2 + 9 = -4t$

48) Find the distance between $(2, 3)$ and $(7, 5)$.

**Multiply and simplify.**

49) $(4 - 6i)(3 - 6i)$

50) Graph $y = -x^2 + 2x + 3$. Identify the vertex.

# Appendix

## Section A.1  Mean, Median, and Mode

### Objectives

1. Find the Mean
2. Find the Weighted Mean
3. Find the Median
4. Find the Mode

One way to analyze data or a list of numbers is to look for a **measure of central tendency.** This is a number that can be used to represent the entire list of numbers. In this section, we will learn about three measures of central tendency: the mean (or average), the median, and the mode. Let's start with the mean.

### 1. Find the Mean

What is the **mean,** or **average,** of a list of numbers?

> **Definition**
>
> The **mean (average)** is the sum of all the values in a list of numbers divided by the number of values in the list.
>
> $$\text{Mean} = \frac{\text{Sum of all values}}{\text{Number of values}}$$

**Example 1**

One night Heidi's bowling scores were 150, 211, 163, and 192. Find her mean (average) score for the night of bowling.

**Solution**

Let's start by using the definition of mean.

$$\text{Mean} = \frac{150 + 211 + 163 + 192}{4} \qquad \begin{array}{l}\text{Sum of her bowling scores} \\ \hline \text{Number of scores}\end{array}$$

$$= \frac{716}{4} \qquad \text{Add.}$$

$$= 179 \qquad \text{Divide.}$$

Heidi's mean or average bowling score for the night was 179. ∎

**Note**

The mean will not necessarily be a value found in the original group of numbers!

**You Try 1**

The numbers of miles that Candace ran each day during a certain week were 1, 3, 6, 2, 1, 2, and 6. Find the mean of the number of miles she ran each day.

## 2. Find the Weighted Mean

Have you ever wondered how a grade point average (GPA) is calculated? It is usually calculated using a *weighted mean.* If some values in a list of numbers appear more than once, then to find a **weighted mean,** each value is *weighted* by multiplying it by the number of times it appears in the list.

Let's look at how to find the weighted mean (or GPA) of a student's grades. The grade earned for the course is the value in the list, and the number of credits assigned to the course is the weight.

**Example 2**

Annanias's grades from last semester are listed in the table below. His school uses a 4-point scale so that an A is 4 points, a B is 3 points, a C is 2 points, a D is 1 point, and an F is 0 points. Find his GPA.

| Course | Grade | Credits |
|---|---|---|
| Biology | B | 4 |
| Math | A | 5 |
| U.S. History | B | 3 |
| Creative Writing | C | 3 |

### Solution

The grade earned for the course is the value in the list, and the number of credits assigned to the course is the number of times the grade is counted, or the *weight.*

For each course, multiply the grade, in points, by the number of credits that course is worth. Then, find the weighted mean, or GPA, by taking the sum of all of these numbers and dividing by the total number of credits.

| Course | Grade | Grade, in Points | Credits | Grade, in Points · Credits |
|---|---|---|---|---|
| Biology | B | 3 | 4 | $3 \cdot 4 = 12$ |
| Math | A | 4 | 5 | $4 \cdot 5 = 20$ |
| U.S. History | B | 3 | 3 | $3 \cdot 3 = 9$ |
| Creative Writing | C | 2 | 3 | $2 \cdot 3 = 6$ |
| Totals | | | 15 | 47 |

$$\text{Weighted mean (GPA)} = \frac{47}{15} \qquad \frac{\text{Total number of grade points}}{\text{Total number of credits}}$$

$$= 3.1\overline{3} \qquad \text{Divide.}$$

Round $3.1\overline{3}$ to the nearest hundredth. Annanias's GPA was 3.13. ∎

**You Try 2**

Determine Arnold's GPA if he received this grade report:

| Course | Grade | Credits |
|---|---|---|
| Music Appreciation | A | 3 |
| Psychology | B | 4 |
| World History | D | 3 |
| Weight Training | A | 1 |
| Personal Health | B | 3 |

### 3. Find the Median

Another way to represent a list of numbers with a single number is to use the *median*. The **median** of an ordered list of numbers is the middle number. To find the median of a list of values, follow this three-step process.

> **Procedure   How to Find the Median of a List of Values**
>
> **Step 1:**  Arrange the values from lowest to highest.
>
> **Step 2:**  Determine whether there is an even or odd number of values in the list.
>
> **Step 3:**  If there is an *odd number* of values, the median is the *middle number*. If there is an *even number* of values, the median is the *mean (average) of the middle two numbers*.

**Example 3**

Find the median of 3, 5, 10, 9, 8, 21, 22, 15, 4.

**Solution**

Follow the three-step process to find the median.

**Step 1:**  Arrange the values from lowest to highest: 3, 4, 5, 8, 9, 10, 15, 21, 22

**Step 2:**  Determine whether there is an even or odd number of values in the list: There are 9 values in the list; therefore there is an odd number of values.

**Step 3:**  Since there is an odd number of values, the median is the middle value, 9.

The median is 9.                                                                                                              ■

**You Try 3**

Find the median number of pages in a row of books on a bookshelf:
245, 312, 197, 208, 255, 385, 449, 362

### 4. Find the Mode

Another measure of central tendency is the *mode*. The **mode** is the number that appears most frequently in a list of numbers. If there are two numbers that appear most often in a list, then the list has two modes and is said to be **bimodal**. The list can also have *no mode* if no number appears more often than any other number.

**Example 4**

Find the mode in each list of numbers.

a)   20, 30, 40, 70, 50, 30, 90, 60, 80, 20, 30, 10

b)   12, 8, 7, 6, 4, 2, 4, 9, 7, 11, 20, 15

c)   1, 2, 3, 4, 5, 6, 7, 8, 9, 10

**Solution**

a)   If the list is long, it is helpful to arrange the numbers from lowest to highest and underline the numbers that are repeated:

10, <u>20, 20</u>, <u>30, 30, 30</u>, 40, 50, 60, 70, 80, 90

Since 30 appears most often, it is the mode.

b)   Write the numbers from smallest to largest, and underline the numbers that are repeated:

$$2, \underline{4}, \underline{4}, 6, \underline{7}, \underline{7}, 8, 9, 11, 12, 15, 20$$

Since 4 and 7 each appear twice, the list is bimodal. The modes are 4 and 7.

c)   1, 2, 3, 4, 5, 6, 7, 8, 9, 10

The list is already in order and has no repeating numbers. There is no mode.  ■

**You Try 4**

Find the mode in the list of numbers.

Number of text messages Jamie sent each day for a week: 92, 124, 153, 124, 138, 176, 182

## Using Technology

We can use a graphing calculator to find the mean, median, and mode of a list of numbers. For example, find the mean, median, and mode of the list {1, 15, 6, 24, 9, 4, 15, 35, 6, 15}. First clear any lists by pressing  2nd  +  4  ENTER  as shown on the left.

Next enter the numbers in the list L1. Press  2nd  STAT  to select the list L1 and then enter each number in the list by typing each number and pressing  ENTER  as shown on the middle screen.

The graphing calculator will find the mean and median in the list. Press  STAT , press the right arrow key to move over to the CALC menu, and press  ENTER  to execute 1-Var Stats. Press  ENTER  to display the results as shown on the screen below right.

The mean of the list of numbers is represented by the variable $\bar{x}$ on the first line of the display, so the mean is 13.

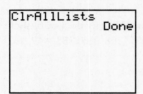

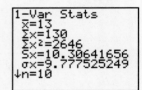

We need to scroll down by pressing the down arrow three times to view the median of the list of data. The median is shown as Med on the screen, so the median is 12 as shown below on the left.

The calculator does not report the mode of the list of the numbers directly. We need to display the list of numbers in ascending order to decide which number occurs most frequently. The data can be sorted in the list L1 by pressing  STAT  2  to select Sort A(, which means sort the list in ascending order. Press  2nd  STAT  to select L1, press  ENTER , and then press  ENTER  to sort the list.

Press  STAT  1  to view the sorted list L1. Scroll through the list by pressing the down arrow key. The number 15 occurs most frequently (3 times), so 15 is the mode.

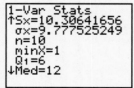

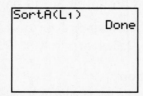

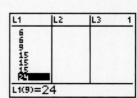

Find the mean, median, and mode of the list of numbers.

1)   2, 8, 22, 7, 9, 12, 46, 21, 5, 8          2)   15, 48, 20, 59, 48, 31, 21, 14, 48, 16

3)   2, 4, −5, 8, −1, 6, 10, 8, 9, 19          4)   17, 28, −12, 35, 5, 19, 17, 31, 24, 6

5)   1, 6, 13, 2, 4, 7, 12, 2, 14, −1          6)   44, 38, 28, 43, 25, 14, 30, 48, 26, 14

**Answers to You Try Exercises**

1)  3 miles     2)  2.86 GPA     3)  283.5 pages     4)  124 text messages

**Answers to Technology Exercises**

1)  mean: 14; median: 8.5; mode: 8     2)  mean: 32; median: 26; mode: 48

3)  mean: 6; median: 7; mode: 8     4)  mean: 17; median: 18; mode: 17

5)  mean: 6; median: 5; mode: 2     6)  mean: 31; median: 29; mode: 14

## A.1 Exercises

**Mixed Exercises: Objectives 1, 3, and 4**

For the following lists of numbers find three measures of central tendency: a) the mean b) the median and c) the mode. Round to the nearest tenth, where appropriate.

1)  19, 14, 29, 34, 24

2)  3, 11, 5, 7, 9

3)  2, 8, 10, 4, 8

4)  19, 17, 11, 18, 19

5)  33, 46, 52, 34, 46, 51

6)  74, 55, 87, 55, 87, 60

7)  24, 38, 33, 38, 24, 27

8)  14, 16, 11, 17, 12, 19

9)  301, 319, 283, 297, 313, 334, 283

10)  321, 132, 231, 123, 231, 213, 312

11)  646, 485, 429, 521, 605, 538, 605, 642

12)  897, 888, 904, 926, 897, 913, 933

13)  Since his start in the NBA in 2003 through the 2008 season, LeBron James has played in 79, 80, 79, 78, 75, and 81 games per season. Find the mean, median, and mode number of games per season. (www.nba.com)

14)  Find the mean, median, and mode of the grams of sugar in a sample of energy drinks: 22, 24, 23, 27, 28, 27, 29, 26, 27.

15)  Find the mean, median, and mode of the first 9 scores to ever win the U.S. Masters golf tournament. (www.masters.com)
$-4, -6, -3, -5, -3, -9, -8, -8, -8$

16)  Find the mean, median, and mode of the 9 scores to win the U.S. Masters golf tournament from 2002 to 2010: $-12, -7, -9, -12, -7, +1, -8, -12, -16$.

17)  Find the mean, median, and mode for the Palmer family's weekly grocery bill for a month: $134, $162, $154, $145, $152.

18)  Find the mean, median, and mode for a hotel's occupancy rate from Sunday through Saturday: 68%, 74%, 76%, 82%, 87%, 85%, 71%.

19)  Find the mode for the cost of a 1-day adult ski lift ticket at the following Colorado resorts for the 2009–2010 season. (www.black-diamond.com)

| Resort | Cost |
|---|---|
| Arapahoe Basin | $65 |
| Beaver Creek | $97 |
| Breckenridge | $92 |
| Copper Mountain | $92 |
| Eldora | $65 |
| Snowmass | $92 |
| Telluride | $92 |

20)  Find the mean cost of tuition for in-state students at the following Illinois state schools for the 2009–2010 academic year. (college websites)

| School | Cost |
|---|---|
| Eastern Illinois Univ. | $7170 |
| Northern Illinois Univ. | $7260 |
| Univ. of Illinois/Chicago | $7260 |
| Univ. of Illinois/Urbana | $9484 |

## Objective 2: Find the Weighted Mean

Find the weighted mean for each list of numbers. Round to the nearest tenth, where appropriate.

21)

| Calories | Frequency |
|----------|-----------|
| 120 | 4 |
| 350 | 2 |
| 400 | 3 |
| 540 | 1 |

22)

| Price of Grocery Item | Frequency |
|-----------------------|-----------|
| $0.99 | 8 |
| $1.49 | 3 |
| $1.99 | 6 |
| $2.59 | 4 |
| $3.99 | 5 |

23)

| Boxes of Girl Scout Cookies Sold | Frequency |
|----------------------------------|-----------|
| 19 | 1 |
| 30 | 3 |
| 50 | 7 |
| 82 | 1 |
| 100 | 2 |
| 144 | 1 |

24)

| Students in Class | Frequency |
|-------------------|-----------|
| 20 | 1 |
| 23 | 3 |
| 25 | 3 |
| 26 | 1 |
| 28 | 4 |
| 29 | 2 |

Find the GPA for each student with the following grade reports. Let A = 4 points, B = 3 points, C = 2 points, D = 1 point, and F = 0 points. Round the answer to the nearest hundredth.

25)

| Course | Grade | Credits |
|--------|-------|---------|
| English Composition | A | 3 |
| Art History | B | 3 |
| Calculus I | A | 5 |
| French | B | 4 |

26)

| Course | Grade | Credits |
|--------|-------|---------|
| Precalculus | B | 4 |
| Child Psychology | A | 4 |
| Educational Psychology | B | 3 |
| American History | C | 3 |
| Children's Literature | A | 2 |

27)

| Course | Grade | Credits |
|--------|-------|---------|
| Technical Writing | B | 3 |
| Biomechanics | A | 4 |
| Immunology | B | 4 |
| Exercise Physiology | C | 3 |
| English Literature | D | 3 |

28)

| Course | Grade | Credits |
|--------|-------|---------|
| Chemistry | D | 4 |
| Chemistry Lab | C | 1 |
| Music Theory | A | 3 |
| Economics | C | 4 |
| Shakespeare | B | 3 |

## Section A.2   Systems of Linear Equations in Three Variables

**Objectives**

1. Define a System of Three Equations in Three Variables
2. Learn the Steps for Solving Systems of Linear Equations in Three Variables
3. *Case 1:* Solve a System of Three Equations in Three Variables When Each Equation Contains Three Variables
4. Solve Special Systems
5. *Case 2:* Solve a System of Three Equations in Three Variables When at Least One Equation Is Missing a Variable
6. Solve an Applied Problem

In Chapter 5, we discussed different ways to solve systems of linear equations in two variables. In this section, we will learn how to solve a system of linear equations in *three* variables.

### 1. Define a System of Three Equations in Three Variables

**Definition**

A **linear equation in three variables** is an equation of the form $Ax + By + Cz = D$ where $A$, $B$, and $C$ are not all zero and where $A$, $B$, $C$, and $D$ are real numbers. Solutions to this type of an equation are **ordered triples** of the form $(x, y, z)$.

An example of a linear equation in three variables is $2x - y + 3z = 12$.
    Some solutions to the equation are

$$(5, 1, 1) \quad \text{since} \quad 2(5) - (1) + 3(1) = 12$$
$$(3, 0, 2) \quad \text{since} \quad 2(3) - (0) + 3(2) = 12$$
$$(6, -3, -1) \quad \text{since} \quad 2(6) - (-3) + 3(-1) = 12$$

There are infinitely many solutions.
    Ordered pairs, like $(2, 5)$, are plotted on a two-dimensional rectangular coordinate system. Ordered triples, however, like $(1, 2, 3)$ and $(3, 0, 2)$, are graphed on a three-dimensional coordinate system.

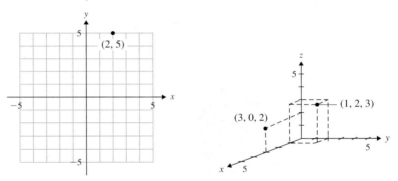

The graph of a linear equation in two variables is a line, but the graph of a linear equation in *three* variables is a *plane*.
    A **solution to a system of linear equations in three variables** is an *ordered triple* that satisfies each equation in the system. Like systems of linear equations in two variables, systems of linear equations in *three* variables can have *one solution* (the system is consistent), *no solution* (the system is inconsistent), or *infinitely many solutions* (the equations are dependent).
    Here is an example of a system of linear equations in three variables:

$$x + 4y + 2z = 10$$
$$3x - y + z = 6$$
$$2x + 3y - z = -4$$

In Section 5.1, we solved systems of linear equations in *two* variables by graphing. Since the graph of an equation like $x + 4y + 2z = 10$ is a *plane*, however, solving a system like the one on the previous page by graphing would not be practical. But let's look at what happens, graphically, when a system of linear equations in three variables has one solution, no solution, or an infinite number of solutions.

**One solution:**

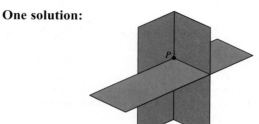

Intersect at point $P$

All three planes intersect at one point; this is the solution of the system.

**No solution:**

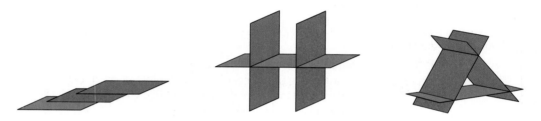

*None* of the planes may intersect or *two* of the planes may intersect, but if there is no solution to the system, *all three planes* do not have a common point of intersection.

**Infinite number of solutions:**

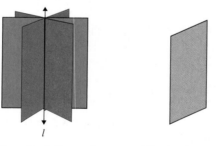

Intersection is the set of points on a line.

All points in common

The three planes may intersect so that they have a line in common—the infinite set of points on the line is the solution to the system. Or, the set of all points in a plane gives the infinite number of solutions.

## 2. Learn the Steps for Solving Systems of Linear Equations in Three Variables

We will discuss how to solve these systems by looking at two different cases.

*Case 1:* Every equation in the system contains three variables.

*Case 2:* At least one of the equations in the system contains only two variables.

We begin with Case 1.

---

**Procedure   Solving a System of Linear Equations in Three Variables**

If each equation contains three variables, then to solve a system of linear equations in three variables, follow these steps.

**Step 1:**  **Label** the equations ①, ②, and ③.

**Step 2:**  **Choose a variable to eliminate. Eliminate** this variable from *two* sets of *two* equations using the elimination method. You will obtain two equations containing the same two variables. Label one of these new equations $\boxed{A}$ and the other $\boxed{B}$.

**Step 3:**  **Use the elimination method to eliminate a variable from equations $\boxed{A}$ and $\boxed{B}$.** You have now found the value of one variable.

**Step 4:**  **Find the value of another variable** by substituting the value found in Step 3 into either equation $\boxed{A}$ or $\boxed{B}$ and solving for the second variable.

**Step 5:**  **Find the value of the third variable** by substituting the values of the two variables found in 3) and 4) into equation ①, ②, or ③.

**Step 6:**  **Check** the solution in each of the original equations, and **write the solution as an ordered triple.**

---

### 3. *Case 1:* Solve a System of Three Equations in Three Variables When Each Equation Contains Three Variables

**Example 1**

Solve.   ① $x + 2y - 2z = 3$
②  $2x + y + 3z = 1$
③   $x - 2y + z = -10$

**Solution**

**Step 1:**  **Label** the equations ①, ②, and ③.

**Step 2:**  **Choose a variable to eliminate:** *y.*

We will eliminate $y$ from *two* sets of *two* equations.

a) *Equations* ① *and* ③. Add the equations to eliminate $y$. Label the resulting equation $\boxed{A}$.

$$
\begin{array}{ll}
① & x + 2y - 2z = 3 \\
③ \;+ & \underline{x - 2y + \phantom{2}z = -10} \\
\boxed{A} & 2x - \phantom{2y} \phantom{+} z = -7
\end{array}
$$

b) *Equations* ② *and* ③. To eliminate $y$, multiply equation ② by 2 and add it to equation ③. Label the resulting equation $\boxed{B}$.

$$
\begin{array}{ll}
2 \times ② & 4x + 2y + 6z = 2 \\
③ \;+ & \underline{\phantom{4}x - 2y + \phantom{6}z = -10} \\
\boxed{B} & 5x + \phantom{2y} 7z = -8
\end{array}
$$

**Note**

Equations $\boxed{A}$ and $\boxed{B}$ contain only *two* variables and they are the *same* variables, *x* and *z*.

**Step 3:**  **Use the elimination method to eliminate a variable from equations $\boxed{A}$ and $\boxed{B}$.**

$$
\begin{array}{l}
2x - \phantom{7}z = -7 \;\boxed{A} \\
5x + 7z = -8 \;\boxed{B}
\end{array}
$$

We will eliminate $z$ from $\boxed{A}$ and $\boxed{B}$. Multiply $\boxed{A}$ by 7 and add it to $\boxed{B}$.

$$
\begin{array}{rl}
7 \times \boxed{A} & 14x - 7z = -49 \\
\boxed{B} \;+ & \underline{5x + 7z = -8} \\
& 19x \qquad\;\; = -57 \\
& \boxed{x = -3} \qquad \text{Divide by 19.}
\end{array}
$$

**Step 4:** **Find the value of another variable** by substituting $x = -3$ into either equation $\boxed{A}$ or $\boxed{B}$.

We will substitute $x = -3$ into $\boxed{A}$ since it has smaller coefficients.

$$
\begin{array}{rll}
\boxed{A} & 2x - z = -7 & \\
& 2(-3) - z = -7 & \text{Substitute } -3 \text{ for } x. \\
& -6 - z = -7 & \text{Multiply.} \\
& -z = -1 & \text{Add 6.} \\
& \boxed{z = 1} & \text{Divide by } -1.
\end{array}
$$

**Step 5:** **Find the value of the third variable** by substituting $x = -3$ and $z = 1$ into either equation, ①, ②, or ③.

We will substitute $x = -3$ and $z = 1$ into ① to solve for $y$.

$$
\begin{array}{rll}
① & x + 2y - 2z = 3 & \\
& -3 + 2y - 2(1) = 3 & \text{Substitute } -3 \text{ for } x \text{ and } 1 \text{ for } z. \\
& -3 + 2y - 2 = 3 & \text{Multiply.} \\
& 2y - 5 = 3 & \text{Combine like terms.} \\
& 2y = 8 & \text{Add 5.} \\
& \boxed{y = 4} & \text{Divide by 2.}
\end{array}
$$

**Step 6:** **Check** the solution, $(-3, 4, 1)$, in each of the original equations, and **write the solution.**

$$
\begin{array}{rll}
① & x + 2y - 2z = 3 & \\
& -3 + 2(4) - 2(1) \stackrel{?}{=} 3 & \\
& -3 + 8 - 2 \stackrel{?}{=} 3 & \\
& 3 = 3 & \checkmark \text{ True}
\end{array}
$$

$$
\begin{array}{rll}
② & 2x + y + 3z = 1 & \\
& 2(-3) + 4 + 3(1) \stackrel{?}{=} 1 & \\
& -6 + 4 + 3 \stackrel{?}{=} 1 & \\
& 1 = 1 & \checkmark \text{ True}
\end{array}
$$

$$
\begin{array}{rll}
③ & x - 2y + z = -10 & \\
& -3 - 2(4) + 1 \stackrel{?}{=} -10 & \\
& -3 - 8 + 1 \stackrel{?}{=} -10 & \\
& -10 = -10 & \checkmark \text{ True}
\end{array}
$$

The solution is $(-3, 4, 1)$. ∎

**You Try 1**

Solve     $x + 2y + 3z = -11$
$3x - y + z = 0$
$-2x + 3y - z = 4$

## 4. Solve Special Systems

**Example 2**

Solve.   ① $-3x + 2y - z = 5$
② $x + 4y + z = -4$
③ $9x - 6y + 3z = -2$

### Solution

Follow the steps.

***Step 1:***   **Label** the equations ①, ②, and ③.

***Step 2:***   **Choose a variable to eliminate.** $z$ will be the easiest. Eliminate $z$ from *two* sets of *two* equations.

a) *Equations* ① *and* ②. Add the equations to eliminate $z$. Label the resulting equation Ⓐ.

$$\begin{array}{r} ①\quad -3x + 2y - z = 5 \\ ② \;+\; \underline{\quad x + 4y + z = -4} \\ \boxed{A}\quad -2x + 6y \phantom{+z}\; = 1 \end{array}$$

b) *Equations* ① *and* ③. To eliminate $z$, multiply equation ① by 3 and add it to equation ③. Label the resulting equation Ⓑ.

$$① \quad -3x + 2y - z = 5 \longrightarrow 3 \times ① \quad \begin{array}{r} -9x + 6y - 3z = 15 \\ ③ \;+\; \underline{\quad 9x - 6y + 3z = -2} \\ \boxed{B}\quad\;\; 0 = 13 \quad \text{False} \end{array}$$

Since the variables drop out and we get the false statement $0 = 13$, equations one and two have no ordered triple that satisfies each equation.
  The system is inconsistent, so there is no solution. The solution set is $\varnothing$.   ∎

**Note**

When the variables drop out and you get a false statement, there is *no solution* to the system. The system is inconsistent, so the solution set is $\varnothing$.

**Example 3**

Solve.   ① $-4x - 2y + 8z = -12$
② $2x + y - 4z = 6$
③ $6x + 3y - 12z = 18$

*Solution*

Follow the steps.

*Step 1:* **Label** the equations ①, ②, and ③.

*Step 2:* **Choose a variable to eliminate.** Eliminate $y$ from *two* sets of *two* equations.

a) *Equations* ① *and* ②. To eliminate $y$, multiply equation ② by 2 and add it to equation 1. Label the resulting equation $\boxed{A}$.

$$
\begin{array}{ll}
2 \times ② & 4x + 2y - 8z = 12 \\
① & +\ -4x - 2y + 8z = -12 \\
\hline
\boxed{A} & \qquad\qquad\quad 0 = 0 \qquad \text{True}
\end{array}
$$

The variables dropped out and we obtained the true statement $0 = 0$. This is because equation ① is a multiple of equation ②.

Notice, also, that equation 3 is a multiple of equation ②. (Equation ③ $= 3 \times$ equation ②.)

The equations are dependent. There are an infinite number of solutions of the form $\{(x, y, z) \mid 2x + y - 4z = 6\}$. The equations all have the same graph. ■

**You Try 2**

Solve each system of equations.

a) $\quad 8x + 20y - 4z = -16$
$\quad\ -6x - 15y + 3z = 12$
$\quad\quad\ 2x + 5y - z = -4$

b) $\quad\quad x + 4y - 3z = 2$
$\quad\quad 2x - 5y + 2z = -8$
$\quad -3x - 12y + 9z = 7$

### 5. *Case 2:* Solve a System of Three Equations in Three Variables When at Least One Equation Is Missing a Variable

**Example 4**

Solve.  ① $5x - 2y = 6$
② $\quad y + 2z = 1$
③ $3x - 4z = -8$

*Solution*

First, notice that while this *is* a system of three equations in three variables, none of the equations contains three variables. Furthermore, each equation is "missing" a different variable.

We will use many of the *ideas* outlined in the steps for solving a system of three equations, but after labeling our equations ①, ②, and ③, we will begin by using *substitution* rather than the elimination method.

*Step 1:* Label the equations ①, ②, ③.

*Step 2:* The goal of Step 2 is to obtain two equations that contain the same two variables. We will modify this step from the way it was outlined on p. A-9.

In order to obtain *two* equations with the same *two* variables, we will use *substitution*.

Look at equation ②. Since $y$ is the only variable in the system with a coefficient of 1, we will solve equation ② for $y$.

$$
\begin{array}{ll}
② \quad y + 2z = 1 & \\
\quad\quad\ y = 1 - 2z & \text{Subtract } 2z.
\end{array}
$$

Substitute $y = 1 - 2z$ into equation ① to obtain an equation containing the variables $x$ and $z$. Simplify. Label the resulting equation $\boxed{A}$.

$$\begin{array}{rl}
① \qquad 5x - 2y = 6 & \\
5x - 2(1 - 2z) = 6 & \text{Substitute } 1 - 2z \text{ for } y. \\
5x - 2 + 4z = 6 & \text{Distribute.} \\
\boxed{A} \qquad 5x + 4z = 8 & \text{Add 2.}
\end{array}$$

**Step 3:** The goal of Step 3 is to solve for one of the variables. Equations $\boxed{A}$ and ③ contain only $x$ and $z$.

We will eliminate $z$ from $\boxed{A}$ and ③. Add the two equations to eliminate $z$, then solve for $x$.

$$\begin{array}{rl}
\boxed{A} \qquad 5x + 4z = 8 & \\
③ \quad \underline{+\ 3x - 4z = -8} & \\
8x \qquad\quad = 0 & \\
\boxed{x = 0} & \text{Divide by 8.}
\end{array}$$

**Step 4:** Find the value of another variable by substituting $x = 0$ into either $\boxed{A}$, ①, or ③. We will substitute $x = 0$ into $\boxed{A}$.

$$\begin{array}{rl}
\boxed{A} \qquad 5x + 4z = 8 & \\
5(0) + 4z = 8 & \text{Substitute 0 for } x. \\
0 + 4z = 8 & \text{Multiply.} \\
4z = 8 & \\
\boxed{z = 2} & \text{Divide by 4.}
\end{array}$$

**Step 5:** Find the value of the third variable by substituting $x = 0$ into ① or $z = 2$ into ②. We will substitute $x = 0$ into ① to find $y$.

$$\begin{array}{rl}
① \qquad 5x - 2y = 6 & \\
5(0) - 2y = 6 & \text{Substitute 0 for } x. \\
0 - 2y = 6 & \text{Multiply.} \\
-2y = 6 & \\
\boxed{y = -3} & \text{Divide by } -2.
\end{array}$$

**Step 6:** Check the solution $(0, -3, 2)$ in each of the original equations.

The check is left to the student. The solution is $(0, -3, 2)$. ■

**You Try 3**

Solve.    $x + 2y = 8$
$2y + 3z = 1$
$3x - z = -3$

---

**Summary  How to Solve a System of Three Linear Equations in Three Variables**

1) *Case 1*: If every equation in the system contains three variables, then follow the steps on p. A-9. See Example 1.

2) *Case 2*: If at least one of the equations in the system contains only two variables, follow the steps on p. A-9 but in a modified form. See Example 4.

## 6. Solve an Applied Problem

To solve applications involving a system of three equations in three variables, we will use the process outlined in Section 5.4.

---

**Example 5**

Write a system of equations and solve.

    The top three gold-producing nations in 2002 were South Africa, the United States, and Australia. Together, these three countries produced 37% of the gold during that year. Australia's share was 2% less than that of the United States, while South Africa's percentage was 1.5 times Australia's percentage of world gold production. Determine what percentage of the world's gold supply was produced by each country in 2002. (Market Share Reporter—2005, Vol. 1, "Mine Product" http://www.gold.org/value/market/supply-demand/min_production.html from World Gold Council)

**Solution**

*Step 1:*  **Read** the problem carefully and identify what we are being asked to find.

    We must determine the percentage of the world's gold produced by South Africa, the United States, and Australia in 2002.

    **Define the unknowns;** assign a variable to represent each unknown quantity.

*Step 2:*  **Choose variables** to represent the unknown quantities.

    $x$ = percentage of world's gold supply produced by South Africa
    $y$ = percentage of world's gold supply produced by the United States
    $z$ = percentage of world's gold supply produced by Australia

*Step 3:*  **Write a system of equations relating the variables.**

    i)  *To write one equation* we will use the information that says *together* the three countries produced 37% of the gold.

$$x + y + z = 37$$

    ii)  *To write a second equation* we will use the information that says Australia's share was 2% less than that of the United States.

$$z = y - 2$$

    iii)  *To write the third equation* we will use the statement that says South Africa's percentage was 1.5 times Australia's percentage.

$$x = 1.5z$$

    The system is  ① $x + y + z = 37$
               ②         $z = y - 2$
               ③         $x = 1.5z$

*Step 4:*  **Solve the system.**

    Since two of the equations contain only two variables, we will solve the system following the steps in some modified form.

    Label the equations ①, ②, and ③. Using substitution as our first step, we can rewrite the first equation in terms of a single variable, $z$, and solve for $z$.

    Solve equation ② for $y$, and substitute $z + 2$ for $y$ in equation ①.

        ②  $z = y - 2$
            $z + 2 = y$    Solve for $y$.

Using equation ③, substitute $1.5z$ for $x$ in equation ①.

$$\begin{array}{ll}
③ & x = 1.5z \\
① & x + y + z = 37 \\
& (1.5z) + (z + 2) + z = 37 \quad \text{Substitute } 1.5z \text{ for } x \text{ and } z + 2 \text{ for } y. \\
& 3.5z + 2 = 37 \quad \text{Combine like terms.} \\
& 3.5z = 35 \quad \text{Subtract 2.} \\
& \boxed{z = 10} \quad \text{Divide by 3.5.}
\end{array}$$

**To solve for $x$,** we can substitute $z = 10$ into equation ③.

$$\begin{array}{ll}
③ & x = 1.5z \\
& x = 1.5(10) \quad \text{Substitute 10 for } z. \\
& \boxed{x = 15} \quad \text{Multiply.}
\end{array}$$

**To solve for $y$,** we can substitute $z = 10$ into equation ②.

$$\begin{array}{ll}
② & z = y - 2 \\
& 10 = y - 2 \quad \text{Substitute 10 for } z. \\
& \boxed{12 = y} \quad \text{Solve for } y.
\end{array}$$

The solution of the system is $(15, 12, 10)$.

***Step 5:*** **Check** the answer and **interpret** the solution as it relates to the problem.

In 2002, South Africa produced 15% of the world's gold, the United States produced 12%, and Australia produced 10%.

The check is left to the student.

**You Try 4**

Write a system of equations and solve.
  Amelia, Bella, and Carmen are sisters. Bella is 5 yr older than Carmen, and Amelia's age is 5 yr less than twice Carmen's age. The sum of their ages is 48. How old is each girl?

**Answers to You Try Exercises**

1) $(2, 1, -5)$    2) a) infinite number of solutions of the form $\{(x, y, z) | 2x + 5y - z = -4\}$
b) $\varnothing$    3) $(-2, 5, -3)$    4) Amelia: 19; Bella: 17; Carmen: 12

## A.2 Exercises

**Objective 1: Define a System of Three Equations in Three Variables**

Determine whether the ordered triple is a solution of the system.

1) $4x + 3y - 7z = -6$
  $x - 2y + 5z = -3$
  $-x + y + 2z = 7$
  $(-2, 3, 1)$

2) $3x + y + 2z = 2$
  $-2x - y + z = 5$
  $x + 2y - z = -11$
  $(1, -5, 2)$

3) $-x + y - 2z = 2$
  $3x - y + 5z = 4$
  $2x + 3y - z = 7$
  $(0, 6, 2)$

4) $6x - y + 4z = 4$
  $-2x + y - z = 5$
  $2x - 3y + z = 2$
  $\left(-\dfrac{1}{2}, -3, 1\right)$

**Mixed Exercises: Objectives 2–4**

Solve each system. See Examples 1–3.

5) $x + 3y + z = 3$
  $4x - 2y + 3z = 7$
  $-2x + y - z = -1$

6) $x - y + 2z = -7$
  $-3x - 2y + z = -10$
  $5x + 4y + 3z = 4$

7) $5x + 3y - z = -2$
  $-2x + 3y + 2z = 3$
  $x + 6y + z = -1$

8) $-2x - 2y + 3z = 2$
  $3x + 3y - 5z = -3$
  $-x + y - z = 9$

9) $3a + 5b - 3c = -4$
  $a - 3b + c = 6$
  $-4a + 6b + 2c = -6$

10) $a - 4b + 2c = -7$
  $3a - 8b + c = 7$
  $6a - 12b + 3c = 12$

11) $a - 5b + c = -4$
$3a + 2b - 4c = -3$
$6a + 4b - 8c = 9$

12) $-a + 2b - 12c = 8$
$-6a + 2b - 8c = -3$
$3a - b + 4c = 4$

13) $-15x - 3y + 9z = 3$
$5x + y - 3z = -1$
$10x + 2y - 6z = -2$

14) $-4x + 10y - 16z = -6$
$-6x + 15y - 24z = -9$
$2x - 5y + 8z = 3$

15) $-3a + 12b - 9c = -3$
$5a - 20b + 15c = 5$
$-a + 4b - 3c = -1$

16) $3x - 12y + 6z = 4$
$-x + 4y - 2z = 7$
$5x + 3y + z = -2$

## Objective 5: Case 2

Solve each system. See Example 4.

17) $5x - 2y + z = -5$
$x - y - 2z = 7$
$4y + 3z = 5$

18) $-x + z = 9$
$-2x + 4y - z = 4$
$7x + 2y + 3z = -1$

19) $a + 15b = 5$
$4a + 10b + c = -6$
$-2a - 5b - 2c = -3$

20) $2x - 6y - 3z = 4$
$-3y + 2z = -6$
$-x + 3y + z = -1$

21) $x + 2y + 3z = 4$
$-3x + y = -7$
$4y + 3z = -10$

22) $-3a + 5b + c = -4$
$a + 5b = 3$
$4a - 3c = -11$

23) $-5x + z = -3$
$4x - y = -1$
$3y - 7z = 1$

24) $a + b = 1$
$a - 5c = 2$
$b + 2c = -4$

VIDEO 25) $4a + 2b = -11$
$-8a - 3c = -7$
$b + 2c = 1$

26) $3x + 4y = -6$
$-x + 3z = 1$
$2y + 3z = -1$

## Mixed Exercises: Objectives 2–5

Solve each system.

27) $6x + 3y - 3z = -1$
$10x + 5y - 5z = 4$
$x - 3y + 4z = 6$

28) $2x + 3y - z = 0$
$x - 4y - 2z = -5$
$-4x + 5y + 3z = -4$

29) $7x + 8y - z = 16$
$-\dfrac{1}{2}x - 2y + \dfrac{3}{2}z = 1$
$\dfrac{4}{3}x + 4y - 3z = -\dfrac{2}{3}$

30) $3a + b - 2c = -3$
$9a + 3b - 6c = -9$
$-6a - 2b + 4c = 6$

31) $2a - 3b = -4$
$3b - c = 8$
$-5a + 4c = -4$

32) $5x + y - 2z = -2$
$-\dfrac{1}{2}x - \dfrac{3}{4}y + 2z = \dfrac{5}{4}$
$x - 6z = 3$

33) $-4x + 6y + 3z = 3$
$-\dfrac{2}{3}x + y + \dfrac{1}{2}z = \dfrac{1}{2}$
$12x - 18y - 9z = -9$

34) $x - \dfrac{5}{2}y + \dfrac{1}{2}z = \dfrac{5}{4}$
$x + 3y - z = 4$
$-6x + 15y - 3z = -1$

35) $a + b + 9c = -3$
$-5a - 2b + 3c = 10$
$4a + 3b + 6c = -15$

36) $2x + 3y = 2$
$-3x + 4z = 0$
$y - 5z = -17$

VIDEO 37) $2x - y + 4z = -1$
$x + 3y + z = -5$
$-3x + 2y = 7$

38) $a + 3b - 8c = 2$
$-2a - 5b + 4c = -1$
$4a + b + 16c = -4$

## Objective 6: Solve an Applied Problem

Write a system of equations and solve.

39) Moe buys two hot dogs, two orders of fries, and a large soda for $9.00. Larry buys two hot dogs, one order of fries, and two large sodas for $9.50, and Curly spends $11.00 on three hot dogs, two orders of fries, and a large

soda. Find the price of a hot dog, an order of fries, and a large soda.

40) A movie theater charges $9.00 for an adult's ticket, $7.00 for a ticket for seniors 60 and over, and $6.00 for a child's ticket. For a particular movie, the theater sold a total of 290 tickets, which brought in $2400. The number of seniors' tickets sold was twice the number of children's tickets sold. Determine the number of adults', seniors', and children's tickets sold.

41) A Chocolate Chip Peanut Crunch Clif Bar contains 4 fewer grams of protein than a Chocolate Peanut Butter Balance Bar Plus. A Chocolate Peanut Butter Protein Plus PowerBar contains 9 more grams of protein than the Balance Bar Plus. All three bars contain a total of 50 g of protein. How many grams of protein are in each type of bar? (www.clifbar.com, www.balance.com, www.powerbar.com)

42) A 1-tablespoon serving size of Hellman's Real Mayonnaise has 55 more calories than the same serving size of Hellman's Light Mayonnaise. Miracle Whip and Hellman's Light have the same number of calories in a 1-tablespoon serving size. If the three spreads have a total of 160 calories in one serving, determine the number of calories in one serving of each. (product labels)

43) The three NBA teams with the highest revenues in 2002–2003 were the New York Knicks, the Los Angeles Lakers, and the Chicago Bulls. Their revenues totaled $428 million. The Lakers took in $30 million more than the Bulls, and the Knicks took in $11 million more than the Lakers. Determine the revenue of each team during the 2002–2003 season. (*Forbes*, Feb. 16, 2004, p. 66)

44) The best-selling paper towel brands in 2002 were Bounty, Brawny, and Scott. Together they accounted for 59% of the market. Bounty's market share was 25% more than Brawny's, and Scott's market share was 2% less than Brawny's. What percentage of the market did each brand hold in 2002? (*USA Today*, Oct. 23, 2003, p. 3B from Information Resources, Inc.)

45) Ticket prices to a Cubs game at Wrigley Field vary depending on whether they are on a value date, a regular date, or a prime date. At the beginning of the 2008 season, Bill, Corrinne, and Jason bought tickets in the bleachers for several games. Bill spent $367 on four value dates, four regular dates, and three prime dates. Corrinne bought tickets for four value dates, three regular dates, and two prime dates for $286. Jason spent $219 on three value dates, three regular dates, and one prime date. How much did it cost to sit in the bleachers at Wrigley Field on a value date, regular date, and prime date in 2008? (http://chicago.cubs.mlb.com)

46) To see the Boston Red Sox play at Fenway Park in 2009, two field box seats, three infield grandstand seats, and five bleacher seats cost $530. The cost of four field box seats, two infield grandstand seats, and three bleacher seats was $678. The total cost of buying one of each type of ticket was $201. What was the cost of each type of ticket during the 2009 season? (http://boston.redsox.mlb.com)

47) The measure of the largest angle of a triangle is twice the middle angle. The smallest angle measures 28° less than the middle angle. Find the measures of the angles of the triangle. (Hint: Recall that the sum of the measures of the angles of a triangle is 180°.)

48) The measure of the smallest angle of a triangle is one-third the measure of the largest angle. The middle angle measures 30° less than the largest angle. Find the measures of the angles of the triangle. (Hint: Recall that the sum of the measures of the angles of a triangle is 180°.)

49) The smallest angle of a triangle measures 44° less than the largest angle. The sum of the two smaller angles is 20° more than the measure of the largest angle. Find the measures of the angles of the triangle.

50) The sum of the measures of the two smaller angles of a triangle is 40° less than the largest angle. The measure of the largest angle is twice the measure of the middle angle. Find the measures of the angles of the triangle.

51) The perimeter of a triangle is 29 cm. The longest side is 5 cm longer than the shortest side, and the sum of the two smaller sides is 5 cm more than the longest side. Find the lengths of the sides of the triangle.

52) The shortest side of a triangle is half the length of the longest side. The sum of the two smaller sides is 2 in. more than the longest side. Find the lengths of the sides if the perimeter is 58 in.

## Section A.3   Quadratic Inequalities

**Objectives**

1. Solve a Quadratic Inequality by Graphing
2. Solve a Quadratic Inequality Using Test Points
3. Solve Quadratic Inequalities with Special Solutions
4. Solve an Inequality of Higher Degree

In Chapter 3, we learned how to solve *linear* inequalities such as $3x - 5 \leq 16$. In this section, we will discuss how to solve *quadratic* and *rational* inequalities.

**Quadratic Inequalities**

### Definition

A **quadratic inequality** can be written in the form

$$ax^2 + bx + c \leq 0 \quad \text{or} \quad ax^2 + bx + c \geq 0$$

where $a$, $b$, and $c$ are real numbers and $a \neq 0$. ($<$ and $>$ may be substituted for $\leq$ and $\geq$.)

### 1. Solve a Quadratic Inequality by Graphing

To understand how to solve a quadratic inequality, let's look at the graph of a quadratic function.

**Example 1**

a)   Graph $y = x^2 - 2x - 3$.

b)   Solve $x^2 - 2x - 3 < 0$.

c)   Solve $x^2 - 2x - 3 \geq 0$.

**Solution**

a)   The graph of the quadratic function $y = x^2 - 2x - 3$ is a parabola that opens upward. Use the vertex formula to confirm that the vertex is $(1, -4)$.
    To find the $y$-intercept, let $x = 0$ and solve for $y$.

$$y = (0)^2 - 2(0) - 3$$
$$y = -3$$

The $y$-intercept is $(0, -3)$.
To find the $x$-intercepts, let $y = 0$ and solve for $x$.

$$0 = x^2 - 2x - 3$$
$$0 = (x - 3)(x + 1) \qquad \text{Factor.}$$
$$x - 3 = 0 \quad \text{or} \quad x + 1 = 0 \qquad \text{Set each factor equal to 0.}$$
$$x = 3 \quad \text{or} \qquad x = -1 \qquad \text{Solve.}$$

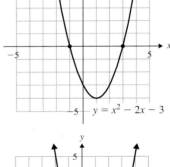

b)   We will use the graph of $y = x^2 - 2x - 3$ to solve the inequality $x^2 - 2x - 3 < 0$. That is, to solve $x^2 - 2x - 3 < 0$ we must ask ourselves, "Where are the *y-values* of the function *less than* zero?"
    The $y$-values of the function are less than zero when the $x$-values are greater than $-1$ and less than 3, as shown at the right.

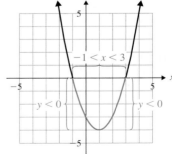

The solution set of $x^2 - 2x - 3 < 0$ (in interval notation) is $(-1, 3)$.

c)  To solve $x^2 - 2x - 3 \geq 0$ means to find the $x$-values for which the $y$-*values* of the function $y = x^2 - 2x - 3$ are *greater than or equal to* zero. (Recall that the $x$-intercepts are where the function equals zero.)

   The $y$-values of the function are greater than or equal to zero when $x \leq -1$ or when $x \geq 3$. The solution set of $x^2 - 2x - 3 \geq 0$ is $(-\infty, -1] \cup [3, \infty)$.

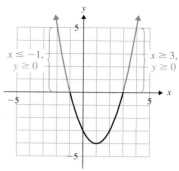

When $x \leq -1$ or $x \geq 3$, the $y$-values are greater than or equal to 0.    ■

## You Try 1

a)  Graph $y = x^2 + 6x + 5$.     b)  Solve $x^2 + 6x + 5 \leq 0$.     c)  Solve $x^2 + 6x + 5 > 0$.

## 2. Solve a Quadratic Inequality Using Test Points

Example 1 illustrates how the $x$-intercepts of $y = x^2 - 2x - 3$ break up the $x$-axis into the three separate intervals: $x < -1$, $-1 < x < 3$, and $x > 3$. We can use this idea of intervals to solve a quadratic inequality without graphing.

## Example 2

Solve $x^2 - 2x - 3 < 0$.

### Solution

Begin by solving the equation $x^2 - 2x - 3 = 0$.

$$x^2 - 2x - 3 = 0$$
$$(x - 3)(x + 1) = 0 \qquad \text{Factor.}$$
$$x - 3 = 0 \quad \text{or} \quad x + 1 = 0 \qquad \text{Set each factor equal to 0.}$$
$$x = 3 \quad \text{or} \qquad x = -1 \qquad \text{Solve.}$$

(These are the $x$-intercepts of $y = x^2 - 2x - 3$.)

### Note

The $<$ indicates that we want to find the values of $x$ that will make $x^2 - 2x - 3 < 0$; that is, find the values of $x$ that make $x^2 - 2x - 3$ a *negative* number.

Put $x = 3$ and $x = -1$ on a number line with the smaller number on the left. This breaks up the number line into three intervals: $x < -1$, $-1 < x < 3$, and $x > 3$.

Choose a test number in each interval and substitute it into $x^2 - 2x - 3$ to determine whether that value makes $x^2 - 2x - 3$ positive or negative. (If one number in the interval makes $x^2 - 2x - 3$ positive, then *all* numbers in that interval will make $x^2 - 2x - 3$ positive.) Indicate the result on the number line.

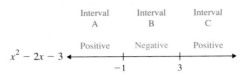

***Interval A:*** $(x < -1)$  As a test number, choose any number less than $-1$. We will choose $-2$. Evaluate $x^2 - 2x - 3$ for $x = -2$.

$$x^2 - 2x - 3 = (-2)^2 - 2(-2) - 3 \qquad \text{Substitute } -2 \text{ for } x.$$
$$= 4 + 4 - 3$$
$$= 8 - 3 = 5$$

When $x = -2$, $x^2 - 2x - 3$ is *positive*. Therefore, $x^2 - 2x - 3$ will be positive for all values of $x$ in this interval. Indicate this on the number line, as seen above.

***Interval B:*** $(-1 < x < 3)$  As a test number, choose any number between $-1$ and 3. We will choose 0. Evaluate $x^2 - 2x - 3$ for $x = 0$.

$$x^2 - 2x - 3 = (0)^2 - 2(0) - 3 \qquad \text{Substitute } 0 \text{ for } x.$$
$$= 0 - 0 - 3 = -3$$

When $x = 0$, $x^2 - 2x - 3$ is *negative*. Therefore, $x^2 - 2x - 3$ will be negative for all values of $x$ in this interval. Indicate this on the number line above.

***Interval C:*** $(x > 3)$  As a test number, choose any number greater than 3. We will choose 4. Evaluate $x^2 - 2x - 3$ for $x = 4$.

$$x^2 - 2x - 3 = (4)^2 - 2(4) - 3 \qquad \text{Substitute } 4 \text{ for } x.$$
$$= 16 - 8 - 3$$
$$= 8 - 3 = 5$$

When $x = 4$, $x^2 - 2x - 3$ is *positive*. Therefore, $x^2 - 2x - 3$ will be positive for all values of $x$ in this interval. Indicate this on the number line.

Look at the number line. The solution set of $x^2 - 2x - 3 < 0$ consists of the interval(s) where $x^2 - 2x - 3$ is *negative*. This is in interval B, $(-1, 3)$.

The graph of the solution set is $\xleftarrow{\hspace{1em}} \substack{+\!+\!+\!+\!+\diamond+\!+\!+\!+\diamond+\!+ \\ -5\ -4\ -3\ -2\ -1\ \ 0\ \ 1\ \ 2\ \ 3\ \ 4\ \ 5} \xrightarrow{\hspace{1em}}$ .

The solution set is $(-1, 3)$. This is the same as the result we obtained in Example 1 by graphing.

**You Try 2**

Solve $x^2 + 5x + 4 \leq 0$. Graph the solution set and write the solution in interval notation.

Next we will summarize how to solve a quadratic inequality.

> **Procedure    How to Solve a Quadratic Inequality**
>
> **Step 1:**  **Write the inequality in the form $ax^2 + bx + c \leq 0$ or $ax^2 + bx + c \geq 0$.**
> ($<$ and $>$ may be substituted for $\leq$ and $\geq 0$.) If the inequality symbol is $<$ or $\leq$, we are looking for a *negative* quantity in the interval on the number line. If the inequality symbol is $>$ or $\geq$, we are looking for a *positive* quantity in the interval.
>
> **Step 2:**  **Solve** the equation $ax^2 + bx + c = 0$.
>
> **Step 3:**  **Put the solutions of $ax^2 + bx + c = 0$ on a number line.** These values break up the number line into intervals.
>
> **Step 4:**  **Choose a test number in each interval** to determine whether $ax^2 + bx + c$ is positive or negative in each interval. Indicate this on the number line.
>
> **Step 5:**  **If the inequality is in the form $ax^2 + bx + c \leq 0$ or $ax^2 + bx + c < 0$, then the solution set contains the numbers in the interval where $ax^2 + bx + c$ is *negative*. If the inequality is in the form $ax^2 + bx + c \geq 0$ or $ax^2 + bx + c > 0$, then the solution set contains the numbers in the interval where $ax^2 + bx + c$ is *positive*.**
>
> **Step 6:**  **If the inequality symbol is $\leq$ or $\geq$, then the endpoints of the interval(s) (the numbers found in Step 3) are included in the solution set.** Indicate this with brackets in the interval notation.
> **If the inequality symbol is $<$ or $>$, then the endpoints of the interval(s) are not included in the solution set.** Indicate this with parentheses in interval notation.

## 3. Solve Quadratic Inequalities with Special Solutions

We should look carefully at the inequality before trying to solve it. Sometimes, it is not necessary to go through all of the steps.

**Example 3**

Solve.

a)  $(y + 4)^2 \geq -5$    b)  $(t - 8)^2 < -3$

**Solution**

a)  The inequality $(y + 4)^2 \geq -5$ says that a squared quantity, $(y + 4)^2$, is greater than or equal to a *negative* number, $-5$. *This is always true.* (A squared quantity will *always* be greater than or equal to zero.) Any real number, $y$, will satisfy the inequality.

The solution set is $(-\infty, \infty)$.

b)  The inequality $(t - 8)^2 < -3$ says that a squared quantity, $(t - 8)^2$, is less than a *negative* number, $-3$. *There is no real number value for t so that $(t - 8)^2 < -3$.*

The solution set is $\varnothing$. ∎

**You Try 3**

Solve.

a)  $(k + 2)^2 \leq -4$    b)  $(z - 9)^2 > -1$

## 4. Solve an Inequality of Higher Degree

Other polynomial inequalities in factored form can be solved in the same way that we solve quadratic inequalities.

**Example 4**

Solve $(c - 2)(c + 5)(c - 4) < 0$.

**Solution**

This is the factored form of a third-degree polynomial. Since the inequality is $<$, the solution set will contain the intervals where $(c - 2)(c + 5)(c - 4)$ is *negative*.

Solve $(c - 2)(c + 5)(c - 4) = 0$.

$$c - 2 = 0 \quad \text{or} \quad c + 5 = 0 \quad \text{or} \quad c - 4 = 0 \qquad \text{Set each factor equal to 0.}$$
$$c = 2 \quad \text{or} \quad c = -5 \quad \text{or} \quad c = 4 \qquad \text{Solve.}$$

Put $c = 2$, $c = -5$, and $c = 4$ on a number line, and test a number in each interval.

| Interval | $c < -5$ | $-5 < c < 2$ | $2 < c < 4$ | $c > 4$ |
|---|---|---|---|---|
| **Test number** | $c = -6$ | $c = 0$ | $c = 3$ | $c = 5$ |
| **Evaluate** $(c - 2)(c + 5)(c - 4)$ | $(-6 - 2)(-6 + 5)(-6 - 4)$ | $(0 - 2)(0 + 5)(0 - 4)$ | $(3 - 2)(3 + 5)(3 - 4)$ | $(5 - 2)(5 + 5)(5 - 4)$ |
| | $= (-8)(-1)(-10)$ | $= (-2)(5)(-4)$ | $= (1)(8)(-1)$ | $= (3)(10)(1)$ |
| | $= -80$ | $= 40$ | $= -8$ | $= 30$ |
| **Sign** | Negative | Positive | Negative | Positive |

We can see that the intervals where $(c - 2)(c + 5)(c - 4)$ is negative are $(-\infty, -5)$ and $(2, 4)$. The endpoints are not included since the inequality is $<$.

The graph of the solution set is

The solution set of $(c - 2)(c + 5)(c - 4) < 0$ is $(-\infty, -5) \cup (2, 4)$.

 **You Try 4**

Solve $(y + 3)(y - 1)(y + 1) \geq 0$. Graph the solution set and write the solution in interval notation.

**Answers to You Try Exercises**

1)   a)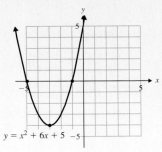
$y = x^2 + 6x + 5$

b) $[-5, -1]$   c) $(-\infty, -5) \cup (-1, \infty)$

2) ←——●——————————→; $[-4, -1]$
  $-5\ -4\ -3\ -2\ -1\quad 0\quad 1\quad 2\quad 3\quad 4\quad 5$

3)   a) $\varnothing$   b) $(-\infty, \infty)$

4) ←————●————●————→; $[-3, -1] \cup [1, \infty)$
  $-5\ -4\ -3\ -2\ -1\quad 0\quad 1\quad 2\quad 3\quad 4\quad 5$

# A.3 Exercises

## Objective 1: Solve a Quadratic Inequality by Graphing

For Exercises 1–4, use the graph of the function to solve each inequality.

1) $y = x^2 + 4x - 5$

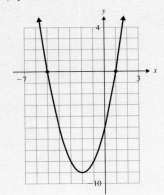

a) $x^2 + 4x - 5 \leq 0$

b) $x^2 + 4x - 5 > 0$

2) $y = x^2 - 6x + 8$

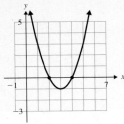

a) $x^2 - 6x + 8 > 0$

b) $x^2 - 6x + 8 \leq 0$

3) $y = -\dfrac{1}{2}x^2 + x + \dfrac{3}{2}$

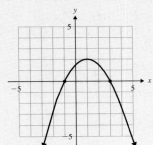

a) $-\dfrac{1}{2}x^2 + x + \dfrac{3}{2} \geq 0$

b) $-\dfrac{1}{2}x^2 + x + \dfrac{3}{2} < 0$

4) $y = -x^2 - 8x - 12$

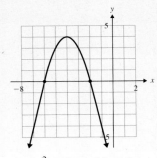

a) $-x^2 - 8x - 12 < 0$

b) $-x^2 - 8x - 12 \geq 0$

## Objective 2: Solve a Quadratic Inequality Using Test Points

Solve each quadratic inequality. Graph the solution set and write the solution in interval notation.

5) $x^2 + 6x - 7 \geq 0$

6) $m^2 - 2m - 24 > 0$

7) $c^2 + 5c < 36$

8) $t^2 + 36 \leq 15t$

VIDEO 9) $r^2 - 13r > -42$

10) $v^2 + 10v < -16$

11) $3z^2 + 14z - 24 \leq 0$

12) $5k^2 + 36k + 7 \geq 0$

13) $7p^2 - 4 > 12p$

14) $4w^2 - 19w < 30$

15) $b^2 - 9b > 0$

16) $c^2 + 12c \leq 0$

17) $4y^2 \leq -5y$

18) $2a^2 \geq 7a$

19) $m^2 - 64 < 0$

20) $p^2 - 144 > 0$

21) $121 - h^2 \leq 0$

22) $1 - d^2 > 0$

VIDEO 23) $144 \geq 9s^2$

24) $81 \leq 25q^2$

## Objective 3: Solve Quadratic Inequalities with Special Solutions

Solve each inequality.

25) $(k + 7)^2 \geq -9$

26) $(h + 5)^2 \geq -2$

27) $(3v - 11)^2 > -20$

28) $(r + 4)^2 < -3$

29) $(2y - 1)^2 < -8$

30) $(4d - 3)^2 > -1$

31) $(n + 3)^2 \leq -10$

32) $(5s - 2)^2 \leq -9$

## Objective 4: Solve an Inequality of Higher Degree

Solve each inequality. Graph the solution set and write the solution in interval notation.

VIDEO 33) $(r + 2)(r - 5)(r - 1) \leq 0$

34) $(b + 2)(b - 3)(b - 12) > 0$

35) $(j - 7)(j - 5)(j + 9) \geq 0$

36) $(m + 4)(m - 7)(m + 1) \leq 0$

37) $(6c + 1)(c + 7)(4c - 3) < 0$

38) $(t + 2)(4t - 7)(5t - 1) \geq 0$

# Answers to Exercises

## Chapter 1

### Section 1.1

1) a) $\dfrac{2}{5}$  b) $\dfrac{2}{3}$  c) 1

3) $\dfrac{1}{2}$

5) a) 1, 2, 3, 6, 9, 18  b) 1, 2, 4, 5, 8, 10, 20, 40  c) 1, 23

7) a) composite  b) composite  c) prime

9) Composite. It is divisible by 2 and has other factors as well.

11) a) $2 \cdot 3 \cdot 3$  b) $2 \cdot 3 \cdot 3 \cdot 3$  c) $2 \cdot 3 \cdot 7$
   d) $2 \cdot 3 \cdot 5 \cdot 5$

13) a) $\dfrac{3}{4}$  b) $\dfrac{3}{4}$  c) $\dfrac{12}{5}$ or $2\dfrac{2}{5}$  d) $\dfrac{3}{7}$

15) a) $\dfrac{6}{35}$  b) $\dfrac{10}{39}$  c) $\dfrac{7}{15}$  d) $\dfrac{12}{25}$  e) $\dfrac{1}{2}$  f) $\dfrac{7}{4}$ or $1\dfrac{3}{4}$

17) She multiplied the whole numbers and multiplied the fractions. She should have converted the mixed numbers to improper fractions before multiplying. Correct answer: $\dfrac{77}{6}$ or $12\dfrac{5}{6}$.

19) a) $\dfrac{1}{12}$  b) $\dfrac{15}{44}$  c) $\dfrac{4}{7}$  d) 7  e) $\dfrac{24}{7}$ or $3\dfrac{3}{7}$  f) $\dfrac{1}{14}$

21) 30  23) a) 30  b) 24  c) 36

25) a) $\dfrac{8}{11}$  b) $\dfrac{3}{5}$  c) $\dfrac{3}{5}$  d) $\dfrac{7}{18}$  e) $\dfrac{29}{30}$  f) $\dfrac{1}{18}$
   g) $\dfrac{71}{63}$ or $1\dfrac{8}{63}$  h) $\dfrac{7}{12}$  i) $\dfrac{7}{5}$ or $1\dfrac{2}{5}$  j) $\dfrac{41}{54}$

27) a) $14\dfrac{7}{11}$  b) $11\dfrac{2}{5}$  c) $6\dfrac{1}{2}$  d) $5\dfrac{9}{20}$  e) $1\dfrac{2}{5}$  f) $4\dfrac{13}{40}$
   g) $11\dfrac{5}{28}$  h) $8\dfrac{9}{20}$

29) four bears; $\dfrac{1}{3}$ yd remaining  31) 50

33) $16\dfrac{1}{2}$ in. by $22\dfrac{5}{8}$ in.  35) $3\dfrac{5}{12}$ cups  37) $5\dfrac{3}{20}$ gal

39) 35  41) $7\dfrac{23}{24}$ in.  43) 240

### Section 1.2

1) a) base: 6; exponent: 4  b) base: 2; exponent: 3
   c) base: $\dfrac{9}{8}$; exponent: 5

3) a) $9^4$  b) $2^8$  c) $\left(\dfrac{1}{4}\right)^3$

5) a) 64  b) 121  c) 16  d) 125  e) 81  f) 144  g) 1
   h) $\dfrac{9}{100}$  i) $\dfrac{1}{64}$  j) 0.09

7) $(0.5)^2 = 0.5 \cdot 0.5 = 0.25$ or $(0.5)^2 = \left(\dfrac{1}{2}\right)^2 = \dfrac{1}{4}$

9) Answers may vary.

11) 19  13) 38  15) 23  17) 17  19) $\dfrac{13}{20}$  21) $\dfrac{19}{18}$ or $1\dfrac{1}{18}$

23) 4  25) 15  27) 19  29) 37  31) 11  33) $\dfrac{5}{6}$

35) $\dfrac{27}{7}$ or $3\dfrac{6}{7}$

### Section 1.3

1) acute  3) straight  5) supplementary; complementary

7) 31°  9) 78°  11) 37°  13) 142°

15) $m\angle A = m\angle C = 149°, m\angle B = 31°$  17) 180

19) 39°; obtuse  21) 39°; right  23) equilateral  25) isosceles

27) true  29) $A = 80$ ft$^2$; $P = 36$ ft

31) $A = 42$ cm$^2$; $P = 29.25$ cm

33) $A = 42.25$ mi$^2$; $P = 26$ mi  35) $A = 162$ in$^2$; $P = 52$ in.

37) a) $A = 25\pi$ in$^2$; $A \approx 78.5$ in$^2$  b) $C = 10\pi$ in; $C \approx 31.4$ in.

39) a) $A = 6.25\pi$ m$^2$; $A \approx 19.625$ m$^2$
   b) $C = 5\pi$ m; $C = 15.7$ m

41) $A = \dfrac{1}{4}\pi$ m$^2$; $C = \pi$ m  43) $A = 49\pi$ ft$^2$; $C = 14\pi$ ft

45) $A = 376$ m$^2$; $P = 86$ m  47) $A = 201.16$ in$^2$; $P = 67.4$ in.

49) 88 in$^2$  51) 25.75 ft$^2$  53) 177.5 cm$^2$  55) 70 m$^3$

57) $288\pi$ in$^3$  59) $\dfrac{500}{3}\pi$ ft$^3$  61) $136\pi$ cm$^3$

63) a) 58.5 ft$^2$  b) No, it would cost $1170 to use this glass.

65) a) 226.08 ft$^3$  b) 1691 gal  67) a) 62.8 in.  b) 314 in$^2$

69) 6395.4 gal

71) No. This granite countertop would cost $2970.00.

73) 28.9 in.  75) a) 44 ft$^2$  b) $752  77) 140.4 in$^2$

79) 1205.76 ft$^3$

## Section 1.4

1) Answers may vary.

3) a) 17   b) 17, 0   c) 17, 0, −25

   d) $17, 3.8, \dfrac{4}{5}, 0, -25, 6.\overline{7}, -2\dfrac{1}{8}$

   e) $\sqrt{10}, 9.721983\ldots$   f) all numbers in the set

5) true   7) false   9) true

11)
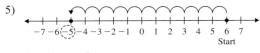

13)
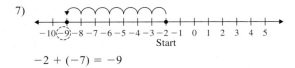

15) the distance of the number from zero   17) −8

19) 15   21) $\dfrac{3}{4}$   23) 10   25) $\dfrac{9}{4}$   27) −14   29) 13

31) $-4\dfrac{1}{7}$   33) $-10, -2, 0, \dfrac{9}{10}, 3.8, 7$

35) $-6.51, -6.5, -5, 2, 7\dfrac{1}{3}, 7\dfrac{5}{6}$   37) true   39) false

41) false   43) false   45) −53   47) 1.4 million

49) −419,000

## Section 1.5

1) Answers may vary.   3) Answers may vary.

5)

$6 - 11 = -5$

7)

$-2 + (-7) = -9$

9) −7   11) −14   13) 23   15) −11   17) −850   19) $\dfrac{1}{6}$

21) $-\dfrac{25}{24}$ or $-1\dfrac{1}{24}$   23) $-\dfrac{8}{45}$   25) 2.7   27) −9.23

29) 15   31) −11   33) −2   35) −19   37) 23

39) $-\dfrac{19}{18}$ or $-1\dfrac{1}{18}$   41) $\dfrac{5}{24}$   43) 12   45) 11   47) −7

49) false   51) false   53) true

55) $-18 + 6 = -12$. His score in the 2005 Masters was −12.

57) $6,110,000 - 5,790,000 = 320,000$. The carbon emissions of China were 320,000 thousand metric tons more than those of the United States.

59) $881,566 + 45,407 = 926,973$. There were 926,973 flights at O' Hare in 2007.

61) a) −3   b) −134   c) −131   d) −283

63) a) −0.7   b) 0.4   c) 0.1   d) −0.2

65) $5 + 7$; 12   67) $10 - 16$; −6   69) $9 - (-8)$; 17

71) $-21 + 13$; −8   73) $-20 + 30$; 10   75) $23 - 19$; 4

77) $(-5 + 11) - 18$; −12

## Section 1.6

1) negative   3) −56   5) 45   7) 84   9) $-\dfrac{2}{15}$   11) 1.4

13) 135   15) −84   17) when $k$ is negative

19) when $k \neq 0$   21) 36   23) −125   25) 9   27) −49

29) −32   31) positive   33) 10   35) −4   37) −8

39) $\dfrac{10}{13}$   41) 0   43) $-\dfrac{3}{2}$ or $-1\dfrac{1}{2}$   45) −33   47) 43

49) 16   51) 16   53) $\dfrac{1}{4}$   55) $-12 \cdot 6$; −72

57) $(-7)(-5) + 9$; 44   59) $\dfrac{63}{-9} + 7$; 0

61) $(-4)(-8) - 19$; 13   63) $\dfrac{-100}{4} - (-7 + 2)$; −20

65) $2[18 + (-31)]$; −26   67) $\dfrac{2}{3}(-27)$; −18

69) $12(-5) + \dfrac{1}{2}(36)$; −42

## Section 1.7

1)

| Term | Coeff. |
|------|--------|
| $7p^2$ | 7 |
| $-6p$ | −6 |
| 4 | 4 |

The constant is 4.

3)

| Term | Coeff. |
|------|--------|
| $x^2y^2$ | 1 |
| $2xy$ | 2 |
| $-y$ | −1 |
| 11 | 11 |

The constant is 11.

5)

| Term | Coeff. |
|------|--------|
| $-2g^5$ | −2 |
| $\dfrac{g^4}{5}$ | $\dfrac{1}{5}$ |
| $3.8g^2$ | 3.8 |
| $g$ | 1 |
| −1 | −1 |

The constant is −1.

7) a) 11  b) $-17$  9) $-17$  11) 0  13) $-3$  15) $\dfrac{3}{2}$

17) No. The exponents are different.  19) Yes. Both are $a^3b$-terms.

21) 1  23) $-5$  25) distributive  27) identity

29) commutative  31) associative  33) $19 + p$

35) $(8 + 1) + 9$  37) $3k - 21$  39) $y$

41) No. Subtraction is not commutative.

43) $2 \cdot 1 + 2 \cdot 9 = 2 + 18 = 20$

45) $-2 \cdot 5 + (-2) \cdot 7 = -10 + (-14) = -24$

47) $4 \cdot 8 - 4 \cdot (3) = 32 - 12 = 20$

49) $-10 + 4 = -6$

51) $8y + 8 \cdot 3 = 8y + 24$  53) $-10z + (-10) \cdot 6 = -10z - 60$

55) $-3x - (-3) \cdot (4y) - (-3) \cdot (6) = -3x + 12y + 18$

57) $8c - 9d + 14$  59) $24p + 7$  61) $-19y^2 + 30$

63) $\dfrac{16}{5}r - \dfrac{2}{9}$  65) $7w + 10$  67) 0  69) $-5g + 2$  71) $-3t$

73) $26x + 37$  75) $\dfrac{11}{10}z + \dfrac{13}{2}$  77) $\dfrac{65}{8}t - \dfrac{19}{16}$

79) $-1.1x - 19.6$  81) $x + 18$  83) $x - 6$  85) $x - 3$

87) $12 + 2x$  89) $(3 + 2x) - 7;\ 2x - 4$

91) $(x + 15) - 5;\ x + 10$

**Chapter 1 Review Exercises**

1) a) 1, 2, 4, 8, 16  b) 1, 37  3) a) $\dfrac{2}{5}$  b) $\dfrac{23}{39}$  5) $\dfrac{3}{10}$

7) 40  9) $\dfrac{3}{14}$  11) $\dfrac{11}{12}$  13) $\dfrac{7}{10}$  15) $\dfrac{19}{56}$  17) $6\dfrac{13}{24}$  19) 81

21) $\dfrac{27}{64}$  23) 10  25) $\dfrac{1}{4}$  27) $102°$

29) $A = 6\dfrac{9}{16}$ mi$^2$; $P = 10\dfrac{3}{4}$ mi  31) $A = 100$ in$^2$; $P = 40$ in.

33) a) $A = 9\pi$ in$^2$; $A \approx 28.26$ in$^2$
   b) $C = 6\pi$ in.; $C \approx 18.84$ in.

35) $360.66$ cm$^2$  37) $1.3\pi$ ft$^3$  39) $\dfrac{125}{8}$ in$^3$ or $15\dfrac{5}{8}$ in$^3$

41) a) $\{-16, 0, 4\}$  b) $\left\{\dfrac{7}{15}, -16, 0, 3.\overline{2}, 8.5, 4\right\}$

   c) $\{4\}$  d) $\{0, 4\}$  e) $\{\sqrt{31}, 6.01832...\}$

43) a) 18  b) $-7$  45) 19  47) $-\dfrac{5}{24}$  49) $-12$  51) 72

53) $-12$  55) $\dfrac{15}{4}$ or $3\dfrac{3}{4}$  57) $-36$  59) 64  61) 27

63) $-9$  65) $-31$  67) $\dfrac{-120}{-3};\ 40$  69) $(-4) \cdot 7 - 15;\ -43$

71)

| Term | Coeff. |
|---|---|
| $5z^4$ | 5 |
| $-8z^3$ | $-8$ |
| $\dfrac{3}{5}z^2$ | $\dfrac{3}{5}$ |
| $-z$ | $-1$ |
| 14 | 14 |

73) $\dfrac{1}{52}$  75) inverse  77) distributive

79) $7 \cdot 3 - 7 \cdot 9 = 21 - 63 = -42$  81) $-15 + 3 = -12$

83) $12m - 10$  85) $17y^2 - 3y - 3$  87) $\dfrac{31}{4}n - \dfrac{9}{2}$

**Chapter 1 Test**

1) $2 \cdot 3 \cdot 5 \cdot 7$  2) a) $\dfrac{5}{8}$  b) $\dfrac{3}{4}$  3) $\dfrac{5}{24}$  4) $\dfrac{23}{36}$  5) $7\dfrac{5}{12}$

6) $\dfrac{1}{27}$  7) $-\dfrac{1}{4}$  8) $-17$  9) 20  10) $-\dfrac{1}{12}$  11) 60

12) $-3.7$  13) $-49$  14) $\dfrac{2}{3}$  15) 15,211 ft

16) a) 125  b) $-16$  c) 43  d) $-37$  17) $149°$

18) $49°$; acute

19) a) $A = 9$ mm$^2$; $P = 14.6$ mm  b) $A = 105$ cm$^2$; $P = 44$ cm
   c) $A = 200$ in$^2$; $P = 68$ in.

20) 9 ft$^3$  21) a) $81\pi$ ft$^2$  b) $254.34$ ft$^2$

22) a) 22, 0  b) 22  c) $\sqrt{43}$, $8.0934...$

   d) $22, -7, 0$  e) $3\dfrac{1}{5}, 22\ -7, 0, 6.2, 1.\overline{5}$

23)

24) a) $-4 + 27;\ 23$  b) $17 - 5(-6);\ 47$

25)

| Term | Coeff. |
|---|---|
| $4p^3$ | 4 |
| $-p^2$ | $-1$ |
| $\dfrac{1}{3}p$ | $\dfrac{1}{3}$ |
| $-10$ | $-10$ |

26) $\dfrac{1}{3}$  27) a) commutative  b) associative

   c) inverse  d) distributive

28) a) $(-4) \cdot 2 + (-4) \cdot 7 = -8 + (-28) = -36$
   b) $3 \cdot 8m - 3 \cdot 3n + 3 \cdot 11 = 24m - 9n + 33$

29) a) $-6k^2 + 4k - 14$  b) $6c - \dfrac{49}{6}$  30) $2x - 9$

# Chapter 2

## Section 2.1A

1) $9^6$   3) $\left(\frac{1}{7}\right)^4$   5) $(-5)^7$   7) $(-3y)^8$   9) base: 6; exponent: 8

11) base: 0.05; exponent: 7   13) base: $-8$; exponent: 5

15) base: $9x$; exponent: 8   17) base: $-11a$; exponent: 2

19) base: $p$; exponent: 4   21) base: $y$; exponent: 2

23) $(3+4)^2 = 49$, $3^2 + 4^2 = 25$. They are not equivalent because when evaluating $(3+4)^2$, first add $3+4$ to get 7, then square the 7.

25) Answers may vary.   27) No. $3t^4 = 3 \cdot t^4$; $(3t)^4 = 3^4 \cdot t^4 = 81t^4$

29) 32   31) 121   33) 16   35) $-81$   37) $-8$   39) $\frac{1}{125}$

41) 32   43) 81   45) 200   47) $\frac{1}{64}$   49) $8^{12}$   51) $5^{11}$

53) $(-7)^8$   55) $b^6$   57) $k^6$   59) $8y^5$   61) $54m^{15}$   63) $-42r^5$

65) $28t^{16}$   67) $-40x^6$   69) $8b^{15}$   71) $y^{12}$   73) $w^{77}$   75) 729

77) $(-5)^6$   79) $\frac{1}{81}$   81) $\frac{36}{a^2}$   83) $\frac{m^5}{n^5}$   85) $10{,}000y^4$

87) $81p^4$   89) $-64a^3b^3$   91) $6x^3y^3$   93) $-9t^4u^4$

95) a) $A = 3w^2$ sq units; $P = 8w$ units
   b) $A = 5k^5$ sq units; $P = 10k^3 + 2k^2$ units

97) $\frac{3}{8}x^2$ sq units

## Section 2.1B

1) operations   3) $k^{24}$   5) $200z^{26}$   7) $-6a^{31}b^7$   9) 121

11) $-64t^{18}u^{26}$   13) $288k^{14}l^4$   15) $\frac{3}{4g^{15}}$   17) $\frac{49}{4}n^{22}$   19) $900h^{28}$

21) $-147w^{45}$   23) $\frac{36x^6}{25y^{10}}$   25) $\frac{d^{18}}{4c^{30}}$   27) $\frac{2a^{36}b^{63}}{9c^2}$   29) $\frac{r^{39}}{242t^5}$

31) $\frac{2}{3}x^{24}y^{14}$   33) $-\frac{1}{10}c^{29}d^{18}$   35) $\frac{125x^{15}y^6}{z^{12}}$   37) $\frac{81t^{16}u^{36}}{16v^{28}}$

39) $\frac{9w^{10}}{x^6y^{12}}$   41) a) $20l^2$ units   b) $25l^4$ sq units

43) a) $\frac{3}{8}x^2$ sq units   b) $\frac{11}{4}x$ units

## Section 2.2A

1) false   3) true   5) 1   7) $-1$   9) 0   11) 2   13) $\frac{1}{36}$

15) $\frac{1}{16}$   17) $\frac{1}{125}$   19) 64   21) 32   23) $\frac{27}{64}$   25) $\frac{49}{81}$

27) $-64$   29) $\frac{64}{9}$   31) $-\frac{1}{64}$   33) $-1$   35) $\frac{1}{16}$

37) $\frac{13}{36}$   39) $\frac{83}{81}$

## Section 2.2B

1) a) $w$   b) $n$   c) $2p$   d) $c$   3) 1   5) $-2$   7) 2   9) $\frac{1}{d^3}$

11) $\frac{1}{p}$   13) $\frac{b^3}{a^{10}}$   15) $\frac{x^5}{y^8}$   17) $\frac{t^5u^3}{8}$   19) $\frac{5m^6}{n^2}$   21) $2t^{11}u^5$

23) $\frac{8a^6c^{10}}{5bd}$   25) $2x^7y^6z^4$   27) $\frac{36}{a^2}$   29) $\frac{q^5}{32n^5}$   31) $\frac{c^2d^2}{144b^2}$

33) $-\frac{9}{k^2}$   35) $\frac{3}{t^3}$   37) $-\frac{1}{m^9}$   39) $z^{10}$   41) $j$   43) $5n^2$   45) $cd^3$

## Section 2.3

1) You must subtract the denominator's exponent from the numerator's exponent; $a^2$.

3) $d^5$   5) $m^4$   7) $8t^7$   9) 36   11) 81   13) $\frac{1}{16}$   15) $\frac{1}{125}$

17) $10d^2$   19) $\frac{2}{3}c^5$   21) $\frac{1}{y^5}$   23) $\frac{1}{x^9}$   25) $\frac{1}{t^3}$   27) $\frac{1}{a^{10}}$   29) $t^3$

31) $\frac{15}{w^8}$   33) $-\frac{6}{k^3}$   35) $a^3b^7$   37) $\frac{2k^3}{3t^8}$   39) $\frac{10}{x^5y^5}$

41) $\frac{w^6}{9v^3}$   43) $\frac{3}{8}c^4d$   45) $(x+y)^7$   47) $(c+d)^6$

## Putting It All Together

1) $\frac{16}{81}$   2) 64   3) 1   4) $-125$   5) $\frac{9}{100}$   6) $\frac{49}{9}$   7) 9

8) $-125$   9) $\frac{1}{100}$   10) $\frac{1}{8}$   11) $\frac{1}{32}$   12) 81   13) $-\frac{27}{125}$

14) 64   15) $\frac{1}{36}$   16) $\frac{13}{36}$   17) $270g^{12}$   18) $56d^9$   19) $\frac{33}{s^{11}}$

20) $\frac{1}{c^5}$   21) $\frac{16}{81}x^{40}y^{24}$   22) $\frac{a^9b^{15}}{1000}$   23) $\frac{n^6}{81m^{16}}$   24) $\frac{r^8s^{24}}{81}$

25) $-b^{15}$   26) $h^{88}$   27) $-27m^{15}n^6$   28) $169a^{12}b^2$

29) $-6z^3$   30) $-9w^9$   31) $\frac{t^{18}}{s^{42}}$   32) $\frac{1}{m^3n^{14}}$   33) $a^{14}b^3c^7$

34) $\frac{4}{9v^{10}}$   35) $\frac{27u^{30}}{64v^{21}}$   36) $\frac{81}{x^6y^8}$   37) $-27t^6u^{15}$

38) $\frac{1}{144}k^{16}m^2$   39) $\frac{1}{h^{18}}$   40) $-\frac{1}{d^{20}}$   41) $\frac{h^4}{16}$   42) $\frac{13}{f^2}$

43) $56c^{10}$   44) $80p^{15}$   45) $\frac{3}{a^5}$   46) $\frac{1}{9r^2s^2}$   47) $\frac{6}{55}r^{10}$

48) $\frac{1}{f^{36}}$   49) $\frac{a^2b^9}{c}$   50) $\frac{x^9y^{24}}{z^3}$   51) $\frac{72n^3}{5m^5}$   52) $\frac{t^7}{100s^7}$   53) 1

54) 1   55) $\frac{9}{49d^6}$   56) $\frac{1}{100x^{10}y^8}$   57) $p^{12c}$   58) $25d^{8t}$

59) $y^{4m}$   60) $x^{4c}$   61) $\frac{1}{t^{3b}}$   62) $\frac{1}{a^{7y}}$   63) $\frac{5}{8c^{7x}}$   64) $-\frac{3}{8y^{8a}}$

## Section 2.4

1) yes  3) no  5) no  7) yes  9) Answers may vary.

11) Answers may vary.  13) 7176.5  15) 0.0406

17) 0.1200006  19) −0.000068  21) −52,600  23) 0.000008

25) 602,196.7  27) 3,000,000  29) −0.000744

31) 24,428,000  33) 0.000000000025 meters

35) $2.1105 \times 10^3$  37) $9.6 \times 10^{-5}$  39) $-7 \times 10^6$

41) $3.4 \times 10^3$  43) $8 \times 10^{-4}$  45) $-7.6 \times 10^{-2}$  47) $6 \times 10^3$

49) $3.808 \times 10^8$ kg  51) $1 \times 10^{-8}$ cm  53) 30,000

55) 690,000  57) −1200  59) −0.06  61) −0.0005

63) 160,000  65) 0.0001239  67) 5,256,000,000 particles

69) 17,000 lb/cow  71) 26,400,000 droplets  73) $6083

75) $1.34784 \times 10^9$ m  77) 20 metric tons

## Chapter 2 Review Exercises

1) a) $8^6$  b) $(-7)^4$  3) a) 32  b) $\dfrac{1}{27}$  c) $7^{12}$  d) $k^{30}$

5) a) $125y^3$  b) $-14m^{16}$  c) $\dfrac{a^6}{b^6}$  d) $6x^2y^2$  e) $\dfrac{25}{3}c^8$

7) a) $z^{22}$  b) $-18c^{10}d^{16}$  c) 125  d) $\dfrac{25t^6}{2u^{21}}$

9) a) 1  b) −1  c) $\dfrac{1}{9}$  d) $-\dfrac{5}{36}$  e) $\dfrac{125}{64}$

11) a) $\dfrac{1}{v^9}$  b) $\dfrac{c^2}{81}$  c) $y^8$  d) $-\dfrac{7}{k^9}$  e) $\dfrac{19a}{z^4}$  f) $\dfrac{20n^5}{m^6}$  g) $\dfrac{k^5}{32j^5}$

13) a) 9  b) $r^8$  c) $\dfrac{3}{2t^5}$  d) $\dfrac{3x^7}{5y}$

15) a) $81s^{16}t^{20}$  b) $2a^{16}$  c) $\dfrac{y^{18}}{z^{24}}$  d) $-36x^{11}y^{11}$  e) $\dfrac{d^{25}}{c^{35}}$

  f) $8m^3n^{12}$  g) $\dfrac{125t^9}{27k^{18}}$  h) 14

17) a) $y^{10k}$  b) $x^{10p}$  c) $z^{7c}$  d) $\dfrac{1}{t^{5d}}$  19) −418.5

21) 0.00067  23) 20,000  25) $5.75 \times 10^{-5}$  27) $3.2 \times 10^7$

29) $1.78 \times 10^5$  31) $9.315 \times 10^{-4}$  33) 0.0000004

35) 3.6  37) 7500  39) 30,000 quills

41) 0.00000000000000299 g  43) 25,740,000

## Chapter 2 Test

1) $(-3)^3$  2) $x^5$  3) 125  4) $\dfrac{1}{x^7}$  5) $8^{36}$  6) $p^5$  7) 81

8) 1  9) $\dfrac{1}{32}$  10) $\dfrac{3}{16}$  11) $-\dfrac{27}{64}$  12) $\dfrac{49}{100}$  13) $125n^{18}$

14) $-30p^{12}$  15) $m^6$  16) $\dfrac{a^4}{b^6}$  17) $-\dfrac{t^{33}}{27u^{27}}$  18) $\dfrac{8}{y^9}$

19) 1  20) $2m + n$  21) $\dfrac{3a^4c^2}{5b^3d^3}$  22) $y^{18}$  23) $t^{13k}$

24) 728,300  25) $1.65 \times 10^{-4}$  26) −50,000

27) 21,800,000  28) 0.00000000000000000182 g

## Cumulative Review for Chapters 1–2

1) $\dfrac{3}{5}$  2) $\dfrac{7}{12}$  3) $\dfrac{7}{25}$  4) 12  5) −28  6) −81  7) −1

8) 42  9) a) $346\dfrac{2}{3}$yd  b) $11,520  10) 62  11) $V = \dfrac{4}{3}\pi r^3$

12) a) −4, 3  b) $\sqrt{11}$  c) 3  d) $-4, 3, -2.1\overline{3}, 2\dfrac{2}{3}$  e) 3

13) 261  14) $\dfrac{9}{2}m - 15n + \dfrac{21}{4}$  15) $-3t^2 + 37t - 25$

16) $\dfrac{1}{2}x - 13$  17) $4^{10}$  18) $\dfrac{y^3}{x^3}$  19) $\dfrac{1}{4x^5}$  20) $-\dfrac{81r^4}{t^{12}}$

21) $-28z^8$  22) $\dfrac{1}{n^7}$  23) $-\dfrac{32b^5}{a^{30}}$  24) $7.29 \times 10^{-4}$

25) 58,280

# Chapter 3

## Section 3.1

1) expression  3) equation  5) No, it is an expression.

7) b, c  9) no  11) yes  13) yes  15) {17}  17) {−6}

19) {−4}  21) {0}  23) {−0.7}  25) {4.8}  27) $\left\{-\dfrac{1}{2}\right\}$

29) $\left\{\dfrac{17}{20}\right\}$  31) Answers may vary.  33) {4}  35) {−7}

37) {12}  39) $\left\{-\dfrac{9}{4}\right\}$  41) {0.23}  43) {14}  45) {−1}

47) {6.5}  49) {48}  51) {−39}  53) $\left\{-\dfrac{9}{2}\right\}$  55) {−45}

57) {84}  59) $\left\{\dfrac{5}{3}\right\}$  61) $\left\{-\dfrac{11}{5}\right\}$  63) {18}  65) {−14}

67) {7}  69) {−1}  71) {0}  73) $\left\{-\dfrac{2}{5}\right\}$  75) $\left\{\dfrac{8}{3}\right\}$

77) {−1}  79) {10}  81) {−5}  83) $\left\{-\dfrac{33}{5}\right\}$  85) $\left\{\dfrac{24}{5}\right\}$

87) $\left\{-\dfrac{3}{2}\right\}$  89) {−18}  91) {2.7}  93) {−6.2}  95) $\left\{-\dfrac{2}{5}\right\}$

97) {9}  99) {54}  101) {−49}  103) {56}  105) {−1.4}

107) {−30}  109) $\left\{\dfrac{16}{5}\right\}$

## Section 3.2

1) Step 1: Clear parentheses and combine like terms on each side of the equation. Step 2: Isolate the variable. Step 3: Solve for the variable. Step 4: Check the solution.

3) Combine like terms; $8x + 11 - 11 = 27 - 11$; Combine like terms; $\dfrac{8x}{8} = \dfrac{16}{8}$; $x = 2$; $\{2\}$

5) $\{4\}$   7) $\{-11\}$   9) $\left\{-\dfrac{5}{2}\right\}$   11) $\{19\}$   13) $\left\{-\dfrac{5}{4}\right\}$

15) $\{3\}$   17) $\{-2\}$   19) $\{-3\}$   21) $\{5\}$   23) $\left\{-\dfrac{7}{4}\right\}$

25) $\{0\}$   27) $\left\{-\dfrac{4}{7}\right\}$   29) $\{6\}$   31) $\{-1\}$   33) $\{2\}$

35) $\left\{\dfrac{1}{4}\right\}$   37) $\{-5\}$   39) $\{1\}$   41) $\left\{-\dfrac{5}{2}\right\}$   43) $\{7\}$

45) $\{-6\}$   47) $\{15\}$   49) $\{-3\}$   51) $\{1\}$   53) $\{0\}$

55) $\left\{\dfrac{1}{2}\right\}$   57) $\{8\}$   59) $\left\{\dfrac{19}{7}\right\}$

## Section 3.3

1) Eliminate the fractions by multiplying both sides of the equation by the LCD of all the fractions in the equation.

3) Multiply both sides of the equation by 8.   5) $\{5\}$   7) $\{4\}$

9) $\left\{-\dfrac{2}{3}\right\}$   11) $\{-8\}$   13) $\left\{\dfrac{20}{9}\right\}$   15) $\{-1\}$   17) $\{3\}$

19) $\{-20\}$   21) $\{2\}$   23) $\{-0.15\}$   25) $\{10\}$   27) $\{600\}$

29) The variable is eliminated, and you get a false statement like $5 = -12$.   31) $\varnothing$   33) {all real numbers}   35) $\varnothing$

37) $\{100\}$   39) $\{25\}$   41) $\{6\}$   43) $\{16\}$   45) $\left\{-\dfrac{2}{3}\right\}$

47) $\{8\}$   49) {all real numbers}   51) $\left\{-\dfrac{41}{10}\right\}$   53) $\{0\}$

55) $\{6000\}$   57) $\varnothing$   59) $\left\{\dfrac{3}{4}\right\}$

61) 1. Read the problem until you understand it.   2. Choose a variable to represent an unknown quantity.   3. Translate the problem from English into an equation.   4. Solve the equation.   5. Check the answer in the original problem and interpret the solution as it relates to the problem.

63) $x + 12 = 5$; $-7$   65) $x - 9 = 12$; 21

67) $2x + 5 = 17$; 6   69) $2x + 11 = 13$; 1

71) $3x - 8 = 40$; 16   73) $\dfrac{3}{4}x = 33$; 44   75) $\dfrac{1}{2}x - 9 = 3$; 24

77) $2x - 3 = x + 8$; 11   79) $\dfrac{1}{3}x + 10 = x - 2$; 18

81) $x - 24 = \dfrac{x}{9}$; 27   83) $x + \dfrac{2}{3}x = 25$; 15

85) $x - 2x = 13$; $-13$

## Section 3.4

1) $c + 14$   3) $c - 37$   5) $\dfrac{1}{2}s$   7) $14 - x$   9) The number of children must be a whole number.   11) It is an even number.

13) 1905: 4.2 inches; 2004: 3.0 inches   15) Lance: 7; Miguel: 5

17) regular: 260 mg; decaf: 20 mg   19) Spanish: 186; French: 124

21) 11 in., 25 in.   23) bracelet: 9.5 in.; necklace: 19 in.

25) Derek: 2 ft; Cory: 3 ft; Tamara: 1 ft   27) 41, 42, 43

29) 18, 20   31) $-15, -13, -11$   33) 107, 108

35) Jimmy: 13; Kelly: 7   37) 5 ft, 11 ft

39) Bonnaroo: 70,000; Lollapalooza: 225,000

41) 57, 58, 59   43) Helen: 140 lb; Tara: 155 lb; Mike: 207 lb

45) 12 in., 24 in., 36 in.   47) Lil Wayne: 2.88 million; Coldplay: 2.15 million; Taylor Swift: 2.11 million

49) 72, 74, 76

## Section 3.5

1) $42.50   3) $20.65   5) $19.60   7) $140.00   9) $17.60

11) $66.80   13) 1800 acres   15) 1015   17) 4400 people

19) $9   21) $6955   23) $315   25) $9000 at 6% and $6000 at 7%   27) $1650 at 6% and $2100 at 5%   29) $3000 at 9.5% and $4500 at 6.5%   31) 3 oz   33) 8.25 mL

35) 16 oz of the 4% acid solution, 8 oz of the 10% acid solution

37) 3 L   39) 2 lb   41) 250 g   43) $1\dfrac{1}{5}$ gallons   45) $8.75

47) 8500   49) CD: $2000; IRA: $4000; mutual fund: $3000

51) $32.00   53) $38,600   55) 9%: 3 oz; 17%: 9 oz

57) peanuts: 7 lb; cashews: 3 lb   59) 4%: $9000; 7%: $11,000

61) 16 oz of orange juice, 60 oz of the fruit drink   63) 2,100,000

## Section 3.6

1) No. The height of a triangle cannot be a negative number.

3) cubic centimeters   5) $\dfrac{11}{4}$   7) 3000   9) 2.5   11) $9.2\pi$

13) 4   15) 4   17) 9   19) 4   21) 10   23) 78 ft   25) 8 in.

27) 314 yd$^2$   29) 67 mph   31) 24 in.   33) 3 ft   35) 2.5%

37) 18 in. $\times$ 28 in.   39) 12 ft $\times$ 19 ft   41) 2 in., 8 in.

43) 1.5 ft, 1.5 ft, 2.5 ft   45) $m\angle A = 35°, m\angle C = 62°$

47) $m\angle A = 26°, m\angle B = 52°$   49) $m\angle A = 44°,$ $m\angle B = m\angle C = 68°$   51) 43°, 43°   53) 172°, 172°

55) 38°, 38°   57) 144°, 36°   59) 120°, 60°   61) 73°, 107°

63) $180 - x$   65) 63°   67) 24°   69) angle: 20°; comp: 70°; supp: 160°   71) 72°   73) 45°   75) a) $x = 21$   b) $x = y - h$   c) $x = c - r$   77) a) $c = 7$   b) $c = \dfrac{d}{a}$   c) $c = \dfrac{v}{m}$

79) a) $x = 44$   b) $a = ry$   c) $a = dw$

81) a) $d = 3$   b) $d = \dfrac{z + a}{k}$   83) a) $h = -\dfrac{2}{3}$   b) $h = \dfrac{n - v}{q}$

85) $m = \dfrac{F}{a}$   87) $c = nv$   89) $\sigma = \dfrac{E}{T^4}$   91) $h = \dfrac{3V}{\pi r^2}$

93) $E = IR$   95) $R = \dfrac{I}{PT}$   97) $l = \dfrac{P - 2w}{2}$ or $l = \dfrac{P}{2} - w$

99) $N = \dfrac{2.5H}{D^2}$   101) $b_2 = \dfrac{2A}{h} - b_1$ or $b_2 = \dfrac{2A - hb_1}{h}$

103) $h^2 = \dfrac{S}{\pi} - \dfrac{c^2}{4}$ or $h^2 = \dfrac{1}{4}\left(\dfrac{4S}{\pi} - c^2\right)$

105) a) $w = \dfrac{P - 2l}{2}$ or $w = \dfrac{P}{2} - l$   b) 3 cm

107) a) $F = \dfrac{9}{5}C + 32$   b) 68°F

## Section 3.7

1) Answers may vary, but some possible answers are $\dfrac{6}{8}, \dfrac{9}{12},$ and $\dfrac{12}{16}$.

3) Yes, a percent can be written as a fraction with a denominator of 100. For example, 25% can be written as $\dfrac{25}{100}$ or $\dfrac{1}{4}$.   5) $\dfrac{4}{3}$

7) $\dfrac{2}{25}$   9) $\dfrac{1}{4}$   11) $\dfrac{2}{3}$   13) $\dfrac{3}{8}$   15) package of 8: $0.786 per battery   17) 48-oz jar: $0.177 per oz

19) 24-oz box: $0.262 per oz   21) A ratio is a quotient of two quantities. A proportion is a statement that two ratios are equal.

23) true   25) false   27) true   29) {2}   31) {40}   33) {18}

35) $\left\{\dfrac{8}{3}\right\}$   37) {−2}   39) {11}   41) {−1}   43) $\left\{\dfrac{5}{2}\right\}$

45) $3.54   47) $\dfrac{1}{2}$ cup   49) 82.5 mg   51) 360   53) 8 lb

55) 35.75 Euros   57) $x = 10$   59) $x = 13$   61) $x = 63$

63) a) $0.70   b) 70¢   65) a) $4.22   b) 422¢

67) a) $2.20   b) 220¢   69) a) $0.25q$   b) $25q$

71) a) $0.10d$   b) $10d$   73) a) $0.01p + 0.05n$   b) $p + 5n$

75) 9 dimes, 17 quarters   77) 11 $5 bills, 18 $1 bills

79) 19 adults, 38 children   81) Marc Anthony: $86; Santana: $66.50   83) miles   85) northbound: 200 mph; southbound: 250 mph   87) 5 hours   89) $1\dfrac{2}{3}$ hr

91) 36 minutes   93) Nick: 14 mph; Scott: 12 mph

95) passenger train: 50 mph; freight train: 30 mph

97) 7735 yen

99) $\dfrac{1}{4}$ hour or 15 min   101) 27 dimes, 16 quarters

103) jet: 400 mph; small plane: 200 mph   105) 240

## Section 3.8

1) Use brackets when there is a ≤ or ≥ symbol.

3) $(-\infty, 4)$   5) $[-3, \infty)$

7)   a) $\{k \mid k \le 2\}$   b) $(-\infty, 2]$

9)   a) $\left\{c \mid c < \dfrac{5}{2}\right\}$   b) $\left(-\infty, \dfrac{5}{2}\right)$

11)
a) $\{a \mid a \ge -4\}$   b) $[-4, \infty)$

13) when you multiply or divide the inequality by a negative number

15)
a) $\{k \mid k \ge -2\}$   b) $[-2, \infty)$

17)
a) $\{c \mid c \le 4\}$   b) $(-\infty, 4]$

19)
a) $\{d \mid d < -1\}$   b) $(-\infty, -1)$

21)
a) $\{z \mid z > 5\}$   b) $(5, \infty)$

23)
a) $\{m \mid m > 3\}$   b) $(3, \infty)$

25)
a) $\left\{x \mid x < -\dfrac{7}{4}\right\}$   b) $\left(-\infty, -\dfrac{7}{4}\right)$

27)
a) $\{b \mid b \ge -8\}$   b) $[-8, \infty)$

29)
a) $\left\{a \mid a > \dfrac{5}{3}\right\}$   b) $\left(\dfrac{5}{3}, \infty\right)$

31)
a) $\{k \mid k \ge -15\}$   b) $[-15, \infty)$

33)
a) $\{c \mid c > 8\}$   b) $(8, \infty)$

35)
$(-\infty, 7]$

37)
$[-1, \infty)$

39)
$(-\infty, 12)$

41) 
$$\left(-\infty, -\frac{3}{2}\right)$$

43) 
$$(-\infty, 5]$$

45) 
$$(0, \infty)$$

47) 
$$(-\infty, 20)$$

49) $[-5, 3]$    51) $(-3, 0]$

53) 
a) $\{y \mid -4 < y < 0\}$    b) $(-4, 0)$

55) 
a) $\{k \mid -3 \le k \le 2\}$    b) $[-3, 2]$

57) 
a) $\left\{n \mid \dfrac{1}{2} < n \le 3\right\}$    b) $\left(\dfrac{1}{2}, 3\right]$

59) 
$[-3, 1]$

61) 
$$\left(-5, \frac{7}{2}\right)$$

63) 
$[2, 5]$

65) 
$(-8, 4)$

67) 
$$\left[\frac{7}{4}, 3\right)$$

69) 
$[5, 9]$

71) 
$(1, 3]$

73) 
$[5, 8)$

75) 17    77) 26    79) at most 6 mi    81) 89 or higher

## Chapter 3 Review Exercises

1) no    3) The variables are eliminated and you get a false statement like $5 = 13$.

5) $\{-19\}$    7) $\{-8\}$    9) $\{36\}$    11) $\left\{\dfrac{15}{2}\right\}$    13) $\{5\}$

15) $\left\{-\dfrac{2}{3}\right\}$    17) $\{0\}$    19) $\{10\}$    21) $\{$all real numbers$\}$

23) $2x - 9 = 25$; 17    25) Thursday: 75; Friday: 51

27) 14 in., 22 in.    29) 6 lb    31) $1500 at 2%, $4500 at 4%

33) 7    35) 7 in.    37) $m\angle A = 55°, m\angle B = 55°, m\angle C = 70°$

39) $61°, 61°$    41) $p = z + n$    43) $b = \dfrac{2A}{h}$

45) Yes. It can be written as $\dfrac{15}{100}$ or $\dfrac{3}{20}$.

47) $\dfrac{4}{5}$    49) $\{12\}$    51) 1125    53) 12 $20 bills, 10 $10 bills

55) Jared: 5 mph; Meg: 6 mph

57) 
$(-3, \infty)$

59) 
$(-\infty, 4]$

61) 
$[-4, -1]$

63) 
$$\left(-2, -\frac{1}{2}\right)$$

65) $\left\{\dfrac{5}{2}\right\}$    67) $\{-42\}$    69) $\varnothing$    71) $\{13\}$    73) $\{-1\}$

75) 120 oz    77) 17, 19    79) 8 cm, 11 cm, 16 cm

81) 360

## Chapter 3 Test

1) $\left\{-\dfrac{7}{9}\right\}$    2) $\{-9\}$    3) $\{6\}$    4) $\{-3\}$    5) $\left\{\dfrac{2}{3}\right\}$

6) $\varnothing$    7) $\{1\}$

8) A ratio is a quotient of two quantities. A proportion is a statement that two ratios are equal.

9) 36, 38, 40    10) 250 mL    11) $31.90    12) 10 in. × 15 in.

13) eastbound: 66 mph; westbound: 72 mph

14) $a = \dfrac{4B}{n}$    15) $h = \dfrac{S - 2\pi r^2}{2\pi r}$ or $h = \dfrac{S}{2\pi r} - r$

16) $m\angle A = 26°, m\angle B = 115°, m\angle C = 39°$

17) 
$(-\infty, -2]$

18) 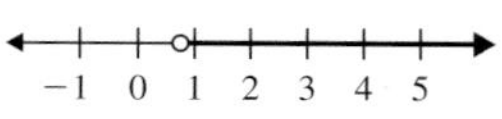

$\left(\frac{3}{4}, \infty\right)$

19) 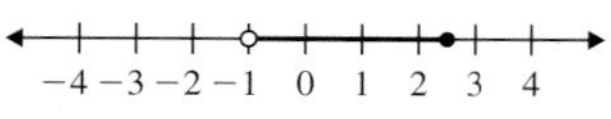

$\left(-1, \frac{5}{2}\right]$

20) at least 77

## Cumulative Review for Chapters 1–3

1) $-\frac{11}{24}$ 2) $\frac{15}{2}$ 3) 54 4) $-87$ 5) $-47$ 6) 27 $\text{cm}^2$

7) $\{-5, 0, 9\}$ 8) $\left\{\frac{3}{4}, -5, 2.5, 0, 0.\bar{4}, 9\right\}$

9) $\{0, 9\}$ 10) distributive

11) No. For example, $10 - 3 \neq 3 - 10$.

12) $17y^2 - 18y$ 13) $\frac{5}{4}r^{12}$ 14) $-72m^{33}$ 15) $-\frac{9}{2z^6}$

16) $\frac{1}{4c^6d^{10}}$ 17) $8.95 \times 10^{-6}$ 18) $\left\{-\frac{23}{2}\right\}$

19) $\{4\}$ 20) {all real numbers} 21) $\left\{\frac{11}{8}\right\}$

22) $\{18\}$ 23) car: 60 mph; train: 70 mph

24) $(-\infty, -6]$ 25) $\left(-1, \frac{5}{2}\right)$

## Chapter 4

### Section 4.1

1) 16.1 gallons 3) 2004 and 2006; 15.9 gallons

5) Consumption was increasing. 7) New Jersey; 86.3%

9) Florida's graduation rate is about 32.4% less than New Jersey's.

11) Answers may vary. 13) yes 15) yes 17) no 19) yes

21) 5 23) $-\frac{7}{2}$ 25) 5

27)

| x | y |
|---|---|
| 0 | $-4$ |
| 1 | $-2$ |
| $-1$ | $-6$ |
| $-2$ | $-8$ |

29)

| x | y |
|---|---|
| 0 | 0 |
| $\frac{1}{2}$ | 2 |
| 3 | 12 |
| $-5$ | $-20$ |

31)

| x | y |
|---|---|
| 0 | $-2$ |
| $-\frac{8}{5}$ | 0 |
| 1 | $-\frac{13}{4}$ |
| $-\frac{12}{5}$ | 1 |

33)

| x | y |
|---|---|
| 0 | $-2$ |
| $-3$ | $-2$ |
| 8 | $-2$ |
| 17 | $-2$ |

35) Answers may vary.

37) $A$: $(-2, 1)$, quadrant II; $B$: $(5, 0)$, no quadrant; $C(-2, -1)$; uadrant III; $D$: $(0, -1)$, no quadrant; $E$: $(2, -2)$; quadrant IV; $F$: $(3, 4)$; quadrant I

39–41) 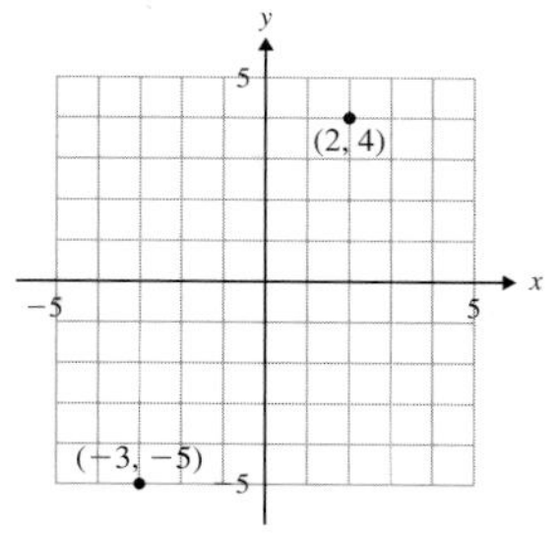


43–45) 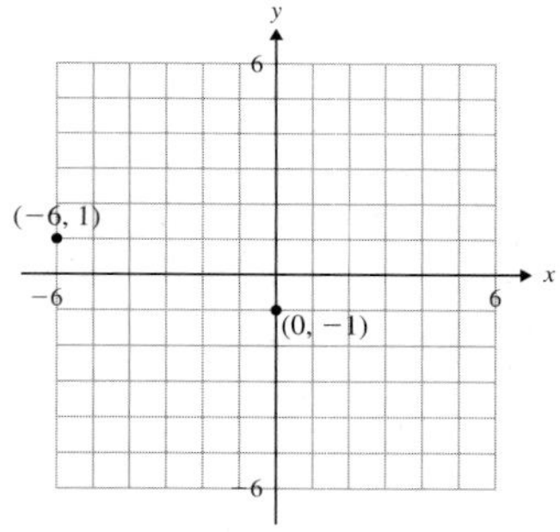


47–49) 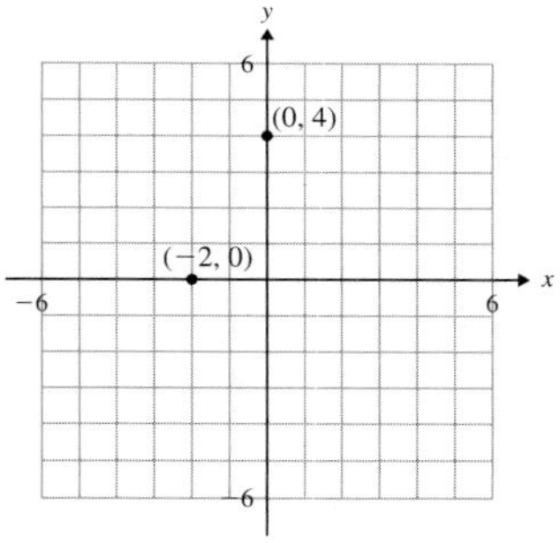


51–53) 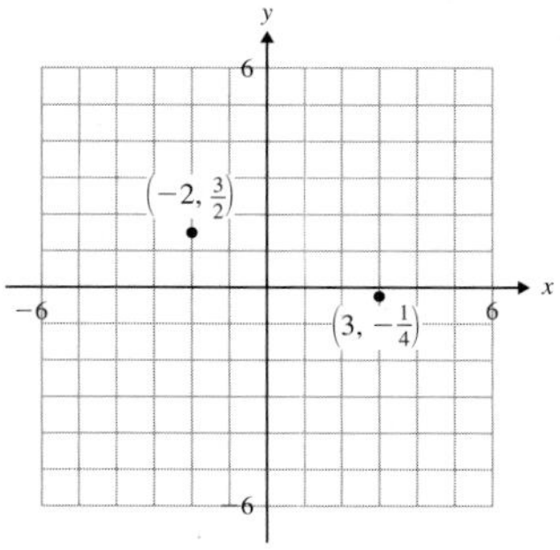


55) 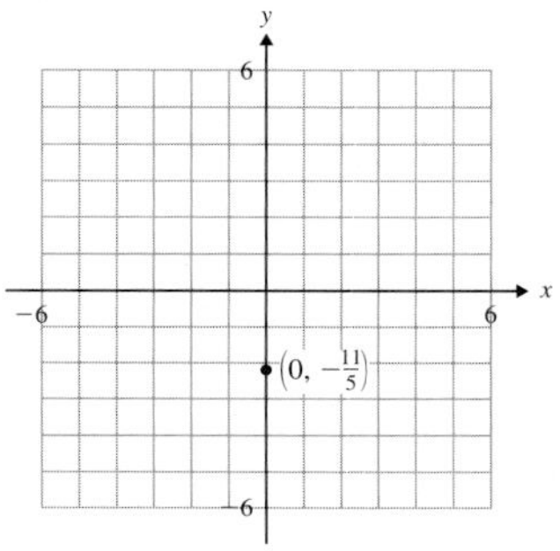

57)

| x | y |
|---|---|
| 0 | 3 |
| $\frac{3}{4}$ | 0 |
| 2 | −5 |
| −1 | 7 |

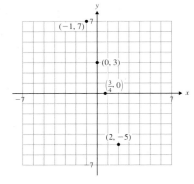

59)

| x | y |
|---|---|
| 0 | 0 |
| −1 | −1 |
| 3 | 3 |
| −5 | −5 |

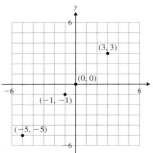

61)

| x | y |
|---|---|
| 0 | 3 |
| 4 | 0 |
| 1 | $\frac{9}{4}$ |
| −4 | 6 |

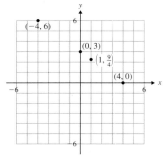

63)

| x | y |
|---|---|
| 0 | −1 |
| 1 | −1 |
| −3 | −1 |
| −1 | −1 |

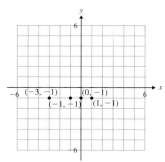

65)

| x | y |
|---|---|
| 0 | 2 |
| −2 | $\frac{3}{2}$ |
| 4 | 3 |
| −1 | $\frac{7}{4}$ |

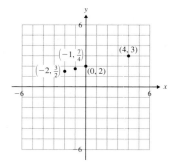

67) a) $(3, -5), (6, -3), (-3, -9)$

b) $\left(1, -\dfrac{19}{3}\right), \left(5, -\dfrac{11}{3}\right), \left(-2, -\dfrac{25}{3}\right)$

c) The x-values in part a) are multiples of the denominator of $\dfrac{2}{3}$. So, when you multiply $\dfrac{2}{3}$ by a multiple of 3 the fraction is eliminated.

69) negative    71) negative    73) positive    75) zero

77) a) x represents the year; y represents the number of visitors
   b) In 2004, there were 37.4 million visitors to Las Vegas.
   c) 38.9 million    d) 2005    e) 2 million    f) (2007, 39.2)

79) a) (1985, 52.9), (1990, 50.6), (1995, 42.4), (2000, 41.4), (2005, 40.0)
   b)

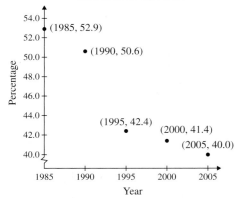

   c) In the year 2000, 41.4% of all fatal highway accidents involved alcohol.

81) a)

| x | y |
|---|---|
| 100.00 | 9.50 |
| 140.00 | 13.30 |
| 210.72 | 20.0184 |
| 250.00 | 23.75 |

   b)

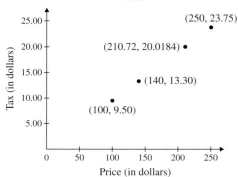

   c) If a bill totals $140.00, the sales tax will be $13.30.
   d) $20.02    e) They lie on a straight line.    f) $200.00

**Section 4.2**

1) line

3)

| x | y |
|---|---|
| 0 | 4 |
| −1 | 6 |
| 2 | 0 |
| 3 | −2 |

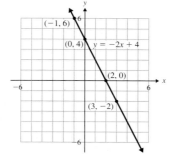

5)

| x | y |
|---|---|
| 0 | 7 |
| 2 | 10 |
| −2 | 4 |
| −4 | 1 |

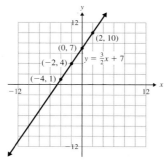

7)

| x | y |
|---|---|
| $\frac{3}{2}$ | 0 |
| 0 | 3 |
| $\frac{1}{2}$ | 2 |
| −1 | 5 |

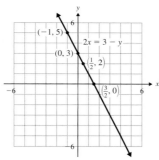

9)

| x | y |
|---|---|
| $-\frac{4}{9}$ | 5 |
| $-\frac{4}{9}$ | 0 |
| $-\frac{4}{9}$ | −1 |
| $-\frac{4}{9}$ | −2 |

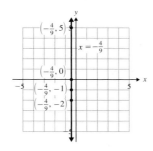

11) It is the point where the graph intersects the y-axis. Let x = 0 in the equation and solve for y.

13) (1, 0), (0, −1), (2, 1)

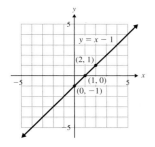

15) (4, 0), (0, −3), (2, $-\frac{3}{2}$)

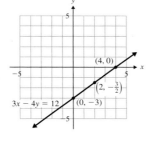

17) (−2, 0), (0, $-\frac{3}{2}$,) (2, −3)

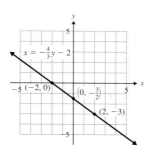

19) (4, 0), (0, −8), (2, −4)

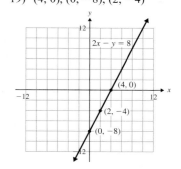

21) (0, 0), (1, −1), (−1, 1)

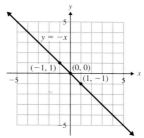

23) (0, 0), (3, 4), (−3, −4)

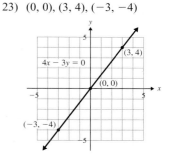

25) (5, 0), (5, 2), (5, −1)

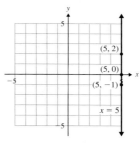

27) (0, 0), (1, 0), (−2, 0)

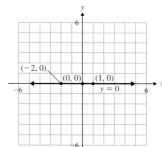

29) ($\frac{4}{3}$, 0), ($\frac{4}{3}$, 1), ($\frac{4}{3}$, −2)

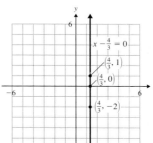

31) ($\frac{9}{4}$, 0), (0, −9), (3, 3)

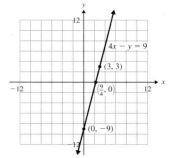

33) (0, 0)

35) a)

| x | y |
|---|---|
| 0 | 0 |
| 4 | 5.16 |
| 7 | 9.03 |
| 12 | 15.48 |

(0, 0), (4, 5.16), (7, 9.03), (12, 15.48)

b) (0, 0): If no songs are purchased, the cost is $0.
(4, 5.16): The cost of downloading 4 songs is $5.16.
(7, 9.03): The cost of downloading 7 songs is $9.03.
(12, 15.48): The cost of downloading 12 songs is $15.48.

c)

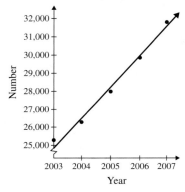

Cost of Downloading
Popular Songs from iTunes

d) 9

37) a) 2004: 26,275; 2007: 31,801
b) 2004: 26,578; 2007: 31,564; yes, they are close.
c)

Number of Science and Engineering
Doctorates Awarded in the U.S.

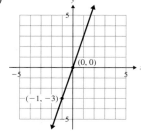

d) The $y$-intercept is 24,916. In 2003, approximately 24,916 science and engineering doctorates were awarded. It looks like it is within about 300 units of the plotted point.
e) 39,874

**Section 4.3**

1) The slope of a line is the ratio of vertical change to horizontal change. It is $\dfrac{\text{Change in } y}{\text{Change in } x}$ or $\dfrac{\text{Rise}}{\text{Run}}$ or $\dfrac{y_2 - y_1}{x_2 - x_1}$, where $(x_1, y_1)$ and $(x_2, y_2)$ are points on the line.

3) It slants upward from left to right.

5) undefined   7) $m = \dfrac{3}{4}$

9) $m = -\dfrac{2}{3}$

11) $m = -3$

13) Slope is undefined.

15)

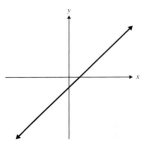

17) 2   19) $-4$   21) $-\dfrac{11}{5}$   23) undefined   25) 0   27) $\dfrac{3}{7}$

29) 0.75   31) $\dfrac{79}{12}$

33) No. The slope of the slide is $0.\overline{6}$. This is more than the recommended slope.

35) Yes. The slope of the driveway is 0.0375. This is less than the maximum slope allowed.

37) $\dfrac{6}{13}$

39) a) 2.89 mil; 2.70 mil
b) negative
c) The number of injuries is decreasing.
d) $m = -0.1$; the number of injuries is decreasing by about 0.1 million, or 100,000 per year.

41)

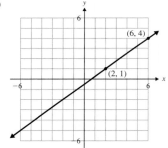

43)

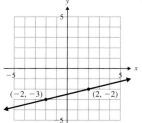

45)

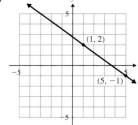

47)

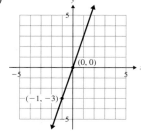

49)

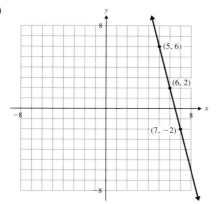

51)

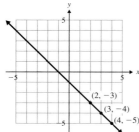

53)

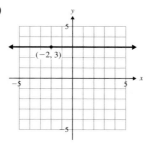

55)

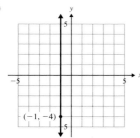

57)

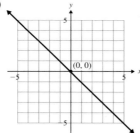

**Section 4.4**

1) The slope is $m$, and the $y$-intercept is $(0, b)$.

3) $m = \dfrac{2}{5}$, $y$-int: $(0, -6)$

5) $m = -\dfrac{3}{2}$, $y$-int: $(0, 3)$

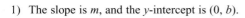
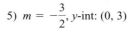
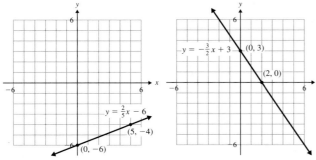
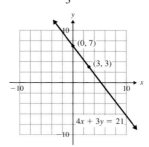

7) $m = \dfrac{3}{4}$, $y$-int: $(0, 2)$

9) $m = -2$, $y$-int: $(0, -3)$

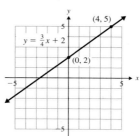
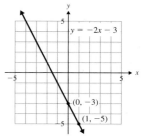

11) $m = 5$, $y$-int: $(0, 0)$

13) $m = -\dfrac{3}{2}$, $y$-int: $\left(0, -\dfrac{7}{2}\right)$

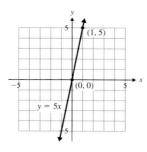
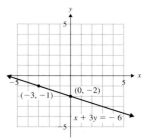

15) $m = 0$, $y$-int: $(0, 6)$

17) $y = -\dfrac{1}{3}x - 2$

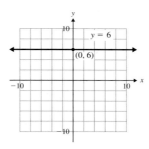
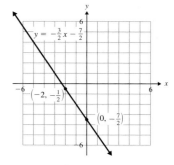

19) $y = -\dfrac{4}{3}x + 7$

21)  This cannot be written in slope-intercept form.

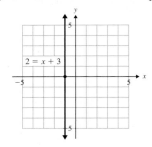

23)  $y = -\dfrac{2}{3}x + 6$          25)  $y = -5$

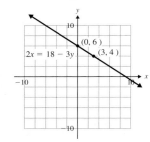          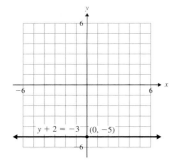

27)  a)  (0, 0); if Kolya works 0 hr, he earns \$0.
     b)  $m = 8.50$; Kolya earns \$8.50 per hour.
     c)  \$102.00

29)  a)  (0, 18); when the joey comes out of the pouch, it
         weighs 18 oz.
     b)  24 oz
     c)  A joey gains 2 oz per week after coming out of its
         mother's pouch.
     d)  7 weeks

31)  a)  (0, 0); \$0 = 0 rupees
     b)  48.2; each American dollar is worth 48.2 rupees.
     c)  3856 rupees
     d)  \$50.00

33)  $y = -4x + 7$    35)  $y = \frac{9}{5}x - 3$    37)  $y = -\frac{5}{2}x - 1$

39)  $y = x + 2$    41)  $y = 0$

43)  Their slopes are negative reciprocals, or one line is vertical
     and one is horizontal.

45)  perpendicular    47)  parallel    49)  neither    51)  neither

53)  perpendicular    55)  parallel    57)  perpendicular

59)  parallel    61)  perpendicular    63)  neither    65)  parallel

67)  parallel    69)  perpendicular

## Section 4.5

1)  $2x + y = -4$    3)  $x - y = 1$    5)  $4x - 5y = -5$

7)  $4x + 12y = -15$

9)  Substitute the slope and $y$-intercept into $y = mx + b$.

11)  $y = -7x + 2$    13)  $4x + y = 6$

15)  $2x - 7y = 21$    17)  $y = -x$

19)  a)  $y - y_1 = m(x - x_1)$
     b)  Substitute the slope and point into the point-slope formula.

21)  $y = x + 2$    23)  $y = -5x + 19$    25)  $4x - y = -7$

27)  $2x - 5y = -50$    29)  $y = -\dfrac{5}{4}x + \dfrac{29}{4}$    31)  $5x - 6y = -15$

33)  Find the slope and use it and one of the points in the
     point-slope formula.

35)  $y = -3x + 4$    37)  $y = 2x - 3$    39)  $y = -\dfrac{1}{3}x + \dfrac{10}{3}$

41)  $x + 3y = -2$    43)  $5x - 3y = 18$    45)  $y = -3.0x + 1.4$

47)  $y = \dfrac{3}{4}x - 1$    49)  $y = -3x - 4$    51)  $y = 3$

53)  $y = -\dfrac{4}{3}x + \dfrac{5}{3}$    55)  $y = x + 2$    57)  $y = 7x + 6$

59)  $x = 3$    61)  $y = 3$    63)  $y = -4x - 4$

65)  $y = -3x + 20$    67)  $y = \dfrac{1}{2}x - 2$

69)  They have the same slopes and different $y$-intercepts.

71)  $y = 4x + 2$    73)  $4x - y = 0$    75)  $x + 2y = 10$

77)  $y = 5x - 2$    79)  $y = \dfrac{3}{2}x - 4$    81)  $x - 5y = 10$

83)  $y = -x - 5$    85)  $3x - y = 10$    87)  $y = -3x - 12$

89)  $y = -x + 11$    91)  $y = 4$    93)  $y = 2$

95)  $y = -\dfrac{2}{7}x + \dfrac{1}{7}$    97)  $y = -\dfrac{3}{2}$

99)  a)  $y = 4603.3x + 81,150$
     b)  The average salary of a mathematician is increasing by
         \$4603.30 per year.
     c)  \$122,579.70

101)  a)  $y = -15,000x + 500,000$
      b)  The budget is being cut by \$15,000 per year.
      c)  \$455,000
      d)  2016

103)  a)  $y = 8x + 100$
      b)  A kitten gains about 8 g per day.
      c)  140 g; 212 g
      d)  23 days

105)  a)  $E = 1.6A + 28.4$          b)  40

## Section 4.6

1)  a)  any set of ordered pairs
    b)  a relation in which each element of the domain
        corresponds to exactly one element of the range
    c)  Answers may vary.

3)  domain: $\{-4, 1, 4, 8\}$; range: $\{-4, -3, 2, 10\}$; function

5)  domain: $\{1, 9, 25\}$; range: $\{-3, -1, 1, 5, 7\}$; not a function

7)  domain: $\{-4, -3, -1, 0\}$; range: $\left\{-2, -\dfrac{1}{2}\right\}$; function

9)  domain: $\{-4.1, -2.3, 3.0\}$; range: $\{-5.7, 3.1, 6.2, 7.8\}$;
    not a function

11)  domain: {Hawaii, New York, Miami}; range: {State, City};
     not a function

13)  domain: $\{-5, -2, 0, 6\}$; range: $\{-11, -5, 4\}$; function

15) domain: $(-\infty, \infty)$; range: $(-\infty, \infty)$; function

17) domain: $[-4, \infty)$; range: $(-\infty, \infty)$; not a function

19) domain: $(-\infty, \infty)$; range: $(-\infty, 6]$; function

21) $(-\infty, \infty)$  23) $(-\infty, \infty)$  25) $(-\infty, \infty)$  27) $(-\infty, \infty)$

29) $(-\infty, 0) \cup (0, \infty)$  31) $(-\infty, 5) \cup (5, \infty)$

33) $(-\infty, -1) \cup (-1, \infty)$  35) $(-\infty, 20) \cup (20, \infty)$

37) $y$ is a function, and $y$ is a function of $x$.

39) a) $y = 11$  b) $f(6) = 11$  41) a) $y = -9$  b) $f(3) = -9$

43) $-3$  45) $-11$  47) $-2$  49) $-1$

51) $\dfrac{1}{4}$  53) a) $f(-2) = 12; f(5) = 4$  55) $f(-2) = 4; f(5) = 2$

57) $f(-2) = -2; f(5) = 2$  59) $f(-2) = -2; f(5) = -2$

61) $-4$  63) $-4$

65) $m = -4$; $y$-int: $(0, -1)$

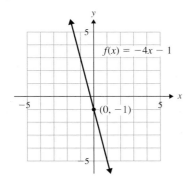

67) $m = -\dfrac{1}{3}$; $y$-int: $(0, -4)$

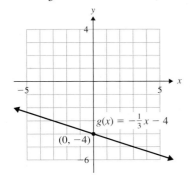

69) $m = 5$; $y$-int: $(0, -1)$

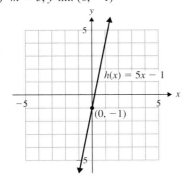

71) $m = \dfrac{3}{2}$; $y$-int: $\left(0, -\dfrac{5}{2}\right)$

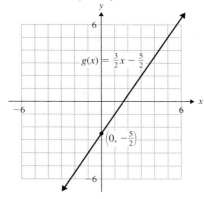

73) a) $36.00  b) $58.50  c) 8.5 hr
   d) $m = 9$; Fiona pays the babysitter $9 per hour.

75) a) $131.00  b) $250.00  c) 7

77) a) 180.9 cm (or about 5 ft 11 in. tall)

79) 167.9 cm (or about 5 ft 6 in. tall)

## Chapter 4 Review Exercises

1) yes  3) yes  5) 14  7) $-9$

9)

| x | y |
|---|---|
| 0 | $-14$ |
| 6 | $-8$ |
| $-3$ | $-17$ |
| $-8$ | $-22$ |

11)

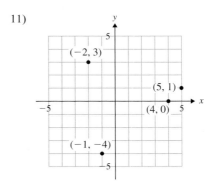

13) a)

| x | y |
|---|---|
| 10 | 50 |
| 18 | 54 |
| 29 | 59.50 |
| 36 | 63 |

(10, 50), (18, 54), (29, 59.50), (36, 63)

b)

Cost of Renting a Pick-Up

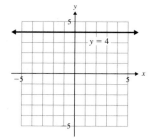

c) The cost of renting the pick-up is $74.00 if it is driven 58 miles.

15)

| x | y |
|---|---|
| 0 | 4 |
| 1 | 2 |
| 2 | 0 |
| 3 | −2 |

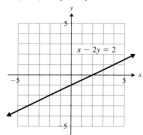

17) (2, 0), (0, −1);
(4, 1) may vary.

19) (2, 0), (0, 1);
(−2, 2) may vary.

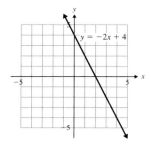

21) (0, 4); (2, 4),
(−1, 4) may vary.

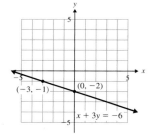

23) $\frac{3}{2}$   25) 5   27) $\frac{2}{3}$   29) −7   31) 0

33) a) $4.00
b) The slope is positive, so the value of the album is increasing over time.
c) $m = 1$; the value of the album is increasing by $1.00 per year.

35)

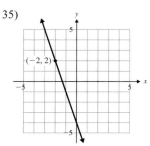

37)

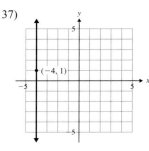

39) $m = -1$, y-int: (0, 5)   41) $m = \frac{2}{5}$, y-int: (0, −6)

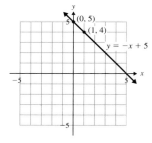

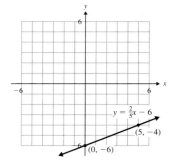

43) $m = -\frac{1}{3}$, y-int: (0, −2)   45) $m = -1$, y-int: (0, 0)

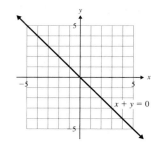

47) a) (0, 197.6); in 2003, the value of the squash crop was about $197.6 million.
b) It has been increasing by $7.9 million per year.
c) $213 million; $213.4 million

49) parallel   51) perpendicular   53) $y = 6x + 10$

55) $y = -\frac{3}{4}x + 7$   57) $y = -2x + 9$   59) $y = 7$

61) $3x - y = 7$   63) $5x - 2y = 8$   65) $4x + y = 0$

67) $x + y = 7$

69) a) $y = 3500x + 62{,}000$
b) Mr. Romanski's salary is increasing by $3500 per year.
c) $72,500   d) 2014

71) $y = -8x + 6$   73) $x - 2y = -18$   75) $y = -\frac{1}{5}x + 10$

77) $y = x - 12$   79) $y = \frac{3}{2}x + 2$   81) $y = 8$

83) domain: {−6, 5, 8, 10}; range: {0, 1, 4}; function

85) domain: {CD, DVD, Blu-ray disc};
range: {Music, Movie, PS3}; not a function

87) domain: $(-\infty, \infty)$; range: $(-\infty, \infty)$; function

89) $(-\infty, \infty)$   91) $(-\infty, 0) \cup (0, \infty)$   93) $(-\infty, \infty)$

95) $f(2) = 7; f(-1) = -8$   97) a) $-1$   b) 23   c) $-16$   d) 4

99)

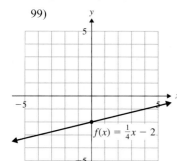

101)

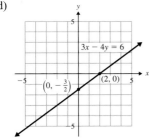

## Chapter 4 Test

1) Yes

2)

| x | y |
|---|---|
| 0 | −2 |
| −2 | −5 |
| 4 | 4 |
| 2 | 1 |

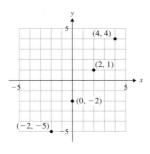

3) positive; negative

4) a) $(2, 0)$   b) $(0, -\frac{3}{2})$   c) Answers may vary.

d)

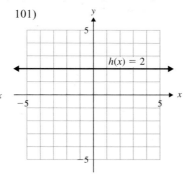

5)

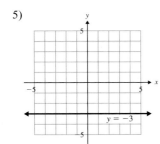

6)

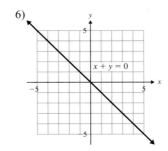

7) a) $-\dfrac{5}{4}$   b) 0

8)

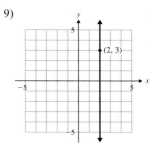

9)

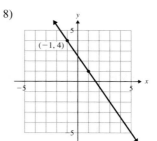

10) $y = \dfrac{3}{2}x - 5$

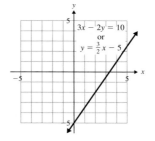

11) $y = 7x - 10$   12) $x + 3y = 12$   13) perpendicular

14) a) $y = -\dfrac{1}{2}x + 7$   b) $y = \dfrac{3}{4}x - \dfrac{1}{4}$

15) a) 399   b) $y = -9x + 419$

c) According to the equation, 401 students attended the school in 2007. The actual number was 399.

d) The school is losing 9 students per year.

e) $(0, 419)$; in 2005, 419 students attended this school.

f) 347

16) It is a relation in which each element of the domain corresponds to exactly one element of the range.

17) domain: $\{0, 1, 16\}$;
range: $\{0, -1, 2, -2\}$;
not a function

18) domain: $(-\infty, \infty)$; range: $[-3, \infty)$;
function

19) $(-\infty, \infty)$   20) $(-\infty, -8) \cup (-8, \infty)$   21) 4   22) 2

23) $-19$   24) 20   25) 11   26) $-3$

27)

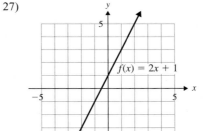

28) a) \$14.91   b) 114 min

## Cumulative Review for Chapters 1–4

1) $\dfrac{14}{33}$    2) 39 in.    3) $-81$    4) $\dfrac{14}{25}$    5) $-\dfrac{13}{8}$    6) 12

7) $2(17) - 9; 25$    8) $-20k^{15}$    9) $\dfrac{1}{16w^{32}}$    10) $\{-15\}$

11) $\{1\}$    12) $\varnothing$    13) $\left[\dfrac{5}{2}, \infty\right)$    14) $\$340{,}000$

15) $m\angle A = 29°, m\angle B = 131°$    16) 10 ft and 14 ft

17) $m = \dfrac{8}{3}$    18)

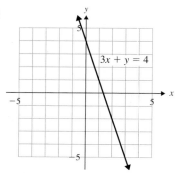

19) $3x - 8y = 8$    20) $y = \dfrac{3}{4}x$    21) $(-\infty, \infty)$    22) 26

23) 8    24) $-28$

25)

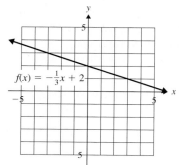

## Chapter 5

### Section 5.1

1) yes    3) no    5) yes    7) no    9) The lines are parallel.

11) $(3, 1)$

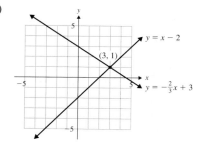

13) $(2, 3)$

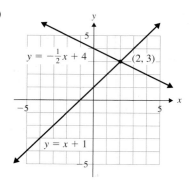

15) $(4, -5)$

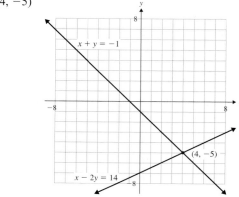

17) $(-1, -4)$

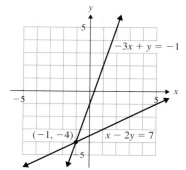

19) $\varnothing$; inconsistent system

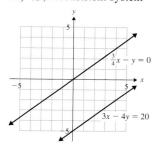

21) infinite number of solutions of the form
$\left\{(x, y) \mid y = \dfrac{1}{3}x - 2\right\}$; dependent equations

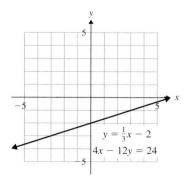

23) (0, 2)

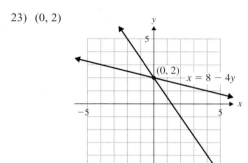

25) infinite number of solutions of the form $\{(x, y) \mid y = -3x + 1\}$; dependent equations

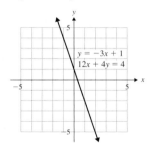

27) (−2, 2)

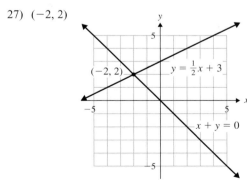

29) (1, −1)

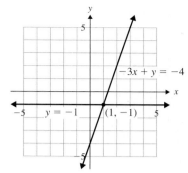

31) ∅; inconsistent system

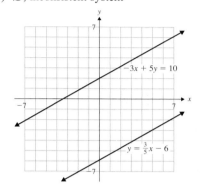

33–37) Answers may vary.    39) C; (−3, 4) is in quadrant II.

41) B; (4.1, 0) is the only point on the positive x-axis.

43) The slopes are different.    45) one solution

47) no solution    49) infinite number of solutions

51) one solution    53) no solution

55) a) 1985–2000
   b) (2000, 1.4); in the year 2000, 1.4% of foreign students
      were from Hong Kong and 1.4% were from Malaysia.
   c) 1985–1990 and 2000–2005
   d) 1985–1990; this line segment has the most negative slope.

57) (3, −4)  59) (4, 1)  61) (−2.25, −1.6)

**Section 5.2**

1) It is the only variable with a coefficient of 1.

3) The variables are eliminated, and you get a false statement.

5) (2, 5)  7) (−3, −2)  9) (1, −2)  11) (0, −7)  13) ∅

15) infinite number of solutions of the form $\{(x, y) \mid x - 2y = 10\}$

17) $\left(-\dfrac{4}{5}, 3\right)$  19) (4, 5)

21) infinite number of solutions of the form $\{(x, y) \mid -x + 2y = 2\}$

23) (−3, 4)  25) $\left(\dfrac{5}{3}, 2\right)$  27) ∅

29) Multiply the equation by the LCD of the fractions to
    eliminate the fractions.

31) (6, 1)  33) (−6, 4)  35) (3, −2)

37) infinite number of solutions of the form $\{(x, y)|y - \frac{5}{2}x = -2\}$

39) $(8, 0)$   41) $(3, 5)$   43) $(1.5, -1)$   45) $(-16, -12)$   47) $\varnothing$

49) $(6, 4)$   51) $(-4, -9)$   53) $(0, -8)$

55) a) A+: $96.00; Rock Bottom: $110.00
   b) A+: $180.00; Rock Bottom: $145.00
   c) $(200, 120)$; if the cargo trailer is driven 200 miles, the cost would be the same from each company: $120.00.
   d) If it is driven less than 200 miles, it is cheaper to rent from A+. If it is driven more than 200 miles, it is cheaper to rent from Rock Bottom Rental. If the trailer is driven exactly 200 miles, the cost is the same from each company.

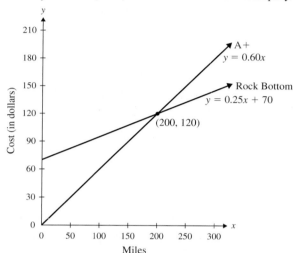

**Section 5.3**

1) Add the equations.   3) $(5, 8)$   5) $(-6, -2)$   7) $(-7, 2)$

9) $(2, 0)$   11) $(7, -1)$   13) $(0, 2)$   15) $\left(-\frac{2}{3}, -5\right)$

17) infinite number of solutions of the form $\{(x, y)|9x - y = 2\}$

19) $(8, 1)$   21) $\left(5, -\frac{3}{2}\right)$   23) $\varnothing$   25) $(9, 5)$   27) $\varnothing$

29) Eliminate the fractions. Multiply the first equation by 4, and multiply the second equation by 24.

31) $\left(\frac{5}{8}, 4\right)$   33) $\left(-\frac{9}{2}, -13\right)$   35) $(-6, 1)$

37) infinite number of solutions of the form $\left\{(x, y)\middle|y = \frac{2}{3}x - 7\right\}$   39) $(1, 1)$

41) $(-7, -4)$   43) $(12, -1)$   45) $\varnothing$

47) $(0.25, 5)$   49) $(4, 3)$   51) $\left(-\frac{3}{2}, 4\right)$   53) $(1, 1)$

55) $\left(-\frac{123}{17}, \frac{78}{17}\right)$   57) $\left(-\frac{203}{10}, \frac{49}{5}\right)$   59) $3$   61) $-8$

63) (a) $5$   b) $c$ can be any real number except 5.

65) a) $3$   b) $a$ can be any real number except 3.

67) $\left(\frac{2}{5}, \frac{2}{b}\right)$   69) $\left(-\frac{1}{4a}, \frac{19}{4b}\right)$

**Chapter 5 Putting It All Together**

1) Elimination method; none of the coefficients is 1 or $-1$; $(5, 6)$.

2) Substitution; the first equation is solved for $x$ and does not contain any fractions; $(3, 5)$.

3) Since the coefficient of $y$ in the second equation is 1, you can solve for $y$ and use substitution. Or, multiply the second equation by 5 and use the elimination method. Either method will work well; $\left(\frac{1}{4}, -3\right)$.

4) Elimination method; none of the coefficients is 1 or $-1$; $\left(0, -\frac{2}{5}\right)$.

5) Substitution; the second equation is solved for $x$ and does not contain any fractions; $(1, -7)$.

6) The second equation is solved for $y$, but it contains two fractions. Multiply this equation by 4 to eliminate the fractions, then write it in the form $Ax + By = C$. Use the elimination method to solve the system; $(6, 4)$.

7) $(6, 0)$   8) $(-2, 2)$   9) $\left(-\frac{2}{3}, \frac{1}{5}\right)$   10) $(1, 4)$   11) $\varnothing$

12) infinite number of solutions of the form $\{(x, y)|y = -6x + 5\}$

13) $\left(-\frac{1}{2}, 1\right)$   14) $\left(\frac{5}{6}, -\frac{3}{2}\right)$   15) $(4, 3)$   16) $\left(\frac{1}{6}, 6\right)$

17) $(9, -7)$   18) $(10, -9)$   19) $(0, 4)$   20) $(-3, -1)$

21) infinite number of solutions of the form $\{(x, y)|3x - y = 5\}$

22) $\varnothing$   23) $(-9, -14)$   24) $\left(2, \frac{4}{3}\right)$   25) $\left(-\frac{68}{41}, \frac{64}{41}\right)$

26) $\left(-\frac{53}{34}, \frac{24}{17}\right)$   27) $\left(\frac{3}{4}, 0\right)$   28) $\left(5, \frac{3}{4}\right)$   29) $(1, 1)$

30) $(-2, 4)$   31) $\left(-\frac{5}{2}, 10\right)$   32) $(-4, -6)$

33) $(2, 2)$

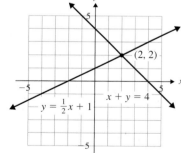

34) $(-1, -2)$

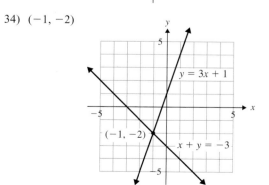

35) $(-4, 4)$

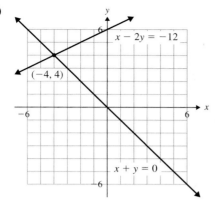

36) $(2, 4)$

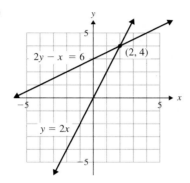

37) $\varnothing$

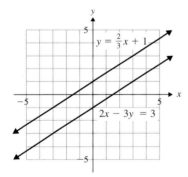

38) infinite number of solutions of the form $\{(x, y)|y = -\frac{5}{2}x - 3\}$

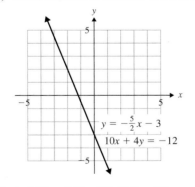

39) $(-1.25, -0.5)$    40) $(7.5, -13)$

## Section 5.4

1) 38 and 49

3) *The Dark Knight:* $67.2 million; *Transformers:* $60.6 million

5) Beyonce: 16; T.I.: 11    7) Urdu: 325,000; Polish: 650,000

9) Gagarin: 108 min; Shepard: 15 min

11) width: 30 in.; height: 80 in.

13) length: 110 mm; width: 61.8 mm

15) width: 34 ft; length: 51 ft

17) $m\angle x = 67.5°$; $m\angle y = 112.5°$

19) T-shirt: $20.00; hockey puck: $8.00

21) bobblehead: $19.00; mug: $12.00

23) hamburger: $0.61; small fries: $1.39

25) wrapping paper: $7.00; gift bags: $8.00

27) 9%: 3 oz; 17%: 9 oz

29) pure acid: 2 L; 25%: 8 L

31) Asian Treasure: 24 oz; Pearadise: 36 oz

33) taco: 330 mg; chalupa: 650 mg

35) 2%: $2500; 4%: $3500    37) $0.44: 12; $0.28: 4

39) Michael: 9 mph; Jan: 8 mph

41) small plane: 240 mph; jet: 400 mph

43) Pam: 8 mph; Jim: 10 mph

45) speed of boat in still water: 6 mph; speed of the current: 1 mph

47) speed of boat in still water: 13 mph; speed of the current: 3 mph

49) speed of jet in still air: 450 mph; speed of the wind: 50 mph

## Section 5.5

1) Answers may vary.    3) Answers may vary.

5) Answers may vary.

7)

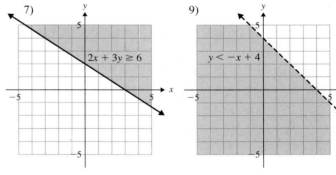

9)

11)

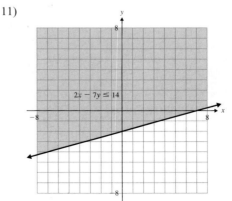

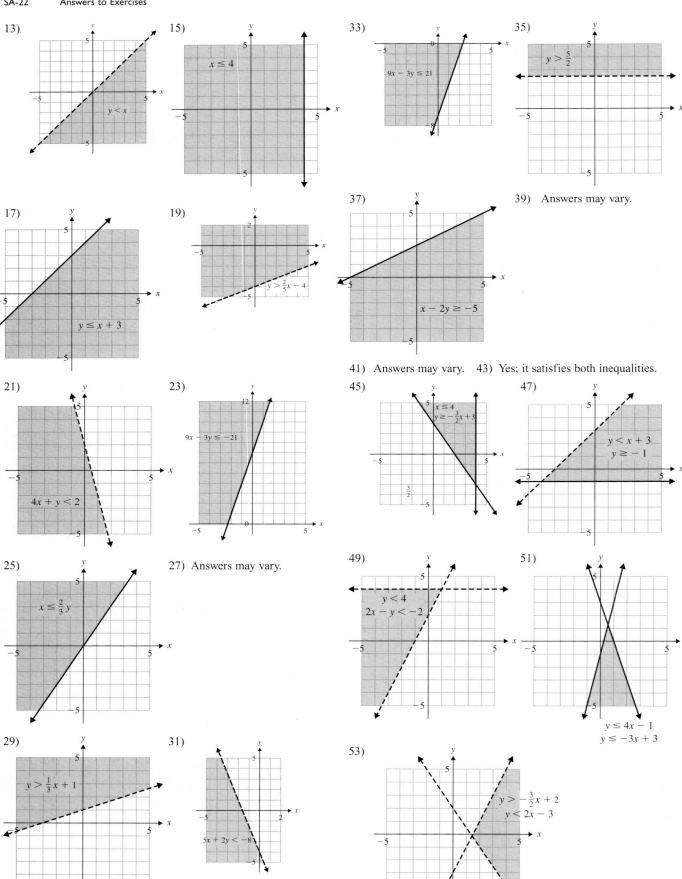

13)  $y < x$

15)  $x \le 4$

33)  $9x - 3y \le 21$

35)  $y > \frac{5}{2}$

17)  $y \le x + 3$

19)  $y > \frac{2}{5}x - 4$

37)  $x - 2y \ge -5$

39)  Answers may vary.

41)  Answers may vary.    43)  Yes; it satisfies both inequalities.

21)  $4x + y < 2$

23)  $9x - 3y \le -21$

45)  $x \le 4$
     $y \ge -\frac{3}{2}x + 3$
     $\frac{3}{2}$

47)  $y < x + 3$
     $y \ge -1$

25)  $x \le \frac{2}{3}y$

27)  Answers may vary.

49)  $y < 4$
     $2x - y < -2$

51)  $y \le 4x - 1$
     $y \le -3x + 3$

29)  $y > \frac{1}{3}x + 1$

31)  $5x + 2y < -8$

53)  $y > -\frac{3}{2}x + 2$
     $y < 2x - 3$

55)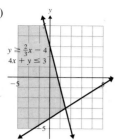

57)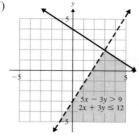

9)  infinite number of solutions    11)  (2, 5)

13)  (5, 3)    15)  (−4, −1)

17)  infinite number of solutions of the form $\{(x, y)|5x − 2y = 4\}$

19)  $\left(\dfrac{33}{43}, −\dfrac{36}{43}\right)$

21)  when one of the variables has a coefficient of 1 or −1

23)  $\left(−\dfrac{5}{3}, 2\right)$    25)  (0, 3)    27)  $\left(\dfrac{3}{4}, 0\right)$    29)  ∅

31)  white: 94; chocolate: 47

33)  Edwin: 8 mph; Camille: 6 mph

35)  length: 12 cm; width: 7 cm

37)  quarters: 35; dimes: 28

39)  hand warmers: $4.50; socks: $18.50    41)  Answers may vary.

59)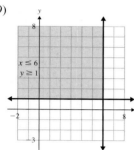

61)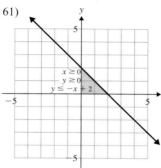

43)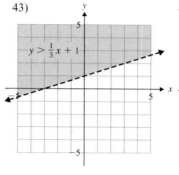

45)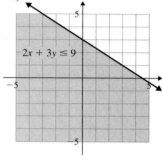

63)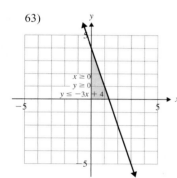

65)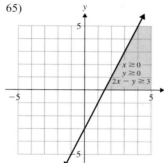

47)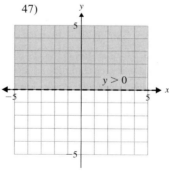

49)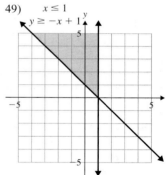

## Chapter 5 Review Exercises

1)  no    3)  The lines are parallel.

5)  ∅

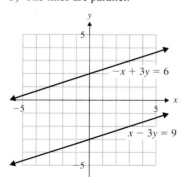

7)  (−3, −1)

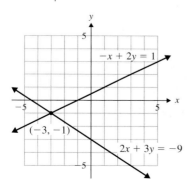

51)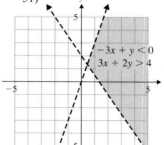

53)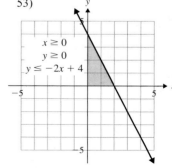

**Chapter 5 Test**

1)  yes

2)  $(4, -2)$

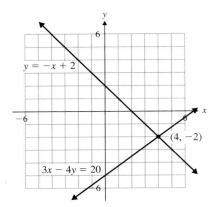

3)  $\varnothing$

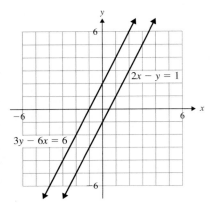

4)  a) 2001; approximately 4.6% of the population was unemployed.

b) (2003, 4.3): in 2003, 4.3% of the population of Hawaii and New Hampshire was unemployed.

c) Hawaii from 2003 to 2005; this means that during this time, Hawaii experienced the largest decrease in the unemployment rate during all years represented on the graph whether for Hawaii or for New Hampshire.

5)  $\left(-5, -\dfrac{1}{2}\right)$   6)  infinite number of solutions of the form $\left\{(x, y) \,\middle|\, y = \dfrac{1}{2}x - 3\right\}$   7)  (3, 1)   8)  (0, 6)   9)  $\varnothing$

10)  $(-2, -8)$   11)  $(2, -4)$   12)  $\left(-\dfrac{4}{3}, 0\right)$

13)  Answers may vary.   14)  Yellowstone: 2.2 mil acres; Death Valley: 3.3 mil acres   15)  adult: \$45.00; child: \$20.00

16)  length: 38 cm; width: 19 cm   17)  12%: 40 mL; 30%: 32 mL

18)  Rory: 42 mph; Lorelai: 38 mph

19)

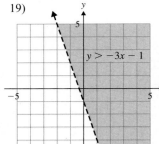

20)

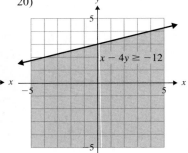

21)

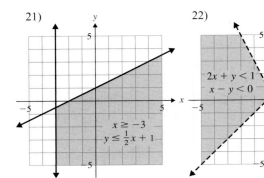

22)

**Cumulative Review for Chapters 1–5**

1)  $\dfrac{41}{30}$   2)  $9\dfrac{1}{3}$   3)  $-29$   4)  30 in$^2$   5)  $-12x^2 - 15x + 3$

6)  $32p^{20}$   7)  $\dfrac{63}{x^4}$   8)  $\dfrac{3n^4}{2m^{10}}$   9)  $7.319 \times 10^{-4}$   10)  $\varnothing$

11)  $\left\{\dfrac{1}{2}\right\}$   12)  (1, 6)   13)  15.8 mpg

14)  a)  $h = \dfrac{2A}{b_1 + b_2}$   b)  6 cm

15)

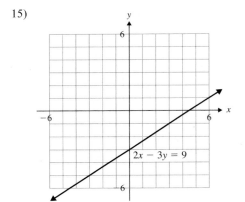

16)  $x$-int: (16, 0); $y$-int: (0, $-2$)   17)  $y = \dfrac{1}{4}x + \dfrac{5}{4}$

18)  perpendicular   19) (4, 10)   20) (3, 1)   21) $\varnothing$

22)  $\{(x, y)|3x + 9y = -2\}$   23) $(-2, -7)$

24)  4-ft boards: 16; 6-ft boards: 32

25)

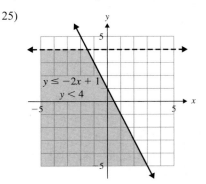

# Chapter 6

## Section 6.1

1) quotient rule; $k^6$   3) power rule for a product; $16h^4$

5) 64   7) $-64$   9) $\frac{1}{6}$   11) 81   13) $\frac{16}{81}$   15) $-31$   17) $\frac{1}{64}$

19) $t^{13}$   21) $-16c^9$   23) $z^{24}$   25) $125p^{30}$   27) $-\frac{8}{27}a^{21}b^3$

29) $f^4$   31) $7v$   33) $\frac{d^4}{6}$   35) $\frac{1}{x^6}$   37) $\frac{1}{m}$   39) $\frac{3}{2k^4}$

41) $20m^8n^{14}$   43) $24y^8$   45) $\frac{1}{49a^8b^2}$   47) $\frac{b^5}{a^3}$   49) $\frac{x^3}{y^{20}}$

51) $\frac{4a^2b^{15}}{3c^{17}}$   53) $\frac{y^{15}}{x^5}$   55) $\frac{64c^6}{a^6b^3}$   57) $\frac{9}{h^8k^8}$

59) $\frac{c^6}{27d^{18}}$   61) $\frac{u^{13}v^4}{32}$

63) $A = 10x^2$ sq units;
P = 14x units

65) $A = \frac{3}{16}p^2$ sq units;
P = 2p units

67) $k^{6a}$   69) $g^{8x}$   71) $x^{3b}$   73) $\frac{1}{8r^{18m}}$

## Section 6.2

1) Yes; the coefficients are real numbers and the exponents are whole numbers.

3) No; one of the exponents is a negative number.

5) No; two of the exponents are fractions.

7) binomial   9) trinomial   11) monomial

13) It is the same as the degree of the term in the polynomial with the highest degree.

15) Add the exponents on the variables.

17)
| Term | Coeff. | Degree |
|------|--------|--------|
| $3y^4$ | 3 | 4 |
| $7y^3$ | 7 | 3 |
| $-2y$ | $-2$ | 1 |
| 8 | 8 | 0 |

Degree of polynomial is 4.

19)
| Term | Coeff. | Degree |
|------|--------|--------|
| $-4x^2y^3$ | $-4$ | 5 |
| $-x^2y^2$ | $-1$ | 4 |
| $\frac{2}{3}xy$ | $\frac{2}{3}$ | 2 |
| $5y$ | 5 | 1 |

Degree of polynomial is 5.

21) a) 1   b) 13   23) 37   25) 146   27) $-23$

29) a) $y = 680$; if he rents the equipment for 5 hours, the cost of building the road will be $680.00.
b) $920.00   c) 8 hours

31) $13z$   33) $-11c^2 + 7c$   35) $-4.3t^6 + 4.2t^2$

37) $6a^2b^2 - 7ab^2$   39) $9n - 5$   41) $-5a^3 + 7a$

43) $12r^2 + 6r + 11$   45) $4b^2 - 3$

47) $\frac{7}{18}w^4 - \frac{1}{2}w^2 - \frac{3}{8}w - \frac{3}{2}$   49) $2m^2 + 3m + 19$

51) $10c^4 - \frac{1}{5}c^3 - \frac{1}{4}c + \frac{25}{9}$   53) $1.2d^3 + 7.7d^2 - 11.3d + 0.6$

55) $12w - 4$   57) $-y + 2$   59) $-2b^2 - 10b + 19$

61) $4f^4 - 14f^3 + 6f^2 - 8f + 9$   63) $5.8r^2 + 6.5r + 11.8$

65) $7j^2 + 9j - 5$   67) $8s^5 - 4s^4 - 4s^2 + 1$

69) $\frac{1}{16}r^2 + \frac{7}{9}r - \frac{5}{6}$   71) Answers may vary.

73) No. If the coefficients of the like terms are opposite in sign, their sum will be zero. Example: $(3x^2 + 4x + 5) + (2x^2 - 4x + 1) = 5x^2 + 6$

75) $-7a^4 - 12a^2 + 14$   77) $7n + 8$   79) $4w^3 + 7w^2 - w - 2$

81) $\frac{4}{3}y^3 - \frac{7}{4}y^2 + 3y - \frac{1}{14}$   83) $m^3 - 7m^2 - 4m - 7$

85) $9p^2 + 2p - 8$   87) $5z^6 + 9z^2 - 4$   89) $-7p^2 + 9p$

91) $4w + 14z$   93) $-5ac + 12a + 5c$   95) $4u^2v^2 + 31uv - 4$

97) $17x^3y^2 - 11x^2y^2 - 20$   99) $6x + 6$ units

101) $10p^2 - 2p - 6$ units

103) a) 16   b) 4   105) a) 32   b) 8   107) $-\frac{2}{3}$   109) 30

111)

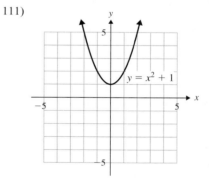

113)

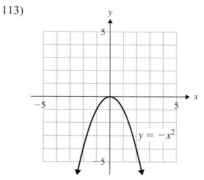

115)

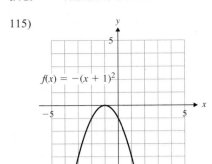

$f(x) = -(x + 1)^2$

117)

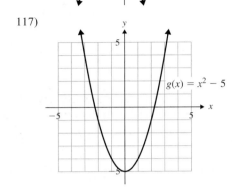

$g(x) = x^2 - 5$

119)

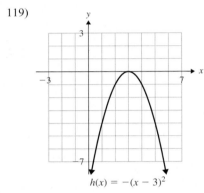

$h(x) = -(x - 3)^2$

## Section 6.3

1) Answers may vary.    3) $24m^8$    5) $-32c^6$

7) $10a^2 - 35a$    9) $-42c^2 - 12c$    11) $6v^5 - 24v^4 - 12v^3$

13) $-36b^5 + 18b^4 + 54b^3 + 81b^2$

15) $3a^3b^3 + 18a^3b^2 - 39a^2b^2 + 21a^2b$

17) $-9k^6 - 12k + \dfrac{9}{5}k^4$    19) $6c^3 + 11c^2 - 45c + 28$

21) $3f^3 - 13f^2 - 14f + 20$

23) $8x^4 - 22x^3 + 17x^2 - 26x - 10$

25) $4y^4 + \dfrac{7}{3}y^3 + 45y^2 + 28y - 36$

27) $s^4 + 3s^3 - 5s^2 + 11s - 6$

29) $-24h^5 + 26h^4 - 53h^3 + 19h^2 - 18h$

31) $15y^3 - 22y^2 + 17y - 6$    33) First, Outer, Inner, Last

35) $w^2 + 12w + 35$    37) $r^2 + 6r - 27$    39) $y^2 - 8y + 7$

41) $3p^2 + p - 14$    43) $21n^2 + 19n + 4$

45) $4w^2 - 17w + 15$    47) $12a^2 + ab - 20b^2$

49) $48x^2 + 74xy + 21y^2$    51) $v^2 + \dfrac{13}{12}v + \dfrac{1}{4}$

53) $\dfrac{1}{3}a^2 + \dfrac{17}{6}ab - 5b^2$

55) a) $4y + 4$ units
    b) $y^2 + 2y - 15$ sq units

57) a) $2m^2 + 2m + 14$ units
    b) $3m^3 - 6m^2 + 21m$ sq units

59) $3n^2 - \dfrac{5}{2}n$ sq units    61) Both are correct.

63) $8n^2 + 14n - 30$    65) $-5z^4 + 50z^3 - 80z^2$

67) $c^3 + 6c^2 + 5c - 12$    69) $3x^3 - 25x^2 + 44x - 12$

71) $2p^5 + 34p^3 + 120p$    73) $y^2 - 25$    75) $a^2 - 49$

77) $9 - p^2$    79) $u^2 - \dfrac{1}{25}$    81) $\dfrac{4}{9} - k^2$    83) $4r^2 - 49$

85) $k^2 - 64j^2$    87) $d^2 + 8d + 16$    89) $n^2 - 26n + 169$

91) $h^2 - 12h + 36$    93) $9u^2 + 6u + 1$

95) $4d^2 - 20d + 25$    97) $9c^2 + 12cd + 4d^2$

99) $\dfrac{9}{4}k^2 + 24km + 64m^2$

101) $4a^2 + 4ab + b^2 + 12a + 6b + 9$

103) $f^2 - 6fg + 9g^2 - 16$

105) No. The order of operations tells us to perform exponents, $(r + 2)^2$, before multiplying by 3.

107) $7y^2 + 28y + 28$    109) $4c^3 + 24c^2 + 36c$

111) $r^3 + 15r^2 + 75r + 125$    113) $g^3 - 12g^2 + 48g - 64$

115) $8a^3 - 12a^2 + 6a - 1$

117) $h^4 + 12h^3 + 54h^2 + 108h + 81$

119) $625t^4 - 1000t^3 + 600t^2 - 160t + 16$

121) No; $(x + 2)^2 = x^2 + 4x + 4$

123) $c^2 - 5c - 84$    125) $-40a^2 + 68a - 24$

127) $10k^3 - 37k^2 - 38k + 9$    129) $\dfrac{1}{36} - h^2$

131) $27c^3 + 27c^2 + 9c + 1$    133) $\dfrac{9}{32}p^{11}$

135) $a^2 + 14ab + 49b^2$    137) $-5z^3 + 30z^2 - 45z$

139) $a^3 + 12a^2 + 48a + 64$ cubic units

141) $\pi k^2 + 10\pi k + 25\pi$ sq units

143) $9c^2 - 15c + 4$ sq units

## Section 6.4

1) dividend: $6c^3 + 15c^2 - 9c$; divisor: $3c$; quotient: $2c^2 + 5c - 3$    3) Answers may vary.

5) $7p^4 + 3p^3 + 4p^2$    7) $3w^2 - 10w - 9$

9) $11z^5 + 7z^4 - 19z^2 + 1$    11) $h^6 + 6h^4 - 12h$

13) $3r^6 - r^3 + \dfrac{1}{2r}$    15) $4d^2 - 6d + 9$

17) $7k^5 + 2k^3 - 11k^2 - 9$    19) $-5d^3 + 1$

21) $\dfrac{5}{3}w^3 + 2w - 1 + \dfrac{1}{3w}$    23) $6k^3 - 2k - \dfrac{15}{2} - \dfrac{3}{2k} + \dfrac{1}{2k^5}$

25) $8p^3q^2 + 10p^2q - 9p + 3$    27) $2s^4t^5 - 4s^3t^3 - \dfrac{1}{7}st^2 + \dfrac{3}{s}$

29) The answer is incorrect. When you divide $5p$ by $5p$, you get 1. The quotient should be $8p^2 - 2p + 1$.

31) dividend: $12w^3 - 2w^2 - 23w - 7$; divisor: $3w + 1$; quotient: $4w^2 - 2w - 7$

33) 2    35) $158\dfrac{1}{6}$    37) $437\dfrac{4}{9}$    39) $g + 4$    41) $a + 6$

43) $k - 6$    45) $x + 8$    47) $2h^2 + 5h + 1$

49) $3p^2 + 5p - 1$    51) $7m + 12 + \dfrac{7}{m - 4}$

53) $4a^2 - 7a + 2 - \dfrac{8}{5a - 2}$    55) $n^2 - 3n + 9$

57) $2r^2 + 4r + 5$    59) $6x^2 - 9x + 5 - \dfrac{11}{2x + 3}$

61) $k^2 + k + 5$    63) $3t^2 - 8t - 6 + \dfrac{2t - 4}{5t^2 - 1}$

65) No. For example, $\dfrac{12x + 8}{3x} = 4 + \dfrac{8}{3x}$. The quotient is not a polynomial because one term has a variable in the denominator.

67) $5a^2b^2 + 3a^2b - \dfrac{1}{10} + \dfrac{1}{5ab}$

69) $-3f^3 + 6f^2 - 2f + 9 + \dfrac{23}{5f - 2}$

71) $8t + 5 + \dfrac{11}{t - 3}$    73) $16p^2 + 12p + 9$    75) $6x^2 + x - 7$

77) $4v^2 - 7v + 3$    79) $5h^2 - 3h - 2 + \dfrac{1}{2h^2 - 9}$

81) $m^2 - 4$    83) $-\dfrac{8}{9}t^3 - 5t + 3 + \dfrac{9}{t}$    85) $\dfrac{1}{2}x + 5$

87) $3p^2 - 5p + 1 + \dfrac{6p + 20}{7p^2 + 2p - 4}$    89) $3x + 7$ units

91) $30n^2 - 36n + 12$ units    93) $3x^2 - 7x + 2$ mph

**Chapter 6 Review Exercises**

1) 32    3) $\dfrac{125}{8}$    5) $p^{28}$    7) $5t^6$    9) $-42c^9$    11) $\dfrac{1}{k^5}$

13) $-\dfrac{48s^4}{r^3}$    15) $\dfrac{9y^{28}}{4x^6}$    17) $\dfrac{1}{m^2n^6}$

19) $A = 12f^2$ sq units; $P = 14f$ units    21) $y^{7a}$    23) $r^{9x}$

25)

| Term | Coeff. | Degree |
|------|--------|--------|
| $7s^3$ | 7 | 3 |
| $-9s^2$ | $-9$ | 2 |
| $s$ | 1 | 1 |
| 6 | 6 | 0 |

Degree of polynomial is 3.

27) 31    29) a) 20  b) $-6$    31) $-2c^2 + c + 5$

33) $9.8j^3 + 4.3j^2 + 3.4j - 0.5$    35) $\dfrac{1}{2}k^2 - k + 6$

37) $-3x^2y^2 + 9x^2y - 5xy + 6$    39) $3m + 4n - 3$

41) $6x^2 - 11x - 28$    43) $4d^2 + 6d + 6$ units

45)

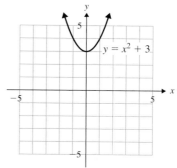

47)

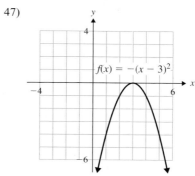

49) $24r^2 - 39r$    51) $-32w^4 - 24w^3 - 8w^2 - 2w + 3$

53) $y^2 - 12y + 27$    55) $6n^2 - 29n + 28$

57) $-a^2 + 3a + 130$    59) $4p^4q^4 + 66p^3q^4 - 6p^2q^3 + 24pq^2$

61) $4x^2 - 16xy - 9y^2$

63) $10x^6 + 50x^5 - 123x^4 - 15x^3 + 42x^2 + 30x - 72$

65) $8f^4 - 76f^3 + 168f^2$    67) $z^3 + 8z^2 + 19z + 12$

69) $\dfrac{1}{7}d^2 - \dfrac{11}{14}d - 24$    71) $c^2 + 8c + 16$

73) $16p^2 - 24p + 9$    75) $x^3 - 9x^2 + 27x - 27$

77) $m^2 - 6m + 9 + 2mn - 6n + n^2$    79) $p^2 - 169$

81) $\dfrac{81}{4} - \dfrac{25}{36}x^2$    83) $9a^2 - \dfrac{1}{4}b^2$    85) $3u^3 + 24u^2 + 48u$

87) a) $2n^2 + 7n - 22$ sq units  b) $6n + 18$ units

89) $4t^2 - 10t - 5$    91) $w + 5$    93) $4r^2 + r - 3$

95) $t + 2 - \dfrac{3}{2t} + \dfrac{10}{7t^2}$    97) $2v - 1 + \dfrac{6}{4v + 9}$

99) $3v^2 - 7v + 8$    101) $c^2 + 2c + 4$

103) $6k^2 - 4k + 7 - \dfrac{18}{3k + 2}$    105) $-\dfrac{5}{3}x^3y^2 - 4xy^2 + 1 - \dfrac{5}{4y^2}$

107) $8a + 2$ units    109) $20c^3 - 12c^2 - 11c + 1$

111) $144 - 49w^2$    113) $-40r^{21}t^{27}$

115) $13a^3b^3 + 7ab^2 - \dfrac{5}{3}b + \dfrac{1}{3ab^2}$

117) $h^3 - 15h^2 + 75h - 125$    119) $2c^2 - 8c + 9 + \dfrac{5}{c + 4}$

121) $\dfrac{y^{12}}{125}$    123) $2p^2 + 3p - 5$

## Chapter 6 Test

1) $\dfrac{64}{27}$   2) 625   3) $-32p^9$   4) $32t^{15}$   5) $\dfrac{g^4}{h^{10}}$   6) $\dfrac{25a^6}{9b^{18}}$

7) a) $-1$ b) 3   8) 9   9) 3   10) $24h^5 - 12h^4 + 4h^3$

11) $12a^3b^2 - 3a^2b^2 - 3ab + 9$   12) $9y^2 - 13y - 7$

13) $-11c^3 - 12c^2 - 19c - 14$   14) $u^2 - 14u + 45$

15) $8g^2 + 10g + 3$   16) $v^2 - \dfrac{4}{25}$   17) $6x^2 - 11xy - 7y^2$

18) $-12n^3 - 8n^2 + 63n - 40$   19) $2y^3 + 24y^2 + 72y$

20) $9m^2 - 24m + 16$   21) $\dfrac{16}{9}x^2 + \dfrac{8}{3}xy + y^2$

22) $t^3 - 6t^2 + 12t - 8$

23)

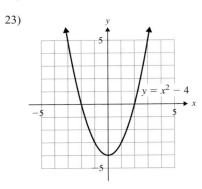

$y = x^2 - 4$

24) $w + 3$

25) $3m^2 - 5m + 1 - \dfrac{3}{4m}$   26) $6p^2 - p + 5 - \dfrac{15}{3p - 7}$

27) $y^2 + 3y + 9$   28) $2r^2 + 3r - 4$

29) a) $3d^2 - 14d - 5$ sq units
    b) $8d - 8$ units

30) $4n + 3$ units

## Cumulative Review for Chapters 1–6

1) a) $\{41, 0\}$
   b) $\{-15, 41, 0\}$
   c) $\left\{\dfrac{3}{8}, -15, 2.1, 41, 0.\overline{52}, 0\right\}$

2) $-87$   3) $\dfrac{75}{31}$ or $2\dfrac{13}{31}$   4) $-32a^{14}$   5) $c^{17}$   6) $\dfrac{64}{p^{21}}$

7) $\left\{-\dfrac{35}{3}\right\}$   8) $\varnothing$   9) $(-\infty, -5]$

10) 40 mL of 15% solution; 30 mL of 8% solution

11) $x$-int: $(8, 0)$; $y$-int: $(0, -3)$

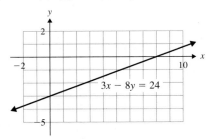

$3x - 8y = 24$

12)

$y = -4$

13) $3x + y = -5$   14) $y = -\dfrac{1}{4}x + 3$   15) $\left(-5, \dfrac{1}{2}\right)$

16) width: 12 cm; length: 35 cm   17) $-2q^2 - 45$

18) $n^2 + n - 56$   19) $9a^2 - 121$

20) $ab^2 - \dfrac{3}{2b} + \dfrac{5}{a^2b} + \dfrac{1}{2a^3b}$

21) $5p^2 + p - 7 - \dfrac{16}{p - 3}$

22) $12n^4 + 4n^3 - 35n^2 - 9n + 18$   23) $5c^3 - 40c^2 + 80c$

24) $4z^2 - 2z + 1$   25) $-7$

## Chapter 7

### Section 7.1

1) 7   3) $6p^2$   5) $4n^6$   7) $5a^2b$   9) $21r^3s^2$   11) $ab$

13) $(k - 9)$   15) Answers may vary.   17) yes   19) no

21) yes   23) $2(w + 5)$   25) $9(2z^2 - 1)$   27) $10m(10m^2 - 3)$

29) $r^2(r^7 + 1)$   31) $\dfrac{1}{5}y(y + 4)$   33) does not factor

35) $5n^3(2n^2 - n + 8)$   37) $8p^3(5p^3 + 5p^2 - p + 1)$

39) $9a^2b(7ab^2 - 4ab + 1)$   41) $-6(5n + 7)$

43) $-4w^3(3w^2 + 4)$   45) $-1(k - 3)$   47) $(t - 5)(u + 6)$

49) $(6x + 1)(y - z)$   51) $(q + 12)(p + 1)$

53) $(9k + 8)(5h^2 - 1)$   55) $(b + 2)(a + 7)$

57) $(3t + 4)(r - 9)$   59) $(2b + 5c)(4b + c^2)$

61) $(g - 7)(f + 4)$   63) $(t - 10)(s - 6)$

65) $(5u + 6)(t - 1)$   67) $(12g^3 + h)(3g - 8h)$

69) Answers may vary.   71) $4(xy + 3x + 5y + 15)$; Group the
    terms and factor out the GCF from each group.; $4(y + 3)(x + 5)$

73) $3(c + 7)(d + 2)$   75) $2p(p - 4)(q - 5)$

77) $(5s - 6)(2t + 1)$   79) $(3a^2 - 2b)(a - 7b)$

81) $2uv(4u + 5)(v + 2)$   83) $(n + 7)(3m + 10)$

85) $8(2b - 3)$   87) $(d + 6)(c - 4)$   89) $2a^3(3a - 4)(b + 2)$

91) $(d + 4)(7c + 3)$   93) $(g - 1)(d + 1)$   95) $x^3y^2(x + 12y)$

97) $4(m + 3)(n + 2)$   99) $-2(3p^2 + 10p - 1)$

## Section 7.2

1) a) $5, 2$   b) $-8, 7$   c) $5, -1$   d) $-9, -4$

3) They are negative.   5) Can I factor out a GCF?

7) Can I factor again?   9) $n + 2$   11) $c - 10$   13) $x + 4$

15) $(g + 6)(g + 2)$   17) $(y + 8)(y + 2)$   19) $(w - 9)(w - 8)$

21) $(b - 4)(b + 1)$   23) prime   25) $(c - 9)(c - 4)$

27) $(m + 10)(m - 6)$   29) $(r - 12)(r + 8)$   31) prime

33) $(x + 8)(x + 8)$ or $(x + 8)^2$   35) $(n - 1)(n - 1)$ or $(n - 1)^2$

37) $(d + 12)(d + 2)$   39) prime   41) $2(k - 3)(k - 8)$

43) $5h(h + 5)(h + 2)$   45) $r^2(r + 12)(r - 11)$

47) $7q(q^2 - 7q - 6)$   49) $3z^2(z + 4)(z + 4)$ or $3z^2(z + 4)^2$

51) $xy(y - 9)(y + 7)$   53) $-(m + 5)(m + 7)$

55) $-(c + 7)(c - 4)$   57) $-(z - 3)(z - 10)$

59) $-(p - 8)(p + 7)$   61) $(x + 4y)(x + 3y)$

63) $(c - 8d)(c + d)$   65) $(u - 5v)(u - 9v)$

67) $(m - 3n)(m + 7n)$   69) $(a + 12b)(a + 12b)$ or $(a + 12b)^2$

71) No; from $(3x + 6)$ you can factor out a 3. The correct answer is $3(x + 2)(x + 5)$.   73) yes   75) $2(x + 5)(x + 3)$

77) $(n - 4)(n - 2)$   79) $(m - 4n)(m + 11n)$   81) prime

83) $4q(q - 3)(q - 4)$   85) $-(k + 9)(k + 9)$ or $-(k + 9)^2$

87) $4h^3(h + 7)(h + 1)$   89) $(k + 12)(k + 9)$

91) $pq(p - 7q)(p - 10q)$   93) prime

95) $(x - 12y)(x - y)$   97) $5v^3(v^2 + 11v - 9)$

99) $6xy^2(x - 9)(x + 1)$   101) $(z - 9)(z - 4)$

103) $(ab + 6)(ab + 7)$   105) $(x + y)(z + 10)(z - 3)$

107) $(a - b)(c - 7)(c - 4)$

109) $(p + q)(r + 12)(r + 12)$ or $(p + q)(r + 12)^2$

## Section 7.3

1) a) $10, -5$   b) $-27, -1$   c) $6, 2$   d) $-12, 6$

3) $(3c + 8)(c + 4)$   5) $(6k - 7)(k - 1)$

7) $(2x - 9y)(3x + 4y)$   9) Can I factor out a GCF?

11) $4k^2 + 17k + 18$   13) $t + 2$   15) $3a + 2$   17) $3x - y$

19) $(2h + 3)(h + 5)$   21) $(7y - 4)(y - 1)$   23) $(5b - 6)(b + 3)$

25) $(3p + 2)(2p - 1)$   27) $(2t + 3)(2t + 5)$

29) $(9x - 4y)(x - y)$   31) because 2 can be factored out of $2x - 4$, but 2 cannot be factored out of $2x^2 + 13x - 24$

33) $(2r + 5)(r + 2)$   35) $(3u - 5)(u - 6)$

37) $(7a - 4)(a + 5)$   39) $(3y + 10)(2y + 1)$

41) $(9w - 7)(w + 3)$   43) $(4c - 3)(2c - 9)$

45) $(2k + 11)(2k + 9)$   47) $(10b + 9)(2b - 5)$

49) $(2r - 3t)(r + 8t)$   51) $(6a - b)(a - 4b)$

53) $(4z - 3)(z + 2)$; the answer is the same.

55) $(3p + 2)(p - 6)$   57) $(4k + 3)(k + 3)$

59) $2w(5w + 1)(3w + 7)$   61) $3(7r - 2)(r - 4)$   63) prime

65) $(7b + 3)(6b - 1)$   67) $(7x - 3y)(x - 2y)$

69) $2(d + 5)(d - 4)$   71) $r^2t^2(6r + 1)(5r + 3)$   73) $(3k - 7)^2$

75) $(n + 9)(2m - 7)(m + 1)$   77) $(u + 4)^2(3v + 4)(2v + 5)$

79) $(2a - 1)^4(5b - 6)(3b - 2)$   81) $-(n + 12)(n - 4)$

83) $-(7a + 3)(a - 1)$   85) $-(5z - 2)(2z - 3)$

87) $-5m(2m + 9)(2m + 3)$   89) $-ab(6a + b)(a - 2b)$

## Section 7.4

1) a) $49$   b) $81$   c) $36$   d) $100$   e) $25$   f) $16$
   g) $121$   h) $\dfrac{1}{9}$   i) $\dfrac{9}{64}$

3) a) $c^2$   b) $3r$   c) $9p$   d) $6m^2$   e) $\dfrac{1}{2}$   f) $\dfrac{12}{5}$

5) $y^2 + 12y + 36$

7) The middle term does not equal $2(2a)(-3)$. It would have to equal $-12a$ to be a perfect square trinomial.

9) $(h + 5)^2$   11) $(b - 7)^2$   13) $(2w + 1)^2$   15) $(3k - 4)^2$

17) $\left(c + \dfrac{1}{2}\right)^2$   19) $\left(k - \dfrac{7}{5}\right)^2$   21) $(a + 4b)^2$   23) $(5m - 3n)^2$

25) $4(f + 3)^2$   27) $5a^2(a + 3)^2$   29) $-4(2y + 5)^2$

31) $3h(25h^2 - 2h + 4)$   33) a) $x^2 - 81$   b) $81 - x^2$

35) $w - 8$   37) $11 - p$   39) $8c - 5b$   41) $(k + 2)(k - 2)$

43) $(c + 5)(c - 5)$   45) prime   47) $\left(x + \dfrac{1}{3}\right)\left(x - \dfrac{1}{3}\right)$

49) $\left(a + \dfrac{2}{7}\right)\left(a - \dfrac{2}{7}\right)$   51) $(12 + v)(12 - v)$

53) $(1 + h)(1 - h)$   55) $\left(\dfrac{6}{5} + b\right)\left(\dfrac{6}{5} - b\right)$

57) $(10m + 7)(10m - 7)$   59) $(13k + 1)(13k - 1)$

61) prime   63) $\left(\dfrac{1}{3}t + \dfrac{5}{2}\right)\left(\dfrac{1}{3}t - \dfrac{5}{2}\right)$   65) $(u^2 + 10)(u^2 - 10)$

67) $(6c + d^2)(6c - d^2)$   69) $(r^2 + 1)(r + 1)(r - 1)$

71) $(r^2 + 9t^2)(r + 3t)(r - 3t)$   73) $5(u + 3)(u - 3)$

75) $2(n + 12)(n - 12)$   77) $3z^2(2z + 5)(2z - 5)$

79) a) $64$   b) $1$   c) $1000$   d) $27$   e) $125$   f) $8$

81) a) $m$   b) $3t$   c) $2b$   d) $h^2$   83) $y^2 - 2y + 4$

85) $(t + 4)(t^2 - 4t + 16)$   87) $(z - 1)(z^2 + z + 1)$

89) $(3m - 5)(9m^2 + 15m + 25)$   91) $(5y - 2)(25y^2 + 10y + 4)$

93) $(10c - d)(100c^2 + 10cd + d^2)$

95) $(2j + 3k)(4j^2 - 6jk + 9k^2)$

97) $(4x + 5y)(16x^2 - 20xy + 25y^2)$

99) $6(c + 2)(c^2 - 2c + 4)$

101) $7(v - 10w)(v^2 + 10vw + 100w^2)$

103) $(h + 2)(h - 2)(h^2 - 2h + 4)(h^2 + 2h + 4)$

105) $7(2d + 1)$   107) $4(2k + 3)(k - 2)$

109) $(r + 1)(r^2 - 7r + 19)$   111) $(c - 1)(c^2 + 13c + 61)$

## Chapter 7 Putting It All Together

1) $(c + 8)(c + 7)$   2) $(r + 10)(r - 10)$   3) $(u + 9)(v + 6)$

4) $(5t + 4)(t - 8)$   5) $(2p - 7)(p - 3)$   6) $(h - 11)^2$

7) $9v^3 (v^2 + 10v - 6)$   8) $(m + 10n)(m - 4n)$

9) $4q (3q + 8)(2q - 1)$   10) $5(k - 2)(k^2 + 2k + 4)$

11) $(g + 5)(g^2 - 5g + 25)$   12) $(x - 9)(y - 1)$

13) $(12 + w)(12 - w)$   14) prime   15) $(3r + 2t)^2$

16) $5(8b - 7)$   17) $7n^2 (n - 10)(n + 1)$

18) $(2x - 3)(2x + 5)$   19) prime   20) $4a (b + 3)(c - 6)$

21) $5(2x - 3)(4x^2 + 6x + 9)$   22) $(7c + 4)^2$

23) $\left( m + \dfrac{1}{10} \right)\left( m - \dfrac{1}{10} \right)$   24) prime

25) $(5y - 6)(2x + 1)(2x - 1)$   26) $4ab(5a^2 + 3b)(5a^2 - 3b)$

27) $(p + 15q)(p + 2q)$   28) $8(k + 2)(k^2 - 2k + 4)$

29) prime   30) $6g^2h (2g^2h^2 + 9g + 5)$   31) $2(5n - 2)^2$

32) $(8a - 9)(a + 1)$   33) $3(12r + 7s)(r + s)$

34) $\left( t + \dfrac{9}{13} \right)\left( t - \dfrac{9}{13} \right)$   35) $(9x^2 + y^2)(3x + y)(3x - y)$

36) $(v - 12)(v - 11)$   37) $2(a - 9)(a + 4)$

38) $(p + 1)(p - 1)(q - 6)$   39) $\left( h - \dfrac{1}{5} \right)^2$

40) $(m - 4)(m^2 + 4m + 16)$   41) $(8u - 5)(2v + 3)$

42) $-9r (3r - 10)(r - 2)$   43) $(4b + 3)(2b - 5)$

44) $3(2b + 3)^2$   45) $4y^2z^2 (2y^2z - 7yz - 10y + 1)$

46) $(7 + p)(7 - p)$   47) prime   48) $6hk (h + k)(h + 8k)$

49) $(4u + 5v)^2$   50) $(b^2 + 4)(b + 2)(b - 2)$

51) $(8k - 3)(3k + 5)$   52) prime

53) $5(s - 4t)(s^2 + 4st + 16t^2)$   54) $12w^3 (3w^3 - 7w + 1)$

55) $(a - 1)(b - 1)$   56) $(d + 8)^2$   57) $7(h + 1)(h - 1)$

58) $(3p - 4q)(3p - 2q)$   59) $6(m - 5)^2$   60) prime

61) $(11z + 13)(11z - 13)$

62) $(4a - 5b)(16a^2 + 20ab + 25b^2)$   63) $-3(4r + 1)(r + 6)$

64) $9(c + 4)(c + 2)$   65) $(n + 1)(n^2 - n + 1)$   66) $(4t + 1)^2$

67) $(9u^2 + v^2)(3u + v)(3u - v)$   68) $2(3u + 7v^2)(2u + v)$

69) $(13h + 2)(h + 1)$   70) $2g (g - 8)(g + 7)$

71) $t^4 (5t^3 - 8)$   72) $\left( m + \dfrac{12}{5} \right)\left( m - \dfrac{12}{5} \right)$

73) $(d - 10)(d + 3)$   74) $(5k - 6)^2$   75) prime

76) $2(3w + 2)(9w^2 - 6w + 4)$   77) $(r + 1)^2$

78) $(b - 12)(b - 7)$   79) $(7n + 10)(7n - 10)$

80) $9y^2 (y + 3)(y - 3)$   81) $(2z + 1)(y + 11)(y - 5)$

82) $(2k - 7)(h + 5)(h - 9)$   83) $(t + 1)(t - 4)$

84) $v(v + 2)$   85) $z(z + 3)$   86) $(3n + 7)(3n - 10)$

87) $4ab$   88) $-4y (x + 2y)$   89) $3(7p - q)(p - q)$

90) $(7s - t)(s + 3t)$   91) $(r + 5)(r^2 + r + 7)$

92) $(d - 3)(d^2 - 12d + 39)$   93) $(k - 8)(k^2 - 13k + 43)$

94) $2(w - 1)(4w^2 + 22w + 49)$   95) $(a + b - 4)(a - b - 4)$

96) $(x + y + 3)(x - y + 3)$   97) $(s + t + 9)(s - t + 9)$

98) $(m + n - 1)(m - n - 1)$

## Section 7.5

1) $ax^2 + bx + c = 0$

3) a) quadratic   b) linear   c) quadratic   d) linear

5) If the product of two quantities equals 0, then one or both of the quantities must be zero.

7) $\{-11, 4\}$   9) $\left\{ \dfrac{3}{2}, 10 \right\}$   11) $\{0, 12\}$   13) $\left\{ -\dfrac{5}{3} \right\}$

15) $\left\{ -\dfrac{1}{2}, -\dfrac{2}{9} \right\}$   17) $\left\{ -\dfrac{1}{4}, \dfrac{2}{5} \right\}$   19) $\{0, 4.6\}$

21) No; the product of the factors must equal zero.

23) $\{-6, -2\}$   25) $\{-10, 11\}$   27) $\left\{ \dfrac{4}{3}, 2 \right\}$   29) $\left\{ -\dfrac{3}{4}, \dfrac{2}{3} \right\}$

31) $\{-12, 5\}$   33) $\{6, 9\}$   35) $\{-8, 8\}$   37) $\left\{ -\dfrac{7}{10}, \dfrac{7}{10} \right\}$

39) $\left\{ -\dfrac{6}{5}, -1 \right\}$   41) $\{0, 4\}$   43) $\{-6, 2\}$   45) $\{7, 12\}$

47) $\{-9, -7\}$   49) $\{8\}$   51) $\left\{ -3, -\dfrac{4}{5} \right\}$

53) No. You cannot divide an equation by a variable because you may eliminate a solution and may be dividing by zero.

55) $\left\{ -6, 0, \dfrac{9}{8} \right\}$   57) $\left\{ -\dfrac{7}{6}, 2, 3 \right\}$   59) $\{0, -7, 7\}$

61) $\{-4, 0, 9\}$   63) $\left\{ 0, \dfrac{5}{2}, 3 \right\}$   65) $\left\{ 0, -\dfrac{2}{9}, \dfrac{2}{9} \right\}$

67) $\left\{ \dfrac{7}{2}, \dfrac{9}{2} \right\}$   69) $\left\{ \dfrac{1}{3} \right\}$   71) $\{-5, 6\}$   73) $\{0, -4, 4\}$

75) $\left\{ -\dfrac{1}{3}, 9 \right\}$   77) $\{-11, 11\}$   79) $\left\{ \dfrac{5}{2}, 3 \right\}$   81) $\left\{ -\dfrac{1}{3}, 11 \right\}$

83) $\left\{ -\dfrac{1}{13}, 1 \right\}$   85) $\left\{ -3, -\dfrac{3}{2}, 4 \right\}$   87) $\left\{ \dfrac{1}{5}, \dfrac{7}{2} \right\}$

89) $\{-8, -1, 1\}$   91) $-8, -2$   93) $\dfrac{5}{2}, 4$   95) $-5, 5$

97) $0, 1, 6$

**Section 7.6**

1) length = 7 in.; width = 4 in.

3) base = 11 cm; height = 8 cm

5) base = 9 in.; height = 4 in.

7) height = 6 in.; width = 9 in.

9) length = 4 ft; width = 2 ft   11) 2.5 ft by 6 ft

13) length = 6 in.; width = 5 in.

15) height = 7 cm; base = 6 cm   17) 8 and 9, or 1 and 2

19) $6, 8, 10$; or $-2, 0, 2$   21) $7, 9, 11$

23) Answers may vary.   25) 9   27) 5   29) 20   31) 3, 4, 5

33) $5, 12, 13$   35) 10 cm   37) 5 ft   39) 5 mi

41) a) 144 ft   b) after 2 sec   c) 3 sec

43) a) 288 ft   b) 117 ft   c) 324 ft   d) 176 ft

45) a) 0 ft   b) after 2 sec and after 4 sec   c) 144 ft
    d) after 6 sec

47) a) $2835   b) $2880   c) $18

**Chapter 7 Review Exercises**

1) 8   3) $5h^3$   5) $9(7t + 5)$   7) $2p^4(p^2 - 10p + 1)$

9) $(m + 8)(n - 5)$   11) $-5r(3r^2 + 8r - 1)$

13) $(a + 9)(b + 2)$   15) $(x - 7)(4y - 3)$   17) $(q + 6)(q + 4)$

19) $(z - 12)(z + 6)$   21) $(m - 3n)(m - 10n)$

23) $4(v - 8)(v + 2)$   25) $9w^2(w - 1)(w + 2)$

27) $(3r - 2)(r - 7)$   29) $(2p - 5)(2p + 1)$

31) $2(3c + 2)(2c + 5)$   33) $(5x - 3y)(2x + 9y)$

35) $(w + 7)(w - 7)$   37) $(8t + 5u)(8t - 5u)$   39) prime

41) $4x(4 + x)(4 - x)$   43) $(r + 6)^2$   45) $5(2k - 3)^2$

47) $(v - 3)(v^2 + 3v + 9)$   49) $(5x + 4y)(25x^2 - 20xy + 16y^2)$

51) $(5z + 4)(2z - 3)$   53) $k^2(3k + 4)(3k - 4)$

55) $(d - 12)(d - 5)$   57) $(3b + 1)(a + 2)(a - 2)$

59) $6(2p - q)(4p^2 + 2pq + q^2)$   61) $(x + y - 1)(x - y + 9)$

63) $(5c - 2)^2$   65) $\left\{-\dfrac{7}{3}, 0\right\}$   67) $\left\{2, \dfrac{9}{2}\right\}$   69) $\{-9, -8\}$

71) $\left\{-\dfrac{11}{9}, \dfrac{11}{9}\right\}$   73) $\{-4, 10\}$   75) $\{-6, -5\}$

77) $\left\{-\dfrac{4}{5}, 1, \dfrac{4}{3}\right\}$   79) $\left\{\dfrac{1}{4}, 4\right\}$   81) $\left\{0, -\dfrac{2}{3}, \dfrac{2}{3}\right\}$

83) base = 9 cm; height = 4 cm   85) base = 6 ft; height = 2 ft

87) 15   89) length = 4 ft; width = 2.5 ft

91) $-1, 0, 1$; or $4, 5, 6$   93) 3 miles

**Chapter 7 Test**

1)  See whether you can factor out a GCF.

2) $(h - 8)(h - 6)$

3) $(6 + v)(6 - v)$   4) $(7p - 8)(p + 2)$

5) $4ab^2(5a^2b^2 + 9ab + 1)$   6) prime

7) $(4t - 3u)(16t^2 + 12tu + 9u^2)$   8) $4z(z + 4)(z + 3)$

9) $(6r - 5)^2$   10) $(n + 7)(n + 2)(n - 2)$

11) $(x + 3y)(x - 6y)$   12) $(9c^2 + d^2)(3c + d)(3c - d)$

13) $(q - 4)^2(p + 15)(p + 2)$   14) $2(8w - 5)(2w + 3)$

15) $k^5(k + 1)(k^2 - k + 1)$   16) $\{-9, 4\}$   17) $\{0, -12, 12\}$

18) $\left\{-\dfrac{4}{7}, \dfrac{4}{7}\right\}$   19) $\{4, 8\}$   20) $\left\{-\dfrac{8}{3}, -\dfrac{1}{2}\right\}$   21) $\left\{-\dfrac{2}{5}, 3\right\}$

22) length = 6 in.; width = 2 in.   23) $9, 11, 13$   24) 9 miles

25) length = 42 ft; width = 6 ft

26) a)   184 ft   b) 544 ft

    c) when $t = 2\dfrac{1}{2}$ sec and when $t = 10$ sec   d) $12\dfrac{1}{2}$ sec

**Cumulative Review for Chapters 1–7**

1) $-\dfrac{5}{18}$   2) $\dfrac{10}{27}$   3) $18p^8q^2$   4) $\dfrac{7a^{11}}{10b^5}$   5) $8.39 \times 10^{-5}$

6) $\{-2\}$   7) $b_2 = \dfrac{2A - hb_1}{h}$   8) $[-12, \infty)$

9)

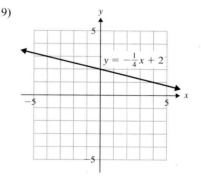

10) $y = \dfrac{5}{3}x + \dfrac{4}{3}$   11) $\varnothing$   12) adults: 120; students: 345

13) $8w^2 - 2w - 21$   14) $9n^2 + 24n + 16$

15) $12z^3 + 32z^2 - 53z + 15$   16) $13v^2 - 13v - 5$

17) $4x^2 + 7x - 2$   18) $3m - 1 + \dfrac{1}{2m}$

19) $3(2c + 9)(c - 2)$   20) prime

21) $(x + 4)(y + 1)(y - 1)$   22) $\left(\dfrac{1}{2} + b\right)\left(\dfrac{1}{2} - b\right)$

23) $(h + 5)(h^2 - 5h + 25)$   24) $\left\{\dfrac{4}{3}, 3\right\}$   25) $\left\{0, -\dfrac{3}{2}, \dfrac{3}{2}\right\}$

# Chapter 8

## Section 8.1

1) when its denominator equals zero

3) a) $\dfrac{5}{17}$   b) $\dfrac{5}{8}$   5) a) $-\dfrac{4}{3}$   b) undefined   7) a) $\dfrac{4}{3}$   b) 0

9) Set the denominator equal to zero and solve for the variable. That value cannot be substituted into the expression because it will make the denominator equal to zero.

11) a) $-4$   b) 0   13) a) $\dfrac{7}{2}$   b) $-\dfrac{1}{4}$

15) a) $0, 11$   b) $\dfrac{9}{5}$   17) a) never equals 0   b) 0

19) a) 0   b) $-4, -5$   21) a) $-20$   b) $\dfrac{3}{2}, -3$

23) a) $-6, -3$   b) 0

25) a) 0
b) never undefined—any real number may be substituted for $y$

27) $\dfrac{7x}{3}$   29) $\dfrac{3}{7g^2}$   31) $\dfrac{4}{5}$   33) $-\dfrac{7}{3}$   35) $\dfrac{13}{10}$   37) $g - 8$

39) $\dfrac{1}{t + 5}$   41) $\dfrac{3c + 4}{c + 2}$   43) $\dfrac{q + 5}{2q + 3}$   45) $\dfrac{w + 5}{5}$

47) $\dfrac{9}{c + 3}$   49) $\dfrac{4(u - 5)}{13}$   51) $\dfrac{m + n}{m^2 + mn + n^2}$   53) $-1$

55) $-1$   57) $-m - 11$   59) $-\dfrac{6}{x + 2}$   61) $-4(b + 2)$

63) $-\dfrac{y^2 + 2}{7}$   65) $\dfrac{3}{2(c - 1)}$   67) $-\dfrac{r^2 + rt + t^2}{r + t}$

69) $\dfrac{b - 6}{4(b + 1)}$   71) $4h^3 - 8h^2 + 1$   73) $-\dfrac{1}{2w^2}$

75) $-\dfrac{5}{(v - 2)(v - 1)}$

77) possible answers:
$\dfrac{-u - 7}{u - 2}, \dfrac{-(u + 7)}{u - 2}, \dfrac{u + 7}{2 - u}, \dfrac{u + 7}{-(u - 2)}, \dfrac{u + 7}{-u + 2}$

79) possible answers:
$\dfrac{-9 + 5t}{2t - 3}, \dfrac{5t - 9}{2t - 3}, \dfrac{-(9 - 5t)}{2t - 3}, \dfrac{9 - 5t}{-2t + 3}, \dfrac{9 - 5t}{3 - 2t}, \dfrac{9 - 5t}{-(2t - 3)}$

81) possible answers:
$-\dfrac{12m}{m^2 - 3}, \dfrac{12m}{-(m^2 - 3)}, \dfrac{12m}{-m^2 + 3}, \dfrac{12m}{3 - m^2}$

83) $4y - 3$   85) $4a^2 - 10a + 25$   87) $3x + 2$   89) $2c + 4$

91) $3k + 1$

## Section 8.2

1) $\dfrac{35}{54}$   3) $\dfrac{5}{21}$   5) $\dfrac{16b^4}{27}$   7) $\dfrac{t^2}{6s^6}$   9) $\dfrac{5}{2}$   11) $\dfrac{1}{2t(3t - 2)}$

13) $\dfrac{2u^2}{3(4u - 5)^2}$   15) 6   17) $\dfrac{3(y - 5)}{2y^2}$   19) $\dfrac{5}{c}$

21) $-7x(x + 11)$   23) 6   25) $\dfrac{28}{3}$   27) $\dfrac{32m^7}{35}$   29) $-\dfrac{20}{3g^3h^2}$

31) $\dfrac{8}{3k^6(k - 2)}$   33) $8q(p + 7)$   35) $\dfrac{5}{q(q - 7)}$

37) $\dfrac{z + 10}{(2z + 1)(z + 8)}$   39) $\dfrac{3}{4(3a + 1)}$   41) $\dfrac{7}{2}$   43) $\dfrac{c + 1}{6c}$

45) Answers may vary.   47) $h^2 + 3h - 10$   49) $7 - 2z$

51) $\dfrac{25}{16}$   53) $\dfrac{1}{18}$   55) $\dfrac{1}{4}$   57) $\dfrac{4}{3r^2}$   59) $\dfrac{a - 5}{12a^8}$

61) $\dfrac{2(4x + 5)}{3x^4}$   63) $\dfrac{(c + 6)(c + 1)}{9(c + 5)}$   65) $\dfrac{5x + 1}{4x}$

67) $\dfrac{k(2k + 3)}{2}$   69) $\dfrac{t^4(3t - 1)}{2(t + 2)}$   71) $-\dfrac{h^4}{3(h + 8)}$

73) $\dfrac{3x^6}{4y^6}$   75) $\dfrac{7}{3}$   77) $-\dfrac{a}{6}$   79) $\dfrac{a^2}{a - 4}$   81) $-\dfrac{3}{4(x + y)}$

83) $\dfrac{m + 2}{16}$   85) $\dfrac{4j - 1}{3j + 2}$   87) $\dfrac{x}{12(x - 9)}$   89) $\dfrac{12x^5}{y^3}$

## Section 8.3

1) 60   3) 120   5) $n^{11}$   7) $28r^7$   9) $36z^5$   11) $110m^4$

13) $24x^3y^2$   15) $11(z - 3)$   17) $w(2w + 1)$

19) Factor the denominators.

21) $10(c - 1)$   23) $3p^5(3p - 2)$   25) $(m - 7)(m - 3)$

27) $(z + 3)(z + 8)(z - 3)$   29) $(t + 6)(t - 6)(t + 3)$

31) $a - 8$ or $8 - a$   33) $x - y$ or $y - x$

35) Answers may vary.   37) $\dfrac{28}{48}$   39) $\dfrac{72}{9z}$   41) $\dfrac{21k^3}{56k^4}$

43) $\dfrac{12t^2u^3}{10t^7u^5}$   45) $\dfrac{7r}{r(3r + 4)}$   47) $\dfrac{4v^6}{16v^5(v - 3)}$

49) $\dfrac{9x^2 - 45x}{(x + 6)(x - 5)}$   51) $\dfrac{z^2 + 5z - 24}{(2z - 5)(z + 8)}$

53) $-\dfrac{5}{p - 3}$   55) $\dfrac{8c}{7 - 6c}$   57) $\dfrac{8}{15} = \dfrac{16}{30}; \dfrac{1}{6} = \dfrac{5}{30}$

59) $\dfrac{4}{u} = \dfrac{4u^2}{u^3}; \dfrac{8}{u^3} = \dfrac{8}{u^3}$   61) $\dfrac{9}{8n^6} = \dfrac{27}{24n^6}; \dfrac{2}{3n^2} = \dfrac{16n^4}{24n^6}$

63) $\dfrac{6}{4a^3b^5} = \dfrac{6a}{4a^4b^5}; \dfrac{6}{a^4b} = \dfrac{24b^4}{4a^4b^5}$

65) $\dfrac{r}{5} = \dfrac{r^2 - 4r}{5(r - 4)}; \dfrac{2}{r - 4} = \dfrac{10}{5(r - 4)}$

67) $\dfrac{3}{d} = \dfrac{3d - 27}{d(d - 9)}; \dfrac{7}{d - 9} = \dfrac{7d}{d(d - 9)}$

69) $\dfrac{m}{m + 7} = \dfrac{m^2}{m(m + 7)}; \dfrac{3}{m} = \dfrac{3m + 21}{m(m + 7)}$

71) $\dfrac{a}{30a - 15} = \dfrac{2a}{30(2a - 1)}; \dfrac{1}{12a - 6} = \dfrac{5}{30(2a - 1)}$

73) $\dfrac{9}{k - 9} = \dfrac{9k + 27}{(k - 9)(k + 3)}; \dfrac{5k}{k + 3} = \dfrac{5k^2 - 45k}{(k - 9)(k + 3)}$

75) $\dfrac{3}{a + 2} = \dfrac{9a + 12}{(a + 2)(3a + 4)}; \dfrac{2a}{3a + 4} = \dfrac{2a^2 + 4a}{(a + 2)(3a + 4)}$

77) $\dfrac{9y}{y^2 - y - 42} = \dfrac{18y^2}{2y(y + 6)(y - 7)};$

$\dfrac{3}{2y^2 + 12y} = \dfrac{3y - 21}{2y(y + 6)(y - 7)}$

79) $\dfrac{c}{c^2 + 9c + 18} = \dfrac{c^2 + 6c}{(c + 6)^2(c + 3)};$

$\dfrac{11}{c^2 + 12c + 36} = \dfrac{11c + 33}{(c + 6)^2(c + 3)}$

81) $\dfrac{11}{g - 3} = \dfrac{11g + 33}{(g + 3)(g - 3)}; \dfrac{4}{9 - g^2} = -\dfrac{4}{(g + 3)(g - 3)}$

83) $\dfrac{4}{3x - 4} = \dfrac{12x + 16}{(3x + 4)(3x - 4)};$

$\dfrac{7x}{16 - 9x^2} = -\dfrac{7x}{(3x + 4)(3x - 4)}$

85) $\dfrac{2}{z^2 + 3z} = \dfrac{6z + 18}{3z(z + 3)^2}; \dfrac{6}{3z^2 + 9z} = \dfrac{6z + 18}{3z(z + 3)^2};$

$\dfrac{8}{z^2 + 6z + 9} = \dfrac{24z}{3z(z + 3)^2}$

87) $\dfrac{t}{t^2 - 13t + 30} = \dfrac{t^2 + 3t}{(t + 3)(t - 3)(t - 10)};$

$\dfrac{6}{t - 10} = \dfrac{6t^2 - 54}{(t + 3)(t - 3)(t - 10)};$

$\dfrac{7}{t^2 - 9} = \dfrac{7t - 70}{(t + 3)(t - 3)(t - 10)}$

89) $-\dfrac{9}{h^3 + 8} = -\dfrac{45}{5(h + 2)(h^2 - 2h + 4)};$

$\dfrac{2h}{5h^2 - 10h + 20} = \dfrac{2h^2 + 4}{5(h + 2)(h^2 - 2h + 4)}$

## Section 8.4

1) $\dfrac{7}{8}$   3) $\dfrac{4}{7}$   5) $-\dfrac{18}{p}$   7) $\dfrac{5}{c}$   9) $\dfrac{z + 6}{z - 1}$   11) 2

13) $\dfrac{5}{t}$   15) $\dfrac{d + 6}{d + 5}$

17) a) $18b^4$

b) Multiply the numerator and denominator of $\dfrac{4}{9b^2}$ by $2b^2$,

and multiply the numerator and denominator of $\dfrac{5}{6b^4}$ by 3.

c) $\dfrac{4}{9b^2} = \dfrac{8b^2}{18b^4}; \dfrac{5}{6b^4} = \dfrac{15}{18b^4}$

19) $\dfrac{31}{40}$   21) $\dfrac{8t + 9}{6}$   23) $\dfrac{6h^2 + 50}{15h^3}$   25) $\dfrac{3 - 14f}{2f^2}$

27) $\dfrac{16y + 9}{y(y + 3)}$   29) $\dfrac{11d + 32}{d(d - 8)}$   31) $\dfrac{3(5c + 18)}{(c - 4)(c + 8)}$

33) $\dfrac{m^2 - 16m - 10}{(3m + 5)(m - 10)}$   35) $\dfrac{3u + 2}{u - 1}$

37) $\dfrac{7g^2 - 53g + 6}{(g + 2)(g + 8)(g - 8)}$   39) $\dfrac{3(a + 1)}{(a - 9)(a - 1)}$

41) $\dfrac{2(x^2 - 5x + 8)}{(x - 4)(x + 5)(x - 3)}$   43) $\dfrac{4b^2 + 28b + 3}{3(b - 4)(b + 3)}$

45) No. If the sum is rewritten as $\dfrac{9}{x - 6} - \dfrac{4}{x - 6}$, then the

LCD $= x - 6$. If the sum is rewritten as $\dfrac{-9}{6 - x} + \dfrac{4}{6 - x}$,

then the LCD is $6 - x$.

47) $\dfrac{6}{q - 4}$ or $-\dfrac{6}{4 - q}$   49) $\dfrac{26}{f - 7}$ or $-\dfrac{26}{7 - f}$

51) $\dfrac{8 - x}{x - 4}$ or $\dfrac{x - 8}{4 - x}$   53) 1   55) $\dfrac{3(1 + 2u)}{2u - 3v}$ or $-\dfrac{3(1 + 2u)}{3v - 2u}$

57) $-\dfrac{2(x - 1)}{(x + 3)(x - 3)}$   59) $\dfrac{3(3a + 4)}{(2a + 3)(2a - 3)}$

61) $\dfrac{-10a^2 + 8a - 11}{a(a - 2)}$   63) $\dfrac{10b^2 + 2b + 15}{b(b + 8)(3b + 1)}$

65) $\dfrac{c^2 - 2c + 20}{(c - 4)^2(c + 3)}$   67) $\dfrac{17a + 23b}{4(a + b)(a - b)}$

69) $\dfrac{2(9v + 2)}{(6v + 1)(3v + 2)(v - 5)}$   71) $\dfrac{7g^2 - 45g + 205}{5g(g - 6)(2g - 5)}$

73) a) $\dfrac{2(k - 4)}{k + 1}$   b) $\dfrac{k^2 - 3k + 28}{2(k + 1)}$

75) a) $\dfrac{6h}{(h + 5)^2(h + 4)}$   b) $\dfrac{2(h^2 + 4h + 6)}{(h + 5)(h + 4)}$   77) $\dfrac{49x + 6}{4x^2}$

## Chapter 8 Putting It All Together

1) a) 0   b) $\dfrac{1}{2}$   2) a) $\dfrac{3}{5}$   b) undefined

3) a) undefined   b) $\dfrac{7}{45}$   4) a) 3   b) $\dfrac{1}{2}$

5) a) $-6, 6$   b) 0   6) a) $-3, -\dfrac{5}{2}$   b) 4

7) a) $-4, 2$   b) $\dfrac{3}{5}$   8) a) $-8, 8$   b) $\dfrac{8}{5}$

9) a) 0   b) never equals 0

10) a) never undefined—any real number may be substituted
    for $t$   b) 15

11) $4w^{11}$   12) $\dfrac{7}{3n^5}$   13) $\dfrac{m + 9}{2(m + 4)}$   14) $\dfrac{1}{j - 5}$   15) $-\dfrac{3}{n + 2}$

16) $-\dfrac{1}{y + 5}$   17) $\dfrac{32}{3}$   18) $\dfrac{2(2f - 11)}{f(f + 11)}$

19) $\dfrac{4j^2 + 27j + 27}{(j + 9)(j - 9)(j + 6)}$   20) $\dfrac{5a^2b}{3}$   21) $\dfrac{y}{3z^2}$

22) $\dfrac{8q^2 - 37q + 21}{(q - 5)(q + 4)(q + 7)}$   23) $\dfrac{x^2 + 2x + 12}{(2x + 1)^2(x - 4)}$

24) $-\dfrac{n - 4}{4}$ or $\dfrac{4 - n}{4}$   25) $-\dfrac{m + 7}{8}$   26) $\dfrac{12}{r - 7}$

27) $\dfrac{19x}{(y + 5)(x^2 - 2x + 4)}$

28) $\dfrac{6}{d^6}$   29) $\dfrac{9d^2 + 8d + 24}{d^2(d + 3)}$

30) $\dfrac{3}{5}$   31) $\dfrac{3k^3(3k + 1)}{2}$   32) $\dfrac{35}{12z}$

33) $-\dfrac{(w - 15)(w - 1)}{(w + 5)(w - 5)(w - 7)}$   34) $\dfrac{3a^3(a^2 + 5)}{10}$

35) $\dfrac{2(7x - 1)}{(x - 8)(x + 3)}$   36) $\dfrac{3y^3 + 16}{12y^4}$   37) $\dfrac{h + 5}{8(2h + 1)}$

38) $(b - 3)(b + 4)$   39) $\dfrac{3m + 20n}{7m - 4n}$   40) $-1$

41) $\dfrac{-2p^2 - 8p + 11}{p(p + 7)(p - 8)}$   42) $\dfrac{5u^2 + 37u - 19}{u(3u - 2)(u + 1)}$   43) $\dfrac{2(t + 1)^2}{t}$

44) $\dfrac{1}{5r^2(3r - 7)}$   45) $\dfrac{c^2}{24}$   46) $\dfrac{9}{2}$   47) $1$

48) $\dfrac{12p^2 - 92p - 15}{(4p + 3)(p + 2)(p - 6)}$   49) $\dfrac{3m + 1}{7(m + 4)}$

50) $\dfrac{4(c^2 + 16)}{3(c - 2)}$   51) $-\dfrac{1}{3k}$   52) $-\dfrac{11}{5w}$

53) a) $\dfrac{6z}{(z + 5)(z + 2)}$   b) $\dfrac{2(z^2 + 8z + 30)}{(z + 5)(z + 2)}$

54) $\dfrac{111n + 8}{36n^2}$

## Section 8.5

1) Method 1: Rewrite it as a division problem, then simplify.

$\dfrac{2}{9} \div \dfrac{5}{18} = \dfrac{2}{9} \cdot \dfrac{\overset{2}{\cancel{18}}}{5} = \dfrac{4}{5}$

Method 2: Multiply the numerator and denominator by 18, the LCD of $\dfrac{2}{9}$ and $\dfrac{5}{18}$. Then simplify.

$\dfrac{\overset{2}{\cancel{18}}\left(\dfrac{2}{9}\right)}{\underset{1}{\cancel{18}}\left(\dfrac{5}{18}\right)} = \dfrac{4}{5}$

3) $\dfrac{20}{63}$   5) $\dfrac{u}{v}$   7) $x^2y$   9) $\dfrac{2m^4}{15n^2}$   11) $\dfrac{18}{m}$   13) $3(g - 6)$

15) $\dfrac{5d^2}{2}$   17) $\dfrac{c - 8}{11}$   19) $\dfrac{1}{28}$   21) $\dfrac{r(r - 4s)}{3r + s}$   23) $\dfrac{8}{r(3t - r^2)}$

25) $\dfrac{8w + 17}{10(w + 1)}$   27) $\dfrac{1}{28}$   29) $\dfrac{r(r - 4s)}{3r + s}$   31) $\dfrac{8}{r(3t - r^2)}$

33) $\dfrac{8w + 17}{10(w + 1)}$   35) Answers may vary.

37) $\dfrac{a}{12}$   39) $-\dfrac{n + 6}{2(3n - 5)}$   41) $\dfrac{6 - w^2}{w + 6}$   43) $2$   45) $\dfrac{t - 5}{t + 4}$

47) $\dfrac{9xy}{2(x + y)}$   49) $\dfrac{3c^2}{5}$   51) $-\dfrac{38}{21}$   53) $\dfrac{4}{35}$   55) $\dfrac{1}{v(3u - 2v^2)}$

57) $a + b$   59) $\dfrac{4(x + 2)}{x - 6}$   61) $\dfrac{z}{y^2}$   63) $\dfrac{11}{m}$

65) $\dfrac{2}{(h - 2)(h + 2)}$   67) $\dfrac{2(v - 9)(v + 2)}{3(v + 3)(v + 1)}$

## Section 8.6

1) Eliminate the denominators.   3) difference; $\dfrac{8r + 15}{6}$

5) equation; $\left\{-\dfrac{2}{5}\right\}$   7) sum; $\dfrac{a^2 + 3a + 33}{a^2(a + 11)}$

9) equation; $\{12\}$   11) $0, 2$   13) $0, 3, -3$   15) $-4, 9$

17) $\{-1\}$   19) $\left\{\dfrac{16}{3}\right\}$   21) $\{-4\}$   23) $\{-11\}$   25) $\{1\}$

27) $\{-6\}$   29) $\left\{\dfrac{7}{3}\right\}$   31) $\varnothing$   33) $\{-15\}$   35) $\{2\}$

37) $\{-6, -2\}$   39) $\{3, 5\}$   41) $\{-8\}$   43) $\{-10, 12\}$

45) $\{3\}$   47) $\left\{\dfrac{21}{5}\right\}$   49) $\varnothing$   51) $\{-1\}$   53) $\{0, -12\}$

55) $\{8\}$   57) $\{3, 1\}$   59) $\varnothing$   61) $\{-5, -3\}$   63) $\{-3\}$

65) $\left\{-\dfrac{2}{3}\right\}$   67) $\{-5, 4\}$   69) $\{5\}$   71) $\{-2, 8\}$   73) $\varnothing$

75) $m = \dfrac{CA}{W}$   77) $b = \dfrac{rt}{2a}$   79) $x = \dfrac{t + u}{3B}$   81) $n = \dfrac{dz - t}{d}$

83) $s = \dfrac{3A - hr}{h}$   85) $y = \dfrac{kx + raz}{r}$   87) $r = \dfrac{st}{s + t}$

89) $z = \dfrac{4xy}{x - 5y}$

## Section 8.7

1) $\{60\}$   3) $\{12\}$   5) 12 ft   7) 4   9) 800

11) length: 48 ft; width: 30 ft

13) used tutoring: 9; did not use tutoring: 24

15) a) 6 mph   b) 10 mph

17) a) $x + 40$ mph   b) $x - 30$ mph   19) 10 mph

21) 225 mph   23) 4 mph   25) 40 mph   27) 60 mph

29) $\dfrac{1}{3}$ homework/hour   31) $\dfrac{1}{t}$ job/hr   33) $2\dfrac{2}{9}$ hr   35) $2\dfrac{2}{5}$ hr

37) 20 min   39) 7.5 hr   41) 2 hr   43) increases   45) direct

47) inverse   49) $A = kw$   51) $x = \dfrac{k}{g}$   53) $C = kd^3$

55) $b = \dfrac{k}{z^2}$   57) a) 9   b) $z = 9x$   c) 54   59) a) 60

b) $T = \dfrac{60}{n}$   c) 12   61) a) 576   b) $u = \dfrac{576}{v^2}$   c) 64

63) 77   65) 9   67) 64   69) 28   71) \$720.00   73) 8 cm

75) 0.96 lb   77) 15 in.   79) 180 lb   81) 62.5 lumens

## Chapter 8 Review Exercises

1) a) 0   b) 0   3) a) 0   b) $-\dfrac{11}{4}$

5) a) never equals 0   b) $-\dfrac{3}{2}, \dfrac{3}{2}$   7) a) $-\dfrac{5}{3}, 2$

b) never undefined—any real number may be substituted for $m$

9) $11k^6$   11) $\dfrac{r-8}{4r}$   13) $\dfrac{1}{2z+1}$   15) $-\dfrac{1}{x+11}$

17) possible answers:
$$\dfrac{-4n-1}{5-3n}, \dfrac{-(4n+1)}{5-3n}, \dfrac{4n+1}{3n-5}, \dfrac{4n+1}{-5+3n}, \dfrac{4n+1}{-(5-3n)}$$

19) $b+3$   21) $\dfrac{24}{35}$   23) $\dfrac{t+2}{2(t+6)}$   25) $\dfrac{x-3}{45}$

27) $\dfrac{1}{2r^2(r-7)}$   29) $\dfrac{q}{35p^2}$   31) $\dfrac{1}{3}$   33) $\dfrac{3}{10}$

35) 30   37) $k^5$   39) $(4x+9)(x-7)$   41) $w-5$ or $5-w$

43) $(c+4)(c+5)(c-7)$   45) $\dfrac{12y^2}{20y^3}$

47) $\dfrac{6z}{z(2z+5)}$   49) $\dfrac{t^2+t-12}{(3t+1)(t+4)}$

51) $\dfrac{8c}{c^2+5c-24} = \dfrac{8c^2-24c}{(c-3)^2(c+8)}$;
$\dfrac{5}{c^2-6c+9} = \dfrac{5c+40}{(c-3)^2(c+8)}$

53) $\dfrac{7}{2q^2-12q} = \dfrac{7q+42}{2q(q+6)(q-6)}$;
$\dfrac{3q}{36-q^2} = -\dfrac{6q^2}{2q(q+6)(q-6)}$;
$\dfrac{q-5}{2q^2+12q} = \dfrac{q^2-11q+30}{2q(q+6)(q-6)}$

55) $\dfrac{4}{3c}$   57) $\dfrac{36u-5v}{40u^3v^2}$   59) $\dfrac{n^2-12n+20}{n(3n-5)}$

61) $\dfrac{4y-37}{(y-3)(y+2)}$   63) $\dfrac{(k-7)(k+2)}{k(k+7)^2}$

65) $\dfrac{t+20}{t-18}$   67) $\dfrac{5w^2-51w+8}{(2w-7)(w+8)(w+3)}$

69) $\dfrac{2b^2+7b+2}{2b(3b+2)(3b-2)}$

71) a) $\dfrac{2}{x(x+2)}$   b) $\dfrac{2x^3+4x+8}{x^2(x+2)}$   73) $\dfrac{y}{x^2}$   75) $\dfrac{p^2+4}{p^2+9}$

77) $\dfrac{1}{5}$   79) $\dfrac{(y+4)(y-9)}{(y-8)(y+6)}$   81) $r+t$   83) $\{4\}$

85) $\{-7, 5\}$   87) $\{-3\}$   89) $\{-1, 5\}$   91) $\varnothing$   93) $D = \dfrac{s+T}{R}$

95) $k = \dfrac{cw-N}{aw}$   97) $R_1 = \dfrac{R_2R_3}{R_2-R_3}$   99) 2 mph

101) $3\dfrac{1}{13}$ hr   103) 24   105) 216 cm$^2$

107) $\dfrac{n^2+6n+3}{(2n-1)(n+2)}$   109) $\dfrac{a+2}{(4a-7)(2a+5)}$

111) $-\dfrac{cd}{(c+d)(c-d)}$   113) $\{1, 3\}$   115) $\{-10\}$

## Chapter 8 Test

1) $-\dfrac{3}{8}$   2) a) $-10$   b) $\dfrac{9}{2}$   3) a) $-4, 9$   b) never equals zero

4) $\dfrac{1}{3t^4u^3}$   5) $\dfrac{h-8}{3h+1}$

6) possible answers: $-\dfrac{m-7}{4m-5}, \dfrac{m-7}{5-4m}, \dfrac{m-7}{-4m+5}$

7) $z(z+6)$   8) $\dfrac{2}{3r}$   9) $\dfrac{7b}{5a^6}$   10) $-\dfrac{13h}{36}$

11) $\dfrac{c^2+20c+30}{(3c+5)(c+2)}$   12) $-\dfrac{k^2+2}{4}$   13) $\dfrac{2d}{5(d+3)}$

14) $\dfrac{t-14}{t-7}$ or $\dfrac{14-t}{7-t}$   15) $\dfrac{-2v^2-2v+39}{(2v-3)(v-2)(v+9)}$

16) $\dfrac{m}{m-2}$   17) $\dfrac{x+y}{4xy}$   18) $\{-2\}$   19) $\varnothing$   20) $\{-7, 3\}$

21) $b = \dfrac{ac}{a-c}$   22) Ricardo: 3 hr; Michael: 6 hr   23) 12 mph

24) 75   25) 10 L

## Cumulative Review Chapters 1–8

1) 45 cm$^2$   2) 63   3) $32p^{15}$   4) $\dfrac{1}{125y^6}$   5) 5 ft by 9 ft

6) $\left(-\infty, \dfrac{7}{4}\right)$   7) $[0, 15]$

8) $x$-int: $\left(\dfrac{3}{2}, 0\right)$; $y$-int: $(0, -2)$

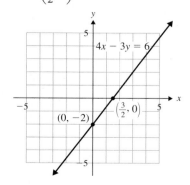

9) $-\dfrac{4}{3}$   10) $\left(3, -\dfrac{5}{2}\right)$   11) $4n^2 - 12n + 9$   12) $64a^2 - b^2$

13) $3h^2 - \dfrac{5}{3}h + 1 - \dfrac{2}{3h^2}$   14) $5k^2 - 2k - 3 + \dfrac{4}{k + 4}$

15) $(2d + 5)(2d - 3)$   16) $3(z^2 + 4)(z + 2)(z - 2)$

17) $(r + 8)(t - 1)$   18) $\{-12, -3\}$

19) a) $0, 6$   b) $-\dfrac{2}{7}$   20) $\dfrac{3(c - 2)}{c - 6}$

21) $\dfrac{2n}{4 - n}$   22) $\dfrac{3(y - 5)}{y(y + 5)}$   23) $\dfrac{r - 6}{r - 11}$   24) $\{6\}$   25) $3$

# Chapter 9

## Section 9.1

1) False; the $\sqrt{\phantom{x}}$ symbol means to find only the positive square root of 100. $\sqrt{100} = 10$

3) False; the square root of a negative number is not a real number.

5) $9$ and $-9$   7) $2$ and $-2$   9) $30$ and $-30$   11) $\dfrac{1}{6}$ and $-\dfrac{1}{6}$

13) $0.5$ and $-0.5$   15) $12$   17) $3$   19) not real   21) $-6$

23) $\dfrac{8}{9}$   25) $\dfrac{2}{3}$   27) $-\dfrac{1}{4}$   29) $0.7$   31) $-0.12$

33) $3.3$

35) $1.4$

37) $5.7$

39) $7.4$

41) $10$   43) $\sqrt{65}$   45) $12$   47) $\sqrt{2}$   49) $\sqrt{55}$ in.   51) $12$ ft

53) $17$ ft   55) $6.6$ in.   57) Ivan is right. You must combine like terms in the radicand before evaluating the square root.

59) $13$   61) $\sqrt{65}$   63) $\sqrt{109}$   65) $50$   67) $5$   69) $\sqrt{13}$

71) $\sqrt{61}$   73) $\sqrt{157}$   75) $\dfrac{5}{8}$   77) True

79) False; the odd root of a negative number is a negative number.

81) $\sqrt[3]{64}$ is the number you cube to get 64. $\sqrt[3]{64} = 4$

83) No; the even root of a negative number is not a real number.

85) $2$   87) $5$   89) $-1$   91) $3$   93) not real   95) $-2$

97) $-2$   99) $3$   101) not real   103) $\dfrac{2}{5}$   105) $7$   107) $5$

109) not real

## Section 9.2

1) $\sqrt{6}$   3) $\sqrt{130}$   5) $\sqrt{15n}$   7) False; 24 contains a factor of 4 which is a perfect square.

9) Factor.; $\sqrt{4} \cdot \sqrt{15}$; Simplify.

11) $2\sqrt{5}$   13) $3\sqrt{10}$   15) simplified   17) $-5\sqrt{3}$

19) $12\sqrt{5}$   21) $-40$   23) $2\sqrt{3}$   25) $6$   27) $2\sqrt{15}$

29) $42\sqrt{2}$   31) $6\sqrt{10}$   33) $\dfrac{3}{4}$   35) $4$   37) $2\sqrt{3}$

39) $-\dfrac{\sqrt{3}}{8}$   41) $\dfrac{2\sqrt{13}}{7}$   43) $8\sqrt{21}$   45) $\dfrac{4\sqrt{10}}{3}$

47) $\dfrac{\sqrt{5}}{4}$   49) $\dfrac{10}{7}$   51) $c$   53) $t^3$   55) $11b^5$   57) $6q^7$

59) $2r^2\sqrt{7}$   61) $3z^6\sqrt{2}$   63) $\dfrac{y^4}{13}$   65) $\dfrac{3\sqrt{11}}{w}$   67) $r^2s^6$

69) $6x^5y$   71) $\dfrac{m^9n^4}{7}$   73) Factor.; $\sqrt{p^8} \cdot \sqrt{p^1}$; Simplify.

75) $h\sqrt{h}$   77) $g^6\sqrt{g}$   79) $10w^2\sqrt{w}$   81) $5t^5\sqrt{3t}$

83) $q^3\sqrt{13q}$   85) $xy^4\sqrt{y}$   87) $2t^4u^2\sqrt{tu}$

89) (i) Its radicand will not contain any factors that are perfect cubes.
  (ii) The radicand will not contain fractions.
  (iii) There will be no radical in the denominator of a fraction.

91) $\sqrt[3]{12}$   93) $\sqrt[4]{9n}$   95) $2\sqrt[3]{3}$   97) $2\sqrt[3]{9}$   99) $3\sqrt[3]{4}$

101) $2\sqrt[4]{3}$   103) $3\sqrt[4]{2}$   105) $\dfrac{1}{4}$   107) $\dfrac{1}{2}$   109) $5\sqrt[3]{2}$

111) $10\sqrt[3]{2}$   113) $r$   115) $w^4$   117) $3d^5$   119) $5a^6b^2$

121) $\dfrac{t^3}{2}$   123) $t\sqrt[3]{t}$   125) $x^3\sqrt[3]{x^2}$   127) $b^5\sqrt[3]{b^2}$   129) $5g^2$

131) $3xy^4$   133) $2k^2\sqrt[3]{3k}$   135) $n$   137) $m\sqrt[4]{m}$

139) $2\sqrt{2}$ in.

## Section 9.3

1) They have the same index and the same radicand.

3) $9\sqrt{11}$   5) $-5\sqrt{2}$   7) $7\sqrt[3]{5}$   9) $18\sqrt[3]{2}$   11) $12 - 7\sqrt{13}$

13) Factor.; $\sqrt{4} \cdot \sqrt{6} + \sqrt{6}$; Simplify.; $3\sqrt{6}$   15) $4\sqrt{3}$

17) $6\sqrt{3}$   19) $-7\sqrt{7}$   21) $10\sqrt{2}$   23) $-5\sqrt{2}$   25) $6\sqrt{10}$

27) $-\sqrt{2}$   29) $11\sqrt{2}$   31) $7\sqrt{6}$   33) $12\sqrt[3]{11}$   35) $-\sqrt[3]{3}$

37) $32\sqrt[3]{2}$   39) $15\sqrt{c}$   41) $-4\sqrt{3a}$   43) $4\sqrt{5b}$   45) $2\sqrt{2n}$

47) $15\sqrt{3z}$   49) $21\sqrt{7v}$   51) $4\sqrt{h} - 4\sqrt{2p}$   53) $17z\sqrt{3}$

55) $14r\sqrt{7}$   57) $13q\sqrt{q}$   59) $3m^3\sqrt{m}$   61) $12w^3\sqrt{w}$

63) $7y\sqrt{xy}$   65) $3uv\sqrt{5u}$   67) $17m\sqrt{3mn}$   69) $-10\sqrt[3]{t^2}$

71) $17\sqrt[3]{x^2}$   73) $-7r^5\sqrt[3]{r}$   75) $25a\sqrt[3]{3a^2}$   77) $-18c^2\sqrt[3]{4c}$

79) $14p^2q\sqrt[3]{11pq^2}$   81) $5x$   83) $5c^3\sqrt[4]{c^3}$

85) $-8\sqrt[3]{y^2} + 12\sqrt[5]{y^2}$   87) $10y^2z^2\sqrt[4]{11z}$   89) $28\sqrt{7}$ in.

## Section 9.4

1) $5m + 15$  3) $7\sqrt{2} + 42$  5) $\sqrt{10} - 8\sqrt{2}$  7) $-30\sqrt{2}$

9) $6\sqrt{5}$  11) $-\sqrt{30}$  13) $7\sqrt{2} + 2\sqrt{3}$  15) $a - 8\sqrt{ab}$

17) $x\sqrt{6y} + 3y\sqrt{3x}$

19) Both are examples of multiplication of two binomials. They can be multiplied using FOIL.

21) $4 \cdot 3 + 4\sqrt{5} + 3\sqrt{5} + \sqrt{5} \cdot \sqrt{5}$; Multiply.; $17 + 7\sqrt{5}$

23) $k^2 + 9k + 18$  25) $19 + 8\sqrt{7}$  27) $-26 + 3\sqrt{2}$

29) $26 - 9\sqrt{15}$  31) $86 - 14\sqrt{3}$

33) $3\sqrt{7} + 3\sqrt{2} + 2\sqrt{35} + 2\sqrt{10}$  35) $m + 6\sqrt{7mn} + 35n$

37) $-2\sqrt{6pq} + 30p - 16q$

39) Both are examples of the square of a binomial. We can multiply them using the formula $(a - b)^2 = a^2 - 2ab + b^2$.

41) $4b^2 - 44b + 121$  43) $4 + 2\sqrt{3}$  45) $18 - 2\sqrt{65}$

47) $38 + 4\sqrt{70}$  49) $38 - 16\sqrt{3}$  51) $k + 2\sqrt{11k} + 11$

53) $x - 2\sqrt{xy} + y$  55) $(a + b)(a - b) = a^2 - b^2$

57) $a^2 - 81$  59) $-4$  61) $34$  63) $46$  65) $-39$  67) $c - d$

69) $36 - t$  71) $64f - g$  73) $79 + 18\sqrt{2}$

75) $7\sqrt{2mn} + n\sqrt{5}$  77) $29$  79) $11 - 4\sqrt{7}$

81) $15a - 4\sqrt{5ab} + b$  83) $\sqrt{10} - 3\sqrt{15} + 2\sqrt{5} - 3\sqrt{30}$

85) $\sqrt[3]{4} - 16$  87) $41$  89) $2\sqrt{10}$

## Section 9.5

1) Eliminate the radical from the denominator.

3) $\dfrac{\sqrt{6}}{6}$  5) $\dfrac{7\sqrt{3}}{3}$  7) $4\sqrt{2}$  9) $\dfrac{9\sqrt{10}}{2}$  11) $\dfrac{\sqrt{2}}{3}$

13) $-5\sqrt{2}$  15) $\dfrac{\sqrt{77}}{7}$  17) $\dfrac{2\sqrt{30}}{3}$  19) $\dfrac{\sqrt{10}}{6}$  21) $\dfrac{4\sqrt{6}}{3}$

23) $\dfrac{\sqrt{30}}{2}$  25) $\dfrac{\sqrt{2}}{2}$  27) $\dfrac{\sqrt{15}}{10}$  29) $\dfrac{\sqrt{10}}{20}$  31) $\dfrac{\sqrt{30}}{3}$

33) $\dfrac{2\sqrt{5}}{5}$  35) $\dfrac{2\sqrt{78}}{39}$  37) $\dfrac{9}{4}$  39) $\dfrac{5\sqrt{k}}{k}$  41) $\dfrac{\sqrt{6d}}{d}$

43) $\dfrac{4\sqrt{a}}{a}$  45) $\dfrac{\sqrt{xy}}{y}$  47) $\dfrac{2p\sqrt{10pq}}{q}$  49) $\dfrac{8v^3\sqrt{5vw}}{5w}$

51) $3h^2$  53) $\dfrac{x^3\sqrt{15y}}{3y}$  55) $\dfrac{\sqrt{2n}}{n}$  57) $\dfrac{v\sqrt{3u}}{3u}$  59) $\dfrac{10\sqrt{5p}}{p^2}$

61) $2^2$ or $4$  63) $3$  65) $c^2$

67) He multiplied incorrectly: $\dfrac{9}{\sqrt[3]{2}} \cdot \dfrac{\sqrt[3]{2}}{\sqrt[3]{2}} = \dfrac{9\sqrt[3]{2}}{\sqrt[3]{4}}$.

The correct way is $\dfrac{9}{\sqrt[3]{2}} \cdot \dfrac{\sqrt[3]{4}}{\sqrt[3]{4}} = \dfrac{9\sqrt[3]{4}}{\sqrt[3]{8}} = \dfrac{9\sqrt[3]{4}}{2}$.

69) $\dfrac{4\sqrt[3]{9}}{3}$  71) $\dfrac{\sqrt[3]{2}}{2}$  73) $7\sqrt[3]{25}$  75) $\dfrac{2\sqrt[3]{3}}{3}$  77) $\dfrac{\sqrt[3]{55}}{5}$

79) $\dfrac{6\sqrt[3]{x^2}}{x}$  81) $\dfrac{7\sqrt[3]{m}}{m}$  83) $\dfrac{10\sqrt[3]{c^2}}{c}$  85) $\dfrac{\sqrt[3]{4w}}{w}$  87) $\dfrac{8\sqrt[3]{4t^2}}{t}$

89) $\dfrac{\sqrt[3]{30ab^2}}{6b}$  91) $\dfrac{7\sqrt[4]{8}}{2}$  93) $\dfrac{2\sqrt[4]{x}}{x}$

95) Change the sign between the two terms.

97) $(4 - \sqrt{3}); 13$  99) $(\sqrt{5} + \sqrt{11}); -6$

101) $(\sqrt{p} + 6); p - 36$

103) Multiply by the conjugate.;
$(a + b)(a - b) = a^2 - b^2; \dfrac{20 + 5\sqrt{3}}{16 - 3}; \dfrac{20 + 5\sqrt{3}}{13}$

105) $\dfrac{18 - 6\sqrt{2}}{7}$  107) $\dfrac{35 + 5\sqrt{3}}{46}$  109) $-\dfrac{10\sqrt{5} + 60}{31}$

111) $2\sqrt{5} + 3\sqrt{2}$  113) $\dfrac{17 - \sqrt{3}}{13}$

115) $6 + 2\sqrt{6} - \sqrt{2} - \sqrt{3}$  117) $\dfrac{12\sqrt{w} - 36}{w - 9}$

119) $\dfrac{45 + 9\sqrt{n}}{25 - n}$  121) $\dfrac{m - \sqrt{mn}}{m - n}$  123) $1 + 4\sqrt{5}$

125) $\dfrac{5 - 11\sqrt{3}}{3}$  127) $\dfrac{\sqrt{5} + 2}{3}$  129) $-6 - 5\sqrt{2}$

131) $\dfrac{\sqrt{22}}{10}$  133) $\dfrac{\sqrt[3]{36}}{3}$  135) $\dfrac{5a\sqrt{30ab}}{6b}$

137) $-\dfrac{37 + 12\sqrt{10}}{71}$  139) $\dfrac{2\sqrt{21}}{7}$  141) $\dfrac{8\sqrt{2y}}{y}$

143) $r = 2\sqrt{3}$ in.

## Section 9.6

1) Sometimes there are extraneous solutions.

3) $\{144\}$  5) $\left\{\dfrac{1}{9}\right\}$  7) $\{20\}$  9) $\{7\}$  11) $\{1\}$  13) $\varnothing$

15) $\left\{-\dfrac{31}{2}\right\}$  17) $\{3\}$  19) $\varnothing$  21) $\{2\}$  23) $\{1\}$  25) $\{63\}$

27) $\{-3\}$  29) $\varnothing$  31) $\{9\}$  33) $\{1, 16\}$  35) $\{9\}$  37) $\varnothing$

39) $\{2, 3\}$  41) $\{-1, 3\}$  43) $y^2 + 14y + 49$

45) $n + 14\sqrt{n} + 49$  47) $19 - 8\sqrt{c + 3} + c$

49) $12p + 4\sqrt{3p - 4} - 15$  51) $\{4\}$  53) $\{49\}$  55) $\{2, 18\}$

57) $\{1\}$  59) $\varnothing$  61) $\left\{\dfrac{1}{4}\right\}$  63) $\{-2, 2\}$  65) $\{3\}$  67) $\{0\}$

69) Cube both sides of the equation.  71) $\{-125\}$

73) $\{-1\}$  75) $\left\{-\dfrac{7}{3}\right\}$  77) $\{-2, 7\}$  79) $\left\{-4, -\dfrac{3}{2}\right\}$

81) $\{71\}$  83) $\{-2, 8\}$  85) $\{-8\}$  87) $25$  89) $4$

91) 6 ft $\times$ 6 ft  93) 62 mph; the car was not speeding.

95) 1600 ft

## Section 9.7

1) The denominator of 2 becomes the index of the radical. $36^{1/2} = \sqrt{36}$

3) 4    5) 10    7) 3    9) $-4$    11) $\dfrac{3}{2}$    13) $\dfrac{5}{4}$    15) $-\dfrac{10}{7}$

17) The denominator of 4 becomes the index of the radical. The numerator of 3 is the power to which we raise the radical expression. $16^{3/4} = (\sqrt[4]{16})^3$

19) 16    21) 27    23) 32    25) $-216$    27) $\dfrac{32}{243}$    29) $-\dfrac{16}{81}$

31) False; the negative exponent does not make the result negative. $121^{-1/2} = \dfrac{1}{11}$

33) $\dfrac{1}{4}$    35) $\dfrac{1}{10}$    37) $\dfrac{1}{5}$    39) $-\dfrac{1}{3}$    41) $\dfrac{1}{32}$    43) $\dfrac{1}{16}$    45) $-\dfrac{1}{9}$

47) $-\dfrac{1}{32}$    49) 125    51) 4    53) $7^{4/5}$    55) 81    57) $\dfrac{1}{125}$

59) 7    61) $6^{1/3}$    63) $k^{7/2}$    65) $\dfrac{1}{m^{7/10}}$    67) $-72v^{11/8}$

69) $n^{1/3}$    71) $\dfrac{13}{36c^{3/2}}$    73) $q^6$    75) $\dfrac{1}{x^{5/3}}$    77) $z^{3/20}$

79) 7    81) 2    83) $\sqrt[3]{t^2}$    85) $\sqrt{h}$    87) $d^2$    89) 1 sec

91) a) 7711    b) 28,284

## Chapter 9 Review Exercises

1) 7    3) $-4$    5) $-5$    7) 3    9) 6    11) 10    13) $3\sqrt{5}$

15) $3\sqrt[3]{3}$    17) $2\sqrt[4]{5}$    19) $2\sqrt{2}$    21) $-2$    23) $h^4$

25) $b^8$    27) $w^7\sqrt{w}$    29) $3d^2\sqrt{7}$    31) $2x^6y^2\sqrt{11y}$

33) $2m^5\sqrt[3]{6m^2}$    35) $\dfrac{25\sqrt{3}}{7}$    37) $\sqrt{30}$    39) $5\sqrt{2}$

41) $\sqrt[3]{30}$    43) $4\sqrt{5} + 5\sqrt{6}$    45) $-14\sqrt[3]{9}$    47) $-25n^2\sqrt{2n}$

49) $7 - \sqrt{35}$    51) $13 + 2\sqrt{2}$    53) $49 - 20\sqrt{6}$    55) 1

57) $6\sqrt{3}$    59) $\dfrac{7\sqrt{2h}}{h}$    61) $5\sqrt[3]{9}$    63) $\dfrac{7\sqrt[3]{4c^2}}{2c}$    65) $\dfrac{6 - 2\sqrt{2}}{11}$

67) $\dfrac{54 + 6\sqrt{x}}{81 - x}$    69) $2 - 3\sqrt{3}$    71) $\{81\}$    73) $\{9\}$

75) $\{-7\}$    77) $\{1, 2\}$    79) $\{1\}$

81) The denominator of the fractional exponent becomes the index on the radical. The numerator is the power to which we raise the radical expression. $8^{2/3} = (\sqrt[3]{8})^2$

83) 8    85) $\dfrac{4}{5}$    87) $-3$    89) 25    91) $\dfrac{100}{9}$    93) $\dfrac{1}{11}$

95) $-\dfrac{1}{27}$    97) 16    99) $\dfrac{1}{16}$    101) $-24p^{1/3}$    103) $\dfrac{1}{m^4}$

105) $\sqrt{n}$    107) $5d\sqrt{d}$    109) $33\sqrt{2}$    111) $t^3$

113) $\dfrac{6\sqrt{5} - 5\sqrt{2}}{26}$    115) $\dfrac{1}{16}$    117) $19 - 8\sqrt{3}$    119) $8p^7\sqrt{2p}$

## Chapter 9 Test

1) 11    2) not real    3) $-10$    4) 3    5) $\sqrt{39}$ cm    6) $\sqrt{5}$

7) $5\sqrt{5}$    8) $2\sqrt[3]{6}$    9) $4\sqrt{2}$    10) $m^3$    11) $h^3\sqrt{h}$    12) $k^4$

13) $b^4\sqrt[3]{b^2}$    14) $3m^2n^4\sqrt{7m}$    15) $\dfrac{x^5y^2\sqrt[3]{y}}{2}$    16) $2w^5\sqrt{15w}$

17) $6y$    18) $z^3\sqrt[3]{z}$    19) $8\sqrt{3} - 3\sqrt{2}$    20) $-5a^2\sqrt{2a}$

21) $\dfrac{4\sqrt{5}}{9}$    22) $5\sqrt{2} - 9\sqrt{5}$

23) $4\sqrt{3} + 36 - 3\sqrt{21} - 27\sqrt{7}$    24) $-11$

25) $49 - 28\sqrt{t} + 4t$    26) $\dfrac{\sqrt{6}}{3}$    27) $\dfrac{33 - 11\sqrt{2}}{7}$    28) $\dfrac{\sqrt{35k}}{5k}$

29) $6\sqrt[3]{2}$    30) $2 - \sqrt{3}$

31) The principal square root of a number cannot equal a negative number.

32) $\left\{\dfrac{5}{2}\right\}$    33) $\{2, 3\}$    34) $\{3\}$    35) 4 in.    36) $-2$

37) $\left\{\dfrac{1}{13}\right\}$

38) We can change $81^{3/4}$ to a radical expression and evaluate. The denominator of the fractional exponent becomes the index on the radical. The numerator is the power to which we raise the radical expression. Therefore, $81^{3/4} = (\sqrt[4]{81})^3 = (3)^3 = 27$.

39) $-20g^{7/8}$    40) $\dfrac{7}{2w^{1/3}}$    41) $\dfrac{y^2}{x^{3/2}}$

## Cumulative Review for Chapters 1–9

1) $\dfrac{15}{14}$    2) $2(-9 + 4); -10$    3) $\dfrac{1}{16}$    4) $2a - \dfrac{9}{2}b + 1$

5) $9.41 \times 10^{-4}$    6) $\left\{\dfrac{5}{3}\right\}$

7)

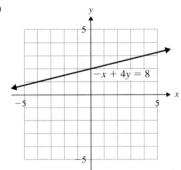

$-x + 4y = 8$

8) $y = -\dfrac{6}{5}x - \dfrac{14}{5}$    9) $\left(-\dfrac{1}{3}, 4\right)$    10) $8t^3 - 42t^2 + 7t + 15$

11) $-n^2 + 9n + 4$    12) $6a^2 + 3a - 7$    13) $(p - 5)^2$

14) $3(2y - 3)(y - 2)$    15) $(h - 9)(h^2 + 4)$    16) $\{-4, 7\}$

17) length = 8 in., width = 5 in.    18) $\dfrac{4m^2 - 9m - 15}{4m(m + 3)}$

19) $-\dfrac{3k(k-9)}{k+8}$   20) $\left\{-\dfrac{5}{2}, -2\right\}$   21) $\dfrac{4\sqrt{15}}{9}$   22) $2x^2\sqrt[3]{9}$

23) $-\dfrac{\sqrt{3}+2\sqrt{2}}{5}$   24) 16   25) 9

## Chapter 10

### Section 10.1

1) factoring and the square root property; $\{-9, 9\}$

3) $\{-4, 4\}$   5) $\{-\sqrt{11}, \sqrt{11}\}$   7) $\varnothing$   9) $\left\{-\dfrac{5}{3}, \dfrac{5}{3}\right\}$

11) $\{-0.2, 0.2\}$   13) $\{-12, 12\}$   15) $\{-\sqrt{19}, \sqrt{19}\}$

17) $\{-3\sqrt{6}, 3\sqrt{6}\}$   19) $\left\{-\dfrac{\sqrt{5}}{8}, \dfrac{\sqrt{5}}{8}\right\}$   21) $\{-\sqrt{14}, \sqrt{14}\}$

23) $\varnothing$   25) $\{-3, 3\}$   27) $\{-2\sqrt{3}, 2\sqrt{3}\}$   29) $\{-2, 2\}$

31) $\varnothing$   33) $\left\{-\dfrac{3}{2}, \dfrac{3}{2}\right\}$   35) $\left\{-\dfrac{\sqrt{10}}{3}, \dfrac{\sqrt{10}}{3}\right\}$

37) $\left\{-\dfrac{\sqrt{55}}{5}, \dfrac{\sqrt{55}}{5}\right\}$   39) $\{-11, -1\}$   41) $\{7, 9\}$

43) $\{-2 - \sqrt{13}, -2 + \sqrt{13}\}$   45) $\{10 - 3\sqrt{5}, 10 + 3\sqrt{5}\}$

47) $\varnothing$   49) $\{-3 - 2\sqrt{17}, -3 + 2\sqrt{17}\}$   51) $\{-1, 2\}$

53) $\left\{-4, \dfrac{3}{2}\right\}$   55) $\left\{\dfrac{10 - 2\sqrt{6}}{3}, \dfrac{10 + 2\sqrt{6}}{3}\right\}$

57) $\left\{\dfrac{-8 - 5\sqrt{5}}{5}, \dfrac{-8 + 5\sqrt{5}}{5}\right\}$

59) $\left\{\dfrac{-3 - 3\sqrt{6}}{2}, \dfrac{-3 + 3\sqrt{6}}{2}\right\}$   61) $\varnothing$   63) $\left\{-\dfrac{2}{5}, \dfrac{6}{5}\right\}$

65) $\left\{-\dfrac{5}{3}, -\dfrac{2}{3}\right\}$   67) $\left\{\dfrac{11 - \sqrt{14}}{2}, \dfrac{11 + \sqrt{14}}{2}\right\}$   69) $\{0, 4\}$

71) 9 cm   73) $\dfrac{1}{2}$ m   75) 3 m   77) 16 m/sec

### Section 10.2

1) a trinomial whose factored form is the square of a binomial; examples may vary.

3) $\dfrac{1}{2}(18) = 9$; $9^2 = 81$; $y^2 + 18y + 81$; $y^2 + 18y + 81$; $(y + 9)^2$

5) $a^2 + 12a + 36$; $(a + 6)^2$   7) $k^2 - 10k + 25$; $(k - 5)^2$

9) $g^2 - 24g + 144$; $(g - 12)^2$   11) $h^2 + 9h + \dfrac{81}{4}$; $\left(h + \dfrac{9}{2}\right)^2$

13) $x^2 - x + \dfrac{1}{4}$; $\left(x - \dfrac{1}{2}\right)^2$   15) Answers may vary.

17) $\{-4, -2\}$   19) $\{5, 9\}$   21) $\varnothing$

23) $\{-2 - \sqrt{15}, -2 + \sqrt{15}\}$   25) $\{5 - 2\sqrt{7}, 5 + 2\sqrt{7}\}$

27) $\varnothing$   29) $\{-8, 5\}$   31) $\{-7, 8\}$

33) $\left\{-\dfrac{9}{2} - \dfrac{\sqrt{33}}{2}, -\dfrac{9}{2} + \dfrac{\sqrt{33}}{2}\right\}$   35) $\varnothing$

37) No, because the coefficient of $x^2$ is not 1.

39) $\left\{-\dfrac{11}{2}, -\dfrac{5}{2}\right\}$   41) $\{5 - 2\sqrt{3}, 5 + 2\sqrt{3}\}$   43) $\{3, 4\}$

45) $\left\{\dfrac{9}{2} - \dfrac{\sqrt{33}}{2}, \dfrac{9}{2} + \dfrac{\sqrt{33}}{2}\right\}$   47) $\left\{\dfrac{1}{4}, \dfrac{3}{4}\right\}$   49) $\varnothing$

51) $\{-7, 5\}$   53) $\{-3 - 4\sqrt{2}, -3 + 4\sqrt{2}\}$   55) $\left\{-\dfrac{2}{3}, 1\right\}$

57) $\left\{-\dfrac{11}{5} - \dfrac{\sqrt{86}}{5}, -\dfrac{11}{5} + \dfrac{\sqrt{86}}{5}\right\}$

59) width = 10 in., length = 17 in.

61) base = 15 cm, height = 8 cm   63) 10, 24, 26

### Section 10.3

1) $x = \dfrac{-b \pm \sqrt{b^2 - 4ac}}{2a}$

3) The equation must be written as $3x^2 - 5x - 4 = 0$ before identifying the values of $a$, $b$, and $c$. $a = 3$, $b = -5$, $c = -4$;
$x = \dfrac{-(-5) \pm \sqrt{(-5)^2 - 4(3)(-4)}}{2(3)}$

5) $\{-4, 2\}$   7) $\left\{\dfrac{1}{2}, \dfrac{2}{3}\right\}$   9) $\left\{\dfrac{5 - \sqrt{17}}{2}, \dfrac{5 + \sqrt{17}}{2}\right\}$

11) $\left\{\dfrac{1 - \sqrt{13}}{3}, \dfrac{1 + \sqrt{13}}{3}\right\}$   13) $\varnothing$   15) $\{-11, 0\}$

17) $\left\{\dfrac{3 - \sqrt{3}}{2}, \dfrac{3 + \sqrt{3}}{2}\right\}$   19) $\{-1 - \sqrt{10}, -1 + \sqrt{10}\}$

21) $\varnothing$   23) $\{3, 5\}$   25) $\left\{\dfrac{1}{4}\right\}$   27) $\{-3 - \sqrt{5}, -3 + \sqrt{5}\}$

29) $\left\{\dfrac{1 - \sqrt{55}}{9}, \dfrac{1 + \sqrt{55}}{9}\right\}$   31) $\{0.25, 0.5\}$

33) $\left\{-\dfrac{\sqrt{3}}{4}, \dfrac{\sqrt{3}}{4}\right\}$   35) $\left\{-\dfrac{2}{3}, \dfrac{2}{3}\right\}$   37) $\left\{\dfrac{3}{2}\right\}$   39) $\varnothing$

41) $\left\{\dfrac{5 - \sqrt{19}}{2}, \dfrac{5 + \sqrt{19}}{2}\right\}$   43) $\dfrac{2 + \sqrt{70}}{3}$ in. or about 3.46 in.

45) a) 2 sec   b) $\dfrac{3 + \sqrt{33}}{4}$ sec or about 2.19 sec

### Putting It All Together

1) $\{-5\sqrt{3}, 5\sqrt{3}\}$   2) $\{4 - 2\sqrt{5}, 4 + 2\sqrt{5}\}$   3) $\{-5, 4\}$

4) $\varnothing$   5) $\{-3 - \sqrt{2}, -3 + \sqrt{2}\}$   6) $\left\{-\dfrac{3}{4}, 2\right\}$

7) $\left\{-5, \dfrac{1}{3}\right\}$   8) $\varnothing$   9) $\varnothing$

10) $\left\{\dfrac{-3 - \sqrt{15}}{2}, \dfrac{-3 + \sqrt{15}}{2}\right\}$   11) $\varnothing$   12) $\varnothing$   13) $\{-4, 10\}$

14) $\{-6, 6\}$   15) $\{3 - \sqrt{14}, 3 + \sqrt{14}\}$   16) $\{-7, -2, 0\}$

17) $\left\{\dfrac{5}{2}, 3\right\}$   18) $\{0, 3\}$   19) $\{3, 6\}$

20) $\left\{\dfrac{-3 - \sqrt{105}}{8}, \dfrac{-3 + \sqrt{105}}{8}\right\}$   21) $\varnothing$

22) $\{-2 - \sqrt{6}, -2 + \sqrt{6}\}$   23) $\left\{-2, \dfrac{1}{2}\right\}$   24) $\left\{-\dfrac{2}{3}, \dfrac{1}{2}\right\}$

25) $\{0, 5\}$   26) $\varnothing$   27) $\left\{-\dfrac{3}{2}, 0, \dfrac{3}{2}\right\}$   28) $\{-8, 4\}$

29) $\left\{\dfrac{9 - \sqrt{57}}{4}, \dfrac{9 + \sqrt{57}}{4}\right\}$   30) $\left\{\dfrac{5 - \sqrt{17}}{4}, \dfrac{5 + \sqrt{17}}{4}\right\}$

**Section 10.4**

1) False   3) True   5) $6i$   7) $4i$   9) $i\sqrt{5}$   11) $3i\sqrt{3}$

13) $3i\sqrt{5}$   15) $4i\sqrt{2}$

17) Add the real parts and add the imaginary parts.

19) $-3 + 12i$   21) $4 - 9i$   23) $-\dfrac{1}{8} - \dfrac{13}{12}i$   25) $2i$

27) Both are products of binomials, so we can multiply both using FOIL.

29) $14 - 12i$   31) $-20 - \dfrac{15}{2}i$   33) $-21 - 6i$   35) $15 + 20i$

37) $-2 + 26i$   39) $14 + 18i$   41) $34 - 58i$   43) $-\dfrac{2}{15} + \dfrac{2}{5}i$

45) conjugate: $3 - 5i$; product: 34

47) conjugate: $-10 + 2i$; product: 104

49) conjugate: $-7 - 9i$; product: 130

51) Multiply the numerator and denominator by the conjugate of the denominator.

53) $\dfrac{2}{37} + \dfrac{12}{37}i$   55) $-\dfrac{12}{17} - \dfrac{20}{17}i$   57) $-\dfrac{2}{3} + \dfrac{4}{3}i$

59) $-\dfrac{74}{85} + \dfrac{27}{85}i$   61) $-\dfrac{38}{29} + \dfrac{21}{29}i$   63) $-7i$   65) $\{-3i, 3i\}$

67) $\{-i\sqrt{11}, i\sqrt{11}\}$   69) $\{-2i\sqrt{7}, 2i\sqrt{7}\}$   71) $\{-5i, 5i\}$

73) $\{-3 - 5i, -3 + 5i\}$   75) $\{8 - 2i\sqrt{5}, 8 + 2i\sqrt{5}\}$

77) $\left\{-\dfrac{1}{4} - \dfrac{\sqrt{3}}{4}i, -\dfrac{1}{4} + \dfrac{\sqrt{3}}{4}i\right\}$

79) $\left\{-\dfrac{1}{6} - \dfrac{\sqrt{23}}{6}i, -\dfrac{1}{6} + \dfrac{\sqrt{23}}{6}i\right\}$

81) $\{-1 - i\sqrt{3}, -1 + i\sqrt{3}\}$   83) $\{2 - 5i, 2 + 5i\}$

85) $\left\{\dfrac{-3 - \sqrt{23}}{2}, \dfrac{-3 + \sqrt{23}}{2}\right\}$

87) $\left\{\dfrac{1}{4} - \dfrac{\sqrt{11}}{4}i, \dfrac{1}{4} + \dfrac{\sqrt{11}}{4}i\right\}$

89) $\left\{-1 - \dfrac{\sqrt{6}}{3}i, -1 + \dfrac{\sqrt{6}}{3}i\right\}$

91) $\left\{\dfrac{6}{5} - \dfrac{3}{5}i, \dfrac{6}{5} + \dfrac{3}{5}i\right\}$   93) $\{2 - i, 2 + i\}$

95) $\{-4 - i\sqrt{13}, -4 + i\sqrt{13}\}$

97) $\{3 - 3i\sqrt{2}, 3 + 3i\sqrt{2}\}$

99) $\left\{-\dfrac{3}{2} - \dfrac{\sqrt{7}}{2}i, -\dfrac{3}{2} + \dfrac{\sqrt{7}}{2}i\right\}$

**Section 10.5**

1) The $x$-coordinate of the vertex is $-\dfrac{b}{2a}$. Substitute that value into the equation to find the $y$-coordinate of the vertex.

3) Vertex: $(0, -4)$

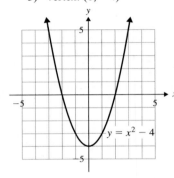

5) Vertex: $(0, -1)$

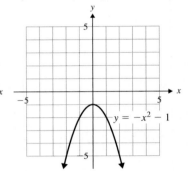

7) Vertex: $(3, 0)$

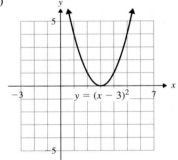

9) Vertex: $(2, -1)$

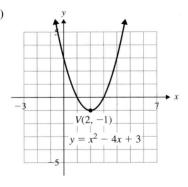

11) Vertex: $(1, 4)$

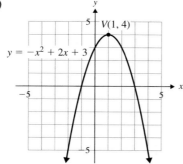

$y = -x^2 + 2x + 3$

$V(1, 4)$

13) Vertex: $(-1, 5)$

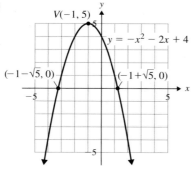

$V(-1, 5)$

$y = -x^2 - 2x + 4$

$(-1-\sqrt{5}, 0)$   $(-1+\sqrt{5}, 0)$

15) Vertex: $(-3, 2)$

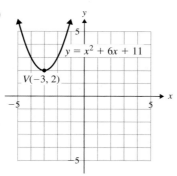

$y = x^2 + 6x + 11$

$V(-3, 2)$

17) Vertex: $(1, 0)$

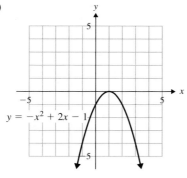

$y = -x^2 + 2x - 1$

19) Vertex: $(1, -2)$

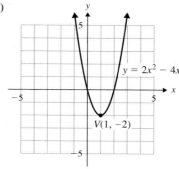

$y = 2x^2 - 4x$

$V(1, -2)$

21) $\{-1, 3\}$   23) $\{-1\}$   25) $\varnothing$   27) $\{0.4, 3\}$

29) $\{-4.5, -0.25\}$   31) $\{-3.65, 1.65\}$

33) Answers may vary. One equation is $y = x^2 - 6x + 8$.

35) Answers may vary. One equation is $y = x^2 + 2x - 3$.

37) $(-\infty, \infty)$   39) $[-2, \infty)$   41) $(-\infty, 0]$

43) $(-\infty, 3]$   45) $[-1, \infty)$   47) a) 4   b) 7

49) a) 9 sec   b) 1296 ft

**Chapter 10 Review Exercises**

1) $\{-9, 9\}$   3) $\{-\sqrt{11}, \sqrt{11}\}$   5) $\varnothing$   7) $\left\{-12, -\dfrac{4}{3}\right\}$

9) $x^2 + 18x + 81; (x + 9)^2$   11) $z^2 - 3z + \dfrac{9}{4}; \left(z - \dfrac{3}{2}\right)^2$

13) $\{-9, -1\}$   15) $\varnothing$   17) $\left\{-\dfrac{5}{2} - \dfrac{\sqrt{17}}{2}, -\dfrac{5}{2} + \dfrac{\sqrt{17}}{2}\right\}$

19) $\left\{-\dfrac{3}{2}, \dfrac{7}{2}\right\}$   21) $x = \dfrac{-b \pm \sqrt{b^2 - 4ac}}{2a}$   23) $\{-4, -3\}$

25) $\varnothing$   27) $\{5 - 3\sqrt{2}, 5 + 3\sqrt{2}\}$

29) $\left\{\dfrac{-2 - \sqrt{10}}{3}, \dfrac{-2 + \sqrt{10}}{3}\right\}$   31) 3 in.   33) $\{-4, 7\}$

35) $\left\{\dfrac{-1 - \sqrt{13}}{3}, \dfrac{-1 + \sqrt{13}}{3}\right\}$   37) $\{9 - \sqrt{5}, 9 + \sqrt{5}\}$

39) $\left\{0, \dfrac{1}{3}, 6\right\}$   41) $\varnothing$   43) $\{0, 4\}$   45) $12i$   47) $2i\sqrt{10}$

49) $-3 + 5i$   51) $-8 - 21i$   53) $-1$   55) $-\dfrac{9}{2} + 9i$

57) $-43 + 23i$   59) conjugate: $-6 + 4i$; product: 52

61) $\dfrac{21}{10} - \dfrac{7}{10}i$   63) $-\dfrac{26}{25} + \dfrac{32}{25}i$

65) The first problem is the sum of two complex numbers, and the second problem is the product of two complex numbers.

67) $4 + 7i$   69) $-24 - 18i$   71) $\dfrac{24}{17} - \dfrac{6}{17}i$   73) $-1 + 23i$

75) $\{-6i, 6i\}$    77) $\{4 - i, 4 + i\}$

79) $\{-2 - 3i\sqrt{3}, -2 + 3i\sqrt{3}\}$    81) $\left\{\dfrac{3}{2} - \dfrac{3}{2}i, \dfrac{3}{2} + \dfrac{3}{2}i\right\}$

83) parabola

85) vertex: $(0, 4)$

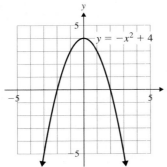

87) vertex: $(-2, -1)$

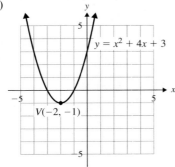

89) vertex: $(3, 2)$

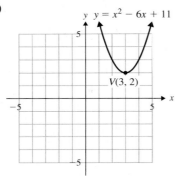

91) $\{-0.5, 0.75\}$    93) domain: $(-\infty, \infty)$; range: $[0, \infty)$

95) domain: $(-\infty, \infty)$; range: $(-\infty, -1]$

**Chapter 10 Test**

1) $\{-\sqrt{6}, \sqrt{6}\}$    2) 4 m/sec    3) $\{6 - \sqrt{10}, 6 + \sqrt{10}\}$

4) $\left\{-\dfrac{5}{2} - \dfrac{\sqrt{13}}{2}, -\dfrac{5}{2} + \dfrac{\sqrt{13}}{2}\right\}$

5) She divided only the 12 in the numerator by the 4 in the denominator. The correct way to simplify is
$$\dfrac{12 \pm 8\sqrt{5}}{4} = \dfrac{4(3 \pm 2\sqrt{5})}{4} = 3 \pm 2\sqrt{5}.$$

6) $\left\{\dfrac{-9 - \sqrt{41}}{4}, \dfrac{-9 + \sqrt{41}}{4}\right\}$    7) $\{7 - 2\sqrt{3}, 7 + 2\sqrt{3}\}$

8) $\{-6, 2\}$    9) $\left\{\dfrac{1 - \sqrt{13}}{4}, \dfrac{1 + \sqrt{13}}{4}\right\}$    10) $\varnothing$

11) a) 3 sec   b) $\dfrac{5 + \sqrt{57}}{4}$ sec or about 3.14 sec    12) $11i$

13) $4i\sqrt{3}$    14) $-20 + 12i$    15) $-4 + i$    16) $43 - 6i$

17) $\dfrac{18}{17} - \dfrac{4}{17}i$    18) $-\dfrac{33}{29} - \dfrac{10}{29}i$

19) $\{-8 - i\sqrt{3}, -8 + i\sqrt{3}\}$    20) $\left\{\dfrac{1}{5} - \dfrac{2}{5}i, \dfrac{1}{5} + \dfrac{2}{5}i\right\}$

21) vertex: $(1, 0)$

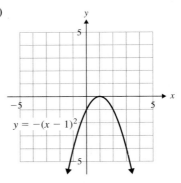

22) vertex: $(-1, -4)$

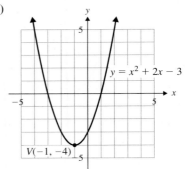

23) vertex: $(2, 3)$

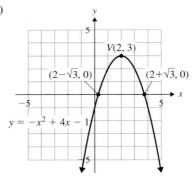

24) $\{2, 4\}$    25) domain: $(-\infty, \infty)$; range: $[-4, \infty)$

**Cumulative Review: Chapter 1–10**

1) $\dfrac{19}{56}$    2) $\dfrac{1}{4}$    3) $A = 29.6$ mm$^2$; $P = 24$ mm

4) a) $\{2\}$   b) $\{-6, 0, 2\}$   c) $\left\{-6, 14.38, \dfrac{3}{11}, 2, 5.\overline{7}, 0\right\}$

d) $\{\sqrt{23}, 9.21743819 \ldots\}$

5) $-17°F$  6) a) 32  b) $\dfrac{1}{27}$  c) $\dfrac{x^{17}}{16y^{12}}$  d) 4

7) a) 1  b) 2  c) $-\dfrac{1}{36}$  d) $\dfrac{1}{16}$

8) a) $81s^{16}t^{20}$  b) $d^{25}$  c) $\dfrac{125t^9}{27k^{18}}$  d) 14   9) $5.75 \times 10^{-5}$

10) 0.0000004   11) {8}   12) {all real numbers}   13) 10

14) \$4500 at 7% and \$7500 at 8%   15) $(-2, 3]$

16)
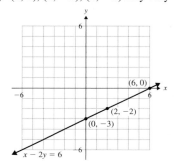

| x | y |
|---|---|
| 2 | 0 |
| 0 | $-\dfrac{4}{3}$ |
| 3 | $\dfrac{2}{3}$ |
| $-4$ | $-4$ |

17) $(6, 0), (0, -3); (2, -2)$ may vary

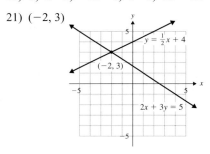

18) $5x + 3y = 6$  19) $y = 2x - 7$

20) a) 8  b) $-27$  c) 5  d) $5c - 22$

21) $(-2, 3)$

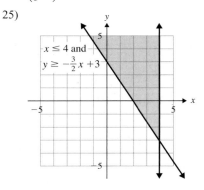

22) $\left(\dfrac{1}{3}, 5\right)$   23) $(2, 1)$   24) car: 60 mph; bus: 48 mph

25)

26) $\dfrac{4q^{30}}{9p^6}$   27) $7.9p^3 + 5.1p^2 + 4.8p - 3.6$

28) $\dfrac{5}{4}k^2 - \dfrac{2}{3}k + 7$   29) $-35w^2 - 45w + 180$

30) $2q - 4 - \dfrac{7}{3q + 7}$   31) $(t + 7u)(t - 9u)$

32) $(3g - 11)(g + 4)$   33) $(a + 8)^2$   34) $\left\{-\dfrac{5}{2}, \dfrac{1}{2}\right\}$

35) length = 7 cm; width = 4 cm   36) $\dfrac{x - 5}{55}$

37) $\dfrac{10p^2 - 67p - 53}{4(p + 1)(p - 7)}$   38) $\dfrac{4 + 9z}{10z^2}$   39) $\dfrac{c - 1}{c - 2}$   40) {$-3, 3$}

41) $10k^8$   42) $6\sqrt{5} - 4\sqrt{3}$   43) $\sqrt{z} - 2$   44) $\dfrac{1}{27}$

45) {12}   46) $\{-3\sqrt{2}, 3\sqrt{2}\}$   47) $\{-2 - i\sqrt{5}, -2 + i\sqrt{5}\}$

48) $\sqrt{29}$   49) $-24 - 42i$

50) Vertex: $(1, 4)$

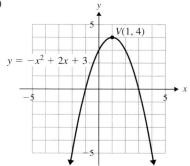

# Appendix

## Section A.1

1) a) 24  b) 24  c) no mode   3) a) 6.4  b) 8  c) 8
5) a) 43.7  b) 46  c) 46   7) a) 30.7  b) 30  c) 24 and 38
9) a) 304.3  b) 301  c) 283   11) a) 558.9  b) 571.5  c) 605
13) 78.7 games; 79 games; 79 games   15) $-6; -6; -8$
17) \$149.40; \$152; no mode   19) \$92   21) 292
23) 59   25) 3.53   27) 2.71

## Section A.2

1) yes   3) no   5) $(-2, 0, 5)$   7) $(1, -1, 4)$   9) $\left(2, -\dfrac{1}{2}, \dfrac{5}{2}\right)$

11) $\varnothing$   13) $\{(x, y, z) \mid 5x + y - 3z = -1\}$

15) $\{(a, b, c) \mid -a + 4b - 3c = -1\}$   17) $(2, 5, -5)$

19) $\left(-4, \dfrac{3}{5}, 4\right)$   21) $(0, -7, 6)$   23) $(1, 5, 2)$

25) $\left(-\dfrac{1}{4}, -5, 3\right)$   27) $\varnothing$   29) $\left(4, -\dfrac{3}{2}, 0\right)$   31) $(4, 4, 4)$

33) $\{(x, y, z) \mid -4x + 6y + 3z = 3\}$   35) $\left(1, -7, \dfrac{1}{3}\right)$

37) $(-3, -1, 1)$   39) hot dog: \$2.00; fries: \$1.50; soda: \$2.00

41) Clif Bar: 11g; Balance Bar: 15 g; PowerBar: 24 g

43) Knicks: \$160 million; Lakers: \$149 million;
Bulls: \$119 million   45) value: \$22; regular: \$36; prime: \$45

47) $104°; 52°, 24°$   49) $80°, 64°, 36°$   51) 12 cm, 10 cm, 7 cm

## Section A.3

1) a) $[-5, 1]$   b) $(-\infty, -5) \cup (1, \infty)$

3) a) $[-1, 3]$   b) $(-\infty, -1) \cup (3, \infty)$

5) $(-\infty, -7] \cup [1, \infty)$

7) $(-9, 4)$

9) $(-\infty, 6) \cup (7, \infty)$

11) $\left[-6, \dfrac{4}{3}\right]$

13) $\left(-\infty, -\dfrac{2}{7}\right) \cup (2, \infty)$

15) $(-\infty, 0) \cup (9, \infty)$

17) $\left[-\dfrac{5}{4}, 0\right]$

19) $(-8, 8)$

21) $(-\infty, -11] \cup [11, \infty)$

23) $[-4, 4]$

25) $(-\infty, \infty)$   27) $(-\infty, \infty)$   29) $\varnothing$   31) $\varnothing$

33) $(-\infty, -2] \cup [1, 5]$

35) $[-9, 5] \cup [7, \infty)$

37) $(-\infty, -7) \cup \left(-\dfrac{1}{6}, \dfrac{3}{4}\right)$

# Photo Credits

**Page 1:** © The McGraw-Hill Companies, Inc./Lars A. Niki, photographer; **p. 15:** © Vol. 61 PhotoDisc/Getty RF; **p. 16:** © BananaStock/Punchstock RF; **p. 105 (left):** © Vol. 16 PhotoDisc RF/Getty; **p. 105 (right):** © Ingram Publishing/Alamy RF; **p. 110:** © Digital Vision RF; **p. 113:** © EP100/PhotoDisc/Getty RF; **p. 142:** Photo by Keith Weller/USDA; **p. 143 (top):** © Big Cheese Photo/Punchstock RF; **p. 143 (bottom):** © Ingram Publishing/Superstock RF; **p. 144:**© PhotoLink/Getty RF; **p. 150:** © BrandX/Punchstock RF; **p. 151:** © Corbis RF; **p. 152 (left):** © Keith Brofsky/Getty Images RF; **p. 152 (right):** © Getty RF; **p. 163:** © Vol. 51 PhotoDisc/Getty RF; **p. 170:** © The McGraw-Hill Companies, Inc./Ken Cavanagh, photographer; **p. 179:** © The McGraw-Hill Companies, Inc./Jill Braaten, photographer; **p. 180:** © John Wilkes Studio/Corbis RF; **p. 181:** © RubberBall Productions RF; **p. 189:** © Getty RF; **p. 191 (top):** © S. Meltzer/PhotoLink/Getty RF; **p. 191 (bottom):** © Dynamic Graphics/Jupiter Images RF; **p. 196:** © Corbis RF; **p. 198:** © Royalty-Free/Corbis; **p. 200:** © Digital Vision/Getty RF; **p. 201:** © The McGraw-Hill Companies, Inc./Jill Braaten, photographer; **p. 213:** © S. Meltzer/PhotoLink/Getty RF; **p. 220:** © Digital Vision RF; **p. 224:** © Royalty-Free/Corbis; **p. 243:** © Brooklyn Productions/Corbis RF; **p. 250:** © Adam Gault/Getty RF; **p. 255:** © Seide Preis/Getty RF; **p. 268:** © Ingram Publishing/Alamy RF; **p. 281:** © DT01/Getty RF; **p. 317:** © The McGraw-Hill Companies, Inc.; **p. 318 (left):** © Burke/Triolo/BrandX; **p. 318 (right):** © BrandX/Corbis RF; **p. 319:** © Digital Vision/Getty RF; **p. 337:** © The McGraw-Hill Companies, Inc./Ken Cavanagh, photographer; **p. 338:** imac/Alamy RF; **p. 339 (left):** © Thinkstock/Jupiter Images RF; **p. 339 (right):** © Sandra Ivany/BrandX/Getty RF; **p. 383:** © The McGraw-Hill Companies, Inc./Jill Braaten, photographer; **p. 433:** © The McGraw-Hill Companies, Inc./John Flournoy, photographer; **p. 445:** © EP046 PhotoDisc/Getty RF; **p. 498:** © Daisuke Morita/Getty RF; **p. 505:** © Dynamic Graphics/Jupiter Images RF; **p. 508:** © Corbis RF; **p. 509:** © Stockbyte/Getty RF; **p. 510:** PhotoAlto RF; **p. 518:** © C Squared Studios/Getty RF; **p. 521:** © PhotoDisc/Getty RF; **p. 571:** © Royalty-Free/Corbis; **p. 577:** © Jonnie Miles/Getty RF; **p. 587:** © Vol. 86 PhotoDisc/Getty RF; **p. 593 (top):** © Philip Coblentz/Brand X Pictures/PictureQuest RF; **p. 593 ( bottom):** © Purestock/Getty RF.

# Index